Sources and Studies
in the History of Mathematics and
Physical Sciences

Managing Editor
J.Z. Buchwald

Associate Editors
J.L. Berggren and J. Lützen

Advisory Board
C. Fraser, T. Sauer, A. Shapiro

For further volumes:
http://www.springer.com/series/4142

The *Liber mahameleth*

A 12th-century mathematical treatise

Part Three
Mathematical Commentary, Bibliography, Index

Jacques Sesiano

Jacques Sesiano
Département de Mathématiques
Ecole polytechnique fédérale
Lausanne, Switzerland

ISSN 2196-8810   ISSN 2196-8829 (electronic)
ISBN 978-3-319-03939-8   ISBN 978-3-319-03940-4 (eBook)
DOI 10.1007/978-3-319-03940-4
Springer Cham Heidelberg New York Dordrecht London

Library of Congress Control Number: 2014930575

© Springer International Publishing Switzerland 2014
This work is subject to copyright. All rights are reserved by the Publisher, whether the whole or part of the material is concerned, specifically the rights of translation, reprinting, reuse of illustrations, recitation, broadcasting, reproduction on microfilms or in any other physical way, and transmission or information storage and retrieval, electronic adaptation, computer software, or by similar or dissimilar methodology now known or hereafter developed. Exempted from this legal reservation are brief excerpts in connection with reviews or scholarly analysis or material supplied specifically for the purpose of being entered and executed on a computer system, for exclusive use by the purchaser of the work. Duplication of this publication or parts thereof is permitted only under the provisions of the Copyright Law of the Publisher's location, in its current version, and permission for use must always be obtained from Springer. Permissions for use may be obtained through RightsLink at the Copyright Clearance Center. Violations are liable to prosecution under the respective Copyright Law.
The use of general descriptive names, registered names, trademarks, service marks, etc. in this publication does not imply, even in the absence of a specific statement, that such names are exempt from the relevant protective laws and regulations and therefore free for general use.
While the advice and information in this book are believed to be true and accurate at the date of publication, neither the authors nor the editors nor the publisher can accept any legal responsibility for any errors or omissions that may be made. The publisher makes no warranty, express or implied, with respect to the material contained herein.

Printed on acid-free paper

Springer is part of Springer Science+Business Media (www.springer.com)

Table of contents
Part Three

Mathematical Commentary

Mathematical Commentary 1137

Book A

Chapter A–I: Numbers 1139
  1. Place of number 1139
  2. Composition of numbers 1141

Chapter A–II: Premisses 1144
Summary 1144
  1. Complements to Euclid's *Elements* 1145
  2. Identities from Book II of Euclid's *Elements* 1153

Chapter A–III: Multiplication of integers 1159
Summary 1159
  1. Various cases of multiplication 1162
  2. Multiplication of digits among themselves 1163
    A. The multiplication table 1163
    B. Rules 1164
  3. Multiplication of units of two orders 1165
    A. Note of a number 1165
    B. Multiplication of two single-figure numbers using the note .. 1166
    C. Multiplication of two single-figure numbers using rules 1169
  4. Multiplication of composite numbers below one thousand 1171
  5. Multiplication of numbers with repeated thousands 1172
    A. Multiplication of single-figure numbers (multiples of limits) . 1173
    B. Multiplication of composite numbers 1175
    C. Multiplication using subtraction 1176
  6. Taking fractions of thousands repeated 1178

Chapter A–IV: Division of integers 1181
Summary 1181
  1. Division of larger by smaller 1183
    A. Division of integers without thousands 1183
    B. Division of integers with thousands 1184
    C. Practical rules for some particular cases 1185
  2. Division of smaller by larger ('denomination') 1188
    A. Premisses on divisibility 1188
    B. Expressing a fraction when the denominator is a product of elementary numbers 1191
    C. Other cases of denomination 1194

Chapter A–V: Multiplication of fractions 1197

Summary 1197
   1. Multiplication of a fractional expression by an integer 1200
      A. Multiplicand a fraction 1201
      B. Multiplicand a fraction of a fraction 1203
      C. Multiplicand a compound fraction 1205
   2. Multiplication of fractional expressions not containing integers . 1207
      A. Multiplicand a fraction 1207
      B. Multiplicand a fraction of a fraction 1209
      C. Multiplicand a compound fraction 1210
   3. Conversion of fractions 1212
      A. Conversion into a fraction 1212
      B. Conversion into a fraction of a fraction 1214
   4. Multiplication of fractional expressions containing integers 1217
   5. Case of irregular fractions 1224
      A. Computation with irregular fractions 1224
      B. Ambiguous formulations 1229
   6. Other problems 1231
      A. A particular problem 1231
      B. Combined expressions 1232

Chapter A–VI: Addition of fractions 1234

Summary 1234
   1. Addition of fractional expressions 1238
      A. Addend a fraction 1238
      B. Addend a compound fraction 1239
      C. Addend a fraction of a fraction 1240
      D. Case of different fractions 1241
   2. Addition of irregular fractions 1243
      A. Addition proper 1243
      B. Ambiguous formulations 1244
   3. Problems on amounts 1245
      A. Known amounts 1245
      B. Unknown amounts 1247
   4. Summing series 1251
      A. Sum required 1251
      B. Last term required 1253

Chapter A–VII: Subtraction of fractions 1255

Summary 1255
   1. Subtraction of fractional expressions 1256
   2. Subtracting irregular fractions 1264
   3. Problems on amounts 1266
      A. Known amounts 1266
      B. Unknown amounts 1267

Chapter A–VIII: Division of fractions 1275

Summary 1275

1. Inverting a fractional expression 1279
   A. Redintegrating a fraction to unity 1279
   B. Reducing an integer with a fraction to unity 1280
   C. Same, by a direct rule 1281
2. Denominating fractional expressions 1282
3. Division of fractions without integers 1283
4. Division of irregular fractions 1286
5. Division of fractional expressions containing integers 1288
   A. Dividend an integer and divisor a fraction 1288
   B. Dividend an integer and divisor an integer with a fractional expression 1290
   C. Dividend an integer with a fraction and divisor a fraction .. 1291
   D. Dividend and divisor integers with a fraction 1293
6. Other problems 1295
   A. An indeterminate problem 1295
   B. Recapitulation: Arithmetical operations with fractions 1295
   C. Combined expressions 1297
7. Sharing amounts of money 1299
   A. Known amounts 1299
   B. Unknown amounts 1301

Chapter A–IX: Roots 1303

Summary 1303
Appendix: Comparison with Abū Kāmil's treatment 1315
1. Approximation of square roots 1316
   A. Approximate roots of non-square integers 1316
   B. Approximate roots of fractional expressions 1316
2. Operations with square roots 1319
   A. Multiplication of square roots 1319
   B. Addition of square roots 1320
   C. Subtraction of square roots 1322
   D. Division of square roots 1323
3. Operations with fourth roots 1327
   A. Multiplication of fourth roots 1327
   B. Addition of fourth roots 1329
   C. Subtraction of fourth roots 1331
   D. Division of fourth roots 1332
4. Square root extraction of binomials and apotomes 1336

Book B

Introduction 1343

Chapter B–I: Buying and selling 1345

Summary 1345
1. The five fundamental problems 1351
   A. One unknown 1351
   B. Two unknowns 1356
2. Problems involving 'things' 1362

viii   Part Three: Mathematical Commentary, Bibliography, Index

    3. Second-degree problems 1365
        A. Two quantities unknown 1366
        B. The four quantities are unknown 1369
        C. Other problems 1377
    4. Corn of various kinds 1386

Chapter B–II: Profit 1395

Summary 1395
    1. The five fundamental problems 1399
        A. One unknown 1399
        B. Two unknowns 1400
    2. The twenty kinds 1403
    3. Variation in capital 1409
        A. Different capital 1410
        B. Sale not completed 1413
    4. Other problems 1418
        A. A particular problem 1418
        B. Data involving roots 1422

Chapter B–III: Profit in partnership 1426

Summary 1426
    1. Three partners 1428
    2. Other cases 1431
        A. Data involving roots 1431
        B. A particular problem 1433

Chapter B–IV: Sharing out according to prescribed parts 1434

Summary 1434

Chapter B–V: Masses 1442

Summary 1442
    1. Constituents of mixed masses 1445
    2. Making smaller pieces 1446
    3. Alloy determination 1446

Chapter B–VI: Drapery 1448

Summary 1448
    1. Cutting a piece 1450
        A. One unknown 1450
        B. Two unknowns 1451
    2. Mixed material 1453

Chapter B–VII: Linen cloths 1457

Summary 1457
    1. Rectangular pieces 1458
        A. One unknown 1458
        B. Two unknowns 1461
    2. Circular pieces 1462

Chapter B–VIII: Grinding 1466

Summary 1466
   1. Payment from each bushel 1469
   2. Payment from another bushel 1469
      A. One unknown 1469
      B. Two unknowns 1471
   3. Increment in volume 1473
      A. Payment from each bushel 1473
      B. Payment from another bushel 1473
   4. Other problems on corn 1474

Chapter B–IX: Boiling must 1476

Summary 1476
   1. One quantity unknown 1480
      A. One overflow 1480
      B. Several overflows 1483
      C. Particular cases of a single overflow 1484
   2. Several quantities unknown 1484
      A. Two quantities unknown 1484
      B. Three quantities unknown 1489
      C. Particular cases 1490

Chapter B–X: Borrowing 1492

Summary 1492

Chapter B–XI: Hiring workers 1495

Summary 1495
   1. One worker 1502
      A. Simple problems 1502
      B. Problems involving 'things' 1503
      C. Solution formulae involving roots 1515
      D. Wage in cash and kind 1524
   2. Two workers (or single lazy worker) 1526
      A. Monthly wages 1527
      B. Daily wages 1534
   3. Three workers 1542

Chapter B–XII: Wages in arithmetical progression 1551

Summary 1551
   1. Number of workers known 1552
   2. Number of workers required 1556

Chapter B–XIII: Hiring a carrier 1563

Summary 1563
   1. One unknown 1566
   2. Two unknowns 1570
   3. Problems involving 'things' 1576

Chapter B–XIV: Hiring stone-cutters 1578

Summary 1578
   1. Cutting stones 1580

2. Kind of work not specified 1585
A. One unknown 1585
B. Two unknowns 1587
3. Other kinds of work............................... 1588

Chapter B–XV: Consumption of lamp-oil 1592
Summary 1592
1. Consumption of one lamp known..................... 1593
2. Consumption of several lamps known 1599

Chapter B–XVI: Consumption by animals 1603
Summary 1603
1. One quantity unknown 1604
A. Consumption of one animal known 1604
B. Consumption of several animals known 1608
2. Two quantities unknown 1612
3. Consumption in arithmetical progression 1615
Appendix: Bushels from different regions 1616
Summary 1616

Chapter B–XVII: Consumption of bread 1618
Summary 1618
Appendix: Other units of capacity 1625
Summary 1625

Chapter B–XVIII: Exchanging moneys 1629
Summary 1629
1. One kind of money exchanged for one kind of money 1635
A. One morabitinus for two types of nummus 1635
B. Morabitini for nummi and solidi 1636
2. Two kinds of money exchanged for one kind of money 1638
A. Final number of nummi not specified 1638
B. Final number of nummi specified 1639
3. More than two kinds of money exchanged for one kind of money 1644
4. Particular problems 1652
A. One morabitinus exchanged 1652
B. Various kinds of morabitinus exchanged 1659
C. Another kind of problem........................... 1661

Chapter B–XIX: Cisterns 1663
Summary 1663
1. Pipes filling a cistern 1664
2. Stone thrown into a cistern 1664
3. Content of a cask................................. 1668

Chapter B–XX: Ladders 1670
Summary 1670
1. The ladder....................................... 1675

    A. One unknown 1675
    B. Two unknowns 1677
  2. The broken tree 1686
  3. The falling tree 1691
    A. Forward motion 1691
    B. Forward and retrograde motion 1692
  4. Planting trees 1693
  5. The two towers 1693
Chapter B–XXI: Bundles 1698
Summary 1698
Chapter B–XXII: Messengers 1700
Summary 1700
  1. Forward motion 1701
  2. Forward and retrograde motion 1702
Chapter B–XXIII: Mutual lending 1704
Summary 1704
  1. Successive doubling 1707
  2. Buying a horse by borrowing from partners 1709
    A. Each borrows from his neighbour 1709
    B. Each borrows from all partners 1710

Bibliography 1715

Index 1721

Mathematical Commentary

Each chapter opens with a summary of what is to follow and the mathematical methods involved. Then comes, for each problem, a full mathematical transcription together, if need be, with complements, references or general remarks. The transcription may, at times, seem lengthy to the modern reader. However, once he has glanced over the translated text, he will realize that there was no other way of making systematically intelligible the verbal reasonings and sometimes abstruse content of the text. In any event, imposing concision on its admittedly verbose author would have meant sacrificing some parts of the text.

Book A
Chapter A–I:
Numbers

The opening chapter of the *Liber mahameleth*, on numbers, is divided into two parts. The first concerns the place of number among all things that exist and distinguishes between its two aspects: number when considered as such and when applied to human use. The second part is devoted to the numeration and verbal expression of the natural integers in the decimal system. Here we shall paraphrase and comment on, rather than summarize, these two parts.

---

1. Place of number.

It is to Avicenna (Ibn Sīnā, 980–1037) that we owe the conciliation of Aristotelian tradition, as influenced by neo-Platonism, with Moslem philosophy and theology. His *Shifā'* (Healing) deals successively with logic, physics and natural sciences, mathematics, psychology and metaphysics. In the introductory chapter (*madkhal*) of the *Logic*, Avicenna presents a classification of the sciences.[†] He begins by making the distinction between theoretical and practical philosophy, the object of the first being the study of things the existence of which is independent of our freewill and actions, whereas the second is concerned with things depending on our freewill and actions. The things of the first kind are further subdivided according to their relationship to motion and matter.[‡] But let us see how Avicenna explains all this (in the twelfth-century Latin translation[*]):

> *Res (al-ashyā') autem que sunt, aut habent esse non ex nostro arbitrio (ikhtiyār) vel opere (fi'l), vel habent esse ex nostro arbitrio et opere (...). Res autem que sunt quarum esse non est ex voluntate (ikhtiyār again) nostra vel opere (...) dividuntur in duo: in res (al-umūr) que commiscentur motui (tukhālit al-haraka), et res que non commiscentur motui.*[‖]

[†] Commented translation of this part in M. Marmura's *Avicenna on the division of the sciences*.

[‡] Motion ($\kappa\acute{\iota}\nu\eta\sigma\iota\varsigma$) in the Aristotelian sense of passing from one state to another (either naturally, as is mostly the case, or by constraint): change in quantity, quality, place, and so on.

[*] Venetian edition of the *Logic*, fol. 2ra (*Capitulum de intrando apud scientias*). I have provided the Arabic words in a few cases; see the Cairo edition of the *Mantiq*, 1 (*al-Madkhal*), pp. 12–13.

[‖] 'like mind and God (the Creator)' (*mithla al-'aql wa'l-bāri'*), the Arabic text adds.

Res autem que commiscentur motui dividuntur in duo: quia aut in res que non habent esse nisi quia possibile est eas admisceri (ḵẖālaṭa again) motui, sicut est humanitas (insānīya) et quadratura (tarbīʿ) et his similia; aut in res que habent esse absque hoc. Ille autem que non habent esse nisi quia possibile est eas admisceri motui iterum dividuntur in duo: quia aut sic sunt quod nec esse nec intelligi possunt absque materia propria (mādda muʿayyana), sicut forma (ṣūra) humana aut asinina, aut sic quod possunt intelligi absque materia —sed non esse— sicut quadratura, ad quam intelligendam non est omnino necesse appropriari eam aliqua specie materie, nec considerari secundum aliquam dispositionem motus. Res autem que commiscentur motui et habent esse sine illo sunt sicut identitas (huwīya) et unitas (waḥda) et multitudo (kaṯẖra) et causalitas.

The similarity to our author's introduction is striking —it has merely been adapted to both the Christian and mathematical context, 'God and mind' being replaced by 'God and an angel' and 'multitude' by 'number'. But let us now turn our attention to the subdivision according to the *Liber mahameleth* (see also Fig. 1).°

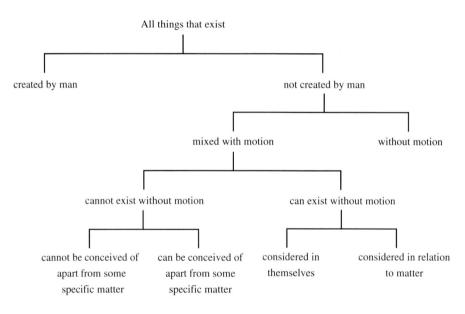

Fig. 1

Of all things that exist, some are created by man and others not, that is, do not depend on his choice or action. Of those things which are not

---

° As mentioned in our General Introduction (Part I, p. lx), similar passages are found in the *De divisione philosophiae* by the Spanish philosopher, translator, and archdeacon of Segovia, Dominicus Gundissalinus (ed. Baur, pp. 10–14 & 90–93) and in the *Opusculum de scientiis* attributed to al-Fārābī (ed. Camerarius, pp. 14–15).

created by man, some —like God and an angel— are not at all, nor can be, subject to motion (change), whereas others are mixed with motion. Of those which are mixed with motion, some can exist only in connection with motion, while others can also have an existence independent of motion; examples of the first kind are humanity and squareness, and examples of the second kind are unity, number, causality. But there is a further difference between the examples of the first kind: humanity cannot be conceived of apart from some specific matter since the existence of man is indissociable from his material body; but squareness (like other usual geometric forms) can be understood independently of any specific kind of substance, even if it will come into existence only through matter.

Number is thus a thing not created by man, mixed with motion but which can also exist independently of motion. Furthermore, it can be considered either in itself, inasmuch as it is number, or in relation to matter. Indeed, on the one hand, a (natural) number can be considered for its *intrinsic properties*, such as parity, divisibility, relation between it and the sum of its divisors (not including itself)*; the study of such properties of natural numbers is the subject of 'theoretical arithmetic' (in Arabic: *'ilm al-'adad al-naẓarī*) —what we today call number theory— elementary examples of which are found in the *Introduction to arithmetic* by Nicomachos (*fl.* A.D. 100), mentioned by our author. On the other hand, number can be considered *in relation to motion and matter*. This is the subject of 'practical arithmetic' (*'ilm al-'adad al-'amalī*), of which there are two aspects, both to do with human use: with on the one hand, the operations of reckoning, namely the four arithmetical ones and that of root extraction, all of them taught, as our author reminds us, in Muḥammad ibn Mūsā al-Khwārizmī's *Arithmetic* (c. 820); and on the other, the application of number to commercial transactions, domestic activities and various kinds of measuring —that is, to problems necessarily involving magnitudes having concrete signification.

This introduction, initially concerned with the place of number, has now quite naturally moved on to presenting the work itself: the subject will be practical arithmetic in its two aspects, Book A being devoted to arithmetical reckoning and Book B to applications.

2. Composition of numbers.

Our author follows the ancient conception: that unit is not itself a number but the generator of (natural) numbers.‡ As asserted in our text, natural numbers proceed to infinity since each time a unit can be added.

---

* The sum may be larger than the number, which is then *superfluus*, or smaller, in which case the number is *diminutus*. Gundissalinus, who (unlike our text) has this example, omits the case of equality, thus of 'perfect' numbers.

‡ See, for instance, Heath, *History of Greek Mathematics*, I, pp. 69–70.

On the other hand, the structural and physiological constraints of language —and the difficulty of having to remember an ever-increasing quantity of words— set limits to the differentiation of numbers by name. Thus we are to solve the problem of expressing an *infinite* quantity of positive integers by means of a *limited* quantity of words.

For this purpose, we shall first introduce, by means of what we call today the ten-scale, a periodicity. Consider, with our author, the decimal system. After a first set, that of the nine digits, which form the first 'order' (*ordo*, Arabic *martaba*), we shall introduce a larger unit, the 'limit' (*limes*, Arabic *'aqd*) of the next order, multiples of which will be counted by means of the digits of the first order; this second order is that of the tens. The same will then be done for the next two orders, those of the hundreds and the thousands. Thus far we have been able to restrict the number of names to twelve: the nine digits and the three names of the limits that we have introduced, namely 'ten', 'hundred', 'thousand'.††

ten 9	8	7	6	5	4	3	2	1	
1	1	1	1	1	1	1	1	1	1
2	2	2	2	2	2	2	2	2	2
3	3	3	3	3	3	3	3	3	3
4	4	4	4	4	4	4	4	4	4
5	5	5	5	5	5	5	5	5	5
6	6	6	6	6	6	6	6	6	6
7	7	7	7	7	7	7	7	7	7
8	8	8	8	8	8	8	8	8	8
9	9	9	9	9	9	9	9	9	9
of the thousand times thousand times thousands	of the hundred thousand times thousands	of the ten thousand times thousands	of the thousand times thousands	of the hundreds of thousands	of the tens of thousands	of the thousands	of the hundreds	of the tens	order of the digits

Fig. 2

This will in fact be enough to express numbers verbally: on the model of the first class, consisting of the first three orders, we shall form further classes, of three orders each, using no other words than the above twelve.

---

†† In Arabic and Latin the names for 11, 12, ... , 19 are indeed simply those of the units combined with that of ten (as in English from 13 on). But in Latin (not in Arabic) 20 has a name of its own; see Part I, p. xviii.

Thus, for the second class, the limit of the fourth order will be a thousand, the limit of the fifth, ten thousand, that of the sixth, a hundred thousand. The first element of the next order, and of the next class, will be a thousand thousands; indeed, where our languages introduce a new word, 'million', Arabic (much as Latin does) repeats the word 'thousand': *alf alf* (and *milies mille*). Again, the first element of the next class, our 10^9, will be *milies milies mille* (*alf alf alf*). Such is the principle of 'repetition' (*iteratio*, Arabic *takrār*): what distinguishes any three-order class from the one before is the addition to its numerals of a (or a further) 'thousand'. That is why the sequence of numeral names in Arabic (and Latin) ends with 'thousand' and no more words are needed.

The table (Fig. 2) shows how to express verbally any natural number ($< 10^{10}$) proposed in number symbols (or conversely): each of its digits, the decimal place of which is indicated in the upper line, is followed by the name of the corresponding order unit, indicated in the bottom line.

Remark. In writing numbers, a certain (decimal) place (*differentia*) is assigned to each order. The purpose of the 'note' (*nota*) is to determine the highest place once the number is expressed verbally. How to determine the note will be taught at the beginning of the chapter on multiplication of integers (A–III). But passages, apparently interpolated, already allude to it at the end of this chapter.

Chapter A–II:
Premisses

Summary

1. Complements to Euclid's *Elements*.

These premisses are not found in Euclid's *Elements* but are required for the subsequent demonstrations. They consist of nine complements proper introduced by the author, designated here by P_1 to P_9, plus one proposition found in Euclid's *Elements* (P'_7), explicitly quoted because of its similarity to one of the complements, and a proposition from Abū Kāmil's *Algebra* (P'_3). We can summarize them as follows.

— If a_1, a_2, a_3 are three numbers, then

$$(a_1 \cdot a_2) \cdot a_3 = a_1 \cdot (a_2 \cdot a_3) \qquad (P_2)$$

$$\frac{a_1}{a_2} \cdot a_3 = \frac{a_1 \cdot a_3}{a_2} \qquad (P_5)$$

$$\frac{\frac{a_1}{a_2}}{a_3} = \frac{a_1}{a_2 \cdot a_3}, \quad \frac{a_3}{\frac{a_1}{a_2}} = \frac{a_2 \cdot a_3}{a_1} \qquad (P_4,\ P_9)$$

$$\frac{a_1}{a_3} \pm \frac{a_2}{a_3} = \frac{a_1 \pm a_2}{a_3}. \qquad (P_8,\ P_7)$$

— If a_1, a_2, a_3, a_4 are four numbers, then

$$(a_1 \cdot a_2)(a_3 \cdot a_4) = (a_1 \cdot a_3)(a_2 \cdot a_4) \qquad (P_3)$$

$$\left(\frac{a_1}{a_2}\right)\left(\frac{a_3}{a_4}\right) = \frac{a_1 \cdot a_3}{a_2 \cdot a_4}. \qquad (P'_3)$$

— If a_1, a_2, a_3, a_4, a_5, a_6 are six numbers such that $a_1 : a_2 = a_3 : a_4$ and $a_5 : a_2 = a_6 : a_4$, then

$$\frac{a_1 \pm a_5}{a_2} = \frac{a_3 \pm a_6}{a_4} \quad (a_1 > a_5,\ a_3 > a_6). \qquad (P'_7,\ P_6)$$

— If a_1, a_2, ..., a_n are in arithmetical progression, then

$$a_1 + a_n = a_i + a_{n-i+1} \qquad (P_1)$$

and this is equal to twice the middle term if there is one, that is, when n is odd.

2. Identities from Book II of Euclid's *Elements*.

Book II of Euclid's *Elements* is frequently said to deal with 'geometrical algebra' since its first ten propositions correspond to algebraic identities. Thus it is not surprising that, after the above premisses, the author wished to make explicit the algebraic content of these formally geometrical theorems. So we find here (designated by PE_1–PE_{10}) an adaptation of *Elements* II.1–10 to numbers instead of to segments as in the *Elements*. Note, though, that whenever such a proposition is used in a later part of his work, the author's reference, if any, is always to the *Elements*; there is just one exception (B.327, p. 1660 below) and an indirect allusion to a second part in A–II (A.200a). Indeed, the author's purpose here is just to show how these theorems apply to numbers.°

These ten propositions rely on the distributivity law or on a few simple identities used in algebraic reckoning since Mesopotamian times. They are, or are reducible to, the four following identities:

$$a \cdot (a_1 + a_2 + \ldots + a_n) = a \cdot a_1 + a \cdot a_2 + \ldots + a \cdot a_n \qquad (PE_1,\ PE_2,\ PE_3)$$

$$(a_1 \pm a_2)^2 = a_1^2 \pm 2\,a_1 \cdot a_2 + a_2^2 \qquad (PE_4,\ PE_7)$$

$$\left(\frac{a_1 + a_2}{2}\right)^2 = a_1 \cdot a_2 + \left(\frac{a_1 - a_2}{2}\right)^2 \qquad (PE_5,\ PE_6,\ PE_8)$$

$$\frac{a_1^2 + a_2^2}{2} = \left(\frac{a_1 + a_2}{2}\right)^2 + \left(\frac{a_1 - a_2}{2}\right)^2. \qquad (PE_9,\ PE_{10})$$

---

1. Complements to Euclid's *Elements*.

FIRST PREMISS (P_1).[†] Let $a_1, a_2, \ldots, a_n$ be integers in arithmetic progression with the common difference δ. Then

$$a_1 + a_n = a_i + a_{n-i+1} \quad (2 \leq i \leq n-1)$$

and in addition, for n odd,

$$a_1 + a_n = a_i + a_{n-i+1} = 2 \cdot a_{\frac{n+1}{2}}.$$

Demonstration.

• Let (Fig. 3) the numbers, in even quantity, be A, BG, DH, $ÇK$, TQ, LM. To prove that

$$A + LM = BG + TQ = DH + ÇK.$$

Put $PG = BG - A$, $FH = DH - BG$, $CK = ÇK - DH$, $ZQ = TQ - ÇK$, $RM = LM - TQ$, with $PG = FH = CK = ZQ = RM = \delta$.

In order to prove that $A + LM = BG + TQ$, we first add $MN = A$ to LM. So, first, $LN = A + LM$. On the other hand, since $RN = BG$

° Even if he does not need PE_8–PE_{10} later on. We read also in Campanus' 13th-century edition of the *Elements*, after II.10: *Hec autem* (Propositio II.10) *et omnes premisse veritatem habent in numeris sicut in lineis.*

† Used in the proof following A.13; see also B.244.

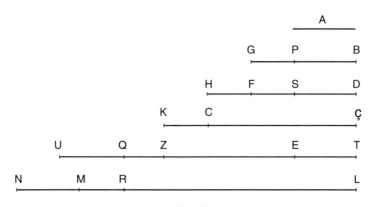

Fig. 3

(for $MN = A = BP$ and $RM = PG$) and $LR = TQ$ (by construction), we shall have, by addition, $LN = BG + TQ$.

We have thus proved that $A + LM = BG + TQ$; by a similar reasoning we could prove that this is also equal to $DH + ÇK$.

• Consider now n odd, that is, with TQ as the last term. We can prove as before that $A + TQ = BG + ÇK$. We must further prove that $A + TQ = 2 \cdot DH$.

We add $QU = A$ to TQ and cut $TE = A$ from TQ; similarly, we cut $DS = A$ from DH. Thus clearly $EQ = 4\delta$ and $SH = 2\delta$, so that $EQ = 2\,SH$, while $TE + QU = 2\,DS$; therefore
$$TU = TE + QU + EQ = 2\,A + 4\,\delta = 2\,(A + 2\,\delta) = 2 \cdot DH,$$
so that $TU = A + TQ = a_1 + a_n$ is indeed equal to $2\,DH = 2 \cdot a_{\frac{n+1}{2}}$.

<u>Second premiss</u> (P_2).* Let a_1, a_2, a_3 be three numbers. Then
$$(a_1 \cdot a_2) \cdot a_3 = a_1 \cdot (a_2 \cdot a_3).$$

Demonstration. Let (Fig. 4) the three numbers be A, B, G; to prove that $(A \cdot B) \cdot G = A \cdot (B \cdot G)$.

Put $A \cdot B = D$, $B \cdot G = H$; to prove that $D \cdot G = A \cdot H$. Now since $A \cdot B = D$ and $B \cdot G = H$, so (*Elements* VII.18) $A : G = D : H$, so that (*Elements* VII.19) $A \cdot H = D \cdot G$.

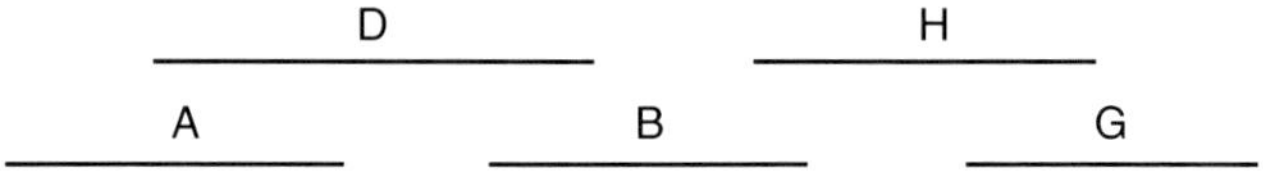

Fig. 4

Remark. According to the statement in the text, we have here a mixture of associativity and commutativity, for it corresponds to $(a_1 \cdot a_2)\,a_3 =$

---

* Used in P_4, A.131b, B.41a, B.42a, B.260c.

$(a_3 \cdot a_2) a_1$. Commutativity is taken for granted (*Elements* VII.16); which is why the multiplication table in A–III presents only the products in and above the diagonal, that is, $k \cdot l$ with $l \geq k$ for k fixed.

THIRD PREMISS (P_3).|| Let a_1, a_2, a_3, a_4 be four numbers. Then
$$(a_1 \cdot a_2)(a_3 \cdot a_4) = (a_1 \cdot a_3)(a_2 \cdot a_4).$$

Demonstrations. There are two: one relies on the same theorems as above (*Elements* VII.18–19) whereas the other uses a proposition from Abū Kāmil's *Algebra*. But the author will subsequently show that Abū Kāmil's proposition may itself be proved by VII.17–19, thus by a demonstration said to be 'much simpler' than Abū Kāmil's own.

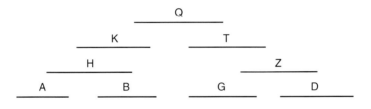

Fig. 5

(**a**) Demonstration by means of *Elements* VII.18–19. Let (Fig. 5) the four numbers be A, B, G, D; to prove that $(A \cdot B)(G \cdot D) = (A \cdot G)(B \cdot D)$.

Put $A \cdot B = H$, $G \cdot D = Z$, $A \cdot G = K$, $B \cdot D = T$; we are thus to prove that $H \cdot Z = K \cdot T$.

Since $A \cdot B = H$ and $B \cdot D = T$, then (*Elements* VII.18) $A : D = H : T$. Again, since $A \cdot G = K$ and $G \cdot D = Z$, then also $A : D = K : Z$. Hence $H : T = K : Z$ and therefore (*Elements* VII.19) $H \cdot Z = K \cdot T$.

(**b**) Demonstration by means of Abū Kāmil's proposition (P_3'), which asserts that:
$$\left(\frac{a_1}{a_2}\right)\left(\frac{a_3}{a_4}\right) = \frac{a_1 \cdot a_3}{a_2 \cdot a_4}.$$

Considering as before that $A \cdot B = H$, $G \cdot D = Z$, $A \cdot G = K$, $B \cdot D = T$, we must prove that
$$(Q =) H \cdot Z = K \cdot T.$$

Now by hypothesis and using Abū Kāmil's above statement, we have
$$A = \frac{H}{B}, \quad G = \frac{Z}{D}, \quad A \cdot G = \frac{H}{B} \cdot \frac{Z}{D} = \frac{H \cdot Z}{B \cdot D},$$

so that $(A \cdot G)(B \cdot D) = H \cdot Z = (A \cdot B)(G \cdot D)$.

|| Used in the proof following A.13, in A.37*b*, A.134*c*, B.219 (quoted).

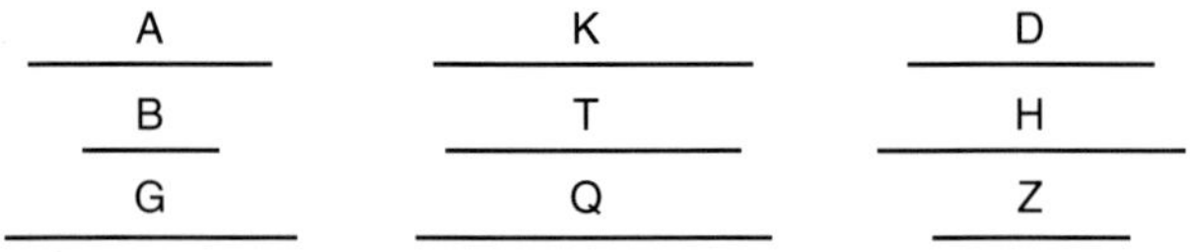

```
         A              K              D
         B              T              H
         G              Q              Z
```

Fig. 6

(*c*) Author's proof of Abū Kāmil's proposition.*

Let (Fig. 6) $\frac{A}{B} = G$, $\frac{D}{H} = Z$, $A \cdot D = K$, $B \cdot H = T$, $G \cdot Z = Q$; to prove that $\frac{K}{T} = Q$, thus $\frac{A \cdot D}{B \cdot H} = \frac{A}{B} \cdot \frac{D}{H}$.

Since $G \cdot B = A$ and $G \cdot Z = Q$, so (*Elements* VII.17) $B : Z = A : Q$. Again, since $B \cdot H = T$ and $Z \cdot H = D$, so (*Elements* VII.18) $B : Z = T : D$. Therefore $A : Q = T : D$, and so

$$A \cdot D = K = Q \cdot T \quad \text{and} \quad \frac{K}{T} = Q.$$

(*d*) Our author does not reproduce Abū Kāmil's proof, which is the following.‡

Since $G \cdot B = A$ and $G \cdot Z = Q$, so $B : Z = A : Q$, then $A \cdot Z = B \cdot Q$. Multiplying both sides by H produces $A \cdot Z \cdot H = H \cdot B \cdot Q$; therefore, since $Z \cdot H = D$ and $B \cdot H = T$, $A \cdot D = T \cdot Q$.

Remark. As we see, the only difference is that our author avoids the common multiplication by once again applying *Elements* VII.17-18. This is insignificant, but has the advantage of using the same Euclidean theorems as before. That is why our author calls his own proof 'much simpler'. A similar instance is found in A.124*a*: a second proof is called 'easier' because it relies on *Elements* VII.17–19 whereas the first uses P_3'.

<u>Fourth premiss</u> (P_4).° Let a_1, a_2, a_3 be three numbers. Then

$$\frac{\frac{a_1}{a_2}}{a_3} = \frac{a_1}{a_2 \cdot a_3}.$$

---

* Abū Kāmil's proposition is used in A.100*a*, A.124*a*, A.134*a*, B.254*d*, B.255*a*, B.264*a* (generalized to six quantities).

‡ See fol. 64^v–65^r of the Arabic manuscript (pp. 128–129 of the printed reproduction), or lines 3393–3430 of the Latin translation. (This is in Book III of Abū Kāmil's *Algebra*, where he deals with quadratic equations with irrational coefficients and solutions: see our *Introduction to the History of Algebra*, pp. 69–72.) For the purpose of easier comparison, we have changed the letters in Abū Kāmil's demonstration by adopting those used in the *Liber mahameleth*.

° Used in A.93*b* (statement repeated), B.141 (quoted), B.145 (interpolated proof), B.282*a*, B.283*a*, B.291*a*, B.292*a*. In the last four occurrences, P_4 is extended (see below).

Demonstration. Let there be three numbers A, B, D, and put $\dfrac{A}{B} = G$ and $\dfrac{G}{D} = H$ (Fig. 7); to prove that
$$H = \frac{A}{B \cdot D}.$$

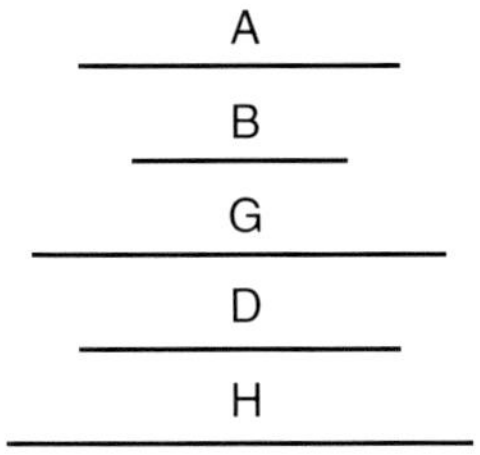

$$\begin{array}{c} \underline{A} \\ \underline{B} \\ \underline{G} \\ \underline{D} \\ \underline{H} \end{array}$$

Fig. 7

From the two given relations we infer that $A = G \cdot B = (H \cdot D) \cdot B$. But since, by P_2, $(H \cdot D) \cdot B = H \cdot (D \cdot B) = H \cdot (B \cdot D)$, so $A = H \cdot (B \cdot D)$, whence the assertion.

Remarks. P_4 and the coming P_5 are related: the first is the case of division, the second of multiplication (see also P_9). Note too that, in relation with the extension of the rule of three in B–XV and B–XVI, P_4 takes the forms

$$\frac{\frac{a_1 \cdot a_2}{a_3}}{a_4} = \frac{a_1 \cdot a_2}{a_3 \cdot a_4}, \qquad \frac{\frac{a_1 \cdot a_2}{a_4} a_3}{a_5} = \frac{\frac{(a_1 \cdot a_2) a_3}{a_4}}{a_5} = \frac{(a_1 \cdot a_2) a_3}{a_4 \cdot a_5}.$$

See Translation, notes 1689, 1693, 1715, 1725.

FIFTH PREMISS (P_5).[††] Let a_1, a_2, a_3 be three numbers. Then
$$\frac{a_1}{a_2} \cdot a_3 = \frac{a_1 \cdot a_3}{a_2}.$$
This is extended to four quantities in B.301a, where a_1 is itself a product, so that P_5 becomes
$$\frac{a_1 \cdot a_2}{a_3} \cdot a_4 = \frac{(a_1 \cdot a_2) a_4}{a_3}.$$
Remark. P_5 is a corollary of Abū Kāmil's proposition (with $a_4 = 1$).

Demonstration. Let the three numbers be (Fig. 8) A, B, D, with $\dfrac{A}{B} = G$ and $G \cdot D = H$; to prove that
$$H = \frac{A \cdot D}{B}.$$

[††] Used in A.46, A.55, A.123b, A.131a, A.247, B introduction (rule *ii*), B.1, B.257b (quoted), B.260b, B.277b (interpolated proof), B.310c. See Translation, notes 487, 854, 1788.

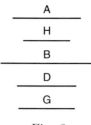

Fig. 8

Since $G \cdot B = A$ while $G \cdot D = H$, then $B : D = A : H$, and therefore $A \cdot D = B \cdot H$, which proves the assertion.

Remark. The demonstration again relies on the group *Elements* VII.17–19.

SIXTH PREMISS (P_6).[‖] If $a_1, a_2, \ldots, a_5, a_6$ are six numbers such that $a_1 : a_2 = a_3 : a_4$ and $a_5 : a_2 = a_6 : a_4$, then

$$\frac{a_1 - a_5}{a_2} = \frac{a_3 - a_6}{a_4} \quad (a_1 > a_5,\ a_3 > a_6).$$

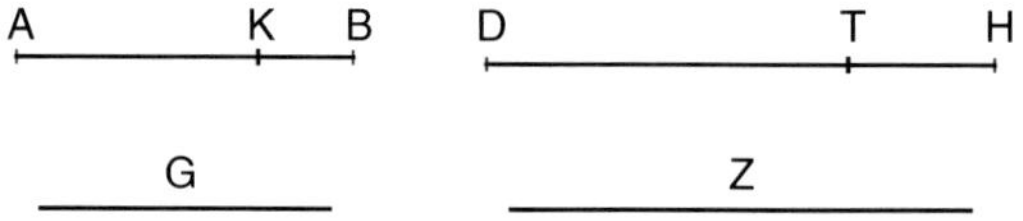

Fig. 9

Demonstration. Let (Fig. 9) the six numbers be AB, G, DH, Z, AK, DT (where $AB > AK$, $DH > DT$), with $\dfrac{AB}{G} = \dfrac{DH}{Z}$, $\dfrac{AK}{G} = \dfrac{DT}{Z}$; to prove that

$$\frac{AB - AK}{G} = \frac{BK}{G} = \frac{DH - DT}{Z} = \frac{HT}{Z}.$$

From the first given relation and the second, inverted, thus from

$$\frac{AB}{G} = \frac{DH}{Z} \text{ and } \frac{G}{AK} = \frac{Z}{DT},$$ we infer, *ex æquali*, that $\dfrac{AB}{AK} = \dfrac{DH}{DT}$.

From the second given relation and by separation of the result just obtained, thus from

$$\frac{AK}{G} = \frac{DT}{Z} \text{ and } \frac{AB - AK}{AK} = \frac{DH - DT}{DT},$$ that is, $\dfrac{BK}{AK} = \dfrac{HT}{DT}$,

we infer, *ex æquali*, that $\dfrac{BK}{G} = \dfrac{HT}{Z}$, which is the desired result.

Remark. The corresponding case for addition is P_7'.

SEVENTH PREMISS (P_7).^{∗∗} Let a_1, a_2, a_3 be three numbers. Then

---

[‖] Used in P_7, A.192a, A.194a, B.90.
^{∗∗} Used in A.200a.

Fig. 10

$$\frac{a_1}{a_3} - \frac{a_2}{a_3} = \frac{a_1 - a_2}{a_3} \quad (a_1 > a_2).$$

Demonstration. It relies on the previous premiss. Let (Fig. 10) $AB = a_1$, $AK = a_2$, thus $KB = a_1 - a_2$, and $a_3 = G$. Let further $\frac{AB}{G} = DH$, $\frac{AK}{G} = DT$, thus $\frac{AB}{G} - \frac{AK}{G} = DH - DT = TH$. To prove that

$$TH = \frac{KB}{G}.$$

Since

$$\frac{AB}{G} = \frac{DH}{1}, \quad \frac{AK}{G} = \frac{DT}{1},$$

we are in the situation of P_6 with $Z = 1$; therefore

$$\frac{AB - AK}{G} = \frac{DH - DT}{1}, \quad \text{that is,} \quad \frac{KB}{G} = TH.$$

Remark. The corresponding case for addition is P_8.

<u>A Euclidean theorem</u> (P_7').[†] If $a_1 : a_2 = a_3 : a_4$ and $a_5 : a_2 = a_6 : a_4$, then

$$\frac{a_1 + a_5}{a_2} = \frac{a_3 + a_6}{a_4}.$$

This is the analogue of P_6 for the additive case. But since it is found in the *Elements* (V.24), the proof is omitted. Indeed, the reader is supposed to know or have the *Elements* to hand.

Remark. Since Euclid does not have P_6, the *Liber mahameleth* introduces it. In his 18th-century English version of the *Elements*, Robert Simson felt the same need and added as Corollary 1 to Proposition V.24 the following statement: *If the same hypothesis be made as in the proposition, the excess of the first and fifth shall be to the second as the excess of the third and sixth to the fourth. The demonstration of this is the same with that of the proposition if division be used instead of composition.*[°]

As P_6 was used to prove P_7, P_7' will serve to prove P_8, which is the analogue of P_7 for the additive case.

<u>Eighth premiss</u> (P_8).[††] Let a_1, a_2, a_3 be three numbers. Then

---

[†] Used in P_8, PE_1, A.151a, A.192a, B.241 (lacuna completed). Extended to three ratios in PE_1 and A.151a.

[°] See Heath's *Euclid*, II, p. 184; 'division' for ratios is synonymous with 'separation'.

[††] Used in A.152a, A.297 (implicitly).

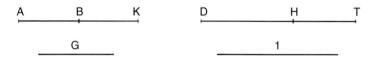

Fig. 11

$$\frac{a_1}{a_3} + \frac{a_2}{a_3} = \frac{a_1 + a_2}{a_3}.$$

Demonstration. Let (Fig. 11) the three numbers be $AB = a_1$, $BK = a_2$, thus $AK = a_1 + a_2$, $G = a_3$. Let further $\frac{AB}{G} = DH$, $\frac{BK}{G} = HT$; then $\frac{AB}{G} + \frac{BK}{G} = DH + HT = DT$. To prove that

$$DT = \frac{AK}{G}.$$

As in the proof of P_7, we have

$$\frac{AB}{G} = \frac{DH}{1}, \quad \frac{BK}{G} = \frac{HT}{1};$$

so we may apply P_7' with $a_4 = 1$, thus

$$\frac{AB + BK}{G} = \frac{DH + HT}{1}, \quad \text{that is,} \quad \frac{AK}{G} = DT.$$

NINTH PREMISS (P_9).* Converse of P_4. Let a_1, a_2, a_3 be three numbers; then

$$\frac{a_3}{\frac{a_1}{a_2}} = \frac{a_3 \cdot a_2}{a_1}.$$

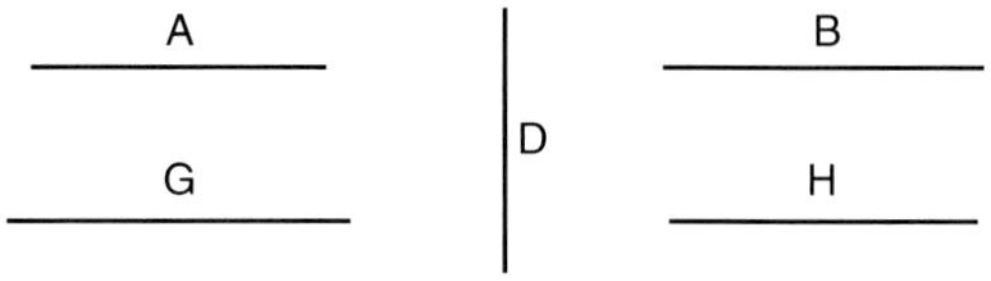

```
        A                          B
      ─────              │       ─────
                         │D
        G                          H
      ─────              │       ─────
```

Fig. 12

Demonstration. Let (Fig. 12) $A = a_1$, $B = a_2$, $D = a_3$, $\frac{A}{B} = G$, $\frac{D}{G} = H$; to prove that

$$H = \frac{D \cdot B}{A}.$$

The proof relies once again on *Elements* VII.17–19. Since $G \cdot B = A$ and $G \cdot H = D$, so $H : B = D : A$, whence $H \cdot A = D \cdot B$, which proves the assertion.

---

* Not used.

2. Identities from Book II of Euclid's *Elements*.

ELEMENTS II.1 (PE_1).‡ Let a, b be two numbers, and $b = b_1 + b_2 + \ldots + b_n$; then
$$a \cdot b = a \cdot b_1 + a \cdot b_2 + \ldots + a \cdot b_n.$$

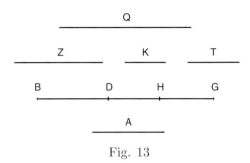

Fig. 13

Demonstration. Let (Fig. 13) $a = A$, $b = BG = BD + DH + HG$; to prove that $A \cdot BG = A \cdot BD + A \cdot DH + A \cdot HG$, or that $Q = Z + K + T$ if we put $A \cdot BG = Q$, $A \cdot BD = Z$, $A \cdot DH = K$, $A \cdot HG = T$.

From the first three of these last relations we infer, by *Elements* VII.17, that
$$\frac{BD}{BG} = \frac{Z}{Q} \quad \text{and} \quad \frac{DH}{BG} = \frac{K}{Q};$$
thus by $P'_7 = $ *Elements* V.24,
$$\frac{BD+DH}{BG} = \frac{BH}{BG} = \frac{Z+K}{Q}.$$
Similarly, from
$$\frac{BH}{BG} = \frac{Z+K}{Q} \quad \text{and} \quad \frac{HG}{BG} = \frac{T}{Q},$$
we find, by P'_7 again, that
$$\frac{BH+HG}{BG} = \frac{BG}{BG} = \frac{Z+K+T}{Q}.$$
Therefore, as required to prove, $Z + K + T = Q$.*

Remark. As seen in our summary of this chapter (p. 1145), the two coming propositions are particular cases of this theorem.

ELEMENTS II.2 (PE_2).° If $a = a_1 + a_2 + \ldots + a_n$, then

---

‡ Used in $PE_2 - PE_6$, A.32, A.45, A.123c, A.124c (interpolated reference), B.48, B.185c, B.232a, B.251a, B.316c (note 1832), B.327.

* The author has explicitly noted that the demonstrations of II.1–10, when applied to numbers, may require the use of *later* propositions from the *Elements*. This is the case here (and the only one). (See Translation, p. 598, note 79.)

° Used in $PE_4 - PE_6$, A.214, B.185c.

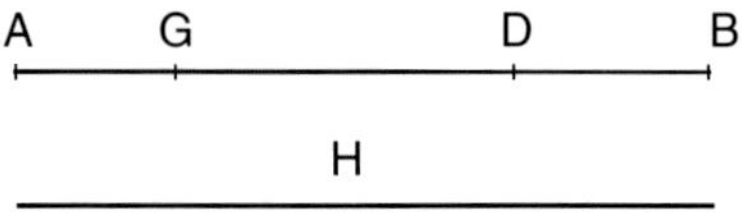

A   G   D   B

H

Fig. 14

$$a^2 = a \cdot a_1 + a \cdot a_2 + \ldots + a \cdot a_n.$$

Demonstration. Let (Fig. 14) $a = AB = AG + GD + DB$; to prove that
$$AB^2 = AB \cdot AG + AB \cdot GD + AB \cdot DB.$$

We provisionally put $AB = H$ and apply PE_1 to $H \cdot (AG + GD + DB)$.

<u>Elements</u> II.3 (PE_3).[†] If $a = a_1 + a_2$, then
$$a \cdot a_2 = a_2^2 + a_2 \cdot a_1.$$

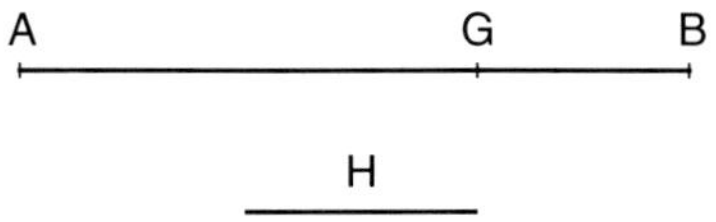

A   G   B

H

Fig. 15

Demonstration. Let (Fig. 15) $a = AB = AG + GB$; to prove that
$$AB \cdot GB = GB^2 + GB \cdot AG.$$

Put here $H = GB$ and apply PE_1 to $H \cdot AB$. Then
$$H \cdot AB = H \cdot AG + H \cdot GB = GB \cdot AG + GB^2,$$
whence the requirement since $H \cdot AB$ equals $AB \cdot GB$.

<u>Elements</u> II.4 (PE_4).[‖] If $a = a_1 + a_2$, then
$$a^2 = a_1^2 + a_2^2 + 2 \cdot a_1 \cdot a_2.$$

A   G   B

Fig. 16

Demonstration. Let (Fig. 16) $AB = AG + GB$; to prove that
$$AB^2 = AG^2 + GB^2 + 2 \cdot AG \cdot GB.$$

Applying PE_2 & PE_1, we find
$$AB^2 = AB \cdot AG + AB \cdot GB = \left(AG^2 + AG \cdot GB\right) + \left(AG \cdot GB + GB^2\right),$$
whence the assertion.

[†] Mentioned in B.351a (interpolated quotation).
[‖] Used in $PE_7 - PE_9$, A.289, A.311, B.344a, B.346, B.351a.

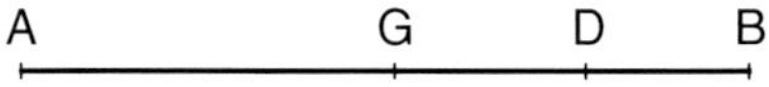

```
A           G    D   B
├───────────┼────┼───┤
```

Fig. 17

Elements II.5 (PE_5).‡ If $a = 2a' = a_1 + a_2$ $(a_1 > a_2)$, then
$$a'^2 = a_1 \cdot a_2 + (a' - a_2)^2.$$

Demonstration. Let (Fig. 17) AB be divided at G into equal, at D into unequal parts; to prove that
$$GB^2 = AD \cdot DB + GD^2.$$

We have successively

$$\begin{aligned}
GB^2 &= GB \cdot GD + GB \cdot DB & &(\text{by } PE_2) \\
&= GB \cdot GD + AG \cdot DB & &(AG = GB) \\
&= GD^2 + DB \cdot GD + AG \cdot DB & &(\text{by } PE_1) \\
&= GD^2 + AD \cdot DB.
\end{aligned}$$

Remark. This proposition is equivalent to
$$\left(\frac{a_1 + a_2}{2}\right)^2 = a_1 \cdot a_2 + \left(\frac{a_1 - a_2}{2}\right)^2,$$
which will be in constant use for solving quadratic problems. (The same holds for the next theorem, reduced to the same identity by writing $a + b = a_1$, $b = a_2$.)

Elements II.6 (PE_6).* If $a = 2a'$ and b are two numbers, then
$$(a' + b)^2 = (a + b) \cdot b + a'^2.$$

```
A        G        B            D
├────────┼────────┼────────────┤
```

Fig. 18

Demonstration. Let (Fig. 18) $AB = a$, G its mid-point, and $BD = b$; to prove that
$$GD^2 = AD \cdot BD + GB^2.$$

We have successively

$$\begin{aligned}
GD^2 &= GD \cdot GB + GD \cdot BD & &(\text{by } PE_2) \\
&= GB^2 + GB \cdot BD + GD \cdot BD & &(\text{by } PE_1) \\
&= GB^2 + AG \cdot BD + GD \cdot BD & &(AG = GB) \\
&= GB^2 + AD \cdot BD.
\end{aligned}$$

---

‡ Used in A.323*b*, B.30, B.185, B.252, B.346, B.348, and whenever an equation of the form $x^2 + q = px$ is solved geometrically.

* Used in A.214, A.298*b*, A.317*a*, A.320*a*, B.31, B.37, B.138, B.185*c*, B.324, B.345, B.349, and whenever an equation of the form $x^2 \pm px = q$ is solved geometrically.

Elements II.7 (PE_7).†† If $a = a_1 + a_2$, then
$$a^2 + a_2^2 = 2 \cdot a \cdot a_2 + a_1^2.$$

Remark. Since $a_1 = a - a_2$, this is used wherever we need a development of the form $(\alpha - \beta)^2$. See, for instance, A.44.

Fig. 19

Demonstration. Let (Fig. 19) $AB = a$ be divided at G; to prove that
$$AB^2 + GB^2 = 2 \cdot AB \cdot GB + AG^2.$$

We have successively

$$\begin{aligned} AB^2 + GB^2 &= \left(AG^2 + GB^2 + 2 \cdot AG \cdot GB\right) + GB^2 &&\text{(by } PE_4\text{)} \\ &= AG^2 + 2 \cdot GB^2 + 2 \cdot AG \cdot GB \\ &= AG^2 + 2 \cdot AB \cdot GB. \end{aligned}$$

Elements II.8 (PE_8).‖ If $a = a_1 + a_2$, then
$$(a + a_1)^2 = 4 \cdot a \cdot a_1 + a_2^2.$$

Fig. 20

Demonstration. Let (Fig. 20) $AB = a$ be divided at G and its part $a_1 = GB = BD$ be added to it; to prove that
$$AD^2 = 4 \cdot AB \cdot BD + AG^2.$$

We have first, by PE_4,
$$AD^2 = AB^2 + BD^2 + 2 \cdot AB \cdot BD.$$

Consider the first two terms on the right side. Since $BD = GB$ and using PE_7,
$$AB^2 + BD^2 = AB^2 + GB^2 = 2AB \cdot GB + AG^2 = 2AB \cdot BD + AG^2.$$
So
$$\begin{aligned} AD^2 &= 2 \cdot AB \cdot BD + AG^2 + 2 \cdot AB \cdot BD \\ &= 4 \cdot AB \cdot BD + AG^2. \end{aligned}$$

---

†† Used in PE_8, A.44, A.214, A.291, B.345.
‖ Not used.

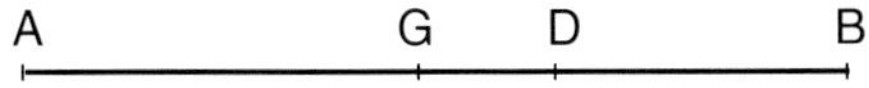

```
A            G   D            B
|------------|---|------------|
```

Fig. 21

ELEMENTS II.9 (PE_9).° If a is a number with $a = 2a' = a_1 + a_2$ $(a_1 > a_2)$, then
$$a_1^2 + a_2^2 = 2 \cdot a'^2 + 2(a' - a_2)^2.$$

Demonstration. Let (Fig. 21) $AB = a$ be divided into equal parts at G and into unequal parts at D; to prove that
$$AD^2 + DB^2 = 2 \cdot AG^2 + 2 \cdot GD^2.$$

Evidently, since $AG = GB$,
$$AB^2 = 2\,AG^2 + 2\,GB^2.$$

Now, by PE_4, $GB^2 = GD^2 + DB^2 + 2\,GD \cdot DB$, so that
$$2\,GB^2 = 2\,GD^2 + 2\,DB^2 + 4\,GD \cdot DB,$$
and therefore
$$AB^2 = 2\,AG^2 + 2\,GD^2 + 2\,DB^2 + 4\,GD \cdot DB.$$

On the other hand, by PE_4 again,
$$AB^2 = AD^2 + DB^2 + 2\,AD \cdot DB.$$

Equate now the last two results:
$$AD^2 + DB^2 + 2\,AD \cdot DB = 2\,AG^2 + 2\,GD^2 + 2\,DB^2 + 4\,GD \cdot DB. \quad (*)$$

Let us consider the last two terms, $2\,DB^2 + 4\,GD \cdot DB$. Since
$$2\,DB^2 + 2\,GD \cdot DB = 2\,(DB + GD)\,DB = 2\,GB \cdot DB = 2\,AG \cdot DB,$$
they become
$$2\,DB^2 + 4\,GD \cdot DB = 2\,AG \cdot DB + 2\,GD \cdot DB = 2\,AD \cdot DB,$$
and the relation $(*)$ will be
$$AD^2 + DB^2 + 2\,AD \cdot DB = 2\,AG^2 + 2\,GD^2 + 2\,AD \cdot DB,$$
whence, as asserted,
$$AD^2 + DB^2 = 2\,AG^2 + 2\,GD^2.$$

ELEMENTS II.10 (PE_{10}).* If $a = 2a'$ and b are two numbers, then
$$(a+b)^2 + b^2 = 2\left[a'^2 + (a'+b)^2\right].$$

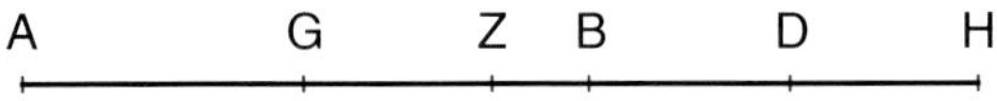

```
A        G   Z B     D    H
|--------|---|-|-----|----|
```

Fig. 22

° Used in PE_{10}.
* Not used.

Demonstration. Let (Fig. 22) $AB = a$, G its mid-point, and $BD = b$; to prove that
$$AD^2 + BD^2 = 2 \cdot [AG^2 + GD^2].$$
Put $DH = GZ = BD$; then
$$AZ = AG + GZ = \tfrac{1}{2} \cdot AB + \tfrac{1}{2} \cdot BH = \tfrac{1}{2} \cdot AH.$$
Thus AH is divided at Z into equal, at D into unequal parts. By PE_9, we have then $AD^2 + DH^2 = 2 \cdot AZ^2 + 2 \cdot ZD^2$.

But, first, $AZ = GD$ (since $AG = GB$ and $GZ = BD$), and so
$$AD^2 + DH^2 = 2\,GD^2 + 2\,ZD^2.$$
Next, $ZD = AG$: since $GZ = BD$, then 'let ZB be (added in) common', whence $GZ + ZB = BD + ZB$, thus $GB = ZD$, and $GB = AG$); so
$$AD^2 + DH^2 = 2\,GD^2 + 2\,AG^2.$$
Finally, since $DH = BD$,
$$AD^2 + BD^2 = 2\,GD^2 + 2\,AG^2.$$

Remark. The demonstration and figure correspond to the situation $a' > b$, that is, $GB > GZ$, whereby Z falls between G and B. The reasoning would remain the same if $b > a'$ (Z between B and D), the only difference being that, in order to infer that $ZD = AG = GB$, we would need to consider $GZ - ZB = BD - ZB$; the text would then have 'let the common (segment) ZB be removed'.

Chapter A–III:
Multiplication of integers

Summary

As the author observes, beginning with addition would be more appropriate, but since the Arabs 'mostly' begin with multiplication, so will he. (This does not change much: addition of integers is in any event not taught in the *Liber mahameleth*.)

1. Various cases of multiplication.

The basic quantities considered for arithmetical operations in the *Liber mahameleth* are of three kinds: integer, fraction, fraction of a fraction — thus

$$N, \quad \frac{k}{l}, \quad \frac{r}{s}\frac{1}{l},$$

where all letters represent natural numbers. Considering all possible multiplications involving in each of the two factors not only one kind but also an additive combination of them, we find twenty-eight cases, listed by the author.[†] But the present chapter, being concerned only with the multiplication of integers, will treat in Sections 2 to 5 multiplication of the various types of integers seen in the second section of A–I: units, tens, hundreds, thousands, and their associations, first without then with thousands repeated.

2. Multiplication of digits among themselves.

We are first taught multiplication of the numbers from 1 to 10, all in words. Expressed in figures, this takes the form of a triangular table, thus without the products below the diagonal; commutativity is therefore taken for granted (*Elements* VII.16). A set of rules follows whereby a connection between some of these products may be established, and thus some computed by means of others. In any event it is said that we are supposed to learn the multiplication table by heart, and this will from now on be considered as done.

3. Multiplication of units of two orders.

The first direct application of this is computation of the products of two numbers each with just the first figure significant, that is, different

---

[†] For a 'fraction and a fraction of a fraction', thus an expression of the form

$$\frac{k}{l} + \frac{r}{s}\frac{1}{l},$$

we shall use the term 'compound fraction'.

from zero. But we then need to know to which order the product belongs, and this will be determined by calculating its 'note' (*nota*, Arabic *uss*).*

A. *Note of a number*.

Suppose we wish to know the order occupied by a given number written in our numerical symbols, say N with the decimal representation $a_k\, a_{k-1} \cdots a_2\, a_1\, a_0$ (thus $N = a_k 10^k + a_{k-1} 10^{k-1} + \ldots + a_2 10^2 + a_1 10 + a_0$ and a_i the digits of N, with $a_k \neq 0$). Since the order of N is indicated by the place of its highest digit, we need only consider $a_k \cdot 10^k$, or even 10^k. For $k = 0$, it will be the first order, for $k \neq 0$ it will be the $(k+1)$-th order. The note of N, which we shall designate by $\nu(N)$, will thus be given by

$$\nu(N) = \nu\big(a_k \cdot 10^k\big) = \nu(10^k) = k + 1,$$

while a_k itself will give the rank (*locus*, Arabic *manzila*) of N among the nine numbers of the same order.

Now in the *Liber mahameleth* numbers are all expressed in words; which, as seen in A–I, requires the word 'thousand' to be repeated. Thus, in order to determine the order from the verbal expression, we are to consider as the highest element of N not $a_k \cdot 10^k$ but rather $a \cdot 10^{3m}$, where m is the number of times 'thousand' is repeated and a the integer preceding all these repetitions (with $1 \leq a \leq 999$). Two problems then arise in connection with the note of N when expressed in words.

(*i*) Find the note of N from its verbal expression. We then have

$$\nu(N) = \nu\big(a \cdot 10^{3m}\big) = 3m + t,$$

with $t = 1$ if a belongs to the units, $t = 2$ if to the tens and $t = 3$ if to the hundreds. This is illustrated by, respectively, A.1 & A.3; A.1′ & A.4; A.1″ & A.5. The rule itself will be stated, and demonstrated, at the end (after A.5).

(*ii*) Inversely, when a note is given, find the corresponding order or, more precisely, the verbal expression of its first term (its 'limit'). Since

$$\frac{\nu(N)}{3} = m + \frac{t}{3},$$

m will be the number of thousands repeated while the remainder t will indicate whether the limit belongs to the units (case $t = 1$) or to the tens (case $t = 2$). For the case $t = 0$, thus when $\nu(N)$ is divisible by 3, we must proceed otherwise. We shall consider that $m - 1$ is the number of repetitions and that $t = 3$ indicates the hundreds. See, for these three cases, A.2′ & A.7, A.2 & A.6, A.2″ & A.8. The rule is given after A.8.

B. *Multiplication of two single-figure numbers using the note*.

Let $N_1 = a_1 \cdot 10^k$ and $N_2 = a_2 \cdot 10^l$, with a_1 and a_2 digits. First form $a_1 \cdot a_2 = a$, the result of which is known from the multiplication table, then compute the note of the whole product. Now since the product is $a \cdot 10^{k+l}$, the note will be, if a is a digit,

---

* A word like 'index' would perhaps have been better were it not already in common mathematical use in connection with roots.

$$\nu(N_1 \cdot N_2) = k + l + 1 = (k+1) + (l+1) - 1 = \nu(N_1) + \nu(N_2) - 1.$$

If a is not a digit but a two-figure number, $\nu(N)$ will indicate the place of its digit. This rule is explained before A.9 and proved after A.13. Illustrative examples are: units by tens (A.9), tens by tens (A.10), tens by hundreds (A.11), and finally products of numbers with repeated thousands (A.12–13).

C. *Multiplication of two single-figure numbers using rules*.

Instead of each time calculating the note by the above rule i, we may just learn to which order belongs the product in question. We are thus taught which rules to follow for various elementary cases, namely: units by tens (A.14), by hundreds (A.15), by thousands (A.16); then tens by tens (A.17), by hundreds (A.18), by thousands (A.19); finally hundreds by hundreds (A.20) and by thousands (A.21).

4. Multiplication of composite numbers below one thousand.

The multiplication of one two-figure number by another is, as inferred from Premiss PE_1, reducible to calculating and then adding four products of the types seen in Section 3 (A.22–25). This is also the case of hundreds when for both the last figure is zero (A.26–27). So far we have been able to work out the result with our hands since they can express four-figure numbers. The case of hundreds comprising each three significant figures is different, and we have to write out the computation, as A.28 asserts. We shall then have: the two factors in two rows, with the intermediate results below and their corresponding places aligned vertically for the subsequent addition.

Simplified computations are possible when one or both factors happen to be just below some multiple of a limit: the factor in question is then replaced by the round number minus the difference (A.23 *seqq*.). Since we are then to multiply differences, the rules of signs have been introduced at the beginning of this section.

5. Multiplication of numbers with repeated thousands.

The examples are in disorder. We are taught the multiplication of limits of thousands (A.30), of units of thousands or ten thousands (A.33–36), of higher thousands (A.31, A.37), all with just one significant figure. Disregarding initially the repetitions, we are left with the multiplication of two single-figure numbers, an operation known from Sections 2 & 3, to the result of which the disregarded thousands are then appended. A.32, A.38, A.39, A.41 involve composite numbers, A.40 is a multiplication of two four-figure numbers. In A.32 the basic principle is demonstrated: each of the two factors is decomposed into a sum and each part of the first factor is multiplied by each part of the other and the results are summed up. This is the distributive law taught in Premiss PE_1.

As before, certain problems can be more conveniently solved using subtraction (A.44, A.45, both with a demonstration). As a preliminary, we are taught how to subtract one number from another when repeated thousands are involved (A.42, A.43).

6. Taking fractions of thousands repeated.

Although in keeping with the subject of repeated thousands, the last section (A.46–51) is in fact a preliminary to the next chapter on division. For it will be applied from A.55 on. The principle in taking fractions of repetitions is always the same: If we are to take the fraction $\frac{p}{q}$ ($p < q \leq 999$) of $a \cdot 1000^k$, where a is the integer preceding the repeated 'thousand', we shall consider

$$\left(\frac{p}{q} \cdot a\right) \cdot 1000^k, \quad \text{or} \quad \left(\frac{p}{q} \cdot a \cdot 1000\right) \cdot 1000^{k-1},$$

according to whether $q < a$ or $q > a$, and so take the fraction of an integer and append the repetitions left out. In all examples q does not divide a, which means that a fraction must also be taken of the remaining repetitions.

---

1. Various cases of multiplication.

The basic quantities considered in the *Liber mahameleth* belong to one of three types:

— integer (*integer* or *integrum*, Arabic ṣaḥīḥ), symbolized in the enumeration of cases below by N;

— fraction $\frac{k}{l}$ (*fractio*, Arabic kasr), written F below;

— fraction of a fraction $\frac{r}{s}\frac{1}{l}$ (*fractio fractionis*, Arabic kasr kasr or, vocalized, kasru kasrin), our $F'F$.

Considering now arithmetical operations applied to expressions formed by one or more of these three terms, including one kind by itself, we then come to the twenty-eight fundamental cases mentioned by the author and alluded to elsewhere (Translation, p. 701, note 483). We have listed them below and indicated the examples treated in the *Liber mahameleth*, with the operation involved (m for multiplication, a for addition, c for conversion of fractions, s for subtraction, d for division and d' for denomination, that is, when the divisor is the larger term).

• Single by same single:

I. $N \circ N$ (subject of this chapter for multiplication, of the next for division).

II. $F \circ F$: A.99–101 (m), A.108–111 (c), A.145–146 (a), A.187–188 (s), A.230–231 (d'), A.236–238 (d).

III. $F'F \circ F'F$: A.104–105 (m), A.117–118 & A.120 (c), A.149 (a), A.196 (s).

• Single by different single:

IV. $N \circ F$: A.90–92 (m), A.215–216 & A.225–226 & A.246–248 (d).

V. $N \circ F'F$: A.93–94 (m), A.217–219 & A.249–250 (d).

VI. $F \circ F'F$: A.102–103 (m), A.115–116 (c), A.239 (d, d').

• Single by pair:

VII. $N \circ (N + F)$: A.221–223 & A.227–228 & A.232–233 (d'), A.254 (d).
VIII. $N \circ (F + F'F)$: A.96 & A.98 (m), A.220 & A.251–252 (d).
IX. $N \circ (N + F'F)$: A.255 (d).
X. $F \circ (N + F)$: A.123 (m), A.193 (s), A.257 (d).
XI. $F \circ (N + F'F)$.
XII. $F \circ (F + F'F)$: A.106 (m), A.112–114 (c), A.147 (a), A.189 (s), A.240 (d, d').
XIII. $F'F \circ (N + F)$: A.197 (s), A.258 (d, d').
XIV. $F'F \circ (N + F'F)$.
XV. $F'F \circ (F + F'F)$: A.150 (a).

- Single by triad:

XVI. $N \circ (N + F + F'F)$: A.224 & A.229 (d'), A.256 (d).
XVII. $F \circ (N + F + F'F)$.
XVIII. $F'F \circ (N + F + F'F)$.

- Pair by pair:

XIX. $(N+F) \circ (N+F)$: A.124–125 (m), A.194 (s), A.234–235 (d'), A.260–261 (d).
XX. $(N + F) \circ (F + F'F)$: A.195 (s), A.259 (d).
XXI. $(N + F) \circ (N + F'F)$: A.263 (d).
XXII. $(N + F'F) \circ (N + F'F)$.[‡]
XXIII. $(N + F'F) \circ (F + F'F)$.
XXIV. $(F+F'F) \circ (F+F'F)$: A.126–127 (m), A.148 (a), A.190 (s), A.241 (d).

- Pair by triad:

XXV. $(N + F) \circ (N + F + F'F)$: A.128 ($m$), A.262 ($d$).
XXVI. $(F + F'F) \circ (N + F + F'F)$: A.137 ($m$), A.191 ($s$).
XXVII. $(N + F'F) \circ (N + F + F'F)$.

- Triad by triad:

XXVIII. $(N + F + F'F) \circ (N + F + F'F)$: A.129 ($m$), A.198 ($s$).

Chapter A–III being exclusively about the multiplication of integers, it will concern case I only. As mentioned in our summary, we shall start by learning how to multiply digits, then higher-order units $a \cdot 10^t$, either using the note or by rules, before going on to composite numbers, of the first three orders to begin with and then from one thousand on.

2. Multiplication of digits among themselves.
A. *The multiplication table*.

[‡] XXII & XXIII follow XXIV in the enumeration.

The table is in words, but does not repeat the products obtained by commuting the factors. It thus corresponds to the following triangular shape, where we have put the multiplicand on the right and the multipliers above:

1	2	3	4	5	6	7	8	9	10	
1	2	3	4	5	6	7	8	9	10	1
	4	6	8	10	12	14	16	18	20	2
		9	12	15	18	21	24	27	30	3
			16	20	24	28	32	36	40	4
				25	30	35	40	45	50	5
					36	42	48	54	60	6
						49	56	63	70	7
							64	72	80	8
								81	90	9
									100	10

As the author notes, it is necessary to know this table before proceeding further. The subsequent rules will help calculate or recalculate, if need be, some of its elements.

B. *Rules*.

Let n, m, n_i, m_i represent digits.

(i) $n^2 = (n-1)(n+1) + 1$,

or, generally,

(i') $n^2 = (n-2)(n+2) + 2^2 = \ldots = [n-(n-1)] \cdot [n+(n-1)] + (n-1)^2$.
Remark. This is just the identity $n^2 - t^2 = (n+t)(n-t)$. It may simplify the computations for larger squares, as e.g. $8^2 = (8-2)(8+2) + 2^2 = 6 \cdot 10 + 4$.

(ii) $n^2 = m_1 \cdot m_2$ if $m_1 : n = n : m_2$.
Remark. Thus n is a mean proportional between m_1 and m_2. This corresponds to *Elements* II.14.

(iii) $n^2 = m_1^2 + m_2^2 + 2 \cdot m_1 \cdot m_2$ if $n = m_1 + m_2$.
Remark. This is PE_4.

(iv) $n^2 = 10 \cdot n - (10-n) \cdot n$.
Remark. With this, as with i', some squares in the lower part of the diagonal ($n > 5$) may be found by means of the products in the upper part.

(v) $n_1 \cdot n_2 = 10 \cdot n_1 - (10 - n_2) \cdot n_1$.
Remark. iv is a particular case of this.

(vi) $n_1 \cdot n_2 = n_1 \cdot (m_1 + m_2 + \ldots + m_k)$ if $n_2 = \sum_1^k m_i$.

Remark. This is PE_1.

3. Multiplication of units of two orders.

A. *Note of a number*.

As seen above (p. 1160), the note $\nu(N)$ of a number $N = \sum_{0}^{k} a_i 10^i$ is given by $\nu(N) = k + 1$ when N is expressed in numerical symbols. Let now N be expressed verbally, by means of m repetitions of 'thousand' and let a be the integer preceding these repetitions (thus $a < 10^3$). This gives rise to the two fundamental problems mentioned in our summary.

(i) Find the note of a number. We have then, as stated and demonstrated after A.5,
$$\nu(N) = 3m + t,$$
with $t = 1$ if only a single digit precedes the repetitions (A.1, A.3), or $t = 2$ if a belongs to the tens (A.1′, A.4), or finally $t = 3$ if a belongs to the hundreds (A.1″, A.5).

Demonstration (follows A.5). Each addition of 'thousand' occurs after a group of three consecutive places occupied by units, tens and hundreds respectively. All these places will be counted when the number of repetitions is multiplied by 3. It thus remains only to add the order of the part which precedes the repetitions, and which has at most three figures.

(A.1) MSS *A, B, C*. Given $N = 1\,000\,000\,000\,000$, find $\nu(N)$.
This is a unit with four repetitions; so $t = 1$, $m = 4$, and
$$\nu(N) = 3 \cdot 4 + 1 = 13 \ \left(= \nu\!\left(1 \cdot 10^{3 \cdot 4}\right)\right).$$

(A.1′) Given $N = 10\,000\,000\,000\,000$, find $\nu(N)$.
This is a ten with four repetitions, so $t = 2$, $m = 4$, and
$$\nu(N) = 3 \cdot 4 + 2 = 14 \ \left(= \nu\!\left(10 \cdot 10^{3 \cdot 4}\right)\right).$$

(A.1″) Given $N = 100\,000\,000\,000\,000$, find $\nu(N)$.
This is a hundred with four repetitions, so $t = 3$, $m = 4$, and
$$\nu(N) = 3 \cdot 4 + 3 = 15 \ \left(= \nu\!\left(100 \cdot 10^{3 \cdot 4}\right)\right).$$

(ii) Find the order unit when the note ν is known. As explained after A.8 (and seen in our summary), we shall have
$$\frac{\nu(N)}{3} = m + \frac{t}{3} \quad (t = 1,\ 2), \qquad \frac{\nu(N)}{3} = m - 1 + \frac{t}{3} \quad (t = 3),$$

with m (or $m-1$) indicating the repetitions and t the order of the number preceding them.

(A.2) MSS $\mathcal{A}$, $\mathcal{B}$, $\mathcal{C}$. Given $\nu(N) = 14$, find N.

Dividing 14 by 3 gives $m = 4$, $t = 2$, that is, we shall have 10 followed by four repetitions; so $N = 10\,000\,000\,000\,000$.

(A.2′) Given $\nu(N) = 13$, find N.

Dividing 13 by 3 gives $m = 4$, $t = 1$, that is, we shall have 1 followed by four repetitions; so $N = 1\,000\,000\,000\,000$.

(A.2″) Given $\nu(N) = 15$, find N.

Dividing 15 by 3 gives $m = 5$, $t = 0$, or $m = 4$ and $t = 3$, that is, we shall have 100 followed by four repetitions; so $N = 100\,000\,000\,000\,000$.

— **Further examples of case i.**

(A.3) MSS $\mathcal{B}$, $\mathcal{C}$ only. Given $N = 3\,000\,000\,000\,000\,000$, find $\nu(N)$.
Since $m = 5$, $t = 1$ (digit 3 irrelevant), so
$$\nu(N) = 3 \cdot 5 + 1 = 16 \; \left(= \nu\!\left(1 \cdot 10^{3 \cdot 5}\right)\right).$$

(A.4) MSS $\mathcal{B}$, $\mathcal{C}$. Given $N = 10\,000\,000\,000\,000\,000\,000\,000\,000$, find $\nu(N)$.
Since $m = 8$, $t = 2$, so
$$\nu(N) = 3 \cdot 8 + 2 = 26 \; \left(= \nu\!\left(10 \cdot 10^{3 \cdot 8}\right)\right).$$

(A.5) MSS $\mathcal{B}$, $\mathcal{C}$. Given $N = 100\,000\,000\,000\,000\,000\,000$, find $\nu(N)$.
Since $m = 6$, $t = 3$, so
$$\nu(N) = 3 \cdot 6 + 3 = 21 \; \left(= \nu\!\left(100 \cdot 10^{3 \cdot 6}\right)\right).$$

— **Further examples of case ii.**

(A.6) MSS $\mathcal{A}$, $\mathcal{B}$, $\mathcal{C}$. Given $\nu(N) = 11$, find N.
Since $11 = 3 \cdot 3 + 2$, so $m = 3$, $t = 2$ and
$$N = 10\,000\,000\,000,$$
or with some other of the tens.

(A.7) MSS $\mathcal{A}$, $\mathcal{B}$, $\mathcal{C}$. Given $\nu(N) = 13$, find N. Same as A.2′.
Since $13 = 3 \cdot 4 + 1$, so $m = 4$, $t = 1$ and
$$N = 1\,000\,000\,000\,000,$$
or with some other of the digits.

(A.8) MSS $\mathcal{A}$, $\mathcal{B}$, $\mathcal{C}$. Given $\nu(N) = 18$, find N.
Since $18 = 5 \cdot 3 + 3$, so $m = 5$, $t = 3$ and
$$N = 100\,000\,000\,000\,000\,000,$$
or with some other of the hundreds.

B. *Multiplication of two single-figure numbers using the note.*

We have already learnt the multiplication table for digits and, after that, how to calculate the note of a verbalized number. We shall now learn how to compute the product of two numbers of which (if they are written) only the first digit is not zero. In other words, they are of the form $N_1 = a_1 \cdot 10^k$ and $N_2 = a_2 \cdot 10^l$, with a_1 and a_2 digits. Since we know from the table the product $a_1 \cdot a_2$, we need only compute the note of the product, for this will tell us which decimal place the units of the product are to occupy.

The rule is given at the outset by the author. Consider the ranks of the numbers in their order (thus a_1 and a_2), and multiply them; this gives $a = a_1 \cdot a_2$ which will be, as we know from the table, a number with at most two figures. Next, compute $\nu(N_1)$ and $\nu(N_2)$ and form $\nu(N_1)+\nu(N_2)-1$, which is the note of the product; this will give the place of the digit of a, the next highest place being for its tens, if any. This enables us to express the result. (Remember that so far none of these numbers is supposed to be written down.)

— **Units by tens.**

(A.9) MSS $\mathcal{A}, \mathcal{B}, \mathcal{C}$. Multiply 3 by 40.

By the table, $3 \cdot 4 = 12$. Since $\nu(3) = 1$, $\nu(40) = 2$, then $\nu(3 \cdot 40) = 2$, which indicates tens. Therefore
$$3 \cdot 40 = 12\,0.$$

— **Tens by tens.**

(A.10) MSS $\mathcal{A}, \mathcal{B}, \mathcal{C}$. Multiply 30 by 50.

By the table, $3 \cdot 5 = 15$. Since $\nu(30) = 2$, $\nu(50) = 2$, then $\nu(30 \cdot 50) = 3$, which indicates hundreds. Therefore
$$30 \cdot 50 = 15\,00.$$

— **Tens by hundreds.**

(A.11) MSS $\mathcal{A}, \mathcal{B}, \mathcal{C}$. Multiply 20 by 500.

By the table, $2 \cdot 5 = 10$. Since $\nu(20) = 2$, $\nu(500) = 3$, then $\nu(20 \cdot 500) = 4$, which indicates thousands. Therefore
$$20 \cdot 500 = 10\,000.$$

— **Numbers with repetitions.**

(A.12) MSS $\mathcal{A}, \mathcal{B}, \mathcal{C}$. Multiply N_1 by N_2 with (when written) $N_1 = 100\,000\,000\,000$, $N_2 = 5\,000\,000\,000\,000$.

By the table, $1 \cdot 5 = 5$. Since $\nu(N_1) = 3 \cdot 3 + 3 = 12$, $\nu(N_2) = 4 \cdot 3 + 1 = 13$, then $\nu(N_1 \cdot N_2) = 24$, whence $m = 7$ and $t = 3$. Since this indicates hundreds with seven repetitions,
$$N_1 \cdot N_2 = 500\,000\,000\,000\,000\,000\,000\,000$$
(or $500 \cdot 1000^7$).

(A.13) MSS $\mathcal{A}, \mathcal{B}, \mathcal{C}$. Multiply N_1 by N_2 with $N_1 = 8\,000\,000\,000\,000$, $N_2 = 400\,000\,000\,000\,000\,000\,000\,000$.

By the table, $8 \cdot 4 = 32$. Since $\nu(N_1) = 4 \cdot 3 + 1 = 13$, $\nu(N_2) = 7 \cdot 3 + 3 = 24$, then $\nu(N_1 \cdot N_2) = 36$, whence $m = 11$ and $t = 3$. Since this indicates hundreds with eleven repetitions,

$$N_1 \cdot N_2 = 3\,200\,000\,000\,000\,000\,000\,000\,000\,000\,000\,000\,000$$

(thus $3200 \cdot 1000^{11}$, or, to be closer to the expression of the text, $3 \cdot 1000^{12} + 200 \cdot 1000^{11}$).

This section on the note concludes with a long demonstration showing that $\nu(N_1 \cdot N_2) = \nu(N_1) + \nu(N_2) - 1$. It first proves (part a below) the case $a_1 = a_2 = 1$; that is, N_1 and N_2 are the so-called limits of their orders, and the author will incidentally show that multiplying two limits always produces a limit, thus the first element of a higher order. Then he considers (b below) the case $a_1, a_2 \neq 1$, and finally (c below) the particular case where $N_1 = N_2 = N$. There is much interpolating.

(**a**) Let the consecutive powers of 10, starting with $A = 1$, be $A, B, G, D, H, Z, K, T, Q, L$. Since this is a geometric progression, we have the following continued proportion:

$$\frac{A}{B} = \frac{B}{G} = \frac{G}{D} = \ldots = \frac{Q}{L}.$$

According to *Elements* IX.11, if in such a sequence of terms in continued proportion one of them is divided by some smaller one, the quotient will be some term of the sequence. Inversely, any product of two terms of the proportion will be some subsequent term of this sequence, thus also the first term of some order. Let $D \cdot K$ produce M, which is thus some subsequent term in the continued proportion. But $D \cdot K = M$ can be written as $M : K = D : 1$, with all four terms belonging to the above sequence since $1 = A$. From *Elements* VIII.8 we know that as many terms of the continued proportion will fall between M and K as between D and A. Now to each of these terms is assigned a note, and these notes themselves form an arithmetical progression with common difference 1. Since M is as far from K as D is from A, then, according to Premiss P_1, the sum of the notes of M and A must equal the sum of the notes of K and D, that is,

$$\nu(M) + \nu(A) = \nu(K) + \nu(D).$$

Therefore

$$\nu(M) = \nu(K) + \nu(D) - \nu(A) = \nu(K) + \nu(D) - 1,$$

or, in our writing,

$$\nu(N_1 \cdot N_2) = \nu(N_1) + \nu(N_2) - 1,$$

which proves the assertion when N_1 and N_2 are the initial terms in their orders.

(**b**) Assume now that D and K, our N_1 and N_2, are no longer first in their orders, so that $N_1 = a_1 \cdot 10^k$, $N_2 = a_2 \cdot 10^l$ with $a_1, a_2 \neq 1$. Then (by P_3)

$$N_1 \cdot N_2 = a_1 \cdot 10^k \cdot a_2 \cdot 10^l = a_1 \cdot a_2 \left(10^k \cdot 10^l\right)$$

and the units of the product will occupy the place determined in part *a*, the tens the next highest.

(**c**) Assume now $N_1 = N_2 = N$. In that case we shall have
$$\nu(N^2) = 2 \cdot \nu(N) - 1.$$

A long interpolation follows. The justification of $\nu(N_1 \cdot N_2) = \nu(N_1) + \nu(N_2) - 1$ by 'another author who puts it more succinctly' is just begging the question. We consider, as before, the sequence of the powers of 10. Since $N_1 \cdot N_2 : N_2 = N_1 : 1$, we have (by *Elements* VII.19) $N_1 \cdot N_2 \cdot 1 = N_1 \cdot N_2$, so $\nu(N_1 \cdot N_2 \cdot 1) = \nu(N_1 \cdot N_2)$. From this *identity* the interpolator infers that $\nu(N_1 \cdot N_2) + \nu(1) = \nu(N_1) + \nu(N_2)$, and, since $\nu(1) = 1$, this concludes his proof. A later reader noted a flaw in the proof, namely that transition from the note of the product to the sum of the notes had not been proved.

In the same vein, the interpolator observes that if the notes were to form an arithmetical progression with constant difference t and first term t as well ($t \neq 1$), we would have (in accordance with P_1)
$$\nu(N_1 \cdot N_2) = \nu(N_1) + \nu(N_2) - t, \quad \nu(N^2) = 2 \cdot \nu(N) - t.$$

C. *Multiplication of two single-figure numbers using rules*.

We now find the explanation, each time with a rule and an example, of multiplication of a digit by a unit of higher order (rules *i–iii*, A.14–16; multiplication of digits has already been taught and is henceforth considered as known); of tens by tens or by a unit of higher order (rules *iv–vi*, A.17–19); of hundreds by hundreds or by a unit of higher order (rules *vii– viii*, A.20–21). Rules proper are *i–ii*, *iv–v*, *vii* since the orders to which belong multiplicand and multiplier are specified; whereas in the others, the multiplier is defined to a repetition.*

— **Units by tens.**

(*i*) When one of the units is multiplied by one of the tens, the units of the result will become tens and the tens, hundreds.

(**A.14**) MSS $\mathcal{A}, \mathcal{B}, \mathcal{C}$. Multiply 7 by 70.

Since, by the table, $7 \cdot 7 = 49$, so
$$7 \cdot 70 = 490.$$

Remark. Rule *i* sees the first occurrence of *figura*, the usual word for the written symbols of numerals (Translation, p. 620).

— **Units by hundreds.**

(*ii*) When one of the units is multiplied by one of the hundreds, the units of the result will become hundreds and the tens, thousands.

(**A.15**) MSS $\mathcal{A}, \mathcal{B}, \mathcal{C}$. Multiply 7 by 300.

Since, by the table, $7 \cdot 3 = 21$, so
$$7 \cdot 300 = 2100.$$

---

* The rules proper will be repeated by an interpolator before A.22.

— **Units by thousands.**

(*iii*) When one of the units is multiplied by one of the thousands, 'thousand' however often repeated, the units of the result will be in the highest place of the multiplier, the articles in the next higher.

(**A.16**) MSS $\mathcal{A}, \mathcal{B}, \mathcal{C}$. Multiply 6 by 30 000.

Since, by the table, $6 \cdot 3 = 18$, put 8 in the same place as 3. So
$$6 \cdot 30\,000 = 180\,000.$$

— **Tens by tens.**

(*iv*) When one of the tens is multiplied by itself or another of the tens, the units of the result will become hundreds and the tens, thousands.

(**A.17**) MSS $\mathcal{A}, \mathcal{B}, \mathcal{C}$. Multiply 30 by 70.

Since, by the table, $3 \cdot 7 = 21$, 1 will be a hundred, so
$$30 \cdot 70 = 2100.$$

— **Tens by hundreds.**

(*v*) When one of the tens is multiplied by one of the hundreds, the units of the result will become thousands and the tens, ten thousands.

(**A.18**) MSS $\mathcal{A}, \mathcal{B}, \mathcal{C}$. Multiply 30 by 500.

Since, by the table, $3 \cdot 5 = 15$,
$$30 \cdot 500 = 15\,000.$$

— **Tens by thousands, repeated or not.**

(*vi*) When one of the tens is multiplied by one of the thousands, 'thousand' however often repeated, the units of the result will be next to the highest place of the multiplier and the tens in the next higher.

(**A.19**) MSS $\mathcal{A}, \mathcal{B}, \mathcal{C}$. Multiply 30 by 4000.

Since, by the table, $3 \cdot 4 = 12$, so
$$30 \cdot 4000 = 120\,000.$$

— **Hundreds by hundreds.**

(*vii*) When one of the hundreds is multiplied by one of the hundreds, the units of the result will become ten thousands and the tens, hundred thousands.

(**A.20**) MSS $\mathcal{A}, \mathcal{B}, \mathcal{C}$. Multiply 300 by 500.

Since, by the table, $3 \cdot 5 = 15$, so
$$300 \cdot 500 = 150\,000.$$

— **Hundreds by thousands, repeated or not.**

(*viii*) When one of the hundreds is multiplied by one of the thousands, 'thousand' however often repeated, the units of the result will be two places after the highest place of the multiplier and the tens in the next higher.

(**A.21**) MSS $\mathcal{A}, \mathcal{B}, \mathcal{C}$. Multiply 200 by 5 000 000.

Since, by the table, $2 \cdot 5 = 10$, so

$$200 \cdot 5\,000\,000 = 1\,000\,000\,000.$$

4. Multiplication of composite numbers below one thousand.

We find as a preliminary three further rules, namely the rules of signs. Two of them will be needed to simplify some multiplications by expressing the factors, or one of them, as a difference (A.23–25, A.27, A.29, A.44–45). These rules are:

(ix) Additive by additive gives additive.

(x) Additive by subtractive gives subtractive, subtractive by additive gives subtractive.

(xi) Subtractive by subtractive gives additive.

Remark. That these rules should receive so little attention from the author is surprising: it looks as if they have been introduced only to facilitate some multiplications in a particular group of problems.

We then come to multiplication of numbers with two digits (A.22–25), then three (A.26–29, A.26–27 reducible to the former case). Decomposition of the numbers and distributivity (PE_1) allow us to reduce these problems to the cases already seen (multiplication of two-digit numbers to multiplications of one-digit numbers, multiplication of three-digit numbers to the cases with one or two). In A.22–27, the number of significant digits in the result is at most four, and they are consecutive; so we can still perform the computations without writing the numbers, since finger-reckoning enables us to signify one two-digit number in each hand. The situation changes with A.28.

(**A.22**) MSS $\mathcal{A}, \mathcal{B}, \mathcal{C}$. Multiply 13 by 14.

(**a**) By decomposition and distributivity,
$$13 \cdot 14 = 10 \cdot 10 + 3 \cdot 10 + 4 \cdot 10 + 3 \cdot 4.$$

(**b**) In this particular case where the article is the same, we may calculate three products only
$$13 \cdot 14 = 10 \cdot 10 + (3+4) \cdot 10 + 3 \cdot 4.$$

(**A.23**) MSS $\mathcal{A}, \mathcal{B}, \mathcal{C}$. Multiply 38 by 40.

(**a**) We can use the previous rules for multiplying units by tens and tens among themselves (rules i, A.14, & iv, A.17); thus
$$38 \cdot 40 = 8 \cdot 40 + 30 \cdot 40.$$

(**b**) But since 38 contains a higher digit we may reduce the computation to a multiplication involving a subtraction and use rule x:
$$38 \cdot 40 = (40 - 2) \cdot 40 = 40 \cdot 40 - 2 \cdot 40 = 1600 - 80 = 1520.$$

(**A.24**) MSS $\mathcal{A}, \mathcal{B}, \mathcal{C}$. Multiply 79 by 32.

Since 79 has a higher digit, use subtraction and rule x:

$$79 \cdot 32 = (80-1) \cdot 32 = 80 \cdot 32 - 32 = 2560 - 32 = 2528.$$

(**A.25**) MSS $\mathcal{A}, \mathcal{B}, \mathcal{C}$. Multiply 79 by 58.

Both having a higher digit, use subtraction twice and rules x & xi:
$$79 \cdot 58 = (80-1) \cdot (60-2) = 80 \cdot 60 - 1 \cdot 60 - 2 \cdot 80 + 1 \cdot 2$$
$$= 4800 - 220 + 2 = 4582.$$

(**A.26**) MSS $\mathcal{A}, \mathcal{B}, \mathcal{C}$. Multiply 320 by 640.

Subtraction is of no help here. So calculate
$$320 \cdot 640 = (300+20) \cdot (600+40) = 300 \cdot 600 + 300 \cdot 40 + 20 \cdot 600 + 20 \cdot 40$$
$$= 180\,000 + 12\,000 + 12\,000 + 800 = 204\,800.$$

(**A.27**) MSS $\mathcal{A}, \mathcal{B}, \mathcal{C}$. Multiply 390 by 390.

(*a*) Proceed according to A.26:
$$390 \cdot 390 = (300+90) \cdot (300+90).$$

(*b*) Proceed using rules x & xi (and the rule seen in A.22*b*)
$$390 \cdot 390 = (400-10) \cdot (400-10) = 400 \cdot 400 - 10 \cdot 400 \cdot 2 + 10 \cdot 10$$
$$= 160\,000 - 8000 + 100 = 152\,100.$$

(**A.28**) MSS $\mathcal{A}, \mathcal{B}, \mathcal{C}$. Multiply 728 by 464.

This is the first instance where we are obliged to write down the intermediate results, each hand being able to signify at most one two-digit number. To perform the operation in writing, put the two factors in two rows, with multiplicand above, then intermediate results below, with the corresponding decimal places aligned vertically for the subsequent addition.[†]

$$728 \cdot 464 = 700 \cdot 400 + 28 \cdot 400 + 700 \cdot 64 + 28 \cdot 64$$
$$= 280\,000 + 11\,200 + 44\,800 + 1792 = \mathit{337\,792}.[‡]$$

(**A.29**) MSS $\mathcal{A}, \mathcal{B}, \mathcal{C}$. Multiply 999 by 999.

(*a*) Proceed as in A.28, that is,
$$999 \cdot 999 = (900+99) \cdot (900+99).$$

(*b*) Use the subtractive way seen in A.23–25 & A.27:
$$999 \cdot 999 = (1000-1) \cdot (1000-1)$$
$$= 1000 \cdot 1000 - 1 \cdot 1000 - 1000 \cdot 1 + 1 \cdot 1$$
$$= 1\,000\,000 - 2000 + 1 = 998\,001.$$

Remark. No use here of the simplification seen in A.22*b* and A.27*b*.

5. Multiplication of numbers with repeated thousands.

[†] In operations with integers, the two numbers are superposed (whereas, as we shall see from A–V on, the two fractional expressions are put side by side with their terms written vertically).

[‡] From now on we shall write results omitted by the author in italics.

As mentioned in our summary, there is some disorder in the coming problems. After A.30 should come A.33–36 (single thousands), then A.31 and A.37 (higher thousands), next A.38–41 and finally A.32 (all with composite numbers).

A. *Multiplication of single-figure numbers (multiples of limits).**

(A.30) MSS $\mathcal{A}, \mathcal{B}, \mathcal{C}$. Multiply 1000, or generally 1000^m, by 1000.

We need only append to one factor the repetitions belonging to the other ('thousand's in words, '000's in numerals). In modern terms, this means adding together the exponents of the powers of 1000.

Remark. $1000 \cdot 1000$ is explained here, although it has already occurred (in the previous problem).

(A.31) MSS $\mathcal{A}, \mathcal{B}, \mathcal{C}$. Multiply 6 000 000 000 by 7 000 000.

Disregarding the repetitions, this reduces to $6 \cdot 7 = 42$ (by the table). Thus, appending the five repetitions,
$$6\,000\,000\,000 \cdot 7\,000\,000 = 42\,000\,000\,000\,000\,000.$$

(A.32) MSS $\mathcal{A}, \mathcal{B}, \mathcal{C}$. Multiply 100 000 020 006 by 50 100 000.

Considering that
$$100\,000\,020\,006 = 100\,000\,000\,000 + 20\,000 + 6,$$
$$50\,100\,000 = 50\,000\,000 + 100\,000,$$
we are to calculate six products, computation of all of which is known from before (A.16–21, A.30–31):

$$50\,000\,000 \cdot 6 = 300\,000\,000$$
$$50\,000\,000 \cdot 20\,000 = 1\,000\,000\,000\,000$$
$$50\,000\,000 \cdot 100\,000\,000\,000 = 5\,000\,000\,000\,000\,000\,000$$
$$100\,000 \cdot 100\,000\,000\,000 = 10\,000\,000\,000\,000\,000$$
$$100\,000 \cdot 20\,000 = 2\,000\,000\,000$$
$$100\,000 \cdot 6 = 600\,000.$$

So we shall have
$$100\,000\,020\,006 \cdot 50\,100\,000 = 5\,010\,001\,002\,300\,600\,000.$$

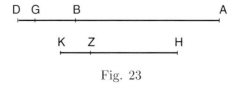

Fig. 23

---

* Not including A.32, misplaced.

Demonstration. Let (Fig. 23) the two factors be AD and HK. Take, on these two segments, $100\,000\,000\,000 = AB$, $20\,000 = BG$, $6 = GD$, $50\,000\,000 = HZ$, $100\,000 = ZK$; to prove that
$$AD \cdot HK = HZ \cdot AB + HZ \cdot BG + HZ \cdot GD$$
$$+ ZK \cdot AB + ZK \cdot BG + ZK \cdot GD.$$
By distributivity (*Elements* II.1 = PE_1) we may write:
$$AD \cdot HK = AD \cdot HZ + AD \cdot ZK$$
$$= (AB + BG + GD)\, HZ + (AB + BG + GD)\, ZK,$$
whence the requirement. This proof, the author adds, can be used in other similar cases (allusion to A.39–41, which are to precede A.32).[†]

(A.33) MSS $\mathcal{A}, \mathcal{B}, \mathcal{C}$. Multiply 4000 by 6000.

Disregarding the repetitions, this reduces to $4 \cdot 6 = 24$ (known from the multiplication table). Then, appending the repetitions, we obtain
$$4000 \cdot 6000 = 24\,000\,000.$$

(A.34) MSS $\mathcal{A}, \mathcal{B}, \mathcal{C}$. Multiply 7000 by 2000.

Disregarding the repetitions, this reduces to $7 \cdot 2 = 14$. Then, appending the repetitions, we obtain
$$7000 \cdot 2000 = 14\,000\,000.$$

(A.35) MSS $\mathcal{A}, \mathcal{B}, \mathcal{C}$. Multiply 10 000 by 10 000.

Disregarding the repetitions, this reduces to $10 \cdot 10 = 100$ (known from the table). Then, appending the repetitions, we obtain
$$10\,000 \cdot 10\,000 = 100\,000\,000.$$

(A.36) MSS $\mathcal{A}, \mathcal{B}, \mathcal{C}$. Multiply 6 000 by 40 000.

Disregarding the repetitions, this reduces to $6 \cdot 40 = 240$ (rule i and A.14). Then, appending the repetitions, we find
$$6\,000 \cdot 40\,000 = 240\,000\,000.$$

(A.37) MSS $\mathcal{A}, \mathcal{B}, \mathcal{C}$. Multiply 100 000 000 000 by 5 000 000 000 000.

(*a*) Disregarding the repetitions, we shall apply rule ii (A.15) and then append the repetitions.

(*b*) Since $100\,000\,000\,000$ is $100 \cdot 1\,000\,000\,000$ while $5\,000\,000\,000\,000$ is $5 \cdot 1\,000\,000\,000\,000$, the multiplication becomes (by virtue of P_3) $5 \cdot 100$ to be multiplied by the $(3+4=)$ seven repetitions. Thus
$$100\,000\,000\,000 \cdot 5\,000\,000\,000\,000 = 500\,000\,000\,000\,000\,000\,000\,000.$$

Remark. This is presented in the extant text as another way. As a matter of fact, it proves the procedure employed since A.33: we shall multiply the two factors preceding the repetitions and append to the result all the repetitions disregarded (which amounts, as seen in A.30, to multiplying by them).

[†] In the *Liber mahameleth* we sometimes find the proof at the end of a group of problems. See A.45, A.55 (misplaced), A.151*a*, A.260*a*.

B. *Multiplication of composite numbers*.

(A.38) MSS $\mathcal{A}, \mathcal{B}, \mathcal{C}$. Multiply 15 000 by 15 000.

Disregarding the repetitions, we compute $15 \cdot 15 = 225$ (A.22); whence
$$15\,000 \cdot 15\,000 = 225\,000\,000.$$

(A.39) MSS $\mathcal{A}, \mathcal{B}, \mathcal{C}$. Multiply 6400 by 3800.

Since $6400 = 6000 + 400$ and $3800 = 3000 + 800$, we shall calculate four products of the kinds seen before (A.15, A.20, A.33):

$$6000 \cdot 3000 = 18\,000\,000$$
$$3000 \cdot 400 = 1\,200\,000$$
$$6000 \cdot 800 = 4\,800\,000$$
$$400 \cdot 800 = 320\,000.$$

Adding this will give the result.

(A.40) MSS $\mathcal{A}, \mathcal{B}, \mathcal{C}$. Multiply 6468 by 4564.

Considering that
$$6468 \cdot 4564 = (6000 + 400 + 68) \cdot (4000 + 500 + 64),$$
we have to compute nine products as taught before (A.15–16, A.18–22a, A.24–25, A.31, A.33):

$$4000 \cdot 6000 = 24\,000\,000$$
$$4000 \cdot 400 = 1\,600\,000$$
$$4000 \cdot 68 = 272\,000$$
$$500 \cdot 6000 = 3\,000\,000$$
$$500 \cdot 400 = 200\,000$$
$$500 \cdot 68 = 34\,000$$
$$64 \cdot 6000 = 384\,000$$
$$64 \cdot 400 = 25\,600$$
$$64 \cdot 68 = 4352.$$

Adding the corresponding places will give the result. Note that in the figure found in the text, no zeros are written and the digits of all intermediate results have been successively inserted from the top (below the final result). The quantities in the rows of the figure are therefore *not* the intermediate results.

(A.41) MSS $\mathcal{A}, \mathcal{B}, \mathcal{C}$. Multiply 300 040 000 by 5 400 000 000.

This is decomposed into
$$(300\,000\,000 + 40\,000) \cdot (5\,000\,000\,000 + 400\,000\,000).$$

We find respectively:

$$1\,500\,000\,000\,000\,000\,000$$
$$120\,000\,000\,000\,000\,000$$
$$200\,000\,000\,000\,000$$
$$16\,000\,000\,000\,000$$

to be then added together.

C. *Multiplication using subtraction*.

A.42 & A.43 teach how to subtract a quantity from repeated thousands. The purpose, as explicitly said at the end of A.43, is to simplify the multiplication by representing one of the factors as a difference. This has already been used in the multiplication of the tens and hundreds (A.23–25, A.27 & A.29). We shall now apply this to the multiplication of higher numbers, either different or the same, both cases being then proved. For the subtraction itself, as stated at the beginning of A.42 and illustrated in A.42 & A.43, the minuend is decomposed from order unit to order unit until we reach the unit of the next higher order to the subtrahend.

(A.42) MSS $\mathcal{A}, \mathcal{B}, \mathcal{C}$. Subtract 1 from $1\,000\,000\,000$.

Decomposition gives

$$1\,000\,000\,000 = 900\,000\,000 + 100\,000\,000$$
$$100\,000\,000 = 99\,000\,000 + 1\,000\,000$$
$$1\,000\,000 = 900\,000 + 100\,000$$
$$100\,000 = 99\,000 + 1000$$
$$1000 = 900 + 100$$
$$100 = 90 + 10,$$

and 10 meets the requirement. This means that, since $1\,000\,000\,000$ is representable as

$$900\,000\,000 + 99\,000\,000 + 900\,000 + 99\,000 + 900 + 90 + 10,$$

we shall perform the subtraction on the last term and add the remainder to the unmodified numbers ('left as they stand'); this gives $999\,999\,999$.

(A.43) MSS $\mathcal{A}, \mathcal{B}, \mathcal{C}$. Subtract $20\,000\,000$ from $30\,000\,000\,000\,000\,000$. We have here

$$30\,000\,000\,000\,000\,000 = 29\,000\,000\,000\,000\,000 + 1\,000\,000\,000\,000\,000$$
$$1\,000\,000\,000\,000\,000 = 900\,000\,000\,000\,000 + 100\,000\,000\,000\,000$$
$$100\,000\,000\,000\,000 = 99\,000\,000\,000\,000 + 1\,000\,000\,000\,000$$
$$1\,000\,000\,000\,000 = 900\,000\,000\,000 + 100\,000\,000\,000$$
$$100\,000\,000\,000 = 99\,000\,000\,000 + 1\,000\,000\,000$$
$$1\,000\,000\,000 = 900\,000\,000 + 100\,000\,000.$$

Performing then $100\,000\,000 - 20\,000\,000$ and adding to the remaining $80\,000\,000$ all the unmodified quantities will give the result, namely $29\,999\,999\,980\,000\,000$.

(A.44) MSS $\mathcal{A}, \mathcal{B}, \mathcal{C}$. Multiply 9999 by 9999.

Proceeding as in A.40 will give nine products. It is simpler to use, as in the group A.22–29, the subtraction:
$$(10\,000 - 1)(10\,000 - 1) = 100\,000\,000 - 20\,000 + 1 = 99\,980\,001.$$

```
A                G  B
|----------------|--|
```

Fig. 24

Demonstration. Let (Fig. 24) $10\,000 = AB$, $1 = GB$, required AG^2. Since (*Elements* II.7=PE_7) $AB^2 + GB^2 = 2 \cdot AB \cdot GB + AG^2$, we can calculate AG^2.

(A.45) MSS $\mathcal{A}, \mathcal{B}, \mathcal{C}$. Multiply the quantities $48\,999\,999\,960\,000\,000\,000$ and $20\,999\,999\,980\,000$.

Proceeding as before (A.32, A.39–40) will be, the author remarks, rather irksome. But the multiplier can be expressed as
$$49\,000\,000\,000\,000\,000\,000 - 40\,000\,000\,000$$
and the multiplicand as
$$21\,000\,000\,000\,000 - 20\,000;$$
so only four multiplications are needed, giving as the final result
$$1\,028\,999\,998\,180\,000\,000\,800\,000\,000\,000\,000.^\dagger$$

Demonstration. Similar to the demonstration in A.44, but with two segments of straight line since the two factors are different. As often, the demonstration is at the end of the section, with the more complicated example.

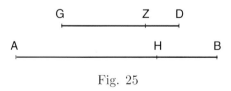

Fig. 25

---

† We recognize in a figure added by the reader in MS $\mathcal{B}$ (see Latin text) the given quantities, the result, the round numbers and the four products, namely

$$1\,029\,000\,000\,000\,000\,000\,000\,000\,000\,000\,000$$
$$-\ 980\,000\,000\,000\,000\,000\,000\,000$$
$$-\ 840\,000\,000\,000\,000\,000\,000\,000$$
$$+\ 800\,000\,000\,000\,000$$

with, each time, the partial sums calculated.

Let (Fig. 25) $49\,000\,000\,000\,000\,000\,000 = AB$, $40\,000\,000\,000 = HB$, $21\,000\,000\,000\,000 = GD$, $20\,000 = ZD$, required $AH \cdot GZ$. By distributivity (*Elements* II.1 = PE_1), or the proof in A.32,

$$AB \cdot GD = (AH + HB)(GZ + ZD)$$
$$= AH \cdot GZ + AH \cdot ZD + HB \cdot GZ + HB \cdot ZD.$$

Adding $HB \cdot ZD$ on both sides gives

$$AB \cdot GD + HB \cdot ZD = AH \cdot GZ + AH \cdot ZD + HB \cdot GZ + 2 \cdot HB \cdot ZD.$$

Since

$$HB \cdot GZ + HB \cdot ZD = HB \cdot GD$$

while

$$AH \cdot ZD + HB \cdot ZD = AB \cdot ZD,$$

so

$$AB \cdot GD + HB \cdot ZD = AH \cdot GZ + AB \cdot ZD + HB \cdot GD.$$

Therefore, for the required product,

$$AH \cdot GZ = AB \cdot GD + HB \cdot ZD - AB \cdot ZD - HB \cdot GD.$$

---

6. Taking fractions of thousands repeated.

The last section, still on the subject of repeated thousands, is actually more of a preliminary to the next chapter on division, where it will be applied from A.55 on. If we are to take the fraction $\frac{p}{q}$ of $a \cdot 1000^k$, where a (≤ 999) is the integer preceding the k 'thousand's, we shall consider

$$\left(\frac{p}{q} \cdot a\right) \cdot 1000^k \quad \text{or} \quad \left(\frac{p}{q} \cdot 1000 \cdot a\right) \cdot 1000^{k-1},$$

according to whether $q < a$ or pa (A.48–51) or $q > a$ or pa (A.46–47). After taking the fraction of the integer we shall append the repetitions left out.[‡] In all examples (A.46–51), q does not divide the integer preceding the repetitions, so that a fraction will also multiply the repetitions. The text considers first aliquot, then non-aliquot fractions.

(A.46) MSS $\mathcal{A}$, $\mathcal{B}$, $\mathcal{C}$. Take $\frac{1}{3}$ of $1\,000\,000\,000$.
Since

$$\frac{1}{3} \cdot 1000 = 333 + \frac{1}{3}$$

we can calculate, iteratively,

$$\frac{1}{3} \cdot 1\,000\,000\,000 = \left(333 + \frac{1}{3}\right) 1\,000\,000 = 333\,000\,000 + \frac{1}{3} \cdot 1\,000\,000$$

$$= 333\,000\,000 + \left(333 + \frac{1}{3}\right) 1000 = 333\,000\,000 + 333\,000 + \frac{1}{3} \cdot 1000$$

[‡] Although multiplication of a fraction by an integer will be taught later (A–V).

$$= 333\,333\,333 + \frac{1}{3}.$$

Demonstration. From A.30 and P_5,

$$\frac{1\,000\,000\,000}{3} = \frac{1000 \cdot 1\,000\,000}{3} = \frac{1000}{3} \cdot 1\,000\,000,$$

and, likewise,
$$\frac{1\,000\,000}{3} = \frac{1000}{3} \cdot 1000.$$

(A.47) MSS $\mathcal{A}, \mathcal{B}, \mathcal{C}$. Take $\frac{1}{9}$ of $1\,000\,000\,000$.

Since
$$\frac{1}{9} \cdot 1000 = 111 + \frac{1}{9}$$

so, iteratively,

$$\frac{1}{9} \cdot 1\,000\,000\,000 = 111\,000\,000 + \frac{1}{9} \cdot 1\,000\,000$$

$$= 111\,111\,000 + \frac{1}{9} \cdot 1000 = 111\,111\,111 + \frac{1}{9}.$$

(A.48) MSS $\mathcal{A}, \mathcal{B}, \mathcal{C}$. Take $\frac{5}{6}$ of $10\,000\,000\,000$.

Since
$$\frac{5}{6} \cdot 10 = 8 + \frac{1}{3}$$

so, iteratively,

$$\frac{5}{6} \cdot 10\,000\,000\,000 = \left(8 + \frac{1}{3}\right) 1\,000\,000\,000$$

$$= 8\,000\,000\,000 + \frac{1}{3} \cdot 1\,000\,000\,000 = 8\,333\,333\,333 + \frac{1}{3},$$

for taking a third of $1\,000\,000\,000$ is known from A.46.

(A.49) MSS $\mathcal{A}, \mathcal{B}, \mathcal{C}$. Take $\frac{3}{5}$ of $8\,000\,000$.

Since
$$\frac{3}{5} \cdot 8 = 4 + \frac{4}{5}$$

so

$$\frac{3}{5} \cdot 8\,000\,000 = \left(4 + \frac{4}{5}\right) 1\,000\,000 = 4\,000\,000 + 800\,000 = 4\,800\,000.$$

(A.50) MSS $\mathcal{A}, \mathcal{B}, \mathcal{C}$. Take $\frac{5}{8}$ of $100\,000\,000\,000\,000$.

Since
$$\frac{5}{8} \cdot 100 = 62 + \frac{1}{2}$$

so
$$\frac{5}{8} \cdot 100\,000\,000\,000\,000 = \left(62 + \frac{1}{2}\right) 1\,000\,000\,000\,000$$
$$= 62\,000\,000\,000\,000 + 500\,000\,000\,000 = 62\,500\,000\,000\,000.$$

(A.51) MSS $\mathcal{A}, \mathcal{B}, \mathcal{C}$. Take $\frac{5}{7}$ of $20\,000\,000\,000$.

We are told to take, as before, the fraction from the part preceding the repetition, and then to multiply by (or append) the repetition, thus

$$\left(\frac{5}{7} \cdot 20\right) 1\,000\,000\,000 = \left(14 + \frac{2}{7}\right) 1\,000\,000\,000.$$

Chapter A–IV:
Division of integers

Summary

In his introduction, the author states some fundamental facts about division. If it involves two quantities numbering objects of the same kind, the purpose is to determine the ratio one quantity has to the other; if the kinds are different, we have the problem of sharing. The operation, though, remains the same and obeys the rule that multiplying the result by the divisor gives the dividend (rule *i*). Next, division takes two forms (rule *ii*): if larger is divided by smaller, it is division proper; if smaller by larger, it is called 'denomination'. This determines the two sections of this chapter.

1. Division of larger by smaller.

Suppose we have to divide a by b, with a and b integers ($a > b$). Then

— either find the quantity c such that $c \cdot b = a$;

— or determine the inverse of b and multiply it by a.

If a and b have a common factor, it will be conveniently left out (rule *iii* and A.54). This will in particular be applied to the division of integers with repeated thousands, the number of repetitions in the divisor being subtracted from that in the dividend (A.55, A.57, A.59). Once this reduction is performed, we may provisionally leave out the repetitions in excess in the dividend, perform the division and reintroduce them at the end. We are already acquainted with that from the last section of the previous chapter (A.46–51).

Remark. Division of integers is obviously supposed to be known: most examples, and the whole subsequent part of this section, are devoted to the particular case of dividing thousands.

The later part of this section presents practical rules which simplify the computations for division *and* multiplication when either multiplicand or dividend has the simple form 500, 2000, 1 000 000. They are conveniently illustrated by means of fixed quantities of gold coins (*morabitini*): 500 represents the number of morabitini in a money-bag, 1 000 000 the number of morabitini in a treasury, thus 2000 is the number of money-bags in a treasury. The problems and the rules are the following:

• Find the number b of money-bags for a given quantity q of morabitini (A.60–61), or, conversely, find q when b is given (A.62–63).

<u>Rule</u>: To divide a given number by 500, remove from it one repetition and multiply the result by 2; to multiply it by 500, divide by 2 and append one repetition.

- Find the number t of treasuries for a given quantity q of morabitini (A.64–65), or, conversely, find q when t is given (A.66–67).

Rule: To divide a given number by $1\,000\,000$, remove from it two repetitions; to multiply it by $1\,000\,000$, append two repetitions.

- Find the number b of money-bags for a given quantity t of treasuries (A.68–69), or, conversely, find t when b is given (A.70–71).

Rule: To multiply a given number by 2000, double it and append one repetition; to divide it by 2000, remove one repetition and divide by 2.

2. Division of smaller by larger ('denomination').

The Arabic language has particular names for 'half', 'third' and so on to 'tenth', which makes it possible to express a proper fraction with denominator ≤ 10 as in English ('six tenths'). In other cases a circumlocution is necessary ('six parts of eleven parts of one' or 'six parts out of eleven'). Whence the preference for expressing a general fraction by means of 'elementary' fractions, that is, those having as their denominators 'elementary' numbers, thus numbers from 2 to 10. This will be possible if the denominator of the general fraction considered is a product of elementary numbers. For this purpose, we are first to consider divisibility by $2, \ldots, 10$ and, more precisely, to find criteria for determining when a given quantity is exactly divisible by these numbers. The *Liber mahameleth* gives twenty-nine rules concerning divisibility by these elementary numbers; they can be classified into four groups:

- Any number not divisible by an elementary number is not divisible by any of that elementary number's multiples (rules *i–iv*; multiples ≤ 10 only considered).

- Any number divisible by an elementary number is divisible by any of that elementary number's divisors (rules *v–x*).

- A number not divisible by an elementary number may be divisible by some of that elementary number's divisors, depending on the remainder (rules *xi–xviii*).

- Finally, we find criteria for divisibility by 10 (rule *xix* & *xxviii*, A.77–78), by 9 (rule *xx*, A.72), by 8 (rule *xxi*, A.73), by 7 (rule *xxii*, A.74–75), by 6 (rule *xxiii*, A.76), by 5 (rule *xxiv* & *xxix*, A.79), by 4 (rule *xxv*), by 3 (rule *xxvi*), by 2 (rule *xxvii*).

As intended, this concludes with the general problem of representing a fraction as a sum of fractions in the form

$$\frac{p}{q} = \frac{p_1}{q_1} + \frac{p_2}{q_1 \cdot q_2} + \frac{p_3}{q_1 \cdot q_2 \cdot q_3} + \ldots + \frac{p_{l-1}}{q_1 \cdot q_2 \cdot \ldots \cdot q_{l-1}} + \frac{p_l}{q_1 \cdot q_2 \cdot \ldots \cdot q_l}.$$

where $2 \leq q_i \leq 10$ and q has only q_i's as its factors (rules *xxx–xxxi'*, A.80).

The section ends with examples of denominating from other numbers: first 1, or any digit, from composite (that is, mixed) numbers (A.81–83) or thousands repeated (A.84–85), then other numbers from composite numbers (A.86–87) or thousands repeated (A.88–89).

1. Division of larger by smaller.

Division in general has its origin in two distinct problems:

• Sharing: A number of things is to be shared equally among a number of persons, and we wish to know how many things each is to receive, for instance ten *nummi* among five men.* The two given quantities are then of a different kind.

• Comparing: We wish to know how many times one quantity is contained in another of the same kind, for instance ten nummi in twenty nummi. This means determining the amount of the ratio between these two quantities.

(i) In either case, the operation is the same. Indeed, the quotient multiplied by the divisor will give the dividend:

$$\frac{a}{b} = c \quad \text{implies} \quad b \cdot c = a.$$

This will be used not only to verify the result but also find it (A.52a & c).

(ii) The operation of division takes two forms: If $a > b$, we have division proper (*divisio*, Arabic *qisma*); if $a < b$, we have denomination (*denominatio*, Arabic *tasmīya*).

(iii) If both the dividend and the divisor contain a common factor, this can be cancelled (A.54).

A. *Division of integers without thousands*.

Suppose we are to divide a by b. The author considers three ways to obtain the result:

— Find the number which when multiplying b produces a (see A.52a, A.53).

— Find the inverse of b (that is, denominate 1 from b) and multiply the result by a (A.52b).

— If the dividend is not exactly divisible by the divisor, find the integral quotient and denominate the remainder from the divisor (A.53).

(A.52) MSS $\mathcal{A}, \mathcal{B}, \mathcal{C}$. Divide 20 by 4.

(**a**) Find the number which multiplying 4 gives 20. By the multiplication table, this is 5.

(**b**) Find the inverse of the divisor and multiply it by the dividend; thus consider

$$\tfrac{1}{4} \cdot 20.$$

Justification. Since, from rule i, $1 : b = c : a$, so

$$\frac{1}{b} \cdot a = c.$$

(**c**) Verification. According to rule i, $b \cdot c$ must give a, and this is the case for $4 \cdot 5$.

---

* The *nummus* will appear throughout the *Liber mahameleth* as basic monetary unit.

Remark. This is the first occurrence of a 'verification' of the result (*experientia*, Arabic *imtiḥān*). Numerous instances in Book B.

(**A.53**) MSS $\mathcal{A}, \mathcal{B}, \mathcal{C}$. Divide 60 by 8.
Find which number multiplied by 8 gives 60.

Since this is a division with remainder, find the largest integer which when multiplied by 8 gives less than 60. This is 7, with the remainder 4 to be denominated from the divisor. Thus
$$\frac{60}{8} = 7 + \frac{4}{8} = 7 + \frac{1}{2}.$$

Remark. We have not yet been taught denomination and barely know fractions (from A–III.6).

(**A.54**) MSS $\mathcal{A}, \mathcal{B}, \mathcal{C}$. Divide 24 by 8. Application of rule *iii*.

Removing the (here: a) common factor, namely 4, leaves the division of 6 by 2.

B. *Division of integers with thousands*.

Considering that in this section the dividend is larger than the divisor, there are two possibilities.

• If only the dividend contains repetitions, we shall disregard some of them so as to leave a number still larger than the divisor, perform the division, and append to the quotient the repetitions left out.

• If both the dividend and the divisor contain repetitions, we shall apply rule *iii* and cancel the repetitions common to both terms; this will leave us with the previous situation.

Accordingly, suppose we are to divide $A \cdot 1000^n$ by $B \cdot 1000^m$ ($A, B < 1000$ and $n > m$); then we shall be left with dividing $A \cdot 1000^{n-m}$ by B. If $A > B$, we divide A by B and multiply the result by 1000^{n-m}, as is done in A.55–57 (proof in A.55); if $A < B$, we divide $1000 \cdot A$ by B and multiply the result by 1000^{n-m-1}, as seen in A.58–59. The same procedure was applied in the last section of A–III.

(**A.55**) MSS $\mathcal{A}, \mathcal{B}, \mathcal{C}$. Divide 100 000 000 000 000 000 by 15 000 000. This problem should rather follow A.57.

This is a direct application of rule *iii*:
$$\frac{100\,000\,000\,000\,000\,000}{15\,000\,000} = \frac{100}{15} 1000^{5-2} = \left(6 + \frac{2}{3}\right) \cdot 1\,000\,000\,000.$$

We are now in a known situation (A.46–51 for the fractional part).

Demonstration. Using general terms, let us consider the division of $A \cdot 1000^n$ by $B \cdot 1000^m$. By *Elements* VII.18,
$$\frac{A \cdot 1000^n}{B \cdot 1000^m} = \frac{A \cdot 1000^{n-m} \cdot 1000^m}{B \cdot 1000^m} = \frac{A \cdot 1000^{n-m}}{B},$$
and using P_5, we indeed find that

$$\frac{A \cdot 1000^{n-m}}{B} = \frac{A}{B} \cdot 1000^{n-m}.$$

(**A.56**) MSS $\mathcal{A}, \mathcal{B}, \mathcal{C}$. Divide 100 000 000 by 12.

$$\frac{100\,000\,000}{12} = \frac{100}{12} \cdot 1\,000\,000 = \left(8 + \frac{1}{3}\right) \cdot 1\,000\,000$$

whence (A.46)

$$\frac{100\,000\,000}{12} = 8\,333\,333 + \frac{1}{3}.$$

(**A.57**) MSS $\mathcal{A}, \mathcal{B}, \mathcal{C}$. Divide 50 000 000 000 000 by 8000.

$$\frac{50\,000\,000\,000\,000}{8000} = \frac{50\,000\,000\,000}{8} = \frac{50}{8} \cdot 1\,000\,000\,000$$
$$= \left(6 + \frac{1}{4}\right) \cdot 1\,000\,000\,000 = 6\,250\,000\,000.$$

(**A.58**) MSS $\mathcal{A}, \mathcal{B}, \mathcal{C}$. Divide 8 000 000 000 by 400.

(*a*) Reduced to

$$\frac{8000}{400} \cdot 1\,000\,000 = 20\,000\,000.$$

(*b*) May be reduced to

$$\frac{1000}{400} \cdot 8\,000\,000 = \left(2 + \frac{1}{2}\right) \cdot 8\,000\,000 = 20\,000\,000.$$

Remark. This seems an odd way to calculate, but it is in accordance with P_5 and also occurs in the next problem and in part C below.

(**A.59**) MSS $\mathcal{A}, \mathcal{B}, \mathcal{C}$. Divide 6 000 000 000 000 by 200 000 000.
Reduced (rule *iii*) to 6 000 000 divided by 200, then (A.58) to

$$\frac{6000}{200} \cdot 1000 \quad \text{or} \quad \frac{1000}{200} \cdot 6000.$$

C. *Practical rules for some particular cases*.

We have seen how, for division as well as for multiplication (A–III), we may separate the repetitions from their multiplicative factors for the computations. This determines the rules to come, which simplify the computations when the multiplier or the dividend takes a simple form, namely 500, 2000 or 1 000 000. Actually, these rules are not given as such but presented in the form of some rather fanciful accounting which involves gold coins (*morabitini*), money-bags (of 500 morabitini each) and treasuries (of 1 000 000 each, thus equivalent to 2000 money-bags). The Moorish and Spanish origin is evident from the vocabulary. The *morabitinus* was a gold coin minted in Christian Spain which came to replace the equivalent in use under the Islamic rulers of Spain before the Christian reconquest, the dynasty of the *al-Murābiṭūn*, or Almoravides —whence its name. Furthermore, the *domus pecunie* is a faithful translation of the Arabic *bayt al-māl*, used to mean the Treasury, later also smaller treasuries such as those of institutions or foundations.† For each of the above conversions there will be a pair of problems, in the first of which the simplified rule is stated and then applied. We thus find the following conversions:

(i) Find the number b of money-bags for a given quantity q of morabitini (A.60–61); or conversely (i') find q when b is given (A.62–63).

(ii) Find the number t of treasuries for a given quantity q of morabitini (A.64–65); or conversely (ii') find q when t is given (A.66–67).

(iii) Find the number b of money-bags for a given quantity t of treasuries (A.68–69); or conversely (iii') find t when b is given (A.70–71).

The corresponding rules are then:

(i) To divide by 500, remove one repetition then multiply by 2; (i') to multiply by 500, divide by 2 then append one repetition.

(ii) To divide by 1 000 000, remove two repetitions; (ii') to multiply by 1 000 000, append two repetitions.

(iii) To multiply by 2000, double the given number then append one repetition; (iii') to divide by 2000, remove one repetition then divide by 2.

(i) MORABITINI AND MONEY-BAGS.

(**A.60**) MSS $\mathcal{A}, \mathcal{B}, \mathcal{C}$. Required the number b of money-bags for the quantity $q = 20\,000\,000$ morabitini.

(**a**) As seen in the section on dividing numbers with repeated thousands (A.58), compute
$$b = \frac{q}{500} = \frac{20\,000\,000}{500} = 40\,000.$$

(**b**) Apply rule i, that is, remove one repetition and take twice the result; then $b = 2 \cdot 20\,000$.

This is justified by the procedure seen in A.58–59: we remove one 'thousand' from the repetitions, divide it by 500, and multiply the quotient, thus 2, by the remaining expression, namely 20 000.

(**A.61**) MSS $\mathcal{A}, \mathcal{B}, \mathcal{C}$. Required the number b of money-bags for the quantity $q = 600\,000\,000\,000\,000$ morabitini.

(**a**) Compute
$$b = \frac{q}{500} = \frac{600\,000\,000\,000\,000}{500}.$$

(**b**) Applying rule i gives $b = 2 \cdot 600\,000\,000\,000 = 1\,200\,000\,000\,000$.

(**A.62**) MSS $\mathcal{A}, \mathcal{B}, \mathcal{C}$. Required now the quantity q of morabitini corresponding to $b = 200\,000$ money-bags.

(**a**) By multiplication: $q = b \cdot 500 = 200\,000 \cdot 500 = 100\,000\,000$.

---

† Another rendering is the *casa scazi* (adaptation of 'Schatzkammer') added at the beginning of A.42 by an early reader referring to this section.

(**b**) By rule i', take first half of b, thus
$$\frac{1}{2} \cdot 200\,000 = 100\,000$$
and add one repetition.

Justification: 500 being half a thousand, we may, instead of multiplying by 500, take half the given number and multiply the result by a thousand.

(**A.63**) MSS $\mathcal{A}, \mathcal{B}, \mathcal{C}$. Required the quantity q of morabitini corresponding to $b = 3\,000\,000$ money-bags.

(**a**) By multiplication: $q = b \cdot 500 = 3\,000\,000 \cdot 500 = 1\,500\,000\,000$.

(**b**) By rule i', take half of b, thus $1\,500\,000$, and add one repetition.

(*ii*) MORABITINI AND TREASURIES.

(**A.64**) MSS $\mathcal{A}, \mathcal{B}, \mathcal{C}$. Required the number t of treasuries corresponding to $q = 10\,000\,000\,000$ morabitini.

(**a**) By division:
$$t = \frac{q}{1\,000\,000} = \frac{10\,000\,000\,000}{1\,000\,000} = 10\,000.$$

(**b**) By rule ii, just remove twice 'thousand' from q.

(**A.65**) MSS $\mathcal{A}, \mathcal{B}, \mathcal{C}$. Required the number t of treasuries corresponding to $q = 100\,000\,000\,000\,000$ morabitini.
By rule ii, removing two repetitions gives $t = 100\,000\,000$.

(**A.66**) MSS $\mathcal{A}, \mathcal{B}, \mathcal{C}$. Required the number q of morabitini corresponding to $t = 10\,000$ treasuries.

(**a**) By multiplication: $q = 10\,000 \cdot 1\,000\,000 = 10\,000\,000\,000$.

(**b**) By rule ii', append to $10\,000$ two repetitions.

(**A.67**) MSS $\mathcal{A}, \mathcal{B}, \mathcal{C}$. Required the number q of morabitini corresponding to $t = 100\,000\,000$ treasuries.

By rule ii', append to t two repetitions, which gives $100\,000\,000\,000\,000$.

(*iii*) MONEY-BAGS AND TREASURIES.

(**A.68**) MSS $\mathcal{A}, \mathcal{B}, \mathcal{C}$. Required the number b of money-bags corresponding to $t = 300$ treasuries.

(**a**) The number of money-bags in a treasury being 2000, so $b = t \cdot 2000$.

(**b**) By rule iii, double the given number and append to the result one repetition.

(**A.69**) MSS $\mathcal{A}, \mathcal{B}, \mathcal{C}$. Required the number b of money-bags in $t = 600\,000$ treasuries.

By rule iii, add one repetition to $2t$, which gives $b = 1\,200\,000\,000$.

(**A.70**) MSS $\mathcal{A}, \mathcal{B}, \mathcal{C}$. Required the number t of treasuries corresponding to $b = 600\,000$ money-bags.

(**a**) By division:
$$t = \frac{b}{2000} = \frac{600\,000}{2000} = 300.$$

(**b**) By rule iii', remove from the given number one repetition and divide the result by 2.

(**A.71**) MSS $\mathcal{A}, \mathcal{B}, \mathcal{C}$. Required the number t of treasuries corresponding to $b = 100\,000\,000$ money-bags.

By rule iii', remove one repetition and divide by 2, which gives $t = 50\,000$.

This section ends with a table showing entries from which we obtain the appropriate rule of multiplication. It corresponds to the table below (Fig. 26).

from \ to	morabitini	money-bags	treasuries
morabitini	●	i'	ii'
money-bags	i	●	iii
treasuries	ii	iii'	●

Fig. 26

2. Division of smaller by larger ('denomination').

A. *Premisses on divisibility*.

We shall now encounter twenty-nine rules concerning the divisibility of integers by what we call 'elementary numbers', that is, the natural numbers from 2 to 10. These rules, preserved only in MS $\mathcal{A}$, can be divided into four categories.

• Any number not divisible by an elementary number is not divisible by any of that elementary number's multiples. Since only the multiples of 2, 3, 4, 5 are concerned, this means that

(i) If N is not divisible by 2, it is not by 4, 6, 8, 10 either.

(ii) If N is not divisible by 3, it is not by 6 or 9 either.

(iii) If N is not divisible by 4, it is not by 8 either.

(iv) If N is not divisible by 5, it is not by 10 either.

• Any number divisible by an elementary number is divisible by any of that elementary number's divisors. Since only the composite numbers up to 10 (thus 10, 9, 8, 6, 4) are concerned, this means that

(v) Any number N divisible by 10 is divisible by 5 and 2.

(*vi*) Any number N divisible by 9 is divisible by 3.

(*vii*) Any number N divisible by 8 is divisible by 4 and 2.

(*viii*) Any number N divisible by 6 is divisible by 3 and 2.

(*ix*) Any number N divisible by 4 is divisible by 2.

(*x*) An odd number is not divisible by an even number.†

• A number not divisible by an elementary number may be divisible by some of this elementary number's divisors or their multiples, depending on the remainder.

(*xi*) Dividing the integer N by 9, which has the divisor 3, we have $N = 9 \cdot t + r$. Then
  — if $r = 0$, $N = 9t$ is divisible by 9 and by 3, if N is even by 6 as well;‡
  — if $r = 6$, $N = 9t + 6$ is divisible by 3, if N is even by 6 as well;*
  — if $r = 3$, $N = 9t + 3$ is divisible by 3.°

(*xii*) Dividing the integer N by 8, which has the divisor 4 (thus also 2), we have $N = 8t + r$. Then
  — if $r = 0$, $N = 8t$ is divisible by 8, 4 and 2 (according to *vii*);
  — if $r = 4$, $N = 8t + 4$ is divisible by 4 (and 2).

(*xiii*) N divisible by 7 has the form $N = 7t$; 7, prime, has no divisors.

(*xiv*) Dividing the integer N by 6, which has the divisor 3 (and 2), we have $N = 6t + r$. Then
  — if $r = 0$, $N = 6t$ is divisible by 6 (and by 3, according to *viii*);
  — if $r = 3$, $N = 6t + 3$ is divisible by 3.

(*xv*) N divisible by 5 has the form $N = 5t$; 5, prime, has no divisors.

(*xvi*) N, divisible by 4, has the form $N = 4t$ (and is thus divisible by 2).

(*xvii*) N, divisible by 3, has the form $N = 3t$.

(*xviii*) All numbers which are not divisible by the digits taken in descending order till 1 will be divisible (at most) by prime factors ≥ 11 (*Elements* VII.31).

• Criteria of divisibility. To find the criterion of divisibility of an integer N by the elementary integer m, the author proceeds, in modern language, as follows. Consider that, with the digits symbolized as a_i, N is written $a_n \, a_{n-1} \cdots a_2 \, a_1 \, a_0$, so that $N = a_n \cdot 10^n + a_{n-1} \cdot 10^{n-1} + \ldots + a_2 \cdot 10^2 + a_1 \cdot 10 + a_0$. Suppose that 10^k divided by m leaves the remainder $r_k < m$. Then N will be divisible by m if
$$r_n \cdot a_n + r_{n-1} \cdot a_{n-1} + \ldots + r_2 \cdot a_2 + r_1 \cdot a_1 + a_0$$
is divisible by m.

---

† This general statement would have been better placed at the beginning.
‡ $N = 18k + 9$ if $t = 2k + 1$ and $N = 18k$ if $t = 2k$.
* $N = 18k + 15$ if $t = 2k + 1$ and $N = 18k + 6$ if $t = 2k$.
° And by 6 and 2 if t is odd (N even).

(xix) A number is divisible by 10 if its last digit (our a_0) is 0. Indeed, all 10^k, k natural, are divisible by 10.

(xx) A number is divisible by 9 if the sum of its figures is divisible by 9. That is, we shall just have to consider the divisibility by 9 of
$$a_n + a_{n-1} + \ldots + a_2 + a_1 + a_0.$$
Indeed, since each power of 10 when divided by 9 leaves 1 as remainder, N will be divisible by 9 if the above sum is.

(A.72) MS $\mathcal{A}$. Examine the divisibility of 154 by 9.

Taking 1 for each 100, 1 for each 10, and 4, gives $1 \cdot 1 + 5 \cdot 1 + 4 = 10$. Since 10 is not divisible by 9, neither is 154.

(xxi) Divisibility by 8. A number is divisible by 8 if the number of units plus twice the number of tens plus four times the number of hundreds is divisible by 8. So we shall consider the divisibility of
$$4 \cdot a_2 + 2 \cdot a_1 + a_0.$$
Indeed, the remainder of 100 is 4, that of 10, 2, while, since 1000 is divisible by 8, higher powers of 10 need not be considered.

(A.73) MS $\mathcal{A}$. Examine the divisibility of 264 by 8.

Since $4 \cdot 2 + 2 \cdot 6 + 4 = 24$ is divisible by 8, so is 264.

($xxii$) Divisibility by 7. The remainders of the successive powers of 10 after division by 7 are, respectively, 3 (for 10), 2 (for 10^2), 6 (for 10^3), 4 (for 10^4), 5 (for 10^5), 1 (for 10^6), generally 1 for 10^{6k}, 3 for $10 \cdot 10^{6k} = 10^{6k+1}$, 2 for $100 \cdot 10^{6k} = 10^{6k+2}$, 6 for 10^{6k+3}, 4 for $10 \cdot 10^{6k+3} = 10^{6k+4}$, 5 for $100 \cdot 10^{6k+3} = 10^{6k+5}$. The divisibility of N by 7 will therefore depend upon the divisibility by 7 of
$$\ldots + \left(\ldots + 2\,a_8 + 3\,a_7 + a_6\right) + \left(5\,a_5 + 4\,a_4 + 6\,a_3 + 2\,a_2 + 3\,a_1 + a_0\right).$$

(A.74) MS $\mathcal{A}$. Examine the divisibility of 2348 by 7.

Since $6 \cdot 2 + 2 \cdot 3 + 3 \cdot 4 + 8 = 38$ is not divisible by 7, neither is 2348.

(A.75) MS $\mathcal{A}$. Other example, with higher thousands: examine the divisibility by 7 of 4 430 000 000.

We observe that 30 and 400 are followed by an even number of repetitions and 4 by an odd number of them; thus the number is, in our symbolism, of the form $4 \cdot 10^{6k+3} + 400 \cdot 10^{6k} + 30 \cdot 10^{6k}$, and we shall consider $6 \cdot 4 + 2 \cdot 4 + 3 \cdot 3 = 41$. Since 41 is not divisible by 7, neither is the proposed number.

($xxiii$) Divisibility by 6. A number is divisible by 6 if its last digit plus four times the sum of the other figures is divisible by 6. So we shall consider the divisibility by 6 of the expression
$$4\left(\ldots + a_3 + a_2 + a_1\right) + a_0.$$
Indeed, any power 10^k ($k \geq 1$) leaves the remainder 4 after division by 6.

(A.76) MS $\mathcal{A}$. Example. Divisibility of 2324 by 6.

Since $4 \cdot 2 + 4 \cdot 3 + 4 \cdot 2 + 4 = 32$ is not divisible by 6, neither is 2324.

(*xxiv*) Divisibility by 5. If the last (significant) figure is not 5, the number will not be divisible by 5, otherwise it will be.

(*xxv*) Divisibility by 4. A number is divisible by 4 if twice the number of tens plus the digit, thus $2\,a_1 + a_0$, is divisible by 4.

Indeed, since 100 is divisible by 4, any multiple of 100 is divisible by 4, so that we may consider only $10\,a_1 + a_0$; but this is $4 \cdot 2\,a_1 + (2\,a_1 + a_0)$, whence the rule given.

(*xxvi*) Divisibility by 3. Consider the divisibility by 3 of
$$a_n + a_{n-1} + \ldots + a_2 + a_1 + a_0,$$
that is, of the sum of the figures representing N.

Indeed, $10^k = 999\ldots9 + 1$ (with 9 repeated k times), so the remainder of the division by 3 is always 1.

(*xxvii*) Divisibility by 2. The last digit must be divisible by 2, thus the number even.

This section concludes with two practical rules for the division by 10 and 5.

(*xxviii*) To find a tenth of N (divisible by 10), move it one place to the right (that is, drop the final zero).

(**A.77–78**) MS $\mathcal{A}$. A tenth of 1200 is 120, a tenth of 120 is 12.

(*xxix*) To find a fifth of N (divisible by 5). If N is divisible by 10, do as before, thus obtaining a tenth, and double it; if N ends with 5, disregard it, do as before and add 1 to the result.

(**A.79**) MS $\mathcal{A}$. A fifth of $25 = 20 + 5$ is $2 \cdot 2 + 1$.

(*xxx*) For the other elementary numbers, just perform the division.

B. *Expressing a fraction when the denominator is a product of elementary numbers*.

Since we now know the criteria of divisibility for the elementary numbers, we will have no difficulty in finding such factors for any given number. This step will now be applied to expressing a fraction $\frac{p}{q}$ as the sum of products of elementary fractions.

(*xxxi*) Take the denominator q and divide it successively by the elementary numbers, beginning with 10 and ending with 2. Since q is supposed to have as factors only digits and 10 (and all factors different), it will finally take the form $q = q_1 \cdot q_2 \cdot \ldots \cdot q_l$ with $q_1 > q_2 > \ldots > q_l$ and $10 \geq q_i \geq 2$.° We may then express the proposed fraction as a sum of fractions with products of the q_i's in the denominators. This is obtained as follows.[†]

The given fraction, which now has the form

$$\frac{p}{q_1 \cdot q_2 \cdot \ldots \cdot q_l} \quad \text{with} \quad p < q_1 \cdot q_2 \cdot \ldots \cdot q_l,$$

° According to the text: $q = q_1^{\alpha_1} \cdot q_2^{\alpha_2} \cdot \ldots \cdot q_l^{\alpha_l}$ (not all factors different).
[†] The procedure would be the same with other q_i's.

may be written as
$$\frac{1}{q_1}\frac{p}{q_2 \cdot q_3 \cdot \ldots \cdot q_l};$$
we shall have for the second factor, if $p > q_2 \cdot q_3 \cdot \ldots \cdot q_l$ (otherwise we do with the next q_i's the same as with q_1),
$$\frac{p}{q_2 \cdot q_3 \cdot \ldots \cdot q_l} = p_1 + \frac{r_1}{q_2 \cdot q_3 \cdot \ldots \cdot q_l} \quad (r_1 < q_2 \cdot q_3 \cdot \ldots \cdot q_l),$$
whence, for the original expression,
$$\frac{p}{q} = \frac{p_1}{q_1} + \frac{1}{q_1} \cdot \frac{r_1}{q_2 \cdot q_3 \cdot \ldots \cdot q_l}.$$

Writing again
$$\frac{r_1}{q_2 \cdot q_3 \cdot \ldots \cdot q_l} = \frac{1}{q_2}\frac{r_1}{q_3 \cdot q_4 \cdot \ldots \cdot q_l},$$
we shall have, if $r_1 > q_3 \cdot q_4 \cdot \ldots \cdot q_l$,
$$\frac{r_1}{q_3 \cdot q_4 \cdot \ldots \cdot q_l} = p_2 + \frac{r_2}{q_3 \cdot q_4 \cdot \ldots \cdot q_l} \quad (r_2 < q_3 \cdot q_4 \cdot \ldots \cdot q_l),$$
whence, for the original expression,
$$\frac{p}{q} = \frac{p_1}{q_1} + \frac{p_2}{q_1}\frac{1}{q_2} + \frac{1}{q_1}\frac{1}{q_2} \cdot \frac{r_2}{q_3 \cdot q_4 \cdot \ldots \cdot q_l}.$$

Let us write once more
$$\frac{r_2}{q_3 \cdot q_4 \cdot \ldots \cdot q_l} = \frac{1}{q_3}\frac{r_2}{q_4 \cdot q_5 \cdot \ldots \cdot q_l},$$
so again, if $r_2 > q_4 \cdot q_5 \cdot \ldots \cdot q_l$,
$$\frac{r_2}{q_4 \cdot q_5 \cdot \ldots \cdot q_l} = p_3 + \frac{r_3}{q_4 \cdot q_5 \cdot \ldots \cdot q_l} \quad (r_3 < q_4 \cdot q_5 \cdot \ldots \cdot q_l),$$
giving
$$\frac{p}{q} = \frac{p_1}{q_1} + \frac{p_2}{q_1}\frac{1}{q_2} + \frac{p_3}{q_1}\frac{1}{q_2}\frac{1}{q_3} + \frac{1}{q_1}\frac{1}{q_2}\frac{1}{q_3} \cdot \frac{r_3}{q_4 \cdot q_5 \ldots \cdot q_l}.$$

Continuing in the same manner, we shall end with
$$\frac{r_{l-3}}{q_{l-1} \cdot q_l} = p_{l-2} + \frac{r_{l-2}}{q_{l-1} \cdot q_l} \quad (r_{l-2} < q_{l-1} \cdot q_l).$$

Writing then
$$\frac{r_{l-2}}{q_{l-1} \cdot q_l} = \frac{1}{q_{l-1}}\frac{r_{l-2}}{q_l},$$

we shall have, if $r_{l-2} > q_l$,

$$\frac{r_{l-2}}{q_l} = p_{l-1} + \frac{r_{l-1}}{q_l} \quad (r_{l-1} = p_l < q_l).$$

So the complete representation of the fraction will be

$$\frac{p}{q} = \frac{p_1}{q_1} + \frac{p_2}{q_1}\frac{1}{q_2} + \frac{p_3}{q_1}\frac{1}{q_2}\frac{1}{q_3} + \ldots + \frac{p_{l-1}}{q_1}\frac{1}{q_2}\frac{1}{q_3} \ldots \frac{1}{q_{l-1}} + \frac{p_l}{q_1}\frac{1}{q_2}\frac{1}{q_3} \ldots \frac{1}{q_l}.$$

Remark. Another way to write this is by means of the ascending fraction

$$\frac{p}{q} = \cfrac{p_1 + \cfrac{p_2 + \cfrac{p_3 + \cfrac{\ldots + \cfrac{p_l}{q_l}}{q_4}}{q_3}}{q_2}}{q_1}.$$

(*xxxi′*) Since we have performed the division by taking the elementary numbers in decreasing order, the larger denominators appear first in the final result.

But we may also change the order of the factors so that, as is customary, the smaller denominators appear first. See the final form of the result in A.80.

(**A.80**) MS $\mathcal{A}$. Example. Find the representation of

$$\frac{p}{q} = \frac{1836}{5040}.$$

First, factorize the denominator:
$$5040 = 10 \cdot 504 = 10 \cdot 9 \cdot 56 = 10 \cdot 9 \cdot 8 \cdot 7.$$

Next, proceed with the successive divisions of the numerator:

$$\frac{1836}{5040} = \frac{1}{10}\frac{1836}{504} = \frac{1}{10}\left(3 + \frac{324}{504}\right) = \frac{3}{10} + \frac{1}{10}\frac{324}{504}$$

$$\frac{1}{10}\frac{324}{504} = \frac{1}{10}\frac{1}{9}\frac{324}{56} = \frac{1}{10}\frac{1}{9}\left(5 + \frac{44}{56}\right) = \frac{5}{10}\frac{1}{9} + \frac{1}{10}\frac{1}{9}\frac{44}{56}$$

$$\frac{1}{10}\frac{1}{9}\frac{44}{56} = \frac{1}{10}\frac{1}{9}\frac{1}{8}\frac{44}{7} = \frac{1}{10}\frac{1}{9}\frac{1}{8}\left(6 + \frac{2}{7}\right) = \frac{6}{10}\frac{1}{9}\frac{1}{8} + \frac{1}{10}\frac{1}{9}\frac{1}{8}\frac{2}{7},$$

and no further reduction is possible. So finally, with the factors arranged as in the text,

$$\frac{1836}{5040} = \frac{3}{10} + \frac{5}{9}\frac{1}{10} + \frac{6}{8}\frac{1}{9}\frac{1}{10} + \frac{2}{7}\frac{1}{8}\frac{1}{9}\frac{1}{10}.$$

Remark. Beginning the computation with the smaller denominators would give
$$\frac{1836}{5040} = \frac{2}{7} + \frac{4}{7}\frac{1}{8} + \frac{3}{7}\frac{1}{8}\frac{1}{9} + \frac{6}{7}\frac{1}{8}\frac{1}{9}\frac{1}{10}.$$
But in other cases this might not work: we could end up with a remainder larger than a divisor already used.

C. *Other cases of denomination*.

— **Denominating 1, or any digit, from a composite.**

(A.81) MSS $\mathcal{A}, \mathcal{B}, \mathcal{C}$. Denominate 1 from 12.

Since $12 = 3 \cdot 4$, so $\frac{1}{4} \cdot 12 = 3$ while $\frac{1}{3} \cdot 3 = 1$, or else $\frac{1}{3} \cdot 12 = 4$ while $\frac{1}{4} \cdot 4 = 1$, so we have in both cases
$$1 = \frac{1}{3}\frac{1}{4} \cdot 12 \quad \text{and} \quad \frac{1}{12} = \frac{1}{3}\frac{1}{4}.$$

Since also $12 = 2 \cdot 6$, so $\frac{1}{6} \cdot 12 = 2$ while $\frac{1}{2} \cdot 2 = 1$, or else $\frac{1}{2} \cdot 12 = 6$ while $\frac{1}{6} \cdot 6 = 1$, so we have
$$1 = \frac{1}{2}\frac{1}{6} \cdot 12 \quad \text{and} \quad \frac{1}{12} = \frac{1}{2}\frac{1}{6}.$$

(A.82) MSS $\mathcal{A}, \mathcal{B}, \mathcal{C}$. Denominate 1 from 13.

Since 13 is prime, no factorization is possible and we have simply
$$1 = \frac{1}{13} \cdot 13.$$

(A.83) MSS $\mathcal{A}, \mathcal{B}, \mathcal{C}$. Denominate 1 from 14.

Since $14 = 2 \cdot 7$, so $\frac{1}{7} \cdot 14 = 2$ while $\frac{1}{2} \cdot 2 = 1$, or else $\frac{1}{2} \cdot 14 = 7$ while $\frac{1}{7} \cdot 7 = 1$, so we have in either case
$$1 = \frac{1}{2}\frac{1}{7} \cdot 14 \quad \text{and} \quad \frac{1}{14} = \frac{1}{2}\frac{1}{7}.$$

— **Denominate 1, or any digit, from repeated thousands.**

(A.84) MSS $\mathcal{A}, \mathcal{B}, \mathcal{C}$. Denominate 1 from 1000.

Since $100 = \frac{1}{10} \cdot 1000$, $10 = \frac{1}{10} \cdot 100$, $1 = \frac{1}{10} \cdot 10$, then
$$1 = \frac{1}{10}\frac{1}{10}\frac{1}{10} \cdot 1000.$$

Likewise, for the next repetition, since $1000 = \frac{1}{10}\frac{1}{10}\frac{1}{10} \cdot 1\,000\,000$ while (from above) $1 = \frac{1}{10}\frac{1}{10}\frac{1}{10} \cdot 1000$, so
$$1 = \frac{1}{10}\frac{1}{10}\frac{1}{10}\frac{1}{10}\frac{1}{10}\frac{1}{10} \cdot 1\,000\,000.$$

Generally, if the denominator has k repetitions of 'thousand', we shall repeat $3k$ times '(of) a tenth' since

$$1 = \left(\frac{1}{10}\frac{1}{10}\frac{1}{10}\right)^k \cdot 10^{3k} = \left(\frac{1}{10}\right)^{3k} \cdot 10^{3k}.$$

(A.85) MSS $\mathcal{A}, \mathcal{B}, \mathcal{C}$. Denominate 1 from 80 000.

The principle for denominating numbers with repetitions is just the same as that seen in taking fractions from repetitions (A.46–51): If the denominator is $a \cdot 1000^k$, where a (≤ 999) is the integer preceding the k 'thousand's, we shall consider

$$\left(\frac{1}{a}\right) \cdot \frac{1}{1000^k};$$

that is, we shall denominate 1 (or the digit taking its place) from a and append $3k$ times '(of) a tenth'. In the present case, where $80\,000 = 80 \cdot 1000$, since $1 = \frac{1}{8}\frac{1}{10} \cdot 80$ while $1 = \frac{1}{10}\frac{1}{10}\frac{1}{10} \cdot 1000$, then

$$1 = \frac{1}{8}\frac{1}{10}\frac{1}{10}\frac{1}{10}\frac{1}{10} \cdot 80\,000.$$

— **Denominating composites.**

(A.86) MSS $\mathcal{A}, \mathcal{B}, \mathcal{C}$. Denominate 12 from 27.

Since $27 = 3 \cdot 9$, so $3 = \frac{1}{9} \cdot 27$. But $12 = 4 \cdot 3$, so

$$12 = \frac{4}{9} \cdot 27.$$

Remark. This is a particular case since the two terms have a common factor.

(A.87) MSS $\mathcal{A}, \mathcal{B}, \mathcal{C}$. Denominate 14 from 45.

Since $45 = 9 \cdot 5$, so $5 = \frac{1}{9} \cdot 45$. But $14 = \left(2 + \frac{4}{5}\right) \cdot 5$, so

$$14 = \left(\frac{2}{9} + \frac{4}{5}\frac{1}{9}\right) \cdot 45.$$

Remark. When he comes to deal with fractions (from A–V on), the author will calculate this as follows:

$$\frac{14}{45} = \frac{14}{5 \cdot 9} = \frac{10 + 4}{5 \cdot 9} = \frac{2}{9} + \frac{4}{5}\frac{1}{9}.$$

— **Denominating from repeated thousands.**

Analogous to the division of numbers with repetitions (A.55–59): if we are to denominate $a \cdot 1000^m$ from $b \cdot 1000^n$ ($a, b < 1000$, $m < n$), we shall denominate a from b if $a < b$ or from $b \cdot 1000$ if $a > b$; then we shall append the repetitions left after removing those which are common; that is to say, we shall append as many times $\frac{1}{10}\frac{1}{10}\frac{1}{10}$ as is either $n - m$ or $n - m - 1$. The two coming problems illustrate these two operations.

(A.88) MSS $\mathcal{A}, \mathcal{B}, \mathcal{C}$. Denominate 5000 from 40 000 000 000.

Removing the common repetition leaves us with denominating 5 from 40 000 000, thus (disregarding two 'thousand') 5 from 40, which is $\frac{1}{8}$. The answer is therefore, transcribing the 'tenth repeated' of the text by an exponent,

$$\frac{5\,000}{40\,000\,000} = \frac{1}{8}\left(\frac{1}{10}\right)^6.$$

(A.89) MSS $\mathcal{A}, \mathcal{B}, \mathcal{C}$. Denominate 400 from 10 000 000.

Denominating 400 from 10 000 gives $\frac{2}{5}\frac{1}{10}$. Appending the repetition gives
$$\frac{400}{10\,000\,000} = \frac{2}{5}\left(\frac{1}{10}\right)^4.$$

Chapter A–V:
Multiplication of fractions

Summary

Multiplication of fractional expressions will involve the four types of quantities encountered at the beginning of Chapter A–III, namely

— integers N

— fractions $\dfrac{k}{l}$ $\quad(k<l)$

— fractions of a fraction $\dfrac{r}{s}\dfrac{1}{l}$

— compound fractions $\dfrac{k}{l}+\dfrac{r}{s}\dfrac{1}{l}$.

Multiplying any of these four types by the first three will produce a result of the same form, at most with the denominators of the multiplier being appended to those of the multiplicand; indeed,

$$N\left(\dfrac{k}{l}+\dfrac{r}{s}\dfrac{1}{l}\right)=\dfrac{N\cdot k}{l}+\dfrac{N\cdot r}{s}\dfrac{1}{l},$$

$$\dfrac{p}{q}\left(\dfrac{k}{l}+\dfrac{r}{s}\dfrac{1}{l}\right)=\dfrac{p\cdot k}{l}\dfrac{1}{q}+\dfrac{p\cdot r}{s}\dfrac{1}{l}\dfrac{1}{q}.$$

This is no longer the case if both factors contain two or more terms added. For then

$$\left(N+\dfrac{p}{q}\right)\left(\dfrac{k}{l}+\dfrac{r}{s}\dfrac{1}{l}\right)=\dfrac{1}{l}\left(N\cdot k+\dfrac{p\cdot k}{q}+\dfrac{N\cdot r}{s}+\dfrac{p\cdot r}{q}\dfrac{1}{s}\right),$$

$$\left(\dfrac{p}{q}+\dfrac{t}{u}\dfrac{1}{q}\right)\left(\dfrac{k}{l}+\dfrac{r}{s}\dfrac{1}{l}\right)=\dfrac{1}{q}\dfrac{1}{l}\left(p\cdot k+\dfrac{t\cdot k}{u}+\dfrac{p\cdot r}{s}+\dfrac{t\cdot r}{u}\dfrac{1}{s}\right),$$

and we have two fractions with different denominators in the parentheses. In order to arrive at the customary form, we shall have to convert one of these fractions into the other.

It is on the basis of this distinction that the major part of the present chapter is arranged. Thus we are first taught multiplications which do not require such conversion (A.90–107, to which may be added A.123); next, the operation of conversion is explained (A.108–122); and then the remaining cases of multiplication, where conversion may or must be performed, are treated (A.124–130, A.136–137). The chapter ends with cases where the fractional expressions involved are of other types: first, so-called irregular fractions (A.131–135 & A.138–139) and then more intricate expressions (A.140–144).

1. Multiplication of a fractional expression by an integer.

Suppose we are to compute
$$\frac{p}{q} \cdot N,$$
where $p \neq 1$ since otherwise it would be the division of two integers. Three ways are used by the author:

1. Multiply the fractional expression by its denominator, then the (integral) product by the integer, and divide the result by the denominator.
$$\frac{p}{q} \cdot N = \left[\left(\frac{p}{q} \cdot q\right) \cdot N\right] \frac{1}{q} = (p \cdot N) \frac{1}{q}.$$

2. Multiply the numerator of the fraction directly by the integer and divide the product by the denominator.
$$\frac{p}{q} \cdot N = \frac{p \cdot N}{q}.$$

3. Divide the integer by the denominator and multiply the quotient by the numerator.
$$\frac{p}{q} \cdot N = \frac{N}{q} \cdot p.$$

These are the three ways explained (with the first demonstrated) in A.90. If, instead of a fraction as in A.90–92, we have a fraction of a fraction (A.93–95) or a compound fraction (A.96–98), the product of the denominators will be formed; otherwise, the fractions to be multiplied may be taken in succession. Note that the main denominator in the results is commonly that of the multiplying fraction (but see, e.g., A.124*b*). These denominators must be kept, even if a simpler form is possible (see remarks in A.97 and A.121, pp. 1206 & 1216 below).

2. Multiplication of fractional expressions not containing integers.

We find successively: fraction by fraction (A.99–101), fraction by fraction of a fraction (A.102–103), fraction of a fraction by fraction of a fraction (A.104–105), fraction by compound fraction (A.106–107; the product of a compound fraction by a fraction of a fraction is included).*

As before, we may first multiply the fractional expression by the product of the denominators or directly multiply the numerators and divide the result by the denominators taken in succession. Furthermore, when one compound fraction occurs, multiplication term by term is possible. Here

---

* Other cases will be treated later on for they require knowledge of Sections 3 and 4: those in A.126–127 (product of two compound fractions) and A.136 (one factor is a sum of 'different fractions', Arabic *kusūr mukhtalifa*).

again, the denominators in the result are those of the second factor, if not stated otherwise (see A.101 b, A.106 c).

3. Conversion of fractions.

The nature of conversion is explained in A.108 and the general rule in A.109. Suppose a whole is divided into l equal parts and then into s equal parts and we wish to know how many of the second parts are equivalent to a given number, say k, of the first parts. Since $k : l = r : s$, with r required, then

$$r = \frac{k \cdot s}{l} = \alpha + \frac{\beta}{l}.$$

A simple fraction will thus mostly be transformed into a compound fraction —even though the text speaks about 'conversion of a fraction into a fraction'.

We are first taught the conversion, into a proposed kind of fraction (one denominator given), of fractions (A.108–111), compound fractions (A.112–114), fractions of a fraction (A.115–116). Next, we consider the conversion, this time into a proposed kind of fraction of a fraction (two denominators given), of fractions of a fraction (A.117–120) and compound fractions (A.121–122).

4. Multiplication of fractional expressions containing integers.

The following cases are considered:

— integer and fraction by fraction (A.123);

— integer and fraction by integer and fraction (A.124–125, A.130);

— integer and compound fraction by integer and fraction (A.128), by integer and compound fraction (A.129), by compound fraction alone (A.137, misplaced).

We may form the product of the denominators, as before, or multiply term by term (*secundum differentias*). The second way will mostly entail converting fractions.

5. Case of irregular fractions.

An irregular fraction (*fractio irregularis*, Arabic *kasr al-'adad*, 'fraction of a number') is of the form

$$\frac{k}{l} \text{ of } M \; \left(\text{thus } \frac{k}{l} \cdot M\right),$$

where M may be an integer or an integer with a fraction.

We consider first the products involving such expressions. Thus an irregular fraction with M an integer is multiplied by an integer (A.131) or by a similar expression (A.134), and an irregular fraction with M an integer with a fraction is successively multiplied by an integer (A.132), by an integer with a fraction (A.133) or by a similar expression (A.135).

With irregular fractions comes the problem of ambiguous formulation. The absence of case-endings in English means that an enunciation such as that of A.138, 'multiply a fourth of five and two fifths of six by a tenth of three and an eighth of four', admits of (e.g.) two interpretations, namely

$$\left(\frac{1}{4}\cdot 5 + \frac{2}{5}\cdot 6\right)\left(\frac{1}{10}\cdot 3 + \frac{1}{8}\cdot 4\right)$$

$$\frac{1}{4}\left(5 + \frac{2}{5}\cdot 6\right)\cdot\frac{1}{10}\left(3 + \frac{1}{8}\cdot 4\right).$$

In other words, it is unclear whether the fraction at the beginning of each of the two expressions multiplies just the first term or both. Dealing with such various interpretations is the object of A.135 –135′ & A.138–139. Here the author is obviously complying with an Arabic tradition: with the Latin case-endings no confusion is possible (Part I, p. xviii).

6. Other problems.

After a particular case (A.140) come four problems which deal with the multiplication of combined expressions where the multiplier or the multiplicand contains *subtractive* fractions (A.141–144). They are found only in MS $\mathcal{B}$, where they follow our A.140. With $\mathcal{B}$ transmitting the disordered text, the question of their actual place in the *Liber mahameleth* might arise. Our decision to leave them after A.140 was guided by the fact that the subtractions involved are elementary and do not require knowledge of the subtraction of fractions taught in A–VII. Furthermore, such problems about *kasr bi'l-istithnā'* also occur at the end of the chapter on multiplication in a contemporary Spanish-Arabic treatise, that by Muḥammad ibn 'Abdallāh al-Ḥaṣṣār.*

---

1. Multiplication of a fractional expression by an integer.

Preliminary remark. From now on until the end of A–VI, the text often contains tables setting out in numerical symbols the quantities involved. It is worth noting that they are mainly just a transcription of the data. Often they do not contain any result. Whatever the case, these tables should *not* be considered as a mean of computation for there is no indication of any operation (or, at most, a small line connecting the quantities to be multiplied, as is commonly seen in mediaeval mathematical manuscripts). In these figures an aliquot fraction $\frac{1}{q}$ is represented either by $\frac{1}{q\cdot}$ or by $q\cdot$, depending on whether the text reads 'one fourth' or 'a fourth';[†] a fraction $\frac{p}{q}$ is represented by $\frac{p}{q\cdot}$ ($\frac{p}{q\cdot}$ means $p + \frac{1}{q}$). The copyists did not understand the meaning of these dots. In MSS $\mathcal{B}$ and $\mathcal{C}$ they are missing in half the

---

* See Suter's summary, p. 28.
† See examples from A.111 on; exceptions are rare (one in A.138).

instances; in addition, in MS $\mathcal{C}$ they frequently occur where they should not (A.138, A.146, A.153). They were obviously often mistaken for the dots commonly used in mediaeval manuscripts to separate numbers written in figures from words (with 2, e.g., appearing as 2. or .2.). The case of MS $\mathcal{A}$ is different: the second hand, who added the figures, made use of the fractional line to transcribe the fractions. At least that is how he started off; later he gave up and just reproduced what he thought he saw. That is why, in the table found in A.135, we find a mediaeval abbreviation for *et* confused with 7, and, in the introduction to Book B, $\frac{1}{2.}$ (meaning $1 + \frac{1}{2}$) is taken to be $\frac{1}{2}$ (further such examples in B.4 and B.6). This mathematical commentary does not include any of these tables, possibly later additions (see Part I, p. lix).

A. *Multiplicand a fraction*.

(A.90) MSS $\mathcal{A}, \mathcal{B}, \mathcal{C}$. Multiplication of a simple fraction by a digit.

$$\frac{3}{4} \cdot 7$$

We find here the three ways mentioned in our summary.

(a) Introducing the denominator as a further multiplier.

$$\frac{3}{4} \cdot 7 = \left[\left(\frac{3}{4} \cdot 4\right) 7\right] \frac{1}{4} = (3 \cdot 7) \frac{1}{4} = \frac{21}{4} = 5 + \frac{1}{4}.$$

Demonstration. The text justifies this with two arguments (or: steps).$^{\|}$

— Since

$$\frac{3}{4} : 1 = \frac{3}{4} \cdot m : m,$$

for any m (*Elements* VII.18), we have on the one hand ($m = 4$)

$$\frac{3}{4} : 1 = \frac{3}{4} \cdot 4 : 4,$$

and on the other ($m = 7$)

$$\frac{3}{4} : 1 = \frac{3}{4} \cdot 7 : 7.$$

Equating the right sides leads us to

$$\frac{3}{4} \cdot 7 = \left[\left(\frac{3}{4} \cdot 4\right) 7\right] \frac{1}{4}.$$

— Since, as mentioned by the author in the multiplication table for integers, multiplying any number (whether integral or fractional) by 1 leaves it unchanged, we have

$$\frac{1}{4} \cdot 1 = \frac{1}{4}, \quad \frac{3}{4} \cdot 1 = \frac{3}{4} = 3 \cdot \frac{1}{4},$$

and, taking m such elements (and using P_5),

$^{\|}$ The extant text seems to be lacunary.

$$\frac{1}{4} \cdot m = \frac{m}{4}, \quad \frac{3}{4} \cdot m = \frac{3 \cdot m}{4},$$

whence in our case

$$\frac{3}{4} \cdot 7 = \frac{21}{4},$$

and performing the division will give the required result, as in b.

(**b**) Directly multiplying the numerator by the integer, then dividing the result by the denominator:

$$\frac{3}{4} \cdot 7 = \frac{3 \cdot 7}{4}.$$

(**c**) Dividing the integer first, then multiplying:*

$$\frac{3}{4} \cdot 7 = \frac{7}{4} \cdot 3 = \left(1 + \frac{3}{4}\right) 3.$$

(**A.91**) MSS *A, B, C.* Multiplication of a simple fraction by a composite integer.

$$\frac{3}{5} \cdot 47$$

(**a**) As in A.90 we may consider the three ways

$$\frac{3}{5} \cdot 47 = \left[\left(\frac{3}{5} \cdot 5\right) 47\right] \frac{1}{5}, \quad \frac{3 \cdot 47}{5}, \quad \frac{47}{5} \cdot 3.$$

(**b**) In this last case we may use the method seen in A.53 and so reduce the problem to A.90 (multiplication of a fraction by a digit):

$$\frac{3}{5} \cdot 47 = \frac{3}{5}(45 + 2) = 3 \cdot 9 + \frac{3}{5} \cdot 2 = 27 + 1 + \frac{1}{5} = 28 + \frac{1}{5}.$$

(**A.92**) MSS *A, B, C.* Multiplication of a non-elementary fraction by a composite integer.†

$$\frac{8}{13} \cdot 46$$

(**a**) Multiplying by the denominator.

$$\frac{8}{13} \cdot 46 = \left[\left(\frac{8}{13} \cdot 13\right) 46\right] \frac{1}{13} = \frac{8 \cdot 46}{13} = \frac{368}{13} = 28 + \frac{4}{13}.$$

---

* Remember (p. 1172, footnote) that computations or results written in italics here are not found in the text.

† An 'elementary' fraction is one with denominator ≤ 10, a 'non-elementary' one thus with denominator > 10. The way of expressing these two classes of fractions is different in Arabic. See above, p. 1182.

(**b**) Directly, multiplying together numerator and multiplier.
$$\frac{8}{13} \cdot 46 = \frac{8 \cdot 46}{13}.$$

(**c**) Proceed as in A.91b, taking the largest integer below 46 divisible by 13:
$$\frac{8}{13} \cdot 46 = \frac{8}{13}(39 + 7) = 8 \cdot 3 + \frac{8 \cdot 7}{13} = 24 + 4 + \frac{4}{13} = 28 + \frac{4}{13}.$$

B. *Multiplicand a fraction of a fraction*.

(**A.93**) MSS $\mathcal{B}, \mathcal{C}; \mathcal{A}$, but omits the demonstration and b. Multiplication of a fraction of a fraction by an integer.
$$\frac{3}{4}\frac{1}{7} \cdot 15$$

(**a**) Forming the product of the denominators: $4 \cdot 7 = 28$; then
$$\frac{3}{4}\frac{1}{7} \cdot 15 = \left[\left(\frac{3}{4}\frac{1}{7} \cdot 28\right) 15\right] \frac{1}{28} = \frac{3 \cdot 15}{28} = \frac{45}{28}$$
$$= \frac{28 + 16 + 1}{4 \cdot 7} = 1 + \frac{4}{7} + \frac{1}{4}\frac{1}{7}.$$

Demonstration. As in A.90a, with two arguments.

— Since, as known from A.90a,
$$\frac{3}{4}\frac{1}{7} 28 : 28 = \frac{3}{4}\frac{1}{7} 15 : 15,$$
we have
$$\frac{3}{4}\frac{1}{7} 15 = \frac{\left(\frac{3}{4}\frac{1}{7} 28\right) 15}{28}.$$

— Considering that
$$\frac{3}{4}\frac{1}{7} \cdot 1 = \frac{3}{4}\frac{1}{7},$$
and taking fifteen times this gives
$$\frac{3}{4}\frac{1}{7} \cdot 15 = \frac{3 \cdot 15}{4}\frac{1}{7} = \frac{3 \cdot 15}{4 \cdot 7} = \frac{45}{4}\frac{1}{7} \quad \text{or} \quad \frac{45}{7}\frac{1}{4}.$$

In other words, we multiply the numerator of the fraction by the integer and divide the result by one denominator, then by the other.

(**b**) Directly, multiplying the numerator by the integer and dividing the result by the product of the denominators:
$$\frac{3}{4}\frac{1}{7} \cdot 15 = \frac{3 \cdot 15}{4 \cdot 7} = \frac{45}{28}.$$

Demonstration. By P_4,

$$\frac{\frac{45}{4}}{7} = \frac{\frac{45}{7}}{4} = \frac{45}{4 \cdot 7} = \frac{45}{28}.$$

(*c*) Case of successive division. Consider

$$\left(\frac{3 \cdot 15}{4}\right)\frac{1}{7} \quad \text{or} \quad \left(\frac{3 \cdot 15}{7}\right)\frac{1}{4}.$$

In the first case we obtain

$$\frac{45}{4}\frac{1}{7} = \left(11 + \frac{1}{4}\right)\frac{1}{7} = 1 + \frac{4}{7} + \frac{1}{4}\frac{1}{7};$$

in the second

$$\frac{45}{7}\frac{1}{4} = \left(6 + \frac{3}{7}\right)\frac{1}{4} = 1 + \frac{2}{4} + \frac{3}{7}\frac{1}{4}.$$

(*d*) Reduction to the former subsection (A.90–92). (This differs only in form from *c*.)

$$\frac{3}{4}\frac{1}{7} \cdot 15 = \left(\frac{3}{4} \cdot 15\right)\frac{1}{7} = \left(11 + \frac{1}{4}\right)\frac{1}{7}.$$

(**A.94**) MSS $\mathcal{B}, \mathcal{C}; \mathcal{A}$, but without *c*. Multiplication of a non-elementary fraction of a fraction by an integer.

$$\frac{3}{7}\frac{1}{11} \cdot 36$$

(*a*) Forming the product of the denominators $7 \cdot 11 = 77$, compute

$$\frac{3}{7}\frac{1}{11} \cdot 36 = \left[\left(\frac{3}{7}\frac{1}{11} \cdot 77\right)36\right]\frac{1}{77} = \frac{108}{77} = \frac{77 + 28 + 3}{7 \cdot 11} = 1 + \frac{4}{11} + \frac{3}{7}\frac{1}{11}.$$

(*b*) Consider as before

$$\frac{3}{7}\frac{1}{11} \cdot 36 = \frac{3 \cdot 36}{7}\frac{1}{11} = \frac{108}{7}\frac{1}{11} \quad \text{or} \quad \frac{108}{11}\frac{1}{7}.$$

(*c*) Or else take

$$\left(\frac{3}{7} \cdot 36\right)\frac{1}{11} = \left(15 + \frac{3}{7}\right)\frac{1}{11}.$$

• Generally, we are told, if we are to multiply an integer by a fraction of a fraction with several terms, we shall multiply the numerator by the integer and divide the result by the product of the denominators:

$$\frac{k}{l_1}\frac{1}{l_2} \cdots \frac{1}{l_m} \cdot N = \frac{k \cdot N}{l_1 \cdot l_2 \cdot \ldots \cdot l_m}.$$

(**A.95**) MSS $\mathcal{A}, \mathcal{B}, \mathcal{C}$. Multiplication of a fraction of a fraction of a fraction by an integer.

$$\frac{3\ 1\ 1}{7\ 5\ 8} \cdot 59$$

From the wording ('for instance'), this problem is supposed to illustrate the above rule; as a matter of fact, the treatment contains just the same intermediary steps as before (using the product of the denominators and dividing in succession). The above rule might have initially been added in the margin (see Translation, p. 660, note 346).

(*a*) Forming the product of the denominators: $7 \cdot 5 \cdot 8 = 280$ and multiplying the fraction by it.

$$\frac{3\ 1\ 1}{7\ 5\ 8} \cdot 59 = \left[\left(\frac{3\ 1\ 1}{7\ 5\ 8} \cdot 280\right) \cdot 59\right] \frac{1}{280} = \frac{3 \cdot 59}{280} = \frac{177}{280}.$$

(*b*) Multiplying together integer and numerator and dividing successively.

$$\frac{3\ 1\ 1}{7\ 5\ 8} \cdot 59 = \left(\frac{177}{5}\right)\frac{1\ 1}{7\ 8} = \left(35 + \frac{2}{5}\right)\frac{1\ 1}{7\ 8} = \left(5 + \frac{2\ 1}{5\ 7}\right)\frac{1}{8} = \frac{5}{8} + \frac{2\ 1\ 1}{5\ 7\ 8}.$$

We may take, the author observes, the fractions in any other order, for instance

$$\frac{3\ 1\ 1}{7\ 5\ 8} \cdot 59 = \frac{177}{7}\frac{1\ 1}{5\ 8} = \left(25 + \frac{2}{7}\right)\frac{1\ 1}{5\ 8} = \frac{5}{8} + \frac{2\ 1\ 1}{7\ 5\ 8}.$$

Remark. As we have already seen in A.93*c* (and shall see later on: A.101*b*), the form of the result may then vary; thus, with 8 as the first divisor, we shall have

$$\frac{177}{8}\frac{1\ 1}{7\ 5} = \left(22 + \frac{1}{8}\right)\frac{1\ 1}{7\ 5} = \left(3 + \frac{1}{7} + \frac{1\ 1}{8\ 7}\right)\frac{1}{5} = \frac{3}{5} + \frac{1\ 1}{7\ 5} + \frac{1\ 1\ 1}{8\ 7\ 5}.$$

Showing the equivalence of such different results requires knowledge of the next section, on converting fractions. But for the successive divisions the denominators will usually be taken in increasing order.

(*c*) Considering the fractions in succession (differs only in form from *b*).

$$\frac{3\ 1\ 1}{7\ 5\ 8} \cdot 59 = \left[\left(\frac{3}{7} \cdot 59\right)\frac{1}{5}\right]\frac{1}{8}.$$

These various ways are applicable whatever the number of fractions in the multiplicand, the author concludes.

C. *Multiplicand a compound fraction*.

(**A.96**) MSS $\mathcal{A}, \mathcal{B}, \mathcal{C}$. Multiplication of a compound fraction by an integer.

$$\left(\frac{5}{7} + \frac{3\ 1}{4\ 7}\right) \cdot 10$$

(**a**) Forming the product of the denominators: $4 \cdot 7 = 28$, so

$$\left(\frac{5}{7} + \frac{3}{4}\frac{1}{7}\right) \cdot 10 = \left[\left(\frac{5}{7} \cdot 28 + \frac{3}{4}\frac{1}{7} \cdot 28\right) 10\right] \frac{1}{28}$$

$$= \frac{230}{28} = \frac{224 + 4 + 2}{28} = 8 + \frac{1}{7} + \frac{1}{2}\frac{1}{7}.$$

The proof for this case (of compound fraction), the author adds, is like the one already seen, which is to mean that seen for a fraction (A.90a) and a fraction of a fraction (A.93a).*

(**b**) Considering the fractions in succession:

$$\left(\frac{5}{7} + \frac{3}{4}\frac{1}{7}\right) \cdot 10 = \left[\left(5 + \frac{3}{4}\right) \cdot 10\right] \frac{1}{7} = \left(57 + \frac{1}{2}\right) \frac{1}{7} = 8 + \frac{1}{7} + \frac{1}{2}\frac{1}{7}.$$

(**A.97**) MSS $\mathcal{A}$, $\mathcal{B}$, $\mathcal{C}$. Multiplication of a compound fraction with several terms by an integer (extension, as in A.95, to several fractions with elementary denominators (≤ 10)).

$$\left(\frac{3}{7}\frac{1}{10} + \frac{1}{4}\frac{1}{7}\frac{1}{10}\right) \cdot 54$$

(**a**) Forming the product of the denominators: $4 \cdot 7 \cdot 10 = 280$, so

$$\left(\frac{3}{7}\frac{1}{10} + \frac{1}{4}\frac{1}{7}\frac{1}{10}\right) \cdot 54 = \left[\left(\frac{3}{7}\frac{1}{10} \cdot 280 + \frac{1}{4}\frac{1}{7}\frac{1}{10} \cdot 280\right) 54\right] \frac{1}{280}$$

$$= \frac{13 \cdot 54}{280} = \frac{702}{280} = \frac{560 + 140 + 2}{4 \cdot 7 \cdot 10} = 2 + \frac{5}{10} + \frac{1}{2}\frac{1}{7}\frac{1}{10}.$$

Remark. And not, since we are to keep the same kind of fraction,

$$\frac{560 + 140 + 2}{4 \cdot 7 \cdot 10} = 2 + \frac{1}{2} + \frac{1}{2}\frac{1}{7}\frac{1}{10}.$$

(**b**) Considering the fractions in succession:

$$\left(\frac{3}{7}\frac{1}{10} + \frac{1}{4}\frac{1}{7}\frac{1}{10}\right) \cdot 54 = \left[\left(3 + \frac{1}{4}\right) \cdot 54\right] \frac{1}{7}\frac{1}{10} = \left(175 + \frac{1}{2}\right) \cdot \frac{1}{7}\frac{1}{10}$$

$$= \left(25 + \frac{1}{2}\frac{1}{7}\right) \cdot \frac{1}{10} = 2 + \frac{5}{10} + \frac{1}{2}\frac{1}{7}\frac{1}{10}.$$

(**A.98**) MSS $\mathcal{B}$, $\mathcal{C}$ only. Multiplication of a non-elementary compound fraction by an integer.

$$\left(\frac{4}{11} + \frac{1}{5}\frac{1}{11}\right) \cdot 36$$

---

* Using the representation of the compound fraction as $\dfrac{5 + \frac{3}{4}}{7}$.

(**a**) Forming the product of the denominators: $5 \cdot 11 = 55$, so

$$\left(\frac{4}{11} + \frac{1}{5}\frac{1}{11}\right) \cdot 36 = \left[\left(\frac{4}{11} \cdot 55 + \frac{1}{5}\frac{1}{11} \cdot 55\right) \cdot 36\right] \frac{1}{55}$$
$$= \frac{21 \cdot 36}{55} = \frac{756}{55} = \frac{715 + 40 + 1}{5 \cdot 11} = 13 + \frac{8}{11} + \frac{1}{5}\frac{1}{11}.$$

(**b**) Considering the fractions in succession

$$\left(\frac{4}{11} + \frac{1}{5}\frac{1}{11}\right) \cdot 36 = \left[\left(4 + \frac{1}{5}\right) \cdot 36\right] \frac{1}{11} = \left(151 + \frac{1}{5}\right) \frac{1}{11}$$
$$= 13 + \frac{8}{11} + \frac{1}{5}\frac{1}{11}.$$

---

2. Multiplication of fractional expressions not containing integers.

A. *Multiplicand a fraction*.

We shall meet successively with the multiplications of aliquot, non-aliquot and non-elementary fractions. The denominator of the multiplier is generally the main one (and thus found in the answer).$^{\|}$

(**A.99**) MSS $\mathcal{A}, \mathcal{B}, \mathcal{C}$. Multiplication of one aliquot fraction by another.

$$\frac{1}{4} \cdot \frac{1}{8}$$

Just as we had $\frac{p}{q} \cdot N$ computed from $\frac{p}{q} \cdot N : N = \frac{p}{q} : 1$, we shall consider here

$$\frac{1}{4} \cdot \frac{1}{8} : \frac{1}{8} = \frac{1}{4} : 1.$$

Since the required quantity will be to $\frac{1}{8}$ as $\frac{1}{4}$ is to 1, it will equal a fourth of an eighth, that is, the two denominations will remain unaltered, one just being attached to the other.

(**A.100**) MSS $\mathcal{A}, \mathcal{B}, \mathcal{C}$. Multiplication of one non-aliquot fraction by another.

$$\frac{3}{4} \cdot \frac{5}{8}$$

(**a**) Forming the product of the denominators: $4 \cdot 8 = 32$. This then reduces the problem to A.90, but applied to two terms, and we shall accordingly split up the product of the denominators (whence the subsequent demonstration).

$$\frac{3}{4} \cdot \frac{5}{8} = \left(\frac{3}{4} \cdot 4 \cdot \frac{5}{8} \cdot 8\right) \frac{1}{32} = \frac{15}{32} = \frac{12 + 3}{4 \cdot 8} = \frac{3}{8} + \frac{3}{4}\frac{1}{8}.$$

$^{\|}$ Other possibility mentioned in A.101*b*, A.102*b*.

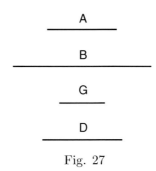

Fig. 27

Demonstration. Let (Fig. 27) $4 = A$, $8 = B$, $\frac{3}{4} \cdot 4 = G$, $\frac{5}{8} \cdot 8 = D$; then (A.90)

$$\frac{G}{A} = \frac{\frac{3}{4} \cdot 4}{4} = \frac{3}{4}, \quad \frac{D}{B} = \frac{\frac{5}{8} \cdot 8}{8} = \frac{5}{8}.$$

Since by P'_3 (Abū Kāmil's proposition)

$$\frac{G}{A} \cdot \frac{D}{B} = \frac{G \cdot D}{A \cdot B},$$

we have proved that indeed

$$\frac{3}{4} \cdot \frac{5}{8} = \frac{\left(\frac{3}{4} \cdot 4\right)\left(\frac{5}{8} \cdot 8\right)}{4 \cdot 8} = \left(\frac{3}{4} 4 \cdot \frac{5}{8} 8\right)\frac{1}{32}.$$

This proof will be referred to several times in similar cases, namely when a fraction is multiplied by a fraction of a fraction (A.102a), or by a compound fraction (A.106a), or when a fraction of a fraction is multiplied by another (A.104a).

(**b**) Forming the product of the numerators and dividing successively.

$$\frac{3}{4} \cdot \frac{5}{8} = \left(\frac{3 \cdot 5}{4}\right)\frac{1}{8} = \frac{15}{4}\frac{1}{8} = \left(3 + \frac{3}{4}\right)\frac{1}{8} = \frac{3}{8} + \frac{3}{4}\frac{1}{8}.$$

Remark. Inverting the divisions, as we are told to do as well, will lead to the less convenient expression

$$\frac{1}{4} + \frac{7}{8}\frac{1}{4}.$$

Demonstration (incomplete; see Translation, p. 662). As in A.90a & A.93a but with a fraction replacing the integer m in the relation

$$\frac{p}{q} \cdot m : m = \frac{p}{q} : 1.$$

(**A.101**) MSS $\mathcal{A}$, $\mathcal{B}$, $\mathcal{C}$. Multiplication of a fraction by a non-elementary fraction.

$$\frac{3}{7} \cdot \frac{9}{13}$$

(**a**) Forming the product of the denominators: $7 \cdot 13 = 91$, then, as before,

$$\frac{3}{7} \cdot \frac{9}{13} = \left(\frac{3}{7} \cdot 7 \cdot \frac{9}{13} 13\right)\frac{1}{91} = \frac{3 \cdot 9}{91} = \frac{27}{91} = \frac{21 + 6}{7 \cdot 13} = \frac{3}{13} + \frac{6}{7}\frac{1}{13}.$$

(**b**) Forming the product of the numerators and dividing by either denominator first.

$$\frac{3}{7} \cdot \frac{9}{13} = \frac{3 \cdot 9}{7} \cdot \frac{1}{13} = \frac{27}{7} \cdot \frac{1}{13} = \left(3 + \frac{6}{7}\right)\frac{1}{13} = \frac{3}{13} + \frac{6}{7}\frac{1}{13},$$

$$\frac{3}{7} \cdot \frac{9}{13} = \frac{3 \cdot 9}{13} \cdot \frac{1}{7} = \frac{27}{13} \cdot \frac{1}{7} = \left(2 + \frac{1}{13}\right)\frac{1}{7} = \frac{2}{7} + \frac{1}{13}\frac{1}{7}.$$

B. *Multiplicand a fraction of a fraction*.

(**A.102**) MSS $\mathcal{A}$, $\mathcal{B}$, $\mathcal{C}$. Multiplication of a fraction of a fraction by a fraction.

$$\frac{3}{4}\frac{1}{5} \cdot \frac{7}{8}$$

(**a**) Taking the denominator of each expression (thus a product for the fraction of a fraction) and multiplying them. They are here $4 \cdot 5 = 20$ and 8, and their product is 160. Then

$$\frac{3}{4}\frac{1}{5} \cdot \frac{7}{8} = \left[\left(\frac{3}{4}\frac{1}{5} \cdot 20\right)\left(\frac{7}{8} \cdot 8\right)\right]\frac{1}{160} = \frac{3 \cdot 7}{160} = \frac{21}{160} = \frac{20+1}{4 \cdot 5 \cdot 8} = \frac{1}{8} + \frac{1}{4}\frac{1}{5}\frac{1}{8}.$$

The proof is clear, the text says (same as in A.100a).

(**b**) Successive computation. Since multiplying the numerators gives 21, we have

$$21 \cdot \frac{1}{4}\frac{1}{5}\frac{1}{8} = \left(5 + \frac{1}{4}\right)\frac{1}{5}\frac{1}{8} = \left(1 + \frac{1}{4}\frac{1}{5}\right)\frac{1}{8} = \frac{1}{8} + \frac{1}{4}\frac{1}{5}\frac{1}{8}.$$

Remark. The author notes here again that the result is the same by whichever denominator we divide first. But, first, the form may not be the same. Second, beginning here with 8 will turn out to be inconvenient.

(**A.103**) MSS $\mathcal{A}$, $\mathcal{B}$, $\mathcal{C}$. Multiplication of a non-elementary fraction of a fraction by a fraction.

$$\frac{4}{5}\frac{1}{11} \cdot \frac{1}{9}$$

(**a**) Forming the product of the denominators of each of the two expressions, $5 \cdot 11$ and 9, thus $55 \cdot 9 = 495$.

$$\frac{4}{5}\frac{1}{11} \cdot \frac{1}{9} = \left[\left(\frac{4}{5}\frac{1}{11} \cdot 55\right)\left(\frac{1}{9} \cdot 9\right)\right]\frac{1}{495} = \frac{4}{495}.$$

(**b**) Successive division of the product of the numerators, here just 4, gives

$$\frac{4}{5}\frac{1}{9}\frac{1}{11} \quad \text{or} \quad \frac{4}{5}\frac{1}{11}\frac{1}{9},$$

for we may mention the fractions in any order.

Remark. Only the verbal expression will differ according to the order in which we append the fractions since the numerator is smaller than, and prime to, each of the denominators.

(A.104) MSS $\mathcal{A}$, $\mathcal{B}$, $\mathcal{C}$. Multiplication of a fraction of a fraction by a fraction of a fraction.

$$\frac{3}{4}\frac{1}{5} \cdot \frac{7}{8}\frac{1}{6}$$

(a) Forming the product of the denominators: $4 \cdot 5 = 20$, $8 \cdot 6 = 48$, thus $20 \cdot 48 = 960$. Then

$$\frac{3}{4}\frac{1}{5} \cdot \frac{7}{8}\frac{1}{6} = \left[\left(\frac{3}{4}\frac{1}{5} \cdot 20\right)\left(\frac{7}{8}\frac{1}{6} \cdot 48\right)\right]\frac{1}{960} = \frac{3 \cdot 7}{960}$$

$$= \frac{21}{960} = \frac{20+1}{4 \cdot 5 \cdot 8 \cdot 6} = \frac{1}{8}\frac{1}{6} + \frac{1}{4}\frac{1}{5}\frac{1}{8}\frac{1}{6}.$$

Here again, we are told, the proof is clear (A.100 *a*).

(b) Direct computation, multiplying the numerators and then taking the fractions in any order we wish. Thus, taking the denominators in increasing order

$$\frac{3}{4}\frac{1}{5} \cdot \frac{7}{8}\frac{1}{6} = \frac{21}{4}\frac{1}{5}\frac{1}{6}\frac{1}{8} = \left(5+\frac{1}{4}\right) \cdot \frac{1}{5}\frac{1}{6}\frac{1}{8} = \left(1+\frac{1}{4}\frac{1}{5}\right) \cdot \frac{1}{6}\frac{1}{8} = \frac{1}{6}\frac{1}{8} + \frac{1}{4}\frac{1}{5}\frac{1}{6}\frac{1}{8}.$$

Remark. Dividing first by 6 would give

$$\frac{21}{6}\frac{1}{4}\frac{1}{5}\frac{1}{8} = \left(3+\frac{1}{2}\right) \cdot \frac{1}{4}\frac{1}{5}\frac{1}{8} = \frac{3}{4}\frac{1}{5}\frac{1}{8} + \frac{1}{2}\frac{1}{4}\frac{1}{5}\frac{1}{8}$$

and not, as erroneously computed by an early reader,

$$\frac{21}{6}\frac{1}{4}\frac{1}{5}\frac{1}{8} = \left(3+\frac{1}{2}\right) \cdot \frac{1}{4}\frac{1}{5}\frac{1}{8} = \left(\frac{7}{2}\frac{1}{4}\right)\frac{1}{5} = \frac{7}{8}\frac{1}{5},$$

who has omitted the last factor one eighth (perhaps because it occurs twice).

(A.105) MSS $\mathcal{A}$, $\mathcal{B}$, $\mathcal{C}$. Multiplication of a fraction of a fraction by a non-elementary fraction of a fraction.

$$\frac{4}{5}\frac{1}{9} \cdot \frac{1}{7}\frac{1}{11}$$

We are told to proceed as before:

(a) forming the product of the denominators;

(b) multiplying the numerators and taking the fractions in successive order.

Remark. As in A.103, the numerator is smaller than, and prime with, each of the denominators, and so the computation becomes a mere putting together of the names of the fractions.

C. *Multiplicand a compound fraction*.

(A.106) MSS $\mathcal{A}$, $\mathcal{B}$, $\mathcal{C}$. Multiplication of a compound fraction by a (non-elementary) fraction.

$$\left(\frac{5}{7}+\frac{3\ 1}{4\ 7}\right)\cdot\frac{10}{11}$$

(*a*) Forming the product of the denominators: $4\cdot 7 = 28$ and 11, thus $28\cdot 11 = 308$. So

$$\left(\frac{5}{7}+\frac{3\ 1}{4\ 7}\right)\cdot\frac{10}{11} = \left[\left(\frac{5}{7}+\frac{3\ 1}{4\ 7}\right)28\cdot\frac{10}{11}11\right]\frac{1}{308} = \frac{23\cdot 10}{308} = \frac{230}{308}$$
$$=\frac{224+6}{4\cdot 7\cdot 11} = \frac{8\cdot 28+6}{4\cdot 7\cdot 11} = \frac{8}{11}+\frac{6}{4\cdot 7\cdot 11} = \frac{8}{11}+\frac{1\ 1}{7\ 11}+\frac{1\ 1\ 1}{2\ 7\ 11}.$$

For the proof, the reader is referred to the one in A.100*a*.

(*b*) Multiplying the numerators and taking the fractions individually.*

$$\left(\frac{5}{7}+\frac{3\ 1}{4\ 7}\right)\frac{10}{11} = \left[\left(5+\frac{3}{4}\right)\cdot 10\right]\frac{1}{7}\cdot\frac{1}{11} = \left(57+\frac{1}{2}\right)\frac{1\ 1}{7\ 11}$$
$$= \left(8+\frac{1}{7}+\frac{1\ 1}{2\ 7}\right)\frac{1}{11} = \frac{8}{11}+\frac{1\ 1}{7\ 11}+\frac{1\ 1\ 1}{2\ 7\ 11}.$$

(*c*) Rearranging the terms.

$$\left(\frac{5}{7}+\frac{3\ 1}{4\ 7}\right)\frac{10}{11} = \left[\left(5+\frac{3}{4}\right)\cdot\frac{10}{11}\right]\frac{1}{7} = \left(\frac{50}{11}+\frac{30}{4}\frac{1}{11}\right)\frac{1}{7}$$
$$= \left(4+\frac{6}{11}+\frac{7}{11}+\frac{1\ 1}{2\ 11}\right)\frac{1}{7} = \left(5+\frac{2}{11}+\frac{1\ 1}{2\ 11}\right)\frac{1}{7}$$
$$= \frac{5}{7}+\frac{2\ 1}{11\ 7}+\frac{1\ 1\ 1}{2\ 11\ 7}.$$

(*d*) Multiplying term by term (individual calculations known from A.100–103).

$$\left(\frac{5}{7}+\frac{3\ 1}{4\ 7}\right)\frac{10}{11} = \left(\frac{5}{7}\cdot\frac{10}{11}\right)+\left(\frac{3\ 1}{4\ 7}\cdot\frac{10}{11}\right) = \frac{49+1}{7\cdot 11}+\frac{28+2}{4\cdot 7\cdot 11}$$
$$= \frac{7}{11}+\frac{1\ 1}{7\ 11}+\frac{1}{11}+\frac{1\ 1\ 1}{2\ 7\ 11} = \frac{8}{11}+\frac{1\ 1}{7\ 11}+\frac{1\ 1\ 1}{2\ 7\ 11}.$$

(**A.107**) MSS $\mathcal{A}, \mathcal{B}, \mathcal{C}$. Extension: compound fraction with several terms.

$$\left(\frac{3\ 1}{5\ 7}+\frac{1\ 1\ 1}{4\ 5\ 7}\right)\cdot\left(\frac{5\ 1}{6\ 8}\right)$$

(*a*) Forming the product of the denominators: $(4\cdot 5\cdot 7)\cdot(6\cdot 8) = 140\cdot 48$.

$$\left(\frac{3\ 1}{5\ 7}+\frac{1\ 1\ 1}{4\ 5\ 7}\right)\cdot\left(\frac{5\ 1}{6\ 8}\right) = \left[\left(\frac{3\ 1}{5\ 7}+\frac{1\ 1\ 1}{4\ 5\ 7}\right)140\cdot\frac{5\ 1}{6\ 8}48\right]\frac{1}{140\cdot 48} = \frac{65}{6720}.$$

---

* Remember that the numerator for a compound fraction is an integer with a fraction (p. 1206, footnote).

(**b**) Multiplying the numerators:

$$\left(\frac{3}{5}\frac{1}{7} + \frac{1}{4}\frac{1}{5}\frac{1}{7}\right) \cdot \left(\frac{5}{6}\frac{1}{8}\right) = \left(3 + \frac{1}{4}\right) 5 \cdot \frac{1}{5}\frac{1}{6}\frac{1}{7}\frac{1}{8} = \left(16 + \frac{1}{4}\right) \cdot \frac{1}{5}\frac{1}{6}\frac{1}{7}\frac{1}{8}.$$

Dividing first by one or the other denominator will give

$$\left(3 + \frac{1}{5} + \frac{1\,1}{4\,5}\right) \cdot \frac{1\,1\,1}{7\,6\,8} = \frac{3\,1\,1}{7\,6\,8} + \frac{1\,1\,1\,1}{5\,7\,6\,8} + \frac{1\,1\,1\,1\,1}{4\,5\,7\,6\,8} = \frac{3\,1\,1}{7\,6\,8} + \frac{1\,1\,1\,1}{4\,7\,6\,8}$$

$$\left(2 + \frac{4}{6} + \frac{1\,1}{4\,6}\right) \cdot \frac{1\,1\,1}{8\,5\,7} = \frac{2\,1\,1}{8\,5\,7} + \frac{4\,1\,1\,1}{6\,8\,5\,7} + \frac{1\,1\,1\,1\,1}{4\,6\,8\,5\,7}$$

$$\left(2 + \frac{2}{7} + \frac{1\,1}{4\,7}\right) \cdot \frac{1\,1\,1}{5\,6\,8} = \frac{2\,1\,1}{5\,6\,8} + \frac{2\,1\,1\,1}{7\,5\,6\,8} + \frac{1\,1\,1\,1\,1}{4\,7\,5\,6\,8}$$

$$\left(2 + \frac{1\,1}{4\,8}\right) \cdot \frac{1\,1\,1}{5\,6\,7} = \frac{2\,1\,1}{5\,6\,7} + \frac{1\,1\,1\,1\,1}{4\,8\,5\,6\,7}$$

of which the text has thus the results of the first and the second.[‡]

---

3. Conversion of fractions.

The operation of converting fractions (Arabic *ṣarf kasr ilā kasr*) was not needed before since we merely appended the names of the denominators. But it is necessary, the author says, for multiplying fractions if we use other rules. For instance, when each side contains an integer and a fraction and we multiply term by term (*secundum differentias*), we shall have

$$\left(N_1 + \frac{p_1}{q_1}\right)\left(N_2 + \frac{p_2}{q_2}\right) = N_1 \cdot N_2 + \frac{p_1 N_2}{q_1} + \frac{p_2 N_1}{q_2} + \frac{p_1 p_2}{q_1 q_2},$$

and one of the two fractional terms in the middle will have to be converted into the other. Another computation also requiring conversion of fractions, which thus could not be treated before either, occurs with fractional expressions containing a sum of 'different fractions' (A.130 and A.136).

A. *Conversion into a fraction*.[°]

In A.108 the signification of conversion is explained and the rule inferred, and in A.109 this rule is repeated: if we are to convert the fractional expression $\frac{k}{l}$ into fractions with given denominator s, calculate

$$\left(\frac{k \cdot s}{l}\right) \frac{1}{s},$$

the term in parentheses, reduced if possible, being then the numerator of the required fraction.

[‡] The first is immediately deduced from what precedes since 5 drops out. This is what the text does.

[°] That is: one denominator of the final fraction given. But, as said in our summary (p. 1199), the result will normally be a compound fraction (and the given denominator thus occurs in its two terms).

— Converting a fraction.

(A.108) MSS $\mathcal{A}, \mathcal{B}, \mathcal{C}$. Conversion of a fraction into an elementary fraction.
$$\frac{3}{4} = \frac{r}{5}$$
We have here $3 : 4 = r : 5$, so
$$r = \frac{3 \cdot 5}{4} = \frac{15}{4} = 3 + \frac{3}{4}, \quad \text{hence} \quad \frac{3}{4} = \frac{3}{5} + \frac{3}{4}\frac{1}{5}.$$

(A.109) MSS $\mathcal{A}, \mathcal{B}, \mathcal{C}$. Another example (the denominators are just twice those above).
$$\frac{3}{8} = \frac{r}{10}$$
According to the rule,
$$\frac{3}{8} = \left(\frac{3 \cdot 10}{8}\right)\frac{1}{10} = \frac{30}{8} \cdot \frac{1}{10} = \left(3 + \frac{3}{4}\right)\frac{1}{10} = \frac{3}{10} + \frac{3}{4}\frac{1}{10}.$$

(A.110) MSS $\mathcal{A}, \mathcal{B}, \mathcal{C}$. Another example.
$$\frac{4}{7} = \frac{r}{6}$$
By the rule,
$$\frac{4}{7} = \left(\frac{4 \cdot 6}{7}\right)\frac{1}{6} = \frac{24}{7} \cdot \frac{1}{6} = \left(3 + \frac{3}{7}\right)\frac{1}{6} = \frac{3}{6} + \frac{3}{7}\frac{1}{6}.$$

(A.111) MSS $\mathcal{A}, \mathcal{B}, \mathcal{C}$. Another example, but of a conversion into a non-elementary fraction.
$$\frac{1}{5} = \frac{r}{13}$$
$$\frac{1}{5} = \left(\frac{1 \cdot 13}{5}\right)\frac{1}{13} = \left(2 + \frac{3}{5}\right)\frac{1}{13} = \frac{2}{13} + \frac{3}{5}\frac{1}{13}.$$

— Converting a compound fraction.

By analogy to the foregoing, we shall consider $k : l = r : s$, but now with k an integer plus a fraction.

(A.112) MSS $\mathcal{A}, \mathcal{B}, \mathcal{C}$. Conversion of a compound fraction into a (non-elementary) fraction.
$$\frac{5}{7} + \frac{2}{3}\frac{1}{7} = \frac{r}{11}$$
Since $k : l = r : s$, with $k = 5 + \frac{2}{3}$, $l = 7$, $s = 11$, the number of elevenths will be
$$r = \frac{\left(5 + \frac{2}{3}\right) 11}{7}.$$

(A.113) MSS $\mathcal{A}, \mathcal{B}, \mathcal{C}$. Another example, this time into an elementary fraction.
$$\frac{3}{8} + \frac{1}{2}\frac{1}{8} = \frac{r}{10}$$

$$\frac{3}{8} + \frac{1}{2}\frac{1}{8} = \left[\left(\frac{3}{8} + \frac{1}{2}\frac{1}{8}\right)10\right]\frac{1}{10} = \left(\frac{30}{8} + \frac{5}{8}\right)\frac{1}{10}$$
$$= \left(4 + \frac{3}{8}\right)\frac{1}{10} = \frac{4}{10} + \frac{3}{8}\frac{1}{10}.$$

(A.114) MSS $\mathcal{A}, \mathcal{B}, \mathcal{C}$. Conversion of a non-elementary compound fraction into a fraction.
$$\frac{4}{11} + \frac{1}{3}\frac{1}{11} = \frac{r}{8}$$

$$\frac{4}{11} + \frac{1}{3}\frac{1}{11} = \left[\left(\frac{4}{11} + \frac{1}{3}\frac{1}{11}\right)8\right]\frac{1}{8} = \left(\frac{32}{11} + \frac{2}{11} + \frac{2}{3}\frac{1}{11}\right)\frac{1}{8}$$
$$= \left(3 + \frac{1}{11} + \frac{2}{3}\frac{1}{11}\right)\frac{1}{8} = \frac{3}{8} + \frac{1}{11}\frac{1}{8} + \frac{2}{3}\frac{1}{11}\frac{1}{8}.$$

— **Converting a fraction of a fraction.**

(A.115) MSS $\mathcal{A}, \mathcal{B}, \mathcal{C}$. Conversion of a fraction of a fraction into a fraction.
$$\frac{3}{4}\frac{1}{10} = \frac{r}{6}$$

$$\frac{3}{4}\frac{1}{10} = \left(\frac{3}{4}\frac{1}{10} \cdot 6\right)\frac{1}{6} = \left(\frac{18}{4} \cdot \frac{1}{10}\right)\frac{1}{6} = \left(\frac{4}{10} + \frac{1}{2}\frac{1}{10}\right)\frac{1}{6} = \frac{4}{10}\frac{1}{6} + \frac{1}{2}\frac{1}{10}\frac{1}{6}.$$

(A.116) MSS $\mathcal{A}, \mathcal{B}, \mathcal{C}$. Conversion of a non-elementary fraction of a fraction into a fraction.
$$\frac{3}{5}\frac{1}{11} = \frac{r}{8}$$

$$\frac{3}{5}\frac{1}{11} = \left(\frac{3}{5}\frac{1}{11} \cdot 8\right)\frac{1}{8} = \left(\frac{24}{5} \cdot \frac{1}{11}\right)\frac{1}{8} = \left(\frac{4}{11} + \frac{4}{5}\frac{1}{11}\right)\frac{1}{8} = \frac{4}{11}\frac{1}{8} + \frac{4}{5}\frac{1}{11}\frac{1}{8}.$$

B. *Conversion into a fraction of a fraction*.[*]

— **Converting a fraction of a fraction.**

(A.117) MSS $\mathcal{A}, \mathcal{B}, \mathcal{C}$. Conversion of a fraction of a fraction into a fraction of a fraction.
$$\frac{3}{7}\frac{1}{8} = \frac{r}{6}\frac{1}{10}$$

---

[*] That is: two or several denominators of the final fraction are given.

(**a**) As before, but the multiplier will be the product of the denominators of the fractions into which we are to convert.

$$\frac{3}{7}\frac{1}{8} = \left(\frac{3}{7}\frac{1}{8} \cdot 6 \cdot 10\right)\frac{1}{6}\frac{1}{10} = \frac{180}{7}\frac{1}{8} \cdot \frac{1}{6}\frac{1}{10} = \frac{168 + 8 + 4}{7 \cdot 8} \cdot \frac{1}{6}\frac{1}{10}$$

$$= \left(3 + \frac{1}{7} + \frac{1}{2}\frac{1}{7}\right)\frac{1}{6}\frac{1}{10} = \frac{3}{6}\frac{1}{10} + \frac{1}{7}\frac{1}{6}\frac{1}{10} + \frac{1}{2}\frac{1}{7}\frac{1}{6}\frac{1}{10}.$$

(**b**) Since we are interested in the numerator of the resulting fraction, we may just calculate the common denominators of each expression and then form

$$r = \frac{3 \cdot 60}{56}$$

which are sixths of tenths.

This leads the author to formulate the general rule whereby the conversion into a fraction of a fraction is reduced to the former case (conversion into a fraction):

(*i*) To convert

$$\frac{k}{l_1}\frac{1}{l_2}\ldots\frac{1}{l_n} \quad \text{into} \quad \frac{r}{s_1}\frac{1}{s_2}\ldots\frac{1}{s_m},$$

form $l_1 \cdot l_2 \cdot \ldots \cdot l_n = L$, $s_1 \cdot s_2 \cdot \ldots \cdot s_m = S$, then consider that, since $k : L = r : S$,

$$r = \frac{k \cdot S}{L}$$

will give the quantity of fractions when converted (to be reduced by considering division by the single l_i).

Remark. This reminds us of the general rule given before A.95 (above, p. 1204): it will not be applied subsequently either. (Anyway it is not of much practical use for our purpose of converting into given denominators.) A.117*b* and rule *i* seem to have been added later.

(**A.118**) MSS $\mathcal{A}$, $\mathcal{B}$, $\mathcal{C}$. Conversion of a fraction of a fraction into a non-elementary fraction of a fraction.

$$\frac{1}{5}\frac{1}{7} = \frac{r}{8}\frac{1}{11}$$

$$\frac{1}{5}\frac{1}{7} = \left(\frac{1}{5}\frac{1}{7} \cdot 88\right)\frac{1}{8}\frac{1}{11} = \left(\frac{70 + 15 + 3}{5 \cdot 7}\right)\frac{1}{8}\frac{1}{11}$$

$$= \left(2 + \frac{3}{7} + \frac{3}{5}\frac{1}{7}\right)\frac{1}{8}\frac{1}{11} = \frac{2}{8}\frac{1}{11} + \frac{3}{7}\frac{1}{8}\frac{1}{11} + \frac{3}{5}\frac{1}{7}\frac{1}{8}\frac{1}{11}.$$

(*ii*) If the same fraction occurs on both sides, it may be omitted from the computations and just appended to the result.

(**A.119**) MSS $\mathcal{A}$, $\mathcal{B}$, $\mathcal{C}$. Conversion of a fraction of a fraction having several terms into a fraction of a fraction with several terms, one term the same.

$$\frac{5}{6}\frac{1}{8}\frac{1}{7} = \frac{r}{11}\frac{1}{5}\frac{1}{8}$$

Consider first the conversion of $\frac{5}{6}\frac{1}{7}$ into elevenths of a fifth, a case seen in A.117–118, and then append the fraction left out.

(A.120) MSS $\mathcal{A}, \mathcal{B}, \mathcal{C}$. Another illustration of rule *ii*.

$$\frac{3}{4}\frac{1}{6} = \frac{r}{6}\frac{1}{7}$$

Consider first the conversion of $\frac{3}{4}$ into sevenths, a case seen in A.108–111:

$$\frac{3}{4} = \left(\frac{3}{4} \cdot 7\right)\frac{1}{7} = \left(5 + \frac{1}{4}\right)\frac{1}{7}; \quad \text{then} \quad \frac{3}{4}\frac{1}{6} = \frac{5}{6}\frac{1}{7} + \frac{1}{4}\frac{1}{6}\frac{1}{7}.$$

— **Converting a compound fraction.**

(A.121) MSS $\mathcal{A}, \mathcal{B}, \mathcal{C}$. Conversion of a compound fraction (with several fractional terms) into a fraction of a fraction.

$$\frac{3}{8}\frac{1}{10} + \frac{1}{2}\frac{1}{8}\frac{1}{10} = \frac{r}{6}\frac{1}{7}$$

Since $6 \cdot 7 = 42$,

$$\frac{3}{8}\frac{1}{10} + \frac{1}{2}\frac{1}{8}\frac{1}{10} = \left[\left(\frac{3}{8}\frac{1}{10} + \frac{1}{2}\frac{1}{8}\frac{1}{10}\right)42\right]\frac{1}{6}\frac{1}{7} = \left(\frac{126 + 21}{8 \cdot 10}\right)\frac{1}{6}\frac{1}{7}$$

$$= \left(\frac{147}{8 \cdot 10}\right)\frac{1}{6}\frac{1}{7} = \left(\frac{80 + 64 + 3}{8 \cdot 10}\right)\frac{1}{6}\frac{1}{7}$$

$$= \left(1 + \frac{8}{10} + \frac{3}{8}\frac{1}{10}\right)\frac{1}{6}\frac{1}{7} = \frac{1}{6}\frac{1}{7} + \frac{8}{10}\frac{1}{6}\frac{1}{7} + \frac{3}{8}\frac{1}{10}\frac{1}{6}\frac{1}{7}.$$

Remark. In the second term of the result, the first fraction is not reduced since this would introduce a new type of fraction.

(A.122) MSS $\mathcal{A}, \mathcal{B}, \mathcal{C}$. Another such example but with a non-elementary fraction and use of rule *ii*.

$$\frac{4}{7}\frac{1}{10} + \frac{1}{2}\frac{1}{7}\frac{1}{10} = \frac{r}{10}\frac{1}{13}$$

Since one fraction is common, we shall first convert $\frac{4}{7} + \frac{1}{2}\frac{1}{7}$ into thirteenths as seen in A.112–114:

$$\frac{4}{7} + \frac{1}{2}\frac{1}{7} = \left[\left(\frac{4}{7} + \frac{1}{2}\frac{1}{7}\right)13\right]\frac{1}{13} = \left(\frac{52}{7} + \frac{13}{2}\frac{1}{7}\right)\frac{1}{13}$$

$$= \left(\frac{58}{7} + \frac{1}{2}\frac{1}{7}\right)\frac{1}{13} = \left(8 + \frac{2}{7} + \frac{1}{2}\frac{1}{7}\right)\frac{1}{13} = \frac{8}{13} + \frac{2}{7}\frac{1}{13} + \frac{1}{2}\frac{1}{7}\frac{1}{13}.$$

Adding the disregarded fraction to both, we obtain finally

$$\frac{4}{7}\frac{1}{10} + \frac{1}{2}\frac{1}{7}\frac{1}{10} = \frac{8}{10}\frac{1}{13} + \frac{2}{7}\frac{1}{10}\frac{1}{13} + \frac{1}{2}\frac{1}{7}\frac{1}{10}\frac{1}{13}.$$

4. Multiplication of fractional expressions containing integers.

We now return to the multiplication of fractional expressions. Of the five problems involving integers and fractions —A.123, A.124, A.125, A.128, A.129 (& A.130)— the first, which does not require conversion, could have come before; if it did not, this was probably to keep together the cases involving integers and fractions. The solution of A.126–127 requires conversion (A.126b) and the multiplication of fractional expressions containing integers (A.126c, A.127).

(A.123) MSS $\mathcal{A}$, $\mathcal{B}$, $\mathcal{C}$. Multiplication of a simple fraction by an integer with a fraction.

$$\frac{5}{7} \cdot \left(6 + \frac{2}{3}\right)$$

(*a*) The product of the denominators being $3 \cdot 7 = 21$,

$$\frac{5}{7}\left(6+\frac{2}{3}\right) = \left[\frac{5}{7}\cdot 21\left(6+\frac{2}{3}\right)\right]\frac{1}{21} = \left[15\left(6+\frac{2}{3}\right)\right]\frac{1}{21} = \frac{100}{21}.$$

Demonstration. Analogous to those seen in A.90a & A.93a for pure fractions.

$$\frac{5}{7}\cdot 21 : 21 = \frac{5}{7}\left(6+\frac{2}{3}\right) : \left(6+\frac{2}{3}\right),$$

whence

$$\frac{5}{7}\left(6+\frac{2}{3}\right) = \left[\frac{5}{7}\cdot 21\left(6+\frac{2}{3}\right)\right]\frac{1}{21},$$

as above.

Remark. Here the product of the denominators is not split up between the two factors (as has been usual from A.100 on).

(*b*) Or directly

$$\frac{5}{7}\left(6+\frac{2}{3}\right) = \frac{5\cdot\left(6+\frac{2}{3}\right)}{7},$$

the proof of which is clear (P_5).

(*c*) Or compute term by term, which leads us to what we have seen in Sections 1 & 2.

$$\frac{5}{7}\left(6+\frac{2}{3}\right) = \frac{5}{7}\cdot 6 + \frac{5}{7}\cdot\frac{2}{3} = \frac{30}{7} + \frac{10}{3}\frac{1}{7}$$
$$= 4 + \frac{2}{7} + \frac{3}{7} + \frac{1}{3}\frac{1}{7} = 4 + \frac{5}{7} + \frac{1}{3}\frac{1}{7},$$

the proof of which is clear (distributivity law PE_1).

(A.124) MSS $\mathcal{A}$, $\mathcal{B}$, $\mathcal{C}$. Multiplication of an integer with a fraction by an integer with a fraction.

$$\left(4+\frac{5}{8}\right)\cdot\left(9+\frac{3}{5}\right)$$

(**a**) Form the product of the denominators $8 \cdot 5 = 40$.

$$\left(4 + \frac{5}{8}\right)\left(9 + \frac{3}{5}\right) = \left[\left(4 + \frac{5}{8}\right) 8 \cdot \left(9 + \frac{3}{5}\right) 5\right] \frac{1}{40} = [(32 + 5) \cdot (45 + 3)] \frac{1}{40}$$

$$= \frac{37 \cdot 48}{40} = \frac{1776}{40} = \frac{1760 + 16}{8 \cdot 5} = 44 + \frac{2}{5}.$$

The procedure is the same as that seen earlier, from A.100 on, for multiplying fractional expressions. We are nonetheless given two proofs.

— The first is the same as in A.100a but our memory is being refreshed here, we are told, since it is also applicable to subsequent cases. Let (Fig. 28) $4 + \frac{5}{8} = A$, $9 + \frac{3}{5} = B$, $8 = G$, $5 = D$. Put $A \cdot G = H$ and $B \cdot D = Z$, so that

$$A = \frac{H}{G}, \qquad B = \frac{Z}{D}.$$

We wish to determine $A \cdot B$. By Abū Kāmil's proposition (P'_3),

$$A \cdot B = \frac{H}{G} \cdot \frac{Z}{D} = \frac{H \cdot Z}{G \cdot D} = \frac{(A \cdot G)(B \cdot D)}{G \cdot D},$$

and this proves the above computation.

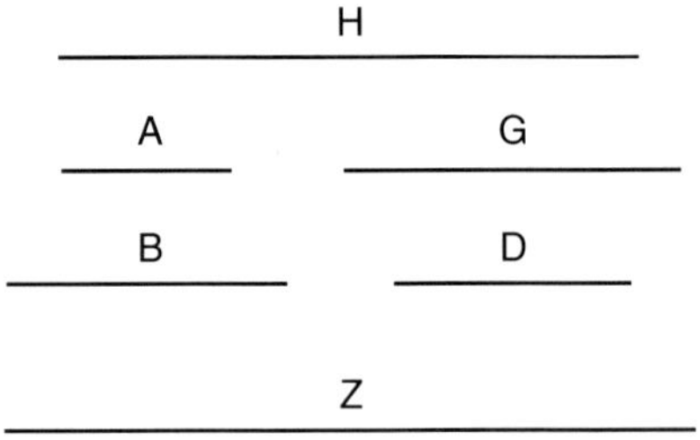

Fig. 28

— Another demonstration is said to be easier; indeed, it relies on the group of theorems *Elements* VII.17–19, by now familiar to the reader after being in constant use in Ch. A–II. Let (Fig. 29) $4 + \frac{5}{8} = A$, $9 + \frac{3}{5} = B$, with $A \cdot B = G$, required quantity; let next $8 = D$ and $\left(4 + \frac{5}{8}\right) 8 = A \cdot D = H$; further, let $5 = K$ and $\left(9 + \frac{3}{5}\right) 5 = B \cdot K = Z$, finally $8 \cdot 5 = D \cdot K = T$.

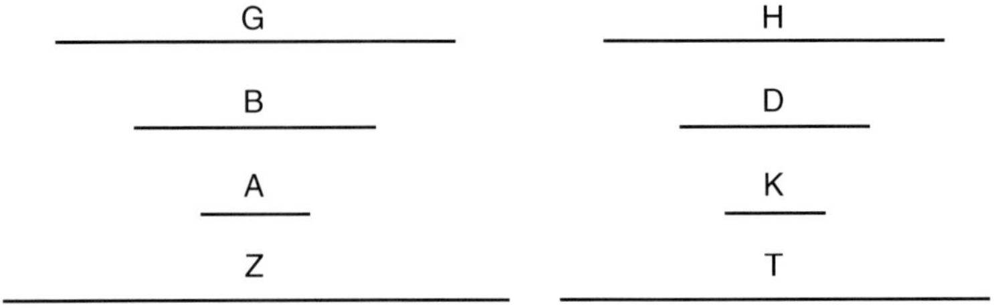

Fig. 29

First, since $A \cdot B = G$ and $A \cdot D = H$, so (*Elements* VII.17) $B : D = G : H$. Second, since $B \cdot K = Z$ and $D \cdot K = T$, so (*Elements* VII.18) $B : D = Z : T$. Therefore, $G : H = Z : T$, thus (*Elements* VII.19) $G \cdot T = H \cdot Z$ and

$$G = \frac{H \cdot Z}{T}, \quad \text{or} \quad \left(4 + \frac{5}{8}\right)\left(9 + \frac{3}{5}\right) = \frac{\left(4 + \frac{5}{8}\right) 8 \cdot \left(9 + \frac{3}{5}\right) 5}{8 \cdot 5}.$$

These two proofs will be referred to in similar cases (A.126a, A.128a, A.129a).

When introducing the conversion of fractions, the author mentioned that this would enable us to use other rules. As we observed at the beginning of Section 3 (p. 1212), this will in particular apply to the multiplication term by term of two fractional expressions which both contain integers. These we shall meet with from now on. (The present problem is, however, not a good example since one type of fraction disappears and thus a conversion is not necessary.)

(**b**) Multiplying term by term.

$$\left(4 + \frac{5}{8}\right)\left(9 + \frac{3}{5}\right) = 4 \cdot 9 + \frac{5}{8} \cdot 9 + \frac{5}{8}\frac{3}{5} + 4 \cdot \frac{3}{5}$$
$$= 36 + \left(5 + \frac{5}{8}\right) + \frac{3}{8} + \left(2 + \frac{2}{5}\right).$$

Converting the only remaining fraction into eighths, we find

$$\frac{2}{5} = \frac{16}{5}\frac{1}{8} = \frac{3}{8} + \frac{1}{5}\frac{1}{8},$$

whence finally

$$\left(4 + \frac{5}{8}\right)\left(9 + \frac{3}{5}\right) = 44 + \frac{3}{8} + \frac{1}{5}\frac{1}{8}.$$

(**c**) Multiplying individually the terms of one expression by the whole of the other leaves us with multiplying an integer with a fraction by first an integer and then a fraction, which procedures are known from Section 1 (integer by fraction) and A.123.

$$\left(4 + \frac{5}{8}\right)\left(9 + \frac{3}{5}\right) = 4 \cdot \left(9 + \frac{3}{5}\right) + \frac{5}{8} \cdot \left(9 + \frac{3}{5}\right)$$
$$= \left(38 + \frac{2}{5}\right) + \left(5 + \frac{5}{8} + \frac{3}{8}\right) = 38 + \frac{2}{5} + 6 = 44 + \frac{2}{5}.$$

We may generalize this way (seen here for the first time) and calculate

$$\left(a + \frac{k_1}{l_1} + \frac{k_2}{l_2} + \ldots + \frac{k_n}{l_n}\right)\left(b + \frac{r_1}{s_1} + \frac{r_2}{s_2} + \ldots + \frac{r_m}{s_m}\right)$$

by considering

$$a \cdot \left(b + \frac{r_1}{s_1} + \ldots + \frac{r_m}{s_m}\right) + \sum_i \frac{k_i}{l_i} \cdot \left(b + \frac{r_1}{s_1} + \ldots + \frac{r_m}{s_m}\right).$$

(A.125) MSS $\mathcal{A}, \mathcal{B}, \mathcal{C}$. Another example, with a non-elementary fraction.
$$\left(7+\frac{2}{5}\right)\cdot\left(8+\frac{4}{11}\right)$$

Proceeding as before (c, and rule):

$$\left(7+\frac{2}{5}\right)\left(8+\frac{4}{11}\right) = 7\cdot\left(8+\frac{4}{11}\right) + \frac{2}{5}\cdot\left(8+\frac{4}{11}\right)$$
$$= \left(56+\frac{28}{11}\right) + \left(\frac{16}{5}+\frac{8}{5}\frac{1}{11}\right) = \left(58+\frac{6}{11}\right) + \left(3+\frac{1}{5}+\frac{1}{11}+\frac{3}{5}\frac{1}{11}\right)$$
$$= 61 + \frac{7}{11} + \frac{11}{5}\frac{1}{11} + \frac{3}{5}\frac{1}{11} = 61 + \frac{9}{11} + \frac{4}{5}\frac{1}{11}.$$

Remark. The text computes the second multiplication as follows:

$$\frac{2}{5}\left(8+\frac{4}{11}\right) = \frac{2}{5}\left(5+3+\frac{4}{11}\right) = 2 + \frac{2}{5}\left(3+\frac{4}{11}\right) = 2 + \frac{2}{5}\cdot\frac{37}{11}$$
$$= 2 + \frac{74}{5}\frac{1}{11} = 2 + \frac{14}{11} + \frac{4}{5}\frac{1}{11} = 3 + \frac{3}{11} + \frac{4}{5}\frac{1}{11}.$$

(A.126) MSS $\mathcal{A}, \mathcal{B}, \mathcal{C}$. Multiplication of a compound fraction by a compound fraction.
$$\left(\frac{5}{8}+\frac{2}{3}\frac{1}{8}\right)\cdot\left(\frac{6}{7}+\frac{3}{4}\frac{1}{7}\right)$$

(*a*) Form the product of the denominators for each side, thus $(3\cdot 8)(4\cdot 7) = 24\cdot 28 = 672$; then consider

$$\left[\left(\frac{5}{8}+\frac{2}{3}\frac{1}{8}\right)24\cdot\left(\frac{6}{7}+\frac{3}{4}\frac{1}{7}\right)28\right]\frac{1}{672} = \frac{17\cdot 27}{672} = \frac{459}{672}.$$

We are told that the two proofs seen in A.124a will apply here. Indeed, the quantities A and B will just signify the two compound fractions.

(*b*) Multiplying one of the two expressions by the terms of the other. We are then to calculate:

$$\left(\frac{5}{8}+\frac{2}{3}\frac{1}{8}\right)\frac{6}{7} = \frac{30}{7}\frac{1}{8} + \frac{12}{3}\frac{1}{7}\frac{1}{8} = \frac{4}{8} + \frac{2}{7}\frac{1}{8} + \frac{4}{7}\frac{1}{8} = \frac{4}{8} + \frac{6}{7}\frac{1}{8}$$

$$\text{or} \quad (*) \quad = \frac{30}{8}\frac{1}{7} + \frac{12}{3}\frac{1}{8}\frac{1}{7} = \frac{3}{7} + \frac{6}{8}\frac{1}{7} + \frac{4}{8}\frac{1}{7} = \frac{4}{7} + \frac{2}{8}\frac{1}{7};$$

$$\left(\frac{5}{8}+\frac{2}{3}\frac{1}{8}\right)\frac{3}{4}\frac{1}{7} = \frac{15}{8}\frac{1}{4}\frac{1}{7} + \frac{6}{3}\frac{1}{4}\frac{1}{8}\frac{1}{7} = \frac{3}{8}\frac{1}{7} + \frac{3}{4}\frac{1}{8}\frac{1}{7} + \frac{1}{2}\frac{1}{8}\frac{1}{7} = \frac{4}{8}\frac{1}{7} + \frac{1}{4}\frac{1}{8}\frac{1}{7}.$$

By adding, we shall arrive either at:
$$\left(\frac{5}{8}+\frac{2}{3}\frac{1}{8}\right)\cdot\left(\frac{6}{7}+\frac{3}{4}\frac{1}{7}\right)=\frac{4}{8}+\frac{6}{8}\frac{1}{7}+\frac{4}{8}\frac{1}{7}+\frac{1}{4}\frac{1}{8}\frac{1}{7}=\frac{5}{8}+\frac{2}{7}\frac{1}{8}+\frac{1}{4}\frac{1}{7}\frac{1}{8},$$
or, if we use the expression (∗), at
$$\left(\frac{5}{8}+\frac{2}{3}\frac{1}{8}\right)\cdot\left(\frac{6}{7}+\frac{3}{4}\frac{1}{7}\right)=\frac{4}{7}+\frac{2}{8}\frac{1}{7}+\frac{4}{8}\frac{1}{7}+\frac{1}{4}\frac{1}{8}\frac{1}{7}=\frac{4}{7}+\frac{6}{8}\frac{1}{7}+\frac{1}{4}\frac{1}{8}\frac{1}{7}.$$

The text is confused. The author, who has partly computed in eighths and in eighths of a seventh, changes these results to sevenths and sevenths of an eighth.[†]

(c) Reduction to the multiplication of an integer with a fraction by an integer with a fraction (A.124–125).

$$\left(\frac{5}{8}+\frac{2}{3}\frac{1}{8}\right)\cdot\left(\frac{6}{7}+\frac{3}{4}\frac{1}{7}\right)=\left(5+\frac{2}{3}\right)\frac{1}{8}\cdot\left(6+\frac{3}{4}\right)\frac{1}{7}$$
$$=\left(38+\frac{1}{4}\right)\frac{1}{7}\frac{1}{8}=\left(5+\frac{3}{7}+\frac{1}{4}\frac{1}{7}\right)\frac{1}{8}=\frac{5}{8}+\frac{3}{7}\frac{1}{8}+\frac{1}{4}\frac{1}{7}\frac{1}{8}.$$

(A.127) MSS 𝒜, ℬ, 𝒞. Multiplication of a compound fraction by a non-elementary compound fraction.
$$\left(\frac{4}{5}+\frac{1}{4}\frac{1}{5}\right)\cdot\left(\frac{7}{11}+\frac{1}{3}\frac{1}{11}\right)$$

Reduce this to the multiplication of an integer and a fraction by an integer and a fraction, as taught just before, namely:
$$\left(4+\frac{1}{4}\right)\frac{1}{5}\cdot\left(7+\frac{1}{3}\right)\frac{1}{11}=\left(31+\frac{1}{6}\right)\frac{1}{5}\frac{1}{11}$$
$$=\left(6+\frac{1}{5}+\frac{1}{6}\frac{1}{5}\right)\frac{1}{11}=\frac{6}{11}+\frac{1}{5}\frac{1}{11}+\frac{1}{6}\frac{1}{5}\frac{1}{11}.$$

(A.128) MSS 𝒜, ℬ, 𝒞. Multiplication of an integer with a compound fraction by an integer with a fraction.

[†] To clarify the text's cumbersome computations, we group here the terms and perform the intermediate additions as the author does:
$$\frac{4}{8}+\frac{6}{7}\frac{1}{8}+\frac{3}{8}\frac{1}{7}+\left(\frac{3}{4}\frac{1}{8}\frac{1}{7}+\frac{1}{2}\frac{1}{8}\frac{1}{7}\right)=\frac{4}{8}+\frac{6}{8}\frac{1}{7}+\frac{3}{8}\frac{1}{7}+\left(\frac{1}{8}\frac{1}{7}+\frac{1}{4}\frac{1}{8}\frac{1}{7}\right)$$
$$=\frac{28}{8}\frac{1}{7}+\frac{7}{8}\frac{1}{7}+\frac{3}{8}\frac{1}{7}+\frac{1}{4}\frac{1}{8}\frac{1}{7}=\frac{35}{8}\frac{1}{7}+\frac{3}{8}\frac{1}{7}+\frac{1}{4}\frac{1}{8}\frac{1}{7}$$
$$=\frac{38}{8}\frac{1}{7}+\frac{1}{4}\frac{1}{8}\frac{1}{7}=\frac{4}{7}+\frac{6}{8}\frac{1}{7}+\frac{1}{4}\frac{1}{8}\frac{1}{7}.$$

$$\left(2+\frac{5}{7}+\frac{2\,1}{3\,7}\right)\cdot\left(4+\frac{3}{8}\right)$$

(**a**) The respective denominators being $3\cdot 7 = 21$ and 8, and their product $21\cdot 8 = 168$, consider

$$\left[\left(2+\frac{5}{7}+\frac{2\,1}{3\,7}\right)21\cdot\left(4+\frac{3}{8}\right)8\right]\frac{1}{168} = \frac{59\cdot 35}{168} = \frac{2065}{168}$$

$$= 12 + \frac{42+6+1}{3\cdot 7\cdot 8} = 12 + \frac{2}{8} + \frac{2\,1}{7\,8} + \frac{1\,1\,1}{3\,7\,8}.$$

This can be proved in the two ways seen in A.124a.

(**b**) Multiplying term by term, thus $2 + \left(\frac{5}{7}+\frac{2\,1}{3\,7}\right)$ by $4 + \frac{3}{8}$.

$2\cdot 4 = 8$

$$\left(\frac{5}{7}+\frac{2\,1}{3\,7}\right)4 = \frac{20}{7} + \frac{8\,1}{3\,7} = 2 + \frac{6}{7} + \frac{2}{7} + \frac{2\,1}{3\,7} = 3 + \frac{1}{7} + \frac{2\,1}{3\,7}$$

$2\cdot\dfrac{3}{8} = \dfrac{6}{8}$

$$\left(\frac{5}{7}+\frac{2\,1}{3\,7}\right)\frac{3}{8} = \frac{15\,1}{7\,8} + \frac{6\,1\,1}{3\,7\,8} = \frac{2}{8} + \frac{1\,1}{7\,8} + \frac{2\,1}{7\,8} = \frac{2}{8} + \frac{3\,1}{7\,8}.$$

Adding this, we find the following:*

$$11 + \frac{6}{8} + \frac{2}{8} + \left(\frac{1}{7}+\frac{2\,1}{3\,7}\right) + \frac{3\,1}{7\,8}.$$

Since

$$\frac{1}{7} + \frac{2\,1}{3\,7} = \frac{8\,1}{7\,8} + \frac{16\,1\,1}{3\,7\,8} = \frac{13\,1}{7\,8} + \frac{1\,1\,1}{3\,7\,8} = \frac{1}{8} + \frac{6\,1}{7\,8} + \frac{1\,1\,1}{3\,7\,8},$$

the above sum becomes

$$11 + \frac{6}{8} + \frac{2}{8} + \frac{1}{8} + \frac{6\,1}{7\,8} + \frac{3\,1}{7\,8} + \frac{1\,1\,1}{3\,7\,8} = 11 + \frac{9}{8} + \frac{1}{8} + \frac{2\,1}{7\,8} + \frac{1\,1\,1}{3\,7\,8}$$

$$= 12 + \frac{2}{8} + \frac{2\,1}{7\,8} + \frac{1\,1\,1}{3\,7\,8}.$$

(**A.129**) MSS $\mathcal{A}, \mathcal{B}, \mathcal{C}$. Multiplication of an integer with a compound fraction by an integer with a (non-elementary) compound fraction.

$$\left(5+\frac{7}{8}+\frac{2\,1}{3\,8}\right)\cdot\left(4+\frac{10}{11}+\frac{1\,1}{2\,11}\right)$$

(**a**) Multiplying the respective denominators $3\cdot 8$ and $2\cdot 11$ gives $24\cdot 22 = 528$; then

$$\left(5+\frac{7}{8}+\frac{2\,1}{3\,8}\right)\cdot\left(4+\frac{10}{11}+\frac{1\,1}{2\,11}\right)$$

$$= \left[\left(5+\frac{7}{8}+\frac{2\,1}{3\,8}\right)24\right]\frac{1\,1}{3\,8}\cdot\left[\left(4+\frac{10}{11}+\frac{1\,1}{2\,11}\right)22\right]\frac{1\,1}{2\,11}$$

$$= \frac{120+21+2}{24}\cdot\frac{88+20+1}{22} = \frac{143}{24}\cdot\frac{109}{22} = \frac{15587}{528}.$$

---

* We delay adding some fractions so that the computations in the text become apparent.

This can again be proved in the two ways seen in A.124a.

(**b**) Multiplying term by term, thus $5 + \left(\dfrac{7}{8} + \dfrac{2}{3}\dfrac{1}{8}\right)$ by $4 + \left(\dfrac{10}{11} + \dfrac{1}{2}\dfrac{1}{11}\right)$.

$5 \cdot 4 = 20$

$\left(\dfrac{7}{8} + \dfrac{2}{3}\dfrac{1}{8}\right) \cdot 4 = \dfrac{28}{8} + \dfrac{8}{3}\dfrac{1}{8} = 3 + \dfrac{6}{8} + \dfrac{2}{3}\dfrac{1}{8}$

$\left(\dfrac{10}{11} + \dfrac{1}{2}\dfrac{1}{11}\right) \cdot 5 = \dfrac{50}{11} + \dfrac{5}{2}\dfrac{1}{11} = 4 + \dfrac{8}{11} + \dfrac{1}{2}\dfrac{1}{11}$

$\left(\dfrac{10}{11} + \dfrac{1}{2}\dfrac{1}{11}\right)\left(\dfrac{7}{8} + \dfrac{2}{3}\dfrac{1}{8}\right) = \left(10 + \dfrac{1}{2}\right) \cdot \left(7 + \dfrac{2}{3}\right)\dfrac{1}{8}\dfrac{1}{11}$

$= \left(80 + \dfrac{1}{2}\right)\dfrac{1}{8}\dfrac{1}{11} = \left(10 + \dfrac{1}{2}\dfrac{1}{8}\right)\dfrac{1}{11} = \dfrac{10}{11} + \dfrac{1}{2}\dfrac{1}{8}\dfrac{1}{11}$.

The text merely gives instructions for computing the final sum: we are to add the fractions by kind, 'beginning with the smallest to the largest', converting where necessary.‡

(**A.130**) MSS $\mathcal{A}$, $\mathcal{B}$, $\mathcal{C}$. Extension: multiplication of an integer with several fractions by an integer with several fractions.

$$\left(6 + \dfrac{1}{5} + \dfrac{1}{3}\right) \cdot \left(8 + \dfrac{5}{6} + \dfrac{1}{4}\right)$$

(**a**) Multiplying the respective denominators 15 and 24 produces 360; then

$$\left(6 + \dfrac{1}{5} + \dfrac{1}{3}\right) \cdot \left(8 + \dfrac{5}{6} + \dfrac{1}{4}\right)$$
$$= \left[\left(6 + \dfrac{1}{5} + \dfrac{1}{3}\right)15 \cdot \left(8 + \dfrac{5}{6} + \dfrac{1}{4}\right)24\right]\dfrac{1}{360} = \dfrac{98 \cdot 218}{360}.$$

Remark. Only the figure (see Translation, p. 683) contains further results:

$$\dfrac{98 \cdot 218}{360} = \dfrac{21\,364}{360} = 59 + \dfrac{124}{360} = 59 + \dfrac{31}{90}.$$

---

‡ We shall find

$20 + 3 + \dfrac{6}{8} + \dfrac{2}{3}\dfrac{1}{8} + 4 + \dfrac{8}{11} + \dfrac{1}{2}\dfrac{1}{11} + \dfrac{10}{11} + \dfrac{1}{2}\dfrac{1}{8}\dfrac{1}{11}$

$= 27 + \dfrac{6}{8} + \dfrac{18}{11} + \dfrac{2}{3}\dfrac{1}{8} + \dfrac{1}{2}\dfrac{1}{11} + \dfrac{1}{2}\dfrac{1}{8}\dfrac{1}{11}$

$= 28 + \dfrac{66}{8}\dfrac{1}{11} + \dfrac{7}{11} + \dfrac{22}{3}\dfrac{1}{8}\dfrac{1}{11} + \dfrac{8}{2}\dfrac{1}{8}\dfrac{1}{11} + \dfrac{1}{2}\dfrac{1}{8}\dfrac{1}{11}$

$= 28 + \dfrac{8}{11} + \dfrac{2}{8}\dfrac{1}{11} + \dfrac{7}{11} + \dfrac{7}{8}\dfrac{1}{11} + \dfrac{1}{3}\dfrac{1}{8}\dfrac{1}{11} + \dfrac{4}{8}\dfrac{1}{11} + \dfrac{1}{2}\dfrac{1}{8}\dfrac{1}{11}$

$= 29 + \dfrac{4}{11} + \dfrac{1}{11} + \dfrac{5}{8}\dfrac{1}{11} + \dfrac{2}{3}\dfrac{1}{2}\dfrac{1}{8}\dfrac{1}{11} + \dfrac{1}{2}\dfrac{1}{8}\dfrac{1}{11}$

$= 29 + \dfrac{5}{11} + \dfrac{5}{8}\dfrac{1}{11} + \dfrac{1}{2}\dfrac{1}{8}\dfrac{1}{11} + \dfrac{2}{3}\dfrac{1}{2}\dfrac{1}{8}\dfrac{1}{11}.$

(*b*) Converting first in order to obtain the same kind of fraction in each of the two expressions.

$$\frac{1}{3} = \frac{5}{3}\frac{1}{5} = \frac{1}{5} + \frac{2}{3}\frac{1}{5}, \quad \text{so} \quad 6 + \frac{1}{5} + \frac{1}{3} = 6 + \frac{2}{5} + \frac{2}{3}\frac{1}{5}$$

$$\frac{1}{4} = \frac{6}{4}\frac{1}{6} = \frac{1}{6} + \frac{1}{2}\frac{1}{6}, \quad \text{so} \quad 8 + \frac{5}{6} + \frac{1}{4} = 9 + \frac{1}{2}\frac{1}{6}.$$

This reduces the problem to A.128–129 just preceding.

Remark. Section 4 is about the multiplication of fractional expressions containing integers, and this example treats the case in which the fraction is replaced by a sum of different fractions. Exactly the same reasoning will apply, the author notes, if in Section 2 (multiplication of fractional expressions without integers) we replace the fractions by a sum of different fractions. He will merely solve an example of the simplest case. This is A.136, misplaced, corresponding to A.99–101 (simple fraction by simple fraction).

---

5. Case of irregular fractions.

A. *Computation with irregular fractions*.

Suppose we have expressions of the form

$$\frac{k}{l} \text{ of } M, \quad \frac{r}{s} \text{ of } M',$$

where M and M' may be integers or integers with fractions. We consider here the products of such expressions. The author's four ways to attain the result are the following.

1. As before, using the denominators and their product (demonstration in A.134*a*).

$$\left[\left(\frac{k}{l} \cdot l\right) \cdot M\right] \cdot \left[\left(\frac{r}{s} \cdot s\right) \cdot M'\right] \frac{1}{l \cdot s}.$$

2. As before, directly (that is, computing successively, according to the wording of the problem: *secundum verba questionis*).

$$\left(\frac{k}{l} \cdot M\right) \cdot \left(\frac{r}{s} \cdot M'\right).$$

This leads us to one of the previous cases of multiplication.

The other two ways just transpose the factors (presumably as is most convenient).

3. Exchanging the multipliers.

$$\left(\frac{k}{l} \cdot M'\right) \cdot \left(\frac{r}{s} \cdot M\right).$$

4. Grouping the factors according to kind (demonstration in A.134c).

$$\left(\frac{k}{l} \cdot \frac{r}{s}\right) \cdot \left(M \cdot M'\right).$$

But before the multiplication of irregular fractions among themselves, the author considers that of irregular fractions by an integral or a fractional quantity (A.131–133).

(A.131) MSS $\mathcal{A}, \mathcal{B}, \mathcal{C}$. Multiplication of an irregular fraction of an integer by an integer.

$$\left(\frac{3}{4} \cdot 5\right) \cdot 7$$

(*a*) First way, with the denominator 4.

$$\left[\left(\frac{3}{4} \cdot 4\right) \cdot 5\right] \frac{1}{4} \cdot 7 = (3 \cdot 5) \frac{1}{4} \cdot 7 = \frac{15 \cdot 7}{4} = \frac{105}{4} = 26 + \frac{1}{4}.$$

The first part of the transformation is known from A.90 while, by P_5,

$$(3 \cdot 5) \frac{1}{4} \cdot 7 = \frac{15}{4} \cdot 7 = \frac{15 \cdot 7}{4}.$$

(*b*) According to the fourth way, which leads us to A.91.

$$\left(\frac{3}{4} \cdot 5\right) \cdot 7 = \frac{3}{4} (5 \cdot 7) = \frac{3}{4} \cdot 35.$$

For, according to P_2,

$$\left(\frac{3}{4} \cdot 5\right) \cdot 7 = \frac{3}{4} \cdot (5 \cdot 7).$$

(*c*) Directly (second way).

$$\left(\frac{3}{4} \cdot 5\right) \cdot 7 = \left(3 + \frac{3}{4}\right) \cdot 7 = 21 + \frac{21}{4} = 21 + 5 + \frac{1}{4} = 26 + \frac{1}{4}.$$

(*d*) Consider, as in the third way,

$$\left(\frac{3}{4} \cdot 7\right) \cdot 5 = \left(5 + \frac{1}{4}\right) \cdot 5.$$

(A.132) MSS $\mathcal{A}, \mathcal{B}, \mathcal{C}$. Multiplication of an irregular fraction of an integer with a fraction by an integer.

$$\frac{4}{5}\left(6 + \frac{1}{3}\right) \cdot 8$$

(*a*) First way, with the denominator 15.

$$\frac{4}{5}\left(6+\frac{1}{3}\right)\cdot 8 = \left[\frac{4}{5}\cdot 15\left(6+\frac{1}{3}\right)\right]\frac{1}{15}\cdot 8$$
$$= \left[12\left(6+\frac{1}{3}\right)\right]\frac{1}{15}\cdot 8 = \frac{76\cdot 8}{15} = \frac{608}{15},$$

this last result being in the figure only.

(*b*) Second way.

$$\left[\frac{4}{5}\left(6+\frac{1}{3}\right)\right]\cdot 8 = \left[\frac{1}{5}\left(25+\frac{1}{3}\right)\right]\cdot 8 = \left(5+\frac{1}{3}\frac{1}{5}\right)\cdot 8.$$

(*c*) Changing the order of the factors, consider

$$\left(\frac{4}{5}\cdot 8\right)\left(6+\frac{1}{3}\right).$$

(*d*) Or else, consider

$$\left[\left(6+\frac{1}{3}\right)8\right]\cdot\frac{4}{5}.$$

(**A.133**) MSS 𝒜, ℬ, 𝒞. Multiplication of an irregular fraction of an integer with a fraction by an integer with a fraction.

$$\frac{4}{7}\left(5+\frac{1}{3}\right)\cdot\left(8+\frac{1}{2}\right)$$

Form the product of the denominators 21 and 2, thus 42.

$$\frac{4}{7}\left(5+\frac{1}{3}\right)\cdot\left(8+\frac{1}{2}\right) = \left[\frac{4}{7}\cdot 21\left(5+\frac{1}{3}\right)\right]\cdot\left[\left(8+\frac{1}{2}\right)2\right]\frac{1}{42}$$
$$= \left[12\left(5+\frac{1}{3}\right)\cdot 17\right]\frac{1}{42} = \frac{64\cdot 17}{42} = \frac{1088}{42},$$

this last result being in the figure only.
The other ways seen in A.132 could be used as well, the author adds.

(**A.134**) MSS 𝒜, ℬ, 𝒞. Multiplication of an irregular fraction of an integer by an irregular fraction of an integer.

$$\left(\frac{2}{3}\cdot 4\right)\cdot\left(\frac{5}{8}\cdot 7\right)$$

(*a*) First way, with the product of the denominators 3 and 8, thus 24. Then

$$\left(\frac{2}{3}\cdot 4\right)\cdot\left(\frac{5}{8}\cdot 7\right) = \left[\left(\frac{2}{3}\cdot 3\right)4\cdot\left(\frac{5}{8}\cdot 8\right)7\right]\frac{1}{24} = \frac{8\cdot 35}{24} = \frac{280}{24} = \frac{35}{3},$$

this last (unreduced) result being in the figure only.
Demonstration. From A.90a (or A.131a) and P_3' respectively:

$$\frac{2}{3}\cdot 4 = \left[\left(\frac{2}{3}\cdot 3\right)4\right]\frac{1}{3} = \frac{8}{3}, \qquad \frac{5}{8}\cdot 7 = \left[\left(\frac{5}{8}\cdot 8\right)7\right]\frac{1}{8} = \frac{35}{8},$$

$$\text{so} \quad \left(\frac{2}{3}\cdot 4\right)\cdot\left(\frac{5}{8}\cdot 7\right) = \frac{8}{3}\cdot\frac{35}{8} = \frac{8\cdot 35}{3\cdot 8}.$$

(*b*) Second way, which reduces the problem to A.124.

$$\left(\frac{2}{3}\cdot 4\right)\cdot\left(\frac{5}{8}\cdot 7\right) = \left(2+\frac{2}{3}\right)\left(4+\frac{3}{8}\right).$$

(*c*) (Our) fourth way, which reduces the problem to A.96.

$$\left(\frac{2}{3}\cdot 4\right)\cdot\left(\frac{5}{8}\cdot 7\right) = \left(\frac{2}{3}\cdot\frac{5}{8}\right)(4\cdot 7) = \frac{10}{3}\frac{1}{8}\cdot 28 = \left(\frac{3}{8}+\frac{1}{3}\frac{1}{8}\right)28.$$

Demonstration. By P_3,

$$\left(\frac{2}{3}\cdot 4\right)\cdot\left(\frac{5}{8}\cdot 7\right) = \left(\frac{2}{3}\cdot\frac{5}{8}\right)(4\cdot 7).$$

(A.135) MSS $\mathcal{A}, \mathcal{B}, \mathcal{C}$. Multiplication of an irregular fraction of an integer with a fraction by an irregular fraction of an integer with a fraction.

$$\frac{2}{3}\left(5+\frac{1}{4}\right)\cdot\frac{2}{7}\left(6+\frac{1}{2}\right)$$

(*a*) First way, with the products of the denominators: $12\cdot 14 = 168$.

$$\frac{2}{3}\left(5+\frac{1}{4}\right)\cdot\frac{2}{7}\left(6+\frac{1}{2}\right) = \left[\frac{2}{3}\cdot 12\left(5+\frac{1}{4}\right)\right]\cdot\left[\frac{2}{7}\cdot 14\left(6+\frac{1}{2}\right)\right]\frac{1}{168}$$

$$= \left[8\left(5+\frac{1}{4}\right)\right]\cdot\left[4\left(6+\frac{1}{2}\right)\right]\frac{1}{168} = \frac{42\cdot 26}{168} = \frac{1092}{168},$$

this last result being in the figure only.
  The proof is as in A.134a, the integer there now being replaced by an integer with a fraction.

(*b*) Second way, thus according to the wording, which reduces the problem to A.124.

$$\frac{2}{3}\left(5+\frac{1}{4}\right)\cdot\frac{2}{7}\left(6+\frac{1}{2}\right) = \left(3+\frac{1}{3}+\frac{2}{3}\frac{1}{4}\right)\left(1+\frac{5}{7}+\frac{1}{7}\right) = \left(3+\frac{1}{2}\right)\left(1+\frac{6}{7}\right).$$

(*c*) Third way, which ends with A.124 again.

$$\frac{2}{3}\left(6+\frac{1}{2}\right)\cdot\frac{2}{7}\left(5+\frac{1}{4}\right) = \left(4+\frac{1}{3}\right)\left(1+\frac{3}{7}+\frac{1}{2}\frac{1}{7}\right) = \left(4+\frac{1}{3}\right)\left(1+\frac{1}{2}\right).$$

(**d**) Fourth way.

$$\left(\frac{2}{3} \cdot \frac{2}{7}\right) \cdot \left[\left(5 + \frac{1}{4}\right) \cdot \left(6 + \frac{1}{2}\right)\right] = \left(\frac{1}{7} + \frac{1}{3}\frac{1}{7}\right)\left(34 + \frac{1}{8}\right)$$

$$= \frac{34}{7} + \frac{1}{7}\frac{1}{8} + \frac{34}{3}\frac{1}{7} + \frac{1}{3}\frac{1}{7}\frac{1}{8} = 4 + \frac{6}{7} + \frac{1}{7}\frac{1}{8} + \frac{11}{7} + \frac{1}{3}\frac{1}{7} + \frac{1}{3}\frac{1}{7}\frac{1}{8}$$

$$= 6 + \frac{3}{7} + \frac{1}{7}\frac{1}{8} + \frac{8}{3}\frac{1}{7}\frac{1}{8} + \frac{1}{3}\frac{1}{7}\frac{1}{8} = 6 + \frac{3}{7} + \frac{4}{7}\frac{1}{8} = 6 + \frac{3}{7} + \frac{1}{2}\frac{1}{7}.$$

(**A.135′**) Multiply $\frac{2}{3}$ of 5 and $\frac{1}{4}$ of 1 by $\frac{2}{7}$ of 6 and $\frac{1}{2}$ of 1.

This problem corresponds to the previous one.* But it can also be interpreted differently, namely as

$$\left(\frac{2}{3} \cdot 5 + \frac{1}{4}\right) \cdot \left(\frac{2}{7} \cdot 6 + \frac{1}{2}\right).$$

Here we meet with the first occurrence of ambiguity in the enunciation. In symbolic mathematics, the use of parentheses overcomes such ambiguity. The case of mediaeval, purely verbal mathematics is different. In Arabic, the use of vocalization signs may avoid any ambiguity; but most copyists omit them from mathematical manuscripts since the meaning is generally clear. The situation of Latin, with its explicit inflexions, is different: there is generally no ambiguity, and thus no confusion possible when numbers are written in words and not in symbols.† Since there is therefore no real reason to consider such cases, our author must be merely following the Arabic tradition. At the beginning of this section, we were told that irregular fractions are (a subject) 'discussed among mathematicians'. If this is not an interpolation, it might have been an attempt by the author to justify treating a topic which is mathematically simple and of interest more to Arab than to Latin readers.

(**a**) First way, with the same product of the denominators as before, namely $168 = 12 \cdot 14$.

$$\left(\frac{2}{3} \cdot 5 + \frac{1}{4}\right) \cdot \left(\frac{2}{7} \cdot 6 + \frac{1}{2}\right)$$

$$= \left(\frac{2}{3} \cdot 12 \cdot 5 + \frac{1}{4} \cdot 12\right)\frac{1}{12} \cdot \left(\frac{2}{7} \cdot 14 \cdot 6 + \frac{1}{2} \cdot 14\right)\frac{1}{14}$$

$$= (40 + 3)\frac{1}{12} \cdot (24 + 7)\frac{1}{14} = \frac{43 \cdot 31}{168}.$$

---

* That the multiplication and the expression involving 'of' are equivalent was pointed out at the beginning of this chapter. See also Translation, p. 655, note 327.

† That is why, in the text itself, ambiguity is mainly confined to the tables, which may thus serve for the various interpretations.

(**b**) Second way, according to the wording, which leads us to a conversion as seen in A.130.

$$\left[\left(\frac{2}{3}\cdot 5\right)+\frac{1}{4}\right]\cdot\left[\left(\frac{2}{7}\cdot 6\right)+\frac{1}{2}\right]=\left(3+\frac{1}{3}+\frac{1}{4}\right)\left(1+\frac{5}{7}+\frac{1}{2}\right)$$

$$=\left(3+\frac{7}{2}\frac{1}{6}\right)\left(1+\frac{17}{2}\frac{1}{7}\right)=\left(3+\frac{3}{6}+\frac{1\,1}{2\,6}\right)\left(2+\frac{1}{7}+\frac{1\,1}{2\,7}\right),$$

to be calculated as in A.129.

(**A.136**) MSS $\mathcal{A}$, $\mathcal{B}$, $\mathcal{C}$. Multiplication of a (non-elementary) fraction by several fractions. This should follow A.130.

$$\frac{10}{11}\left(\frac{2}{9}+\frac{2}{7}\right)$$

(**a**) Change the second expression into one kind of fraction; this will give, the text says, a fraction and a fraction of a fraction (thus a compound fraction). We shall obtain in the two cases, respectively,

$$\frac{2}{9}+\frac{2}{7}=\frac{14}{9}\frac{1}{7}+\frac{2}{7}=\frac{3}{7}+\frac{5}{9}\frac{1}{7}$$

$$\frac{2}{9}+\frac{2}{7}=\frac{2}{9}+\frac{18}{7}\frac{1}{9}=\frac{4}{9}+\frac{4}{7}\frac{1}{9},$$

to be multiplied by the first term as taught before (A.106).

(**b**) Another way is to multiply term by term, which leaves the conversion to the end.

(**A.137**) MSS $\mathcal{A}$, $\mathcal{B}$, $\mathcal{C}$. Multiplication of a compound fraction by an integer with a compound fraction. This problem, too, is misplaced, and should come before A.130.

$$\left(\frac{5}{8}+\frac{2}{3}\frac{1}{8}\right)\cdot\left(6+\frac{2}{7}+\frac{3}{4}\frac{1}{7}\right)$$

$$\left(\frac{5}{8}+\frac{2}{3}\frac{1}{8}\right)\cdot\left(6+\frac{2}{7}+\frac{3}{4}\frac{1}{7}\right)=\left(\frac{5}{8}+\frac{2}{3}\frac{1}{8}\right)6+\left(\frac{5}{8}+\frac{2}{3}\frac{1}{8}\right)\left(\frac{2}{7}+\frac{3}{4}\frac{1}{7}\right),$$

to be calculated as taught in A.96 & A.126–127.

This problem (presumably belonging to the fourth section) concludes the multiplication of proper fractional expressions. As the author says before A.136, acquaintance with the previous methods and understanding their proofs (and familiarity with the conversion of fractions) will enable the reader to solve any problem of that kind.

B. *Ambiguous formulations*.

As he has announced at the end of A.135–135′, the author adds some problems which are (supposedly) ambiguously formulated.

(A.138) MSS 𝒜, ℬ, 𝒞. Multiply 'a fourth of five and two fifths of six by a tenth of three and an eighth of four'.

In the present case, two interpretations are given:

$$(a) \quad \left(\frac{1}{4} \cdot 5 + \frac{2}{5} \cdot 6\right)\left(\frac{1}{10} \cdot 3 + \frac{1}{8} \cdot 4\right)$$

$$(b) \quad \frac{1}{4}\left(5 + \frac{2}{5} \cdot 6\right) \cdot \frac{1}{10}\left(3 + \frac{1}{8} \cdot 4\right).$$

Remark. Note first that the author does not consider mixed cases, with interpretation *a* for the first expression and *b* for the second. Note, too, that in actual fact the Latin text admits of the first interpretation only.

(a) Forming the product of the individual denominators, thus $20 \cdot 80 = 1600$, we shall have

$$\left(\frac{1}{4} \cdot 5 + \frac{2}{5} \cdot 6\right) \cdot \left(\frac{1}{10} \cdot 3 + \frac{1}{8} \cdot 4\right)$$

$$= \left[\left(\frac{1}{4} \cdot 20 \cdot 5 + \frac{2}{5} \cdot 20 \cdot 6\right)\left(\frac{1}{10} \cdot 80 \cdot 3 + \frac{1}{8} \cdot 80 \cdot 4\right)\right] \frac{1}{1600}$$

$$= \left[(25 + 48)(24 + 40)\right] \frac{1}{1600} = \frac{73 \cdot 64}{1600}.$$

(b) Take again the product of the denominators $20 \cdot 80 = 1600$. Then

$$\frac{1}{4}\left(5 + \frac{2}{5} \cdot 6\right) \cdot \frac{1}{10}\left(3 + \frac{1}{8} \cdot 4\right)$$

$$= \left(\frac{1}{4} \cdot 20 \cdot 5 + \frac{1}{4} \cdot \frac{2}{5} \cdot 20 \cdot 6\right)\left(\frac{1}{10} \cdot 80 \cdot 3 + \frac{1}{10}\frac{1}{8} \cdot 80 \cdot 4\right)\frac{1}{1600}$$

$$= \left[(25 + 12)(24 + 4)\right] \frac{1}{1600} = \frac{37 \cdot 28}{1600}.$$

Remark. Problem *b* would be expressed in Latin as: *Si volueris quartam de quinque et duabus quintis* (or: *et duarum quartarum*) *de sex multiplicare in decimam trium et octave de quatuor.* Any confusion is thus excluded.

(A.139) MSS 𝒜, ℬ, 𝒞. Multiply 'two thirds of seven and a half and two fifths of six and a third by two sevenths of four and a tenth and three fourths of nine and a ninth'. Same kind as A.138, but with integers with fractions instead of integers.* This can be understood to mean:

$$(a) \quad \left[\frac{2}{3}\left(7 + \frac{1}{2}\right) + \frac{2}{5}\left(6 + \frac{1}{3}\right)\right] \cdot \left[\frac{2}{7}\left(4 + \frac{1}{10}\right) + \frac{3}{4}\left(9 + \frac{1}{9}\right)\right]$$

---

* Thus, as the author himself says, the first two interpretations here correspond to the two already seen.

(b) $\frac{2}{3}\left[\left(7+\frac{1}{2}\right)+\frac{2}{5}\left(6+\frac{1}{3}\right)\right]\cdot\frac{2}{7}\left[\left(4+\frac{1}{10}\right)+\frac{3}{4}\left(9+\frac{1}{9}\right)\right]$

(c) $\left[\left(\frac{2}{3}\cdot 7+\frac{1}{2}\right)+\left(\frac{2}{5}\cdot 6+\frac{1}{3}\right)\right]\cdot\left[\left(\frac{2}{7}\cdot 4+\frac{1}{10}\right)+\left(\frac{3}{4}\cdot 9+\frac{1}{9}\right)\right]$

(d) $\left[\frac{2}{3}\cdot 7+\left(\frac{1}{2}+\frac{2}{5}\right)\left(6+\frac{1}{3}\right)\right]\cdot\left[\frac{2}{7}\cdot 4+\left(\frac{1}{10}+\frac{3}{4}\right)\left(9+\frac{1}{9}\right)\right].$

As a glossator notes at the end, other cases could have been considered. He must have meant mixed cases, with the first expression in brackets of one of the types above and the second of another; or even, involving a mixture within the brackets, with for example as first term

$$\frac{2}{3}\cdot 7+\frac{1}{2}+\frac{2}{5}\left(6+\frac{1}{3}\right) \quad \text{or} \quad \frac{2}{3}\cdot 7+\left(\frac{1}{2}+\frac{2}{5}\right)6+\frac{1}{3}.$$

Remark. Here again, the author's four distinctions are incompatible with the Latin expression, which admits of just one interpretation; indeed, *et dimidio* (and *et tertia*, and so on) excludes cases *c* and *d* while *duas quintas* (and *tres quartas*) excludes case *b*.

---

6. Other problems.

A. *A particular problem*.

(A.140) MSS $\mathcal{A}$, $\mathcal{B}$. Compute

$$\left(\frac{3\,4\,5\,6\,7}{4\,5\,6\,7\,8}\right)\cdot\left(\frac{4\,5\,6\,7\,8\,9}{5\,6\,7\,8\,9\,10}\right).$$

Preliminary remark. Calculating as done so far would lead to

$$\left(\frac{3\,4\,5\,6\,7}{4\,5\,6\,7\,8}\cdot 6720\right)\left(\frac{4\,5\,6\,7\,8\,9}{5\,6\,7\,8\,9\,10}\cdot 151200\right)\frac{1}{1\,016\,064\,000}.$$

On the other hand, the computation becomes quite elementary with the fractions written in numerals, since each expression reduces to the quotient of its first numerator and its last denominator; which is what the author will end with.

The author first observes that forming the denominating numbers for each side as before would be too long. Therefore, he adds, we shall proceed otherwise. Take of each expression the last denominator and form their product, namely $8\cdot 10=80$. Then, calculate

$$\left(\frac{3\,4\,5\,6\,7}{4\,5\,6\,7\,8}\right)\cdot\left(\frac{4\,5\,6\,7\,8\,9}{5\,6\,7\,8\,9\,10}\right)=\left(\frac{3\,4\,5\,6\,7}{4\,5\,6\,7\,8}\cdot 8\right)\cdot\left(\frac{4\,5\,6\,7\,8\,9}{5\,6\,7\,8\,9\,10}\cdot 10\right)\frac{1}{80}$$

$$=\left(\frac{3\,4\,5\,6}{4\,5\,6\,7}\cdot 7\right)\cdot\left(\frac{4\,5\,6\,7\,8}{5\,6\,7\,8\,9}\cdot 9\right)\frac{1}{80}=\ldots=$$

$$=\frac{3\cdot 4}{80}=\frac{12}{80}=\frac{8+4}{8\cdot 10}=\frac{1}{10}+\frac{1}{2}\frac{1}{10}.$$

B. *Combined expressions*.

As said in our summary (above, p. 1200), the subsequent group of problems is found only in manuscript B. Similar problems for the division are A.265–266.

(A.141) MS B. Difference of fractions by integer and fraction.
$$\left(\frac{3}{4}-\frac{1}{6}\right)\left(5+\frac{1}{3}\right)$$
Multiplying the denominators of the first expression, $4 \cdot 6 = 24$, we have
$$\left(\frac{3}{4}-\frac{1}{6}\right) 24 = 18 - 4 = 14, \quad \text{thus}$$
$$\left(\frac{3}{4}-\frac{1}{6}\right)\left(5+\frac{1}{3}\right) = \left[14\left(5+\frac{1}{3}\right)\right]\frac{1}{24}.$$

(A.142) MS B. Differences of fractions by integer and fraction.
$$\left[\frac{3}{4}-\left(\frac{1}{3}-\frac{1}{8}\right)\right]\left(7+\frac{1}{2}\right)$$
Multiplying the *first two* denominators of the first expression, $4 \cdot 3 = 12$, we have for this expression
$$\left[\frac{3}{4}\cdot 12 - \left(\frac{1}{3}\cdot 12 - \frac{1}{8}\cdot 12\right)\right]\frac{1}{12} = \left[9-\left(4-\left(1+\frac{1}{2}\right)\right)\right]\frac{1}{12} = \left(6+\frac{1}{2}\right)\frac{1}{12},$$
thus, for the proposed multiplication,
$$\left[\frac{3}{4}-\left(\frac{1}{3}-\frac{1}{8}\right)\right]\left(7+\frac{1}{2}\right) = \left(6+\frac{1}{2}\right)\left(7+\frac{1}{2}\right)\frac{1}{12} = \left(48+\frac{3}{4}\right)\frac{1}{12} = 4 + \frac{1}{2}\frac{1}{8}.$$

(A.143) MS B. Multiplication of composite expressions.
$$\frac{3}{4}\left[\frac{5}{6}\left(7+\frac{1}{5}\right)\right] \cdot \left[\frac{9}{10}\cdot 4 - \frac{1}{2}\frac{1}{8}\cdot\left(9+\frac{3}{5}\right)\right]$$
Direct computation. Since
$$\frac{3}{4}\left[\frac{5}{6}\left(7+\frac{1}{5}\right)\right] = \frac{3}{4}\cdot 6 = 4 + \frac{1}{2} \quad \text{and}$$
$$\frac{9}{10}\cdot 4 - \frac{1}{2}\frac{1}{8}\cdot\left(9+\frac{3}{5}\right) = 3 + \frac{3}{5} - \frac{48}{16\cdot 5} = 3, \quad \text{so}$$
$$\frac{3}{4}\left[\frac{5}{6}\left(7+\frac{1}{5}\right)\right] \cdot \left[\frac{9}{10}\cdot 4 - \frac{1}{2}\frac{1}{8}\cdot\left(9+\frac{3}{5}\right)\right] = \left(4+\frac{1}{2}\right)\cdot 3 = 13 + \frac{1}{2}.$$

(A.144) MS B. Other multiplication of composite expressions.
$$\left[\frac{9}{10}-\frac{1}{2}\frac{1}{7}\cdot 11\right] \cdot \left[\left(\frac{5}{9}\cdot\frac{1}{4}\cdot 3\right)\left(\frac{1}{5}\cdot 12\right)\right]$$

In the preceding problem, the second expression became considerably simpler. In the present case, it even reduces to 1, and so the result will just equal the value computed for the first expression.

Since 70 is divisible by 10 and $2 \cdot 7$, take, for the first expression, the common denominator 70. Then

$$\frac{9}{10} - \frac{1}{2}\frac{1}{7} \cdot 11 = \left[\frac{9}{10} \cdot 70 - \frac{1}{2}\frac{1}{7} \cdot 70 \cdot 11\right]\frac{1}{70} = \frac{63 - 55}{70} = \frac{8}{70} = \frac{1}{10} + \frac{1}{7}\frac{1}{10},$$

while for the second expression:[†]

$$\left(\frac{5}{9} \cdot \frac{1}{4} \cdot 3\right)\left(\frac{1}{5} \cdot 12\right) = \frac{5}{9} \cdot \frac{1}{4} \cdot 3 \cdot \left(2 + \frac{2}{5}\right) = \frac{5}{9} \cdot \frac{1}{4} \cdot \left(7 + \frac{1}{5}\right) = \frac{5}{9}\left(1 + \frac{4}{5}\right) = 1.$$

Remark. For computing the first expression the common denominator chosen is not, as usual, the product of all the denominators, but the lowest common denominator. Other instances where the common denominator chosen is smaller than the product are A.158, A.243. (The situation seen in A.142 is different since the chosen denominator is not common.)

[†] According to the computations in the text. We would just cancel the common factors and have the answer immediately. As we have noted in A.140, this reduction becomes evident with the use of numerical symbols.

Chapter A–VI:
Addition of fractions

Summary

1. Addition of fractional expressions.

This first section (A.145–151) treats the addition of fractional expressions: first, fraction to fraction (A.145–146); secondly, compound fraction to fraction (A.147) and to compound fraction (A.148); thirdly, fraction of a fraction to fraction of a fraction (A.149) and to compound fraction (A.150); it ends with the addition of several (different) fractions (A.151). Thus only fraction to fraction of a fraction is left out or, rather, does not receive a separate treatment. As for the case of fractional expressions containing integers, there is no need to consider it; for, as an early reader observed at the end of this section, these integers will just be added separately, then the result to those resulting from adding the fractions.

That makes this section much shorter than the corresponding part on multiplication, which covered three sections (A–V.1–2 & 4). The computations do not require much explanation either: the treatment is similar to what we have seen with multiplication and, furthermore, many problems ended with the addition of fractional expressions. For the computation itself, we may first form the common denominator (here the product of all denominators), which will leave us with calculating a quotient of two integers; or else we may convert one of the two given expressions into the kind of fractions of the other. In the latter case it is preferable, as we are told at the end of this section, to convert into the fractions of the smallest kind, thus with the largest denominator (*genus minoris fractionis*, *ultimum genus fractionis*).

2. Addition of irregular fractions.

As we know from the previous chapter, fifth section, irregular fractions are expressions of the form

$$\frac{k}{l} \text{ of } M$$

where M is an integer or an integer with a fraction. Just three problems form this section: we are taught how to add irregular fractions, first with M an integer (A.152) then an integer with a fraction (A.153), a case which admits of another interpretation (A.154). Considering what we have seen in the previous chapter, this short presentation is quite sufficient.

3. Problems on amounts.

A. *Known amounts*.

A.155–158, on known amounts, are a kind of extension of the problems involving irregular fractions: we are now to calculate a sum of fractions of a known 'amount' (*pecunia*, Arabic *māl*), say S, more specifically we are to determine the value of expressions of the form

(i) $\sum \dfrac{k_i}{l_i} \cdot S,$

(ii) $\sum \dfrac{k_i}{l_i} \cdot S + \dfrac{1}{r} \sum \dfrac{k_i}{l_i} \cdot S,$

(iii) $\sum \dfrac{k_i}{l_i} \cdot S + \dfrac{1}{r}\left(S - \sum \dfrac{k_i}{l_i} \cdot S\right).$

A.155 is of type *i*, A.156 & A.158 of type *ii* (with S respectively an integer and an integer with a fraction). A.156′ and A.157 are of type *iii*, thus (so-called) problems involving a 'remainder' (*residuum*, Arabic *bāqin*). This remainder is supposed to be positive, so that

$$\sum \dfrac{k_i}{l_i} < 1,$$

as stated in A.156′ (even though the value of the whole expression will be positive anyway since $r > 1$ in all problems presented).

The treatment is known from before: calculating L, the product of the denominators, then multiplying the fractional terms by it, next the whole by S and dividing the result by L. After A.155 the given value of S is initially disregarded so that the treatment turns out to be the method of one false position: replace S by L, find then the value M_L of the whole expression, and determine the true value M_S from the proportion $M_S : S = M_L : L$. This will appear from the demonstration in A.155, which makes the transition between the two methods.

B. *Unknown amounts*.

In the following problems, this time on unknown amounts, the value of the proposed expression is a given constant C and S is required. We can, though, again proceed using a false position if the left-hand expression is of type *i* (A.159) or *iii* (A.165).* But if the given expression contains additive or subtractive constants not multiplied by S, as in

(iv) $\sum \left(\dfrac{k_i}{l_i} \cdot S + t_i\right) + \left[\dfrac{1}{r}\left\{S - \sum \left(\dfrac{k_i}{l_i} \cdot S + t_i\right)\right\} + u\right] = C,$

we shall first have to bring them all to the right side in order to arrive at the former situation (A.160–164, A.166). The use of a false position S_0 then gives a value C_0, and we shall infer S from $C_0 : S_0 = C : S$.

In the closing problems, we find other cases, either of the second degree but reducible to the first (A.167–168) or (A.169) indeterminate of the first degree.

This section on unknown amounts is thus our first encounter with equations of the first degree and first-degree problems, which will be fully treated in Book B. But they are not misplaced, as appears from the similar location of such problems in the Spanish-Arabic treatise on arithmetic by

---

* No example of type *ii*.

al-Ḥaṣṣār.‡ Furthermore, there is no question here of algebra for determining the unknown, whereas there will be in Book B.

4. Summing series.

This is indeed a particular section of addition but, as it concerns integers, one which we might have expected to be treated earlier. There is, however, no chapter in the *Liber mahameleth* devoted to the addition of integers; moreover, our book is obviously once again following a tradition: summing series occupies a similar position, at the end of the multiplication of fractions, in al-Ḥaṣṣār's treatise.° As in the former section, we have first (part A) to compute a sum, then (part B) to compute one of the terms when the sum is known.

A. *Sum required*.

We encounter successively the summation formulae for the sequence of natural numbers starting with 1 or not, then the same for odd and even natural numbers; next, the summation formulae for squares of consecutive natural numbers starting with 1 or not, then the same for odd and even numbers; finally we compute the sum of cubes of consecutive natural numbers, then the same for odd and even numbers. The summation formulae known to the author (sometimes slightly different in form from our modern ones) are as follows.

(*i*) Sum of natural numbers (A.170–171):

$$\sum_{1}^{n} k = \frac{n(n+1)}{2}$$

$$\sum_{m}^{n} k = \frac{n(n+1)}{2} - \frac{(m-1)m}{2}$$

(*ii*) Sum of odd natural numbers (A.172–173):

$$\sum_{1}^{n}(2k-1) = \left(\frac{(2n-1)+1}{2}\right)^2$$

$$\sum_{m}^{n}(2k-1) = \left(\frac{(2n-1)+1}{2}\right)^2 - \left(\frac{(2m-1)-1}{2}\right)^2$$

the first of which is equivalent to

$$\sum_{1}^{n}(2k-1) = n^2.$$

(*iii*) Sum of even natural numbers (A.174–175):

---

‡ Already mentioned at the end of our summary to A–V (p. 1200). See Suter, p. 31.
° Suter's summary, pp. 31–34.

$$\sum_{1}^{n} 2k = \frac{2n+2}{2} \cdot \frac{2n}{2}$$

$$\sum_{m}^{n} 2k = \frac{2n+2}{2} \cdot \frac{2n}{2} - \frac{2m-2}{2} \cdot \frac{2m}{2}$$

the first of which is equivalent to

$$\sum_{1}^{n} 2k = n(n+1).$$

(*iv*) Sum of the squares of natural numbers (A.176, A.179):

$$\sum_{1}^{n} k^2 = \frac{n(n+1)}{2} \cdot \left(\frac{2}{3} \cdot n + \frac{1}{3}\right), \quad \text{or} \quad \frac{n(n+1)}{2} \cdot \left(\frac{2}{3}(n-1) + 1\right)$$

$$\sum_{m}^{n} k^2 = \frac{n(n+1)}{2} \cdot \left(\frac{2}{3} \cdot n + \frac{1}{3}\right) - \frac{(m-1)m}{2} \cdot \left(\frac{2}{3} \cdot (m-1) + \frac{1}{3}\right)$$

the first of which is equivalent to

$$\sum_{1}^{n} k^2 = \frac{n(n+1)(2n+1)}{6}.$$

(*v*) Sum of the squares of odd numbers (A.177):

$$\sum_{1}^{n} (2k-1)^2 = \left((2n-1)+2\right) \cdot \frac{(2n-1)+1}{2} \cdot \frac{2n-1}{3}$$

equivalent to

$$\sum_{1}^{n} (2k-1)^2 = \frac{(2n+1)\,2n\,(2n-1)}{6}.$$

(*vi*) Sum of the squares of even numbers (A.178):

$$\sum_{1}^{n} (2k)^2 = \frac{2n+2}{2} \cdot \frac{2n}{2} \cdot \left(\frac{2}{3} \cdot 2n + \frac{2}{3}\right)$$

equivalent to

$$\sum_{1}^{n} (2k)^2 = \frac{(n+1)\,n\,(4n+2)}{3}.$$

(*vii*) Sum of the cubes of natural numbers (A.180–181):

$$\sum_{1}^{n} k^3 = \left(\frac{n(n+1)}{2}\right)^2 \quad \left(= \left(\sum_{1}^{n} k\right)^2\right)$$

$$\sum_{m}^{n} k^3 = \left(\frac{n(n+1)}{2}\right)^2 - \left(\frac{(m-1)m}{2}\right)^2.$$

(*viii*) Sum of the cubes of odd numbers (A.182):

$$\sum_{1}^{n}(2k-1)^3 = \left(\frac{(2n-1)+1}{2}\right)^2 \cdot \left[2 \cdot \left(\frac{(2n-1)+1}{2}\right)^2 - 1\right]$$

equivalent to

$$\sum_{1}^{n}(2k-1)^3 = n^2(2n^2-1).$$

(*ix*) Sum of the cubes of even numbers (A.183):

$$\sum_{1}^{n}(2k)^3 = \left(\frac{2n+2}{2} \cdot \frac{2n}{2}\right) \cdot \left[2 \cdot \frac{2n+2}{2} \cdot \frac{2n}{2}\right]$$

equivalent to

$$\sum_{1}^{n}(2k)^3 = 2 \cdot (n(n+1))^2.$$

Remark. We do not find any demonstration of these formulae. That of a general arithmetical progression will be justified in Book B (B–XII).

B. *Last term required*.

As did Section 3, the present one ends with three problems which, although closely related to what precedes, require quite different treatments (A.184–186): given the sum, find the last term. In all three, square roots occur; in two of them the formula used (but not explained) corresponds to the solution of a quadratic equation.

---

1. Addition of fractional expressions.

A. *Addend a fraction*.

(**A.145**) MSS $\mathcal{A}, \mathcal{B}, \mathcal{C}$.   Addition of a fraction to a fraction.

$$\frac{3}{8} + \frac{4}{5}$$

(*a*) Form the product of the denominators: $8 \cdot 5 = 40$; then

$$\frac{3}{8} + \frac{4}{5} = \left(\frac{3}{8} \cdot 40 + \frac{4}{5} \cdot 40\right)\frac{1}{40} = \frac{15+32}{40}$$

$$= \frac{47}{40} = \frac{40+5+2}{5 \cdot 8} = 1 + \frac{1}{8} + \frac{2}{5}\frac{1}{8}.$$

(*b*) Convert one fraction into the kind of the other.

$$\frac{3}{8} + \frac{4}{5} = \frac{3}{8} + \frac{32}{5}\frac{1}{8} = \frac{3}{8} + \frac{6}{8} + \frac{2}{5}\frac{1}{8} = 1 + \frac{1}{8} + \frac{2}{5}\frac{1}{8}.$$

(**A.146**) MSS $\mathcal{A}, \mathcal{B}, \mathcal{C}$.   Addition of a fraction to a non-elementary fraction.

$$\frac{3}{7} + \frac{10}{11}$$

(**a**) Form the product of the denominators: $7 \cdot 11 = 77$; then

$$\frac{3}{7} + \frac{10}{11} = \left(\frac{3}{7} \cdot 77 + \frac{10}{11} \cdot 77\right)\frac{1}{77} = \frac{33 + 70}{77} = \frac{103}{77}.$$

(**b**) Convert the first fraction into the kind of the second.

$$\frac{3}{7} + \frac{10}{11} = \frac{33}{7}\frac{1}{11} + \frac{10}{11} = \frac{4}{11} + \frac{5}{7}\frac{1}{11} + \frac{10}{11} = 1 + \frac{3}{11} + \frac{5}{7}\frac{1}{11}.$$

(**c**) Convert the second fraction into the kind of the first.

$$\frac{3}{7} + \frac{10}{11} = \frac{3}{7} + \frac{70}{11}\frac{1}{7} = 1 + \frac{2}{7} + \frac{4}{11}\frac{1}{7}.$$

B. *Addend a compound fraction*.

(**A.147**) MSS *A, B, C.* Addition of a compound fraction to a fraction.

$$\left(\frac{3}{5} + \frac{1}{4}\frac{1}{5}\right) + \frac{5}{8}$$

(**a**) Form the product of the denominators: $(5 \cdot 4) \cdot 8 = 20 \cdot 8 = 160$; then

$$\left(\frac{3}{5} + \frac{1}{4}\frac{1}{5}\right) + \frac{5}{8} = \left[\left(\frac{3}{5} + \frac{1}{4}\frac{1}{5}\right) \cdot 160 + \frac{5}{8} \cdot 160\right]\frac{1}{160}$$

$$= (96 + 8 + 100)\frac{1}{160} = \frac{104 + 100}{160} = \frac{204}{160}.$$

(**b**) Convert the first expression into the kind of the second.

$$\left(\frac{3}{5} + \frac{1}{4}\frac{1}{5}\right) + \frac{5}{8} = \frac{24}{5}\frac{1}{8} + \frac{8}{4}\frac{1}{5}\frac{1}{8} + \frac{5}{8} = \frac{4}{8} + \frac{4}{5}\frac{1}{8} + \frac{2}{5}\frac{1}{8} + \frac{5}{8}$$

$$= \frac{5}{8} + \frac{1}{5}\frac{1}{8} + \frac{5}{8} = 1 + \frac{2}{8} + \frac{1}{5}\frac{1}{8}.$$

(**c**) Convert the second fraction into the kind of the first (that is, into fifths).

$$\left(\frac{3}{5} + \frac{1}{4}\frac{1}{5}\right) + \frac{5}{8} = \frac{3}{5} + \frac{1}{4}\frac{1}{5} + \frac{25}{8}\frac{1}{5} = \frac{3}{5} + \frac{1}{4}\frac{1}{5} + \frac{3}{5} + \frac{1}{8}\frac{1}{5} = 1 + \frac{1}{5} + \frac{3}{8}\frac{1}{5}.$$

(**A.148**) MSS *A, B, C.* Addition of a compound fraction to a compound fraction.

$$\left(\frac{2}{5} + \frac{3}{4}\frac{1}{5}\right) + \left(\frac{2}{7} + \frac{2}{3}\frac{1}{7}\right)$$

(*a*) Form the product of the denominators: $(5 \cdot 4)(7 \cdot 3) = 20 \cdot 21 = 420;$*
then

$$\left(\frac{2}{5} + \frac{3}{4}\frac{1}{5}\right) + \left(\frac{2}{7} + \frac{2}{3}\frac{1}{7}\right) = \left[\left(\frac{2}{5} + \frac{3}{4}\frac{1}{5}\right) \cdot 420 + \left(\frac{2}{7} + \frac{2}{3}\frac{1}{7}\right) \cdot 420\right]\frac{1}{420}$$

$$= \left[(168 + 63) + (120 + 40)\right]\frac{1}{420} = \frac{231 + 160}{420} = \frac{391}{420}.$$

(*b*) Convert the first expression into the kind of the second.

$$\frac{2}{5} + \frac{3}{4}\frac{1}{5} = \frac{14}{5}\frac{1}{7} + \frac{21}{4}\frac{1}{5}\frac{1}{7} = \frac{2}{7} + \frac{4}{5}\frac{1}{7} + \frac{1}{7} + \frac{1}{4}\frac{1}{5}\frac{1}{7} = \frac{3}{7} + \frac{4}{5}\frac{1}{7} + \frac{1}{4}\frac{1}{5}\frac{1}{7}.$$

Adding this to the second expression gives:

$$\left(\frac{3}{7} + \frac{4}{5}\frac{1}{7} + \frac{1}{4}\frac{1}{5}\frac{1}{7}\right) + \left(\frac{2}{7} + \frac{2}{3}\frac{1}{7}\right) = \frac{5}{7} + \frac{4}{5}\frac{1}{7} + \frac{1}{4}\frac{1}{5}\frac{1}{7} + \frac{2}{3}\frac{1}{7}.$$

Adding now the last three terms, we obtain first

$$\frac{4}{5}\frac{1}{7} + \frac{1}{4}\frac{1}{5}\frac{1}{7} + \frac{2}{3}\frac{1}{7} = \frac{4}{5}\frac{1}{7} + \frac{1}{4}\frac{1}{5}\frac{1}{7} + \frac{10}{3}\frac{1}{5}\frac{1}{7}$$

$$= \frac{4}{5}\frac{1}{7} + \frac{1}{4}\frac{1}{5}\frac{1}{7} + \frac{3}{5}\frac{1}{7} + \frac{1}{3}\frac{1}{5}\frac{1}{7} = \frac{1}{7} + \frac{2}{5}\frac{1}{7} + \frac{1}{3}\frac{1}{5}\frac{1}{7} + \frac{1}{4}\frac{1}{5}\frac{1}{7}$$

and then for the last two terms

$$\frac{1}{3}\frac{1}{5}\frac{1}{7} + \frac{1}{4}\frac{1}{5}\frac{1}{7} = \frac{6}{3}\frac{1}{6}\frac{1}{5}\frac{1}{7} + \frac{6}{4}\frac{1}{6}\frac{1}{5}\frac{1}{7}$$

$$= \frac{2}{6}\frac{1}{5}\frac{1}{7} + \frac{1}{6}\frac{1}{5}\frac{1}{7} + \frac{2}{4}\frac{1}{6}\frac{1}{5}\frac{1}{7} = \frac{3}{6}\frac{1}{5}\frac{1}{7} + \frac{1}{2}\frac{1}{6}\frac{1}{5}\frac{1}{7}.$$

Putting together all the terms obtained (in boldface type) gives the final result, namely

$$\left(\frac{2}{5} + \frac{3}{4}\frac{1}{5}\right) + \left(\frac{2}{7} + \frac{2}{3}\frac{1}{7}\right) = \frac{6}{7} + \frac{2}{5}\frac{1}{7} + \frac{3}{6}\frac{1}{5}\frac{1}{7} + \frac{1}{2}\frac{1}{6}\frac{1}{5}\frac{1}{7}.$$

C. *Addend a fraction of a fraction*.

(**A.149**) MSS 𝒜, ℬ, 𝒞. Addition of a fraction of a fraction to a fraction of a fraction.

$$\frac{5}{6}\frac{1}{8} + \frac{4}{7}\frac{1}{11}$$

(*a*) Form the product of the denominators: $(6 \cdot 8)(7 \cdot 11) = 48 \cdot 77 = 3696$. So

$$\frac{5}{6}\frac{1}{8} + \frac{4}{7}\frac{1}{11} = \left(\frac{5}{6}\frac{1}{8} \cdot 3696 + \frac{4}{7}\frac{1}{11} \cdot 3696\right)\frac{1}{3696} = \frac{385 + 192}{3696} = \frac{577}{3696}.$$

---

* Some of the values not found in the text are in the table.

(**b**) Convert one fraction, say the first, into the kind of the other (thus into sevenths of elevenths).

$$\frac{5\ 1}{6\ 8} = \frac{385}{6}\frac{1\ 1\ 1}{8\ 7\ 11} = \frac{384+1}{6\cdot 8}\frac{1\ 1}{7\ 11} = \frac{8\ 1}{7\ 11} + \frac{1\ 1\ 1\ 1}{6\ 8\ 7\ 11},$$

so that

$$\frac{5\ 1}{6\ 8} + \frac{4\ 1}{7\ 11} = \frac{8\ 1}{7\ 11} + \frac{1\ 1\ 1\ 1}{6\ 8\ 7\ 11} + \frac{4\ 1}{7\ 11} = \frac{1}{11} + \frac{5\ 1}{7\ 11} + \frac{1\ 1\ 1\ 1}{6\ 8\ 7\ 11}.$$

(**A.150**) MSS $\mathcal{A}, \mathcal{B}, \mathcal{C}$. Addition of a compound fraction to a fraction of a fraction.

$$\left(\frac{4}{5} + \frac{1\ 1}{3\ 5}\right) + \frac{5\ 1}{6\ 11}$$

(**a**) Form the product of the denominators: $(5\cdot 3)(6\cdot 11) = 15\cdot 66 = 990^{\dagger}$. Then we shall consider that

$$\left(\frac{4}{5} + \frac{1\ 1}{3\ 5}\right) + \frac{5\ 1}{6\ 11} = \left[\left(\frac{4}{5} + \frac{1\ 1}{3\ 5}\right)\cdot 990 + \frac{5\ 1}{6\ 11}\cdot 990\right]\frac{1}{990}.$$

(**b**) Convert one fraction, say the first expression, into the kind of the other, namely elevenths (and sixths of elevenths). Then, first,

$$\frac{4}{5} + \frac{1\ 1}{3\ 5} = \frac{44}{5}\frac{1}{11} + \frac{11\ 1\ 1}{3\ 5\ 11} = \frac{8}{11} + \frac{4\ 1}{5\ 11} + \frac{3\ 1}{5\ 11} + \frac{2\ 1\ 1}{3\ 5\ 11}$$

$$= \frac{9}{11} + \frac{2\ 1}{5\ 11} + \frac{2\ 1\ 1}{3\ 5\ 11}.$$

Now, since

$$\frac{2\ 1}{5\ 11} + \frac{2\ 1\ 1}{3\ 5\ 11} = \frac{12\ 1\ 1}{5\ 6\ 11} + \frac{4\ 1\ 1}{5\ 6\ 11} = \frac{3\ 1}{6\ 11} + \frac{1\ 1\ 1}{5\ 6\ 11},$$

we shall have altogether

$$\left(\frac{4}{5} + \frac{1\ 1}{3\ 5}\right) + \frac{5\ 1}{6\ 11} = \frac{9}{11} + \frac{3\ 1}{6\ 11} + \frac{1\ 1\ 1}{5\ 6\ 11} + \frac{5\ 1}{6\ 11}$$

$$= \frac{9}{11} + \frac{1}{11} + \frac{2\ 1}{6\ 11} + \frac{1\ 1\ 1}{5\ 6\ 11} = \frac{10}{11} + \frac{2\ 1}{6\ 11} + \frac{1\ 1\ 1}{5\ 6\ 11}.$$

D. *Case of different fractions*.

(**A.151**) MSS $\mathcal{A}, \mathcal{B}, \mathcal{C}$. Addition of several fractions.

$$\frac{2}{9} + \frac{3}{8} + \frac{10}{11}$$

---

† This value in the table only.

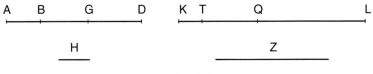

Fig. 30

(**a**) The product of the denominators here is $9 \cdot 8 \cdot 11 = 792$. Therefore

$$\frac{2}{9} + \frac{3}{8} + \frac{10}{11} = \left(\frac{2}{9} \cdot 792 + \frac{3}{8} \cdot 792 + \frac{10}{11} \cdot 792\right) \frac{1}{792}$$

$$= \frac{176 + 297 + 720}{792} = \frac{1193}{792}.$$

Demonstration. Let first (Fig. 30)

$$\frac{2}{9} = AB, \qquad \frac{3}{8} = BG, \qquad \frac{10}{11} = GD,$$

and $1 = H$; let next $792 = Z$ with

$$\frac{2}{9} Z = 176 = KT, \qquad \frac{3}{8} Z = 297 = TQ, \qquad \frac{10}{11} Z = 720 = QL.$$

To prove that

$$AB + BG + GD = \frac{KT + TQ + QL}{Z},$$

which is the previous computation.
Since, from the data, $AB \cdot Z = KT$, $BG \cdot Z = TQ$, $GD \cdot Z = QL$, so

$$\frac{AB}{H} = \frac{KT}{Z}, \qquad \frac{BG}{H} = \frac{TQ}{Z}, \qquad \frac{GD}{H} = \frac{QL}{Z},$$

then (by P'_7, repeated)

$$\frac{AD}{H} = \frac{KL}{Z}, \quad \text{so} \quad AD \cdot Z = KL \cdot H = KL \quad \text{and} \quad AD = \frac{KL}{Z}.$$

This proof is said (by the author himself or by an early reader) to be valid for all previous cases on adding fractions. Indeed, it is the basis of the treatment involving the product of the denominators. In any case, the right place for it is here since we now have three addends instead of two (a compound fraction being considered *de facto* as a single term).

(**b**) Convert the given fractions into one kind. Converting into eighths, we find first

$$\frac{2}{9} = \frac{16 \; 1}{9 \; 8} = \frac{1}{8} + \frac{7 \; 1}{9 \; 8} \quad \text{and} \quad \frac{10}{11} = \frac{80 \; 1}{11 \; 8} = \frac{7}{8} + \frac{3 \; 1}{11 \; 8},$$

so that

$$\frac{2}{9} + \frac{3}{8} + \frac{10}{11} = 1 + \frac{3}{8} + \frac{7 \; 1}{9 \; 8} + \frac{3 \; 1}{11 \; 8}.$$

Next, since

$$\frac{7}{9}\frac{1}{8} + \frac{3}{11}\frac{1}{8} = \frac{77}{11}\frac{1}{9}\frac{1}{8} + \frac{27}{11}\frac{1}{9}\frac{1}{8} = \frac{104}{11}\frac{1}{9}\frac{1}{8} = \frac{99+5}{11 \cdot 9 \cdot 8} = \frac{1}{8} + \frac{5}{11}\frac{1}{9}\frac{1}{8},$$

the required result is

$$\frac{2}{9} + \frac{3}{8} + \frac{10}{11} = 1 + \frac{3}{8} + \frac{1}{8} + \frac{5}{11}\frac{1}{9}\frac{1}{8} = 1 + \frac{4}{8} + \frac{5}{11}\frac{1}{9}\frac{1}{8}.$$

As the text notes, it is more appropriate to convert into the fraction of the smallest kind, that is, with the largest denominator.*

We have already mentioned (p. 1234) that the absence of a problem involving integers plus fractions led an early reader to allude to this (banal) case. With this remark (prompted perhaps by A.152b and A.153b) ends the first section.

---

2. Addition of irregular fractions.

A. *Addition proper*.

(A.152) MSS $\mathcal{A}$, $\mathcal{B}$, $\mathcal{C}$.   Addition of irregular fractions of integers.

$$\frac{3}{5} \cdot 6 + \frac{7}{8} \cdot 9$$

(***a***) Multiplying the individual denominators: $5 \cdot 8 = 40$. Then

$$\frac{3}{5} \cdot 6 + \frac{7}{8} \cdot 9 = \left(\frac{3}{5} 40 \cdot 6 + \frac{7}{8} 40 \cdot 9\right) \frac{1}{40}$$

$$= (24 \cdot 6 + 35 \cdot 9) \frac{1}{40} = \frac{144 + 315}{40} = \frac{459}{40}.$$

Demonstration. We are to prove that

$$\frac{3}{5} \cdot 6 + \frac{7}{8} \cdot 9 = \left(\frac{3}{5} Q \cdot 6 + \frac{7}{8} Q \cdot 9\right) \frac{1}{Q}$$

with Q arbitrary, but suitably chosen, that is, reducing the expression in parentheses to an integer. Now we already know (A.90a) that, individually,

$$\frac{3}{5} \cdot 6 = \left(\frac{3}{5} q \cdot 6\right) \frac{1}{q}, \qquad \frac{7}{8} \cdot 9 = \left(\frac{7}{8} q' \cdot 9\right) \frac{1}{q'}.$$

---

* If we do so here, we shall find

$$\frac{2}{9} + \frac{3}{8} + \frac{10}{11} = 1 + \frac{5}{11} + \frac{5}{9}\frac{1}{11} + \frac{1}{8}\frac{1}{9}\frac{1}{11}.$$

So we shall conveniently choose any q divisible by 5, any q' divisible by 8, and form $Q = q \cdot q'$. With the simplest choice $q = 5$, $q' = 8$, so $q \cdot q' = 40$, we have, as in the present example,

$$\frac{3}{5} \cdot 6 + \frac{7}{8} \cdot 9 = \left(\frac{3}{5} 40 \cdot 6\right) + \left(\frac{7}{8} 40 \cdot 9\right) \frac{1}{40} = \frac{144}{40} + \frac{315}{40} = \frac{144 + 315}{40},$$

this last step, as we are told, by P_8.

(**b**) Computing each term, we have

$$\frac{3}{5} \cdot 6 + \frac{7}{8} \cdot 9 = \left(3 + \frac{3}{5}\right) + \left(7 + \frac{7}{8}\right),$$

which reduces to an addition of fractions (A.145).

(**A.153**) MSS $\mathcal{A}$, $\mathcal{B}$, $\mathcal{C}$. Addition of irregular fractions of integers with fractions.

$$\frac{2}{5}\left(4 + \frac{1}{2}\right) + \frac{3}{8}\left(6 + \frac{1}{3}\right)$$

(**a**) Multiplying the individual denominators: $(5 \cdot 2) \cdot (8 \cdot 3) = 240$, then

$$\frac{2}{5}\left(4 + \frac{1}{2}\right) + \frac{3}{8}\left(6 + \frac{1}{3}\right) = \left[\frac{2}{5} 240 \cdot \left(4 + \frac{1}{2}\right) + \frac{3}{8} 240 \cdot \left(6 + \frac{1}{3}\right)\right]\frac{1}{240}$$

$$= \left[96 \cdot \left(4 + \frac{1}{2}\right) + 90 \cdot \left(6 + \frac{1}{3}\right)\right]\frac{1}{240} = \frac{432 + 570}{240}.$$

This is proved, we are told, as above (the presence of a fraction with the integer makes no difference).

(**b**) Computing each term, which leads us to A.145.

$$\frac{2}{5}\left(4 + \frac{1}{2}\right) + \frac{3}{8}\left(6 + \frac{1}{3}\right) = \left(\frac{8}{5} + \frac{1}{5}\right) + \left(\frac{18}{8} + \frac{1}{8}\right) = \left(1 + \frac{4}{5}\right) + \left(2 + \frac{3}{8}\right).$$

B. *Ambiguous formulations*.

(**A.154**) MSS $\mathcal{A}$, $\mathcal{B}$, $\mathcal{C}$. Add $\frac{2}{5}$ of 4 and $\frac{1}{2}$ of 1 to $\frac{3}{8}$ of 6 and $\frac{1}{3}$ of 1.

This, presented as an independent problem, is in fact another 'reading' (at least, in Arabic) of the previous problem, considering that the fractions are taken of the integers only.||

$$\left(\frac{2}{5} \cdot 4 + \frac{1}{2}\right) + \left(\frac{3}{8} \cdot 6 + \frac{1}{3}\right)$$

|| In the progenitor of MS $\mathcal{C}$ this point has been noted by a reader, for he adds to the enunciation of A.153 that it can be understood in two ways.

(a) Multiplying the individual denominators, thus with the same result as before, we consider

$$\left(\frac{2}{5} \cdot 4 + \frac{1}{2}\right) + \left(\frac{3}{8} \cdot 6 + \frac{1}{3}\right)$$

$$= \left[\left(\frac{2}{5} 240 \cdot 4 + \frac{1}{2} 240\right) + \left(\frac{3}{8} 240 \cdot 6 + \frac{1}{3} 240\right)\right] \frac{1}{240}$$

$$= \left[(384 + 120) + (540 + 80)\right] \frac{1}{240} = \frac{1124}{240}.$$

(b) Computing each term:

$$\left(\frac{2}{5} \cdot 4 + \frac{1}{2}\right) + \left(\frac{3}{8} \cdot 6 + \frac{1}{3}\right) = \left(1 + \frac{3}{5} + \frac{1}{2}\right) + \left(2 + \frac{2}{8} + \frac{1}{3}\right).$$

---

3. Problems on amounts.

A. *Known amounts*.

(**A.155**) MSS $\mathcal{A}, \mathcal{B}, \mathcal{C}$. Calculate

$$\frac{4}{5} \cdot 9 + \frac{3}{4} \cdot 9$$

Multiply the individual denominators: $5 \cdot 4 = 20$; then

$$\frac{4}{5} \cdot 9 + \frac{3}{4} \cdot 9 = \left(\frac{4}{5} \cdot 20 + \frac{3}{4} \cdot 20\right) \frac{1}{20} \cdot 9 = \frac{31 \cdot 9}{20}.$$

Demonstration. As we know (A.93a), for any two quantities Q, S,

$$\sum \frac{k_i}{l_i} \cdot Q : Q = \sum \frac{k_i}{l_i} \cdot S : S$$

where, in this case, S is given and Q, the product of the denominators, has been calculated. Knowing the two terms on the left side, we are now to determine the third term of a proportion from the three others (all integral). It is then pointed out that we may calculate the result as

$$\frac{31 \cdot 9}{20}, \quad \frac{31}{20} \cdot 9, \quad \frac{9}{20} \cdot 31.$$

This (for us futile) change will be emphasized at the beginning of Book B (pp. 1343–1344 below), and constantly applied there for determining the unknown from a proportion. See also below, A.202 (p. 1266).

(**A.156**) MSS $\mathcal{A}, \mathcal{B}, \mathcal{C}$. Calculate

$$\left(\frac{4}{5} \cdot 9 + \frac{3}{4} \cdot 9\right) + \frac{1}{2}\left(\frac{4}{5} \cdot 9 + \frac{3}{4} \cdot 9\right)$$

The product of the denominators being now $5 \cdot 4 \cdot 2 = 40$, and since

$$\left(\frac{4}{5} \cdot 40 + \frac{3}{4} \cdot 40\right) + \frac{1}{2}\left(\frac{4}{5} \cdot 40 + \frac{3}{4} \cdot 40\right) = 62 + \frac{1}{2} \cdot 62 = 93,$$

we shall have $93 : 40 = \text{(required sum)} : 9$, so that

$$\text{required sum} = \frac{93 \cdot 9}{40}.$$

Remark. The solution of this problem could have been deduced immediately from A.155, with no need to form another denominator. Only the analogy between it and the next problem may account for its presence and full treatment.

(A.156') We could not propose, the author says, to add half the remainder, that is, to consider

$$\left(\frac{4}{5} \cdot 9 + \frac{3}{4} \cdot 9\right) + \frac{1}{2}\left[9 - \left(\frac{4}{5} \cdot 9 + \frac{3}{4} \cdot 9\right)\right],$$

since the sum of the fractions taken from 9 is larger than 1.

Remark. What is called the 'remainder' must thus be positive, although the whole expression will in fact not be negative: since this remainder is multiplied by a fraction $\frac{p}{q} < 1$, we have, with S positive,

$$\left(\sum \frac{k_i}{l_i} \cdot S\right) + \frac{p}{q}\left[S - \left(\sum \frac{k_i}{l_i} \cdot S\right)\right] = \frac{p}{q} S + \left(1 - \frac{p}{q}\right)\left(\sum \frac{k_i}{l_i} \cdot S\right) > 0.$$

In accordance with this requirement, the sum of the fractions will be less than 1 in the coming problem.

(A.157) MSS $\mathcal{A}, \mathcal{B}, \mathcal{C}$. Calculate

$$\left(\frac{2}{5} \cdot 9 + \frac{1}{4} \cdot 9\right) + \frac{1}{3}\left[9 - \left(\frac{2}{5} \cdot 9 + \frac{1}{4} \cdot 9\right)\right]$$

The product of the denominators being $5 \cdot 4 \cdot 3 = 60$ and since

$$\left(\frac{2}{5} \cdot 60 + \frac{1}{4} \cdot 60\right) + \frac{1}{3}\left[60 - \left(\frac{2}{5} \cdot 60 + \frac{1}{4} \cdot 60\right)\right] = 39 + \frac{1}{3} \cdot 21 = 46,$$

we have $46 : 60 = \text{(required sum)} : 9$, so

$$\text{required sum} = \frac{46 \cdot 9}{60}.$$

(A.158) MSS $\mathcal{A}, \mathcal{B}, \mathcal{C}$. Same type as A.156, but the given amount now comprises a fraction. We are to calculate

$$\left[\frac{5}{8}\left(7 + \frac{1}{2}\right) + \frac{2}{3}\left(7 + \frac{1}{2}\right)\right] + \frac{1}{4}\left[\frac{5}{8}\left(7 + \frac{1}{2}\right) + \frac{2}{3}\left(7 + \frac{1}{2}\right)\right]$$

The product of the denominators being $8 \cdot 3 \cdot 4 = 96$ (disregarding the fraction in the amount) and since

$$\left(\frac{5}{8} \cdot 96 + \frac{2}{3} \cdot 96\right) + \frac{1}{4}\left(\frac{5}{8} \cdot 96 + \frac{2}{3} \cdot 96\right) = 60 + 64 + 31 = 155,$$

then $155 : 96 = $ (required sum) $: \left(7 + \frac{1}{2}\right)$, whence the required sum.

B. *Unknown amounts*.

In A.159–169 we are to determine an amount when the sum of given fractions of it is known. We find thus here for the first time the determination of an unknown from (first-degree) equations. The reasoning is as before, but the required quantity is the fourth term of the proportion instead of the third.

(A.159) MSS $\mathcal{A}, \mathcal{B}, \mathcal{C}$. Required the amount S such that

$$\frac{1}{3} \cdot S + \frac{1}{4} \cdot S = 10$$

Take $3 \cdot 4 = 12$, for which

$$\frac{1}{3} \cdot 12 + \frac{1}{4} \cdot 12 = 7.$$

Then $7 : 12 = 10 : S$, whence

$$S = \frac{12 \cdot 10}{7} = 17 + \frac{1}{7}.$$

(A.160) MSS $\mathcal{A}, \mathcal{B}, \mathcal{C}$. Required the amount S such that

$$\left(\frac{1}{3} \cdot S + \frac{1}{4} \cdot S\right) + 2 = 10$$

This is equivalent to solving

$$\frac{1}{3} \cdot S + \frac{1}{4} \cdot S = 8,$$

to be treated as above (A.159).

(A.161) MSS $\mathcal{A}, \mathcal{B}, \mathcal{C}$. Required the amount S such that

$$\left(\frac{1}{3} \cdot S + \frac{1}{4} \cdot S\right) - 2 = 10$$

This is equivalent to solving

$$\frac{1}{3} \cdot S + \frac{1}{4} \cdot S = 12,$$

to be treated as above. Hence the general rule given in the text:

• Add to the number on the right side the (absolute) values of all the constants on the left side which are subtractive and subtract the values of all those which are additive. This will lead to the simple case A.159.

Remarks. This is banal for the coming A.162, but caution will be required for problems 'with remainders' where fractions of the constants are taken. Note too that, since the above rule is not applied in A.162, it might be a later addition (possibly by the author himself).[†] We have already noted the presence of general rules which are never put to use (pp. 1205, 1215).

(A.162) MSS $\mathcal{A}, \mathcal{B}, \mathcal{C}$. Required the amount S such that
$$\left(\frac{1}{3} \cdot S + 1\right) + \left(\frac{1}{4} \cdot S - 3\right) = 10$$
This reduces to
$$\frac{1}{3} \cdot S + \frac{1}{4} \cdot S - 2 = 10,$$
which is A.161.

(A.163) MSS $\mathcal{A}, \mathcal{B}, \mathcal{C}$. Required the amount S such that
$$\left[\left(\frac{1}{3} \cdot S + 2\right) + \left(\frac{1}{4} \cdot S - 1\right)\right] + \left[\frac{1}{2}\left\{S - \left[\left(\frac{1}{3} \cdot S + 2\right) + \left(\frac{1}{4} \cdot S - 1\right)\right]\right\} + 4\right] = 10$$

The sum of the constants in the first part is $+1$, thus -1 in the remainder, which when halved reduces to $-\frac{1}{2}$; then the total, with the last term, is $4 + \frac{1}{2}$; we are then to solve
$$\left(\frac{1}{3} \cdot S + \frac{1}{4} \cdot S\right) + \frac{1}{2}\left[S - \left(\frac{1}{3} \cdot S + \frac{1}{4} \cdot S\right)\right] + \left(4 + \frac{1}{2}\right) = 10, \quad \text{or}$$
$$\left(\frac{1}{3} \cdot S + \frac{1}{4} \cdot S\right) + \frac{1}{2}\left[S - \left(\frac{1}{3} \cdot S + \frac{1}{4} \cdot S\right)\right] = 5 + \frac{1}{2},$$
to be treated 'as we have shown before'.

Remark. This alludes to A.165, which must therefore be misplaced and should precede this problem.

(A.164) MSS $\mathcal{A}, \mathcal{B}, \mathcal{C}$. Required the amount S such that
$$\left[\left(\frac{1}{3} \cdot S - 5\right) + \left(\frac{1}{4} \cdot S + 2\right)\right] + \left[\frac{1}{2}\left\{S - \left[\left(\frac{1}{3} \cdot S - 5\right) + \left(\frac{1}{4} \cdot S + 2\right)\right]\right\} - 1\right] = 10$$

The constants are, respectively, -3 for the first part, thus 3 in the remainder, becoming when halved $1 + \frac{1}{2}$, and the additional -1; the sum of these constants being $-(2 + \frac{1}{2})$, the problem becomes
$$\left(\frac{1}{3} \cdot S + \frac{1}{4} \cdot S\right) + \frac{1}{2}\left[S - \left(\frac{1}{3} \cdot S + \frac{1}{4} \cdot S\right)\right] - \left(2 + \frac{1}{2}\right) = 10, \quad \text{or}$$
$$\left(\frac{1}{3} \cdot S + \frac{1}{4} \cdot S\right) + \frac{1}{2}\left[S - \left(\frac{1}{3} \cdot S + \frac{1}{4} \cdot S\right)\right] = 12 + \frac{1}{2},$$

[†] The subsequent allusion to it (A.207) is surely by an early reader.

to be treated 'as has been shown before' (A.165).

Many, more complicated, problems of this kind can be devised, we are told —and reduced in the same way.

(A.165) MSS $\mathcal{A}$, $\mathcal{B}$, $\mathcal{C}$.   Required the amount S such that
$$\left(\frac{1}{3} \cdot S + \frac{1}{5} \cdot S\right) + \frac{1}{4}\left[S - \left(\frac{1}{3} \cdot S + \frac{1}{5} \cdot S\right)\right] = 20$$

As already noted, this problem, involving a remainder but no additional constant, should have appeared before.

The product of the denominators being $3 \cdot 5 \cdot 4 = 60$, we shall consider
$$\left(\frac{1}{3} \cdot 60 + \frac{1}{5} \cdot 60\right) + \frac{1}{4}\left[60 - \left(\frac{1}{3} \cdot 60 + \frac{1}{5} \cdot 60\right)\right] = 32 + \frac{1}{4} \cdot 28 = 39,$$

so that $39 : 60 = 20 : S$, whence $S = \dfrac{60 \cdot 20}{39}$.

(A.166) MSS $\mathcal{A}$, $\mathcal{B}$, $\mathcal{C}$.   Required the amount S such that
$$\left(\frac{1}{5} \cdot S + 2\right) + \left\{\frac{1}{2}\left[S - \left(\frac{1}{5} \cdot S + 2\right)\right] + 4\right\} = 10$$

(a) Direct computation.
$$\left(\frac{1}{5} \cdot S + 2\right) + \left\{\frac{1}{2}\left[S - \left(\frac{1}{5} \cdot S + 2\right)\right] + 4\right\}$$
$$= \left(\frac{1}{5} \cdot S + 2\right) + \left\{\frac{1}{2}\left[\frac{4}{5} \cdot S - 2\right] + 4\right\}$$
$$= \left(\frac{1}{5} \cdot S + 2\right) + \left\{\left[\frac{2}{5} \cdot S - 1\right] + 4\right\}$$
$$= \left(\frac{1}{5} \cdot S + 2\right) + \left(\frac{2}{5} \cdot S + 3\right) = \frac{3}{5} \cdot S + 5 = 10,$$

thus $\dfrac{3}{5} \cdot S = 5$, whence S.

Remark. Since the final determination is not performed, the reader of MS $\mathcal{C}$ has calculated S, in three different ways. He takes the final equation and multiplies it by 25 to obtain $15\,S = 125$; then he multiplies it by 5 to obtain $3\,S = 25$; finally he computes from the beginning and performs the successive additions and subtractions. (He starts by calling our unknown S *unus*, then drops the name, and by the end it has become *res*).

(b) As before, we may first group together the constants, namely $+2$, $\frac{1}{2}(-2)$, $+4$, the sum of which is $+5$, and solve
$$\frac{1}{5} \cdot S + \left[\frac{1}{2}\left(S - \frac{1}{5} \cdot S\right) + 5\right] = 10.$$

— **Other cases.**

The two coming problems A.167–168 fit here well enough for the subject but not for the computations: the first involves a root, not yet considered, the second requires inverting a fraction, not yet considered either (and to be studied later, with the division of fractions).

(A.167) MSS $\mathcal{A}, \mathcal{B}, \mathcal{C}$. Required the amount S such that
$$\left(\frac{1}{3} \cdot S + \frac{1}{4} \cdot S\right)^2 = 49$$

This reduces to
$$\frac{1}{3} \cdot S + \frac{1}{4} \cdot S = 7,$$

thus to A.159.

(A.168) MSS $\mathcal{A}, \mathcal{B}, \mathcal{C}$. Required now the amount S such that
$$\frac{\left(\frac{1}{3} \cdot S + \frac{1}{4} \cdot S\right)^2}{S} = 4 + \frac{1}{2}\frac{1}{6}.$$

With our symbolic algebra the treatment is straightforward; in our text it turns out to be rather awkward.

The author treats it in two steps, for he first computes the result and then justifies his procedure. His computation is as follows. Considering that
$$S = \left(\frac{1}{\frac{1}{3} + \frac{1}{4}}\right)^2 \left(4 + \frac{1}{2}\frac{1}{6}\right) \quad \text{where} \quad \frac{1}{3} + \frac{1}{4} = \frac{3}{6} + \frac{1}{2}\frac{1}{6},$$

we shall have
$$S = \left(\frac{1}{\frac{3}{6} + \frac{1}{2}\frac{1}{6}}\right)^2 \left(4 + \frac{1}{2}\frac{1}{6}\right) = \left(1 + \frac{5}{7}\right)^2 \left(4 + \frac{1}{2}\frac{1}{6}\right).$$

Demonstration. Since, as just seen,
$$\frac{1}{3} \cdot S + \frac{1}{4} \cdot S = \left(\frac{1}{3} + \frac{1}{4}\right) S = \left(\frac{3}{6} + \frac{1}{2}\frac{1}{6}\right) S,$$

we are to solve, according to the data,
$$\left[\left(\frac{3}{6} + \frac{1}{2}\frac{1}{6}\right) S\right]^2 = S \left(4 + \frac{1}{2}\frac{1}{6}\right).$$

This means that
$$\frac{S}{\left(\frac{3}{6} + \frac{1}{2}\frac{1}{6}\right) S} = \frac{\left(\frac{3}{6} + \frac{1}{2}\frac{1}{6}\right) S}{4 + \frac{1}{2}\frac{1}{6}}.$$

Since, for a continued proportion $a : b = b : c$,
$$\frac{a}{b} \cdot \frac{b}{c} = \frac{a}{c} = \left(\frac{a}{b}\right)^2,$$

we shall have here

$$\frac{S}{4+\frac{1}{2}\frac{1}{6}} = \left[\frac{S}{(\frac{3}{6}+\frac{1}{2}\frac{1}{6})S}\right]^2 = \left[\frac{1}{\frac{3}{6}+\frac{1}{2}\frac{1}{6}}\right]^2$$

$$= \left(1+\frac{5}{7}\right)^2 = 1 + \frac{10}{7} + \frac{25}{7}\frac{1}{7} = 2 + \frac{6}{7} + \frac{4}{7}\frac{1}{7},$$

so that

$$S = \left(4 + \frac{1}{2}\frac{1}{6}\right)\left(2 + \frac{6}{7} + \frac{4}{7}\frac{1}{7}\right) = 12.$$

(A.169) MSS $\mathcal{A}$, $\mathcal{B}$. Required now S_1 and S_2 such that

$$\frac{1}{3} \cdot S_1 + \frac{1}{4} \cdot S_2 = 10$$

with $S_1 \neq S_2$ (otherwise this would be A.159). The problem is, as stated, indeterminate (*questio interminata*, Arabic *mas'ala sayyāla*).

(*a*) *Particular solution*. Take for, say, S_1 the product of the denominators, thus $3 \cdot 4 = 12$; then

$$\frac{1}{4} \cdot S_2 = 10 - \frac{1}{3} S_1 = 10 - 4 = 6,$$

and the two numbers $S_1 = 12$, $S_2 = 24$ fulfil the condition.

(*b*) '*More general*' rule. Take any positive pair a_1, a_2 with $a_1 + a_2 = 10$, and, the text (perhaps not the author) says, $a_1 \neq a_2$; putting

$$\frac{1}{3} \cdot S_1 = a_1, \qquad \frac{1}{4} \cdot S_2 = a_2,$$

we shall have $S_1 = 3a_1$, $S_2 = 4a_2$.

Remark. We must not impose $a_1 \neq a_2$, but (considering that $S_1 \neq S_2$) $a_1 : a_2 \neq 4 : 3$.

---

4. Summing series.

A. *Sum required*.

Remark. As is common in mediaeval texts, when half the product of an odd and an even number must be taken, it is usually the even factor which is divided by 2 in the (verbal) formula. We have kept this particularity in the computations below. Note that none of these problems are in MS $\mathcal{C}$.

— **Summing natural numbers.**

(A.170) MSS $\mathcal{A}$, $\mathcal{B}$. Sum of the consecutive natural numbers from 1 to 20.

$$1 + 2 + 3 + \ldots + 20 = (20+1) \cdot \frac{20}{2} = 210.$$

(A.171) MSS $\mathcal{A}$, $\mathcal{B}$. Sum of the consecutive natural numbers from 9 to 20.

$$9 + 10 + 11 + \ldots + 20 = (20+1) \cdot \frac{20}{2} - \frac{9-1}{2} \cdot 9 = 210 - 36 = 174.$$

(A.172) MSS $\mathcal{A}$, $\mathcal{B}$. Sum of the consecutive odd numbers from 1 to 19.

$$1 + 3 + 5 + \ldots + 19 = \left(\frac{19+1}{2}\right)^2 = 100.$$

(A.173) MSS $\mathcal{A}$, $\mathcal{B}$. Sum of the consecutive odd numbers from 9 to 29.

$$9 + 11 + 13 + \ldots + 29 = \left(\frac{29+1}{2}\right)^2 - \left(\frac{9-1}{2}\right)^2 = 225 - 16 = 209.$$

(A.174) MSS $\mathcal{A}$, $\mathcal{B}$. Sum of the consecutive even numbers from 2 to 20.

$$2 + 4 + 6 + \ldots + 20 = \frac{20+2}{2} \cdot \frac{20}{2} = \mathit{110}.$$

(A.175) MSS $\mathcal{A}$, $\mathcal{B}$. Sum of the consecutive even numbers from 10 to 30.

$$10 + 12 + 14 + \ldots + 30 = \frac{30+2}{2} \cdot \frac{30}{2} - \frac{10-2}{2} \cdot \frac{10}{2} = 240 - 20 = 220.$$

— **Summing natural squares.**

(A.176) MSS $\mathcal{A}$, $\mathcal{B}$. Sum of the squares of consecutive natural numbers from 1^2 to 10^2.

(*a*) First way.

$$1^2 + 2^2 + 3^2 + \ldots + 10^2 = (10+1) \cdot \frac{10}{2} \cdot \left(\frac{2}{3} \cdot 10 + \frac{1}{3}\right) = 55 \cdot 7 = 385.$$

(*b*) Second way.

$$1^2 + 2^2 + 3^2 + \ldots + 10^2 = (10+1) \cdot \frac{10}{2} \cdot \left(\frac{2}{3} \cdot (10-1) + 1\right) = \mathit{55} \cdot 7.$$

(A.177) MSS $\mathcal{A}$, $\mathcal{B}$. Sum of the squares of consecutive odd natural numbers from 1^2 to 9^2.

$$1^2 + 3^2 + 5^2 + \ldots + 9^2 = (9+2) \cdot \frac{9+1}{2} \cdot \frac{9}{3} = 11 \cdot 5 \cdot 3 = 165.$$

(A.178) MSS $\mathcal{A}$, $\mathcal{B}$. Sum of the squares of consecutive even natural numbers from 2^2 to 10^2.

$$2^2 + 4^2 + 6^2 + \ldots + 10^2 = \frac{10+2}{2} \cdot \frac{10}{2} \cdot \left(\frac{2}{3} \cdot 10 + \frac{2}{3}\right) = 30 \cdot \left(7 + \frac{1}{3}\right) = 220.$$

(A.179) MSS $\mathcal{A}$, $\mathcal{B}$. Sum of the squares of consecutive natural numbers from 4^2 to 10^2.

$$4^2 + 5^2 + 6^2 + \ldots + 10^2$$
$$= (10+1) \cdot \frac{10}{2} \cdot \left(\frac{2}{3} \cdot 10 + \frac{1}{3}\right) - \frac{4-1}{2} \cdot 4 \cdot \left(\frac{2}{3} \cdot (4-1) + \frac{1}{3}\right)$$
$$= 11 \cdot 5 \cdot 7 - 3 \cdot 2 \cdot \left(2 + \frac{1}{3}\right) = 385 - 14.$$

Remark. We have reason to doubt the genuineness of this problem. First, it is out of place here and should follow A.176. Secondly, the expression is sometimes inappropriate (see Translation, p. 709, notes 523–525). If this problem was in the text originally, it must have been reworded since.

— **Summing natural cubes.**

(A.180) MSS $\mathcal{A}$, $\mathcal{B}$. Sum of the cubes of consecutive natural numbers from 1^3 to 10^3.

$$1^3 + 2^3 + 3^3 + \ldots + 10^3 = \left((10+1) \cdot \frac{10}{2}\right)^2 = 55^2 = 3025.$$

(A.181) MSS $\mathcal{A}$, $\mathcal{B}$. Sum of the cubes of consecutive natural numbers from 5^3 to 10^3.

$$5^3 + 6^3 + 7^3 + \ldots + 10^3 = \left((10+1) \cdot \frac{10}{2}\right)^2 - \left(\frac{5-1}{2} \cdot 5\right)^2$$
$$= 55^2 - 10^2 = 3025 - 100 = 2925.$$

(A.182) MSS $\mathcal{A}$, $\mathcal{B}$. Sum of the cubes of consecutive odd numbers from 1^3 to 9^3.

$$1^3 + 3^3 + 5^3 + \ldots + 9^3 = \left(\frac{9+1}{2}\right)^2 \cdot \left[2\left(\frac{9+1}{2}\right)^2 - 1\right] = 25 \cdot 49 = 1225.$$

(A.183) MSS $\mathcal{A}$, $\mathcal{B}$. Sum of the cubes of consecutive even numbers from 2^3 to 10^3.

$$2^3 + 4^3 + 6^3 + \ldots + 10^3 = \left(\frac{10+2}{2} \cdot \frac{10}{2}\right) \cdot \left[2\left(\frac{10+2}{2} \cdot \frac{10}{2}\right)\right] = 30 \cdot 60 = 1800.$$

B. *Last term required.*

Even though roots are involved, these last problems are in the right place.* As with those preceding, there is no justification of the rules given.

---

* Similar problems in al-Ḥaṣṣār's treatise; see Suter's summary, pp. 33–34.

(A.184) MSS $\mathcal{A}$, $\mathcal{B}$. Given $1+2+3+\cdots+n = 55$, required the last term.

$$n = \sqrt{2\cdot 55 + \frac{1}{4}} - \frac{1}{2} = \sqrt{110 + \frac{1}{4}} - \frac{1}{2} = \left(10 + \frac{1}{2}\right) - \frac{1}{2} = 10.$$

Remark. Since (A.170, rule *i* in our summary, p. 1236)

$$1+2+3+\ldots+n = \frac{n(n+1)}{2} = S,$$

with S given, we have $n^2 + n = 2S$, with the positive root

$$n = -\frac{1}{2} + \sqrt{\left(\frac{1}{2}\right)^2 + 2S}.$$

(A.185) MSS $\mathcal{A}$, $\mathcal{B}$. Given $1+3+\cdots+(2n-1) = 100$, required the last term.

$$2n - 1 = 2\cdot\sqrt{100} - 1 = 19.$$

Remark. Since (A.172, rule *ii* p. 1236 above)

$$1+3+\ldots+(2n-1) = \left(\frac{(2n-1)+1}{2}\right)^2 = S,$$

then $2n - 1 = 2\cdot\sqrt{S} - 1$.

(A.186) MSS $\mathcal{A}$, $\mathcal{B}$. Given $2+4+6+\cdots+2n = 110$, required the last term.

$$2n = \sqrt{4\cdot 110 + 1} - 1 = \sqrt{441} - 1 = 21 - 1.$$

Remark. Since (A.174, rule *iii* pp. 1236–1237 above)

$$2+4+6+\ldots+2n = \frac{2n+2}{2}\cdot\frac{2n}{2} = S,$$

we have $(2n)^2 + 2\cdot 2n = 4S$, with the positive root $2n = -1 + \sqrt{1+4S}$.

Chapter A–VII:
Subtraction of fractions

Summary

1. Subtraction of fractional expressions.

   This is the subject of A.187–199. There are thus twice as many problems as in the corresponding section on addition; for now both minuend and subtrahend will also contain integers. We find examples where a fraction is diminished by a fraction (A.187–188), by a compound fraction (A.189), and by different fractions (A.192, with a proof); a fraction of a fraction is diminished by a fraction of a fraction (A.196); a compound fraction is diminished by a compound fraction (A.190). Problems involving integers are: an integer and a fraction diminished by a fraction (A.193), by a fraction of a fraction (A.197), by a compound fraction (A.195), by an integer and a fraction (A.194; with a proof), and by different fractions (A.199); finally, we find the cases of an integer and a compound fraction minus a compound fraction (A.191) or minus an integer and a compound fraction (A.198). Considering the list at the beginning of A–III, not included are two combinations for pure fractional expressions, while only 6 cases out of 21 involving integers in the fractional expressions are treated.

   What is presented is, however, more than enough to acquaint the reader with the subtraction of fractions, especially since the three ways used are just like those seen for the addition. Usually the product of all the denominators occurring in the given expressions is formed and minuend and subtrahend are multiplied by this product; both expressions then become integers, which brings us back to a subtraction between integers, the result then being divided by the product. But we have the two other possibilities of converting either the subtrahend into the fractions of the minuend or the minuend into the fractions of the subtrahend; both may be presented in the same problem, although the first transformation is more common. Occasionally, if there is an integer in the minuend and the subtrahend is smaller than it, the subtrahend is subtracted from the integer at the outset (A.191, A.193, A.194). This seems easier, but it leaves us, as before, with the addition of fractions with different denominators, which is the longer part.

2. Subtracting irregular fractions.

   This just corresponds to what we have seen for the addition: we find first the case of two irregular fractions of an integer (A.200), then that of two irregular fractions of an integer and a fraction (A.201), which lends itself to another interpretation (A.201′).

   For the computation, we have two possibilities: form the product of the denominators and continue as before, or proceed according to the wording, that is, term by term, which will leave the conversion into the fractions of either the minuend or the subtrahend for the end. These two ways are just the same as those taught in the previous chapter.

3. Problems on amounts.

A. *Known amounts*.

In the problems on known amounts we find expressions of the form

(i) $\quad S - \sum \dfrac{k_i}{l_i} \cdot S$

(ii) $\quad S - \sum \dfrac{k_i}{l_i} \cdot S \pm \dfrac{1}{r}\left[S - \sum \dfrac{k_i}{l_i} \cdot S\right].$

Since in A.202 (case *i*) and A.203 & 203′ (case *ii*) S is a given integer, we have just a computation involving an integer multiplied by fractions, as seen before. The author solves it by means of a proportion, using first, as for the addition, a false position: we shall compute the product of the denominators, L, and calculate the resulting value M_L with L taking the place of S (with L only integers occur); the required quantity M_S will then be found from the proportion $M_L : L = M_S : S$. The condition for a (positive) solution concludes this set of problems (A.203″).

B. *Unknown amounts*.

In the problems on unknown amounts, the value of the proposed expression is a given constant C, and S is required. We can, though, proceed in the same way if the left-hand expression is of type *i* or *ii* (A.204, A.205 & A.205′), the unknown being then the fourth term of the proportion.

As already seen in the chapter on addition, this is no longer the case if constants occur which are not multiplied by S. The two types we encounter are of the form

(iii) $\quad S - \left[\sum \left(\dfrac{k_i}{l_i} \cdot S + t_i\right)\right] = C$

(iv) $\quad S - \left[\sum \left(\dfrac{k_i}{l_i} \cdot S + t_i\right)\right] - \left(\dfrac{1}{r}\left\{S - \sum \left(\dfrac{k_i}{l_i} \cdot S + t_i\right)\right\} + u\right) = C.$

For types *iii* (A.206–209) and *iv* (A.210–212, and the more complicated A.213 solved using a geometrical figure), we must first put all constants on the right side, as seen for the addition.

The chapter on addition ended with second-degree problems (A.184–186). Here there is a single instance (A.214), but with the formula giving the solution fully demonstrated. This is the first instance of a construction often to be met with in Book B.

---

1. Subtraction of fractional expressions.

(**A.187**) MSS $\mathcal{A}$, $\mathcal{B}$, $\mathcal{C}$. Subtraction of a fraction from a fraction. The same quantities are added in A.145.

$$\frac{4}{5} - \frac{3}{8}$$

(**a**) Form the product of the denominators: $5 \cdot 8 = 40$. Then

$$\frac{4}{5} - \frac{3}{8} = \left(\frac{4}{5} \cdot 40 - \frac{3}{8} \cdot 40\right)\frac{1}{40} = \frac{32 - 15}{40}$$

$$= \frac{17}{40} = \frac{15 + 2}{5 \cdot 8} = \frac{3}{8} + \frac{2\ 1}{5\ 8}.$$

(**b**) Convert the minuend into the fractions of the subtrahend.

$$\frac{4}{5} = \frac{32\ 1}{5\ 8} = \frac{30 + 2}{5 \cdot 8} = \frac{6}{8} + \frac{2\ 1}{5\ 8}, \quad \text{so}$$

$$\frac{4}{5} - \frac{3}{8} = \frac{6}{8} + \frac{2\ 1}{5\ 8} - \frac{3}{8} = \frac{3}{8} + \frac{2\ 1}{5\ 8}.$$

(**c**) Convert the subtrahend into the fractions of the minuend.

$$\frac{3}{8} = \frac{15\ 1}{8\ 5} = \frac{8 + 7}{8 \cdot 5} = \frac{1}{5} + \frac{7\ 1}{8\ 5}, \quad \text{so}$$

$$\frac{4}{5} - \frac{3}{8} = \frac{4}{5} - \left(\frac{1}{5} + \frac{7\ 1}{8\ 5}\right) = \frac{3}{5} - \frac{7\ 1}{8\ 5} = \frac{2}{5} + \frac{1\ 1}{8\ 5}.$$

(**A.188**) MSS $\mathcal{A}, \mathcal{B}, \mathcal{C}$. Subtraction of a fraction from a non-elementary fraction. The same quantities are added in A.146.

$$\frac{10}{11} - \frac{3}{7}$$

(**a**) Form the product of the denominators: $11 \cdot 7 = 77$. Then

$$\frac{10}{11} - \frac{3}{7} = \left(\frac{10}{11} \cdot 77 - \frac{3}{7} \cdot 77\right)\frac{1}{77} = \frac{70 - 33}{77}$$

$$= \frac{37}{77} = \frac{35 + 2}{7 \cdot 11} = \frac{5}{11} + \frac{2\ 1}{7\ 11}.$$

(**b**) Convert the subtrahend into the fractions of the minuend.

$$\frac{3}{7} = \frac{33\ 1}{7\ 11} = \frac{28 + 5}{7}\frac{1}{11} = \frac{4}{11} + \frac{5\ 1}{7\ 11}, \quad \text{so}$$

$$\frac{10}{11} - \frac{3}{7} = \frac{10}{11} - \left(\frac{4}{11} + \frac{5\ 1}{7\ 11}\right) = \frac{6}{11} - \frac{5\ 1}{7\ 11} = \frac{5}{11} + \frac{2\ 1}{7\ 11}.$$

(**c**) Convert the minuend into the fractions of the subtrahend.

$$\frac{10}{11} = \frac{70\ 1}{11\ 7} = \frac{6}{7} + \frac{4\ 1}{11\ 7}, \quad \text{so}$$

$$\frac{10}{11} - \frac{3}{7} = \frac{6}{7} + \frac{4\ 1}{11\ 7} - \frac{3}{7} = \frac{3}{7} + \frac{4\ 1}{11\ 7}.$$

(A.189) MSS $\mathcal{A}$, $\mathcal{B}$, $\mathcal{C}$. Subtraction of a compound fraction from a fraction.

$$\frac{6}{7} - \left(\frac{3}{8} + \frac{1\,1}{2\,8}\right)$$

(a) Form the product of the denominators: $7 \cdot 8 \cdot 2 = 112$.

$$\frac{6}{7} - \left(\frac{3}{8} + \frac{1\,1}{2\,8}\right) = \left[\frac{6}{7} \cdot 112 - \left(\frac{3}{8} + \frac{1\,1}{2\,8}\right) \cdot 112\right] \frac{1}{112}$$

$$= (6 \cdot 16 - 3 \cdot 14 - 7) \frac{1}{112} = \frac{96 - 49}{112} = \frac{47}{112}.$$

(b) Convert the subtrahend into the fractions of the minuend.

$$\frac{3}{8} + \frac{1\,1}{2\,8} = \frac{21\,1}{8\,7} + \frac{7\,1\,1}{2\,8\,7} = \frac{2}{7} + \frac{5\,1}{8\,7} + \frac{3\,1}{8\,7} + \frac{1\,1\,1}{2\,8\,7} = \frac{3}{7} + \frac{1\,1\,1}{2\,8\,7}, \quad \text{and}$$

$$\frac{6}{7} - \left(\frac{3}{7} + \frac{1\,1\,1}{2\,8\,7}\right) = \frac{3}{7} - \frac{1\,1\,1}{2\,7\,8} = \frac{2}{7} + \frac{15\,1\,1}{2\,8\,7} = \frac{2}{7} + \frac{7\,1}{8\,7} + \frac{1\,1\,1}{2\,8\,7}.$$

Remark. Note the second term in this last result: reducing would leave us with different kinds of fractions.

(A.190) MSS $\mathcal{A}$, $\mathcal{B}$, $\mathcal{C}$. Subtraction of a compound fraction from a compound fraction.

$$\frac{7}{8} + \frac{1\,1}{2\,8} - \left(\frac{3}{5} + \frac{2\,1}{3\,5}\right)$$

(a) Form the product of the denominators: $(8 \cdot 2)(5 \cdot 3) = 240$. Then compute

$$\frac{7}{8} + \frac{1\,1}{2\,8} - \left(\frac{3}{5} + \frac{2\,1}{3\,5}\right) = \left[\left(\frac{7}{8} + \frac{1\,1}{2\,8}\right) \cdot 240 - \left(\frac{3}{5} + \frac{2\,1}{3\,5}\right) \cdot 240\right] \frac{1}{240}.$$

(b) Convert the subtrahend into the fractions of the minuend.

$$\frac{3}{5} + \frac{2\,1}{3\,5} = \frac{24\,1}{5\,8} + \frac{16\,1\,1}{3\,5\,8} = \frac{4}{8} + \frac{4\,1}{5\,8} + \frac{1}{8} + \frac{1\,1\,1}{3\,5\,8} = \frac{5}{8} + \frac{4\,1}{5\,8} + \frac{1\,1\,1}{3\,5\,8}.$$

Now

$$\frac{7}{8} - \left(\frac{5}{8} + \frac{4\,1}{5\,8} + \frac{1\,1\,1}{3\,5\,8}\right) = \frac{2}{8} - \left(\frac{4\,1}{5\,8} + \frac{1\,1\,1}{3\,5\,8}\right)$$

$$= \frac{1}{8} + \frac{4\,1}{5\,8} + \frac{3\,1\,1}{3\,5\,8} - \left(\frac{4\,1}{5\,8} + \frac{1\,1\,1}{3\,5\,8}\right) = \frac{1}{8} + \frac{2\,1\,1}{3\,5\,8},$$

so finally

$$\frac{7}{8} + \frac{1\,1}{2\,8} - \left(\frac{3}{5} + \frac{2\,1}{3\,5}\right) = \frac{1}{8} + \frac{1\,1}{2\,8} + \frac{2\,1\,1}{3\,5\,8}.$$

(A.191) MSS $\mathcal{A}$, $\mathcal{B}$, $\mathcal{C}$. Subtraction of a compound fraction from an integer with a compound fraction. Problem now involving an integer, whence the new way c.

$$1 + \frac{4}{11} + \frac{1}{2}\frac{1}{11} - \left(\frac{3}{5} + \frac{2}{3}\frac{1}{5}\right)$$

(**a**) Form the product of the denominators: $11 \cdot 2 \cdot 5 \cdot 3$, multiply it by the two given expressions and divide the remainder by the product.

(**b**) Convert the subtrahend into elevenths and subtract it from $\frac{15}{11} + \frac{1}{2}\frac{1}{11}$.

(**c**) Subtract the subtrahend from the integer in the minuend.

$$1 - \left(\frac{3}{5} + \frac{2}{3}\frac{1}{5}\right) = \frac{4}{5} + \frac{3}{3}\frac{1}{5} - \left(\frac{3}{5} + \frac{2}{3}\frac{1}{5}\right) = \frac{1}{5} + \frac{1}{3}\frac{1}{5},$$

then add it to the fractional part of the minuend (A.148).

(**A.192**) MSS *A*, *B*, *C*. Subtraction of several fractions from a (non-elementary) fraction. This problem should be with A.188.

$$\frac{10}{11} - \left(\frac{2}{7} + \frac{3}{10}\right)$$

(**a**) Form the product of the denominators: $11 \cdot 7 \cdot 10 = 770$. Then compute

$$\left[\frac{10}{11} \cdot 770 - \left(\frac{2}{7} \cdot 770 + \frac{3}{10} \cdot 770\right)\right]\frac{1}{770}.$$

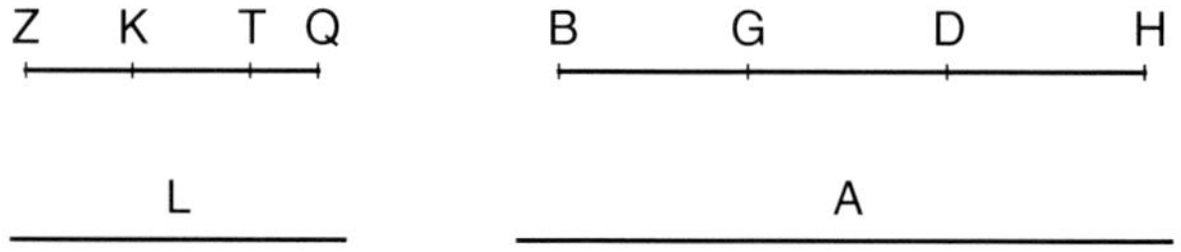

Fig. 31

Demonstration. Analogous to that in A.151 (for addition). Let (Fig. 31) the product of the denominators be A. Put next

$$\frac{2}{7} \cdot A = BG, \quad \frac{3}{10} \cdot A = GD, \quad \frac{10}{11} \cdot A = BH,$$

whence

$$\frac{10}{11} \cdot A - \left(\frac{2}{7} \cdot A + \frac{3}{10} \cdot A\right) = DH;$$

then put, for the fractions themselves,

$$\frac{2}{7} = ZK, \quad \frac{3}{10} = KT, \quad \frac{10}{11} = ZQ,$$

whence

$$\frac{10}{11} - \left(\frac{2}{7} + \frac{3}{10}\right) = TQ.$$

We have to prove that $TQ = \dfrac{DH}{A}$.

Let $1 = L$. Since $ZK : L = BG : A$ and $KT : L = GD : A$, so first, by P'_7, $ZT : L = BD : A$. But we have also $ZQ : L = BH : A$. Therefore, by P_6,

$$\frac{ZQ - ZT}{L} = \frac{BH - BD}{A} \quad \text{thus} \quad \frac{TQ}{L} = \frac{DH}{A},$$

so that $L \cdot DH = DH = TQ \cdot A$, whence the requirement, that is,

$$TQ = \frac{10}{11} - \left(\frac{2}{7} + \frac{3}{10}\right) = \frac{DH}{A} = \left[\frac{10}{11} \cdot A - \left(\frac{2}{7} \cdot A + \frac{3}{10} \cdot A\right)\right]\frac{1}{A}.$$

(b) Convert the subtrahend into the fractions of the minuend.

$$\frac{2}{7} + \frac{3}{10} = \frac{22}{7}\frac{1}{11} + \frac{33}{10}\frac{1}{11} = \frac{3}{11} + \frac{1}{7}\frac{1}{11} + \frac{3}{11} + \frac{3}{10}\frac{1}{11} = \frac{6}{11} + \frac{1}{7}\frac{1}{11} + \frac{3}{10}\frac{1}{11};$$

then, to add the last two terms, either convert the first into tenths of elevenths or the second into sevenths of elevenths. Choosing the first option, the author finds

$$\frac{1}{7}\frac{1}{11} + \frac{3}{10}\frac{1}{11} = \left(\frac{1}{7} + \frac{3}{10}\right)\frac{1}{11} = \left(\frac{10}{7}\frac{1}{10} + \frac{3}{10}\right)\frac{1}{11}$$

$$= \frac{1}{10}\frac{1}{11} + \frac{3}{7}\frac{1}{10}\frac{1}{11} + \frac{3}{10}\frac{1}{11} = \frac{4}{10}\frac{1}{11} + \frac{3}{7}\frac{1}{10}\frac{1}{11}.$$

We have therefore

$$\frac{10}{11} - \left(\frac{2}{7} + \frac{3}{10}\right) = \frac{10}{11} - \left(\frac{6}{11} + \frac{4}{10}\frac{1}{11} + \frac{3}{7}\frac{1}{10}\frac{1}{11}\right)$$

$$= \frac{3}{11} + \frac{1}{11} - \left(\frac{4}{10}\frac{1}{11} + \frac{3}{7}\frac{1}{10}\frac{1}{11}\right)$$

$$= \frac{3}{11} + \frac{9}{10}\frac{1}{11} + \frac{7}{7}\frac{1}{10}\frac{1}{11} - \left(\frac{4}{10}\frac{1}{11} + \frac{3}{7}\frac{1}{10}\frac{1}{11}\right)$$

$$= \frac{3}{11} + \frac{5}{10}\frac{1}{11} + \frac{4}{7}\frac{1}{10}\frac{1}{11}.$$

(A.193) MSS $\mathcal{A}$, $\mathcal{B}$, $\mathcal{C}$. Subtraction of a (non-elementary) fraction from an integer with a fraction.

$$1 + \frac{3}{8} - \frac{8}{11}$$

(a) Form the product of the denominators: $8 \cdot 11 = 88$. Then compute

$$\left(1 + \frac{3}{8}\right) - \frac{8}{11} = \left[\left(1 + \frac{3}{8}\right)88 - \frac{8}{11} \cdot 88\right]\frac{1}{88} = \frac{121 - 64}{88} = \frac{57}{88}.$$

(b) Convert the subtrahend into the fractions of the minuend.

$$\frac{8}{11} = \frac{64}{11}\frac{1}{8} = \frac{5}{8} + \frac{9}{11}\frac{1}{8}, \quad \text{so}$$

$$\left(1+\frac{3}{8}\right) - \frac{8}{11} = \frac{11}{8} - \left(\frac{5}{8}+\frac{9}{11}\frac{1}{8}\right)$$
$$= \frac{10}{8} + \frac{11}{11}\frac{1}{8} - \left(\frac{5}{8}+\frac{9}{11}\frac{1}{8}\right) = \frac{5}{8} + \frac{2}{11}\frac{1}{8}.$$

(c) Convert the minuend into the fractions of the subtrahend.

$$\frac{3}{8} = \frac{33}{8}\frac{1}{11} = \frac{4}{11} + \frac{1}{8}\frac{1}{11}, \quad \text{so}$$

$$\left(1+\frac{3}{8}\right) - \frac{8}{11} = \left(\frac{15}{11}+\frac{1}{8}\frac{1}{11}\right) - \frac{8}{11} = \frac{7}{11} + \frac{1}{8}\frac{1}{11}.$$

(d) Subtract the subtrahend from the integer in the minuend.

$$\left(1+\frac{3}{8}\right) - \frac{8}{11} = 1 - \frac{8}{11} + \frac{3}{8} = \frac{3}{11} + \frac{3}{8},$$

to be transformed into elevenths or into eighths.

(A.194) MSS $\mathcal{A}, \mathcal{B}, \mathcal{C}.$ Subtraction of an integer with a fraction from an integer with a fraction.

$$5+\frac{8}{9} - \left(2+\frac{3}{4}\right)$$

(a) Form the product of the denominators: $9 \cdot 4 = 36$. Then compute

$$\left[\left(5+\frac{8}{9}\right) \cdot 36 - \left(2+\frac{3}{4}\right) \cdot 36\right]\frac{1}{36}.$$

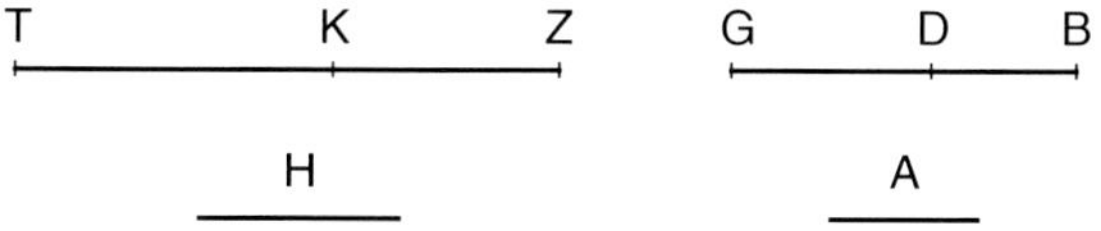

```
T           K           Z      G      D   B
|_____|_____|_____|_____|___|
            |_____|      |_____|
                 H                A
```

Fig. 32

Demonstration. Similar to that in A.192a. Let (Fig. 32) 1 be A. Let then the two terms be

$$2+\frac{3}{4} = DB, \quad 5+\frac{8}{9} = GB, \quad \text{so} \quad \left(5+\frac{8}{9}\right) - \left(2+\frac{3}{4}\right) = GD.$$

Let further the product of the denominators be H and multiply it by the given fractions:

$$\left(2+\frac{3}{4}\right) \cdot H = KZ, \quad \left(5+\frac{8}{9}\right) \cdot H = TZ, \quad \text{so} \quad \left(5+\frac{8}{9}\right) \cdot H - \left(2+\frac{3}{4}\right) \cdot H = TK.$$

To prove that $GD = \dfrac{TK}{H}.$

Just as in the proof in A.192a, we can apply P_6:

$$\frac{GB}{A} = \frac{TZ}{H}, \quad \frac{DB}{A} = \frac{KZ}{H}, \quad \text{whence} \quad \frac{GB - DB}{A} = \frac{TZ - KZ}{H}.$$

Therefore $GD : A = TK : H$, so $GD \cdot H = TK \cdot A = TK$, whence the requirement, that is,

$$GD = \left(5 + \frac{8}{9}\right) - \left(2 + \frac{3}{4}\right) = \frac{TK}{H} = \left[\left(5 + \frac{8}{9}\right) \cdot H - \left(2 + \frac{3}{4}\right) \cdot H\right] \frac{1}{H}.$$

(**b**) As in A.193d, subtracting the subtrahend from the integer in the minuend.

$$5 - \left(2 + \frac{3}{4}\right) + \frac{8}{9} = 2 + \frac{1}{4} + \frac{8}{9};$$

now, since

$$\frac{1}{4} + \frac{8}{9} = \frac{9}{4}\frac{1}{9} + \frac{8}{9} = \frac{2}{9} + \frac{1}{4}\frac{1}{9} + \frac{8}{9} = 1 + \frac{1}{9} + \frac{1}{4}\frac{1}{9},$$

so finally

$$\left(5 + \frac{8}{9}\right) - \left(2 + \frac{3}{4}\right) = 3 + \frac{1}{9} + \frac{1}{4}\frac{1}{9}.$$

(**A.195**) MSS $\mathcal{A}, \mathcal{B}, \mathcal{C}$. Subtraction of a compound fraction from an integer with a (non-elementary) fraction.

$$1 + \frac{4}{13} - \left(\frac{4}{5} + \frac{1}{3}\frac{1}{5}\right)$$

(**a**) Form the product of the denominators.

(**b**) Convert the subtrahend into thirteenths and subtract it from $\frac{17}{13}$.[†]

(**A.196**) MSS $\mathcal{A}, \mathcal{B}, \mathcal{C}$. Subtraction of a fraction of a fraction from a fraction of a fraction.

$$\frac{5}{7}\frac{1}{6} - \frac{3}{5}\frac{1}{8}$$

(**a**) Form the product of the denominators: $(7 \cdot 6)(5 \cdot 8) = 42 \cdot 40 = 1680$. Then

$$\frac{5}{7}\frac{1}{6} - \frac{3}{5}\frac{1}{8} = \left(\frac{5}{7}\frac{1}{6} \cdot 1680 - \frac{3}{5}\frac{1}{8} \cdot 1680\right)\frac{1}{1680} = \frac{200 - 126}{1680} = \frac{74}{1680}.$$

[†] The conversion gives

$$\frac{4}{5} + \frac{1}{3}\frac{1}{5} = \frac{11}{13} + \frac{1}{5}\frac{1}{13} + \frac{1}{3}\frac{1}{5}\frac{1}{13}.$$

(b) Converting the subtrahend.

$$\frac{3}{5}\frac{1}{8} = \frac{126}{5\cdot 8\cdot 7\cdot 6} = \frac{120+5+1}{5\cdot 8\cdot 7\cdot 6} = \frac{3}{7}\frac{1}{6} + \frac{1}{8}\frac{1}{7}\frac{1}{6} + \frac{1}{5}\frac{1}{8}\frac{1}{7}\frac{1}{6},$$ so

$$\frac{5}{7}\frac{1}{6} - \frac{3}{5}\frac{1}{8} = \frac{5}{7}\frac{1}{6} - \left(\frac{3}{7}\frac{1}{6} + \frac{1}{8}\frac{1}{7}\frac{1}{6} + \frac{1}{5}\frac{1}{8}\frac{1}{7}\frac{1}{6}\right)$$

$$= \frac{1}{7}\frac{1}{6} + \frac{7}{8}\frac{1}{7}\frac{1}{6} + \frac{5}{5}\frac{1}{8}\frac{1}{7}\frac{1}{6} - \left(\frac{1}{8}\frac{1}{7}\frac{1}{6} + \frac{1}{5}\frac{1}{8}\frac{1}{7}\frac{1}{6}\right)$$

$$= \frac{1}{7}\frac{1}{6} + \frac{6}{8}\frac{1}{7}\frac{1}{6} + \frac{4}{5}\frac{1}{8}\frac{1}{7}\frac{1}{6}.$$

Remark. We would expect this problem after A.189.

(A.197) MSS $\mathcal{A}$, $\mathcal{B}$, $\mathcal{C}$. Subtraction of a fraction of a fraction from an integer with a (non-elementary) fraction.

$$1 + \frac{1}{11} - \frac{3}{4}\frac{1}{5}$$

(a) Form the product of the denominators: $11 \cdot 4 \cdot 5 = 220$. Then

$$1 + \frac{1}{11} - \frac{3}{4}\frac{1}{5} = \left[\left(1+\frac{1}{11}\right)\cdot 220 - \frac{3}{4}\frac{1}{5}\cdot 220\right]\frac{1}{220} = \frac{240-33}{220} = \frac{207}{220}.$$

(b) Convert the subtrahend into elevenths.

$$\frac{3}{4}\frac{1}{5} = \frac{33}{4}\frac{1}{5}\frac{1}{11} = \frac{20+12+1}{4\cdot 5\cdot 11} = \frac{1}{11} + \frac{3}{5}\frac{1}{11} + \frac{1}{4}\frac{1}{5}\frac{1}{11},$$ so

$$1 + \frac{1}{11} - \frac{3}{4}\frac{1}{5} = \frac{12}{11} - \left(\frac{1}{11} + \frac{3}{5}\frac{1}{11} + \frac{1}{4}\frac{1}{5}\frac{1}{11}\right) = \frac{11}{11} - \left(\frac{3}{5}\frac{1}{11} + \frac{1}{4}\frac{1}{5}\frac{1}{11}\right)$$

$$= \frac{10}{11} + \frac{4}{5}\frac{1}{11} + \frac{4}{4}\frac{1}{5}\frac{1}{11} - \left(\frac{3}{5}\frac{1}{11} + \frac{1}{4}\frac{1}{5}\frac{1}{11}\right) = \frac{10}{11} + \frac{1}{5}\frac{1}{11} + \frac{3}{4}\frac{1}{5}\frac{1}{11}.$$

(A.198) MSS $\mathcal{A}$, $\mathcal{B}$, $\mathcal{C}$. Subtraction of an integer with a compound fraction from an integer with a compound fraction.

$$\left(2 + \frac{5}{8} + \frac{2}{3}\frac{1}{8}\right) - \left(2 + \frac{2}{7} + \frac{3}{4}\frac{1}{7}\right)$$

This is a particular case since the integers disappear. We are then left with

$$\left(\frac{5}{8} + \frac{2}{3}\frac{1}{8}\right) - \left(\frac{2}{7} + \frac{3}{4}\frac{1}{7}\right)$$

to be treated as 'we have taught above' (A.190).

(A.199) MSS $\mathcal{A}$, $\mathcal{B}$, $\mathcal{C}$. Subtraction of several ('different') fractions from an integer with a fraction.

$$\left(1 + \frac{7}{8}\right) - \left(\frac{3}{5} + \frac{4}{7}\right)$$

(**a**) Form the product of the denominators: $8 \cdot 5 \cdot 7 = 280$. Then

$$\left(1 + \frac{7}{8}\right) - \left(\frac{3}{5} + \frac{4}{7}\right) = \left[\left(1 + \frac{7}{8}\right) \cdot 280 - \left(\frac{3}{5} \cdot 280 + \frac{4}{7} \cdot 280\right)\right]\frac{1}{280}$$

$$= \left[(280 + 245) - (168 + 160)\right]\frac{1}{280} = \frac{525 - 328}{280} = \frac{197}{280}.$$

(**b**) Convert the subtrahend into eighths. (We follow closely the computations of the text.)

$$\frac{3}{5} + \frac{4}{7} = \frac{24}{5}\frac{1}{8} + \frac{32}{7}\frac{1}{8} = \frac{4}{8} + \frac{4}{5}\frac{1}{8} + \frac{4}{8} + \frac{4}{7}\frac{1}{8} = 1 + \frac{28 + 20}{5 \cdot 7 \cdot 8}$$

$$= 1 + \frac{35 + 10 + 3}{5 \cdot 7 \cdot 8} = 1 + \frac{1}{8} + \frac{2}{7}\frac{1}{8} + \frac{3}{5}\frac{1}{7}\frac{1}{8}, \quad \text{so}$$

$$\left(1 + \frac{7}{8}\right) - \left(\frac{3}{5} + \frac{4}{7}\right) = \left(1 + \frac{7}{8}\right) - \left(1 + \frac{1}{8} + \frac{2}{7}\frac{1}{8} + \frac{3}{5}\frac{1}{7}\frac{1}{8}\right)$$

$$= \frac{6}{8} - \left(\frac{2}{7}\frac{1}{8} + \frac{3}{5}\frac{1}{7}\frac{1}{8}\right) = \frac{5}{8} + \frac{7}{7}\frac{1}{8} - \left(\frac{2}{7}\frac{1}{8} + \frac{3}{5}\frac{1}{7}\frac{1}{8}\right)$$

$$= \frac{5}{8} + \frac{5}{7}\frac{1}{8} - \frac{3}{5}\frac{1}{7}\frac{1}{8} = \frac{5}{8} + \frac{4}{7}\frac{1}{8} + \frac{5}{5}\frac{1}{7}\frac{1}{8} - \frac{3}{5}\frac{1}{7}\frac{1}{8}$$

$$= \frac{5}{8} + \frac{4}{7}\frac{1}{8} + \frac{2}{5}\frac{1}{7}\frac{1}{8}.$$

---

2. Subtracting irregular fractions.

As said in our summary (pp. 1255–1256), we have here two possibilities:
— either form the product of the denominators and proceed as before;
— or compute each of the two given expressions, then convert into same fractions.

(**A.200**) MSS $\mathcal{A}, \mathcal{B}, \mathcal{C}$. Subtracting irregular fractions of integers.

$$\frac{5}{7} \cdot 6 - \frac{2}{3} \cdot 4$$

(**a**) Form the product of the denominators: $7 \cdot 3 = 21$. Then compute

$$\frac{\left(\frac{5}{7} \cdot 21\right)6 - \left(\frac{2}{3} \cdot 21\right)4}{21} = \frac{90 - 56}{21} = \frac{34}{21}.$$

Demonstration. We know from A.90a (and A.152a) that

$$\left[\left(\frac{2}{3} \cdot 21\right)4\right]\frac{1}{21} = \frac{2}{3} \cdot 4, \quad \left[\left(\frac{5}{7} \cdot 21\right)6\right]\frac{1}{21} = \frac{5}{7} \cdot 6, \quad \text{so (by } P_7\text{)}$$

$$\frac{5}{7} \cdot 6 - \frac{2}{3} \cdot 4 = \frac{\left(\frac{5}{7} \cdot 21\right)6}{21} - \frac{\left(\frac{2}{3} \cdot 21\right)4}{21} = \frac{\left(\frac{5}{7} \cdot 21\right)6 - \left(\frac{2}{3} \cdot 21\right)4}{21}.$$

Remark. We find here in the text the only allusion to the division of the chapter on premisses (A–II) into two parts (the second being the adaptation to numbers of theorems II.1–10 of the *Elements*).

(*b*) Following the wording, that is, calculating the terms individually:

$$\frac{5}{7} \cdot 6 - \frac{2}{3} \cdot 4 = \left(4 + \frac{2}{7}\right) - \left(2 + \frac{2}{3}\right)$$

to be treated according to A.194.

Remark. Another version of *b* found in MSS $\mathcal{B}$ and $\mathcal{C}$ explicitly computes

$$\left(4 + \frac{2}{7}\right) - \left(2 + \frac{2}{3}\right) = \left(4 - \left(2 + \frac{2}{3}\right)\right) + \frac{2}{7} = 1 + \frac{1}{3} + \frac{2}{7}.$$

(**A.201**) MSS $\mathcal{A}, \mathcal{B}, \mathcal{C}$. Subtract $\frac{2}{5}$ of 4 and $\frac{1}{2}$ from $\frac{6}{7}$ of 8 and $\frac{1}{3}$.

The enunciation being ambiguous, the author is going to consider two possibilities. We have already seen such cases (A.135, A.138). Consider first

$$\frac{6}{7} \cdot \left(8 + \frac{1}{3}\right) - \frac{2}{5} \cdot \left(4 + \frac{1}{2}\right)$$

(*a*) Form the product of the denominators: $7 \cdot 3 \cdot 5 \cdot 2 = 210$. Then compute

$$\frac{6}{7}\left(8 + \frac{1}{3}\right) - \frac{2}{5}\left(4 + \frac{1}{2}\right) = \left[\frac{6}{7} \cdot 210 \left(8 + \frac{1}{3}\right) - \frac{2}{5} \cdot 210 \left(4 + \frac{1}{2}\right)\right] \frac{1}{210}$$

$$= \frac{1500 - 378}{210} = \frac{1122}{210}.$$

The proof is, we are told, the same as in A.200. Indeed, as in A.152–153, the integers have just been replaced by integers with fractions.

(*b*) Considering the terms individually:

$$\frac{6}{7}\left(8 + \frac{1}{3}\right) - \frac{2}{5}\left(4 + \frac{1}{2}\right) = \left(7 + \frac{1}{7}\right) - \left(1 + \frac{4}{5}\right)$$

to be treated as in A.194.

(**A.201′**) Subtract $\frac{2}{5}$ of 4 and $\frac{1}{2}$ of 1 from $\frac{6}{7}$ of 8 and $\frac{1}{3}$ of 1.

Consider then

$$\left(\frac{6}{7} \cdot 8 + \frac{1}{3}\right) - \left(\frac{2}{5} \cdot 4 + \frac{1}{2}\right)$$

(*a*) Form the product of the denominators: $7 \cdot 3 \cdot 5 \cdot 2 = 210$. Then compute

$$\left[\left(\frac{6}{7} 210 \cdot 8 + \frac{1}{3} 210\right) - \left(\frac{2}{5} 210 \cdot 4 + \frac{1}{2} 210\right)\right] \frac{1}{210}.$$

(**b**) Considering the terms individually, and converting some of the fractions:
$$\left(\frac{6}{7}\cdot 8+\frac{1}{3}\right)-\left(\frac{2}{5}\cdot 4+\frac{1}{2}\right)=\left(7+\frac{1}{7}+\frac{1}{3}\frac{1}{7}\right)-\left(2+\frac{1}{2}\frac{1}{5}\right).$$

Remark. As noted for similar problems in A–V and A–VI, the Latin case-endings make it possible to distinguish between the two interpretations. (Further clarification is supposed to be provided by the author's including, in the formulation of the second case, 'of 1'.)

---

3. Problems on amounts.
A. *Known amounts*.

(**A.202**) MSS *A*, *B*, *C*.   Compute
$$6-\left(\frac{1}{7}+\frac{1}{9}\right)6$$

Form the product of the denominators: $7\cdot 9=63$. Then
$$63-\left(\frac{1}{7}+\frac{1}{9}\right)63=63-16=47$$

whence
$$6-\left(\frac{1}{7}+\frac{1}{9}\right)6=\frac{6\cdot 47}{63}=\frac{282}{63}=\frac{252+27+3}{9\cdot 7}=4+\frac{3}{7}+\frac{1}{3}\frac{1}{7}.$$

Demonstration. The demonstration is essentially the same as in A.155. The result of operating on the common denominator 63, thus 47, is to 63 itself as the required quantity is to 6.

We may also compute, the text adds,
$$\frac{47}{63}\cdot 6=\frac{42+5}{7\cdot 9}\cdot 6=\left(\frac{6}{9}+\frac{5}{7}\frac{1}{9}\right)\cdot 6,\quad\text{or}\quad\frac{6}{63}\cdot 47=\frac{2}{3}\frac{1}{7}\cdot 47.$$

As we have already observed (p. 1245), attention is often drawn to such banal changes in problems determining one unknown from a proportion. This is again mentioned in the coming problems A.203 & A.203′, and also found in the first two problems on unknown amounts (A.204–205).

(**A.203**) MSS *A*, *B*, *C*.   Compute now
$$\left[9-\left(\frac{1}{5}+\frac{2}{7}\right)9\right]+\frac{1}{2}\left[9-\left(\frac{1}{5}+\frac{2}{7}\right)9\right]$$

Form the product of the denominators: $5\cdot 7\cdot 2=70$. Then, as above, operate first on this number and then deduce the actual quantity by a proportion.
$$\left[70-\left(\frac{1}{5}+\frac{2}{7}\right)70\right]+\frac{1}{2}\left[70-\left(\frac{1}{5}+\frac{2}{7}\right)70\right]$$

$$= (70-34) + \frac{1}{2}(70-34) = 36+18 = 54,$$

then $54:70 =$ (required quantity) $:9$, so required quantity $= \dfrac{9 \cdot 54}{70}$.

(A.203′) Compute

$$\left[9 - \left(\frac{1}{5}+\frac{2}{7}\right)9\right] - \frac{1}{3}\left[9 - \left(\frac{1}{5}+\frac{2}{7}\right)9\right]$$

The product of the denominators being this time $5 \cdot 7 \cdot 3 = 105$, we find

$$\left[105 - \left(\frac{1}{5}+\frac{2}{7}\right)105\right] - \frac{1}{3}\left[105 - \left(\frac{1}{5}+\frac{2}{7}\right)105\right]$$

$$= (105-51) - \frac{1}{3}(105-51) = 54-18 = 36,$$

so $36:105 =$ (required quantity) $:9$, then required quantity $= \dfrac{9 \cdot 36}{105}$.

Remark. The computation in the text is faulty. Instead of calculating $105-(21+30) = 54$, the author appears to have subtracted 21 once more, thus obtaining $54-21=33$ and then $33-\frac{1}{3}33 = 22$. Hence his proportion $22:105 =$ (required quantity) $:9$.

(A.203″) The same will hold if there are more than two fractions. But

$$8 - \left(\frac{2}{3}+\frac{2}{5}\right)8$$

is not solvable, for the sum of the fractions in the subtrahend is greater than 1.

Remark. See similar remark (though less appropriate) in A.156′ (p. 1246).

B. *Unknown amounts*.

Some of these problems are closely related to those in the chapter on addition; indeed, the sum of the fractions taken of the amount is now subtracted from it. Compare A.204, A.206, A.207, A.208, A.210, A.212 with A.159, A.160, A.161, A.162, A.163, A.164.

(A.204) MSS $\mathcal{A}, \mathcal{B}, \mathcal{C}$. Required the amount S such that

$$S - \left(\frac{1}{3}+\frac{1}{4}\right)S = 10$$

Forming the product $3 \cdot 4 = 12$, we compute (A.202)

$$12 - \left(\frac{1}{3}+\frac{1}{4}\right)12 = 5, \quad \text{whence} \quad 5:12 = 10:S, \quad \text{and} \quad S = \frac{10 \cdot 12}{5} = 24.$$

(A.205) MSS $\mathcal{A}, \mathcal{B}, \mathcal{C}$. Required the amount S such that

$$\left[S - \left(\frac{1}{3} + \frac{1}{4}\right)S\right] - \frac{1}{2}\left[S - \left(\frac{1}{3} + \frac{1}{4}\right)S\right] = 10$$

(*a*) Forming the product $3 \cdot 4 \cdot 2 = 24$, we compute (A.203)

$$\left[24 - \left(\frac{1}{3} + \frac{1}{4}\right)24\right] - \frac{1}{2}\left[24 - \left(\frac{1}{3} + \frac{1}{4}\right)24\right]$$

$$= (24 - 14) - \frac{1}{2} \cdot (24 - 14) = 10 - \frac{1}{2} \cdot 10 = 5;$$

then, since $5 : 24 = 10 : S$, $S = \dfrac{10 \cdot 24}{5}$.

(*b*) Or directly. Subtracting the half at the outset leaves

$$\frac{1}{2}\left[S - \left(\frac{1}{3} + \frac{1}{4}\right)S\right] = 10, \quad \text{so} \quad S - \left(\frac{1}{3} + \frac{1}{4}\right)S = 20,$$

which will be treated as before (A.204).

(**A.205′**) Required the amount S such that

$$\left[S - \left(\frac{1}{3} + \frac{1}{4}\right)S\right] - \frac{1}{3}\left[S - \left(\frac{1}{3} + \frac{1}{4}\right)S\right] = 10$$

Then, as in *b* before,

$$\frac{2}{3}\left[S - \left(\frac{1}{3} + \frac{1}{4}\right)S\right] = 10 \quad \text{and} \quad S - \left(\frac{1}{3} + \frac{1}{4}\right)S = 15.$$

(**A.206**) MSS $\mathcal{A}$, $\mathcal{B}$, $\mathcal{C}$. Required the amount S such that

$$S - \left[\left(\frac{1}{3} + \frac{1}{4}\right)S + 2\right] = 10$$

Reduced to A.204 since we have

$$S - \left(\frac{1}{3} + \frac{1}{4}\right)S = 12.$$

(**A.207**) MSS $\mathcal{A}$, $\mathcal{B}$, $\mathcal{C}$. Required the amount S such that

$$S - \left[\left(\frac{1}{3} + \frac{1}{4}\right)S - 2\right] = 10$$

Reduced to A.204 since we have

$$S - \left(\frac{1}{3} + \frac{1}{4}\right)S = 8.$$

• Generally, if we are given, together with the fractions, positive constants t_i, namely if

$$S - \sum\left(\frac{k_i}{l_i} \cdot S \pm t_i\right) = r, \quad \text{with} \quad \sum \frac{k_i}{l_i} < 1 \quad (\text{A.203″}),$$

we shall consider
$$S - \sum \frac{k_i}{l_i} S = r \pm \sum t_i,$$
which is A.204. That is, as asserted in the text, the quantities accompanying the fractions are transferred, with signs unchanged, to the right-hand side.

Remark. In the case of addition (A.161, p. 1247), we were told to change the signs, thus
$$\sum \left(\frac{k_i}{l_i} S \pm t_i \right) = r \quad \text{became} \quad \sum \frac{k_i}{l_i} \cdot S = r \mp \sum t_i.$$
(An early reader has pointed out this difference.) But in neither case is the rule ever applied, the constants being always summed up while still on the left-hand side.

(A.208) MSS $\mathcal{A}, \mathcal{B}, \mathcal{C}$. Required the amount S such that
$$S - \left[\left(\frac{1}{3} \cdot S + 1 \right) + \left(\frac{1}{4} \cdot S - 3 \right) \right] = 10$$
This reduces to A.207 and A.204:
$$S - \left[\left(\frac{1}{3} + \frac{1}{4} \right) S - 2 \right] = 10, \qquad S - \left(\frac{1}{3} + \frac{1}{4} \right) S = 8.$$

(A.209) MSS $\mathcal{A}, \mathcal{B}, \mathcal{C}$. Required the amount S such that
$$S - \left[\left(\frac{1}{3} \cdot S - 2 \right) + \left(\frac{1}{4} \cdot S + 3 \right) \right] = 10$$
Treated as A.206 since it reduces to
$$S - \left[\left(\frac{1}{3} + \frac{1}{4} \right) S + 1 \right] = 10.$$

(A.210) MSS $\mathcal{A}, \mathcal{B}, \mathcal{C}$. Required the amount S such that
$$\left\{ S - \left[\left(\frac{1}{3} \cdot S + 2 \right) + \left(\frac{1}{4} \cdot S - 1 \right) \right] \right\} - \left(\frac{1}{2} \left\{ S - \left[\left(\frac{1}{3} \cdot S + 2 \right) + \left(\frac{1}{4} \cdot S - 1 \right) \right] \right\} + 4 \right) = 10$$

The successive steps found in the text correspond to the following reductions:
$$\left\{ S - \left[\left(\frac{1}{3} + \frac{1}{4} \right) S + 1 \right] \right\} - \left(\frac{1}{2} \left\{ S - \left[\left(\frac{1}{3} + \frac{1}{4} \right) S + 1 \right] \right\} + 4 \right)$$
$$= \left\{ S - \left[\left(\frac{1}{3} + \frac{1}{4} \right) S + 1 \right] \right\} - \left(\frac{1}{2} \left\{ S - \left(\frac{1}{3} + \frac{1}{4} \right) S \right\} + 3 + \frac{1}{2} \right)$$
$$= \left\{ S - \left(\frac{1}{3} + \frac{1}{4} \right) S \right\} - \left(\frac{1}{2} \left\{ S - \left(\frac{1}{3} + \frac{1}{4} \right) S \right\} + 4 + \frac{1}{2} \right) = 10.$$

Removing the constant from the left side, we finally obtain

$$\left\{S - \left(\frac{1}{3} + \frac{1}{4}\right)S\right\} - \frac{1}{2}\left\{S - \left(\frac{1}{3} + \frac{1}{4}\right)S\right\} = 14 + \frac{1}{2},$$

which has the form of A.205.

(A.211) MSS $\mathcal{A}$, $\mathcal{B}$, $\mathcal{C}$. Required the amount S such that

$$\left\{S - \left[\left(\frac{1}{3}\cdot S + 2\right) + \left(\frac{1}{4}\cdot S - 1\right)\right]\right\}$$
$$- \left(\frac{1}{2}\left\{S - \left[\left(\frac{1}{3}\cdot S + 2\right) + \left(\frac{1}{4}\cdot S - 1\right)\right]\right\} - 3\right) = 10$$

Proceeding as taught in A.205b, thus subtracting the second expression directly, and continuing as in the problems seen subsequently, the transformations are successively:

$$\frac{1}{2}\left\{S - \left[\left(\frac{1}{3}\cdot S + 2\right) + \left(\frac{1}{4}\cdot S - 1\right)\right]\right\} + 3 = 10,$$

$$\frac{1}{2}\left\{S - \left[\left(\frac{1}{3}\cdot S + 2\right) + \left(\frac{1}{4}\cdot S - 1\right)\right]\right\} = 7,$$

$$S - \left[\left(\frac{1}{3}\cdot S + 2\right) + \left(\frac{1}{4}\cdot S - 1\right)\right] = 14,$$

$$S - \left[\left(\frac{1}{3} + \frac{1}{4}\right)S + 1\right] = 14.$$

Then $S - \left(\frac{1}{3} + \frac{1}{4}\right)S = 15$, which is A.204 with a different constant on the right side, whence $(5 : 12 = 15 : S)$

$$S = \frac{15 \cdot 12}{5} = 36.$$

(A.212) MSS $\mathcal{A}$, $\mathcal{B}$, $\mathcal{C}$. Required the amount S such that

$$\left\{S - \left[\left(\frac{1}{3}\cdot S - 5\right) + \left(\frac{1}{4}\cdot S + 2\right)\right]\right\}$$
$$- \left(\frac{1}{2}\left\{S - \left[\left(\frac{1}{3}\cdot S - 5\right) + \left(\frac{1}{4}\cdot S + 2\right)\right]\right\} - 1\right) = 10$$

The successive transformations of the left side are:

$$\left\{S-\left[\left(\tfrac{1}{3}+\tfrac{1}{4}\right)S-3\right]\right\}-\left(\tfrac{1}{2}\left\{S-\left[\left(\tfrac{1}{3}+\tfrac{1}{4}\right)S-3\right]\right\}-1\right)$$

$$=\left\{S-\left[\left(\tfrac{1}{3}+\tfrac{1}{4}\right)S-3\right]\right\}-\left(\tfrac{1}{2}\left\{S-\left(\tfrac{1}{3}+\tfrac{1}{4}\right)S\right\}+1+\tfrac{1}{2}-1\right)$$

$$=\left\{S-\left[\left(\tfrac{1}{3}+\tfrac{1}{4}\right)S-3\right]\right\}-\left(\tfrac{1}{2}\left\{S-\left(\tfrac{1}{3}+\tfrac{1}{4}\right)S\right\}+\tfrac{1}{2}\right)$$

$$=\left\{S-\left(\tfrac{1}{3}+\tfrac{1}{4}\right)S\right)\right\}-\left(\tfrac{1}{2}\left\{S-\left(\tfrac{1}{3}+\tfrac{1}{4}\right)S\right\}-\left(2+\tfrac{1}{2}\right)\right)$$

which equals 10. Continuing (unlike the text) we find

$$\tfrac{1}{2}\left\{S-\left(\tfrac{1}{3}+\tfrac{1}{4}\right)S\right\}+\left(2+\tfrac{1}{2}\right)=10,\quad S-\left(\tfrac{1}{3}+\tfrac{1}{4}\right)S=15,$$

the solution to which is known from A.211.

(A.213) MSS $\mathcal{A}$. $\mathcal{B}$, $\mathcal{C}$. Required the amount S such that

$$\left(\left\{S-\left(\tfrac{1}{3}\cdot S+2\right)\right\}-\left[\tfrac{1}{2}\left\{S-\left(\tfrac{1}{3}\cdot S+2\right)\right\}+5\right]\right)$$
$$-\left[\tfrac{2}{5}\left(\left\{S-\left(\tfrac{1}{3}\cdot S+2\right)\right\}-\left[\tfrac{1}{2}\left\{S-\left(\tfrac{1}{3}\cdot S+2\right)\right\}+5\right]\right)-1\right]=11$$

```
  B       KT Z H              D G              A
  |-------|-|-|-|-------------|-|--------------|
```

Fig. 33

Geometrical computation. This indeed facilitates the computations, with the successive subtractions thus arranged on a segment of a straight line. Let us put (Fig. 33):

$S = AB,$

$\tfrac{1}{3}\cdot S = AG,$

$2 = GD,$ thus

$S-\left(\tfrac{1}{3}\cdot S+2\right)=DB,$

$\tfrac{1}{2}\left\{S-\left(\tfrac{1}{3}\cdot S+2\right)\right\}=DH,$

$5 = HZ$, thus

$$\left\{S - \left(\tfrac{1}{3}\cdot S + 2\right)\right\} - \left[\tfrac{1}{2}\left\{S - \left(\tfrac{1}{3}\cdot S + 2\right)\right\} + 5\right] = ZB,$$

$$\tfrac{2}{5}\left(\left\{S - \left(\tfrac{1}{3}\cdot S + 2\right)\right\} - \left[\tfrac{1}{2}\left\{S - \left(\tfrac{1}{3}\cdot S + 2\right)\right\} + 5\right]\right) = ZK,$$

$1 = KT$.

Required the whole segment AB.

Since $TB = 11$, so $TB - KT = BK = 10$; but $BK = \tfrac{3}{5}\cdot ZB$, therefore $ZB = 16 + \tfrac{2}{3}$.

Now, $HZ = 5$, so $HB = 21 + \tfrac{2}{3}$ and, since $HB = \tfrac{1}{2}\cdot DB$, $DB = 43 + \tfrac{1}{3}$.

Now, $GD = 2$, so $BG = 45 + \tfrac{1}{3}$ and, since $BG = \tfrac{2}{3}\cdot AB$, $AB = 68 = S$.

(A.214) MSS $\mathcal{A}$, $\mathcal{B}$, $\mathcal{C}$. Find the amount S such that

$$\left[S - \left(\tfrac{1}{3}\cdot S + 2\right)\right]^2 = S + 24$$

Since this becomes

$$\left[S - \left(\tfrac{1}{3}\cdot S + 2\right)\right]^2 = \left[\tfrac{2}{3}\cdot S - 2\right]^2 = \left(\tfrac{2}{3}\cdot S\right)^2 - 4\cdot\left(\tfrac{2}{3}\cdot S\right) + 4 = S + 24,$$

the author puts (implicitly) $\tfrac{2}{3}\cdot S = S'$; this gives

$$S'^2 - 4S' + 4 = \left(1 + \tfrac{1}{2}\right)\cdot S' + 24, \quad \text{or}$$

$$S'^2 = \left(5 + \tfrac{1}{2}\right)S' + 20,$$

whence

$$S' = 2 + \tfrac{3}{4} + \sqrt{\left(2 + \tfrac{3}{4}\right)^2 + 20} = 2 + \tfrac{3}{4} + \sqrt{7 + \tfrac{4}{8} + \tfrac{1}{2}\tfrac{1}{8} + 20}$$

$$= 2 + \tfrac{3}{4} + \sqrt{27 + \tfrac{4}{8} + \tfrac{1}{2}\tfrac{1}{8}} = 2 + \tfrac{3}{4} + 5 + \tfrac{1}{4} = 8,$$

so that $S = 12$.

In the text itself, there are no explanations, just a sequence of instructions for calculating the various, intermediate and final, numerical results. But the reasoning will be detailed in the subsequent demonstration.

<u>Demonstration.</u>

(α) *Preliminary remark.*

The equation seen above for S' reduces to $S'\left[S' - \left(5 + \frac{1}{2}\right)\right] = 20$. Putting $S' = u$, $S' - \left(5 + \frac{1}{2}\right) = v$, we are to solve
$$\begin{cases} u - v = 5 + \frac{1}{2} \\ u \cdot v = 20, \end{cases}$$
that is, we are to determine two quantities of which we know difference and product.

This is solved by means of two identities. The first,
$$\left(\frac{u+v}{2}\right)^2 = \left(\frac{u-v}{2}\right)^2 + u \cdot v = \left(\frac{5 + \frac{1}{2}}{2}\right)^2 + 20,$$
enables us to determine the half-sum, while with the second,
$$u = \frac{u-v}{2} + \frac{u+v}{2} = \frac{1}{2}\left(5 + \frac{1}{2}\right) + \sqrt{\left(\frac{5 + \frac{1}{2}}{2}\right)^2 + 20}$$
$$= 2 + \frac{3}{4} + \sqrt{27 + \frac{4}{8} + \frac{1}{2}\frac{1}{8}},$$
we can calculate the main unknown. This is the customary way of solving quadratic equations, as numerous examples in Book B will show. The geometrical demonstration below corresponds to this.

```
B       G D H   Z   A
├───────┼─┼─┼───┼───┤
```

Fig. 34

(β) *Demonstration in the text.* Let (Fig. 34)

$S = AB$,

$\frac{1}{3} \cdot S = \frac{1}{3} \cdot AB = BG$, thus $AG = \frac{2}{3} \cdot AB$,

$2 = GD$, thus $AD = AB - (BG + GD) = AB - \left(\frac{1}{3} \cdot AB + 2\right)$.

According to the statement, we must have
$$AD^2 = AB + 24 = \left(1 + \frac{1}{2}\right)AG + 24.$$

Adding $2 \cdot AG \cdot GD$ gives
$$AD^2 + 2 \cdot AG \cdot GD = \left(1 + \frac{1}{2}\right)AG + 2 \cdot AG \cdot GD + 24.$$

But (*Elements* II.7 = PE_7) $AD^2 + 2 \cdot AG \cdot GD = AG^2 + GD^2$, so that we now have

$$AG^2 + GD^2 = \left(1 + \frac{1}{2}\right) \cdot AG + 2 \cdot AG \cdot GD + 24.$$

But $GD = 2$, so the equation reduces to

$$AG^2 = \left(5 + \frac{1}{2}\right) \cdot AG + 20.$$

Since $AG > 5 + \frac{1}{2}$,* so cut AG at H so as to make $AH = 5 + \frac{1}{2}$; the equation becomes

$$AG^2 = AH \cdot AG + 20;$$

but, since $AG^2 = AG(AH + HG) = AG \cdot AH + AG \cdot HG$, the above expression takes the reduced form

$$AG \cdot HG = 20.$$

So we now have

$$\begin{cases} AG - HG = 5 + \frac{1}{2} \\ AG \cdot HG = 20, \end{cases}$$

that is, we know the difference and the product of the two segments AG and HG which we can now determine individually. For this, we shall bisect AH, the difference of the two required segments, at Z; then $AZ = ZH = 2 + \frac{3}{4}$. So AH is a line bisected at Z to which is added HG. We can then apply *Elements* II.6 = PE_6:

$$AG \cdot HG + ZH^2 = ZG^2.$$

But $AG \cdot HG = 20$ while $ZH = 2 + \frac{3}{4}$; so

$$ZG^2 = 20 + 7 + \frac{1}{2} + \frac{1}{2}\frac{1}{8} = 27 + \frac{1}{2} + \frac{1}{2}\frac{1}{8}, \quad \text{then}$$

$$ZG = 5 + \frac{1}{4}, \quad AG = AZ + ZG = 2 + \frac{3}{4} + 5 + \frac{1}{4} = 8 = \frac{2}{3} \cdot AB,$$

whence $AB = 12$ for the required quantity.

---

* For $AG\left(AG - \left(5 + \frac{1}{2}\right)\right) > 0$.

Chapter A–VIII:
Division of fractions

Summary

1. Inverting a fractional expression.

This section may be divided into three parts. First, we are taught how to find the inverse of a fraction, thus how to divide unity by the proposed fraction, an operation called 'redintegration' of a fraction into 1 (al-Ḥaṣṣār calls it *al-jabr*).[‡] We successively consider this for a fraction (A.215–216), a fraction of a fraction (A.217–219) and a compound fraction (A.220); the required inverse is then an integer or an integer with a fraction. Secondly, if the fractional expression to be inverted contains an integer, the operation will be 'finding the (amount of the) ratio' of 1 to the expression considered, the result being a proper fraction (al-Ḥaṣṣār calls this operation *al-khaṭṭ*). A.221–224 illustrate this.

These first two parts, the separation into which originates in the difference between division and denomination, are found only in MSS $\mathcal{B}$ and $\mathcal{C}$. What MS $\mathcal{A}$ has, in common with $\mathcal{B}$ and $\mathcal{C}$, is the third part, which explains the treatment of both cases by the same method, namely that of multiplying numerator and denominator by the denominator of the fraction considered (A.226–229; A.225 may be interpolated).

Taking the inverse of a fractional expression will later have a twofold application: in dividing two expressions, we may multiply the dividend by the inverse of the divisor; when treating quadratic equations, it will allow us to reduce them to the standard forms, with 1 as the coefficient of x^2.

2. Denominating fractional expressions.

This is the subject of A.230–235 (though A.232, and possibly A.233, would fit better in the previous section). The method seen in the last problems of Section 1 forms the basis of denomination as well as division; accordingly, a division involving fractions is reduced to a division involving integers. But now, since the dividend is no longer 1, we shall consider for the common multiplication the denominator in the dividend as well; that is, we shall multiply both terms by the product of the denominators. Finally, it should be noted that, unlike in the previous arithmetical operations, the denominators in the resulting fractions do not need to be the same as in the given expressions.

3. Division of fractions without integers.

[‡] Al-Ḥaṣṣār, author of a contemporary Hispano-Arabic treatise, has already been mentioned (pp. 1200, 1236, 1253).

Suppose we are to divide the fraction $\frac{p}{q}$ by another (smaller) fraction $\frac{r}{s}$. We have three possibilities:

(*i*) As with denomination, we can multiply both by the product of the denominators $q \cdot s$. Then

$$\frac{\frac{p}{q}}{\frac{r}{s}} = \frac{\frac{p}{q} q \cdot s}{\frac{r}{s} q \cdot s} = \frac{p \cdot s}{r \cdot q},$$

which leads to a division involving integers only. This way of operating is found in all problems of this section (A.236–241; proof, for more general cases, in A.244*a* & A.260*a*).

(*ii*) Another possibility is to multiply both by s to have an integer just in the divisor. Then

$$\frac{\frac{p}{q}}{\frac{r}{s}} = \frac{\frac{p}{q} s}{\frac{r}{s} s} = \frac{\frac{p \cdot s}{q}}{r},$$

which is the division of a fraction (or an integer with a fraction) by an integer. See A.239, A.240, A.241.

(*iii*) Another possibility is to find the inverse of $\frac{r}{s}$ (Section 1) and multiply the dividend by it. Then

$$\frac{\frac{p}{q}}{\frac{r}{s}} = \frac{p}{q}\left(a + \frac{k}{l}\right),$$

which brings us back to the multiplication of fractional expressions. See A.237–240'.

Considering now the fractional expressions involved, we find division of a fraction by a fraction (A.236–238), by a fraction of a fraction and conversely (A.239–239'), by a compound fraction and conversely (A.240–240') and, finally, division of two compound fractions (A.241). Dividing a fraction of a fraction by a fraction of a fraction is left out, but by now the treatment is clear enough.

4. Division of irregular fractions.

We find division of two irregular fractions of an integer alone (A.242, A.244) and of an integer with a fraction (A.243, A.245). There is, in the latter case, no mention of ambiguous formulations.

5. Division of fractional expressions containing integers.

This is the largest part, and a very complete one considering the variety of cases treated. For we may have as dividend either an integer alone or an integer with a fraction, and as divisor a fraction or an integer with a fraction. Now all possibilities are considered. Thus an integer alone is divided by a fraction (A.246–248), a fraction of a fraction (A.249–250), a compound fraction (A.251–253); then an integer is divided by an integer with, first, a fraction (A.254), then a fraction of a fraction (A.255), finally a compound fraction (A.256). The same six cases are then examined when the dividend is an integer with a fraction;[∥] for it is successively divided by a fraction (A.257), a fraction of a fraction (A.258), a compound fraction (A.259), then by an integer having with it a fraction (A.260–261), or a fraction of a fraction (A.263), or a compound fraction (A.262).

The ways of solving such problems are those seen above. (If the dividend is an integer, there are just two since i and ii are then equivalent.) For the last way (iii), the text sometimes refers to the chapter on multiplication of fractional expressions (A.255, A.257, A.262).

6. Other problems.

As before, the chapter ends with a few particular problems. The first is indeterminate (A.264) and there are two problems involving more complicated expressions (A.265–266; similar cases for the multiplication are A.141–144). Between these two groups, we find a set of rules which is a kind of recapitulation, telling us how to perform the four arithmetical operations with fractional expressions. The way explained is that most commonly met with before, using the product of the denominators; the required quantity is then, for all four operations, the result of the division of two integers. With this recapitulation, the calculations are extended to fractional expressions containing several different fractions.

Remark. It is certainly useful to recapitulate all four operations. Still, we are left with the impression that this was added later, quite possibly by the author himself despite some novelty in the designations. Some general rules seen before left us with the same impression.*

7. Sharing amounts of money.

At the beginning of the division of integers (see above, p. 1183), we were told about the two aspects of division: the first is determining how many times a quantity is contained in another of the same kind, and the second is sharing a quantity among a certain number of persons in order to determine the part of each. The second aspect is the subject of this last section. It begins with known amounts (A.267–272) and continues with unknown ones (A.273–274).

A. *Known amounts*.

After a particular example of the successive division of a given quantity (A.267), we find three problems illustrating general rules. We are to share a given sum S among a given number n of persons under the condition that consecutive shares s_i and s_{i-1} shall differ by given quantities d_{i-1}. Required are these shares or, rather, the first s_1 since the others will be inferred from it.

∥ Only restriction: this fraction is a simple one.
* See pp. 1204–1205, 1215, 1247–1248, 1269.

(i) The difference of consecutive shares is constant: $d_1 = d_2 = \ldots = d_{n-1} = d$. Then

$$S = s_1 + s_2 + \ldots + s_n$$
$$= s_1 + (s_1 + d) + \ldots + (s_1 + (n-1)d)$$
$$= n \cdot s_1 + (1 + 2 + \ldots + (n-1)) \cdot d = n \cdot s_1 + d \cdot \frac{(n-1)n}{2},$$

whence s_1. See A.268, with $d = 1$. The case $d \neq 1$ is included by the author in the subsequent case.

Remark. There is no reference to the summation formula seen in A.170.

(ii) The difference between consecutive shares is the same except between the first two: $d_1 = k$, $d_2 = d_3 = \ldots = d_{n-1} = d$.

$$S = s_1 + s_2 + \ldots + s_n$$
$$= s_1 + (s_1 + k) + (s_1 + k + d) + \ldots + (s_1 + k + (n-2)d)$$
$$= n \cdot s_1 + k \cdot (n-1) + d \cdot \frac{(n-2)(n-1)}{2},$$

whence s_1. See A.269, with $k = 3$ and $d = 2$.

(iii) In the general case, we shall have $s_i = s_{i-1} + d_{i-1}$ and

$$S = s_1 + s_2 + \ldots + s_n$$
$$= s_1 + (s_1 + d_1) + (s_2 + d_2) + \ldots + (s_{n-1} + d_{n-1})$$
$$= s_1 + (s_1 + d_1) + (s_1 + d_1 + d_2) + \ldots + (s_1 + d_1 + \ldots + d_{n-1})$$
$$= n \cdot s_1 + d_1 + (d_1 + d_2) + (d_1 + d_2 + d_3) + \ldots + (d_1 + \ldots + d_{n-1}),$$

whence s_1. See A.270.

In A.271 & A.272 the parts of S are equal; we know neither their value nor their number (nor therefore the number of recipients) but are given the sum or difference of these two quantities. This leads in both cases to a quadratic equation (with only the formula given).

B. *Unknown amounts*.

A.273–273′ are about dividing up an unknown amount S into equal parts, but successively between n and $n + d$ persons, where n is unknown; we are given d, and the relation (involving a root) between the two parts. An example of the so-called Chinese remainder problem closes the chapter on division. The reader will learn little from this particular problem, solved without clear explanations.

1. Inverting a fractional expression.

As said in our summary, this first section can be divided into three parts. Although these parts differ in aspect, the mode of operating remains much the same. That may be why MS $\mathcal{A}$ omits the first two parts.

A. *Redintegrating a fraction to unity*.

The examples proposed are solved by what we shall call rule i: if the given fraction is $\frac{p}{q}$, so consider that the unit will contain q parts and the given fraction, p. Then $\frac{q}{p}$ will be the required redintegrating factor.

(**A.215**) MSS $\mathcal{B}, \mathcal{C}$. Inverting an aliquot fraction.

$$1 : \frac{1}{3} \quad \text{or} \quad \frac{1}{2}, \frac{1}{4}, \frac{1}{6}, \quad \text{generally} \quad \frac{1}{q}$$

The given fraction corresponds to one part and the unit to 3; thus 3 (or q) such fractions will make up the unit.

(**A.216**) MSS $\mathcal{B}, \mathcal{C}$. Inverting a non-aliquot fraction.

$$1 : \frac{2}{3} \quad \text{or} \quad \frac{3}{5}, \frac{5}{7}, \frac{8}{11}, \quad \text{generally} \quad \frac{p}{q} \quad \text{with} \quad p < q$$

The given fraction corresponds to 2 parts (or p) and the unit to 3 (or q); since $\frac{3}{2}$ multiplying 2 produces 3, the required factor is

$$\frac{3}{2} = 1 + \frac{1}{2},$$

and those of the others, respectively,

$$1 + \frac{2}{3}, \quad 1 + \frac{2}{5}, \quad 1 + \frac{3}{8}, \quad \text{generally} \quad \frac{q}{p}.$$

(**A.217**) MSS $\mathcal{B}, \mathcal{C}$. Inverting an aliquot fraction of a fraction.

$$1 : \frac{1}{2}\frac{1}{6} \quad \text{or} \quad \frac{1}{2}\frac{1}{8}, \frac{1}{4}\frac{1}{7}, \quad \text{generally} \quad \frac{1}{r}\frac{1}{s}$$

As in A.215, but the required factor is $2 \cdot 6 = 12$ (respectively 16, 28, generally $r \cdot s$).

(**A.218**) MSS $\mathcal{B}, \mathcal{C}$. Inverting a non-aliquot fraction of a fraction.

$$1 : \frac{3}{4}\frac{1}{5}$$

As in A.216, but we have 3 out of 20 parts of the unit and the factor multiplying 3 to give 20 is

$$\frac{1}{\frac{3}{4}\frac{1}{5}} = \frac{20}{3} = 6 + \frac{2}{3}.$$

(**A.219**) MSS $\mathcal{B}, \mathcal{C}$. Inverting a non-elementary fraction of a fraction.

$$1 \; : \; \frac{3}{4}\frac{1}{11}$$

$$\frac{1}{\frac{3}{4}\frac{1}{11}} = \frac{44}{3} = 14 + \frac{2}{3}.$$

(**A.220**) MSS $\mathcal{B}, \mathcal{C}$. Inverting a compound fraction.

$$1 \; : \; \frac{2}{7} + \frac{1}{2}\frac{1}{7}$$

We have 5 out of 14 parts of the unit, so the factor will be

$$\frac{1}{\frac{2}{7}+\frac{1}{2}\frac{1}{7}} = \frac{14}{5} = 2 + \frac{4}{5}.$$

B. *Reducing an integer with a fraction to unity*.

To find the amount of the ratio between the unit and the given expression, convert both the unit and the given expression to the latter's denominator (or product of the denominators), and form the quotient of the numerators. This is the meaning of rule *ii* in the text, analogous to *i* before (Section A above) but taking the integer into account: since

$$a + \frac{p}{q} = \frac{aq+p}{q} \quad \text{while} \quad 1 = \frac{q}{q}$$

denominating q from $aq+p$ will give the answer.

(**A.221**) MSS $\mathcal{B}, \mathcal{C}$. Inverting an integer with a fraction.

$$1 \; : \; 1 + \frac{1}{2}$$

See A.216. We have here 3 out of 2 parts of the unit, so

$$\frac{1}{1+\frac{1}{2}} = \frac{2}{3}.$$

(**A.222**) MSS $\mathcal{B}, \mathcal{C}$. Another example.

$$1 \; : \; 2 + \frac{1}{4}$$

We have here 9 out of 4 parts of the unit, so

$$\frac{1}{2+\frac{1}{4}} = \frac{4}{9}.$$

(**A.223**) MSS $\mathcal{B}, \mathcal{C}$. Another example, with a non-elementary fraction.

$$1 \; : \; 3 + \frac{3}{11}$$

We have here 36 out of 11 parts of the unit, so

$$\frac{1}{3+\frac{3}{11}} = \frac{11}{36} = \frac{8+3}{4 \cdot 9} = \frac{2}{9} + \frac{3}{4}\frac{1}{9}.$$

(A.224) MSS $\mathcal{B}, \mathcal{C}$. Inverting an integer with a compound fraction.
$$1 \;:\; 2 + \frac{5}{7} + \frac{1\,1}{2\,7}$$
We have here 39 out of 14 parts of the unit, so
$$\frac{1}{2 + \frac{5}{7} + \frac{1\,1}{2\,7}} = \frac{14}{39} = \frac{12 + 2}{3 \cdot 13} = \frac{4}{13} + \frac{2}{3}\frac{1}{13}.$$

C. *Same, by a direct rule*.

The above examples considered the conversions of 1 and of the given expression into same parts. We shall now be taught a direct computation rule. (In fact, two rules for determining the inverse of a numerical fractional expression are given; however, only the second deserves the name.)

(*iii′*) Divide 1 by the given expression, or denominate if the fractional expression is larger than 1.

(*iii″*) Multiply both 1 and the given expression by the latter's denominator (or product of the denominators) and divide the first product by the second (or denominate).

(A.225) MSS $\mathcal{A}, \mathcal{B}, \mathcal{C}$. Inverting a fraction.
$$1 \;:\; \frac{2}{8}$$
Dividing 1 by the given fraction:
$$\frac{1}{\frac{2}{8}} = 4.$$

Remark. This use of an unreduced fraction is odd, and makes (together with the use of 'rule' *iii′*) the genuineness of this problem questionable.

(A.226) MSS $\mathcal{A}, \mathcal{B}, \mathcal{C}$. Another example of inverting a fraction.
$$1 \;:\; \frac{3}{5}.$$
Using this time the other rule (*iii″*), with the multiplier 5:
$$\frac{1}{\frac{3}{5}} = \frac{5}{\frac{3}{5} \cdot 5} = \frac{5}{3} = 1 + \frac{2}{3}.$$

(A.227) MSS $\mathcal{A}, \mathcal{B}, \mathcal{C}$. Inverting an integer with an aliquot fraction.
$$1 \;:\; 1 + \frac{1}{3}$$
With the multiplier 3:
$$\frac{1}{1 + \frac{1}{3}} = \frac{3}{\left(1 + \frac{1}{3}\right) \cdot 3} = \frac{3}{4}.$$

(A.228) MSS $\mathcal{A}, \mathcal{B}, \mathcal{C}$. Inverting an integer with a fraction, namely
$$1 : 2 + \frac{4}{7}$$
With the multiplier 7:
$$\frac{1}{2+\frac{4}{7}} = \frac{7}{(2+\frac{4}{7}) \cdot 7} = \frac{7}{18} = \frac{6+1}{3 \cdot 6} = \frac{2}{6} + \frac{1}{3}\frac{1}{6}.$$

(A.229) MSS $\mathcal{A}, \mathcal{B}, \mathcal{C}$. Inverting an integer with a compound fraction (with given expression much the same as in A.224).
$$1 : 2 + \frac{3}{7} + \frac{1}{2}\frac{1}{7}$$
With the multiplier $2 \cdot 7 = 14$:
$$\frac{1}{2+\frac{3}{7}+\frac{1}{2}\frac{1}{7}} = \frac{14}{(2+\frac{3}{7}+\frac{1}{2}\frac{1}{7}) \cdot 14} = \frac{14}{35} = \frac{2}{5}.$$

---

2. Denominating fractional expressions.

Since the dividend is no longer 1, rule *iii''* takes the new form: Multiply both fractional expressions by the product of the denominators, and denominate. Note, however, that in A.232–233 the dividend is an integer.

(A.230) MSS $\mathcal{A}, \mathcal{B}, \mathcal{C}$. Denominating an aliquot fraction from an aliquot fraction.
$$\frac{1}{4} : \frac{1}{3}$$
Form the product of the denominators: $4 \cdot 3 = 12$; then
$$\frac{\frac{1}{4}}{\frac{1}{3}} = \frac{\frac{1}{4} \cdot 12}{\frac{1}{3} \cdot 12} = \frac{3}{4}.$$

(A.231) MSS $\mathcal{A}, \mathcal{B}, \mathcal{C}$. Denominating a fraction from a fraction.
$$\frac{5}{6} : \frac{6}{7}$$
Form the product of the denominators: $6 \cdot 7 = 42$; then
$$\frac{\frac{5}{6}}{\frac{6}{7}} = \frac{\frac{5}{6} \cdot 42}{\frac{6}{7} \cdot 42} = \frac{35}{36} = \frac{32+3}{4 \cdot 9} = \frac{8}{9} + \frac{3}{4}\frac{1}{9}.$$

(A.232) MSS $\mathcal{A}, \mathcal{B}, \mathcal{C}$. Denominating an integer from an integer with an aliquot fraction.
$$1 : 2 + \frac{1}{2}$$

Remark. Such problems with dividend 1 were the subject of the previous section (A.221–223, A.227–228).

Taking 2 as the multiplier,

$$\frac{1}{2+\frac{1}{2}} = \frac{2}{\left(2+\frac{1}{2}\right)\cdot 2} = \frac{2}{5}.$$

(A.233) MSS $\mathcal{A}$, $\mathcal{B}$, $\mathcal{C}$. Denominating an integer from an integer with a fraction.

$$2 \;:\; 6+\frac{2}{3}$$

Taking 3 as the multiplier,

$$\frac{2}{6+\frac{2}{3}} = \frac{2\cdot 3}{\left(6+\frac{2}{3}\right)\cdot 3} = \frac{6}{20} = \frac{3}{10}.$$

(A.234) MSS $\mathcal{A}$, $\mathcal{B}$. Denominating an integer with an aliquot fraction from an integer with an aliquot fraction.

$$1+\frac{1}{2} \;:\; 3+\frac{1}{3}$$

Form the product of the denominators: $2\cdot 3 = 6$; then

$$\frac{1+\frac{1}{2}}{3+\frac{1}{3}} = \frac{\left(1+\frac{1}{2}\right)\cdot 6}{\left(3+\frac{1}{3}\right)\cdot 6} = \frac{9}{20} = \frac{8+1}{2\cdot 10} = \frac{4}{10} + \frac{1}{2}\frac{1}{10}.$$

(A.235) MSS $\mathcal{A}$, $\mathcal{B}$. Denominating an integer with a fraction from an integer with a fraction.

$$3+\frac{3}{4} \;:\; 4+\frac{3}{10}$$

Form the product of the denominators: $4\cdot 10 = 40$; then

$$\frac{3+\frac{3}{4}}{4+\frac{3}{10}} = \frac{\left(3+\frac{3}{4}\right)\cdot 40}{\left(4+\frac{3}{10}\right)\cdot 40} = \frac{150}{172} = \frac{75}{86} = \frac{74+1}{2\cdot 43} = \frac{37}{43} + \frac{1}{2}\frac{1}{43}.$$

---

3. Division of fractions without integers.

Remark. Since in two problems the given fractional expressions exchange their rôles, we also find denomination (A.239′, A.240′).

(A.236) MS $\mathcal{B}$. Division of a fraction by a fraction.

$$\frac{4}{5} \;:\; \frac{3}{4}$$

Since this is the first problem of its kind, we are explained what it means: given that $\frac{4}{5}$ is attributed to $\frac{3}{4}$ of a thing, we are to find out what corresponds to the whole thing (that is, if f is the answer, $\frac{4}{5} : \frac{3}{4} = f : 1$). There is a similar introductory explanation to the two chapters on division (A–IV and this one). See also A.246.

Form the product of the denominators, thus $4 \cdot 5 = 20$; then

$$\frac{\frac{4}{5}}{\frac{3}{4}} = \frac{\frac{4}{5} \cdot 20}{\frac{3}{4} \cdot 20} = \frac{16}{15} = 1 + \frac{1}{3}\frac{1}{5}.$$

(**A.237**) MSS $\mathcal{A}, \mathcal{B}, \mathcal{C}$. Another example (this time in all three MSS), with an aliquot fraction in the divisor.

$$\frac{3}{5} : \frac{1}{3}$$

(**a**) Form the product of the denominators $5 \cdot 3 = 15$; then

$$\frac{\frac{3}{5}}{\frac{1}{3}} = \frac{\frac{3}{5} \cdot 15}{\frac{1}{3} \cdot 15} = \frac{9}{5}.$$

(**b**) Finding first the inverse of the divisor and multiplying by it the fraction to be divided. Here the inverse is 3, so

$$\frac{\frac{3}{5}}{\frac{1}{3}} = \frac{3}{5} \cdot 3 = 1 + \frac{4}{5}.$$

(**A.238**) MSS $\mathcal{A}, \mathcal{B}, \mathcal{C}$. Division of a fraction by a non-elementary fraction.

$$\frac{5}{6} : \frac{8}{11}$$

(**a**) As before (multiplying dividend and divisor by $6 \cdot 11 = 66$).

(**b**) Finding first the inverse of the divisor.

$$\frac{1}{\frac{8}{11}} = 1 + \frac{3}{8}, \quad \text{so calculate} \quad \left(1 + \frac{3}{8}\right) \cdot \frac{5}{6}.$$

Remark. All three MSS have, for the inverse, $1 + \frac{3}{5}$.

(**A.239**) MSS $\mathcal{A}, \mathcal{B}, \mathcal{C}$. Division of a fraction by a fraction of a fraction.

$$\frac{10}{13} : \frac{3}{4}\frac{1}{5}$$

(**a**) Form the product of the denominators $13 \cdot 4 \cdot 5 = 260$; then*

$$\frac{\frac{10}{13}}{\frac{3}{4}\frac{1}{5}} = \frac{\frac{10}{13} \cdot 260}{\frac{3}{4}\frac{1}{5} \cdot 260} = \frac{200}{39}.$$

(**b**) Consider only the denominators in the divisor. Since $4 \cdot 5 = 20$,

$$\frac{\frac{10}{13}}{\frac{3}{4}\frac{1}{5}} = \frac{\frac{10}{13} \cdot 20}{\frac{3}{4}\frac{1}{5} \cdot 20}.$$

---

* The quantities in italics are found in A.239'.

(c) Take the inverse of the divisor.

$$\frac{1}{\frac{3\,1}{4\,5}} = 6 + \frac{2}{3}, \quad \text{so compute} \quad \left(6 + \frac{2}{3}\right) \cdot \frac{10}{13}.$$

(**A.239′**) MSS $\mathcal{B}, \mathcal{C}$. Division of a fraction of a fraction by a fraction. Converse problem.

$$\frac{3\,1}{4\,5} : \frac{10}{13}$$

Remark. The copyist of MS $\mathcal{C}$, perhaps following the progenitor, wanted to delete this problem. Indeed, this problem, being the converse of A.239, involves denomination, which is the subject of the previous section.

(**a**) Form the product of the denominators as before (= 260). Then

$$\frac{\frac{3\,1}{4\,5}}{\frac{10}{13}} = \frac{\frac{3\,1}{4\,5} \cdot 260}{\frac{10}{13} \cdot 260} = \frac{39}{200}.$$

(**b**) Take the inverse of the divisor.

$$\frac{1}{\frac{10}{13}} = 1 + \frac{3}{10}, \quad \text{thus}$$

$$\left(1 + \frac{3}{10}\right) \cdot \frac{3\,1}{4\,5} = \frac{3 + \frac{9}{10}}{4} \cdot \frac{1}{5} = \left(\frac{3}{4} + \frac{9\ 1}{10\ 4}\right) \frac{1}{5} = \frac{3\,1}{4\,5} + \frac{9\,1\,1}{10\,4\,5}.$$

(**A.240**) MSS $\mathcal{A}, \mathcal{B}, \mathcal{C}$. Division of a fraction by a compound fraction.

$$\frac{4}{7} : \frac{1}{8} + \frac{2\,1}{3\,8}$$

(**a**) Form the product of the denominators: $7 \cdot 8 \cdot 3 = 168$; then

$$\frac{\frac{4}{7}}{\frac{1}{8} + \frac{2\,1}{3\,8}} = \frac{\frac{4}{7} \cdot 168}{\left(\frac{1}{8} + \frac{2\,1}{3\,8}\right) \cdot 168} = \frac{96}{35}.$$

Remark. The text has 64 instead of 96, an error which is confirmed by the computation in the next, converse problem.

(**b**) Make only the divisor an integer. Since $8 \cdot 3 = 24$,

$$\frac{\frac{4}{7}}{\frac{1}{8} + \frac{2\,1}{3\,8}} = \frac{\frac{4}{7} \cdot 24}{\left(\frac{1}{8} + \frac{2\,1}{3\,8}\right) \cdot 24} = \frac{\frac{96}{7}}{5} = \frac{13 + \frac{5}{7}}{5}.$$

Remark. Performing the division will lead to $2 + \frac{3}{5} + \frac{5\,1}{7\,5}$ (*sic*) or $2 + \frac{5}{7} + \frac{1\,1}{5\,7}$.

(**c**) Take the inverse of the divisor (A.220).

$$\frac{1}{\frac{1}{8} + \frac{2\,1}{3\,8}} = 4 + \frac{4}{5}, \quad \text{then compute} \quad \left(4 + \frac{4}{5}\right) \cdot \frac{4}{7}.$$

(A.240′) Division of a compound fraction by a fraction. Converse problem, thus involving denomination.†

$$\frac{1}{8} + \frac{2}{3}\frac{1}{8} : \frac{4}{7}$$

(a) Form the product of the denominators as before ($= 168$); then

$$\frac{\frac{1}{8} + \frac{2}{3}\frac{1}{8}}{\frac{4}{7}} = \frac{\left(\frac{1}{8} + \frac{2}{3}\frac{1}{8}\right) \cdot 168}{\frac{4}{7} \cdot 168} = \frac{35}{96}.$$

Remark. 64 instead of 96 in the text.

(b) Take the inverse of the divisor.

$$\frac{1}{\frac{4}{7}} = 1 + \frac{3}{4}, \quad \text{then compute} \quad \left(1 + \frac{3}{4}\right)\left(\frac{1}{8} + \frac{2}{3}\frac{1}{8}\right).$$

(A.241) MSS $\mathcal{A}, \mathcal{B}, \mathcal{C}$. Division of a compound fraction by a compound fraction.

$$\frac{5}{8} + \frac{3}{4}\frac{1}{8} : \frac{2}{7} + \frac{1}{2}\frac{1}{7}$$

(a) Form the product of the denominators $(8 \cdot 4) \cdot (7 \cdot 2) = 32 \cdot 14 = 448$; then

$$\frac{\frac{5}{8} + \frac{3}{4}\frac{1}{8}}{\frac{2}{7} + \frac{1}{2}\frac{1}{7}} = \frac{\left(\frac{5}{8} + \frac{3}{4}\frac{1}{8}\right) \cdot 448}{\left(\frac{2}{7} + \frac{1}{2}\frac{1}{7}\right) \cdot 448} = \frac{322}{160}.$$

(b) Making the divisor an integer, thus with the multiplier 14.

$$\frac{\frac{5}{8} + \frac{3}{4}\frac{1}{8}}{\frac{2}{7} + \frac{1}{2}\frac{1}{7}} = \frac{\left(\frac{5}{8} + \frac{3}{4}\frac{1}{8}\right) \cdot 14}{\left(\frac{2}{7} + \frac{1}{2}\frac{1}{7}\right) \cdot 14} = \frac{10 + \frac{1}{2}\frac{1}{8}}{5} = 2 + \frac{1}{2}\frac{1}{8}\frac{1}{5}.$$

---

4. Division of irregular fractions.

(A.242) MSS $\mathcal{B}, \mathcal{C}$. Division of irregular fractions of integers.

$$\frac{3}{4} \cdot 6 : \frac{2}{5} \cdot 4$$

Form the product of the denominators $4 \cdot 5 = 20$; then

$$\frac{\frac{3}{4} \cdot 6}{\frac{2}{5} \cdot 4} = \frac{\left(\frac{3}{4} \cdot 20\right) 6}{\left(\frac{2}{5} \cdot 20\right) 4} = \frac{15 \cdot 6}{8 \cdot 4} = \frac{90}{32} = \frac{64 + 24 + 2}{32} = 2 + \frac{6}{8} + \frac{1}{2}\frac{1}{8}.$$

(A.243) MSS $\mathcal{B}, \mathcal{C}$. Division of irregular fractions of an integer with a fraction.

$$\frac{3}{4} \cdot \left(3 + \frac{1}{5}\right) : \frac{2}{5} \cdot \left(2 + \frac{1}{2}\right)$$

---

† Unlike A.239′, this is in MS $\mathcal{A}$.

We have, since $4 \cdot 5 = 20$ (and $5 \cdot 2 = 10$)

$$\frac{\frac{3}{4}\left(3+\frac{1}{5}\right)}{\frac{2}{5}\left(2+\frac{1}{2}\right)} = \frac{\frac{3}{4} \cdot 20\left(3+\frac{1}{5}\right)}{\frac{2}{5} \cdot 20\left(2+\frac{1}{2}\right)} = \frac{15\left(3+\frac{1}{5}\right)}{8\left(2+\frac{1}{2}\right)} = \frac{48}{20} = 2 + \frac{2}{5}.$$

Remark. Here the common denominator taken is the same as in the previous, related problem (with identical multiplying fractions). Normally, the author just takes the product of all denominators. See also above, p. 1233, remark.

(A.244) MSS $\mathcal{A}, \mathcal{B}, \mathcal{C}$. Another example like A.242. (A.242–243 are not in MS $\mathcal{A}$.)

$$\frac{7}{8} \cdot 6 \;:\; \frac{2}{3} \cdot 5$$

(*a*) Form the product of the denominators $8 \cdot 3 = 24$. Then

$$\frac{\frac{7}{8} \cdot 6}{\frac{2}{3} \cdot 5} = \frac{\left(\frac{7}{8} \cdot 24\right) 6}{\left(\frac{2}{3} \cdot 24\right) 5}.$$

Demonstration. Since (as seen in A.90)

$$\frac{7}{8} \cdot 6 = \frac{\left(\frac{7}{8} \cdot 24\right) \cdot 6}{24} = \frac{126}{24}, \quad \frac{2}{3} \cdot 5 = \frac{\left(\frac{2}{3} \cdot 24\right) \cdot 5}{24} = \frac{80}{24},$$

we shall have

$$\frac{\frac{7}{8} \cdot 6}{\frac{2}{3} \cdot 5} = \frac{\frac{126}{24}}{\frac{80}{24}}.$$

But $\frac{126}{24} : \frac{80}{24} = \frac{126}{24} \cdot t : \frac{80}{24} \cdot t$ by *Elements* VII.17, whence

$$\frac{\frac{126}{24}}{\frac{80}{24}} = \frac{\frac{126}{24} \cdot 24}{\frac{80}{24} \cdot 24} = \frac{126}{80}.$$

This proves that

$$\frac{\frac{7}{8} \cdot 6}{\frac{2}{3} \cdot 5} = \frac{\left(\frac{7}{8} \cdot Q\right) \cdot 6}{\left(\frac{2}{3} \cdot Q\right) \cdot 5}$$

with any (convenient) Q.

(*b*) 'According to the wording of the problem', that is, computing the terms individually (A.90*b*) and then dividing.

$$\frac{\frac{7}{8} \cdot 6}{\frac{2}{3} \cdot 5} = \frac{5+\frac{1}{4}}{3+\frac{1}{3}} = \frac{\left(5+\frac{1}{4}\right) 12}{\left(3+\frac{1}{3}\right) 12} = \frac{63}{40} = \frac{40+20+3}{40} = 1 + \frac{5}{10} + \frac{3}{4}\frac{1}{10}.$$

Remark. This (and A.245*a*) in fact brings us to the division of fractional expressions containing integers, which is the subject of the next section.

(A.245) MSS $\mathcal{A}, \mathcal{B}, \mathcal{C}$. Another example like A.243.

$$\frac{7}{8} \cdot \left(6 + \frac{2}{3}\right) \; : \; \frac{2}{5} \cdot \left(4 + \frac{1}{2}\right)$$

(*a*) Following the wording, thus calculating the terms individually:

$$\frac{\frac{7}{8} \cdot \left(6+\frac{2}{3}\right)}{\frac{2}{5} \cdot \left(4+\frac{1}{2}\right)} = \frac{\frac{42}{8} + \frac{14}{24}}{\frac{8}{5} + \frac{1}{5}} = \frac{5 + \frac{2}{8} + \frac{12+2}{3 \cdot 8}}{1 + \frac{4}{5}} = \frac{5 + \frac{6}{8} + \frac{2}{3}\frac{1}{8}}{1 + \frac{4}{5}}.$$

(*b*) Form the product of the denominators $8 \cdot 3 \cdot 5 \cdot 2 = 240$. Then

$$\frac{\frac{7}{8} \cdot \left(6 + \frac{2}{3}\right)}{\frac{2}{5} \cdot \left(4 + \frac{1}{2}\right)} = \frac{\frac{7}{8} \cdot 240 \cdot \left(6 + \frac{2}{3}\right)}{\frac{2}{5} \cdot 240 \cdot \left(4 + \frac{1}{2}\right)}.$$

This is proved, we are told, as in A.244*a*. Indeed, as in A.153 & A.201, the integer is just replaced by an integer with a fraction.

---

5. Division of fractional expressions containing integers.

There are three ways to solve such problems, or two if the dividend is an integer, namely those mentioned in our summary of the third section.

A. *Dividend an integer and divisor a fraction*.

(A.246) MSS $\mathcal{A}$, $\mathcal{B}$, $\mathcal{C}$.   Division of an integer by an aliquot fraction.

$$10 \; : \; \frac{1}{4}$$

As in A.236, the meaning of the problem is explained first: given that 10 corresponds to $\frac{1}{4}$, find out which quantity f will correspond to 1 (thus $10 : \frac{1}{4} = f : 1$).

Remark. The same kind of explanation is found in the next problem (which in MSS $\mathcal{B}$ and $\mathcal{C}$ does not come next).

(*a*) Taking 4 as the multiplier,

$$\frac{10}{\frac{1}{4}} = \frac{10 \cdot 4}{\frac{1}{4} \cdot 4} = 40.$$

(*b*) Or take the inverse of the divisor, thus 4, and multiply it by 10.

(A.247) MSS $\mathcal{A}$, $\mathcal{B}$, $\mathcal{C}$.   Division of an integer by a non-aliquot fraction.

$$20 \; : \; \frac{3}{4}$$

Taking 4 as the multiplier,

$$\frac{20}{\frac{3}{4}} = \frac{20 \cdot 4}{\frac{3}{4} \cdot 4} = \frac{20 \cdot 4}{3}.$$

Or compute, the text adds, $\frac{4}{3} \cdot 20$, equivalent to the above (by P_5).

(A.248) MSS $\mathcal{B}, \mathcal{C}$. Division of a composite integer by a non-aliquot fraction.
$$15 : \frac{4}{7}$$

(a) Taking 7 as the multiplier,
$$\frac{15}{\frac{4}{7}} = \frac{15 \cdot 7}{\frac{4}{7} \cdot 7} = \frac{15 \cdot 7}{4}.$$

(b) Find the inverse of the divisor and multiply it by the dividend.
$$\frac{1}{\frac{4}{7}} = 1 + \frac{3}{4}, \quad \text{then compute} \quad \left(1 + \frac{3}{4}\right) \cdot 15.$$

(A.249) MSS $\mathcal{A}, \mathcal{B}, \mathcal{C}$. Division of an integer by an (aliquot) fraction of a fraction.
$$8 : \frac{1}{3}\frac{1}{5}$$

(a) Form the product of the denominators: $3 \cdot 5 = 15$. Then
$$\frac{8}{\frac{1}{3}\frac{1}{5}} = \frac{8 \cdot 15}{\frac{1}{3}\frac{1}{5} \cdot 15} = \frac{8 \cdot 15}{1} = 120.$$

(b) The inverse of the divisor being 15, multiply it by 8.

(A.250) MSS $\mathcal{A}, \mathcal{B}, \mathcal{C}$. Division of an integer by a non-elementary fraction of a fraction.
$$15 : \frac{4}{5}\frac{1}{11}$$

(a) Form the product of the denominators: $5 \cdot 11 = 55$. Then
$$\frac{15}{\frac{4}{5}\frac{1}{11}} = \frac{15 \cdot 55}{\frac{4}{5}\frac{1}{11} \cdot 55} = \frac{15 \cdot 55}{4}.$$

(b) Find the inverse of the divisor and multiply it by 8.
$$\frac{1}{\frac{4}{5}\frac{1}{11}} = 13 + \frac{3}{4}, \quad \text{then compute} \quad \left(13 + \frac{3}{4}\right) \cdot 15.$$

(A.251) MSS $\mathcal{A}, \mathcal{B}, \mathcal{C}$. Division of an integer by a compound fraction.
$$10 : \frac{3}{8} + \frac{1}{2}\frac{1}{8}$$

(a) Form the product of the denominators $2 \cdot 8 = 16$. Then
$$\frac{10}{\frac{3}{8} + \frac{1}{2}\frac{1}{8}} = \frac{10 \cdot 16}{\left(\frac{3}{8} + \frac{1}{2}\frac{1}{8}\right) \cdot 16} = \frac{160}{7}.$$

(**b**) Find the inverse of the divisor and multiply it by 10.

$$\frac{1}{\frac{3}{8}+\frac{1}{2}\frac{1}{8}} = 2+\frac{2}{7}, \quad \text{then compute} \quad \left(2+\frac{2}{7}\right) \cdot 10.$$

(**A.252**) MSS $\mathcal{A}$, $\mathcal{B}$, $\mathcal{C}$. Division of an integer by a non-elementary compound fraction.

$$10 \;:\; \frac{5}{11} + \frac{1}{3}\frac{1}{11}$$

(**a**) Form the product of the denominators $3 \cdot 11 = 33$. Then

$$\frac{10}{\frac{5}{11}+\frac{1}{3}\frac{1}{11}} = \frac{10 \cdot 33}{\left(\frac{5}{11}+\frac{1}{3}\frac{1}{11}\right) \cdot 33}.$$

(**b**) Find the inverse of the divisor and multiply it by 10.

$$\frac{1}{\frac{5}{11}+\frac{1}{3}\frac{1}{11}} = \frac{33}{16} = 2+\frac{1}{2}\frac{1}{8}, \quad \text{then compute} \quad \left(2+\frac{1}{2}\frac{1}{8}\right) \cdot 10.$$

(**A.253**) MSS $\mathcal{A}$, $\mathcal{B}$, $\mathcal{C}$. Division of an integer by a non-elementary compound fraction with several terms.

$$8 \;:\; \frac{4}{5}\frac{1}{11} + \frac{2}{3}\frac{1}{5}\frac{1}{11}$$

(**a**) Form the product of the denominators: $5 \cdot 11 \cdot 3 = 165$. Then

$$\frac{8}{\frac{4}{5}\frac{1}{11}+\frac{2}{3}\frac{1}{5}\frac{1}{11}} = \frac{8 \cdot 165}{\left(\frac{4}{5}\frac{1}{11}+\frac{2}{3}\frac{1}{5}\frac{1}{11}\right) \cdot 165} = \frac{8 \cdot 165}{14}.$$

(**b**) Find the inverse of the divisor,

$$\frac{1}{\frac{4}{5}\frac{1}{11}+\frac{2}{3}\frac{1}{5}\frac{1}{11}} = \frac{165}{14} = 11 + \frac{11}{14} = 11 + \frac{5}{7} + \frac{1}{2}\frac{1}{7},$$

and multiply it by the dividend.

B. *Dividend an integer and divisor an integer with a fractional expression.*

(**A.254**) MSS $\mathcal{A}$, $\mathcal{B}$, $\mathcal{C}$. Division of an integer by an integer with a fraction.

$$20 \;:\; 2+\frac{2}{3}$$

(**a**) Taking 3 as the multiplier.

$$\frac{20}{2+\frac{2}{3}} = \frac{20 \cdot 3}{\left(2+\frac{2}{3}\right) \cdot 3} = \frac{60}{8}.$$

(**b**) Find the inverse of the divisor and multiply it by the dividend.

$$\frac{1}{2+\frac{2}{3}} = \frac{3}{8}, \quad \text{then} \quad \frac{3}{8} \cdot 20 = 7 + \frac{1}{2}.$$

(**A.255**) MSS $\mathcal{A}, \mathcal{B}, \mathcal{C}$. Division of an integer by an integer with a fraction of a fraction.

$$30 \; : \; 4 + \frac{1}{2}\frac{1}{6}$$

(**a**) Form the product of the denominators: $2 \cdot 6 = 12$. Then

$$\frac{30}{4+\frac{1}{2}\frac{1}{6}} = \frac{30 \cdot 12}{\left(4+\frac{1}{2}\frac{1}{6}\right) \cdot 12} = \frac{360}{49}.$$

(**b**) Take the inverse of the divisor,

$$\frac{1}{4+\frac{1}{2}\frac{1}{6}} = \frac{12}{49} = \frac{1}{7} + \frac{5}{7}\frac{1}{7},$$

and multiply it by the dividend.

(**A.256**) MSS $\mathcal{A}, \mathcal{B}, \mathcal{C}$. Division of an integer by an integer with a compound fraction.

$$45 \; : \; 3 + \frac{4}{11} + \frac{1}{3}\frac{1}{11}$$

(**a**) Form the product of the denominators: $3 \cdot 11 = 33$. Then

$$\frac{45}{3+\frac{4}{11}+\frac{1}{3}\frac{1}{11}} = \frac{45 \cdot 33}{\left(3+\frac{4}{11}+\frac{1}{3}\frac{1}{11}\right) \cdot 33} = \frac{1485}{112}.$$

(**b**) Take the inverse of the divisor,

$$\frac{1}{3+\frac{4}{11}+\frac{1}{3}\frac{1}{11}} = \frac{33}{112} = \frac{28+4+1}{2 \cdot 7 \cdot 8} = \frac{2}{8} + \frac{2}{7}\frac{1}{8} + \frac{1}{2}\frac{1}{7}\frac{1}{8},$$

and multiply it by the dividend.

C. Dividend an integer with a fraction and divisor a fraction.

Since we now have a fraction in the dividend, there are three possible ways for the solution.

(**A.257**) MSS $\mathcal{A}, \mathcal{B}, \mathcal{C}$. Division of an integer with a fraction by a fraction.

$$30 + \frac{2}{3} \; \left(\text{or } 30 + \frac{1}{3}\right) \; : \; \frac{4}{5}$$

Remark. The dividend is said to be $30 + \frac{2}{3}$, both in the statement and in c, but the computation in a & b is performed with $30 + \frac{1}{3}$.

(a) Form the product of the denominators: $3 \cdot 5 = 15$. Then

$$\frac{30 + \frac{1}{3}}{\frac{4}{5}} = \frac{\left(30 + \frac{1}{3}\right) \cdot 15}{\frac{4}{5} \cdot 15} = \frac{455}{12}.$$

(b) Making only the divisor an integer.

$$\frac{30 + \frac{1}{3}}{\frac{4}{5}} = \frac{\left(30 + \frac{1}{3}\right) \cdot 5}{\frac{4}{5} \cdot 5} = \frac{151 + \frac{2}{3}}{4} = 37 + \frac{10 + 1}{12} = 37 + \frac{5}{6} + \frac{1}{2}\frac{1}{6}.$$

(c) Take the inverse of the divisor,

$$\frac{1}{\frac{4}{5}} = 1 + \frac{1}{4}, \quad \text{then compute} \quad \left(1 + \frac{1}{4}\right)\left(30 + \frac{2}{3}\right).$$

(A.258) MSS $\mathcal{A}$, $\mathcal{B}$, $\mathcal{C}$. Division of an integer with a fraction by a fraction of a fraction.

$$23 + \frac{3}{4} \; : \; \frac{2}{3}\frac{1}{5}$$

(a) Form the product of the denominators: $4 \cdot 3 \cdot 5 = 60$. Then

$$\frac{23 + \frac{3}{4}}{\frac{2}{3}\frac{1}{5}} = \frac{\left(23 + \frac{3}{4}\right) \cdot 60}{\frac{2}{3}\frac{1}{5} \cdot 60} = \frac{\left(23 + \frac{3}{4}\right) \cdot 60}{8}.$$

(b) Making the divisor an integer:

$$\frac{23 + \frac{3}{4}}{\frac{2}{3}\frac{1}{5}} = \frac{\left(23 + \frac{3}{4}\right) \cdot 15}{\frac{2}{3}\frac{1}{5} \cdot 15} = \frac{345 + \frac{45}{4}}{2} = \frac{356 + \frac{1}{4}}{2}.$$

(c) Take the inverse of the divisor,

$$\frac{1}{\frac{2}{3}\frac{1}{5}} = 7 + \frac{1}{2}, \quad \text{then compute} \quad \left(7 + \frac{1}{2}\right)\left(23 + \frac{3}{4}\right).$$

(A.259) MSS $\mathcal{A}$, $\mathcal{B}$, $\mathcal{C}$. Division of an integer with a fraction by a compound fraction.

$$26 + \frac{3}{5} \; : \; \frac{4}{7} + \frac{1}{2}\frac{1}{7}$$

(a) Making the divisor an integer:

$$\frac{26 + \frac{3}{5}}{\frac{4}{7} + \frac{1}{2}\frac{1}{7}} = \frac{\left(26 + \frac{3}{5}\right) \cdot 14}{\left(\frac{4}{7} + \frac{1}{2}\frac{1}{7}\right) \cdot 14} = \frac{364 + \frac{42}{5}}{9} = \frac{372 + \frac{2}{5}}{9}.$$

(**b**) Take the inverse of the divisor,

$$\frac{1}{\frac{4}{7}+\frac{1}{2}\frac{1}{7}} = \frac{14}{9} = 1 + \frac{5}{9}, \quad \text{then compute} \quad \left(1+\frac{5}{9}\right)\left(26+\frac{3}{5}\right).$$

D. *Dividend and divisor integers with a fraction*.

(**A.260**) MSS $\mathcal{A}, \mathcal{B}, \mathcal{C}$. Division of an integer with a fraction by an integer with a fraction.

$$12 + \frac{3}{4} \ : \ 1 + \frac{2}{7}$$

(**a**) Form the product of the denominators: $4 \cdot 7 = 28$. Then

$$\frac{12+\frac{3}{4}}{1+\frac{2}{7}} = \frac{\left(12+\frac{3}{4}\right)\cdot 28}{\left(1+\frac{2}{7}\right)\cdot 28}.$$

Remark. Here the reader of MS $\mathcal{C}$ observes that 28 is the *communis* and $\left(1+\frac{2}{7}\right)28$ the *prelatus*.

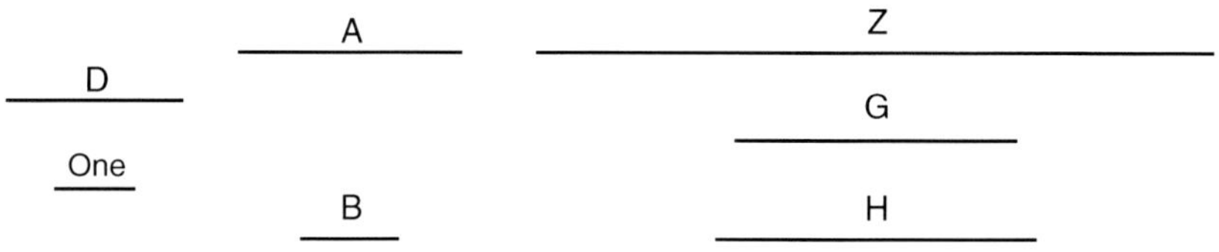

Fig. 35

Demonstration. Let (Fig. 35) $12 + \frac{3}{4} = A$, $1 + \frac{2}{7} = B$, $28 = G$, with $\frac{A}{B} = D$. Let further $A \cdot G = Z$, $B \cdot G = H$. To prove that

$$\frac{Z}{H} = D.$$

Since $A : B = D : 1$, thus $B : A = 1 : D$, while on the other hand (by *Elements* VII.18) $B : A = H : Z$, then $1 : D = H : Z$ and therefore $1 \cdot Z = Z = D \cdot H$. We have thus proved that

$$\frac{A}{B} = D = \frac{Z}{H} = \frac{A \cdot G}{B \cdot G}, \quad \text{that is, that} \quad \frac{12+\frac{3}{4}}{1+\frac{2}{7}} = \frac{\left(12+\frac{3}{4}\right)\cdot 28}{\left(1+\frac{2}{7}\right)\cdot 28}.$$

Remark. Since $A \cdot G = Z$ and $B \cdot G = H$, the result is immediately inferred by *Elements* VII.18. But we are by now used to seeing such recourse to formal proofs, which in addition justify a procedure in principle already used long before (A.226).[†]

[†] 'in principle': in the disordered text, A.260 is the very first problem on division of fractions.

(b) Making the divisor an integer.

$$\frac{12+\frac{3}{4}}{1+\frac{2}{7}} = \frac{\left(12+\frac{3}{4}\right)\cdot 7}{\left(1+\frac{2}{7}\right)\cdot 7} = \frac{\left(12+\frac{3}{4}\right)\cdot 7}{9}.$$

As said in the text, the proof is the same as before (with $G = 28$ becoming $G = 7$).

(A.261) MSS $\mathcal{B}$, $\mathcal{C}$. Another example (not in MS $\mathcal{A}$; similar situation in A.244).

$$20+\frac{3}{4} \;:\; 2+\frac{1}{3}$$

(a) Form the product of the denominators: $4\cdot 3 = 12$. Then

$$\frac{20+\frac{3}{4}}{2+\frac{1}{3}} = \frac{\left(20+\frac{3}{4}\right)\cdot 12}{\left(2+\frac{1}{3}\right)\cdot 12} = \frac{249}{28}.$$

(b) Making the divisor an integer:

$$\frac{20+\frac{3}{4}}{2+\frac{1}{3}} = \frac{\left(20+\frac{3}{4}\right)\cdot 3}{\left(2+\frac{1}{3}\right)\cdot 3} = \frac{62+\frac{1}{4}}{7}.$$

(c) Take the inverse of the divisor,

$$\frac{1}{2+\frac{1}{3}} = \frac{3}{7}, \quad \text{then compute} \quad \frac{3}{7}\cdot\left(20+\frac{3}{4}\right).$$

(A.262) MSS $\mathcal{A}$, $\mathcal{B}$, $\mathcal{C}$. Division of an integer with a fraction by an integer with a compound fraction.

$$17+\frac{10}{11} \;:\; 3+\frac{7}{8}+\frac{1}{2}\frac{1}{8}$$

(a) Form the product of the denominators: $11\cdot 2\cdot 8 = 176$. Then

$$\frac{17+\frac{10}{11}}{3+\frac{7}{8}+\frac{1}{2}\frac{1}{8}} = \frac{\left(17+\frac{10}{11}\right)\cdot 176}{\left(3+\frac{7}{8}+\frac{1}{2}\frac{1}{8}\right)\cdot 176} = \frac{3152}{693}.$$

(b) Making the divisor an integer:

$$\frac{17+\frac{10}{11}}{3+\frac{7}{8}+\frac{1}{2}\frac{1}{8}} = \frac{\left(17+\frac{10}{11}\right)\cdot 16}{\left(3+\frac{7}{8}+\frac{1}{2}\frac{1}{8}\right)\cdot 16} = \frac{\left(17+\frac{10}{11}\right)\cdot 16}{63}.$$

(c) Take the inverse of the divisor,

$$\frac{1}{3+\frac{7}{8}+\frac{1}{2}\frac{1}{8}} = \frac{16}{63} = \frac{14+2}{63} = \frac{2}{9}+\frac{2}{7}\frac{1}{9},$$

and multiply it by the dividend.

(A.263) MSS $\mathcal{A}, \mathcal{B}, \mathcal{C}$. Division of an integer with a fraction by an integer with a fraction of a fraction.
$$25 + \frac{4}{5} \; : \; 4 + \frac{2\;1}{3\;7}$$

(a) Form the product of the denominators, $5 \cdot 3 \cdot 7 = 105$, and operate as before.

(b) Make the divisor an integer by means of the multiplier $3 \cdot 7 = 21$.

(c) Take the inverse of the divisor,
$$\frac{1}{4 + \frac{2\;1}{3\;7}} = \frac{21}{86} = \frac{10}{43} + \frac{1}{2}\frac{1}{43},$$
to be multiplied by the dividend.

---

6. Other problems.

A. *An indeterminate problem*.

(A.264) MSS $\mathcal{A}$ (fragmentary), $\mathcal{B}$. Represent a given integer as the product of two factors, each an integer with a fraction.
$$\left(a + \frac{k}{l}\right)\left(b + \frac{r}{s}\right) = 40$$

This is just the division of an integer by an integer and a fraction. For with the problem being indeterminate, we may choose arbitrarily one of the factors and then divide 40 by it. This is what is done in the three examples presented, with the way of solving being repeated at the end.

$$\text{Taking} \quad 4 + \frac{3}{8} \quad \text{gives} \quad \frac{40}{4 + \frac{3}{8}} = \frac{40 \cdot 8}{35} = \frac{64}{7} = 9 + \frac{1}{7}.$$

$$\text{Taking} \quad 5 + \frac{3}{5} \quad \text{gives} \quad \frac{40}{5 + \frac{3}{5}} = \frac{40 \cdot 5}{28} = \frac{50}{7} = 7 + \frac{1}{7}.$$

$$\text{Taking} \quad 3 + \frac{3}{4} \quad \text{gives} \quad \frac{40}{3 + \frac{3}{4}} = \frac{40 \cdot 4}{15} = \frac{32}{3} = 10 + \frac{2}{3}.$$

(A.264′) Same, but this time one of the factors is supposed to be a proper fraction (which shortens the computations).

$$\text{Taking} \quad \frac{3}{4} \quad \text{gives} \quad \frac{40}{\frac{3}{4}} = \frac{160}{3} = 53 + \frac{1}{3}.$$

B. *Recapitulation: Arithmetical operations with fractions*.

This (MSS $\mathcal{A}$ and $\mathcal{B}$ only) summarizes, and extends to the case of several fractions, what we have seen for clearing away all fractions in the two given expressions. Let these expressions be

$$A = a + \frac{k_1}{l_1} + \frac{k_2}{l_2} + \ldots + \frac{k_n}{l_n}, \qquad B = b + \frac{r_1}{s_1} + \frac{r_2}{s_2} + \ldots + \frac{r_m}{s_m}.$$

— First, form the product of the denominators for each side, namely

$$L = l_1 \cdot l_2 \cdot \ldots \cdot l_n, \qquad S = s_1 \cdot s_2 \cdot \ldots \cdot s_m,$$

which are called the (respective) 'quantities' (*summe**). It is further specified that if there is just one fraction on one side, with denominator, say, l_1, then we shall simply take $L = l_1$.†

— Secondly, multiply these two quantities, thus producing

$$D = L \cdot S = l_1 \cdot l_2 \cdot \ldots \cdot l_n \cdot s_1 \cdot s_2 \cdot \ldots \cdot s_m,$$

which is the 'principal' (*prelatus*‖).

— Thirdly, form each 'intermediate result' (*servatus*, 'kept (in mind)'; Arabic *maḥfūẓ*).‡

   (*i*) For the multiplication, they are the (integral) products $A \cdot L$ and $B \cdot S$.

   (*ii*) For the other three operations, they are the (also integral) products $A \cdot D$ and $B \cdot D$.

— Finally, calculate the required quantity:

- In the case of a multiplication, consider

$$A \cdot B = \frac{(A \cdot L)(B \cdot S)}{D},$$

  that is, divide the product of the intermediate results by the principal.

- In the case of a division, take

$$\frac{A}{B} = \frac{A \cdot D}{B \cdot D},$$

  that is, form the quotient of the intermediate results.

- In the case of an addition or a subtraction, consider

$$A \pm B = \frac{A \cdot D \pm B \cdot D}{D},$$

  that is, divide the sum or difference of the intermediate results by the principal.

---

   * What the text called 'denominating numbers' (*numeri denominationum*) earlier.

   † If there is no fraction on one side, which thus consists of an integer only, we shall take $L = 1$; an early reader notes this case further on in the text.

   ‖ What the text called 'common number' (*numerus communis*) earlier.

   ‡ See Translation, p. 613, note 147.

Remark. In these rules, the *prelatus* is D, thus the product of all denominators, as is the *communis*. But otherwise *prelatus* is used to designate the divisor in the final expression, thus D for multiplication, addition and subtraction but $B \cdot D$ for division (Translation, p. 731, note 610, & p. 735, note 633). These rules may therefore —like others (see our remark above, p. 1248)— have been added later.

For the calculations, a sort of table may be used, which according to the text would be as shown below. We indeed meet in the text with instances of such tables, but only for multiplication (A.90–139) and addition (A.145–153); complete ones as those below are, however, exceptional (A.124a, A.130a).

- **Multiplication**

given expressions	A		B
quantities	L		S
principal		$D = L \cdot S$	
intermediate results	$A \cdot L$		$B \cdot S$
final result		$\dfrac{(A \cdot L)(B \cdot S)}{D}$	

- **Division**

given expressions	A		B
quantities	L		S
principal		$D = L \cdot S$	
intermediate results	$A \cdot D$		$B \cdot D$
final result		$\dfrac{A \cdot D}{B \cdot D}$	

- **Addition and subtraction**

given expressions	A		B
quantities	L		S
principal		$D = L \cdot S$	
intermediate results	$A \cdot D$		$B \cdot D$
final result		$\dfrac{A \cdot D \pm B \cdot D}{D}$	

C. *Combined expressions*.

The next two problems, involving combined expressions, recall A.143–144 in the multiplication, where each of the terms was an intricate expression which had to be computed first.

(A.265) MSS $\mathcal{A}$, $\mathcal{B}$. Calculate

$$10 \;:\; \left(\frac{7}{8}\left(5+\frac{1}{3}\right) - \frac{1}{3}\left(2+\frac{1}{4}\right)\right)$$

Consider the divisor. Since

$$\frac{7}{8}\left(5+\frac{1}{3}\right) = \frac{35}{8} + \frac{7}{24} = 4 + \frac{16}{24} = 4 + \frac{2}{3} \quad \text{while} \quad \frac{1}{3}\left(2+\frac{1}{4}\right) = \frac{3}{4},$$

the divisor becomes

$$4 + \frac{2}{3} - \frac{3}{4} = 3 + \frac{5}{3} - \frac{3}{4} = 3 + \frac{10+1}{12} = 3 + \frac{5}{6} + \frac{1}{2}\frac{1}{6}.$$

So we are now to calculate

$$\frac{10}{3 + \frac{5}{6} + \frac{1}{2}\frac{1}{6}},$$

and this will be treated as in A.256.

(A.266) MSS $\mathcal{A}$, $\mathcal{B}$. Calculate

$$\frac{\frac{3}{4} \cdot \frac{3}{5} \cdot 9}{\frac{3}{10}} \;:\; \left(\frac{2}{3} \cdot 7 - \frac{2+\frac{1}{8}}{\frac{7}{8} + \frac{1}{2}\frac{1}{8}}\right)$$

• Consider first the dividend. Forming the product of the denominators for its upper part, $4 \cdot 5 = 20$, we shall have

$$\frac{3}{4}\frac{3}{5} \cdot 9 = \frac{\left(\frac{3}{4} \cdot \frac{3}{5} \cdot 20\right) 9}{20} = \frac{9 \cdot 9}{20} = \frac{81}{20},$$

while, for its other part,

$$\frac{1}{\frac{3}{10}} = 3 + \frac{1}{3}.$$

So the dividend of the proposed expression becomes

$$\left(3+\frac{1}{3}\right)\frac{81}{20} = \frac{270}{20} = 13 + \frac{1}{2}.$$

• Second, consider the divisor. Its first (additive) term is

$$\frac{2}{3} \cdot 7 = 4 + \frac{2}{3}.$$

Next, take the inverse of the divisor in the subtractive term; with the product of the denominators $2 \cdot 8 = 16$, we find

$$\frac{1}{\frac{7}{8} + \frac{1}{2}\frac{1}{8}} = \frac{16}{15} = 1 + \frac{1}{3}\frac{1}{5},$$

so that the subtractive term will be

$$\frac{2+\frac{1}{8}}{\frac{7}{8}+\frac{1}{2}\frac{1}{8}} = \left(1+\frac{1}{3}\frac{1}{5}\right)\left(2+\frac{1}{8}\right) = \frac{16}{3\cdot 5}\cdot\frac{17}{8} = \frac{34}{3\cdot 5} = 2+\frac{1}{5}+\frac{1}{3}\frac{1}{5}.$$

The divisor of the proposed expression therefore becomes

$$\left(4+\frac{2}{3}\right) - \left(2+\frac{1}{5}+\frac{1}{3}\frac{1}{5}\right) = 2 + \frac{10}{15} - \frac{4}{15} = 2 + \frac{2}{5}.$$

- The result of the division is finally, with the multiplier $2 \cdot 5 = 10$,

$$\frac{13+\frac{1}{2}}{2+\frac{2}{5}} = \frac{135}{24} = \frac{45}{8} = 5 + \frac{5}{8}.$$

7. Sharing amounts of money.

The treatment of these problems is rather uneven: whereas the formulae used in A.268–270 are carefully described, those used for solving A.271–273, which involve second-degree equations and roots, are just applied without any explanation. This reminds us of the group A.184–186.

A. *Known amounts*.

(A.267) MSS $\mathcal{A}$, $\mathcal{B}$. Compute the final share after a series of divisions.

$$\left\{\left[(100:3):7\right]:4\right\}:5$$

Form the product of the denominators $3 \cdot 7 \cdot 4 \cdot 5 = 420$. So

$$\left\{\left[\left(\frac{100}{3}\right)\frac{1}{7}\right]\frac{1}{4}\right\}\frac{1}{5} = \frac{100}{420} = \frac{5}{21} = \frac{1}{7} + \frac{2}{3}\frac{1}{7}.$$

(A.268) MSS $\mathcal{A}$, $\mathcal{B}$. Share ninety nummi among nine men in such a way that, for the parts, $s_i - s_{i-1} = 1$.

Since, as seen in our summary (case i),

$$S = n \cdot s_1 + d \cdot \frac{(n-1)\,n}{2},$$

we shall have, for $S = 90$, $n = 9$, $d = 1$, the formula applied in the text

$$s_1 = \frac{1}{n} \cdot \left[S - d \cdot \frac{(n-1)\,n}{2}\right] = \frac{1}{9} \cdot \left[90 - \frac{8}{2} \cdot 9\right] = \frac{90-36}{9} = \frac{54}{9} = 6,$$

whence $s_i = 6 + (i-1)$.

Remark. We might have expected, for the summation formula, a reference to A.170.

(A.269) MSS $\mathcal{A}$, $\mathcal{B}$. Share a hundred nummi among eight men in such a way that, for the parts, $s_2 = s_1 + k$, with $k = 3$ and, for $i > 2$, $s_i - s_{i-1} = d$ with $d = 2$.

Since, as seen in our summary (case ii),

$$S = n \cdot s_1 + k \cdot (n-1) + d \cdot \frac{(n-2)(n-1)}{2},$$

we shall have

$$s_1 = \frac{1}{n} \cdot \left[S - k \cdot (n-1) - d \cdot \frac{(n-2)(n-1)}{2} \right].$$

As computed in the problem (and explained by the corresponding rule in the text), we find (with $S = 100$, $n = 8$, $k = 3$, $d = 2$)

$$s_1 = \frac{S - [d \cdot (n-2) + k \cdot 2] \frac{n-1}{2}}{n} = \frac{100 - [2 \cdot 6 + 3 \cdot 2] \cdot \frac{7}{2}}{8}$$

$$= \frac{100 - 18 \cdot \frac{7}{2}}{8} = \frac{100 - 63}{8} = \frac{37}{8} = 4 + \frac{5}{8}.$$

Remark. The particular case $k = d$ corresponds to a general arithmetical progression with common difference d. This is included in the rule stated (case ii in the text).

(A.270) MSS $\mathcal{A}$, $\mathcal{B}$. Share eighty nummi among five men in such a way that, for the parts, $s_i - s_{i-1} = d_{i-1}$ with $d_1 = 3$, $d_2 = 1$, $d_3 = 2$, $d_4 = 6$. As explained in our summary (case iii), and in the text,

$$S = n \cdot s_1 + d_1 + (d_1 + d_2) + (d_1 + d_2 + d_3) + \ldots + (d_1 + d_2 + \ldots + d_{n-1}),$$

so we have, for $S = 80$, $n = 5$ and the prescribed differences,

$$s_1 = \frac{1}{n} \left(S - \left[d_1 + (d_1 + d_2) + (d_1 + d_2 + d_3) + (d_1 + d_2 + d_3 + d_4) \right] \right)$$

$$= \frac{80 - [3 + (3+1) + (3+1+2) + (3+1+2+6)]}{5}$$

$$= \frac{80 - [3 + 4 + 6 + 12]}{5} = \frac{80 - 25}{5} = \frac{55}{5} = 11.$$

(A.271) MSS $\mathcal{A}$, $\mathcal{B}$. Share eighteen nummi equally among an unknown number of men given that the number of men plus the share is nine.

Since $S = 18$ and the number of men plus the individual share makes 9, we shall have for n men

$$n + \frac{S}{n} = n + \frac{18}{n} = 9, \quad \text{thus} \quad n^2 + 18 = 9n,$$

from which we infer the formula used (but not explained) in the text

$$n = \frac{9}{2} + \sqrt{\left(\frac{9}{2}\right)^2 - 18} = 4 + \frac{1}{2} + \sqrt{20 + \frac{1}{4} - 18}$$

$$= 4 + \frac{1}{2} + \sqrt{2 + \frac{1}{4}} = 4 + \frac{1}{2} + 1 + \frac{1}{2} = 6.$$

Remark. We thus have six men with three nummi each. But the second positive solution of this equation, which exchanges the values, is also acceptable.

(A.272) MSS $\mathcal{A}$, $\mathcal{B}$. Share forty nummi among an unknown number of men given that the number of men minus the share is three.

Since $S = 40$ and the number of men minus the individual share leaves 3, we shall have for n men

$$n - \frac{S}{n} = n - \frac{40}{n} = 3, \quad \text{thus} \quad n^2 = 3\,n + 40,$$

from which we infer the formula used in the text

$$n = \frac{3}{2} + \sqrt{\left(\frac{3}{2}\right)^2 + 40} = 1 + \frac{1}{2} + \sqrt{2 + \frac{1}{4} + 40}$$
$$= 1 + \frac{1}{2} + \sqrt{42 + \frac{1}{4}} = 1 + \frac{1}{2} + 6 + \frac{1}{2} = 8.$$

B. *Unknown amounts*.

(A.273) MSS $\mathcal{A}$, $\mathcal{B}$. Share S nummi among n men, where S and n are unknown, then among $n+2$ men; the second share is then the root of the first.

We have then

$$\frac{S}{n+2} = \sqrt{\frac{S}{n}}, \quad \text{thus} \quad \frac{S^2}{(n+2)^2} = \frac{S}{n}.$$

Since this problem is indeterminate, we may choose n. Taking $n = 4$ (a square) gives

$$S = \frac{(n+2)^2}{n} = \frac{36}{4} = 9.$$

(A.273') Same, but the second share is (for instance) three times the root of the first. The problem is then of the form

$$\frac{S}{n+2} = k \cdot \sqrt{\frac{S}{n}},$$

whence

$$S = k^2 \cdot \frac{(n+2)^2}{n},$$

as explained in the text.

(A.274) MSS $\mathcal{A}$, $\mathcal{B}$. Find a number N which, when divided by 2, 3, 4, 5, 6 leaves 1 but when divided by 7 leaves no remainder.

The (concise) reasoning of the text corresponds to what follows. Since $60 = 3 \cdot 4 \cdot 5$ is a common (and the least common) multiple of 2, 3, 4, 5, 6, so 60 will leave the remainder 0 when divided by these numbers, and therefore 61 the remainder 1. The same holds, respectively, for all integers of the form $k \cdot 60$ and $k \cdot 60 + 1$, k natural, so that the quantities belonging to the latter set will satisfy the first condition of the problem.

Now $k \cdot 60 + 1 = (k-1)\,60 + 61$, and, when divided by 7 , 61 leaves the remainder 5 and (therefore) 60 the remainder 4. We shall thus seek a number such that $4(k-1) + 5$, or $4 \cdot k + 1$, is divisible by 7. The first possibility is seen to be $k = 5$, and any value $k = 5 + 7 \cdot t$, t natural, will give another solution, so that the general solution is

$$N_k = 60 \cdot k + 1 \quad \text{with} \quad k = 5 + 7 \cdot t,$$

that is,

$$N_t = 301 + 420 \cdot t \quad (t \text{ integer} \geq 0),$$

where 420 is the least common multiple of 2, 3, 4, 5, 6, 7. Thus, as asserted in the text, one solution is $N_0 = 301$ (for $t = 0$, $k = 6$) and another ($t = 1$, $k = 12$) $N_1 = 301 + 420 = 721$.

Remark. The reasoning in the text is in keeping with those seen for the criteria of divisibility in A–IV (above, p. 1189). We do not need to suppose any knowledge in our text of solving generally the indeterminate equation of the first degree $ax - by = c$ in integers.

Chapter A-IX:
Roots
Summary

The notion of 'root' has occurred already before, namely in A.167, A.184–186, A.214, A.271–273. But what we shall learn here, after methods for approximating square roots, is how to apply the four arithmetical operations to both square and fourth roots and how to determine the root of a binomial (with at least one term a quadratic surd). The author does not consider operations with third roots, for we remain in the domain of quadratic (and reducible biquadratic) equations, which is also that of Euclid's ruler-and-compass geometry.°

The chapter on roots may be considered as being the most difficult in Book A. As our author implies, this chapter requires a thorough knowledge of Book X of Euclid's *Elements*, generally considered as the least accessible. Indeed, there are frequent references, in the proofs as well as in the computations, to Book X. Some knowledge of Abū Kāmil's *Algebra* appears to be assumed as well, more specifically its Book I, which contains the study of roots to be applied later to problems in Books III–IV. As is seen from his introduction to this chapter, our author has some reservations about Abū Kāmil's treatment of the subject, said to be somewhat unclear. We shall deal with that at the end of our summary. Note finally that for this chapter we must rely on MS $\mathcal{A}$ only, since $\mathcal{B}$ contains only the beginning of the first section.

1. Approximation of square roots.

No exact method for extracting roots is taught; it is merely mentioned that extracting the roots of integral squares is an easy matter —a somewhat puzzling remark since we would have expected either a description of the algorithm for extracting square roots, as is found in the *Liber algorismi**, or reference to a set of tables, just as elsewhere (B.143′) the reader is referred to a treatise on mensuration for the various formulae in area calculations. For without these aids the procedures taught for successive approximations are of no use: their application presupposes determination of the closest square integer on either side of the given (integral or fractional) quantity.

A. *Approximate roots of non-square integers*.

Suppose first we are to find an approximate value of $\sqrt{N}$, where N is a non-square integer.

° We shall refer throughout to 'fourth roots'; but in the text they are 'roots of a root'.

* See Boncompagni's edition, pp. 75–77, 78–81, with examples.

(i) Consider $a^2 < N < (a+1)^2$, with a integral, so that a and $a+1$ are the integers closest to $\sqrt{N}$; then:
- If N is closer to $(a+1)^2$ than to a^2, form

$$\sqrt{N} \cong (a+1) - \frac{(a+1)^2 - N}{2(a+1)}.$$

See A.277, A.281b. It is further used in B.65 and B.261.
- If N is closer to a^2 than to $(a+1)^2$, form

$$\sqrt{N} \cong a + \frac{N - a^2}{2a}.$$

See A.275, A.281–283, B.350, B.359 (but see the remark there, p. 1694 below).†

Remark. Both types of approximation reduce to the method commonly used in antiquity, namely (with $a^2 < N$ or $> N$)

$$\sqrt{N} \cong \frac{1}{2}\left(a + \frac{N}{a}\right),$$

which is found in Mesopotamian texts and in Heron's *Metrica*. But the distinction made in the text for their use is appropriate: if N is of the form $(a+1)^2 - 1 = a^2 + 2a$, the second (and more common) rule will give $\sqrt{N} \cong a+1$; thus, in B.261 ($N = 48$), the author applies the first rule.

(ii) We may, instead of considering N itself, take any square integer t^2 and calculate, by one of the above methods,

$$\sqrt{N} = \frac{1}{t} \cdot \sqrt{N \cdot t^2}.$$

We can arbitrarily increase precision since we now approximate the root of a larger integer (the larger the better, as stated in the text). This is used in A.276, A.281b, A.282b.

B. *Approximate roots of fractional expressions*.

This second way is also used by the author to approximate roots of fractional expressions by changing a fractional radicand into an integer. For if we are to calculate the root of, say $a + \frac{p}{q}$, we shall multiply it by q^2 (or any square which is an integral multiple of it); we thus obtain

$$\sqrt{a + \frac{p}{q}} = \frac{1}{q} \cdot \sqrt{\left(a + \frac{p}{q}\right)q^2} = \frac{1}{q} \cdot \sqrt{a \cdot q^2 + p \cdot q},$$

which leaves us with approximating the root of an integer in the manner explained above. See A.278–283.

---

† The same rule is found in the *Liber algorismi*, pp. 77–78.

2. Operations with square roots.

A. *Multiplication of square roots*.

The fundamental rule, demonstrated in A.284, is

$$\sqrt{a} \cdot \sqrt{b} = \sqrt{a \cdot b}.$$

Evidently, the result will be rational or not according to whether the product $a \cdot b$ is a square or not, thus also (dividing this product by b^2) to whether the amount of the ratio $a : b$ is a square or not. Hence, as the text states and demonstrates (A.288), if $a : b = \alpha^2 : \beta^2$, thus if the ratio of the given numbers a and b is like the ratio of two squares, the result of the multiplication will be rational. This is to play a central rôle in the addition of roots.

For the multiplication of square roots by rational numbers we shall consider, as is done in A.285–287, the particular case of the above rule:

$$a \cdot \sqrt{b} = \sqrt{a^2} \cdot \sqrt{b} = \sqrt{a^2 \cdot b}.$$

B. *Addition of square roots*.

The fundamental relation, demonstrated in A.289, is

$$\sqrt{a} + \sqrt{b} = \sqrt{(\sqrt{a} + \sqrt{b})^2} = \sqrt{a + b + 2 \cdot \sqrt{a} \cdot \sqrt{b}}.$$

Two possible cases arise, for the resulting expression will take either the form $\sqrt{A}$ or the form $\sqrt{A + \sqrt{B}}$. In the first case, we shall say that the two roots can be added, in the second not since the resulting expression does not bring any simplification. To determine which case it is, we shall consider the ratio $a : b$.

(*i*) If $a : b = \alpha^2 : \beta^2$, then $\sqrt{a} + \sqrt{b} = \sqrt{c}$ (c rational), that is, the sum will be the square root of one number (A.289). Indeed, we have $\sqrt{a} : \sqrt{b} = \alpha : \beta = t$, thus $\sqrt{a} = t \cdot \sqrt{b}$, and consequently

$$\sqrt{a} + \sqrt{b} = (t+1)\sqrt{b} = \sqrt{(t+1)^2 b}, \quad \text{or else}$$

$$\sqrt{a} + \sqrt{b} = \sqrt{a + b + 2 \cdot \sqrt{a} \cdot \sqrt{b}} = \sqrt{a + b + 2 \cdot t \cdot b},$$

of the form $\sqrt{A}$.

(*ii*) If $a : b = \alpha : \beta$, then the roots of a and b cannot be conveniently added (A.290).[‖] For if $a = t \cdot b$, then $\sqrt{a} = \sqrt{t} \cdot \sqrt{b}$, and $2 \cdot \sqrt{a} \cdot \sqrt{b} = 2 \cdot \sqrt{t} \cdot b$, so that

$$\sqrt{a} + \sqrt{b} = \sqrt{a + b + 2 \cdot \sqrt{a} \cdot \sqrt{b}} = \sqrt{a + b + 2 \cdot \sqrt{t \cdot b^2}},$$

[‖] We would say today that *i* & *ii* express a single, necessary and sufficient condition.

of the form $\sqrt{A} + \sqrt{B}$. In this case, we shall just leave the given expression unchanged.

(i') As the author mentions, the first thing to consider is the product $a \cdot b$. For if $a \cdot b$ is a square, then $a : b = \alpha^2 : \beta^2$. This will serve as a criterion to decide if addition is (in the above sense) possible or not.

C. *Subtraction of square roots*.

The fundamental relation, demonstrated in A.291, is

$$\sqrt{a} - \sqrt{b} = \sqrt{a + b - 2 \cdot \sqrt{a \cdot b}}.$$

As with addition, the form of the result depends on whether $a : b = \alpha^2 : \beta^2$ or not, that is, whether $a \cdot b$ is a square or not. The author gives one example of each case (A.291–292).

D. *Division of square roots*.

(1) One term in the divisor (A.293–296).

The fundamental relation, mentioned in A.293, is

$$\frac{\sqrt{a}}{\sqrt{b}} = \sqrt{\frac{a}{b}}.$$

As for multiplication, the case of one term rational is also considered. For a sum of terms in the dividend, we shall divide each separately by the divisor (A.297).

(2) Two terms in the divisor (A.298–302).

By 'binomial', Euclid means the sum of two incommensurable terms, here algebraically of the form $a + \sqrt{b}$ ($a > \sqrt{b}$) or $\sqrt{b} + a$ ($\sqrt{b} > a$) where a and b are rational and b is not a square, or $\sqrt{a} + \sqrt{b}$ with $\sqrt{a}$ and $\sqrt{b}$ incommensurable (*Elements* X.36); in other words, the binomial is not reducible to just one term. To each binomial corresponds what is called an 'apotome' (the Greek ἀποτομή) of the respective forms $a - \sqrt{b}$, $\sqrt{b} - a$ or $\sqrt{a} - \sqrt{b}$ (here $a > b$). A fundamental theorem is that multiplying a binomial by its apotome gives rational (A.298). This is the key to eliminating the irrational part from the divisor. Thus, with c rational or a square root,

$$\frac{c}{\sqrt{a} \pm \sqrt{b}} = \frac{c\left(\sqrt{a} \mp \sqrt{b}\right)}{a - b},$$

$$\frac{c}{a \pm \sqrt{b}} = \frac{c\left(a \mp \sqrt{b}\right)}{a^2 - b}, \qquad \frac{c}{\sqrt{a} \pm b} = \frac{c\left(\sqrt{a} \mp b\right)}{a - b^2}.$$

A.298–300 have the three kinds of binomial in the divisor ($a + \sqrt{b}$, $\sqrt{a} + b$, $\sqrt{a} + \sqrt{b}$), while A.301–302 involve apotomes ($a - \sqrt{b}$, $\sqrt{a} - \sqrt{b}$).

(3) Three terms in the divisor (A.320–322).

Suppose we have a trinomial of the form $\sqrt{a}+\sqrt{b}+\sqrt{c}$ in the divisor. By the fundamental relation

$$(\sqrt{a}+\sqrt{b}+\sqrt{c})(\sqrt{a}+\sqrt{b}-\sqrt{c}) = (\sqrt{a}+\sqrt{b})^2 - c$$
$$= (a+b-c) + \sqrt{4ab},$$

we shall have just one root left in the divisor, and there are three possibilities (A.320):
— if $c < a+b$, the expression takes the form $\sqrt{e}+d$;
— if $c > a+b$, the expression takes the form $\sqrt{e}-d$;
— if $c = a+b$, the expression takes the form $\sqrt{e}$.
Whichever the case, we shall find ourselves in the situation (1) or (2) above.

3. Operations with fourth roots.

A. *Multiplication of fourth roots*.

The fundamental rule, demonstrated in A.303, is

$$\sqrt[4]{a} \cdot \sqrt[4]{b} = \sqrt[4]{a \cdot b},$$

with a and b rational. Three cases are possible, for the result may be a rational number, a square root or a fourth root.

The criterion for determining the case is much the same as for the multiplication of square roots, except that there we considered the ratio $a:b$ whereas here it will be $\sqrt{a}:\sqrt{b}$, with, as before, a and b supposed to be rational non-square numbers. Now the amount of this ratio may be a square root, a rational number or a square, whence the three subsequent possibilities.

(i) Evidently, for multiplication of the two fourth roots to produce a rational number, the product $a \cdot b$ must be a fourth power. This will be the case if $\sqrt{a} : \sqrt{b} = \alpha : \beta = t$ and $t \cdot b = r^2$. For then $\sqrt{a} = t \cdot \sqrt{b}$, thus $\sqrt[4]{a} = \sqrt{t} \cdot \sqrt[4]{b}$, and

$$\sqrt[4]{a} \cdot \sqrt[4]{b} = \sqrt{t} \cdot \sqrt{b} = \sqrt{t \cdot b} = r.$$

See example in A.307.

(ii) Clearly, for multiplication of the two fourth roots to produce a square root, the product $a \cdot b$ must be a square. This will be the case in two situations:

• (ii') either, as before, $\sqrt{a} : \sqrt{b} = \alpha : \beta = t$ but this time $t \cdot b$ is not a square. See A.308 and the demonstration there.

• (ii'') or $\sqrt{a} : \sqrt{b} = \alpha^2 : \beta^2 = t^2$; for then $\sqrt{a} = t^2 \cdot \sqrt{b}$, thus $\sqrt[4]{a} = t \cdot \sqrt[4]{b}$, and

$$\sqrt[4]{a} \cdot \sqrt[4]{b} = t \cdot \sqrt{b} = \sqrt{t^2 \cdot b}.$$

See A.310 (and preceding demonstration).

(*iii*) Multiplication of the two fourth roots will produce a fourth root if the product $a \cdot b$ is just a rational (non-square) number. This will be the case if $\sqrt{a} : \sqrt{b} = \sqrt{t}$. For then $\sqrt{a} = \sqrt{t} \cdot \sqrt{b}$, thus $\sqrt[4]{a} = \sqrt[4]{t} \cdot \sqrt[4]{b}$, and

$$\sqrt[4]{a} \cdot \sqrt[4]{b} = \sqrt[4]{t} \cdot \sqrt[4]{b} = \sqrt[4]{t \cdot b^2}.$$

See A.309 (and A.303).

Before explaining these more theoretical distinctions, the text treats a few examples, first of the general case (application and, as said, demonstration of the fundamental rule, A.303), then with products of the type $\sqrt{a} \cdot \sqrt[4]{b}$ (A.304), $a \cdot \sqrt[4]{b}$ (A.305), $k\sqrt[4]{a} \cdot l\sqrt[4]{b}$ (A.306); they are just analogous to the problems already seen for the multiplication of square roots.

B. *Addition of fourth roots*.

For addition and subtraction, the fundamental relation, demonstrated in A.311, is

$$\sqrt[4]{a} \pm \sqrt[4]{b} = \sqrt{\left(\sqrt[4]{a} \pm \sqrt[4]{b}\right)^2} = \sqrt{\sqrt{a} + \sqrt{b} \pm 2 \cdot \sqrt[4]{a} \cdot \sqrt[4]{b}}.$$

According to what has been seen in the multiplication the author is led to different forms the result may take. These various possibilities, in order of increasing complexity in the expression obtained, are:

(*i*) $\sqrt[4]{a} \pm \sqrt[4]{b} = \sqrt[4]{A}$.
(*ii*) $\sqrt[4]{a} \pm \sqrt[4]{b} = \sqrt{\sqrt{A} \pm B}$.
(*iii*) $\sqrt[4]{a} \pm \sqrt[4]{b} = \sqrt{\sqrt{A} \pm \sqrt{B}}$.
(*iv*) $\sqrt[4]{a} \pm \sqrt[4]{b} = \sqrt{\sqrt{A} \pm \sqrt[4]{B}}$.

Whereas the first three forms bring some kind of simplification, the last one does not. To determine which will occur, consider, as before, the quantity of the ratio of $\sqrt{a}$ to $\sqrt{b}$.

(*i*) If $\sqrt{a} : \sqrt{b} = \alpha^2 : \beta^2 = t^2$, thus (as in case ii'' of the multiplication) $\sqrt[4]{a} = t \cdot \sqrt[4]{b}$, we shall have

$$\sqrt[4]{a} \pm \sqrt[4]{b} = \sqrt{\sqrt{a} + \sqrt{b} \pm 2 \cdot \sqrt[4]{a} \cdot \sqrt[4]{b}} = \sqrt{(t^2 + 1) \cdot \sqrt{b} \pm 2 \cdot t \cdot \sqrt{b}}$$

and therefore

$$\sqrt[4]{a} \pm \sqrt[4]{b} = \sqrt{(t \pm 1)^2 \cdot \sqrt{b}} = \sqrt[4]{(t \pm 1)^4 \, b},$$

which thus reduces to the form $\sqrt[4]{A}$. See example in A.311 and preceding explanation.

(*ii*) If $\sqrt{a} : \sqrt{b} = \alpha : \beta = t$ with $\sqrt{a} \cdot \sqrt{b} = t \cdot b = r^2$ a square (as in case *i* of the multiplication), we shall have

$$\sqrt[4]{a} \pm \sqrt[4]{b} = \sqrt{\sqrt{a} + \sqrt{b} \pm 2 \cdot \sqrt[4]{a} \cdot \sqrt[4]{b}} = \sqrt{(t + 1) \cdot \sqrt{b} \pm 2 \cdot \sqrt{t \cdot b}}$$

and therefore
$$\sqrt[4]{a} \pm \sqrt[4]{b} = \sqrt{\sqrt{(t+1)^2 \cdot b} \pm 2 \cdot \sqrt{t \cdot b}} = \sqrt{\sqrt{(t+1)^2 \cdot b} \pm 2 \cdot r},$$
of the form $\sqrt{\sqrt{A} \pm B}$. See A.312 and preceding explanation.

(iii) If $\sqrt{a} : \sqrt{b} = \alpha : \beta = t$ but $\sqrt{a} \cdot \sqrt{b} = t \cdot b$ not a square (as in case ii' of the multiplication), we shall have
$$\sqrt[4]{a} \pm \sqrt[4]{b} = \sqrt{\sqrt{(t+1)^2 \cdot b} \pm 2 \cdot \sqrt{t \cdot b}} = \sqrt{\sqrt{(t+1)^2 \cdot b} \pm \sqrt{4t \cdot b}},$$
of the form $\sqrt{\sqrt{A} \pm \sqrt{B}}$. See example in A.313 and assertion immediately preceding.

(iv) If $a : b = \alpha : \beta = t$, thus $\sqrt{a} : \sqrt{b} = \sqrt{t}$ (as in case iii of the multiplication), we shall have
$$\sqrt[4]{a} \pm \sqrt[4]{b} = \sqrt{\sqrt{a} + \sqrt{b} \pm 2 \cdot \sqrt[4]{a} \cdot \sqrt[4]{b}} = \sqrt{(\sqrt{t}+1) \cdot \sqrt{b} \pm 2 \cdot \sqrt[4]{t} \cdot \sqrt{b}}$$
and therefore
$$\sqrt[4]{a} \pm \sqrt[4]{b} = \sqrt{\sqrt{(\sqrt{t}+1)^2 \cdot b} \pm \sqrt[4]{16 t \cdot b^2}},$$
with the resulting expression taking the inconvenient form $\sqrt{\sqrt{A} \pm \sqrt[4]{B}}$. Thus, the author remarks, better leave the sum in the form given.

C. *Subtraction of fourth roots*.

The fundamental relation is:
$$\sqrt[4]{a} - \sqrt[4]{b} = \sqrt{\sqrt{a} + \sqrt{b} - 2 \cdot \sqrt[4]{a} \cdot \sqrt[4]{b}}.$$

The conditions are exactly the same as for addition, the binomials being changed into apotomes. See above, negative sign.

D. *Division of fourth roots*.

(1) One term in the divisor.

The fundamental relation is:
$$\frac{\sqrt[4]{a}}{\sqrt[4]{b}} = \sqrt[4]{\frac{a}{b}}.$$

This is proved in the first example (A.314). A.315–316 involve rational quantities, either factors or divisor.

(2) Two terms in the divisor.

Since the case of two terms in the dividend is, as for square roots, reducible to the above situation, let us consider them to be in the divisor.

There are three possibilities: the divisor can be of the form $a + \sqrt[4]{b}$ (A.317), $\sqrt{a} + \sqrt[4]{b}$ (A.319 —difference, 'apotome'), $\sqrt[4]{a} + \sqrt[4]{b}$ (A.318). Proceeding as seen for square roots, we shall consider respectively, according to which term is the greater,

(i) $$\frac{c}{a \pm \sqrt[4]{b}} = \frac{c\left(a \mp \sqrt[4]{b}\right)}{\left(a + \sqrt[4]{b}\right)\left(a - \sqrt[4]{b}\right)} = \frac{ca \mp \sqrt[4]{c^4 b}}{a^2 - \sqrt{b}}$$

$$\frac{c}{\sqrt[4]{b} \pm a} = \frac{c\left(\sqrt[4]{b} \mp a\right)}{\left(\sqrt[4]{b} + a\right)\left(\sqrt[4]{b} - a\right)} = \frac{\sqrt[4]{c^4 b} \mp ca}{\sqrt{b} - a^2}$$

(ii) $$\frac{c}{\sqrt{a} \pm \sqrt[4]{b}} = \frac{c\left(\sqrt{a} \mp \sqrt[4]{b}\right)}{\left(\sqrt{a} + \sqrt[4]{b}\right)\left(\sqrt{a} - \sqrt[4]{b}\right)} = \frac{\sqrt{c^2 a} \mp \sqrt[4]{c^4 b}}{a - \sqrt{b}}$$

$$\frac{c}{\sqrt[4]{b} \pm \sqrt{a}} = \frac{c\left(\sqrt[4]{b} \mp \sqrt{a}\right)}{\left(\sqrt[4]{b} + \sqrt{a}\right)\left(\sqrt[4]{b} - \sqrt{a}\right)} = \frac{\sqrt[4]{c^4 b} \mp \sqrt{c^2 a}}{\sqrt{b} - a}$$

(iii) $$\frac{c}{\sqrt[4]{a} \pm \sqrt[4]{b}} = \frac{c\left(\sqrt[4]{a} \mp \sqrt[4]{b}\right)}{\left(\sqrt[4]{a} + \sqrt[4]{b}\right)\left(\sqrt[4]{a} - \sqrt[4]{b}\right)} = \frac{\sqrt[4]{c^4 a} \mp \sqrt[4]{c^4 b}}{\sqrt{a} - \sqrt{b}} \quad (a > b).$$

In A.317, the author proves (but just in one case, *iii* here above) that the expression resulting in the divisor is an apotome, thus a difference between two incommensurable terms. He then goes on to examine this same third case, which is the only one possibly admitting of simplification, namely if $\sqrt{a}$ and $\sqrt{b}$ are commensurable; for then, as seen in case *i* in the addition of square roots, $\sqrt{a} - \sqrt{b} = \sqrt{d}$, and so the given expression will be the sum or difference of two fourth roots:

$$\frac{c}{\sqrt[4]{a} \pm \sqrt[4]{b}} = \frac{\sqrt[4]{c^4 a} \mp \sqrt[4]{c^4 b}}{\sqrt{d}} = \sqrt[4]{\frac{c^4 a}{d^2}} \mp \sqrt[4]{\frac{c^4 b}{d^2}}.$$

4. Square root extraction of binomials and apotomes.

Suppose now we wish to extract the square root of the sum or difference of two incommensurable terms, which we shall represent here as $\sqrt{A} \pm \sqrt{B}$, with A, B rational (not both squares) and $\sqrt{A} > \sqrt{B}$. We must determine u and v $(u > v)$ such that

$$\sqrt{\sqrt{A} \pm \sqrt{B}} = \sqrt{u} \pm \sqrt{v}.$$

Then two questions arise: How do we calculate u and v and will the expression thus obtained be simpler than what we already have?

(1) <u>Calculation of u and v.</u>

Squaring the above expression, we obtain

$$\sqrt{A} \pm \sqrt{B} = u + v \pm 2 \cdot \sqrt{u \cdot v},$$

and, since $A > B$ and always $u + v > 2 \cdot \sqrt{u \cdot v}$ for $u \neq v$, u and v will be determined by solving
$$\begin{cases} u + v = \sqrt{A} \\ u \cdot v = \frac{1}{4} B. \end{cases}$$
Using the identity
$$\left(\frac{u-v}{2}\right)^2 = \left(\frac{u+v}{2}\right)^2 - u \cdot v = \frac{1}{4} A - \frac{1}{4} B$$
we find
$$\frac{u-v}{2} = \sqrt{\frac{1}{4} A - \frac{1}{4} B} = \frac{1}{2} \sqrt{A - B}$$
and we now know
$$u, v = \frac{u+v}{2} \pm \frac{u-v}{2} = \frac{1}{2} \sqrt{A} \pm \frac{1}{2} \sqrt{A - B},$$
namely
$$u = \frac{1}{2} \sqrt{A} + \frac{1}{2} \sqrt{A - B}, \qquad v = \frac{1}{2} \sqrt{A} - \frac{1}{2} \sqrt{A - B}.$$
Therefore
$$\sqrt{\sqrt{A} \pm \sqrt{B}} = \sqrt{u} \pm \sqrt{v} = \sqrt{\frac{\sqrt{A} + \sqrt{A - B}}{2}} \pm \sqrt{\frac{\sqrt{A} - \sqrt{A - B}}{2}},$$
which is the fundamental identity on which part of Book X of the *Elements* is based.

(2) Possible simplification.

We thus know that the root of a binomial or an apotome is representable as the sum or difference of two roots, in which the radicands themselves are a sum or a difference of two roots. It remains to be seen if, and under what conditions, the expression obtained will be simpler than the one proposed.

Consider thus the identity
$$\sqrt{\sqrt{A} \pm \sqrt{B}} = \sqrt{\frac{\sqrt{A} + \sqrt{A - B}}{2}} \pm \sqrt{\frac{\sqrt{A} - \sqrt{A - B}}{2}}$$
where, as said, $A > B$ while A and B are rational. We already know (by hypothesis) that $\sqrt{A}$ and $\sqrt{B}$ are not commensurable. But $\sqrt{A}$ and $\sqrt{A - B}$ might be, and this is the situation to consider first.

- Assume thus $\sqrt{A}$ and $\sqrt{A - B}$ to be commensurable. We have then $\sqrt{A - B} = t \sqrt{A}$, t rational positive fraction. Consequently,
$$\sqrt{\sqrt{A} \pm \sqrt{B}} = \sqrt{\frac{\sqrt{A} + \sqrt{A - B}}{2}} \pm \sqrt{\frac{\sqrt{A} - \sqrt{A - B}}{2}}$$

$$= \sqrt{\frac{\sqrt{A}+t\sqrt{A}}{2}} \pm \sqrt{\frac{\sqrt{A}-t\sqrt{A}}{2}} = \sqrt{\frac{(1+t)\sqrt{A}}{2}} \pm \sqrt{\frac{(1-t)\sqrt{A}}{2}}.$$

Thus, in the situation where $\sqrt{A}$ and $\sqrt{A-B}$ are commensurable, the general formula takes the form

$$\sqrt{\sqrt{A} \pm \sqrt{B}} = \sqrt[4]{\frac{(1+t)^2 A}{4}} \pm \sqrt[4]{\frac{(1-t)^2 A}{4}} \qquad (*)$$

whereby it appears that the root of the binomial will be equal to the sum of two fourth roots of rational quantities, and to their difference for an apotome. The final result is indeed simpler than the original expression.

- Consider next the situation where $\sqrt{A}$ and $\sqrt{A-B}$ are not commensurable. We have then $\sqrt{A-B} = \sqrt{t}\sqrt{A}$, t not a square. Consequently,

$$\sqrt{\sqrt{A} \pm \sqrt{B}} = \sqrt{\frac{\sqrt{A}+\sqrt{A-B}}{2}} \pm \sqrt{\frac{\sqrt{A}-\sqrt{A-B}}{2}}$$

$$= \sqrt{\frac{\sqrt{A}+\sqrt{t}\sqrt{A}}{2}} \pm \sqrt{\frac{\sqrt{A}-\sqrt{t}\sqrt{A}}{2}}.$$

Thus, in the situation in which $\sqrt{A}$ and $\sqrt{A-B}$ are not commensurable, the general formula takes the form

$$\sqrt{\sqrt{A} \pm \sqrt{B}} = \sqrt{\sqrt{\frac{A}{4}}+\sqrt{\frac{tA}{4}}} \pm \sqrt{\sqrt{\frac{A}{4}}-\sqrt{\frac{tA}{4}}}. \qquad (**)$$

Unlike before, there is no simplification since we now have two terms each taking the same form as the original expression.

By definition, in binomials and apotomes $\sqrt{A}\pm\sqrt{B}$ the quantities $\sqrt{A}$ and $\sqrt{B}$ are incommensurable; both terms may be square roots but one of the two terms may also be rational, thus A or B a square. Combining these three possibilities with the above two situations gives rise to six cases for the binomials and six for the apotomes. They are studied by Euclid in Book X of the *Elements* and listed in the *Liber mahameleth* at the beginning of A.323 for the binomials and A.324 for the apotomes. They are the following.

- Cases *i–iii*: $\sqrt{A}$ and $\sqrt{A-B}$ are commensurable.

(*i*) The given expression $\sqrt{A} \pm \sqrt{B}$ is a *first binomial* or a *first apotome* when, in addition to $\sqrt{A-B} = t\sqrt{A}$, $\sqrt{A}$ (not $\sqrt{B}$) is rational. Accordingly, put $a = \sqrt{A}$ and $b = B$. Since then $\sqrt{A-B} = \sqrt{a^2-b} = t \cdot a$, the above formula (*), namely

$$\sqrt{\sqrt{A} \pm \sqrt{B}} = \sqrt[4]{\frac{(1+t)^2 A}{4}} \pm \sqrt[4]{\frac{(1-t)^2 A}{4}},$$

becomes, expressed with the rational numbers a and b,

$$\sqrt{a \pm \sqrt{b}} = \sqrt{\frac{(1+t)\,a}{2}} \pm \sqrt{\frac{(1-t)\,a}{2}}.$$

Therefore, when A is a square, the required root can be expressed as the sum, or the difference, of two square roots, and Euclid indeed demonstrates that the root of the given expression, thus of a first binomial or a first apotome, is a binomial or an apotome (*Elements* X.54 & X.91). See A.323 (first binomial) and A.324 (first apotome).

Example.[†]

$$6 \pm \sqrt{20}, \quad \text{thus} \quad \sqrt{6 \pm \sqrt{20}} = \sqrt{\frac{6+4}{2}} \pm \sqrt{\frac{6-4}{2}} \ (=\sqrt{5} \pm 1).$$

(*ii*) The given expression $\sqrt{A} \pm \sqrt{B}$ is called a *second binomial* or a *second apotome* when, in addition to $\sqrt{A-B} = t\sqrt{A}$, $\sqrt{B}$ (not $\sqrt{A}$) is rational. Put thus $a = A$ and $b = \sqrt{B}$. Since then $\sqrt{A-B} = \sqrt{a-b^2} = t \cdot \sqrt{a}$, formula (∗), namely

$$\sqrt{\sqrt{A} \pm \sqrt{B}} = \sqrt[4]{\frac{(1+t)^2\,A}{4}} \pm \sqrt[4]{\frac{(1-t)^2\,A}{4}},$$

becomes, with a and b rational,

$$\sqrt{\sqrt{a} \pm b} = \sqrt[4]{\frac{(1+t)^2\,a}{4}} \pm \sqrt[4]{\frac{(1-t)^2\,a}{4}}.$$

The root of the given expression can then be expressed as the sum, or the difference, of two fourth roots, and Euclid indeed shows it to be what he calls a *first bimedial* or a *first apotome of a medial* (*Elements* X.55 & X.92). See A.325 (second binomial).

Example.

$$\sqrt{12} \pm 3, \quad \text{thus} \quad \sqrt{\sqrt{12} \pm 3} = \sqrt[4]{6 + \frac{3}{4}} \pm \sqrt[4]{\frac{3}{4}}.$$

(*iii*) The given expression $\sqrt{A} \pm \sqrt{B}$ is a *third binomial* or a *third apotome* when once again $\sqrt{A-B} = t\sqrt{A}$ but neither $\sqrt{A}$ nor $\sqrt{B}$ is rational; the root will then be a *second bimedial* or a *second apotome of a medial* (*Elements* X.56 & X.93). Putting this time simply $a = A$, $b = B$, thus $\sqrt{A-B} = \sqrt{a-b} = t \cdot \sqrt{a}$, formula (∗), namely

$$\sqrt{\sqrt{A} \pm \sqrt{B}} = \sqrt[4]{\frac{(1+t)^2\,A}{4}} \pm \sqrt[4]{\frac{(1-t)^2\,A}{4}},$$

[†] These numerical examples are the ones given by Muḥammad ibn ʿAbd al-Bāqī al-Baghdādī (ca 1100) in his Commentary on Book X (p. 239 of Suter's summary). See also *Anaritii commentarii*, pp. 342–344.

becomes
$$\sqrt{\sqrt{a}\pm\sqrt{b}} = \sqrt[4]{\frac{(1+t)^2\,a}{4}} \pm \sqrt[4]{\frac{(1-t)^2\,a}{4}}.$$

This is again a sum, or a difference, of two fourth roots.

Example.
$$\sqrt{8}\pm\sqrt{6}, \quad \text{thus} \quad \sqrt{\sqrt{8}\pm\sqrt{6}} = \sqrt[4]{4+\frac{1}{2}} \pm \sqrt[4]{\frac{1}{2}}.$$

- Case *iv–vi*: the roots $\sqrt{A}$ and $\sqrt{A-B}$ are incommensurable.

These three cases do not, as already said, admit of any simplification.

(*iv*) The given expression $\sqrt{A}\pm\sqrt{B}$ is a *fourth binomial* or a *fourth apotome* when $\sqrt{A}$ is rational, as in case *i*, but this time not commensurable with $\sqrt{A-B}$, thus $\sqrt{A-B} = \sqrt{t}\cdot\sqrt{A}$. Then, with $a = \sqrt{A}$, $b = B$, the above formula (**), namely

$$\sqrt{\sqrt{A}\pm\sqrt{B}} = \sqrt{\sqrt{\frac{A}{4}}+\sqrt{\frac{tA}{4}}} \pm \sqrt{\sqrt{\frac{A}{4}}-\sqrt{\frac{tA}{4}}},$$

becomes, with a and b rational,

$$\sqrt{a\pm\sqrt{b}} = \sqrt{\frac{a}{2}+\sqrt{\frac{t\cdot a^2}{4}}} \pm \sqrt{\frac{a}{2}-\sqrt{\frac{t\cdot a^2}{4}}},$$

and the root of the given expression is then what Euclid calls respectively a *major* and a *minor* straight line (*Elements* X.57 & X.94). See A.326 (fourth binomial).

Example.
$$6\pm\sqrt{12}, \quad \text{thus} \quad \sqrt{6\pm\sqrt{12}} = \sqrt{3+\sqrt{6}} \pm \sqrt{3-\sqrt{6}}.$$

(*v*) The given expression is a *fifth binomial* or a *fifth apotome* when, as in the second case, $\sqrt{B}$ is rational, but $\sqrt{A}$ is no longer commensurable with $\sqrt{A-B}$. Then, with $a = A$, $b = \sqrt{B}$, formula (**), that is

$$\sqrt{\sqrt{A}\pm\sqrt{B}} = \sqrt{\sqrt{\frac{A}{4}}+\sqrt{\frac{tA}{4}}} \pm \sqrt{\sqrt{\frac{A}{4}}-\sqrt{\frac{tA}{4}}},$$

becomes

$$\sqrt{\sqrt{a}\pm b} = \sqrt{\sqrt{\frac{a}{4}}+\sqrt{\frac{t\cdot a}{4}}} \pm \sqrt{\sqrt{\frac{a}{4}}-\sqrt{\frac{t\cdot a}{4}}}.$$

The root of the given expression is then a quantity respectively *producing* (when multiplied by itself) *a rational plus a medial* (*Elements* X.58) and *producing with a rational a medial* (*Elements* X.95); indeed, considering the given expression itself, $\sqrt{a}+b$ is, in Euclidean terminology, the sum of a rational and a medial while $\sqrt{a}-b$ is a medial minus a rational.

Example.
$$\sqrt{12} \pm 2, \quad \text{thus} \quad \sqrt{\sqrt{12} \pm 2} = \sqrt{\sqrt{3} + \sqrt{2}} \pm \sqrt{\sqrt{3} - \sqrt{2}}.$$

(*vi*) We finally have a *sixth binomial* and a *sixth apotome* when $\sqrt{A-B}$ is not commensurable with $\sqrt{A}$ since $\sqrt{A-B} = \sqrt{t} \cdot \sqrt{A}$, and neither $\sqrt{A}$ nor $\sqrt{B}$ is rational. Putting this time $a = A$, $b = B$, the formula

$$\sqrt{\sqrt{A} \pm \sqrt{B}} = \sqrt{\sqrt{\frac{A}{4}} + \sqrt{\frac{tA}{4}}} \pm \sqrt{\sqrt{\frac{A}{4}} - \sqrt{\frac{tA}{4}}},$$

becomes

$$\sqrt{\sqrt{a} \pm \sqrt{b}} = \sqrt{\sqrt{\frac{a}{4}} + \sqrt{\frac{t \cdot a}{4}}} \pm \sqrt{\sqrt{\frac{a}{4}} - \sqrt{\frac{t \cdot a}{4}}}.$$

The proposed root is then, as seen in the left-hand expression, simply a quantity *producing two medials* or *producing with a medial a medial* (*Elements* X.59 & X.96).

Example.
$$\sqrt{20} \pm \sqrt{8}, \quad \text{thus} \quad \sqrt{\sqrt{20} \pm \sqrt{8}} = \sqrt{\sqrt{5} + \sqrt{3}} \pm \sqrt{\sqrt{5} - \sqrt{3}}.$$

Appendix: Comparison with Abū Kāmil's treatment.

In the introduction to this section on roots, the author referred to the *Algebra* of Abū Kāmil, who was said to have treated some of this topic, including the proofs, in a not altogether clear manner. Let us examine briefly what Abū Kāmil expounds. The theory of calculation with roots occurs in Book I of the *Algebra*.[†] We find there successively:

— computation (and demonstration) of $k\sqrt{a}$, with k integral or fractional (lines 799–857 of the Latin translation), a passage alluded to in the *Liber mahameleth* (B.205*b*);

— computation of $\sqrt{a} \cdot \sqrt{b}$ (with demonstration) and $k\sqrt{a} \cdot l\sqrt{b}$ (lines 858–914);

— computation of $\frac{\sqrt{a}}{\sqrt{b}}$ (with the demonstration of $\sqrt{\frac{a}{b}} = \frac{\sqrt{a}}{\sqrt{b}}$) and $\frac{k\sqrt{a}}{l\sqrt{b}}$ (lines 931–964), preceded by a demonstration of $\frac{a}{b} \cdot b = a$;

— computation (and demonstration) of $\sqrt{a} \pm \sqrt{b}$ (lines 965–1057), with the condition of possibility $a \cdot b = $ square; the two examples are $\sqrt{18} \pm \sqrt{8}$ (possible, see A.291) and $\sqrt{10} + \sqrt{2}$ (not possible).

It is therefore obvious that, compared with our author's, Abū Kāmil's instructions leave much to be desired; there is no method taught for approximating square roots of numbers (though, admittedly, this belongs

---

[†] Fol. 17^r–21^v of the Arabic text, lines 799–1057 of the printed Latin translation.

more in a treatise on arithmetic than in one on pure algebra), nor is there any consideration of roots of binomials; finally, nothing is explained about operations involving fourth roots.

1. Approximation of square roots.

A. *Approximate roots of non-square integers.*
The text first explains the approximation rules (seen in our summary):
(i) If $a^2 < N < (a+1)^2$, then
$$\sqrt{N} \cong a + \frac{N - a^2}{2a}, \qquad \sqrt{N} \cong (a+1) - \frac{(a+1)^2 - N}{2(a+1)}.$$
depending on whether a^2 or $(a+1)^2$ is closer to N.
(ii) With any square integer t^2,
$$\sqrt{N} = \frac{1}{t} \cdot \sqrt{N \cdot t^2}.$$

(A.275) MS A. Find the approximate square root of 5.
Since $4 < 5 < 9$, take $a^2 = 4$ and apply the appropriate formula i. Then
$$\sqrt{5} \cong 2 + \frac{5-4}{2 \cdot 2} = 2 + \frac{1}{4}.$$

(A.276) MS A. Find the approximate square root of 2.
Choosing $t = 100$, we find, by rule ii,
$$\sqrt{2} = \frac{1}{100} \cdot \sqrt{2 \cdot 10\,000} = \frac{1}{100} \cdot \sqrt{20\,000}.$$

Remark. The text does not pursue the computation. With $a = 141$ ($a^2 < N$), applying formula i will give us
$$\sqrt{2} \cong \frac{141}{100} + \frac{119}{100 \cdot 282} \cong 1.4142198,$$
correct to five decimal places (whereas the two formulae i would give 1.5).

(A.277) MS A. Find the approximate square root of 14.
Since $9 < 14 < 16$, take $(a+1)^2 = 16$ and apply the appropriate formula i. Then
$$\sqrt{14} \cong 4 - \frac{16 - 14}{2 \cdot 4} = 4 - \frac{1}{4} = 3 + \frac{3}{4}.$$

B. *Approximate roots of fractional expressions.*
— **Square roots of square fractional expressions.**

As mentioned in our summary, with rule *ii* we can approximate roots of fractional expressions. We shall meet first the simpler case where the expression given is a square. Suppose the expression

$$a + \frac{p}{q^2}$$

to be a square; we shall then consider

$$\sqrt{a + \frac{p}{q^2}} = \frac{1}{q} \cdot \sqrt{a \cdot q^2 + p},$$

which will give an exact root as well since we have multiplied a square by a square.

(**A.278**) MS *A*. Find the square root of $6 + \frac{1}{4}$.

$$\sqrt{6 + \frac{1}{4}} = \frac{1}{2} \cdot \sqrt{\left(6 + \frac{1}{4}\right) \cdot 4} = \frac{1}{2} \cdot \sqrt{25} = \frac{5}{2} = 2 + \frac{1}{2}.$$

(**A.279**) MS *A*. Find the square root of $5 + \frac{4}{9}$.

$$\sqrt{5 + \frac{4}{9}} = \frac{1}{3} \cdot \sqrt{\left(5 + \frac{4}{9}\right) \cdot 9} = \frac{1}{3} \cdot \sqrt{49} = \frac{7}{3} = 2 + \frac{1}{3}.$$

(**A.280**) MS *A*. Find the square root of $\frac{6}{8} + \frac{1}{8}\frac{1}{8}$.

$$\sqrt{\frac{6}{8} + \frac{1}{8}\frac{1}{8}} = \frac{1}{8} \cdot \sqrt{\left(\frac{6}{8} + \frac{1}{8}\frac{1}{8}\right) \cdot 64} = \frac{1}{8} \cdot \sqrt{49} = \frac{7}{8}.$$

— **Square roots of non-square fractional expressions.**

Suppose next we are given the expression

$$a + \frac{p}{q};$$

we shall then consider

$$\sqrt{a + \frac{p}{q}} = \frac{1}{q} \cdot \sqrt{a \cdot q^2 + p \cdot q},$$

or, if we wish to obtain a closer value by increasing the radicand,

$$\sqrt{a + \frac{p}{q}} = \frac{1}{t \cdot q} \cdot \sqrt{a \cdot q^2 \cdot t^2 + p \cdot q \cdot t^2},$$

both to be calculated by one of the approximation methods taught earlier.

(A.281) MS $\mathcal{A}$. Find the approximate square root of $2 + \tfrac{1}{2}$.

$$\sqrt{2+\tfrac{1}{2}} = \tfrac{1}{2} \cdot \sqrt{\left(2+\tfrac{1}{2}\right) \cdot 4} = \tfrac{1}{2} \cdot \sqrt{10}.$$

(a) Proceeding by method i ($9 < 10 < 16$), then

$$\sqrt{10} \cong 3 + \frac{10-9}{6} = 3 + \frac{1}{6}, \quad \text{thus}$$

$$\sqrt{2+\tfrac{1}{2}} \cong \tfrac{1}{2} \cdot \left(3 + \tfrac{1}{6}\right) = 1 + \tfrac{1}{2} + \tfrac{1}{2}\tfrac{1}{6}.$$

(b) Or else, proceed by method ii with $t = 10$, giving

$$\sqrt{10} = \tfrac{1}{10} \cdot \sqrt{1000} \cong \tfrac{1}{10} \cdot \left(32 - \tfrac{24}{64}\right) = \tfrac{1}{10} \cdot \left(32 - \tfrac{3}{8}\right) = \tfrac{1}{10} \cdot \left(31 + \tfrac{5}{8}\right)$$

and therefore

$$\sqrt{2+\tfrac{1}{2}} = \tfrac{1}{2} \cdot \sqrt{10} \cong \tfrac{1}{20} \cdot \left(31 + \tfrac{5}{8}\right) = 1 + \tfrac{1}{2} + \tfrac{1}{2}\tfrac{1}{10} + \tfrac{5}{8}\tfrac{1}{2}\tfrac{1}{10}.$$

Remark. The second hand in MS $\mathcal{A}$ (see I, pp. xxii–xxiv) has computed $\sqrt{1000}$ using the lower limit (although $\sqrt{1000}$ is closer to 32 than to 31). He thus obtains

$$\sqrt{10} = \tfrac{1}{10} \cdot \sqrt{1000} \cong \tfrac{1}{10} \left(31 + \tfrac{39}{62}\right)$$

and therefore

$$\sqrt{2+\tfrac{1}{2}} = \tfrac{1}{2} \cdot \sqrt{10} \cong \tfrac{1}{20} \cdot \left(31 + \tfrac{39}{62}\right) = 1 + \tfrac{1}{2} + \tfrac{1}{2}\tfrac{1}{20} + \tfrac{39}{62}\tfrac{1}{20}.$$

(A.282) MS $\mathcal{A}$. Find the approximate square root of $1 + \tfrac{3}{5}$.

$$\sqrt{1+\tfrac{3}{5}} = \tfrac{1}{5} \cdot \sqrt{\left(1+\tfrac{3}{5}\right) \cdot 25} = \tfrac{1}{5} \cdot \sqrt{40}.$$

(a) According to method i ($36 < 40 < 49$), or doubling the result in A.281a,

$$\sqrt{40} \cong 6 + \frac{40-36}{12} = 6 + \tfrac{1}{3}$$

and therefore

$$\sqrt{1+\tfrac{3}{5}} \cong \tfrac{1}{5} \cdot \left(6 + \tfrac{1}{3}\right) = 1 + \tfrac{1}{5} + \tfrac{1}{3}\tfrac{1}{5}.$$

(**b**) According to method *ii*, consider

$$\sqrt{40} = \frac{1}{10} \cdot \sqrt{40 \cdot 100} = \frac{1}{10} \cdot \sqrt{4000}, \quad \text{so}$$

$$\sqrt{1 + \frac{3}{5}} = \frac{1}{50} \cdot \sqrt{4000} \cong \frac{1}{50} \cdot \left(63 + \frac{31}{126}\right).$$

(**A.283**) MS $\mathcal{A}$. Find the approximate square root of $\dfrac{3}{13}$ (non-elementary fraction).

Since

$$\sqrt{\frac{3}{13}} = \frac{1}{13} \cdot \sqrt{\frac{3}{13} \cdot 13^2} = \frac{1}{13} \cdot \sqrt{39} \quad \text{and} \quad \sqrt{39} \cong 6 + \frac{39 - 36}{12} = 6 + \frac{1}{4},$$

we have

$$\sqrt{\frac{3}{13}} \cong \frac{1}{13} \cdot \left(6 + \frac{1}{4}\right) = \frac{6}{13} + \frac{1}{4}\frac{1}{13}.$$

This section on the extraction of square roots ends with the remark that we cannot determine exactly (that is, in rational numbers) the root of a non-square number.

---

2. Operations with square roots.

A. *Multiplication of square roots*.

The fundamental relation

$$\sqrt{a} \cdot \sqrt{b} = \sqrt{a \cdot b}$$

will be demonstrated in the first problem (A.284). Another demonstration, in A.288, shows that the result will be rational if $a : b = \alpha^2 : \beta^2$, thus if the ratio of the given numbers a and b is like the ratio of two squares. Between these two demonstrations we shall be taught the multiplication of square roots by rational numbers.

(**A.284**) MS $\mathcal{A}$. Multiply $\sqrt{10}$ by $\sqrt{6}$.

According to the formula to be demonstrated below,

$$\sqrt{10} \cdot \sqrt{6} = \sqrt{10 \cdot 6} = \sqrt{60}.$$

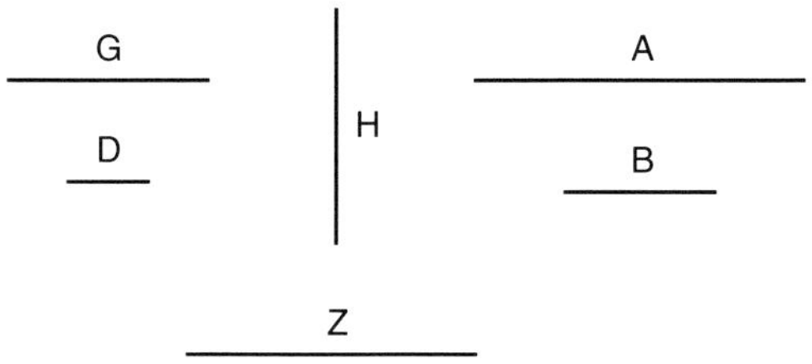

Fig. 36

Demonstration. Let (Fig. 36) $10 = A$, $\sqrt{10} = \sqrt{A} = B$, $6 = G$, $\sqrt{6} = \sqrt{G} = D$; we thus look for the product $B \cdot D$. Let $B \cdot D = H$, $A \cdot G = Z = 60$. To prove that $H = \sqrt{Z}$, that is, that $\sqrt{A} \cdot \sqrt{G} = \sqrt{A \cdot G}$.

Since $B^2 = A$ and $B \cdot D = H$, so (*Elements* VII.17) $B : D = A : H$. Similarly, since $D \cdot B = H$ and $D^2 = G$, so $B : D = H : G$. Therefore $A : H = H : G$, so $A \cdot G = H^2$, that is, $Z = H^2$, and therefore $H = \sqrt{Z}$.

(**A.285**) MS $\mathcal{A}$. Multiply 8 by $\sqrt{10}$. Reduced to the previous problem.

$$8 \cdot \sqrt{10} = \sqrt{64} \cdot \sqrt{10} = \sqrt{640}.$$

(**A.286**) MS $\mathcal{A}$. Find the number to which corresponds 'three roots of ten'.

This will be treated as above, since 'three roots of 10' is $3 \cdot \sqrt{10}$, thus $\sqrt{90}$.

(**A.287**) MS $\mathcal{A}$. Multiply $3\sqrt{6}$ by $5\sqrt{10}$.
By A.285,
$$(3\sqrt{6}) \cdot (5\sqrt{10}) = \sqrt{54} \cdot \sqrt{250} = \sqrt{13\,500}.$$

(**A.288**) MS $\mathcal{A}$. Multiply $\sqrt{8}$ by $\sqrt{18}$. This problem gives, and proves, the condition of rationality.

Since $8 : 18\, (= 4 : 9) = $ square : square, the product of their roots will be rational:
$$\sqrt{8} \cdot \sqrt{18} = \sqrt{144} = 12.$$

Demonstration. Generally, if $a : b = \alpha^2 : \beta^2$, then (*Elements* VIII.26) a and b are 'similar plane numbers' (that is, of the form $a = t \cdot a_1^2$, $b = t \cdot b_1^2$). The product of two such numbers being a square (*Elements* IX.1), the root of this product will be rational.

Remark. What *Elements* VIII.26 actually states is that 'similar plane numbers have to one another the ratio which a square number has to a square number'. The converse, not found in the *Elements*, was added in later Greek times.*

B. Addition of square roots.

The fundamental relation is

$$\sqrt{a} + \sqrt{b} = \sqrt{\left(\sqrt{a} + \sqrt{b}\right)^2} = \sqrt{a + b + 2 \cdot \sqrt{a \cdot b}},$$

which will be demonstrated in the first problem (A.289). Before that, the text considers the two possible cases: (*i*) If $\sqrt{a} + \sqrt{b}$ reduces to the form $\sqrt{A}$, we shall say that the two roots can be added. But (*ii*) if $\sqrt{a} + \sqrt{b}$ leads to the form $\sqrt{A + \sqrt{B}}$, we shall say that the two roots cannot be

---

* See Heath's commented translation of the *Elements*, II, p. 383 and III, pp. 30–31.

added since the resulting expression does not bring any simplification. The given expression will then be left in its original form.

The author successively presents the condition for their addition, its proof, a criterion, then an example of each case.

(*i*) If $a : b = \alpha^2 : \beta^2$, then $\sqrt{a} + \sqrt{b} = \sqrt{c}$ (*c* rational), that is, the sum is the square root of some (rational) number.

Indeed, if two numbers a and b are in such a relation, we have for their roots $\sqrt{a} : \sqrt{b} = \alpha : \beta = t$, that is, $\sqrt{a} = t \cdot \sqrt{b}$, so that $\sqrt{a}$ and $\sqrt{b}$ are commensurable (*Elements* X.9). Consequently, the sum of these roots, namely $(t+1)\sqrt{b}$, is commensurable with each (*Elements* X.15). So their sum is, as they themselves are, rational in square, that is, is the root of a rational number. Indeed,

$$\sqrt{a} + \sqrt{b} = (t+1)\sqrt{b} = \sqrt{(t+1)^2 \, b}, \quad \text{or else}$$

$$\sqrt{a} + \sqrt{b} = \sqrt{a + b + 2 \cdot \sqrt{a \cdot b}} = \sqrt{a + b + 2 \cdot t \cdot b}.$$

(*ii*) But if $a : b \neq \alpha^2 : \beta^2$ (thus $a : b = \alpha : \beta$), then the roots of a and b cannot be conveniently added.

The author then adds that:

(*i'*) If $a \cdot b$ is a square, then (and only then) $a : b = \alpha^2 : \beta^2$.

This will serve as a criterion to decide whether a and b, not squares themselves, are in the ratio of two squares or not, that is, whether the addition is possible or not, and this will be the first thing to consider. For the proof, the author refers to what was seen before (A.288).

(A.289) MS $\mathcal{A}$. Add together $\sqrt{2}$ and $\sqrt{8}$.

Since $2 \cdot 8$ is a square, addition is possible (that is, can be performed). Then

$$\sqrt{2} + \sqrt{8} = \sqrt{2 + 8 + 2 \cdot \sqrt{2 \cdot 8}} = \sqrt{18}.$$

```
         A       B              G
         |───────|──────────────|
```

Fig. 37

Demonstration (of the formula). Let (Fig. 37) $\sqrt{2} = AB$, $\sqrt{8} = BG$; required AG. By *Elements* II.4 ($= PE_4$), $AG^2 = AB^2 + BG^2 + 2 \cdot AB \cdot BG$, and taking the square root on both sides proves the formula. Since $AB \cdot BG = \sqrt{AB^2} \cdot \sqrt{BG^2} = \sqrt{AB^2 \cdot BG^2}$ (A.284), we shall find AG by calculating

$$AG^2 = (AB + BG)^2 = AB^2 + BG^2 + 2 \cdot \sqrt{AB^2 \cdot BG^2}.$$

(A.290) MS $\mathcal{A}$. Add together $\sqrt{6}$ and $\sqrt{10}$.

Since $6 \cdot 10 = 60$ is not a square, these roots cannot be added. Indeed, the formula gives

$$\sqrt{6} + \sqrt{10} = \sqrt{16 + \sqrt{240}},$$

and it is thus simpler to express (in words) their sum as $\sqrt{6} + \sqrt{10}$.

Remark. Indeed, in *verbal* mathematics expressions of the form $\sqrt{a} + \sqrt{b}$ and $\sqrt{a+b}$ are difficult to distinguish: saying 'the root of a and (or: plus) the root of b' is ambiguous. That is why the text must specify here, where a root occurs within a root, 'the root *of the sum* of a and the root of b'. See below, A.292, and remark to B.33, pp. 1373–1374.

C. Subtraction of square roots.

The fundamental relation

$$\sqrt{a} - \sqrt{b} = \sqrt{a + b - 2 \cdot \sqrt{a \cdot b}}$$

is demonstrated in the first problem (A.291). As with addition, the form of the result depends upon whether $a : b = \alpha^2 : \beta^2$ or not, that is, whether $a \cdot b$ is a square or not, which is the first thing to examine. The author gives one example of each case, and notes in the latter the inconvenience (thus unacceptability) of the result.

(A.291) MS $\mathcal{A}$. Subtract $\sqrt{8}$ from $\sqrt{18}$ (same values as in A.288). Subtraction is possible ($18 : 8 = 9 : 4$). Then

$$\sqrt{18} - \sqrt{8} = \sqrt{18 + 8 - 2 \cdot \sqrt{18 \cdot 8}} = \sqrt{26 - 2 \cdot \sqrt{144}} = \sqrt{26 - 24} = \sqrt{2}.$$

A   G   B

Fig. 38

Demonstration (of the formula). Let (Fig. 38) $\sqrt{18} = AB$, $\sqrt{8} = BG$; required AG. By *Elements* II.7 $(= PE_7)$, $AB^2 + BG^2 = 2 \cdot AB \cdot BG + AG^2$, so

$$AG^2 = AB^2 + BG^2 - 2 \cdot AB \cdot BG,$$

and taking the square root on both sides proves the formula.

(A.292) MS $\mathcal{A}$. Subtract $\sqrt{6}$ from $\sqrt{10}$ (same values as in A.290). Applying the rule, we find

$$\sqrt{10} - \sqrt{6} = \sqrt{10 + 6 - 2 \cdot \sqrt{10 \cdot 6}} = \sqrt{16 - 2 \cdot \sqrt{60}} = \sqrt{16 - \sqrt{240}}.$$

It is therefore easier to express it as $\sqrt{10} - \sqrt{6}$: saying 'the root of ten minus the root of six' is preferable to saying 'the root of the remainder from sixteen after subtracting from it the root of two hundred and forty', the text specifies.

D. *Division of square roots*.
— One term in the divisor.

The fundamental relation, mentioned in the first problem (A.293), is

$$\frac{\sqrt{a}}{\sqrt{b}} = \sqrt{\frac{a}{b}}.$$

The coming problems are just analogous to those about multiplication.

(A.293) MS $\mathcal{A}$. Divide $\sqrt{10}$ by $\sqrt{3}$.

$$\frac{\sqrt{10}}{\sqrt{3}} = \sqrt{\frac{10}{3}}.$$

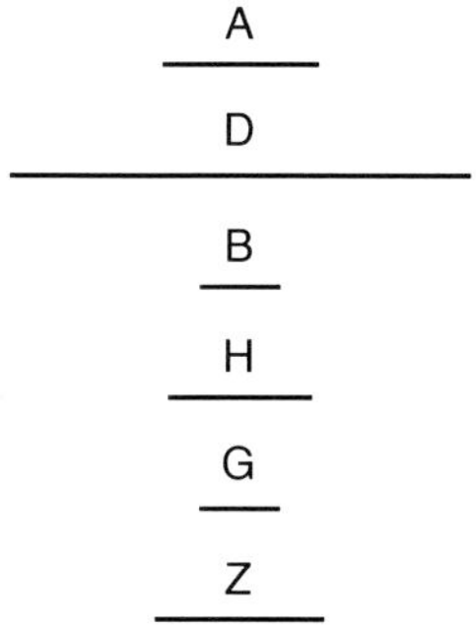

Fig. 39

Demonstration (interpolated). Let (Fig. 39) $(\sqrt{10} =) A$, $(\sqrt{3} =) B$, $\frac{A}{B} = G$, $A^2 = D$, $B^2 = H$, $G^2 = Z$. The text proves that $\frac{D}{H} = Z$.

Since $\frac{A}{B} = G$, so $1 : G = B : A$, thus $1^2 : G^2 = B^2 : A^2$.[†] Therefore $1 : Z = H : D$, and

$$\frac{D}{H} = Z, \quad \text{that is,} \quad \frac{a}{b} = \left(\frac{\sqrt{a}}{\sqrt{b}}\right)^2.$$

Remark. As a proof, the non-interpolated sentence in the text suffices. If ever a formal proof were needed, it should take the following form. Let $\sqrt{a} = A$, $\sqrt{b} = B$, $\frac{A}{B} = G$; $A^2 = D$, $B^2 = H$, $\frac{D}{H} = Z$.

Since $A = B \cdot G$ and $A^2 = D = Z \cdot H = Z \cdot B^2$, that is (A.284), $A = \sqrt{Z} \cdot B$, then

---

[†] That if $a : b = c : d$ then $a^2 : b^2 = c^2 : d^2$ is not proved in the *Elements* but used in X.9; it can, however, be inferred from Book VI (see Heath's Euclid, III, pp. 30–31 & Vitrac's Euclide, 3, p. 119).

$$\frac{A}{B} = \sqrt{Z}, \quad \text{that is,} \quad \frac{\sqrt{a}}{\sqrt{b}} = \sqrt{\frac{a}{b}}.$$

Abū Kāmil's has a full proof, which is as follows.° Let $\sqrt{a} = A$, $\sqrt{b} = B$, $\frac{A}{B} = G$, $A^2 = D$, $B^2 = H$, $Z = \frac{D}{H}$, $G^2 = T$; to prove that $G = \sqrt{Z}$.
Since $\frac{D}{H} = Z$, so $\frac{Z}{D} = \frac{1}{H}$; since $\frac{A}{B} = G$, so $\frac{G}{A} = \frac{1}{B}$ and, squared, $\frac{T}{D} = \frac{1}{H}$. So $\frac{Z}{D} = \frac{T}{D}$, $Z = T = G^2$, $G = \sqrt{Z}$, which means that

$$\frac{\sqrt{a}}{\sqrt{b}} = \sqrt{\frac{a}{b}}.$$

Let us finally note, in our text, the absence of a geometrical demonstration for the analogous case of fourth roots (whereas multiplication and addition are, as here, demonstrated by a geometrical figure).

(A.294) MS $\mathcal{A}$. Divide 10 by $\sqrt{5}$.
As in the case of multiplication (A.285), change 10 into $\sqrt{100}$. Then

$$\frac{10}{\sqrt{5}} = \frac{\sqrt{100}}{\sqrt{5}} = \sqrt{20}.$$

(A.295) MS $\mathcal{A}$. Divide $\sqrt{10}$ by 2.
Same situation, so

$$\frac{\sqrt{10}}{2} = \frac{\sqrt{10}}{\sqrt{4}} = \sqrt{2 + \frac{1}{2}}.$$

(A.296) MS $\mathcal{A}$. Divide $2 \cdot \sqrt{10}$ by $3 \cdot \sqrt{6}$.
Since $2 \cdot \sqrt{10} = \sqrt{40}$ and $3 \cdot \sqrt{6} = \sqrt{54}$,

$$\frac{2 \cdot \sqrt{10}}{3 \cdot \sqrt{6}} = \frac{\sqrt{40}}{\sqrt{54}} = \sqrt{\frac{6}{9} + \frac{2}{3}\frac{1}{9}}.$$

— **Two terms in the dividend or the divisor.**

We have just learned how to divide roots and, before that, to add or subtract them —and if this is possible. We shall now be taught how to divide when either the dividend or the divisor is the sum or difference of two terms at least one of which is a root. The eight possibilities are the following (to which may be added the case $\sqrt{b} \pm a$ in each of the last pairs):

(1) $\dfrac{\sqrt{a} \pm \sqrt{b}}{\sqrt{c}}$, (2) $\dfrac{\sqrt{a} \pm \sqrt{b}}{c}$, (3) $\dfrac{a \pm \sqrt{b}}{\sqrt{c}}$, (4) $\dfrac{a \pm \sqrt{b}}{c}$,

(5) $\dfrac{\sqrt{c}}{\sqrt{a} \pm \sqrt{b}}$, (6) $\dfrac{c}{\sqrt{a} \pm \sqrt{b}}$, (7) $\dfrac{\sqrt{c}}{a \pm \sqrt{b}}$, (8) $\dfrac{c}{a \pm \sqrt{b}}$.

° *Algebra*, fol. 19^v–20^r (pp. 38–39 of the printed reproduction), lines 945–964 of the Latin translation.

- TWO (OR MORE) TERMS IN THE DIVIDEND (CASES 1–4).

Only case (1) is treated (in A.297). Cases (2)–(4) are not considered because they will be solved in the same way. Indeed, all these four cases reduce to two divisions of the type seen in A.293–295. The first two cases may even reduce to just one division if, as the author remarks in A.297, the two terms in the dividend can be transformed into a single root. In the general case of any number of terms, mentioned in the same problem, we shall have either
$$\frac{\sqrt{a_1} + \sqrt{a_2} + \ldots + \sqrt{a_n}}{\sqrt{c}} = \frac{\sqrt{A}}{\sqrt{c}}$$
or
$$\frac{\sqrt{a_1} + \sqrt{a_2} + \ldots + \sqrt{a_n}}{\sqrt{c}} = \frac{\sqrt{A'} + \sum \sqrt{a_k}}{\sqrt{c}}$$
with $\sqrt{A'}$ the sum of those $\sqrt{a_i}$ which can be added and $\sum \sqrt{a_k}$ the sum of the others. Each term in the dividend is then divided by $\sqrt{c}$.

- TWO TERMS IN THE DIVISOR (CASES 5–8).

The situation is less simple when the sum or difference of two (incommensurable) terms stands in the divisor. To make it rational, both dividend and divisor will be multiplied by the difference or the sum of the two terms. (The rationality of the result is proved in A.298b.) Case (5) is treated in A.300 & A.302, case (7) in A.299, case (8) in A.298 & A.301; case (6) would be treated just like case (5). As for the presence of more than two terms in the divisor, this is the subject of the (apparently misplaced) group A.320–322.

- TWO TERMS IN THE DIVIDEND.

(A.297) MS $\mathcal{A}$. Divide $\sqrt{6} + \sqrt{10}$ by $\sqrt{3}$. Since $\sqrt{6}$ and $\sqrt{10}$ cannot be added (A.290), divide the terms individually:
$$\frac{\sqrt{6} + \sqrt{10}}{\sqrt{3}} = \frac{\sqrt{6}}{\sqrt{3}} + \frac{\sqrt{10}}{\sqrt{3}}.$$

Remark. The generalization to more terms concluding this problem is vitiated by the interpolations of a reader under the impression that if the roots could not be added to begin with, they might be after division.

- TWO TERMS IN THE DIVISOR.

(A.298) MS $\mathcal{A}$. Divide 10 by $2 + \sqrt{3}$.
(a) Since $(2 + \sqrt{3})(2 - \sqrt{3}) = 1$, we shall have
$$\frac{10}{2 + \sqrt{3}} = \frac{10(2 - \sqrt{3})}{(2 + \sqrt{3})(2 - \sqrt{3})} = 10(2 - \sqrt{3}) = 20 - \sqrt{300}.$$

In the text, this is partly computed, partly treated geometrically. Let (Fig. 40) $10 = A$, $2 + \sqrt{3} = B$, required $G = \frac{A}{B}$. Put $2 - \sqrt{3} = H$,

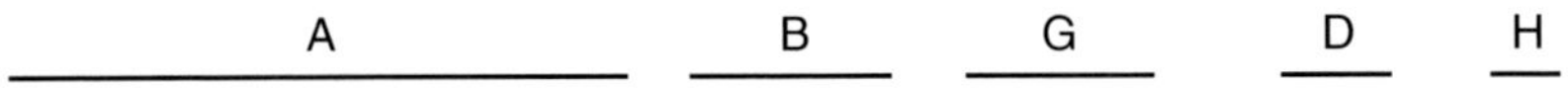

```
  A              B        G        D        H
─────────────  ──────  ──────  ──────  ──
```

Fig. 40

$B \cdot H = D$. Since $A = G \cdot B$ and $D = B \cdot H$, so (*Elements* VII.18) $A : D = G : H$ and, since $A = 10 \cdot D$, then $G = 10 \cdot H = 10\,(2 - \sqrt{3})$.

(**b**) Our author now proves that the product of a binomial and its apotome is rational (in our sense).

Let (Fig. 41) the binomial $\sqrt{a} + \sqrt{b}$ be AB, its two terms AG and GB, with AG^2 and GB^2 rational and, say, $AG = \sqrt{a} > GB = \sqrt{b}$. Take $DG = GB$; then $AD = AG - DG = \sqrt{a} - \sqrt{b}$ is the corresponding apotome. To prove that $AB \cdot AD$ is rational.

Since G is the mid-point of DB while AD is added to DB, we have by *Elements* II.6 ($= PE_6$) $AD \cdot AB + DG^2 = AG^2$, thus $AD \cdot AB = AG^2 - DG^2$, which, being the difference between two rational numbers, is rational.

```
  A                    D   G    B
 ├──────────────────┼──┼────┤
```

Fig. 41

(**A.299**) MS $\mathcal{A}$. Divide $\sqrt{10}$ by $2 + \sqrt{6}$.
The apotome of $2 + \sqrt{6}$ is $\sqrt{6} - 2$, and $(\sqrt{6}+2)(\sqrt{6}-2) = 2$. So

$$\frac{\sqrt{10}}{2+\sqrt{6}} = \frac{\sqrt{10}\,(\sqrt{6}-2)}{(\sqrt{6}+2)(\sqrt{6}-2)} = \frac{\sqrt{10}}{2}\cdot(\sqrt{6}-2)$$

$$= \sqrt{2+\frac{1}{2}}\cdot(\sqrt{6}-2) = \sqrt{15}-\sqrt{10}.$$

(**A.300**) MS $\mathcal{A}$. Divide $\sqrt{10}$ by $\sqrt{2}+\sqrt{3}$.
The apotome of $\sqrt{2}+\sqrt{3}$ is $\sqrt{3}-\sqrt{2}$, and $(\sqrt{2}+\sqrt{3})(\sqrt{3}-\sqrt{2}) = 1$. Therefore

$$\frac{\sqrt{10}}{\sqrt{2}+\sqrt{3}} = \frac{\sqrt{10}\,(\sqrt{3}-\sqrt{2})}{(\sqrt{2}+\sqrt{3})(\sqrt{3}-\sqrt{2})} = \sqrt{30}-\sqrt{20}.$$

(**A.301**) MS $\mathcal{A}$. Divide 10 by $2-\sqrt{3}$. Like A.299, but with an apotome as the divisor.

The binomial corresponding to $2-\sqrt{3}$ is $2+\sqrt{3}$ and their product is rational and equals 1 (A.298). Therefore

$$\frac{10}{2-\sqrt{3}} = \frac{10\,(2+\sqrt{3})}{(2-\sqrt{3})(2+\sqrt{3})} = 20+\sqrt{300}.$$

(A.302) MS A. Divide $\sqrt{10}$ by $\sqrt{5} - \sqrt{3}$. Like A.300, but with an apotome as the divisor.

Since the binomial of $\sqrt{5} - \sqrt{3}$ is $\sqrt{5} + \sqrt{3}$ and their product is 2, so

$$\frac{\sqrt{10}}{\sqrt{5} - \sqrt{3}} = \frac{\sqrt{10}\left(\sqrt{5} + \sqrt{3}\right)}{\left(\sqrt{5} - \sqrt{3}\right)\left(\sqrt{5} + \sqrt{3}\right)} = \frac{\sqrt{10}}{2} \cdot \left(\sqrt{5} + \sqrt{3}\right)$$

$$= \sqrt{2 + \frac{1}{2}} \left(\sqrt{5} + \sqrt{3}\right) = \sqrt{12 + \frac{1}{2}} + \sqrt{7 + \frac{1}{2}}.$$

---

3. Operations with fourth roots.

A. *Multiplication of fourth roots.*

The fundamental rule

$$\sqrt[4]{a} \cdot \sqrt[4]{b} = \sqrt[4]{a \cdot b}$$

will be demonstrated in the first problem (A.303). As seen in our summary, the result may be (with a and b rational)
— a rational number (if $a \cdot b$ is a fourth power)
— a square root (if $a \cdot b$ is a square)
— a fourth root (if $a \cdot b$ is a non-square number).
The author examines these three possibilities by considering the various forms the ratio $\sqrt{a} : \sqrt{b}$ may take (supposing a and b non-square numbers). So the amount of the ratio $\sqrt{a} : \sqrt{b}$ may be a rational number, a square root or a square.

(*i*) As seen in our summary, we shall obtain from the multiplication a *rational number* if $\sqrt{a} : \sqrt{b} = \alpha : \beta = t$ (t rational) with $t \cdot b = r^2$, for then

$$\sqrt[4]{a} \cdot \sqrt[4]{b} = \sqrt{t} \cdot \sqrt{b} = \sqrt{t \cdot b} = r \quad (a \cdot b = r^4).$$

See example in A.307.

(*ii*) We shall obtain a *square root* in two cases: (*ii'*) if either, as before, $\sqrt{a} : \sqrt{b} = \alpha : \beta = t$, rational, but this time $t \cdot b$ is not a square (see A.308 and demonstration before A.307); (*ii''*) if $\sqrt{a} : \sqrt{b} = \alpha^2 : \beta^2 = t^2$, a square (see A.310 & demonstration there).

(*iii*) We shall obtain a *fourth root* if $a : b = \alpha : \beta = t$, that is, $\sqrt{a} : \sqrt{b} = \sqrt{t}$, thus a square root (see A.309 & A.303).

Before explaining these more theoretical distinctions, the author performs some computations, first with the general case (application and demonstration of the fundamental rule, A.303), then with mixed products of the type $\sqrt{a} \cdot \sqrt[4]{b}$ (A.304), $a \cdot \sqrt[4]{b}$ (A.305), $k\sqrt[4]{a} \cdot l\sqrt[4]{b}$ (A.306); they are just analogous to the problems seen before for the multiplication of square roots.

(A.303) MS A. Multiply $\sqrt[4]{7}$ by $\sqrt[4]{10}$ (case *iii*, producing a fourth root).

Since $7 \cdot 10 = 70$,
$$\sqrt[4]{7} \cdot \sqrt[4]{10} = \sqrt[4]{70}.$$

Demonstration. Let (Fig. 42) $10 = A$, $\sqrt{A} = B$, $\sqrt{B} = G = \sqrt[4]{A}$, $7 = D$, $\sqrt{D} = H$, $\sqrt{H} = Z = \sqrt[4]{D}$. Let further $A \cdot D = K$, $B \cdot H = T$, $G \cdot Z = Q$. To prove that $Q = G \cdot Z = \sqrt[4]{A} \cdot \sqrt[4]{D}$ equals $\sqrt[4]{K} = \sqrt[4]{A \cdot D}$.

According to what we have seen in the multiplication of square roots (A.284), we know that

$$T = B \cdot H = \sqrt{A} \cdot \sqrt{D} = \sqrt{A \cdot D} = \sqrt{K} \quad \text{and}$$
$$Q = G \cdot Z = \sqrt{B} \cdot \sqrt{H} = \sqrt{B \cdot H} = \sqrt{T}.$$

Since $Q = \sqrt{T} = \sqrt{\sqrt{K}} = \sqrt[4]{K}$ while, by definition, $Q = \sqrt[4]{A} \cdot \sqrt[4]{D}$, we see that $\sqrt[4]{A} \cdot \sqrt[4]{D} = \sqrt[4]{A \cdot D}$.

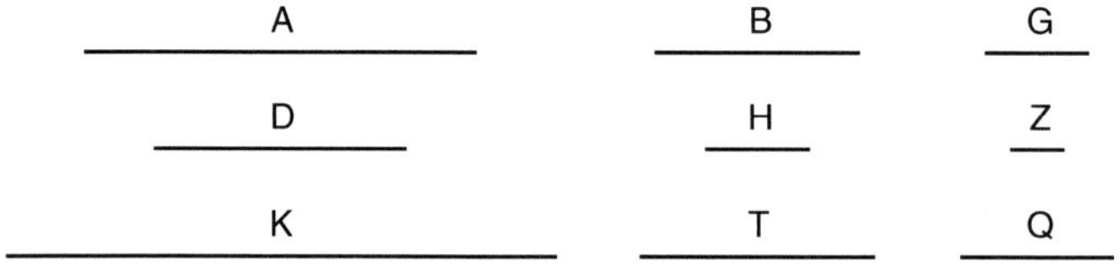

A	B	G
D	H	Z
K	T	Q

Fig. 42

(A.304) MS $\mathcal{A}$. Multiply $\sqrt[4]{10}$ by $\sqrt[4]{30}$.
Since $\sqrt{10} = \sqrt[4]{100}$, we have

$$\sqrt[4]{10} \cdot \sqrt[4]{30} = \sqrt[4]{100} \cdot \sqrt[4]{30} = \sqrt[4]{3000}.$$

(A.305) MS $\mathcal{A}$. Multiply 5 by $\sqrt[4]{10}$.
Since $5 = \sqrt[4]{625}$, we have

$$5 \cdot \sqrt[4]{10} = \sqrt[4]{625} \cdot \sqrt[4]{10} = \sqrt[4]{6250}.$$

(A.306) MS $\mathcal{A}$. Multiply $3 \cdot \sqrt[4]{10}$ by $2 \cdot \sqrt[4]{6}$.
Since (A.305) $3 \cdot \sqrt[4]{10} = \sqrt[4]{810}$ and $2 \cdot \sqrt[4]{6} = \sqrt[4]{96}$, we have (A.303)

$$\left(3 \cdot \sqrt[4]{10}\right) \cdot \left(2 \cdot \sqrt[4]{6}\right) = \sqrt[4]{810} \cdot \sqrt[4]{96} = \sqrt[4]{810 \cdot 96}.$$

The author then mentions the various forms the results can take. If $\sqrt{a} : \sqrt{b} = t$, the product of the two fourth roots will give either (*i*) a number or (*ii'*) a square root. His proof is as follows.

Let the two quantities be $\sqrt[4]{a}$, $\sqrt[4]{b}$. Suppose that their squares are commensurable in length, thus that $\sqrt{a} = t \cdot \sqrt{b}$, t rational; the fourth roots are then $\sqrt[4]{a} = \sqrt{t} \cdot \sqrt[4]{b}$ and $\sqrt[4]{b}$, which are incommensurable in length but commensurable in square. Now *Elements* X.25 tells us that the area formed by the product of two such quantities (which are 'medial' and commensurable in square) is either rational or medial. Indeed, we have $\sqrt[4]{a} \cdot \sqrt[4]{b} = \sqrt{t} \cdot \sqrt{b} = \sqrt{t \cdot b}$, which is either rational (if the radicand $t \cdot b$ is a square) or medial (otherwise). This will be illustrated in the next two problems.

Remark. The use of Euclidean terminology in this context is confusing. Euclid describes a rectangular area as 'medial' when its sides are medial, and rational if its sides are commensurable in square.

(A.307) MS $\mathcal{A}$. Multiply $\sqrt[4]{3}$ by $\sqrt[4]{27}$ (case i).

These are two medials commensurable in square (for $\sqrt{3} : \sqrt{27} = \sqrt{3} \cdot \sqrt{3} : \sqrt{27} \cdot \sqrt{3} = 3 : 9$) but not in length. The result of their multiplication is rational:
$$\sqrt[4]{3} \cdot \sqrt[4]{27} = \sqrt[4]{81} = 3.$$

(A.308) MS $\mathcal{A}$. Multiply $\sqrt[4]{8}$ by $\sqrt[4]{18}$ (case ii').

These are two medials again commensurable in square only, for $\sqrt{8} : \sqrt{18} = \sqrt{8} \cdot \sqrt{18} : \sqrt{18} \cdot \sqrt{18} = 12 : 18 = 6 : 9$. This time the product is 'medial', that is, a square root:
$$\sqrt[4]{8} \cdot \sqrt[4]{18} = \sqrt[4]{144} = \sqrt{12}.$$

Remark. The text does not explain the difference between the two results obtained, that is, that $a \cdot b$ is once a fourth power, once a square only.

We now proceed with two examples of ii'' and iii, the cases themselves being explained by the author before each problem.

(A.309) MS $\mathcal{A}$. Multiply $\sqrt[4]{10}$ by $\sqrt[4]{15}$ (case iii).

Since the two given numbers are not to one another as two squares, for their product is not a square, the result of the multiplication of their square roots will not be a square; thus the result of the multiplication of their fourth roots will be a fourth root.
$$\sqrt[4]{10} \cdot \sqrt[4]{15} = \sqrt[4]{150}.$$

(A.310) MS $\mathcal{A}$. Multiply $\sqrt[4]{2}$ by $\sqrt[4]{32}$ (case ii'').

Consider $\sqrt[4]{a}$ and $\sqrt[4]{b}$. If $\sqrt{a} : \sqrt{b}$ is like the ratio of two squares, then $\sqrt[4]{a} : \sqrt[4]{b}$ is like the ratio of two numbers. These fourth roots are therefore commensurable. Now the multiplication of two commensurable medial quantities (fourth roots) produces, according to *Elements* X.24, a medial result (a square root). This applies to the present example.
$$\sqrt[4]{2} \cdot \sqrt[4]{32} = \sqrt[4]{64} = \sqrt{8}.$$

B. *Addition of fourth roots.*

For addition and subtraction, the fundamental relation is
$$\sqrt[4]{a} \pm \sqrt[4]{b} = \sqrt{\left(\sqrt[4]{a} \pm \sqrt[4]{b}\right)^2} = \sqrt{\sqrt{a} + \sqrt{b} \pm 2 \cdot \sqrt[4]{a} \cdot \sqrt[4]{b}}.$$

This is demonstrated, for the additive case, in the first example (A.311). On account of what was seen in our summary, we know that we may have

(i) $\sqrt[4]{a} \pm \sqrt[4]{b} = \sqrt[4]{A}$ (A.311).
(ii) $\sqrt[4]{a} \pm \sqrt[4]{b} = \sqrt{\sqrt{A} \pm B}$ (A.312).
(iii) $\sqrt[4]{a} \pm \sqrt[4]{b} = \sqrt{\sqrt{A} \pm \sqrt{B}}$ (A.313).
(iv) $\sqrt[4]{a} \pm \sqrt[4]{b} = \sqrt{\sqrt{A} \pm \sqrt[4]{B}}$ (no example).

Having dealt with this above (pp. 1308–1309), we now need only briefly consider how the author treats the three cases where an acceptable change in form is possible.

(i) If $\sqrt{a} : \sqrt{b} = \alpha^2 : \beta^2$, so $\sqrt[4]{a} : \sqrt[4]{b} = \alpha : \beta$ (*Elements* X.9) and therefore $\sqrt[4]{a}$ and $\sqrt[4]{b}$ are commensurable in length (*Elements* X.6). Consequently, their sum will be commensurable with each (*Elements* X.15). Each being medial, and a quantity commensurable with a medial quantity being itself medial (*Elements* X.23), the sum will be medial (fourth root).

(**A.311**) MS $\mathcal{A}$. Add together $\sqrt[4]{3}$ and $\sqrt[4]{243}$.

The text begins by establishing the formula, which is then applied to the example. Let (Fig. 43) $\sqrt[4]{3} = AB$, $\sqrt[4]{243} = BG$, with AG required. By *Elements* II.4 (=PE_4),

$$AG^2 = AB^2 + BG^2 + 2 \cdot AB \cdot BG.$$

```
A           B                    G
|-----------|--------------------|
```

Fig. 43

So far for the formula. In the proposed example, the first two terms on the right side are $AB^2 = \sqrt{3}$ and $BG^2 = \sqrt{243}$. Since they satisfy the situation of case *i* for the addition of *square* roots ($\sqrt{243} : \sqrt{3} = \alpha : \beta = t$), they can be added. We thus obtain

$$AB^2 + BG^2 = \sqrt{3} + \sqrt{243} = \sqrt{246 + 2 \cdot \sqrt{729}} = \sqrt{246 + 2 \cdot 27} = \sqrt{300}.$$

Next, we find (A.303)

$$2 \cdot AB \cdot BG = 2 \cdot \sqrt[4]{3} \cdot \sqrt[4]{243} = 2 \cdot \sqrt[4]{729} = 2 \cdot \sqrt{27} = \sqrt{108}.$$

Consequently, $AG^2 = \sqrt{300} + \sqrt{108}$. Now $\sqrt{300}\ (= 10 \cdot \sqrt{3})$ and $\sqrt{108}\ (= 6 \cdot \sqrt{3})$ are commensurable. As the author says, this must be so. Indeed, the two quantities $\sqrt[4]{a}$ and $\sqrt[4]{b}$ are commensurable, so that $\sqrt[4]{a} = t \cdot \sqrt[4]{b}$; therefore their sum, $\sqrt[4]{a} + \sqrt[4]{b} = (t+1) \sqrt[4]{b}$, is commensurable with each (*Elements* X.15); since the sum of their squares is $\sqrt{a} + \sqrt{b} = (t^2 + 1) \cdot \sqrt{b}$, and twice their product is $2 \cdot \sqrt[4]{a} \cdot \sqrt[4]{b} = 2 \cdot t \cdot \sqrt{b}$, sum and twice the product will clearly be commensurable (*Elements* X.19). The ratio of these commensurable quantities being a rational number, we are in the situation of case *i* for the addition of square roots. We then find

$$\sqrt{300} + \sqrt{108} = \sqrt{300 + 108 + 2 \cdot \sqrt{300 \cdot 108}}$$

$$= \sqrt{408 + 2 \cdot 180} = \sqrt{768}.$$

This is $AG^2 = \left(\sqrt[4]{3} + \sqrt[4]{243}\right)^2$. Therefore

$$\sqrt[4]{3} + \sqrt[4]{243} = \sqrt[4]{768}.$$

(*ii*) If $\sqrt{a} : \sqrt{b} = \alpha : \beta = t$, so $\sqrt{a} + \sqrt{b} = (t+1)\sqrt{b} = \sqrt{(t+1)^2 \cdot b}$; if next $\sqrt{a} \cdot \sqrt{b}$ is a square, then $\sqrt[4]{a} \cdot \sqrt[4]{b} = \sqrt{t \cdot b}$ is rational. Therefore

$$\sqrt[4]{a} + \sqrt[4]{b} = \sqrt{\left(\sqrt[4]{a} + \sqrt[4]{b}\right)^2} = \sqrt{\sqrt{(t+1)^2 \cdot b} + 2 \cdot \sqrt{t \cdot b}} = \sqrt{\sqrt{A} + B},$$

and this is a quantity 'producing a rational and a medial' (*Elements* X.40): the square on the line $\sqrt{\sqrt{A} + B}$ is the area $\sqrt{A} + B$, which is indeed the sum of a rational and a medial.

(A.312) MS $\mathcal{A}$. Add together $\sqrt[4]{3}$ and $\sqrt[4]{27}$ (case *ii*). Since $\sqrt{3} + \sqrt{27} = \sqrt{30 + 2 \cdot 9} = \sqrt{48}$ and $\sqrt[4]{3} \cdot \sqrt[4]{27} = 3$, so

$$\sqrt[4]{3} + \sqrt[4]{27} = \sqrt{\sqrt{48} + 2 \cdot 3} = \sqrt{\sqrt{48} + 6}.$$

(*iii*) If $\sqrt{a} : \sqrt{b} = \alpha : \beta = t$ but $\sqrt{a} \cdot \sqrt{b}$ is not a square, so

$$\sqrt[4]{a} + \sqrt[4]{b} = \sqrt{\sqrt{(t+1)^2 \cdot b} + \sqrt{4t \cdot b}} = \sqrt{\sqrt{A} + \sqrt{B}},$$

which is a quantity 'producing two medials' (*Elements* X.41) since its square is the sum of two medial quantities.

(A.313) MS $\mathcal{A}$. Add together $\sqrt[4]{8}$ and $\sqrt[4]{18}$. Since $\sqrt{8} + \sqrt{18} = \sqrt{26 + 2 \cdot 12} = \sqrt{50}$ and $\sqrt[4]{8} \cdot \sqrt[4]{18} = \sqrt{12}$, so

$$\sqrt[4]{8} + \sqrt[4]{18} = \sqrt{\sqrt{50} + 2 \cdot \sqrt{12}} = \sqrt{\sqrt{50} + \sqrt{48}}.$$

(*iv*) If $a : b = \alpha : \beta = t$, t rational not a square, so

$$\sqrt[4]{a} + \sqrt[4]{b} = \sqrt{\sqrt{(\sqrt{t}+1)^2 \cdot b} + \sqrt[4]{16 \cdot t \cdot b}} = \sqrt{\sqrt{A} + \sqrt[4]{B}},$$

which will be better left in its original, simpler form.

C. *Subtraction of fourth roots*.

Fundamental relation:

$$\sqrt[4]{a} - \sqrt[4]{b} = \sqrt{\sqrt{a} + \sqrt{b} - 2 \cdot \sqrt[4]{a} \cdot \sqrt[4]{b}}.$$

The conditions and the results are exactly the same as for addition, the binomials being merely changed into apotomes.

D. *Division of fourth roots*.

— **One term in the divisor.**

Fundamental relation:
$$\frac{\sqrt[4]{a}}{\sqrt[4]{b}} = \sqrt[4]{\frac{a}{b}}.$$

This is proved in the first example.

(A.314) MS $\mathcal{A}$. Divide $\sqrt[4]{10}$ by $\sqrt[4]{5}$.

Using the formula demonstrated immediately below,
$$\frac{\sqrt[4]{10}}{\sqrt[4]{5}} = \sqrt[4]{\frac{10}{5}} = \sqrt[4]{2}.$$

Demonstration. We shall just apply twice the rule seen in the division of square roots (A.293), that is,
$$\frac{\sqrt[4]{a}}{\sqrt[4]{b}} = \frac{\sqrt{\sqrt{a}}}{\sqrt{\sqrt{b}}} = \sqrt{\left(\frac{\sqrt{a}}{\sqrt{b}}\right)} = \sqrt{\sqrt{\frac{a}{b}}} = \sqrt[4]{\frac{a}{b}}.$$

The coming examples consider the division of $k \cdot \sqrt[4]{a}$ by $l \cdot \sqrt[4]{a}$, then of $\sqrt[4]{a}$ by l.

(A.315) MS $\mathcal{A}$. Divide $2 \cdot \sqrt[4]{10}$ by $3 \cdot \sqrt[4]{8}$.

As seen in A.305,
$$\frac{2 \cdot \sqrt[4]{10}}{3 \cdot \sqrt[4]{8}} = \frac{\sqrt[4]{160}}{\sqrt[4]{648}} = \sqrt[4]{\frac{160}{648}} = \sqrt[4]{\frac{2}{9} + \frac{2}{9}\frac{1}{9}}.$$

(A.316) MS $\mathcal{A}$. Divide $\sqrt[4]{10}$ by 2.
$$\frac{\sqrt[4]{10}}{2} = \frac{\sqrt[4]{10}}{\sqrt[4]{16}} = \sqrt[4]{\frac{10}{16}} = \sqrt[4]{\frac{5}{8}}.$$

— **Two terms in the divisor.**

As seen in our summary (p. 1310), three types occur:

$$(i) \quad \frac{c}{a \pm \sqrt[4]{b}} \quad \left(\text{or} \quad \frac{c}{\sqrt[4]{b} \pm a}\right) \quad (\text{A.317})$$

$$(ii) \quad \frac{c}{\sqrt{a} \pm \sqrt[4]{b}} \quad \left(\text{or} \quad \frac{c}{\sqrt[4]{b} \pm \sqrt{a}}\right) \quad (\text{A.319})$$

$$(iii) \quad \frac{c}{\sqrt[4]{a} \pm \sqrt[4]{b}} \quad \text{with} \quad a > b \quad (\text{A.318}).$$

The principle of the transformation is explained in the first of these problems.

Remark. The author does not explicitly consider in the theory (though he does in the examples) further reduction, for this just means applying what we have seen in the section on square roots, for instance

$$\frac{c}{\sqrt{a}+\sqrt[4]{b}} = \frac{\sqrt{c^2 a}-\sqrt[4]{c^4 b}}{a-\sqrt{b}} = \frac{\left(\sqrt{c^2 a}-\sqrt[4]{c^4 b}\right)\left(a+\sqrt{b}\right)}{a^2-b}.$$

(A.317) MS $\mathcal{A}$. Divide 10 by $2+\sqrt[4]{3}$ (case *i*).

Before treating the example, two theoretical points are established.

(a) Multiplying a binomial involving at least one fourth root by its apotome produces an apotome in the usual sense, that is, a difference between two incommensurable terms which are rational in square. Indeed,

$$(i) \quad \left(a+\sqrt[4]{b}\right)\cdot\left(\pm[a-\sqrt[4]{b}]\right) = \pm\left(a^2-\sqrt{b}\right)$$

$$(ii) \quad \left(\sqrt{a}+\sqrt[4]{b}\right)\cdot\left(\pm[\sqrt{a}-\sqrt[4]{b}]\right) = \pm\left(a-\sqrt{b}\right)$$

$$(iii) \quad \left(\sqrt[4]{a}+\sqrt[4]{b}\right)\cdot\left(\sqrt[4]{a}-\sqrt[4]{b}\right) = \sqrt{a}-\sqrt{b} \quad (a>b).$$

The proof, presented only for the last multiplication, is said to be also valid for the two others (the only difference being that in the result one square root is replaced by a rational term). *Mutatis mutandis*, this is the proof seen in A.298b.

Let (Fig. 44) $\sqrt[4]{a} = AB$, $\sqrt[4]{b} = BG$ ($b<a$, so $BG<AB$), thus AG is the binomial $\sqrt[4]{a}+\sqrt[4]{b}$; putting $DB = BG = \sqrt[4]{b}$, then $AD = AB-DB = \sqrt[4]{a}-\sqrt[4]{b}$ will be the corresponding apotome. To prove that $AG \cdot AD$ is an apotome.

Since B is the mid-point of DG while AD is added to DG, we have by Elements II.6 (= PE_6) $AG \cdot AD + DB^2 = AB^2$, so $AG \cdot AD = AB^2 - DB^2 = \sqrt{a}-\sqrt{b}$, which is an apotome.

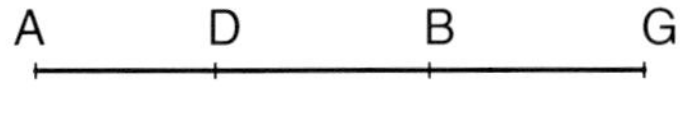

```
A       D       B       G
|-------|-------|-------|
```

Fig. 44

Next, of the three kinds of apotome obtained, namely

$$\pm\left(a^2-\sqrt{b}\right), \quad \pm\left(a-\sqrt{b}\right), \quad \sqrt{a}-\sqrt{b},$$

only the third admits of reduction, provided that $\sqrt{a}$ and $\sqrt{b}$ are commensurable; for then (case *i* of the addition and subtraction of square roots) we shall have $\sqrt{a}-\sqrt{b} = \sqrt{d}$, thus a single root. See A.318.

(**b**) The present example is treated as follows. First,
$$\frac{10}{2+\sqrt[4]{3}} = \frac{10}{(2+\sqrt[4]{3})(2-\sqrt[4]{3})} \cdot (2-\sqrt[4]{3}) = \frac{10}{4-\sqrt{3}} \cdot (2-\sqrt[4]{3});$$
next, dividing 10 by $4 - \sqrt{3}$ as taught before (A.301),
$$\frac{10}{4-\sqrt{3}} = \frac{10\,(4+\sqrt{3})}{(4-\sqrt{3})(4+\sqrt{3})} = \frac{40+\sqrt{300}}{13}.$$
Thus we find
$$\frac{10}{2+\sqrt[4]{3}} = \frac{40+\sqrt{300}}{13} \cdot (2-\sqrt[4]{3}).$$

(**A.318**) MS $\mathcal{A}$. Divide 10 by $\sqrt[4]{3} + \sqrt[4]{12}$ (case *iii*, reducible).
$$\frac{10}{\sqrt[4]{3}+\sqrt[4]{12}} = \frac{10}{(\sqrt[4]{3}+\sqrt[4]{12})(\sqrt[4]{12}-\sqrt[4]{3})} \cdot (\sqrt[4]{12}-\sqrt[4]{3})$$
$$= \frac{10}{\sqrt{12}-\sqrt{3}} \cdot (\sqrt[4]{12}-\sqrt[4]{3}) = \frac{10}{\sqrt{3}} \cdot (\sqrt[4]{12}-\sqrt[4]{3})$$
$$= \frac{\sqrt{100}}{\sqrt{3}} \cdot (\sqrt[4]{12}-\sqrt[4]{3}) = \sqrt{33+\frac{1}{3}} \cdot (\sqrt[4]{12}-\sqrt[4]{3}).$$

(**A.319**) MS $\mathcal{A}$. Divide 10 by $\sqrt{5} - \sqrt[4]{2}$ (case *ii*).
$$\frac{10}{\sqrt{5}-\sqrt[4]{2}} = \frac{10}{(\sqrt{5}-\sqrt[4]{2})(\sqrt{5}+\sqrt[4]{2})} \cdot (\sqrt{5}+\sqrt[4]{2})$$
$$= \frac{10}{5-\sqrt{2}} \cdot (\sqrt{5}+\sqrt[4]{2}) = \frac{10\,(5+\sqrt{2})}{23} \cdot (\sqrt{5}+\sqrt[4]{2}).$$

— **Three terms in the divisor.**

Remark. This section would better fit after A.302 (accordingly, we have discussed it at the end of Section 2 in our summary, p. 1307).

Suppose we have a trinomial of the form $\sqrt{a} + \sqrt{b} + \sqrt{c}$ in the divisor. Since, as seen in the summary,
$$(\sqrt{a}+\sqrt{b}+\sqrt{c})(\sqrt{a}+\sqrt{b}-\sqrt{c}) = (a+b-c) + \sqrt{4\,a\cdot b},$$
there are three possibilities, detailed in A.320.[‡]
(*i*) If $c < a+b$, we have the binomial $\sqrt{e}+d$;
(*ii*) if $c > a+b$, we have the apotome $\sqrt{e}-d$;
(*iii*) if $c = a+b$, we have the quantity rational in square $\sqrt{e}$.

(**A.320**) MS $\mathcal{A}$. Divide 10 by $2 + \sqrt{3} + \sqrt{10}$.

A   B   H G   D

Fig. 45

[‡] The particular case $a\cdot b$ a square is of course omitted.

(**a**) The three above cases are asserted, and proved as follows. Let (Fig. 45) the trinomial be $\sqrt{a} + \sqrt{b} + \sqrt{c} = AD$, with $\sqrt{a} = AB$, $\sqrt{b} = BG$, $\sqrt{c} = GD$ (and supposing here $AB + BG > GD$). Let $HG = GD$, thus $AH = \sqrt{a} + \sqrt{b} - \sqrt{c}$, and let us consider the product $AH \cdot AD$.

Since G is the mid-point of HD while AH is added to HD, by *Elements* II.6 (= PE_6),

$$AH \cdot AD + HG^2 = AG^2, \quad \text{thus} \quad AH \cdot AD = AG^2 - HG^2.$$

Consider first AG^2. Since this is the product of a binomial into itself, it is a first binomial (*Elements* X.54; see also below, A.323a, case i); that is, AG^2 is a number plus a square root, the number being $AB^2 + BG^2$ and the square root $2 \cdot AB \cdot BG$. As for the subtracted quantity HG^2, this is a number. Therefore

(*i*) if $HG^2 < AB^2 + BG^2$, $AH \cdot AD$ is a binomial;

(*ii*) if $HG^2 > AB^2 + BG^2$, $AH \cdot AD$ is an apotome;

(*iii*) if $HG^2 = AB^2 + BG^2$, $AH \cdot AD$ is rational in square.

(**b**) *Computation*.

$$\frac{10}{2+\sqrt{3}+\sqrt{10}} = \frac{10}{(2+\sqrt{3}+\sqrt{10})(2+\sqrt{3}-\sqrt{10})} \cdot (2+\sqrt{3}-\sqrt{10})$$

$$= \frac{10}{(2+\sqrt{3})^2 - 10} \cdot (2+\sqrt{3}-\sqrt{10}) = \frac{10}{4\cdot\sqrt{3}-3} \cdot (2+\sqrt{3}-\sqrt{10})$$

$$= \frac{10}{\sqrt{48}-3} \cdot (2+\sqrt{3}-\sqrt{10}).$$

We thus end up with the cases seen in A.298–301.

(**A.321**) MS $\mathcal{A}$. Divide $\sqrt{10}$ by $\sqrt{6}+\sqrt{7}+\sqrt{8}$.

$$\frac{\sqrt{10}}{\sqrt{6}+\sqrt{7}+\sqrt{8}} = \frac{\sqrt{10}}{(\sqrt{6}+\sqrt{7}+\sqrt{8})(\sqrt{6}+\sqrt{7}-\sqrt{8})} \cdot (\sqrt{6}+\sqrt{7}-\sqrt{8})$$

$$= \frac{\sqrt{10}}{(\sqrt{6}+\sqrt{7})^2 - 8} \cdot (\sqrt{6}+\sqrt{7}-\sqrt{8})$$

$$= \frac{\sqrt{10}}{(13 + 2\sqrt{42}) - 8} \cdot (\sqrt{6}+\sqrt{7}-\sqrt{8})$$

$$= \frac{\sqrt{10}}{5+\sqrt{168}} \cdot (\sqrt{6}+\sqrt{7}-\sqrt{8}) = \frac{\sqrt{10}(\sqrt{168}-5)}{143} \cdot (\sqrt{6}+\sqrt{7}-\sqrt{8}).$$

(**A.322**) MS $\mathcal{A}$. Divide $10 + \sqrt{50}$ by $\sqrt{2}+\sqrt{3}+\sqrt{5}$. As seen in A.297,

$$\frac{10+\sqrt{50}}{\sqrt{2}+\sqrt{3}+\sqrt{5}} = \frac{10}{\sqrt{2}+\sqrt{3}+\sqrt{5}} + \frac{\sqrt{50}}{\sqrt{2}+\sqrt{3}+\sqrt{5}},$$

which is then treated as seen in the two previous problems.

---

4. Square root extraction of binomials and apotomes.

(A.323) MS $\mathcal{A}$. Extract the root of $8 + \sqrt{60}$. The given expression is a first binomial (case i in our summary, with $a = 8$, thus $A = 64$, and $b = B = 60$).

(a) The author begins by enumerating the six cases of binomials and indicates the form their root will take.

(i) If the given expression is a first binomial, that is, of the form $a + \sqrt{b}$ with a and b rational and $\sqrt{a^2 - b} = t \cdot a$, t rational, then its root will be a binomial, as proved in *Elements* X.54.

(ii) If the given expression is a second binomial, that is, of the form $\sqrt{a} + b$ with a and b rational and $\sqrt{a - b^2} = t \cdot \sqrt{a}$, t rational, then its root will be a first bimedial, as proved in *Elements* X.55.

(iii) If the given expression is a third binomial, that is, of the form $\sqrt{a} + \sqrt{b}$ with a and b rational and $\sqrt{a - b} = t \cdot \sqrt{a}$, t rational, then its root will be a second bimedial, as proved in *Elements* X.56.

(iv) If the given expression is a fourth binomial, that is, of the form $a + \sqrt{b}$ with a and b rational and $\sqrt{a^2 - b} = \sqrt{t} \cdot a$, t rational, then its root will be a major, as proved in *Elements* X.57.

(v) If the given expression is a fifth binomial, that is, of the form $\sqrt{a} + b$ with a and b rational and $\sqrt{a - b^2} = \sqrt{t} \cdot \sqrt{a}$, t rational, then its root will be a quantity producing, when squared, a rational plus a medial, as proved in *Elements* X.58.

(vi) If the given expression is a sixth binomial, that is, of the form $\sqrt{a} + \sqrt{b}$ with a and b rational and $\sqrt{a - b} = \sqrt{t} \cdot \sqrt{a}$, t rational, then its root will be a quantity producing, when squared, two medials, as proved in *Elements* X.59.

(b) The present problem, the author says, will exemplify the way to be followed in all six cases.

We know that here the given expression is a first binomial; indeed (*Elements*, def. 1 following X.47), we have first $8 > \sqrt{60}$, with 8 rational while $\sqrt{60}$ is not, and secondly $8^2 = 60 + 2^2$, that is, $\sqrt{8^2 - 60} = 2$ is commensurable with 8.

The computation of the root of this first binomial is performed geometrically, thus combining theory and numerical example. We shall describe it first generally and then proceed with the actual problem.

The quantity $a + \sqrt{b}$ being given, we wish to determine its square root. From $\sqrt{a + \sqrt{b}} = \sqrt{u} + \sqrt{v}$, we obtain, by squaring,

$$a + \sqrt{b} = u + v + 2\sqrt{u \cdot v}.$$

Equating separately the rational and the irrational parts gives

$$\begin{cases} u+v = a \\ u \cdot v = \tfrac{1}{4} \cdot b \end{cases}$$

with a and b known. By eliminating u, we obtain

$$(a-v)v = \tfrac{1}{4} \cdot b = av - v^2 \quad \left(\text{that is, } v^2 + \tfrac{1}{4} \cdot b = av\right).$$

How to solve this *geometrically* is taught in the *Elements*. This is performed in three steps.

(α) We take the segment of straight line a, of known length, and construct on it, as taught in *Elements* VI.28, a rectangle of known area falling short of the whole rectangle by a square area. That is (Fig. 46 α), we construct a rectangle with length $a - v = u$, height v and area equal to the given quantity $\tfrac{1}{4} \cdot b$, which leaves, when compared to the rectangle equal in height but covering the whole segment a, thus with area $a \cdot v$, a square area v^2. With this construction we have a rectangle with sides u and v; that is, u and v are now individually known as segments of straight line.

(β) The next step is to transform these segments into rectangular areas. So we construct a rectangle with length u and height 1, and another with length v and height 1 (Fig. 46 β). Together, their areas equal a.

(γ) These rectangles can now be individually transformed, by *Elements* II.14, into squares with areas u and v and therefore sides $\sqrt{u}$ and $\sqrt{v}$. This being done, we shall arrange these two squares in such a way that their sides are parallel and they themselves touch at one angle (the smaller square being outside the larger for binomials). We complete the figure by adding two rectangles, each of which has its area equal to $\sqrt{u \cdot v} = \tfrac{1}{2}\sqrt{b}$. It appears (Fig. 46 γ) that the area of the completed square will then be $\left(\sqrt{u} + \sqrt{v}\right)^2 = a + \sqrt{b}$; therefore, its side is the quantity sought.

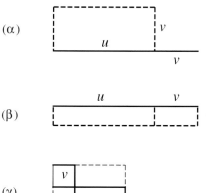

Fig. 46

In the proposed example, let (Fig. 47) $AB = a = 8$, $\sqrt{b} = BG = \sqrt{60}$, and, perpendicularly, $AZ = 1$. So we now have an area, ZG, equal to

$a+\sqrt{b} = 8+\sqrt{60}$. We are therefore to construct the square with $\sqrt{8+\sqrt{60}}$, or $\sqrt{u} + \sqrt{v}$, as its side.

(α) To do so, we first bisect $BG = \sqrt{b} = \sqrt{60}$ at D; thus $GD = DB = \frac{1}{2}\sqrt{b} = \sqrt{15}$. By *Elements* VI.28, we next construct, on the straight line $AB = a$, a rectangle equal in area to DB^2 and falling short by a square equal to BH^2. In the figure, this rectangle is $HAA'H'$ and the square on the remaining part of AB, namely on BH, is $BHH'B'$; the required point H is thus determined, and so therefore are the segments $BH = u$ and $HA = v$.

To find now numerically the lengths of these segments, we have the pair of equations
$$\begin{cases} BH + HA = BA = a \quad (=8) \\ BH \cdot HA = \frac{1}{4} \cdot b \quad (=15). \end{cases}$$

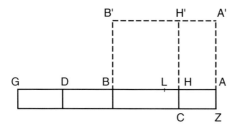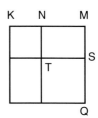

Fig. 47

We proceed as usual with a system of the form
$$\begin{cases} u + v = a \\ u \cdot v = \frac{1}{4} b, \end{cases}$$
namely, we consider the identity
$$\left(\frac{u+v}{2}\right)^2 - u \cdot v = \left(\frac{u-v}{2}\right)^2$$

two terms of which are known. Determination of the unknown term is performed geometrically in the usual way, by *Elements* II.5: we bisect AB at L, thus $BL = LA \; (= 4)$, and AB is now divided into equal parts at L and unequal parts at H, so that
$$BH \cdot HA + LH^2 = LA^2$$

which gives
$$LH^2 = LA^2 - BH \cdot HA = \frac{1}{4} a^2 - \frac{1}{4} b = 16 - 15 = 1,$$

whence
$$LH = \frac{\sqrt{a^2 - b}}{2} = 1.$$

We then find

$$HA = LA - LH = \frac{a - \sqrt{a^2 - b}}{2} = 3 = v$$

$$BH = BL + LH = \frac{a + \sqrt{a^2 - b}}{2} = 5 = u.$$

We thus know that $\sqrt{u} + \sqrt{v} = \sqrt{5} + \sqrt{3}$, but we are still left with the geometrical representation; that is, representing $a + \sqrt{b}$ as a square with side $\sqrt{u} + \sqrt{v}$.

(β) Consider now the rectangular areas with height $AZ = 1$. We have first

$$ZH = HA \cdot AZ = \frac{a - \sqrt{a^2 - b}}{2} = 3$$

$$CB = BH \cdot HC = \frac{a + \sqrt{a^2 - b}}{2} = 5,$$

and then, as known from the data, the two rectangles on GB, each equal to $\sqrt{15}$. We now know all the elements of the rectangle ZAG, which is equal to $a + \sqrt{b}$.

(γ) To transform this whole rectangle into a square area, let us first construct (by *Elements* II.14) two squares respectively equal to the two areas ZH and CB, namely KT and TQ. Since $KT = 3$ and $TQ = 5$, their respective sides are $KN = \sqrt{3}$ and $TS = \sqrt{5}$. Furthermore, the two complementary rectangles are equal to the two rectangles on GB. The completed area KQ, by *Elements* X.54, is then equal to the whole area ZG, thus to $a + \sqrt{b} = 8 + \sqrt{60}$, the root of which is

$$KM = \sqrt{\frac{a + \sqrt{a^2 - b}}{2}} + \sqrt{\frac{a - \sqrt{a^2 - b}}{2}} = \sqrt{5} + \sqrt{3}.$$

This is the required binomial representing the root of $8 + \sqrt{60}$.

(**A.324**) MS $\mathcal{A}$. Extract the root of $225 - \sqrt{50\,000}$. The given expression, of the form $a - \sqrt{b}$, is a first apotome (case i below).

(**a**) The author begins by enumerating, as he did in the previous problem, the different types of root according to the kind of apotome given. They are the following.

(i) If the given expression is a first apotome, that is, of the form $a - \sqrt{b}$ with a and b rational and $\sqrt{a^2 - b} = t \cdot a$, t rational, then its root will be an apotome, as proved in *Elements* X.91.

(ii) If the given expression is a second apotome, that is, of the form $\sqrt{a} - b$ with a and b rational and $\sqrt{a - b^2} = t \cdot \sqrt{a}$, t rational, then its root will be a first apotome of a medial, as proved in *Elements* X.92.

(iii) If the given expression is a third apotome, that is, of the form $\sqrt{a} - \sqrt{b}$ with a and b rational and $\sqrt{a - b} = t \cdot \sqrt{a}$, t rational, then its root will be a second apotome of a medial, as proved in *Elements* X.93.

(*iv*) If the given expression is a fourth apotome, that is, of the form $a - \sqrt{b}$ with a and b rational and $\sqrt{a^2 - b} = \sqrt{t} \cdot a$, t rational, then its root will be a minor, as proved in *Elements* X.94.

(*v*) If the given expression is a fifth apotome, that is, of the form $\sqrt{a} - b$ with a and b rational and $\sqrt{a - b^2} = \sqrt{t} \cdot \sqrt{a}$, t rational, then its root will be a quantity producing, when squared, with a rational a medial, as proved in *Elements* X.95.

(*vi*) If the given expression is a sixth apotome, that is, of the form $\sqrt{a} - \sqrt{b}$ with a and b rational and $\sqrt{a - b} = \sqrt{t} \cdot \sqrt{a}$, t rational, then its root will be a quantity producing, when squared, with a medial a medial, as proved in *Elements* X.96.

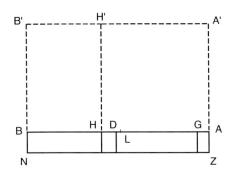
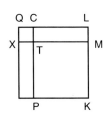

Fig. 48

(**b**) Given thus $a - \sqrt{b} = 225 - \sqrt{50\,000}$, with $\sqrt{a^2 - b} = \sqrt{625} = 25$ which is, like a, rational. The result is computed with a geometric illustration and we shall just, as before, follow the reasoning. Let (Fig. 48) $a = AB$ ($= 225$), $\sqrt{b} = GB$ ($= \sqrt{50\,000}$), so $GA = a - \sqrt{b}$. Taking $AZ = 1$, form the areas $ZB = a$ and $ZG = a - \sqrt{b}$. Let GB be bisected at D, thus $BD = DG = \frac{1}{2} \cdot \sqrt{b}$ ($= \sqrt{12\,500}$). By *Elements* VI.28, construct on AB a rectangle equal in area to DG^2 and falling short by a square; let it be the rectangle $BHH'B'$, with area $BH \cdot HA = DG^2 = 12\,500$, which leaves on HA the square area $HAA'H'$.

Remark. This is again the geometric solution of the equation $v^2 + \frac{1}{4} \cdot b = av$, or $(a - v)v = \frac{1}{4} \cdot b$, where $a = BA$, $\frac{1}{4} \cdot b = GD^2$, $v = BH$, so $a - v = HA$.

We must now determine the numerical value of the segments BH and HA. We know that

$$\begin{cases} BH + HA = AB = a \quad (= 225) \\ BH \cdot HA = \frac{1}{4} \cdot b \quad (= 12\,500). \end{cases}$$

We shall just proceed as before (which is why the author gives the value of BH directly). Bisecting AB at, say, L (not in the author's figure since this step is omitted), we have $BL = LA$ ($= 112 + \frac{1}{2}$) and, using *Elements* II.5,

$$BH \cdot HA + HL^2 = BL^2, \quad \text{so}$$

$$HL^2 = BL^2 - BH \cdot HA = 12\,656 + \frac{1}{4} - 12\,500 = 156 + \frac{1}{4},$$

thus $HL = 12 + \frac{1}{2}$ and

$$BH = BL - HL = \frac{a - \sqrt{a^2 - b}}{2} = 100 = v$$

$$HA = LA + HL = \frac{a + \sqrt{a^2 - b}}{2} = 125 = u$$

whence also for the rectangular areas with height equal to 1:

$$NH = BH \cdot BN = \frac{a - \sqrt{a^2 - b}}{2} = 100$$

$$ZH = HA \cdot AZ = \frac{a + \sqrt{a^2 - b}}{2} = 125.$$

Let us now construct (by *Elements* II.14) the two square areas $QK = ZH = u$ and $TK = NH = v$. Therefore (as seen in *Elements* X.54 and in the preceding problem) area $QT = $ area ZG. Since ZG equals $ZB - GN = a - \sqrt{b}$, the side CT of QT, thus $\sqrt{a - \sqrt{b}} = \sqrt{u} - \sqrt{v}$ will give us the required root. But the side of TK is $TP = \sqrt{100}$ and that of QK is $CP = \sqrt{125}$; therefore

$$CT = CP - TP = \sqrt{\frac{a + \sqrt{a^2 - b}}{2}} - \sqrt{\frac{a - \sqrt{a^2 - b}}{2}} = \sqrt{125} - 10.$$

We have thus found that $\sqrt{225 - \sqrt{50\,000}} = \sqrt{125} - 10$.

Having taught the rule for extracting the root of a first binomial and a first apotome, the author says that we shall treat the remaining cases likewise. One example will show this for a second binomial (A.325); a further example, for a fourth binomial (A.326), is not reducible.

(A.325) MS $\mathcal{A}$. Extract the root of $10 + \sqrt{180}$. The given expression, of the form $\sqrt{a} + b$ ($\sqrt{180} > 10$), is a second binomial (case ii in A.324a and in our summary, p. 1313). The treatment is similar to that in A.323. But only the computations are described in the text here.

Putting $\sqrt{\sqrt{a} + b} = \sqrt{u} + \sqrt{v}$, hence $\sqrt{\sqrt{180} + 10} = \sqrt{u} + \sqrt{v}$, and, squaring, we find $\sqrt{180} + 10 = u + v + 2\sqrt{uv}$, whence (since always, for $u \neq v$, $u + v > 2\sqrt{uv}$, and $\sqrt{180} > 10$)

$$\begin{cases} u + v = \sqrt{180} \\ u \cdot v = \frac{1}{4} \cdot 100 = 25. \end{cases}$$

Applying the identity

$$\left(\frac{u - v}{2}\right)^2 = \left(\frac{u + v}{2}\right)^2 - u \cdot v = \left(\frac{\sqrt{180}}{2}\right)^2 - 25 = 45 - 25 = 20,$$

we now know

$$u, v = \frac{u+v}{2} \pm \frac{u-v}{2} = \sqrt{45} \pm \sqrt{20},$$

whence

$$u = \sqrt{45} + \sqrt{20} = 3 \cdot \sqrt{5} + 2 \cdot \sqrt{5} = 5 \cdot \sqrt{5} = \sqrt{125},$$

$$v = \sqrt{45} - \sqrt{20} = 3 \cdot \sqrt{5} - 2 \cdot \sqrt{5} = \sqrt{5},$$

and therefore

$$\sqrt{\sqrt{180} + 10} = \sqrt{u} + \sqrt{v} = \sqrt[4]{125} + \sqrt[4]{5}.$$

The root of the given expression, which is a second binomial, must be, as seen in case *ii*, a first bimedial; and indeed $\sqrt[4]{125} \cdot \sqrt[4]{5} = 5$ is rational, as stated in *Elements* X.37.

As asserted by the author, the geometric justification would be the same as for case *i* (A.323). We would also calculate the root of a third binomial in the same way. Furthermore, the root of a second apotome would be obtained by analogy to the root of a first apotome (A.324). The same method would be used for a third apotome.

But for a fourth, fifth or sixth binomial, and for the corresponding apotomes, there is no simplification, on the contrary. Thus, as we have seen before and as the text states here, the answer should just be like the question. The last example illustrates this for a fourth binomial.

(A.326) MS *A*. Extract the root of $10 + \sqrt{80}$. The proposed quantity is thus of the form $a + \sqrt{b}$ and therefore corresponds to case *iv* in A.323a and in our summary ($\sqrt{a^2 - b} = \sqrt{20}$ not commensurable with $a = 10$). The rule will give

$$\sqrt{10 + \sqrt{80}} = \sqrt{\frac{10 + \sqrt{20}}{2}} + \sqrt{\frac{10 - \sqrt{20}}{2}} = \sqrt{5 + \sqrt{5}} + \sqrt{5 - \sqrt{5}}$$

and the proposed problem is expressed more conveniently in its original form (on the left); indeed, the right side will contain two terms like the given one.

Book B

As mentioned by the author at the end of his general introduction (Translation, pp. 581–582), among the applications of the science of number are, on the one hand, arithmetical reckoning (which is the subject of Book A) and, on the other, various practical ones such as that of buying and selling. This, then, is the subject of the first chapter of Book B, which opens with a short introduction.

★ ★ ★

Introduction

As a preliminary, two rules are noted, or rather recalled since they have in fact already been used several times in Book A, either for demonstrations (mostly in the Premisses) or in the computations (see, e.g., A.91a, A.155–159, A.165, A.202–205). What is new is that they will be applied to the determination of an unknown term in a proportion in which all terms have a concrete signification. The first rule (Rule of Three) is concerned with calculating this unknown term; the second gives equivalent ways of doing so.

(i) As known from Book A (and *Elements* VII.19), if

$$\frac{a_1}{a_2} = \frac{a_3}{a_4}, \quad \text{then} \quad a_1 \cdot a_4 = a_2 \cdot a_3.$$

The two terms a_1, a_4 are called 'associates' (or 'conjugates', Arabic *muqtarinān*), as are the two other terms a_2, a_3. When one of the four terms is required, we shall divide one term of the other pair by the associate of the required term and multiply the quotient by the remaining term; or else, as the author adds, we shall form the product of the other pair of associates and divide it by the associate of the unknown term, which is equivalent to the previous way (see below, *ii*). Thus

$$a_1 = \frac{a_2}{a_4} \cdot a_3 = \frac{a_3}{a_4} \cdot a_2 = \frac{a_2 \cdot a_3}{a_4}, \quad a_2 = \frac{a_1}{a_3} \cdot a_4 = \frac{a_4}{a_3} \cdot a_1 = \frac{a_1 \cdot a_4}{a_3};$$

$$a_3 = \frac{a_1}{a_2} \cdot a_4 = \frac{a_4}{a_2} \cdot a_1 = \frac{a_1 \cdot a_4}{a_2}, \quad a_4 = \frac{a_2}{a_1} \cdot a_3 = \frac{a_3}{a_1} \cdot a_2 = \frac{a_2 \cdot a_3}{a_1}.$$

All this is illustrated in the text, taking as an example the four numbers 4, 10, 6, 15 (4 : 10 = 6 : 15), each of which is successively determined by means of the three others.

Remark. We are thus told how to compute each of the four terms, including the case in which a_1 or a_2 is unknown. In what follows, however, the first two terms are usually given.

(*ii*) The second rule justifies the equivalence of those three calculations of the required number. Indeed, for any three numbers a_1, a_2, a_3, we have (by Premiss P_5)
$$\frac{a_1 \cdot a_2}{a_3} = \frac{a_1}{a_3} \cdot a_2 = \frac{a_2}{a_3} \cdot a_1.$$
To illustrate this, the author again takes the numbers 4, 6, 10 (with 4 as divisor). The manuscripts contain marginal illustrations of all these examples. Note that these 'three ways', which symbolism makes self-evident, will constantly recur as alternative treatments in the solution of problems.[‡]

[‡] See Index below (p. 1734), 'ways'.

Chapter B–I:
Buying and selling

Summary

Profit (the subject of B–II) does not appear in this first chapter; therefore, when we are told about quantities and prices, the price per unit is the same whether buying or selling. Even if in the same problem the author speaks about bushels bought and sold (B.23, B.27, see Translation) or given and received (B.24, B.25), the unit price remains unchanged.

Remark. The units of hollow measure are the *sextarius* and the *modius*. There is, however, no clear quantitative difference, at least according to the prices indicated; see, for example, B.5 and B.9.

Suppose a quantity q_1 is bought (or sold) for p_1 and a quantity q_2 for p_2. We then have the relation

$$\frac{\text{quantity}}{\text{price}} = \frac{\text{quantity}}{\text{price}}, \quad \text{or} \quad \frac{q_1}{p_1} = \frac{q_2}{p_2}.$$

All the problems in the present chapter, except those at the end, rely on applications of this simple formula. But they range from banal to fairly intricate cases. In the simplest, three quantities are given and the remaining one is required. Less simple is the case where the two quantities on the right side are required and their sum, difference or product is given. A further case arises when two analogous quantities, namely the prices, are unknown but both are expressed as a linear function of another unknown quantity, the price of a 'thing'; this may be solved by transforming the proportion, as in the previous case of two unknowns, or using algebra of the first degree. The more difficult cases lead to the second degree: q_1 and p_2, thus the opposite terms in the proportion (the 'associates' of the introduction), are the two unknowns of which product and sum or difference are given, or all four are unknown but we are given $q_1 \cdot p_1$, $q_2 \cdot p_2$, as well as the sum, difference or product of $q_1 + p_1$ and $q_2 + p_2$.

This is our first application of the transformation of proportions in order to solve a problem and our first encounter with algebra.[†] Indeed, we are taught here some fundamental algebraic operations which will be constantly referred to later on. Determining an unknown was indeed mentioned in the introduction to the *Liber mahameleth* as occurring in the problems on commercial transactions.

Remark. The various transformations of proportions are not explained here, the reader being supposed to be acquainted with the fundamentals of Euclid's *Elements*.

[†] The problems A.159–A.166, A.204–A.213 ('on unknown amounts') were linear problems, but without algebraic designation of the unknown.

1. The five fundamental problems.

The first two cases of these five problems are a plain application of the above proportion (rule of three) since only one of the two terms on the right side of the proportion is required: we know either q_1, p_1, q_2 or q_1, p_1, p_2. In the other three cases, a transformation of the proportion is required since the two terms on the right side are unknown but their sum, difference or product is given: we know q_1, p_1, q_2+p_2, or q_1, p_1, p_2-q_2, or q_1, p_1, $q_2 \cdot p_2$.*

A. *One unknown.*

— **First case** (B.1, B.3–6, B.11–12). Given q_1, p_1, q_2, required the corresponding price p_2; thus the fourth term of the proportion is unknown. This is the problem called 'what is the price?', for we are to find the cost of a known quantity; it is solved by computing p_2 according to the three aforementioned ways (p. 1344):

$$p_2 = \frac{p_1 \cdot q_2}{q_1} = \frac{q_2}{q_1} \cdot p_1 = \frac{p_1}{q_1} \cdot q_2.$$

— **Second case** (B.2, B.7–10, B.13–14). Given q_1, p_1, p_2, required the quantity of bushels q_2 bought or sold for the price p_2; thus the third term of the proportion is unknown. This is the problem 'what shall I have?' or 'what is due to me?', for we are to find the quantity corresponding to a given amount of money; it is solved by computing q_2 as

$$q_2 = \frac{q_1 \cdot p_2}{p_1} = \frac{q_1}{p_1} \cdot p_2 = \frac{p_2}{p_1} \cdot q_1.$$

B. *Two unknowns.*

— **First case** (B.15). Along with p_1, q_1, we know q_2+p_2. Starting with the fundamental proportion

$$\frac{q_1}{p_1} = \frac{q_2}{p_2},$$

we find, by composition,

$$\frac{q_1}{q_1+p_1} = \frac{q_2}{q_2+p_2}, \quad \frac{q_1+p_1}{p_1} = \frac{q_2+p_2}{p_2}$$

so that the two required quantities will be

$$q_2 = \frac{q_1(q_2+p_2)}{q_1+p_1}, \quad p_2 = \frac{p_1(q_2+p_2)}{q_1+p_1}.$$

---

* Giving the quotient is pointless here since this value is already known from q_1 and p_1.

— **Second case** (B.16). We know $p_2 - q_2$. Since also $p_1 > q_1$, we obtain by separation and by conversion, respectively

$$\frac{q_1}{p_1 - q_1} = \frac{q_2}{p_2 - q_2}, \qquad \frac{p_1 - q_1}{p_1} = \frac{p_2 - q_2}{p_2},$$

and therefore

$$q_2 = \frac{q_1 (p_2 - q_2)}{p_1 - q_1}, \qquad p_2 = \frac{p_1 (p_2 - q_2)}{p_1 - q_1}.$$

— **Third case** (B.17). We know $q_2 \cdot p_2$. The transformations

$$\frac{q_1}{p_1} = \frac{q_2}{q_2} \cdot \frac{q_2}{p_2} = \frac{q_2^2}{q_2 \cdot p_2}, \qquad \frac{q_1}{p_1} = \frac{q_2}{p_2} \cdot \frac{p_2}{p_2} = \frac{q_2 \cdot p_2}{p_2^2},$$

lead to

$$q_2 = \sqrt{\frac{q_1 (q_2 \cdot p_2)}{p_1}}, \qquad p_2 = \sqrt{\frac{p_1 (q_2 \cdot p_2)}{q_1}}.$$

B.18–20 are of the same kind, except that the data involve the roots of the two unknowns.

2. Problems involving 'things'.

The fundamental relation in B.21–29 is again

$$\frac{q_1}{p_1} = \frac{q_2}{p_2}.$$

Here we know q_1, q_2 while p_1 and p_2 are linear functions of x, the price of a 'thing'; for even if *res*, translation of the Arabic *shay'*, corresponds as a rule to the algebraic unknown x, here it mainly represents the unknown value of an unspecified object. All these problems end with a linear equation. To reach it, two basic approaches are employed. The first transforms the proportion so as to obtain directly x (or some multiple of it, ax) equal to a number. The second uses the unmodified proportion and, since x occurs on both sides, involves algebraic reckoning, that is, involves the two operations of *jabr* and *muqābala* (see below).

— **Transformation of the proportion.**

This treatment will be applied only if the two prices contain the same number of x's, that is, if $p_1 = a_1 x + b_1$ and $p_2 = a_2 x + b_2$ with $|a_1| = |a_2|$. For, as said above, the purpose of transforming the proportion is to eliminate the x's from one side, thus leaving a linear expression in x equal to a number.

• Case $a_1 = a_2$, thus $p_1 = ax + b_1$ and $p_2 = ax + b_2$ (B.21–25). The unknown x will then be eliminated from one of the sides by separation of the ratio. For

$$\frac{p_1}{p_2} = \frac{q_1}{q_2} \quad \text{gives} \quad \frac{p_1 - p_2}{p_2} = \frac{q_1 - q_2}{q_2}, \quad \text{so}$$

$$ax + b_2 = p_2 = \frac{q_2}{q_1 - q_2} \cdot (p_1 - p_2) = \frac{q_2}{q_1 - q_2} \cdot (b_1 - b_2),$$

and therefore
$$ax = \frac{q_2}{q_1 - q_2} \cdot (b_1 - b_2) - b_2.$$

- Case $a_1 = -a_2$, thus $p_1 = ax + b_1$ and $p_2 = -ax + b_2$ (B.27–29). The unknown x will be eliminated from one of the sides by composition of the ratio. For
$$\frac{p_1}{p_2} = \frac{q_1}{q_2} \quad \text{gives} \quad \frac{p_1 + p_2}{p_2} = \frac{q_1 + q_2}{q_2}, \quad \text{so}$$

$$-ax + b_2 = p_2 = \frac{q_2}{q_1 + q_2} \cdot (p_1 + p_2) = \frac{q_2}{q_1 + q_2} \cdot (b_1 + b_2),$$

and therefore
$$ax = b_2 - \frac{q_2}{q_1 + q_2} \cdot (b_1 + b_2).$$

In both cases, the quantity ax obtained must be positive.

— **Using the unmodified proportion.**

This is what the text calls 'general rule' (*regula generalis*, B.27b & B.28b), because it does not depend on the values of the coefficients of x, and we just determine x by equating the product of the associates ($q_2 \cdot p_1 = q_1 \cdot p_2$). We then obtain

$$a_1 x + b_1 = p_1 = \frac{q_1}{q_2} \cdot p_2 = \frac{q_1}{q_2} \cdot a_2 x + \frac{q_1}{q_2} \cdot b_2,$$

or
$$a_2 x + b_2 = p_2 = \frac{q_2}{q_1} \cdot p_1 = \frac{q_2}{q_1} \cdot a_1 x + \frac{q_2}{q_1} \cdot b_1.$$

We shall then usually end with x's on both sides, whereupon the two operations of *jabr* and *muqābala*, or one of them, may need to be applied in order to reach the final equation. The first means adding to both sides the amount of the subtracted terms, which will leave us with an equation containing additive terms only. The second means removing from both sides the common quantities, which will leave us with an equation containing one term for each power of the unknown —in the present case only two since all equations are linear. The problems have been chosen, or so it seems, as examples of the various possible combinations. Thus the forms of the unreduced equations are the following (with α_i, β_i positive):

$\alpha_1 x + \beta_1 = \alpha_2 x$ (B.21a; subsequent use of *muqābala* only, $\alpha_2 > \alpha_1$)

$\alpha_1 x + \beta_1 = \alpha_2 x + \beta_2$ (B.22a; *muqābala* only, twice)

$\alpha_1 x + \beta_1 = \alpha_2 x - \beta_2$ (B.23b, B.26; *jabr* and *muqābala*, $\alpha_2 > \alpha_1$)

$\beta_1 - \alpha_1 x = \alpha_2 x$ (B.27b; *jabr* only)

$\alpha_1 x + \beta_1 = \beta_2 - \alpha_2 x$ (B.28b; *jabr* and *muqābala*, $\beta_2 > \beta_1$)

$\alpha_1 x - \beta_1 = \beta_2 - \alpha_2 x$ (B.29; *jabr* only, twice).[‡]

When for a linear problem the author later on refers to what he has taught 'in algebra', this will be to one of these cases. Or else it will be to solving the three cases of quadratic equations with positive coefficients b and c, namely $x^2 + c = bx$, $x^2 = bx + c$, $x^2 + bx = c$, each of which is to be examined in the next section (B.30, B.31, B.38).

3. Second-degree problems.

B.30–42 are of a higher level; indeed, some of them are the most difficult in the work, even if the fundamental relation remains

$$\frac{q_1}{p_1} = \frac{q_2}{p_2}.$$

A. *Two quantities unknown*.

In the first three problems B.30–31′ the data link the pairs of opposite terms of the proportion, thus the 'associate' terms as defined in the introduction to Book B. For we have in the first two cases $q_1 \cdot p_2 = c$ and $p_2 \pm q_1 = b$, with b and c given ($q_1 - p_2 = b$ in B.31′). The calculation of the solution is based on the identities involving half-sum and half-difference, as usual in the *Liber mahameleth* and as the geometrical proofs make evident. We have already met that in Book A (A.214, see p. 1273). See also below.

B. *The four quantities are unknown*.

B.32–36 are the difficult problems mentioned above. Along with the values of $q_1 \cdot p_1$, $q_2 \cdot p_2$ we are also given $(q_1+p_1)+(q_2+p_2)$ in B.32–33, $(q_2+p_2)-(q_1+p_1)$ in B.34, $(q_1+p_1)\cdot(q_2+p_2)$ in B.35–36. These problems are solved as follows.
Since

$$\frac{q_1}{q_2} = \frac{p_1}{p_2}, \quad \text{then} \quad \frac{q_1^2}{q_2^2} = \frac{q_1}{q_2} \cdot \frac{q_1}{q_2} = \frac{q_1}{q_2} \cdot \frac{p_1}{p_2} = \frac{q_1 \cdot p_1}{q_2 \cdot p_2},$$

so we find first that

$$\frac{q_1}{q_2} = \sqrt{\frac{q_1 \cdot p_1}{q_2 \cdot p_2}}.$$

Next, since

$$\frac{q_1}{p_1} = \frac{q_2}{p_2}, \quad \text{so} \quad \frac{q_1}{q_1+p_1} = \frac{q_2}{q_2+p_2},$$

we also find that

$$\frac{q_1}{q_2} = \frac{q_1+p_1}{q_2+p_2}.$$

Therefore we have

$$\frac{q_1+p_1}{q_2+p_2} = \sqrt{\frac{q_1 \cdot p_1}{q_2 \cdot p_2}},$$

[‡] The cases in which x occurs on one side only are found, as said, when transformation of the proportion is involved.

whence
$$q_1 + p_1 = \sqrt{\frac{q_1 \cdot p_1}{q_2 \cdot p_2}} \, (q_2 + p_2),$$

from which we infer respectively that
$$(q_1 + p_1)^2 = \left(\sqrt{\frac{q_1 \cdot p_1}{q_2 \cdot p_2}}\right)(q_1 + p_1) \cdot (q_2 + p_2),$$

$$(q_2 + p_2) \pm (q_1 + p_1) = \left(\sqrt{\frac{q_2 \cdot p_2}{q_1 \cdot p_1}} \pm 1\right) \cdot (q_1 + p_1).$$

The second formula enables us to calculate $q_1 + p_1$ when $q_1 \cdot p_1$, $q_2 \cdot p_2$, $(q_2 + p_2) \pm (q_1 + p_1)$ are given, and so does the first when $q_1 \cdot p_1$, $q_2 \cdot p_2$, $(q_1 + p_1) \cdot (q_2 + p_2)$ are known.

In order now to compute q_1 and p_1 separately, we shall use, as in A.214 & A.323–325, the procedure well known since Mesopotamian times, which relies on the two identities involving half-sum and half-difference. In
$$\left(\frac{q_1 + p_1}{2}\right)^2 - q_1 \cdot p_1 = \left(\frac{q_1 - p_1}{2}\right)^2,$$

we know both terms on the left side and can thus compute the term on the right side. After taking its root, we shall know half the sum and half the difference of the two unknowns, and they themselves can then be computed by the second identity, namely
$$q_1, p_1 = \frac{q_1 + p_1}{2} \pm \frac{q_1 - p_1}{2}.$$

We shall determine q_2 and p_2 in the very same manner.

C. *Other problems*.

In this third part we continue mostly with the second degree. In the first two problems, there are roots in the data, with a single quantity of bushels q being a fraction of the square root of its price p, and we know $p \pm q$; we end in this case by solving equations of the types $x^2 = bx + c$ and $x^2 + bx = c$ (B.37–38), the unknown being the root of the required price. We return to pairs of unknowns p_i in the next two problems, where 'things' occur since $p_1 = x \pm \beta$ and $p_2 = k \cdot \sqrt{p_1}$ (β and k known) with q_1 and q_2 given (B.39–40). Finally, in B.41–42, two 'things' occur: we know $x \cdot y$, q, and we must find p from $p = x + y$ and $\frac{p}{q} = \alpha x + \beta$ (α, β known); we end, after reducing the problem, with an equation of the form $x^2 + bx = c$ ($b = 0$ in B.41).

4. Corn of various kinds.

The chapter closes with the case of bushels (*modii*, then *sextarii*) at different prices (B.43–53). Thus we do not have simply, as before, $q_1 : q_2 = p_1 : p_2$ which was applicable to a single type of corn at a single unit price. Here we have corn of n different kinds, q_i bushels of the ith kind, altogether Q bushels, bought for P, and the ith kind costs $p_0^{(i)}$ per bushel. The fundamental relations are then

$$\sum_1^n p_0^{(i)} \cdot q_i = P, \quad \text{with} \quad \sum q_i = Q.$$

This is our first encounter with another algebraic topic, namely solving pairs of linear equations. Here we shall briefly see the condition of solution for $n = 2$ (B.46–47) and an indeterminate case ($n = 3$, B.51); this subject will be thoroughly treated in the chapter on exchanging moneys (B–XVIII). Finally, in the last two problems (B.52–53), the prices $p_0^{(i)}$ form an arithmetical progression.

1. The five fundamental problems.

As said in our summary of this chapter, the initial problems (B.1–14) are concerned with determining, by means of the proportion

$$\frac{q_1}{p_1} = \frac{q_2}{p_2},$$

one unknown on the right side, either p_2 ('what is the price?') or q_2 ('how much shall I have?'). A simple example of each is followed by further examples, mostly involving fractions in the data.

A. *One unknown*.

— **First case: 'What is the price?'**

(B.1) MSS $\mathcal{A}, \mathcal{B}, \mathcal{C}$. The price of three sextarii is ten nummi; what is the price of fourteen sextarii?

Given thus $q_1 = 3$, $p_1 = 10$, $q_2 = 14$. The 'three ways' explained in the introduction to Book B (above, p. 1344) are used to calculate p_2 from the proportion $q_1 : p_1 = q_2 : p_2$.

$$p_2 = \frac{p_1 \cdot q_2}{q_1} = \frac{10 \cdot 14}{3} = \frac{10}{3} \cdot 14 = \frac{14}{3} \cdot 10 = 46 + \frac{2}{3}.$$

Remark. Remember that the results we put in italics do not appear in the text.

— **Second case: 'How much shall I have?'**

(B.2) MSS $\mathcal{A}, \mathcal{B}, \mathcal{C}$. The price of three sextarii is ten nummi; what shall I have for sixty nummi?

Given thus $q_1 = 3$, $p_1 = 10$, $p_2 = 60$, and we can calculate q_2 from the proportion $q_1 : p_1 = q_2 : p_2$ using the 'three ways'.∥

$$q_2 = \frac{q_1 \cdot p_2}{p_1} = \frac{3 \cdot 60}{10} = \frac{180}{10} = \frac{3}{10} \cdot 60 = \frac{60}{10} \cdot 3 = 18.$$

∥ Result calculated by the reader of MS $\mathcal{C}$.

— **Further examples of the first case ('What is the price?')**

B.3–6 deal with the first case, but involve fractions in the given quantities (except for B.5, found in MSS $\mathcal{B}$ & $\mathcal{C}$ only). In the first two, $q_1 = 1$, that is, we know the price p_1 per bushel —as usual when a commodity is proposed: see the remark in the text, at the end of B.4.

(B.3) MSS $\mathcal{A}$, $\mathcal{B}$, $\mathcal{C}$. The price of one sextarius is five nummi and a third; what is the price of ten sextarii?

Given thus $q_1 = 1$, $p_1 = 5 + \frac{1}{3}$, $q_2 = 10$, required p_2. Because $q_1 = 1$, we have simply

$$p_2 = p_1 \cdot q_2 = \left(5 + \frac{1}{3}\right) 10 = 53 + \frac{1}{3}.$$

(B.4) MSS $\mathcal{A}$, $\mathcal{B}$, $\mathcal{C}$. The price of one sextarius is six nummi and a third; what is the price of ten sextarii and a fourth?

Given $q_1 = 1$, $p_1 = 6 + \frac{1}{3}$, $q_2 = 10 + \frac{1}{4}$, required p_2. Still the simple form, but involving two fractions this time.

$$p_2 = p_1 \cdot q_2 = \left(6 + \frac{1}{3}\right)\left(10 + \frac{1}{4}\right),$$

computed according to the way taught before, the text says (A.124).

(B.5) MSS $\mathcal{B}$, $\mathcal{C}$. The price of three sextarii is ten nummi; what is the price of thirteen?

Given $q_1 = 3$, $p_1 = 10$, $q_2 = 13$, required p_2. Here we again encounter the 'three ways' since $q_1 \neq 1$.

$$p_2 = \frac{p_1 \cdot q_2}{q_1} = \frac{10 \cdot 13}{3} = \frac{10}{3} \cdot 13 = \frac{13}{3} \cdot 10.$$

Remark. As noted in the Translation, the genuineness of this problem, so similar to B.1, is questionable. See also B.9, remark.

(B.6) MSS $\mathcal{A}$, $\mathcal{B}$, $\mathcal{C}$. The price of two thirds of a sextarius is two nummi and a fourth; what is the price of ten sextarii and a half?

Given $q_1 = \frac{2}{3}$, $p_1 = 2 + \frac{1}{4}$, $q_2 = 10 + \frac{1}{2}$, required p_2.

Remark. A fraction now occurs in the divisor, and this will lead to further ways of solving. For multiplying dividend and divisor by some appropriate factor f will enable us to clear some or all the fractions and thus simplify multiplication and division; conveniently, f will be the product of all denominators, or of some only, as illustrated below. But the procedure is well known from Book A (see in particular A–VIII, Section 3 —above, p. 1276).

(a) Without auxiliary factor ('three ways'):

$$p_2 = \frac{p_1 \cdot q_2}{q_1} = \frac{\left(2 + \frac{1}{4}\right)\left(10 + \frac{1}{2}\right)}{\frac{2}{3}},$$

$$p_2 = \frac{p_1}{q_1} \cdot q_2 = \left(3 + \frac{3}{8}\right)\left(10 + \frac{1}{2}\right), \qquad p_2 = \frac{q_2}{q_1} \cdot p_1.$$

(**b**) Take $f = 3 \cdot 4 = 12$:

$$p_2 = \frac{(f \cdot p_1) q_2}{f \cdot q_1} = \frac{27 \cdot \left(10 + \frac{1}{2}\right)}{8}.$$

We have thus replaced q_1, p_1 by $q'_1 = 8$, $p'_1 = 27$.

(**c**) Take $f = 3 \cdot 2 = 6$, which will again make the divisor an integer.

$$p_2 = \frac{p_1 (f \cdot q_2)}{f \cdot q_1} = \frac{\left(2 + \frac{1}{4}\right) \cdot 63}{4}.$$

We have thus replaced q_1, q_2 by $q'_1 = 4$, $q'_2 = 63$.

(**d**) In the last part we are told, in order to clear all the fractions, to multiply *each* of the three known terms by the product of the denominators (thus by $f = 3 \cdot 4 \cdot 2 = 24$), which is of course wrong since f will occur twice in the dividend. This last part must be interpolated, for it not only is not applied in the following problems but obviously does not fit in with the recapitulating rules which come after the last problem in this group (B.14). Furthermore, the correct way of clearing the fractions is described, and proved, in B.11.

— **Further examples of the second case ('How much shall I have?')**

B.7–10 deal with the second case. As in the former group, we first consider $q_1 = 1$, but in this case the division is maintained since

$$q_2 = \frac{q_1 \cdot p_2}{p_1}.$$

(**B.7**) MS $\mathcal{C}$ only (presumably interpolated). The price of one sextarius is six nummi; how many shall I have for fifty nummi?

We are thus given $q_1 = 1$, $p_1 = 6$, $p_2 = 50$, and q_2 is required. Then

$$q_2 = \frac{p_2}{p_1} = \frac{50}{6}.$$

(**B.8**) MSS $\mathcal{A}$, $\mathcal{B}$, $\mathcal{C}$. The price of one sextarius is six nummi and a half; what shall I have for forty-four nummi and a third?

Given thus $q_1 = 1$, $p_1 = 6 + \frac{1}{2}$, $p_2 = 44 + \frac{1}{3}$, required q_2.

$$q_2 = \frac{p_2}{p_1} = \frac{44 + \frac{1}{3}}{6 + \frac{1}{2}} = 6 + \frac{10}{13} + \frac{2}{3} \cdot \frac{1}{13},$$

computed according to the way taught for dividing fractions (A.260–261), the text says.

(B.9) MSS $\mathcal{B}$, $\mathcal{C}$. The price of three modii is ten nummi; what shall I have for sixty-four?

Given thus $q_1 = 3$, $p_1 = 10$, $p_2 = 64$, required q_2. Since $q_1 \neq 1$, we again have the three ways.

$$q_2 = \frac{q_1 \cdot p_2}{p_1} = \frac{192}{10} = \frac{3}{10} \cdot 64 = \frac{64}{10} \cdot 3 = 19 + \frac{1}{5}.$$

Remark. As was the case for B.5, an offshoot of B.1, this problem is an offshoot of B.2. MS $\mathcal{A}$ omits both B.5 and B.9.

(B.10) MSS $\mathcal{A}$, $\mathcal{B}$, $\mathcal{C}$. The price of one modius and two thirds is ten nummi and a half; what shall I have for thirty-four nummi and four fifths? Given thus $q_1 = 1 + \frac{2}{3}$, $p_1 = 10 + \frac{1}{2}$, $p_2 = 34 + \frac{4}{5}$, required q_2.

(a) As before, computing

$$q_2 = \frac{q_1 \cdot p_2}{p_1} = \frac{q_1}{p_1} \cdot p_2 = \frac{p_2}{p_1} \cdot q_1.$$

(b) As in B.6, clear the fractions in two terms, one of which must be the divisor, by multiplying dividend and divisor by a quantity containing some of the denominators:

$$q_2 = \frac{(f \cdot q_1) p_2}{f \cdot p_1} \quad \text{or} \quad q_2 = \frac{q_1 (f' \cdot p_2)}{f' \cdot p_1}.$$

Then (e.g. with $f = 3 \cdot 2$ and $f' = 5 \cdot 2$) q_1 and p_1 or p_2 and p_1 are replaced by integers. The proof of this procedure is found in B.13.

* * *

(B.11) MSS $\mathcal{A}$, $\mathcal{B}$, $\mathcal{C}$. The price of five sextarii and a half is seven nummi and a third; what is the price of ten sextarii and a fifth?

This and the next problem, which are examples of the first case, should probably follow B.6. Given thus $q_1 = 5 + \frac{1}{2}$, $p_1 = 7 + \frac{1}{3}$, $q_2 = 10 + \frac{1}{5}$, required p_2. As before,

$$p_2 = \frac{p_1 \cdot q_2}{q_1}.$$

(a) Choosing as f the product of all the denominators and multiplying by f both dividend and divisor will produce a dividend and a divisor which are integral. Thus, with $f = 2 \cdot 3 \cdot 5 = 30$,

$$p_2 = \frac{(f \cdot p_1) q_2}{f \cdot q_1} = \frac{30 \left(7 + \frac{1}{3}\right)\left(10 + \frac{1}{5}\right)}{30 \left(5 + \frac{1}{2}\right)} = \frac{2244}{165} = 13 + \frac{3}{5}.$$

Proof. The validity of this procedure is based on the fact that a ratio, or here a quotient, remains the same when its two terms are multiplied by the same quantity (*Elements* VII.17–18). Thus, with $q_1 : p_1 = q_2 : p_2$,

$$\frac{q_1}{p_1} = \frac{5 + \frac{1}{2}}{7 + \frac{1}{3}} = \frac{f \cdot q_1}{f \cdot p_1} = \frac{30 \left(5 + \frac{1}{2}\right)}{30 \left(7 + \frac{1}{3}\right)} = \frac{165}{220} = \frac{q_2}{p_2} = \frac{10 + \frac{1}{5}}{p_2},$$

that is,
$$p_2 = \frac{220\left(10+\frac{1}{5}\right)}{165} = \frac{(f \cdot p_1)\, q_2}{f \cdot q_1}.$$

(*b*) Direct computation:
$$p_2 = \frac{p_1 \cdot q_2}{q_1} = \frac{\left(7+\frac{1}{3}\right)\left(10+\frac{1}{5}\right)}{5+\frac{1}{2}}.$$

(*c*) To avoid having a divisor with a fraction, we may choose as f the denominator of its fraction, thus changing here $5+\frac{1}{2}$ to 11 and multiplying the dividend by 2.

(*d*) In *b* we may always use the two 'other ways' (dividing one of the two factors).

(**B.12**) MSS *A*, *B*, *C*. The price of five eighths of a sextarius is three fourths of a nummus; what is the price of ten elevenths of a sextarius?

Similar to the preceding, but with pure fractions (one non-elementary, see A.92). Given thus $q_1 = \frac{5}{8}$, $p_1 = \frac{3}{4}$, $q_2 = \frac{10}{11}$, required p_2.

(*a*) With $f = 8 \cdot 4 \cdot 11 = 352$,
$$p_2 = \frac{(f \cdot p_1)\, q_2}{f \cdot q_1} = \frac{240}{220} = 1 + \frac{1}{11}.$$

(*b*) Choose $f = 8$, thus $f \cdot q_1 = 5$ (no more fraction in the divisor).

(*c*) Direct way, considering either
$$p_2 = \frac{p_1 \cdot q_2}{q_1} \quad \text{or} \quad p_2 = \frac{p_1}{q_1} \cdot q_2, \quad p_2 = \frac{q_2}{q_1} \cdot p_1.$$

* * *

(**B.13**) MSS *A*, *B*, *C*. The price of two sextarii and a third is seven and a half; what shall I have for six and three sevenths?

Given thus $q_1 = 2+\frac{1}{3}$, $p_1 = 7+\frac{1}{2}$, $p_2 = 6+\frac{3}{7}$, required q_2. Since we return to the second case, and in view of the likelihood that B.11 & B.12 are misplaced, this should follow B.10.

(*a*) With $f = 3 \cdot 2 \cdot 7 = 42$,
$$q_2 = \frac{(f \cdot q_1)\, p_2}{f \cdot p_1} = \frac{42\left(2+\frac{1}{3}\right)\left(6+\frac{3}{7}\right)}{42\left(7+\frac{1}{2}\right)} = \frac{98\left(6+\frac{3}{7}\right)}{315} = \frac{630}{315} = 2.$$

Proof. The validity of this procedure is shown as in B.11*a*, except that now the third term of the proportion is required. Since
$$\frac{q_1}{p_1} = \frac{2+\frac{1}{3}}{7+\frac{1}{2}} = \frac{f \cdot q_1}{f \cdot p_1} = \frac{42\left(2+\frac{1}{3}\right)}{42\left(7+\frac{1}{2}\right)} = \frac{98}{315} = \frac{q_2}{p_2} = \frac{q_2}{6+\frac{3}{7}},$$

then
$$q_2 = \frac{98\left(6+\frac{3}{7}\right)}{315} = \frac{(f \cdot q_1) p_2}{f \cdot p_1}.$$

(*b*) Direct way:
$$q_2 = \frac{\left(2+\frac{1}{3}\right)\left(6+\frac{3}{7}\right)}{7+\frac{1}{2}}, \quad \text{or} \quad q_2 = \frac{q_1}{p_1} \cdot p_2, \quad q_2 = \frac{p_2}{p_1} \cdot q_1.$$

(*c*) Choose $f = 2$, which gives an integer as the divisor ($f \cdot p_1 = 15$).

(**B.14**) MSS $\mathcal{A}$, $\mathcal{B}$, $\mathcal{C}$. The price of four fifths of a sextarius is two thirds of a nummus; what shall I have for seven eighths?

Given thus $q_1 = \frac{4}{5}$, $p_1 = \frac{2}{3}$, $p_2 = \frac{7}{8}$, required q_2. Similar to B.13 but, like B.12, with pure fractions.

(*a*) Choose $f = 5 \cdot 3 \cdot 8 = 120$, thus

$$q_2 = \frac{(f \cdot q_1) p_2}{f \cdot p_1} = \frac{\left(120 \cdot \frac{4}{5}\right)\frac{7}{8}}{120 \cdot \frac{2}{3}} = \frac{96 \cdot \frac{7}{8}}{80} = \frac{84}{80} = 1 + \frac{1}{2} \cdot \frac{1}{10}.$$

(*b*) Choose $f = 3$, which gives 2 as the new divisor.

(*c*) Direct way:
$$q_2 = \frac{\frac{4}{5} \cdot \frac{7}{8}}{\frac{2}{3}} = \frac{q_1}{p_1} \cdot p_2 = \frac{p_2}{p_1} \cdot q_1.$$

* * *

We have now come to the end of the problems on buying and selling solved by a simple proportion (rule of three). The author recapitulates what the two questions are, how they are solved, and how the divisor can be transformed into an integer (the 'principal'). Consider the proportion $q_1 : p_1 = q_2 : p_2$. In 'what is the price?', where q_1, p_1, q_2 are the known terms and p_2 is required, the divisor is q_1, which is the first of the three given terms, and the denominator of its fraction may be chosen as our f. In 'what shall I have?', where q_1, p_1, p_2 are the known terms and q_2 is required, the divisor is p_1, which is the second of the three given terms, and the denominator of its fraction may become our f.

Remark. Here, elimination of the fraction in the divisor only is considered. We saw the same in the recapitulation following A.264 (above, pp. 1295–1297), which we suggested might be a later addition (though still the author's).

B. *Two unknowns*.

Here we have the three problems with q_1 and p_1 given (as usual) but now also the sum, difference or product of q_2 and p_2 (B.15–17; in B.18–20 the roots of q_2 and p_2 occur). Similar problems, with two quantities unknown (what the text calls *de ignoto*, which we have translated by 'involving unknowns') and requiring transformation of proportions, will constantly recur in later chapters.‡

(B.15) MSS $\mathcal{A}$, $\mathcal{B}$, $\mathcal{C}$. The price of three modii is thirteen nummi; what is the quantity of modii which, added to their price, makes sixty?

Given $q_1 = 3$, $p_1 = 13$, $q_2 + p_2 = 60$, required q_2 (and p_2).

The formulae used are those seen in our introduction to this chapter, namely
$$q_2 = \frac{q_1(q_2 + p_2)}{q_1 + p_1} = \frac{3 \cdot 60}{3 + 13} = \frac{3 \cdot 60}{16},$$
$$p_2 = \frac{p_1(q_2 + p_2)}{q_1 + p_1} = \frac{13 \cdot 60}{3 + 13} = \frac{13 \cdot 60}{16}.$$

After some kind of proof (see remark following B.16 below), mention is made of the 'two ways'
$$p_2 = \frac{p_1}{q_1 + p_1}(q_2 + p_2), \qquad p_2 = \frac{q_2 + p_2}{q_1 + p_1} \cdot p_1.$$

(B.16) MSS $\mathcal{A}$, $\mathcal{B}$, $\mathcal{C}$. The price of three modii is thirteen nummi; what is the quantity of modii which, subtracted from their price, leaves sixty?

Given thus $q_1 = 3$, $p_1 = 13$, $p_2 - q_2 = 60$, required q_2 (and p_2).

The formulae used are those seen in our introduction to this chapter, namely
$$q_2 = \frac{q_1(p_2 - q_2)}{p_1 - q_1} = \frac{3 \cdot 60}{13 - 3} = \frac{3 \cdot 60}{10},$$
$$p_2 = \frac{p_1(p_2 - q_2)}{p_1 - q_1} = \frac{13 \cdot 60}{13 - 3} = \frac{13 \cdot 60}{10}.$$

After a proof similar to that above, the author once again mentions the possibility of using the two banal 'other ways'.

Remark. As to the origin of the formulae, it cannot have been altogether clear to the reader, for both in B.15 and B.16 the author just states the proportion and how to calculate the two required quantities, without any comment about the transformation of the ratio. The next problem will have no justification at all. For proper proofs, we shall have to await the next chapter (B.56–58). In the *Liber mahameleth* there often seems to have been no rereading to check for such inadequacies.

(B.17) MSS $\mathcal{A}$, $\mathcal{B}$, $\mathcal{C}$. The price of three modii is eight nummi; what is the quantity of modii which, multiplied by their price, produces two hundred and sixteen?

---

‡ See B.56–58, B.62–64, B.68–70, B.74–76, B.80–82, B.95–99, B.106–108, B.166–168, B.192–194, B.257–259, B.268–270.

Given thus $q_1 = 3$, $p_1 = 8$, $q_2 \cdot p_2 = 216$, required q_2 (and p_2).
The formulae used are

$$q_2 = \sqrt{\frac{q_1 \, (q_2 \cdot p_2)}{p_1}} = \sqrt{\frac{3 \cdot 216}{8}} = \sqrt{81} = 9,$$

$$p_2 = \sqrt{\frac{p_1 \, (q_2 \cdot p_2)}{q_1}} = \sqrt{\frac{8 \cdot 216}{3}} = \sqrt{576} = 24.$$

Should the radicands not be squares, the author adds, we shall consider approximate roots, calculated as explained at the beginning of A–IX.

Remark. Here, in MSS $\mathcal{B}$ and $\mathcal{C}$, we see traces of the problem

$$q_1 = 4, \quad p_1 = 9, \quad \sqrt[4]{q_2} \cdot \sqrt[4]{p_2} = 13 + \frac{1}{2}.$$

Since this reduces to

$$q_1 = 4, \quad p_1 = 9, \quad \sqrt{q_2} \cdot \sqrt{p_2} = 182 + \frac{1}{4},$$

it is of the same kind as B.20, to which it could once have been an addition.

★ ★ ★

The only difference between the groups B.15–17 and B.18–20 is the occurrence in the latter of roots of the required quantities. We have here a complication of form rather than of content since all these roots happen to be rational.

(B.18) MSS $\mathcal{A}$, $\mathcal{B}$. The price of four sextarii is nine nummi; what is the quantity of sextarii which, when its root is added to the root of their price, makes seven and a half?

Given thus $q_1 = 4$, $p_1 = 9$, $\sqrt{q_2} + \sqrt{p_2} = 7 + \frac{1}{2}$, required q_2 (and p_2). Similar to B.15, but with roots.

(a) By analogy to B.15,

$$\sqrt{q_2} = \frac{\sqrt{q_1}\left(\sqrt{q_2} + \sqrt{p_2}\right)}{\sqrt{q_1} + \sqrt{p_1}} = \frac{\sqrt{4}\left(7 + \frac{1}{2}\right)}{\sqrt{4} + \sqrt{9}} = 2 \cdot \frac{7 + \frac{1}{2}}{5} = 2\left(1 + \frac{1}{2}\right) = 3;$$

$$\sqrt{p_2} = \frac{\sqrt{p_1}\left(\sqrt{q_2} + \sqrt{p_2}\right)}{\sqrt{q_1} + \sqrt{p_1}} = 3 \cdot \frac{7 + \frac{1}{2}}{5} = 3\left(1 + \frac{1}{2}\right) = 4 + \frac{1}{2},$$

whence $q_2 = 9$ and $p_2 = 20 + \frac{1}{4}$.

(b) The quantity q_2 is also computed by the equivalent formula

$$\sqrt{q_2} = \frac{\sqrt{q_2} + \sqrt{p_2}}{\sqrt{\frac{p_1}{q_1}} + 1} = \frac{7 + \frac{1}{2}}{\sqrt{2 + \frac{1}{4}} + 1} = \frac{7 + \frac{1}{2}}{2 + \frac{1}{2}}.$$

(*c*) Another calculation presented is

$$\sqrt{q_2} = \sqrt{\frac{\left(\sqrt{q_2}+\sqrt{p_2}\right)^2}{\frac{p_1}{q_1}-1} + \left(\frac{1}{2}\cdot\frac{2\left(\sqrt{q_2}+\sqrt{p_2}\right)}{\frac{p_1}{q_1}-1}\right)^2} - \frac{1}{2}\cdot\frac{2\left(\sqrt{q_2}+\sqrt{p_2}\right)}{\frac{p_1}{q_1}-1}$$

$$= \sqrt{\frac{\left(7+\frac{1}{2}\right)^2}{1+\frac{1}{4}} + \left(\frac{1}{2}\cdot\frac{15}{1+\frac{1}{4}}\right)^2} - \frac{1}{2}\cdot\frac{15}{1+\frac{1}{4}}$$

$$= \sqrt{\frac{56+\frac{1}{4}}{1+\frac{1}{4}} + \left(\frac{1}{2}\cdot 12\right)^2} - \frac{1}{2}\cdot 12 = \sqrt{45+36} - 6 = 3,$$

whence q_2 and $p_2 = \frac{p_1}{q_1}\cdot q_2$.

Remark. There is no explanation. This formula may be obtained as follows. Since

$$\left(\sqrt{q_2}+\sqrt{p_2}\right)^2 = q_2 + p_2 + 2\sqrt{q_2\cdot p_2} = q_2 + \frac{p_1}{q_1}\cdot q_2 + 2\sqrt{q_2\cdot p_2}$$

$$= 2q_2 + \frac{p_1}{q_1}\cdot q_2 - q_2 + 2\sqrt{q_2\cdot p_2} = \left(\frac{p_1}{q_1}-1\right)q_2 + 2\sqrt{q_2}\left(\sqrt{q_2}+\sqrt{p_2}\right),$$

we have

$$\left(\frac{p_1}{q_1}-1\right)q_2 + 2\sqrt{q_2}\left(\sqrt{q_2}+\sqrt{p_2}\right) = \left(\sqrt{q_2}+\sqrt{p_2}\right)^2,$$

that is,

$$q_2 + \frac{2\left(\sqrt{q_2}+\sqrt{p_2}\right)}{\frac{p_1}{q_1}-1}\cdot\sqrt{q_2} = \frac{\left(\sqrt{q_2}+\sqrt{p_2}\right)^2}{\frac{p_1}{q_1}-1},$$

where p_1, q_1 and $\sqrt{q_2}+\sqrt{p_2}$ are known. Now this is an equation of the second degree for $\sqrt{q_2}$, of which the above formula is the (single) positive solution. Writing it as

$$\sqrt{q_2}\left(\sqrt{q_2} + \frac{2\left(\sqrt{q_2}+\sqrt{p_2}\right)}{\frac{p_1}{q_1}-1}\right) = \frac{\left(\sqrt{q_2}+\sqrt{p_2}\right)^2}{\frac{p_1}{q_1}-1},$$

we are then to solve the system

$$\begin{cases}\left(\sqrt{q_2} + \dfrac{2\left(\sqrt{q_2}+\sqrt{p_2}\right)}{\frac{p_1}{q_1}-1}\right) - \sqrt{q_2} = \dfrac{2\left(\sqrt{q_2}+\sqrt{p_2}\right)}{\frac{p_1}{q_1}-1}\\[2ex] \sqrt{q_2}\left(\sqrt{q_2} + \dfrac{2\left(\sqrt{q_2}+\sqrt{p_2}\right)}{\frac{p_1}{q_1}-1}\right) = \dfrac{\left(\sqrt{q_2}+\sqrt{p_2}\right)^2}{\frac{p_1}{q_1}-1},\end{cases}$$

that is, we are led to find two quantities of which we know difference and product. We have seen such a determination in A.214 and shall be meeting, from B.30 on, other examples of it.

Still, the purpose of this third treatment is not clear (to eliminate the root in the divisor?). Whatever the reason, the solution here is needlessly complicated and does not properly account for the occurrence of the single terms; furthermore, the origin of this solution will remain unintelligible to the reader —and not just because he is not yet familiar with second-degree algebra. Were it not so complex, and *mutatis mutandis* repeated in the next problem, we would have considered it the work of some interpolator.

(B.19) MSS $\mathcal{A}$, $\mathcal{B}$. The price of four sextarii is nine nummi; what is the quantity of sextarii which, when its root is subtracted from the root of their price, leaves one and a half?

Given thus $q_1 = 4$, $p_1 = 9$ (thus $p_1 > q_1$ and so $p_2 > q_2$), $\sqrt{p_2} - \sqrt{q_2} = 1 + \frac{1}{2}$, required q_2 (and p_2). Analogous to B.16, but with roots.

(a) By analogy to B.18a,

$$\sqrt{q_2} = \frac{\sqrt{q_1}\left(\sqrt{p_2} - \sqrt{q_2}\right)}{\sqrt{p_1} - \sqrt{q_1}} = \sqrt{4} \cdot \frac{1+\frac{1}{2}}{\sqrt{9}-\sqrt{4}} = 2 \cdot \frac{1+\frac{1}{2}}{1} = 3,$$

$$\sqrt{p_2} = \frac{\sqrt{p_1}\left(\sqrt{p_2} - \sqrt{q_2}\right)}{\sqrt{p_1} - \sqrt{q_1}} = 3 \cdot \frac{1+\frac{1}{2}}{1} = 4 + \frac{1}{2},$$

whence $q_2 = 9$ and $p_2 = 20 + \frac{1}{4}$.

(b) As in B.18b,

$$\sqrt{q_2} = \frac{\sqrt{p_2} - \sqrt{q_2}}{\sqrt{\frac{p_1}{q_1}} - 1} = \frac{1+\frac{1}{2}}{\sqrt{2+\frac{1}{4}} - 1} = \frac{1+\frac{1}{2}}{\frac{1}{2}} = 3.$$

(c) As in B.18c,

$$\sqrt{q_2} = \sqrt{\frac{\left(\sqrt{p_2}-\sqrt{q_2}\right)^2}{\frac{p_1}{q_1}-1} + \left(\frac{1}{2} \cdot \frac{2\left(\sqrt{p_2}-\sqrt{q_2}\right)}{\frac{p_1}{q_1}-1}\right)^2} + \frac{1}{2} \cdot \frac{2\left(\sqrt{p_2}-\sqrt{q_2}\right)}{\frac{p_1}{q_1}-1}$$

$$= \sqrt{\frac{\left(1+\frac{1}{2}\right)^2}{1+\frac{1}{4}} + \left(\frac{1}{2} \cdot \frac{3}{1+\frac{1}{4}}\right)^2} + \frac{1}{2} \cdot \frac{3}{1+\frac{1}{4}}$$

$$= \sqrt{\frac{2+\frac{1}{4}}{1+\frac{1}{4}} + \left(\frac{1}{2}\left(2+\frac{2}{5}\right)\right)^2} + \frac{1}{2}\left(2+\frac{2}{5}\right)$$

$$= \sqrt{1+\frac{4}{5} + \left(1+\frac{1}{5}\right)^2} + 1 + \frac{1}{5} = 3,$$

whence q_2 and $p_2 = \dfrac{p_1}{q_1} \cdot q_2$.

Remark. There is no explanation here either. This formula can be obtained as before. Since

$$\left(\sqrt{p_2} - \sqrt{q_2}\right)^2 = p_2 + q_2 - 2\sqrt{p_2 \cdot q_2} = \frac{p_1}{q_1} \cdot q_2 + q_2 - 2\sqrt{p_2 \cdot q_2}$$

$$= \frac{p_1}{q_1} \cdot q_2 + 2q_2 - q_2 - 2\sqrt{p_2 \cdot q_2} = \left(\frac{p_1}{q_1} - 1\right)q_2 - 2\sqrt{q_2}\left(\sqrt{p_2} - \sqrt{q_2}\right),$$

we have

$$\left(\frac{p_1}{q_1} - 1\right)q_2 = 2\sqrt{q_2}\left(\sqrt{p_2} - \sqrt{q_2}\right) + \left(\sqrt{p_2} - \sqrt{q_2}\right)^2$$

where p_1, q_1 and $\sqrt{p_2} - \sqrt{q_2}$ are known. This is again an equation of the second degree for $\sqrt{q_2}$, and the above formula is its (single) positive solution; it can be obtained by determining two quantities of which difference and product are known.

(B.20) MSS $\mathcal{A}$, $\mathcal{B}$. The price of four sextarii is nine nummi; what is the quantity of sextarii which, when its root is multiplied by the root of their price, produces twenty-four?

Given $q_1 = 4$, $p_1 = 9$, $\sqrt{q_2} \cdot \sqrt{p_2} = 24$, required q_2 (and p_2). Similar to B.17, but with roots.

(a) By analogy to B.17 (but with dividend and divisor multiplied by $\sqrt{q_1}$ and $\sqrt{p_1}$ respectively),

$$q_2 = \frac{q_1\left(\sqrt{q_2} \cdot \sqrt{p_2}\right)}{\sqrt{q_1} \cdot \sqrt{p_1}} = 4 \cdot \frac{24}{6} = 16,$$

$$p_2 = \frac{p_1\left(\sqrt{q_2} \cdot \sqrt{p_2}\right)}{\sqrt{q_1} \cdot \sqrt{p_1}} = 9 \cdot \frac{24}{6} = 36.$$

(b) An equivalent form of the relation for q_2 is:

$$q_2 = \frac{\sqrt{q_2} \cdot \sqrt{p_2}}{\sqrt{\frac{p_1}{q_1}}} = \frac{24}{1 + \frac{1}{2}}.$$

(c) The third way consists in raising $\sqrt{q_2} \cdot \sqrt{p_2} = 24$ to the square. The problem then becomes $q_1 = 4$, $p_1 = 9$, $q_2 \cdot p_2 = 576$, which brings us back to B.17. A similar reduction is found in B.95.

Remark. The fragment found in B.17 must be a reader's gloss to this problem (since we are told there to proceed as taught°).

° *Tu, fac sicut supra docui.* The expression 'proceed as *I* have taught above' is no proof of genuineness.

2. Problems involving 'things'.

In Section B.21-29 there are again two unknowns but, unlike before, of the same kind (here they are the prices). As a reader will remark at the end of this section, the problems presented will be easier for anyone with some knowledge of Euclid (for transforming the proportions) or of algebra. This is indeed our first encounter with algebra, which is thus not treated, at least in the extant text, in a separate chapter, or theoretically, but appears in solving practical examples. This part thus serves as an introduction to the algebraic use of an unknown, the 'thing'. In all these problems, the thing occurs in linear expressions for p_1 and p_2, while q_1 and q_2 are given numbers; the resulting equations will therefore be linear. Note too that the 'thing' we have here is not just an abstract algebraic unknown but some undefined object of unknown value. This is quite in keeping with other mediaeval mathematical treatises where payment is sometimes partly in cash and partly in kind, with the worker receiving food, a ring or some clothing.* A true algebraic meaning of 'thing' will be encountered in B.41b.

(B.21) MSS $\mathcal{A}$, $\mathcal{B}$, $\mathcal{C}$. The price of three modii is ten nummi plus one thing, and the price of one modius is one thing; what is the thing worth?

Given thus $q_1 = 3$, $p_1 = 10 + x$, $q_2 = 1$, $p_2 = x$, required x.

(a) We have
$$p_1 = 10 + x = \frac{q_1}{q_2} \cdot p_2 = 3p_2 = 3x,$$
whence, subtracting the common quantity x, $2x = 10$ and $x = 5$. We have thus used the unmodified proportion here.

(b) Transforming the proportion. By separating the proportion $p_2 : p_1 = q_2 : q_1$, we find
$$\frac{p_2}{p_1 - p_2} = \frac{q_2}{q_1 - q_2}, \quad \text{thus} \quad p_2 = x = \frac{q_2}{q_1 - q_2} \cdot (p_1 - p_2) = \frac{1}{2} \cdot 10.$$

(B.22) MSS $\mathcal{A}$, $\mathcal{B}$, $\mathcal{C}$. The price of four modii is twenty nummi plus two things, and the price of one modius and a half is two things plus three nummi; what is the thing worth?

Given thus $q_1 = 4$, $p_1 = 20 + 2x$, $q_2 = 1 + \frac{1}{2}$, $p_2 = 2x + 3$, required x.

(a) Using the unmodified proportion. Since $q_1 : q_2 = p_1 : p_2$, so
$$p_1 = \frac{q_1}{q_2} \cdot p_2 = \frac{4}{1 + \frac{1}{2}} \cdot p_2 = \left(2 + \frac{2}{3}\right)(2x + 3) = \left(5 + \frac{1}{3}\right)x + 8,$$

---

* Examples in: Abū Kāmil's *Algebra*, fol. $107^v - 108^r$ (pp. 214–215 in the facsimile edition); al-Karajī's *Kāfī* (Hochheim, III, p. 14) and *Fa<u>kh</u>rī* (p. 77 in Woepcke's *Extrait*); Abū'l-Wafā's *Arithmetic* (p. 352 in Saidan's edition); Rebstock, *Rechnen*, p. 137; also in the Byzantine 14th-century manual edited by K. Vogel, No 95.

and therefore
$$p_1 = 20 + 2x = \left(5 + \frac{1}{3}\right)x + 8.$$

Subtracting the common quantities (here x's and units) gives
$$\left(3 + \frac{1}{3}\right)x = 12, \quad \text{whence} \quad x = 3 + \frac{3}{5}.$$

(*b*) Transforming the proportion.
$$p_2 = 2x + 3 = \frac{q_2}{q_1 - q_2} \cdot (p_1 - p_2) = \frac{1 + \frac{1}{2}}{2 + \frac{1}{2}} \cdot 17 = \frac{3}{5} \cdot 17 = 10 + \frac{1}{5},$$

thus $2x = 7 + \frac{1}{5}$ and $x = 3 + \frac{3}{5}$. At the end of the problem, it is pointed out that this way of solving is valid (that is, convenient) when the coefficients of x in p_1 and p_2 are equal.

(**B.23**) MSS $\mathcal{A}$, $\mathcal{B}$, $\mathcal{C}$. The price of eight modii is twenty nummi plus one thing, and the price of two modii is this thing minus one nummus; what is the thing worth?

Given thus $q_1 = 8$, $p_1 = 20 + x$, $q_2 = 2$, $p_2 = x - 1$, required x.

(*a*) Transforming the proportion.
$$p_2 = x - 1 = \frac{q_2}{q_1 - q_2} \cdot (p_1 - p_2) = \frac{1}{3} \cdot 21 = 7$$

giving $x = 8$. We have ended here with x directly equal to a number, and the way of solving, as the author observes, would be the same if the coefficient of x in the data were some other constant, by which the result found would then have to be divided (same remark in B.27–28, B.39–40).

(*b*) Using the unmodified proportion.
$$p_1 = 20 + x = \frac{q_1}{q_2} \cdot p_2 = 4(x - 1) = 4x - 4,$$

whence $3x = 24$ and $x = 8$.

Remark. As already observed in our summary (p. 1345), these problems become progressively more elaborate in terms of algebraic reckoning. In B.21–22, the coefficients are all positive and therefore only removal of the common terms, the *muqābala* or 'reduction', is required (the same holds for B.25, possibly misplaced). B.23 goes one step further since we also encounter the other algebraic operation, the *jabr* or 'restoration': the final equation contains a subtractive term (here a constant, in B.27*b* the unknown) which, by adding its amount in common, we eliminate and are thus led to an equation containing, as before, positive terms only.

(**B.24**) MS $\mathcal{A}$. The price of six sextarii is ten nummi plus a thing, and the price of two sextarii is this thing; what is the thing worth?

Given thus $q_1 = 6$, $p_1 = 10 + x$, $q_2 = 2$, $p_2 = x$, required x. This and the next, related problem, found only in MS $\mathcal{A}$, would be better placed after B.21. As a matter of fact, B.24 is almost identical with B.21 since the q_i's (this time sextarii) are just doubled and the p_i's are the same.

Transforming the proportion,

$$p_2 = x = \frac{q_2}{q_1 - q_2} \cdot (p_1 - p_2) = \frac{1}{2} \cdot 10 = 5.$$

(B.25) MS $\mathcal{A}$. The price of six sextarii is ten nummi plus one thing, and the price of two sextarii is this thing plus one nummus; what is the thing worth?

Given thus $q_1 = 6$, $p_1 = 10 + x$, $q_2 = 2$, $p_2 = x + 1$, required x. Closely related to the previous one.

Transforming the proportion,

$$p_2 = x + 1 = \frac{q_2}{q_1 - q_2} \cdot (p_1 - p_2) = \frac{1}{2} \cdot 9 = 4 + \frac{1}{2}, \quad x = 3 + \frac{1}{2}.$$

(B.26) MSS $\mathcal{A}, \mathcal{B}, \mathcal{C}$. The price of three modii is twenty nummi plus one thing, and the price of half a modius is two thirds of this thing minus two nummi; what is the thing worth?

Given thus $q_1 = 3$, $p_1 = 20 + x$, $q_2 = \frac{1}{2}$, $p_2 = \frac{2}{3}x - 2$, required x. Similar to B.23, but with fractions.

Using the unmodified proportion,

$$p_1 = 20 + x = \frac{q_1}{q_2} \cdot p_2 = 6 \left(\frac{2}{3} x - 2 \right) = 4x - 12,$$

so $3x = 32$ and $x = 10 + \frac{2}{3}$.

(B.27) MSS $\mathcal{A}, \mathcal{B}$. The price of six modii is ten nummi minus one thing, and the price of two modii is this thing; what is the thing worth?

Given thus $q_1 = 6$, $p_1 = 10 - x$, $q_2 = 2$, $p_2 = x$, required x. Same as B.24 but for the sign of x in p_1, and thus here solvable by composition of the ratio.

(a) Transforming the proportion,

$$p_2 = x = \frac{q_2}{q_1 + q_2} \cdot (p_1 + p_2) = \frac{1}{4} \cdot 10 = 2 + \frac{1}{2}.$$

As noted by the author, this applies as long as the prices are of the form $p_1 = a_1 - bx$, $p_2 = a_2 + bx$ (with the result then being divided by b if $b \neq 1$). This remark, repeated in the next problem, is analogous to the one in B.22.

(b) General rule: using the unmodified proportion $p_1 : p_2 = q_1 : q_2$.

$$p_1 = 10 - x = \frac{q_1}{q_2} \cdot p_2 = 3x,$$

whence $4x = 10$ and $x = 2 + \frac{1}{2}$.

(**B.28**) MSS $\mathcal{A}$, $\mathcal{B}$. The price of four modii is eight nummi minus one thing, and the price of two modii is this thing plus one nummus; what is the thing worth?

Given thus $q_1 = 4$, $p_1 = 8 - x$, $q_2 = 2$, $p_2 = x + 1$, required x.

(***a***) As before, transforming the proportion (this time p_2 contains an additive constant).

$$p_2 = x + 1 = \frac{q_2}{q_1 + q_2} \cdot (p_1 + p_2) = \frac{1}{3} \cdot 9 = 3, \quad \text{thus} \quad x = 2.$$

(***b***) General rule, using the unmodified proportion.

$$p_1 = 8 - x = \frac{q_1}{q_2} \cdot p_2 = 2(x+1) = 2x + 2,$$

whence x, after restoration and reduction.

(**B.29**) MSS $\mathcal{A}$, $\mathcal{B}$, $\mathcal{C}$. The price of four modii is twenty minus two things, and the price of one and a half modii is two things minus three nummi; what is the thing worth?

Given thus $q_1 = 4$, $p_1 = 20 - 2x$, and $q_2 = 1 + \frac{1}{2}$, $p_2 = 2x - 3$, required x. Same as B.22, but for the signs. By the general rule,

$$p_1 = 20 - 2x = \frac{q_1}{q_2} \cdot p_2 = \frac{4}{1 + \frac{1}{2}}(2x - 3) = \left(5 + \frac{1}{3}\right)x - 8, \quad \text{whence}$$

$$\left(7 + \frac{1}{3}\right)x = 28 \quad \text{and} \quad x = 3 + \frac{9}{11}.$$

Remark. There are some oddities in this problem, the last of the section. First, there is no solution by composition of the ratio, although it could be used just as in B.27–28. Second, and more surprising, the computation is faulty. For $\left(5 + \frac{1}{3}\right)x - 8 = 20 - 2x$ becomes $\left(5 + \frac{1}{3}\right)x = 32$, the $-2x$ having been taken as $+4$ nummi. Nevertheless, the value of x computed corresponds to $\left(7 + \frac{1}{3}\right)x = 32$. Still, this problem brings something new since we have restoration of both x's and units.

A remark by an early reader closes this first introductory section on first-degree algebra. As he says, understanding it requires knowledge of algebra or how to work with ratios as taught in Euclid's *Elements*.

The next section will introduce the solution formulae for quadratic equations and justify them by a geometrical reasoning.

3. Second-degree problems.

Since B.15 we have been solving problems 'involving unknowns' (*de ignoto*, see above, pp. 1356–1357): first the two unknowns were q_2 and p_2, then, in the problems involving things, p_1 and p_2. In the present section, the unknowns will be first q and the price per unit p_0 (B.30–31), next all four quantities p_i, q_i will be required (B.32–36); the subsequent problems, first with p and q then with p_1 and p_2, involve roots (B.37–40); finally, the unknown p is of the form $x + y$ (B.41–42)—thus two unknowns must be designated by different names if the treatment is algebraic. We shall be taught here how to solve algebraically or geometrically second-degree equations of the three forms with positive coefficients: $x^2 + c = bx$ (B.30), $x^2 = bx + c$ (B.31) and $x^2 + bx = c$ (B.38).

A. *Two quantities unknown.*

(B.30) MSS *A*, *B*. The price of a certain quantity of sextarii is ninety-three, and this quantity plus the price of one sextarius makes thirty-four; what is this quantity?

Given thus $p = 93 = q \cdot p_0$ and $q + p_0 = 34$, required q.

Computation.

(α) *Preliminary remark.* This amounts to solving the equation

$$x^2 + q \cdot p_0 = (q + p_0) x, \quad \text{or} \quad x^2 + 93 = 34\,x,$$

with its solution giving the two required quantities. Indeed, an equation of the form $x^2 + c = bx$ (b, $c > 0$ with $b^2 > 4c$) has the two positive solutions

$$x = \frac{b}{2} \pm \sqrt{\left(\frac{b}{2}\right)^2 - c},$$

and the values found will be q and p_0.

(β) *Computation in the text.*

$$\frac{34}{2} + \sqrt{\left(\frac{34}{2}\right)^2 - 93} = 17 + \sqrt{289 - 93} = 17 + \sqrt{196} = 17 + 14 = 31$$

then $34 - 31 = 3$.

Now we know that $q + p_0 = 34$. Therefore:
- Assuming $q > p_0$, we shall take $q = 31$, thus $p_0 = 3$.
- Assuming $p_0 > q$, we shall take $q = 3$, thus $p_0 = 31$.

Demonstration.

(α) *Preliminary remark.* Although the previous computation seems to solve a quadratic equation, the subsequent demonstration makes it clear that the author has employed the usual identities. Since we are given $q + p_0$ and $p = q \cdot p_0$, we can calculate

$$\left(\frac{q - p_0}{2}\right)^2 = \left(\frac{q + p_0}{2}\right)^2 - q \cdot p_0, \qquad (1)$$

and we thus know

$$\frac{q+p_0}{2} \quad \text{and} \quad \frac{|q-p_0|}{2} = \sqrt{\left(\frac{q+p_0}{2}\right)^2 - q \cdot p_0}.$$

- If we consider $q > p_0$, then

$$\frac{q+p_0}{2} \pm \sqrt{\left(\frac{q+p_0}{2}\right)^2 - q \cdot p_0} = \frac{q+p_0}{2} \pm \frac{q-p_0}{2}, \tag{2}$$

which gives q and p_0 respectively.

- If we consider $p_0 > q$, then

$$\frac{q+p_0}{2} \pm \sqrt{\left(\frac{q+p_0}{2}\right)^2 - q \cdot p_0} = \frac{q+p_0}{2} \pm \frac{p_0-q}{2}, \tag{3}$$

which gives p_0 and q respectively.

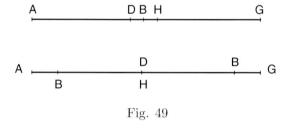

Fig. 49

(β) *Demonstration according to the text.* The geometric illustration of the solution is based on *Elements* II.5 (Premiss PE_5 in A–II), involving a segment of straight line divided into two equal and two unequal parts. Let (Fig. 49, according to MS $\mathcal{A}$ and MS $\mathcal{B}$, respectively) $q = AB$, $p_0 = BG$, thus $AG = 34$ and $AB \cdot BG = 93 = p$. We are then to solve

$$\begin{cases} AB + BG = q + p_0 = 34 \\ AB \cdot BG = q \cdot p_0 = 93, \end{cases}$$

that is, we shall calculate half the difference according to identity (1) above. Now bisecting AG, the section will fall either at D within AB if $AB > BG$ (thus $AD = DG = 17$), or at H within BG if $BG > AB$ (thus $AH = HG = 17$).

- Case $AB > BG$ ($q > p_0$). By Euclid's theorem, $AB \cdot BG + DB^2 = DG^2$. Since $DG^2 = 289$ and $AB \cdot BG = 93$, so $DB^2 = 196$, thus $DB = 14$, hence $AB = AD + DB = 31 = q$ and $BG = DG - DB = 3 = p_0$. This corresponds to (2) above.
- Case $BG > AB$ ($p_0 > q$). Then $AB \cdot BG + BH^2 = AH^2$, and, since $AB \cdot BG = 93$ and $AH^2 = 289$, so $BH^2 = 196$, thus $BH = 14$, whence $BG = BH + HG = 31 = p_0$, $AB = AH - BH = 3 = q$. This corresponds to (3) above.

(B.31) MSS A, B. The price of a certain quantity of sextarii is ninety-three, and subtracting this quantity from the price of one sextarius leaves twenty-eight; what is this quantity?

Given thus $p = 93 = q \cdot p_0$ and $p_0 - q = 28$ (therefore $p_0 > q$), required q.

Computation.

(α) *Preliminary remark.* This amounts to solving the equation

$$p_0^2 = (p_0 - q) p_0 + q \cdot p_0, \quad \text{or} \quad p_0^2 = 28 p_0 + 93,$$

of the form $x^2 = bx + c$, thus with the single positive solution

$$x = \frac{b}{2} + \sqrt{\left(\frac{b}{2}\right)^2 + c}.$$

(β) *Computation in the text.*

$$p_0 = \frac{28}{2} + \sqrt{\left(\frac{28}{2}\right)^2 + 93} = 14 + \sqrt{196 + 93} = 14 + \sqrt{289}$$

$$= 14 + 17 = 31, \qquad q = p_0 - 28 = 3.$$

Demonstration.

(α) *Preliminary remark.* We are given this time $p_0 - q$ and $p = q \cdot p_0$. Since

$$\left(\frac{p_0 + q}{2}\right)^2 = \left(\frac{p_0 - q}{2}\right)^2 + q \cdot p_0, \qquad (*)$$

we have

$$\frac{p_0 + q}{2} = \sqrt{\left(\frac{p_0 - q}{2}\right)^2 + q \cdot p_0},$$

whence

$$p_0 = \frac{p_0 + q}{2} + \frac{p_0 - q}{2} = \sqrt{\left(\frac{p_0 - q}{2}\right)^2 + q \cdot p_0} + \frac{p_0 - q}{2},$$

$$q = \frac{p_0 + q}{2} - \frac{p_0 - q}{2} = \sqrt{\left(\frac{p_0 - q}{2}\right)^2 + q \cdot p_0} - \frac{p_0 - q}{2}.$$

```
A                    D                G B
|————————————————————|————————————————|-|
```

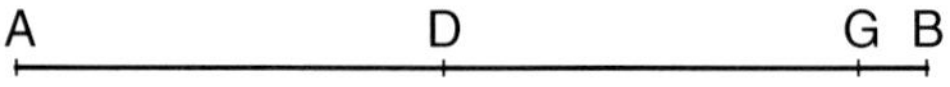

Fig. 50

(β) *Demonstration according to the text.* Based on *Elements* II.6 (PE_6), involving a segment divided into two equal parts to which another segment is appended. Let (Fig. 50) $p_0 = AB$, $q = GB$, thus $AG = p_0 - q = 28$, $AB \cdot GB = 93$. So we are to solve

$$\begin{cases} AB - GB = 28 \\ AB \cdot GB = 93, \end{cases}$$

that is, we shall determine the half-sum (designated below by DB) using the above relation (∗). Bisect AG at D (thus $AD = DG = 14$). Then, by Euclid's theorem, $AB \cdot GB + DG^2 = DB^2$, therefore $DB^2 = 289$, so $DB = 17$, thus $GB = DB - DG = 3 = q$, $AB = AD + DB = 31 = p_0$.

(**B.31**′) If the second condition were instead $q - p_0 = 28$, we would 'operate in the same manner' and find $q = 31$, $p_0 = 3$, and the demonstration would remain 'the same'. Indeed, we shall just put $AB = q$ and $GB = p_0$, thus the rôles of q and p_0 are exchanged; this leads to the equation $q^2 = 28q + 93$, thus the same equation as before, but this time for q.

B. *The four quantities are unknown.*

We now return (B.32–36) to a pair of quantities and a pair of prices. We know the products $q_1 \cdot p_1$ and $q_2 \cdot p_2$ and further, as in the three classical problems *de ignoto*, the sum, difference or product of $q_1 + p_1$ and $q_2 + p_2$.

(**B.32**) MSS $\mathcal{A}$, $\mathcal{B}$. A first quantity of sextarii multiplied by its price produces six, a second quantity of such sextarii multiplied by its price produces twenty-four, and the sum of quantities and prices makes fifteen; what are the two quantities and their prices?

Here, q_1 is the first quantity of sextarii, p_1 their price, q_2 the second quantity of sextarii and p_2 their price; we are thus given $q_1 \cdot p_1 = 6$, $q_2 \cdot p_2 = 24$, $(q_1 + p_1) + (q_2 + p_2) = 15$ and asked for q_1, p_1, q_2, p_2. Computation.

$$q_1 + p_1 = \frac{q_1 + p_1 + q_2 + p_2}{\sqrt{\frac{q_2 \cdot p_2}{q_1 \cdot p_1}} + 1} = \frac{15}{\sqrt{\frac{24}{6}} + 1} = \frac{15}{\sqrt{4} + 1} = \frac{15}{3} = 5$$

and next

$$\frac{|q_1 - p_1|}{2} = \sqrt{\left(\frac{q_1 + p_1}{2}\right)^2 - q_1 \cdot p_1} = \sqrt{\left(2 + \frac{1}{2}\right)^2 - 6} = \sqrt{6 + \frac{1}{4} - 6} = \frac{1}{2}$$

whence

$$\frac{q_1 + p_1}{2} \pm \frac{|q_1 - p_1|}{2} = 2 + \frac{1}{2} \pm \frac{1}{2}$$

which gives 3 and 2 as the values of q_1 and p_1.
- Assuming $q_1 > p_1$, we shall take $q_1 = 3$, $p_1 = 2$.
- Assuming $q_1 < p_1$, we shall choose $q_1 = 2$, $p_1 = 3$.

The computation of q_2 and p_2 is analogous. We shall first calculate

$$q_2 + p_2 = \frac{q_1 + p_1 + q_2 + p_2}{\sqrt{\frac{q_1 \cdot p_1}{q_2 \cdot p_2}} + 1} = \frac{15}{\sqrt{\frac{6}{24}} + 1} = \frac{15}{\sqrt{\frac{1}{4}} + 1} = \frac{15}{1 + \frac{1}{2}} = 10,$$

or, as the text adds at the end of this problem, we shall infer the value of $q_2 + p_2$ directly from that of $q_1 + p_1$ since we know the sum $(q_1 + p_1) + (q_2 + p_2)$. Both ways are used in B.33.

We shall then calculate

$$\frac{|q_2 - p_2|}{2} = \sqrt{\left(\frac{q_2 + p_2}{2}\right)^2 - q_2 \cdot p_2} = \sqrt{\left(\frac{10}{2}\right)^2 - 24} = 1$$

whence

$$\frac{q_2 + p_2}{2} \pm \frac{|q_2 - p_2|}{2} = 5 \pm 1$$

which gives 6 and 4 as the values of q_2 and p_2.

- If we have chosen $q_1 = 3$, $p_1 = 2$, we shall have $q_2 = 6$, $p_2 = 4$.
- If we have taken $q_1 = 2$, $p_1 = 3$, the values will be $q_2 = 4$, $p_2 = 6$.

Demonstration.

(α) *Preliminary remark.* How the solution formulae are obtained is shown geometrically below. It has been seen in the summary to what this corresponds algebraically. Let us repeat the main steps to make the parallel with the author's demonstration evident. We are given $q_1 \cdot p_1$, $q_2 \cdot p_2$, $q_1 + p_1 + q_2 + p_2$. First, since

$$\frac{q_2}{q_1} = \frac{p_2}{p_1}, \quad \text{then} \quad \frac{q_2 \cdot p_2}{q_1 \cdot p_1} = \left(\frac{q_2}{q_1}\right)^2 \quad \text{and} \quad \frac{q_2}{q_1} = \sqrt{\frac{q_2 \cdot p_2}{q_1 \cdot p_1}}; \quad (1)$$

we have thus determined the amount of the ratio $q_2 : q_1 = p_2 : p_1$. Second, starting from the same relation, we obtain successively

$$\frac{q_2}{p_2} = \frac{q_1}{p_1}, \quad \frac{q_2}{q_2 + p_2} = \frac{q_1}{q_1 + p_1}, \quad \frac{q_2}{q_1} = \frac{q_2 + p_2}{q_1 + p_1}. \quad (2)$$

Using the results of (1) and (2), we find

$$q_2 + p_2 = \sqrt{\frac{q_2 \cdot p_2}{q_1 \cdot p_1}} \cdot (q_1 + p_1), \quad (3)$$

whence finally

$$q_1 + p_1 + q_2 + p_2 = \left(\sqrt{\frac{q_2 \cdot p_2}{q_1 \cdot p_1}} + 1\right)(q_1 + p_1), \quad (4)$$

which enables us to determine the unknown term $q_1 + p_1$ on the right. Since we already know the product $q_1 \cdot p_1$, the remainder of the determination is as usual. The same will be done for p_2 and q_2.

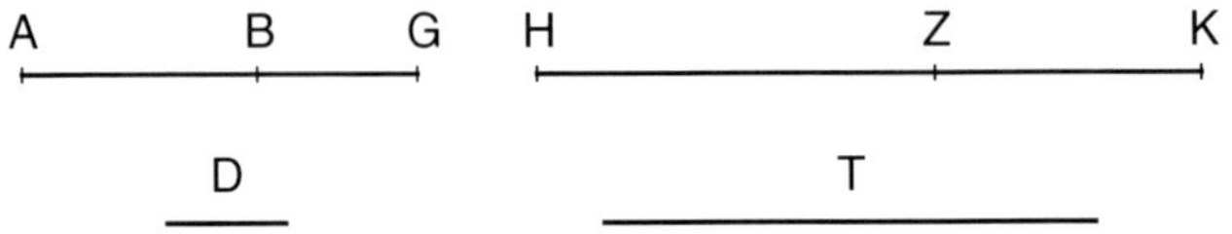

Fig. 51

(β) *Demonstration for p_1 and q_1 according to the text.* Let (Fig. 51) $q_1 = AB$, $p_1 = BG$, with $AB \cdot BG = D = 6$, then $q_2 = HZ$, $p_2 = ZK$, with $HZ \cdot ZK = T = 24$. Since $q_1 : q_2 = p_1 : p_2$,

$$\frac{AB}{HZ} = \frac{BG}{ZK}.$$

Therefore D and T are similar plane figures (*Elements* VI.14).[†] Since two such areas are in a ratio which is compounded of the ratios of the sides (*Elements* VI.23),

$$\frac{T}{D} = \frac{HZ \cdot ZK}{AB \cdot BG} = \frac{HZ}{AB} \cdot \frac{ZK}{BG} = \left(\frac{HZ}{AB}\right)^2. \quad (1)$$

But $T = 4D$, so $HZ = 2AB$. From $HZ : AB = ZK : BG$, we shall have (*Elements* V.12, or operating on the ratios)

$$\frac{HZ}{AB} = \frac{HZ + ZK}{AB + BG} = \frac{HK}{AG}, \quad (2)$$

thus

$$HK = 2AG; \quad (3)$$

next, adding AG, we obtain

$$AG + HK = 3AG. \quad (4)$$

Since $AG + HK = (q_1 + p_1) + (q_2 + p_2) = 15$, then $AG = 5 = q_1 + p_1$.

Let us now calculate q_1 and p_1 individually. We know that $AG = 5$ is divided into two parts AB, BG with $AB \cdot BG = 6$. Thus we must solve

$$\begin{cases} AB + BG = 5 = q_1 + p_1 \\ AB \cdot BG = 6 = q_1 \cdot p_1. \end{cases}$$

The demonstration as it stands in the text does not show most of the subsequent steps, for the procedure is known from B.30. The demonstration once completed is as follows. If we designate the middle of AG by L, either on AB if $AB > BG$ or on BG if $AB < BG$, then AG is divided into two equal segments by L and two unequal segments by B; by *Elements* II.5, $AB \cdot BG + LB^2 = AL^2$, whence

$$LB^2 = AL^2 - AB \cdot BG = \left(2 + \frac{1}{2}\right)^2 - 6 = \frac{1}{4},$$

thus

$$LB = \frac{1}{2} \ \left(= \sqrt{\left(\frac{q_1 + p_1}{2}\right)^2 - q_1 \cdot p_1} = \frac{|q_1 - p_1|}{2}\right),$$

and so $AB = AL + LB = 3 = q_1$, $BG = 2 = p_1$.

[†] D and T are indeed drawn as rectangles in MS $\mathcal{B}$ (same in B.35).

This enables us to determine the other two quantities $HZ = q_2$ and $ZK = p_2$. For since, as seen before, $HZ : AB = ZK : BG = 2$, then $HZ = 2 \cdot AB$ and $ZK = 2 \cdot BG$. If then, as considered in the text, $AB = q_1 > BG = p_1$ (L on AB), so $HZ = 6 = q_2$, $ZK = 4 = p_2$.

For the other possibility (not considered in the extant text of the proof) we shall consider $AB < BG$ (L on BG), and the previous values are just reversed: $AB = 2$, $BG = 3$, $HZ = 4$, $ZK = 6$.

(B.33) MSS $\mathcal{A}$, $\mathcal{B}$. A first quantity of sextarii multiplied by its price produces ten, a second quantity of such sextarii multiplied by its price produces thirty, and the sum of quantities and prices makes twenty; what are the two quantities and their prices?

Same problem as before, but the resulting roots are irrational. This time we are given $q_1 \cdot p_1 = 10$, $q_2 \cdot p_2 = 30$, $(q_1 + p_1) + (q_2 + p_2) = 20$. We find as above (and treating the root as in A.298)

$$q_1 + p_1 = \frac{q_1 + p_1 + q_2 + p_2}{\sqrt{\frac{q_2 \cdot p_2}{q_1 \cdot p_1}} + 1} = \frac{20}{\sqrt{\frac{30}{10}} + 1} = \frac{20}{\sqrt{3} + 1}$$

$$= \frac{20(\sqrt{3} - 1)}{(\sqrt{3} + 1)(\sqrt{3} - 1)} = \frac{20(\sqrt{3} - 1)}{2} = \sqrt{300} - 10.$$

As said in B.32, we may calculate $q_2 + p_2$ either by subtracting this from the given sum $q_1 + p_1 + q_2 + p_2 = 20$, so

$$q_2 + p_2 = 20 - (\sqrt{300} - 10) = 30 - \sqrt{300},$$

or —according to a later (seemingly author's) addition— by calculating

$$q_2 + p_2 = \frac{q_1 + p_1 + q_2 + p_2}{\sqrt{\frac{q_1 \cdot p_1}{q_2 \cdot p_2}} + 1} = \frac{20}{\sqrt{\frac{10}{30}} + 1} = \frac{20}{\sqrt{\frac{1}{3}} + 1}$$

$$= \frac{20\sqrt{3}}{\sqrt{3} + 1} = \frac{20\sqrt{3}(\sqrt{3} - 1)}{2} = 30 - \sqrt{300}.$$

• Let us now calculate q_1 and p_1 separately. We have

$$\begin{cases} q_1 + p_1 = \sqrt{300} - 10 \\ q_1 \cdot p_1 = 10, \end{cases}$$

that is, we know sum and product of the two required quantities. Therefore, as seen several times before,

$$\left(\frac{q_1 - p_1}{2}\right)^2 = \left(\frac{q_1 + p_1}{2}\right)^2 - q_1 \cdot p_1 = \left(\frac{\sqrt{300} - 10}{2}\right)^2 - 10$$

$$= (\sqrt{75} - 5)^2 - 10 = (100 - \sqrt{7500}) - 10 = 90 - \sqrt{7500}$$

so that
$$\frac{|q_1 - p_1|}{2} = \sqrt{90 - \sqrt{7500}}$$
which we leave in this form since it is a fourth binomial (see A.326). We find thus
$$\frac{q_1 + p_1}{2} \pm \frac{|q_1 - p_1|}{2} = \sqrt{75} - 5 \pm \sqrt{90 - \sqrt{7500}}.$$
Therefore, if $q_1 > p_1$,
$$q_1 = \sqrt{75} - 5 + \sqrt{90 - \sqrt{7500}}, \qquad p_1 = \sqrt{75} - 5 - \sqrt{90 - \sqrt{7500}},$$
whereas, if $q_1 < p_1$,
$$q_1 = \sqrt{75} - 5 - \sqrt{90 - \sqrt{7500}}, \qquad p_1 = \sqrt{75} - 5 + \sqrt{90 - \sqrt{7500}}.$$

- The individual values of q_2 and p_2 are computed in the same way. Since
$$\begin{cases} q_2 + p_2 = 30 - \sqrt{300} \\ q_2 \cdot p_2 = 30, \end{cases}$$
$$\left(\frac{q_2 - p_2}{2}\right)^2 = \left(\frac{q_2 + p_2}{2}\right)^2 - q_2 \cdot p_2 = \left(\frac{30 - \sqrt{300}}{2}\right)^2 - 30$$
$$= \left(15 - \sqrt{75}\right)^2 - 30 = \left(300 - \sqrt{67\,500}\right) - 30 = 270 - \sqrt{67\,500},$$
thus
$$\frac{q_2 + p_2}{2} \pm \frac{|q_2 - p_2|}{2} = 15 - \sqrt{75} \pm \sqrt{270 - \sqrt{67\,500}}$$
whence, if $q_1 < p_1$ and therefore $q_2 < p_2$,
$$q_2 = 15 - \sqrt{75} - \sqrt{270 - \sqrt{67\,500}}, \qquad p_2 = 15 - \sqrt{75} + \sqrt{270 - \sqrt{67\,500}},$$
whereas, if $q_1 > p_1$ and therefore $q_2 > p_2$,
$$q_2 = 15 - \sqrt{75} + \sqrt{270 - \sqrt{67\,500}}, \qquad p_2 = 15 - \sqrt{75} - \sqrt{270 - \sqrt{67\,500}}.$$

The author says at the end that the demonstrations are the same as before. As already noted, the difference between this and the previous problem is that here we end up with irrational values (which does not affect the demonstration). The use of irrationals was of course deliberate, as in a subsequent case (B.36).

Remark. The text of this problem illustrates one of the main difficulties encountered in verbal mathematics: indicating where an operation is to stop. Indeed, expressions like $\sqrt{a} + \sqrt{b}$ and $\sqrt{a + \sqrt{b}}$ would *a priori* both be expressed as 'the root of a and the root of b'. Our author distinguishes the second case by saying 'a and the root of b the root of which is taken' (he could also have said 'the root of the sum of a and the root of b'). But if further roots were involved, the situation would become inextricable. A nice illustration of this difficulty in dealing verbally with overlapping radical signs appears in the Latin translation of Abū Kāmil's *Algebra*, preserved in the translator's original version, where we can follow his hesitations and successive corrections.* Among the most noticeable advances made at the time of the introduction of algebraic symbolism, towards the end of the fifteenth century and in the sixteenth, we may mention those of Nicolas Chuquet (in his *Triparty* of 1484), who underlines the expression which is to follow his root sign R^2 (with more underlining when radical signs overlap), and of Rafael Bombelli, who (in his *Algebra* of 1572) encloses the expression following the root sign in a kind of brackets ($\lfloor$ and $\rfloor$).

(B.34) MSS $\mathcal{A}$, $\mathcal{B}$. A first quantity of sextarii multiplied by its price produces six, a second quantity of such sextarii multiplied by its price produces twenty-four, and subtracting the first quantity plus its price from the second quantity plus its price leaves five; what are the two quantities and their prices?

We are now given $q_1 \cdot p_1 = 6$, $q_2 \cdot p_2 = 24$, $(q_2 + p_2) - (q_1 + p_1) = 5$ (difference instead of sum known).

We compute then, by analogy with the above,

$$q_1 + p_1 = \frac{(q_2 + p_2) - (q_1 + p_1)}{\sqrt{\frac{q_2 \cdot p_2}{q_1 \cdot p_1}} - 1} = \frac{5}{\sqrt{\frac{24}{6}} - 1} = \frac{5}{\sqrt{4} - 1} = 5$$

whence $q_2 + p_2 = 5 + (q_1 + p_1) = 10$. The problem thus becomes identical with B.32.

(B.35) MSS $\mathcal{A}$, $\mathcal{B}$. A first quantity of sextarii multiplied by its price produces six, a second quantity of such sextarii multiplied by its price produces twenty-four, and multiplying the first quantity plus its price by the second quantity plus its price produces fifty; what are the two quantities and their prices?

Given thus $q_1 \cdot p_1 = 6$, $q_2 \cdot p_2 = 24$, $(q_1 + p_1) \cdot (q_2 + p_2) = 50$ (the product is now known).

Computation.

$$q_1 + p_1 = \sqrt{\frac{(q_1 + p_1) \cdot (q_2 + p_2)}{\sqrt{\frac{q_2 \cdot p_2}{q_1 \cdot p_1}}}} = \sqrt{\frac{50}{\sqrt{\frac{24}{6}}}} = \sqrt{\frac{50}{\sqrt{4}}} = 5,$$

$$q_2 + p_2 = \frac{(q_1 + p_1) \cdot (q_2 + p_2)}{q_1 + p_1} = \frac{50}{5} = 10.$$

---

* See pp. 319–320 of the edition of this Latin translation.

The required values are again the same as in B.32.

Demonstration.

(α) *Preliminary remark*. As seen in our summary, we infer from

(1) $\quad q_2 + p_2 = \sqrt{\dfrac{q_2 \cdot p_2}{q_1 \cdot p_1}} \cdot (q_1 + p_1) \quad \left(q_1 + p_1 = \sqrt{\dfrac{q_1 \cdot p_1}{q_2 \cdot p_2}} \cdot (q_2 + p_2) \right)$

that

(2) $\quad \sqrt{\dfrac{q_2 \cdot p_2}{q_1 \cdot p_1}} (q_1 + p_1)^2 = (q_1 + p_1)(q_2 + p_2) \quad \left(= \sqrt{\dfrac{q_1 \cdot p_1}{q_2 \cdot p_2}} (q_2 + p_2)^2 \right)$

whence

$$q_1 + p_1 = \sqrt{\dfrac{(q_1 + p_1)(q_2 + p_2)}{\sqrt{\dfrac{q_2 \cdot p_2}{q_1 \cdot p_1}}}} \quad \left(q_2 + p_2 = \sqrt{\dfrac{(q_1 + p_1)(q_2 + p_2)}{\sqrt{\dfrac{q_1 \cdot p_1}{q_2 \cdot p_2}}}} \right).$$

The q_i's and p_i's will then be determined individually as known from before (B.32–33).

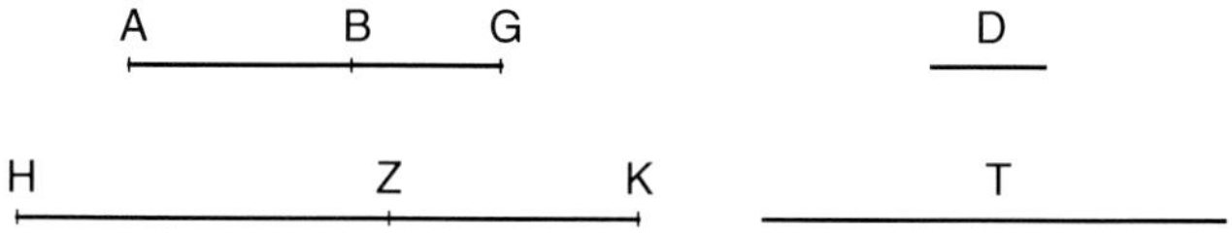

```
A    B   G                    D
|————|———|                    |—|

H         Z        K              T
|—————————|————————|          |———|
```

Fig. 52

(β) *Demonstration according to the text*. Let (Fig. 52) $q_1 = AB$, $p_1 = BG$, with $q_1 \cdot p_1 = AB \cdot BG = D = 6$, then $q_2 = HZ$, $p_2 = ZK$, with $q_2 \cdot p_2 = HZ \cdot ZK = T = 24$, and $(q_1 + p_1) \cdot (q_2 + p_2) = AG \cdot HK = 50$. Thus, as before (B.32),

$$\dfrac{T}{D} = \dfrac{q_2 \cdot p_2}{q_1 \cdot p_1} = \dfrac{q_2^2}{q_1^2} = \left(\dfrac{HZ}{AB}\right)^2.$$

Hence, since $T = 4D$, $HZ = 2AB$, and therefore, by operating on the ratios as in B.32, $HK = 2AG$, which is our (1) above.

Therefore, and this is (2), $AG \cdot HK = AG \cdot 2AG = 50$, whence $AG^2 = 25$, so $AG = 5 = q_1 + p_1$ and $HK = 10 = q_2 + p_2$.

(B.36) MSS $\mathcal{A}$, $\mathcal{B}$. A first quantity of sextarii multiplied by its price produces ten, a second quantity of such sextarii multiplied by its price produces twenty, and multiplying the first quantity plus its price by the second plus its price produces the root of five thousand seven hundred and sixty; what are the two quantities and their prices?

Same problem as before, except that an irrational value occurs: $q_1 \cdot p_1 = 10$, $q_2 \cdot p_2 = 20$, $(q_1 + p_1) \cdot (q_2 + p_2) = \sqrt{5760}$.

Using the same formula as in B.35,

$$q_1 + p_1 = \sqrt{\frac{(q_1+p_1)\cdot(q_2+p_2)}{\sqrt{\frac{q_2\cdot p_2}{q_1\cdot p_1}}}} = \sqrt{\frac{\sqrt{5760}}{\sqrt{\frac{20}{10}}}} = \sqrt{\frac{\sqrt{5760}}{\sqrt{2}}} = \sqrt[4]{2880}.$$

Since $q_1 \cdot p_1 = 10$, we shall calculate

$$\left(\frac{q_1-p_1}{2}\right)^2 = \left(\frac{q_1+p_1}{2}\right)^2 - q_1\cdot p_1 = \left(\frac{\sqrt[4]{2880}}{2}\right)^2 - 10$$

$$= \left(\sqrt[4]{180}\right)^2 - 10 = \sqrt{180} - 10,$$

whence

$$\frac{|q_1-p_1|}{2} = \sqrt{\sqrt{180}-10}.$$

This being a second apotome, we may transform it as taught at the end of Book A (A.325):

$$\frac{|q_1-p_1|}{2} = \sqrt{\frac{\sqrt{180}+\sqrt{180-100}}{2}} - \sqrt{\frac{\sqrt{180}-\sqrt{180-100}}{2}}$$

$$= \sqrt{\sqrt{45}+\sqrt{20}} - \sqrt{\sqrt{45}-\sqrt{20}}$$

$$= \sqrt{3\sqrt{5}+2\sqrt{5}} - \sqrt{3\sqrt{5}-2\sqrt{5}} = \sqrt[4]{125} - \sqrt[4]{5}.$$

Therefore

$$q_1, p_1 = \frac{q_1+p_1}{2} \pm \frac{|q_1-p_1|}{2} = \sqrt[4]{180} \pm \left(\sqrt[4]{125}-\sqrt[4]{5}\right).$$

Thus, if $q_1 > p_1$,

$$q_1 = \sqrt[4]{180} + \sqrt[4]{125} - \sqrt[4]{5} \quad \text{and} \quad p_1 = \sqrt[4]{180} + \sqrt[4]{5} - \sqrt[4]{125},$$

and the reverse if $q_1 < p_1$.[||]

Similarly,

$$q_2 + p_2 = \frac{(q_1+p_1)\cdot(q_2+p_2)}{q_1+p_1} = \frac{\sqrt{5760}}{\sqrt[4]{2880}} = \sqrt[4]{11\,520}$$

whence, because of $q_2 \cdot p_2 = 20$,

$$\left(\frac{q_2-p_2}{2}\right)^2 = \left(\frac{q_2+p_2}{2}\right)^2 - q_2\cdot p_2 = \left(\frac{\sqrt[4]{11\,520}}{2}\right)^2 - 20$$

[||] In verbal algebra (and reckoning) the positive terms come first and are followed by the negative ones.

$$= \left(\sqrt[4]{720}\right)^2 - 20 = \sqrt{720} - 20.$$

Applying the transformation taught in A.325, we obtain

$$\frac{|q_2 - p_2|}{2} = \sqrt{\sqrt{720} - 20} = \sqrt{\frac{\sqrt{720} + \sqrt{320}}{2}} - \sqrt{\frac{\sqrt{720} - \sqrt{320}}{2}}$$

$$= \sqrt{\sqrt{180} + \sqrt{80}} - \sqrt{\sqrt{180} - \sqrt{80}}$$

$$= \sqrt{6\sqrt{5} + 4\sqrt{5}} - \sqrt{6\sqrt{5} - 4\sqrt{5}} = \sqrt[4]{500} - \sqrt[4]{20}.$$

Therefore

$$q_2, p_2 = \frac{q_2 + p_2}{2} \pm \frac{|q_2 - p_2|}{2} = \sqrt[4]{720} \pm \left(\sqrt[4]{500} - \sqrt[4]{20}\right).$$

Then, if we have taken $q_1 > p_1$, we shall have

$$q_2 = \sqrt[4]{720} + \sqrt[4]{500} - \sqrt[4]{20}, \qquad p_2 = \sqrt[4]{720} + \sqrt[4]{20} - \sqrt[4]{500},$$

and the reverse if $q_1 < p_1$.

C. *Other problems*.

In B.37–38, which involve roots in the data, we return to a single quantity of sextarii, as at the beginning of the section (B.30–31). The next two problems, B.39–40, involve 'things' in the data, as did B.21–29. Finally, in B.41–42, we have two things of different values. This is not the first time (see A.169, p. 1251), but here the treatment is algebraic, and thus each unknown is to receive its own designation.

(B.37) MSS *A*, *B*. A certain quantity of sextarii when subtracted from its price leaves thirty-four, and the root of the price is thrice the quantity; what is this quantity?

We thus know that $\sqrt{p} = 3q$, $p - q = 34$.

Computation.

$$\sqrt{p} = \frac{1}{2} \cdot \frac{1}{3} + \sqrt{\left(\frac{1}{2} \cdot \frac{1}{3}\right)^2 + 34} = \frac{1}{6} + \sqrt{\left(\frac{1}{6}\right)^2 + 34} = \frac{1}{6} + 5 + \frac{5}{6} = 6,$$

whence $p = 36$ and $q = 2$.

Remark. This is the solution of $p = \frac{1}{3}\sqrt{p} + 34$, which is a quadratic equation of the form $x^2 = bx + c$ with the single positive solution

$$x = \frac{b}{2} + \sqrt{\left(\frac{b}{2}\right)^2 + c}.$$

Demonstration.

(α) *Preliminary remark.* Since $\sqrt{p} = 3q$ and $p - q = 34$,

$$p - q = p - \frac{1}{3} \cdot \sqrt{p} = \sqrt{p}\left(\sqrt{p} - \frac{1}{3}\right) = 34. \qquad (1)$$

We are then to solve

$$\begin{cases} \sqrt{p} - \left(\sqrt{p} - \frac{1}{3}\right) = \frac{1}{3} \\ \sqrt{p}\left(\sqrt{p} - \frac{1}{3}\right) = 34, \end{cases}$$

and this will be done in the usual way. First,

$$\left(\frac{\sqrt{p} + \left(\sqrt{p} - \frac{1}{3}\right)}{2}\right)^2 = \left(\frac{\sqrt{p} - \left(\sqrt{p} - \frac{1}{3}\right)}{2}\right)^2 + \sqrt{p}\left(\sqrt{p} - \frac{1}{3}\right), \qquad (2)$$

so that

$$\frac{\sqrt{p} + \left(\sqrt{p} - \frac{1}{3}\right)}{2} = \sqrt{\left(\frac{1}{6}\right)^2 + 34}. \qquad (3)$$

Since

$$\sqrt{p} = \frac{\sqrt{p} + \left(\sqrt{p} - \frac{1}{3}\right)}{2} + \frac{\sqrt{p} - \left(\sqrt{p} - \frac{1}{3}\right)}{2},$$

we find

$$\sqrt{p} = \sqrt{\left(\frac{1}{6}\right)^2 + 34} + \frac{1}{6}. \qquad (4)$$

The subsequent demonstration corresponds to this exactly.

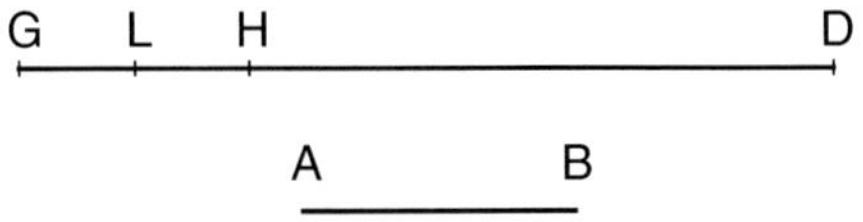

Fig. 53

(β) *Demonstration according to the text.* Let (Fig. 53) $q = AB$, $\sqrt{p} = GD$, with $GD = 3AB$, and further, by hypothesis,

$$p - q = GD^2 - AB = 34 = GD^2 - \frac{1}{3} \cdot GD = GD\left(GD - \frac{1}{3}\right). \qquad (1)$$

Take $GH = \frac{1}{3}$ (which must be on GD since $GD - \frac{1}{3} > 0$). We have then

$$GD(GD - GH) = GD \cdot HD = 34, \qquad (1')$$

that is, we are to solve

$$\begin{cases} GD - HD = \frac{1}{3} \\ GD \cdot HD = 34, \end{cases}$$

Let L be the middle of GH; so $GL = LH = \frac{1}{6}$. By *Elements* II.6,

$$GD \cdot HD + LH^2 = LD^2 = 34 + \left(\frac{1}{6}\right)^2, \qquad (2)$$

so that
$$LD = 5 + \frac{5}{6}. \qquad (3)$$

Therefore
$$GD = GL + LD = 6 = \sqrt{p} \qquad (4)$$

whence $p = 36$, $AB = q = \frac{1}{3} \cdot GD = 2$.

(B.38) MSS $\mathcal{A}$, $\mathcal{B}$. A certain quantity of sextarii plus its price makes eighteen, and the root of the price is twice this quantity; what is this quantity?

Given thus $\sqrt{p} = 2q$, $p + q = 18$, required p and q.

Computation.

$$\sqrt{p} = \sqrt{\left(\frac{1}{2} \cdot \frac{1}{2}\right)^2 + 18} - \frac{1}{2} \cdot \frac{1}{2} = \sqrt{\frac{1}{16} + 18} - \frac{1}{4} = 4 + \frac{1}{4} - \frac{1}{4} = 4,$$

whence $p = 16$, $q = \frac{1}{2} \cdot \sqrt{p} = 2$.

Remark. This solves $p + \frac{1}{2}\sqrt{p} = 18$, which is of the form $x^2 + bx = c$ and has the single positive solution

$$x = -\frac{b}{2} + \sqrt{\left(\frac{b}{2}\right)^2 + c}.$$

Demonstration.

(α) *Preliminary remark.* Since $\sqrt{p} = 2q$ and $p + q = 18$,

$$p + q = p + \frac{1}{2}\sqrt{p} = \sqrt{p}\left(\sqrt{p} + \frac{1}{2}\right) = 18. \qquad (1)$$

We are then to solve

$$\begin{cases} \left(\sqrt{p} + \frac{1}{2}\right) - \sqrt{p} = \frac{1}{2} \\ \left(\sqrt{p} + \frac{1}{2}\right)\sqrt{p} = 18, \end{cases}$$

which is treated as usual:

$$\left(\frac{\left(\sqrt{p} + \frac{1}{2}\right) + \sqrt{p}}{2}\right)^2 = \left(\frac{\left(\sqrt{p} + \frac{1}{2}\right) - \sqrt{p}}{2}\right)^2 + \left(\sqrt{p} + \frac{1}{2}\right)\sqrt{p},$$

so that
$$\frac{\left(\sqrt{p} + \frac{1}{2}\right) + \sqrt{p}}{2} = \sqrt{\left(\frac{1}{4}\right)^2 + 18}. \qquad (2)$$

Therefore,

$$\sqrt{p} + \frac{1}{2} = \frac{\left(\sqrt{p}+\frac{1}{2}\right)+\sqrt{p}}{2} + \frac{\left(\sqrt{p}+\frac{1}{2}\right)-\sqrt{p}}{2}$$

$$= \sqrt{\left(\frac{1}{4}\right)^2 + 18} + \frac{1}{4} = 4 + \frac{1}{4} + \frac{1}{4} = 4 + \frac{1}{2},$$

whence

$$\sqrt{p} = 4 + \frac{1}{2} - \frac{1}{2} = 4. \qquad (3)$$

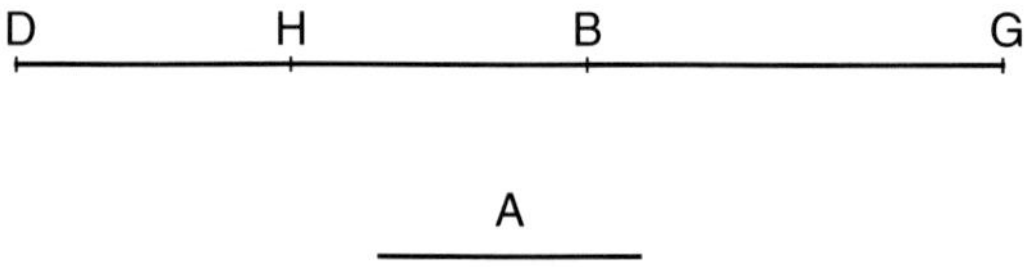

```
D           H              B                     G
|-----------|--------------|---------------------|

                          A
                     |---------|
```

Fig. 54

(β) *Demonstration according to the text.* Let (Fig. 54) $q = A$, $\sqrt{p} = BG$, thus $BG = 2A$. Since $BG^2 + A = 18$, so

$$BG^2 + \frac{1}{2}BG = BG\left(BG + \frac{1}{2}\right) = 18. \qquad (1)$$

Let BG be extended by $DB = \frac{1}{2}$. Then

$$BG(BG + DB) = BG \cdot DG = 18. \qquad (1')$$

We are then to solve

$$\begin{cases} DG - BG = \frac{1}{2} \\ DG \cdot BG = 18. \end{cases}$$

Bisect DB and continue as before, the text says. Indeed, if H is the point of section (not in the text), then $DH = HB = \frac{1}{4}$. Applying *Elements* II.6 gives $BG \cdot DG + HB^2 = HG^2$, thus

$$HG^2 = 18 + \frac{1}{2}\frac{1}{8}, \qquad HG = 4 + \frac{1}{4}, \qquad (2)$$

and

$$BG = HG - HB = 4 = \sqrt{p}, \qquad (3)$$

hence $A = 2 = q$.

★ ★ ★

(B.39) MSS $\mathcal{A}$, $\mathcal{B}$. The price of six modii is four nummi plus a thing, and the price of two modii is three times the root of the previous price; what is the thing worth?

Designating the thing by x, we have in this first problem $q_1 = 6$, $p_1 = x + 4$, $q_2 = 2$, $p_2 = 3\sqrt{x+4}$. Using the unmodified proportion (B.21a),

$$p_1 = x + 4 = \frac{q_1}{q_2} \cdot p_2 = \frac{6}{2} \cdot 3\sqrt{x+4} = 9\sqrt{x+4},$$

hence $\sqrt{x+4} = 9$, $x + 4 = 81$ and $x = 77$.

Operate in the same way, the author adds, if $p_1 = mx + 4$ with $m \neq 1$.

(B.40) MSS $\mathcal{A}$, $\mathcal{B}$. The price of six modii is one thing minus four nummi, and the price of two modii is three times the root of the previous price; what is the thing worth?

Similar problem, with $q_1 = 6$, $p_1 = x - 4$, $q_2 = 2$, $p_2 = 3\sqrt{x-4}$. As before,

$$p_1 = x - 4 = \frac{q_1}{q_2} \cdot p_2 = \frac{6}{2} \cdot 3\sqrt{x-4} = 9\sqrt{x-4},$$

hence $\sqrt{x-4} = 9$, $x - 4 = 81$ and x (or mx) $= 85$.

(B.41) MSS $\mathcal{A}$, $\mathcal{B}$. The price of three sextarii is the sum of two unequal things the product of which is twenty-one, and the price of one sextarius is ten ninths of the lesser thing; what are these things worth?

Here things of different values occur. We have $q = 3$, $p = x + y$ (with $y > x$), $x \cdot y = 21$, p_0 (unit price) $= \left(1 + \frac{1}{9}\right) \cdot x$ (with $p_0 = \frac{1}{3}p$).

(a) *Computation.*

$$x^2 = \frac{21}{\frac{3}{1}\left(1 + \frac{1}{9}\right) - 1} = \frac{21}{3 + \frac{1}{3} - 1} = \frac{21}{2 + \frac{1}{3}} = 9,$$

so $x = 3$, whence $y = 7$. As in the two previous problems, the background of the computations remains unexplained. We are reminded of Mesopotamian treatments of problems, incomprehensible to those not already familiar with the method. In this particularly abstruse problem, the proof will provide a justification.

Demonstration.

(α) *Preliminary remark.* Taking general quantities q and q_0 with the respective prices p and p_0, and knowing that $p = x + y$, $p_0 = ax$, $x \cdot y = b$ (a, b given), we have, since $q : q_0 = p : p_0$, the following relations (with each second one corresponding to the particular case of this problem, namely $q_0 = 1$).

$$p = \frac{p_0 \cdot q}{q_0} = x \cdot \frac{a\,q}{q_0}, \quad \text{or} \quad p = x \cdot a\,q. \qquad (1)$$

Next, since $p = x + y$,

$$y = p - x = x\left(\frac{a\,q}{q_0} - 1\right), \quad \text{or} \quad y = x\,(a\,q - 1); \qquad (2)$$

while, since $xy = b$,
$$xy = x^2 \left(\frac{aq}{q_0} - 1\right) = b, \quad \text{or} \quad x^2 (aq - 1) = b. \quad (3)$$

We thus find
$$x = \sqrt{\frac{b}{\frac{aq}{q_0} - 1}}, \quad \text{or} \quad x = \sqrt{\frac{b}{aq - 1}}, \quad (4)$$

with, in our case, $q_0 = 1$, $q = 3$, $a = 1 + \frac{1}{9}$, $b = 21$.

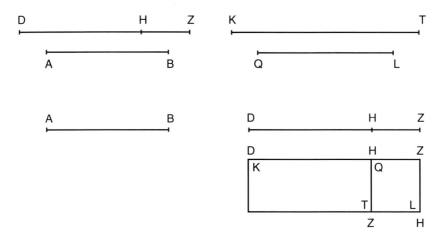

Fig. 55

(β) *Demonstration according to the text.* Let (Fig. 55; the upper one in MS $\mathcal{A}$, the other in $\mathcal{B}^\dagger$) $q = AB = 3$, $y = DH$, $x = HZ$ ($DH > HZ$), with $DH \cdot HZ = KT = 21$ (our b above). Thus

$$p = DH + HZ = DZ = 3\left(1 + \frac{1}{9}\right) HZ = \left(3 + \frac{1}{3}\right) HZ. \quad (1)$$

Then, since $p = x + y$ and $x = HZ$,

$$y = DH = DZ - HZ = \left(2 + \frac{1}{3}\right) HZ, \quad (2)$$

whence
$$DH \cdot HZ = \left(2 + \frac{1}{3}\right) HZ^2 = KT = 21. \quad (3)$$

Let $HZ^2 = QL$. We then have $\left(2 + \frac{1}{3}\right) QL = KT = 21$ and so

$$QL = \frac{KT}{2 + \frac{1}{3}} = \frac{21}{2 + \frac{1}{3}} = 9 = HZ^2, \quad HZ = \sqrt{\frac{21}{2 + \frac{1}{3}}} = 3. \quad (4)$$

---

† In the latter the recurrence of Z and H is confusing.

(**b**) *Algebraic treatment.* Here we find the first clear reference to an algebraic way of solving a problem and, indeed, we are left in no doubt that 'thing' is now attributed to an unknown and not to an object of unknown value (see Translation, p. 809, notes 1000 & 1001). This is also the first time that we encounter the square of the unknown (*census*, Arabic *māl*). Apart from that, the reader will learn little from this algebraic treatment, which is a sequence of instructions without any explanation or reference to former problems.

Following the Arabic tradition, in particular Abū Kāmil's *Algebra* with which he was well acquainted, our author puts as the two unknowns, respectively, *res* ('thing', our x) and *dragma* ('drachma', the Greek coin, Arabic *dirham*, our y).* His subsequent computation is then equivalent to considering that, since $p = x + y$,

$$p_0 = \left(1 + \frac{1}{9}\right)x = \frac{1}{3}p = \frac{1}{3}(x+y), \quad \text{so} \quad \frac{7}{9}x = \frac{1}{3}y,$$

whence $y = \left(2 + \frac{1}{3}\right)x$, and therefore

$$x \cdot y = 21 = \left(2 + \frac{1}{3}\right)x^2,$$

so that $x^2 = 9$, thus $x = 3$, and $y = 7$.

(**B.42**) MSS $\mathcal{A}$, $\mathcal{B}$. The price of five sextarii is the sum of two unequal things the product of which is one hundred and forty-four, and the price of one sextarius is a third of the lesser thing plus two nummi; what are these things worth?

A similar problem, with $q = 5$, $p = x + y$ $(y > x)$, $x \cdot y = 144$, $p_0 = \frac{1}{3}x + 2$ (with $p_0 = \frac{1}{5}p$).

(**a**) Computation. Here again, we have a sequence of computations which will make sense for the reader only if he already has the key to the solution or is able to work it out from the proof.

$$x = \sqrt{\left(\frac{1}{2} \cdot \frac{\frac{5}{1} \cdot 2}{\frac{2}{3}}\right)^2 + \frac{144}{\frac{1}{3} \cdot 5 - 1}} - \frac{1}{2} \cdot \frac{\frac{5}{1} \cdot 2}{\frac{2}{3}}$$

$$= \sqrt{\left(\frac{1}{2} \cdot \frac{10}{\frac{2}{3}}\right)^2 + \frac{144}{\frac{2}{3}}} - \frac{1}{2} \cdot \frac{10}{\frac{2}{3}} = \sqrt{\left(\frac{1}{2} \cdot 15\right)^2 + 216} - \frac{1}{2} \cdot 15$$

$$= \sqrt{56 + \frac{1}{4} + 216} - \left(7 + \frac{1}{2}\right) = \sqrt{272 + \frac{1}{4}} - \left(7 + \frac{1}{2}\right)$$

$$= 16 + \frac{1}{2} - \left(7 + \frac{1}{2}\right) = 9, \quad y = \frac{144}{9} = 16.$$

---

* About this use of a coin's name as unknown by Abū Kāmil, see our *Introduction to the History of Algebra*, p. 76.

Demonstration.

(α) *Preliminary remark.* Considering a pair of quantities q and q_0 with the respective prices p and p_0, and knowing that $p = x + y$, $p_0 = a\,x + c$, $x \cdot y = b$ (a, b, c given), we find, since $q : q_0 = p : p_0$,

$$p = \frac{p_0 \cdot q}{q_0} = \frac{a\,q \cdot x + c\,q}{q_0},$$

whence, since $p = x + y$, and for $q_0 \neq 1$ and $q_0 = 1$ respectively,

$$a\,q \cdot x + c\,q = q_0\,(x + y), \quad \text{or} \quad a\,q \cdot x + c\,q = x + y. \tag{1}$$

Therefore

$$x\,(a\,q - q_0) + c\,q = q_0 \cdot y, \quad \text{or} \quad x\,(a\,q - 1) + c\,q = y, \tag{2}$$

whence, multiplying by x and dividing by q_0,

$$x^2\,\frac{a\,q - q_0}{q_0} + \frac{c\,q}{q_0}\,x = xy, \quad \text{or} \quad x^2\,(a\,q - 1) + c\,q \cdot x = xy, \tag{3}$$

and finally, since xy is given ($= b$),

$$x^2 + \frac{c\,q}{a\,q - q_0}\,x = \frac{b\,q_0}{a\,q - q_0}, \quad \text{or} \quad x^2 + \frac{c\,q}{a\,q - 1}\,x = \frac{b}{a\,q - 1}, \tag{4}$$

that is, with the values given, $x^2 + 15\,x = 216$.

Solving it in our way corresponds to the calculations given in the text. But, as usual, it results in reality from the use of identities. Putting (4) in the form

$$x\left(x + \frac{c\,q}{a\,q - q_0}\right) = \frac{b\,q_0}{a\,q - q_0} \quad \left(\text{thus } x\,(x + 15) = 216\right), \tag{5}$$

we are to solve

$$\begin{cases} \left(x + \dfrac{c\,q}{a\,q - q_0}\right) - x = \dfrac{c\,q}{a\,q - q_0} = 15 \\[1ex] \left(x + \dfrac{c\,q}{a\,q - q_0}\right) \cdot x = \dfrac{b\,q_0}{a\,q - q_0} = 216, \end{cases}$$

and it appears that we have two unknown quantities of which we know the product and the difference. We may then determine the square of half their sum,

$$\left[\frac{1}{2}\left(2x + \frac{c\,q}{a\,q - q_0}\right)\right]^2 = \left[\frac{1}{2} \cdot \frac{c\,q}{a\,q - q_0}\right]^2 + \frac{b\,q_0}{a\,q - q_0},$$

whence

$$x + \frac{1}{2} \cdot \frac{c\,q}{a\,q - q_0} = \sqrt{\left[\frac{1}{2} \cdot \frac{c\,q}{a\,q - q_0}\right]^2 + \frac{b\,q_0}{a\,q - q_0}},$$

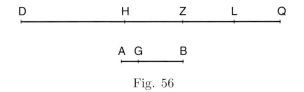

Fig. 56

and therefore

$$x = \sqrt{\left[\frac{1}{2} \cdot \frac{cq}{aq-q_0}\right]^2 + \frac{bq_0}{aq-q_0}} - \frac{1}{2} \cdot \frac{cq}{aq-q_0},$$

that is,

$$x = \sqrt{\left(\frac{15}{2}\right)^2 + 216} - \frac{15}{2}.$$

(β) *Demonstration according to the text.* It corresponds to the above steps. Let (Fig. 56) $q = AB = 5$, $x = HZ$, $y = DH$, thus $DZ = x+y = p$ while $HZ \cdot DH = x \cdot y = 144$, 1 (sextarius) $= q_0 = AG$. Since $p_0 = \frac{1}{3}x + 2$ and $q_0 : q = p_0 : p$,

$$\frac{AG}{AB} = \frac{\frac{1}{3} \cdot HZ + 2}{DZ}, \quad \text{and} \quad \frac{AG}{AB} = \frac{1}{5}, \quad \text{so} \quad \frac{1}{3} \cdot HZ + 2 = \frac{1}{5} \cdot DZ,$$

whence, multiplying by 5,

$$\left(1 + \frac{2}{3}\right) \cdot HZ + 10 = DZ. \qquad (1)$$

Then, subtracting HZ,

$$\frac{2}{3} \cdot HZ + 10 = DH = y. \qquad (2)$$

Since $x \cdot y = 144$ and $x = HZ$,

$$\frac{2}{3} \cdot HZ^2 + 10 \cdot HZ = 144, \qquad (3)$$

and, multiplying by $\frac{3}{2}$,

$$HZ^2 + 15 \cdot HZ = 216. \qquad (4)$$

Putting $ZQ = 15$ (on DZ produced):

$$HZ^2 + ZQ \cdot HZ = HZ(HZ + ZQ) = HZ \cdot HQ = 216. \qquad (5)$$

We are then to solve

$$\begin{cases} HQ - HZ = 15 \\ HQ \cdot HZ = 216. \end{cases}$$

We bisect ZQ at L (so $ZL = LQ = 7 + \frac{1}{2}$) and continue as before (B.31, 37–38). Indeed, we would find (*Elements* II.6) $HZ \cdot HQ + ZL^2 = HL^2$, thus
$$HL^2 = 216 + \left(7 + \frac{1}{2}\right)^2, \qquad HL = 16 + \frac{1}{2},$$
hence
$$HZ = HL - ZL = 9 = x, \qquad DH = \frac{HZ \cdot DH}{HZ} = \frac{144}{9} = 16 = y.$$

(**b**) *Algebraic treatment.* Here too x as *res* and y as *dragma* appear as unknowns. We have, from the data, $p = x + y$ and $p_0 = \frac{1}{3}x + 2$, so
$$p_0 = \frac{p}{q} = \frac{x+y}{q} = \frac{1}{5}x + \frac{1}{5}y = \frac{1}{3}x + 2.$$

Therefore
$$\frac{1}{5}y = \frac{2}{3}\frac{1}{5}x + 2, \qquad y = \frac{2}{3}x + 10,$$
whence finally
$$x \cdot y = \frac{2}{3}x^2 + 10x = 144,$$
the solution of which is 'as we have taught in algebra' (B.38).

---

4. Corn of various kinds.

We now consider corn of n different kinds, q_i of the ith kind, altogether Q modii, bought for P, and the ith kind costs $p_0^{(i)}$ per modius (sextarius in B.48–53). The formulae used are then

$$\sum q_i = Q, \qquad \sum_1^n p_0^{(i)} \cdot q_i = P, \quad \text{and} \quad q_k = \frac{P - \sum_{i \neq k} p_0^{(i)} \cdot q_i}{p_0^{(k)}}.$$

Of particular mathematical interest are the linear systems with two or three unknowns (B.45–47, B.51) and those involving arithmetical progressions (B.52–53).

(**B.43**) MSS $\mathcal{A}$, $\mathcal{B}$, $\mathcal{C}$. The prices of three kinds of corn are respectively six, eight, nine nummi per modius, and with thirty nummi were bought one and a fourth modii of the first and one and two thirds modii of the second; what was bought of the third?

Given thus $p_0^{(1)} = 6$, $p_0^{(2)} = 8$, $p_0^{(3)} = 9$, $P = 30$, $q_1 = 1 + \frac{1}{4}$, $q_2 = 1 + \frac{2}{3}$, required q_3. The computation corresponds to the use of the last formula above.

$$q_3 = \frac{P - p_0^{(1)} \cdot q_1 - p_0^{(2)} \cdot q_2}{p_0^{(3)}} = \frac{30 - 6\left(1 + \frac{1}{4}\right) - 8\left(1 + \frac{2}{3}\right)}{9}$$

$$= \frac{30 - \left(7 + \frac{1}{2}\right) - \left(13 + \frac{1}{3}\right)}{9} = \frac{30 - \left(20 + \frac{5}{6}\right)}{9} = \frac{9 + \frac{1}{6}}{9} = 1 + \frac{1}{6}\frac{1}{9}.$$

This, the text adds, can be generalized to n kinds of corn (with $n-1$ of the q_i's being known).

(B.44) MSS $\mathcal{A}$, $\mathcal{B}$, $\mathcal{C}$. The prices of three kinds of corn are respectively six, eight, ten nummi per modius, and an equal quantity of each was bought for eighteen nummi; how much was it?

Given thus $p_0^{(1)} = 6$, $p_0^{(2)} = 8$, $p_0^{(3)} = 10$, $P = 18$, $q_1 = q_2 = q_3 = q$ (then $Q = 3q$), required q.

$$\left(p_0^{(1)} + p_0^{(2)} + p_0^{(3)}\right) \cdot q = P, \quad \text{so} \quad 24q = 18, \quad \text{and} \quad q = \frac{3}{4}.$$

Remark. See B.48, similar (even for some values: $p_0^{(i)}$ halved).

(B.45) MSS $\mathcal{A}$, $\mathcal{B}$, $\mathcal{C}$. The prices of two kinds of corn are respectively six and eight nummi per modius, and altogether three modii were bought and the same amount was paid for each; what were the two quantities?

Given thus $p_0^{(1)} = 6$, $p_0^{(2)} = 8$, $Q = 3$, $p_1 = p_2$, required q_1, q_2. Here too we have a problem the basis for the solution of which remains unexplained by the author.

(a) First way of solving (introduction of auxiliary quantities).

(α) *Preliminary remark.* Earlier in this chapter we were considering the proportion valid for one type of corn (thus a single unit price p_0), namely

$$\frac{q_1}{q_2} = \frac{p_1}{p_2} = \frac{p_0 \cdot q_1}{p_0 \cdot q_2}.$$

For two kinds of corn at different prices such a proportion will be valid for each kind separately. In this problem, though, we have by hypothesis, since $p_1 = p_2$, that $p_0^{(1)} \cdot q_1 = p_0^{(2)} \cdot q_2$, thus

$$\frac{q_1}{q_2} = \frac{p_0^{(2)}}{p_0^{(1)}}.$$

This leads the author to introduce new quantities in order to return to a direct proportionality.* Let us put

$$c_1 = \frac{f}{p_0^{(1)}}, \quad c_2 = \frac{f}{p_0^{(2)}},$$

with f some optional number. With this we obtain

$$\frac{q_1}{q_2} = \frac{c_1}{c_2} \quad \text{then} \quad \frac{q_1}{q_1 + q_2} = \frac{c_1}{c_1 + c_2}, \quad \frac{q_1 + q_2}{q_2} = \frac{c_1 + c_2}{c_2},$$

---

* He will do the same in other instances. See B.241, B.320a.

with $q_1 + q_2 = Q$, whence

$$q_1 = \frac{c_1}{c_1 + c_2} \cdot Q, \qquad q_2 = \frac{c_2}{c_1 + c_2} \cdot Q.$$

It is thus, we are told, like a problem of partnership, q_i being the profits received by the two partners for a total profit Q according to their invested capitals c_i. These are indeed the formulae used in partnership.

(β) *Computation in the text.* Choosing $f = 24$, we find the two capitals $c_1 = 4$, $c_2 = 3$. Then

$$q_1 = \frac{c_1}{c_1 + c_2} \cdot Q = \frac{4}{7} \cdot 3 = 1 + \frac{5}{7}, \quad p_1 = p_0^{(1)} \cdot q_1 = 6\left(1 + \frac{5}{7}\right) = 10 + \frac{2}{7},$$

$$q_2 = \frac{c_2}{c_1 + c_2} \cdot Q = \frac{3}{7} \cdot 3 = 1 + \frac{2}{7}, \quad p_2 = p_0^{(2)} \cdot q_2 = 8\left(1 + \frac{2}{7}\right) = 10 + \frac{2}{7}.$$

Remark. This problem, also on buying and selling modii, certainly belongs in the present chapter, but its reduction to one on profit in partnership is odd, for that is the subject of a subsequent chapter (B–III). From the manuscripts it does not appear that the solution is interpolated.

(**b**) <u>Second way of solving</u> (does not introduce new quantities).

(α) *Preliminary remark.* From the two forms of the proportion

$$\frac{q_1}{q_2} = \frac{p_0^{(2)}}{p_0^{(1)}} \quad \text{and} \quad \frac{q_2}{q_1} = \frac{p_0^{(1)}}{p_0^{(2)}}$$

we infer that

$$\frac{q_1}{q_1 + q_2} = \frac{q_1}{Q} = \frac{p_0^{(2)}}{p_0^{(2)} + p_0^{(1)}} = \frac{1}{1 + \frac{p_0^{(1)}}{p_0^{(2)}}} = \frac{1}{\frac{p_0^{(1)}}{p_0^{(1)}} + \frac{p_0^{(1)}}{p_0^{(2)}}},$$

$$\frac{q_2}{q_2 + q_1} = \frac{q_2}{Q} = \frac{p_0^{(1)}}{p_0^{(1)} + p_0^{(2)}} = \frac{1}{1 + \frac{p_0^{(2)}}{p_0^{(1)}}} = \frac{1}{\frac{p_0^{(2)}}{p_0^{(2)}} + \frac{p_0^{(2)}}{p_0^{(1)}}},$$

whence

$$q_1 = \frac{Q}{\frac{p_0^{(1)}}{p_0^{(1)}} + \frac{p_0^{(1)}}{p_0^{(2)}}}, \qquad q_2 = \frac{Q}{\frac{p_0^{(2)}}{p_0^{(2)}} + \frac{p_0^{(2)}}{p_0^{(1)}}}.$$

(β) *Computations in the text.* They are, accordingly,

$$q_1 = \frac{3}{\frac{6}{6} + \frac{6}{8}}, \qquad q_2 = \frac{3}{\frac{8}{8} + \frac{8}{6}}.$$

(γ) *Generalization.* These two procedures, the author concludes, may be applied to a greater number of kinds of corn. This means that, if we are given $Q\left(=\sum_1^n q_i\right)$ and the $p_0^{(i)}$, and we must find q_i with $p_0^{(1)} \cdot q_1 = p_0^{(2)} \cdot q_2 = \ldots = p_0^{(n)} \cdot q_n$, we may solve it in two manners:

- According to part a, we put (with f optional)

$$c_i = \frac{f}{p_0^{(i)}}, \quad \text{whence} \quad q_k = \frac{c_k}{\sum_{i=1}^n c_i} \cdot Q.$$

- According to part b, we have, since $p_0^{(1)} \cdot q_1 = p_0^{(i)} \cdot q_i$,

$$Q = q_1 + q_2 + \ldots + q_n = q_1\left(1 + \frac{p_0^{(1)}}{p_0^{(2)}} + \ldots + \frac{p_0^{(1)}}{p_0^{(n)}}\right),$$

or, since generally $p_0^{(k)} \cdot q_k = p_0^{(i)} \cdot q_i$,

$$Q = q_k \cdot \sum_{i=1}^n \frac{p_0^{(k)}}{p_0^{(i)}},$$

whence

$$q_k = \frac{Q}{\sum_{i=1}^n \frac{p_0^{(k)}}{p_0^{(i)}}}.$$

(B.46) MSS $\mathcal{A}$, $\mathcal{B}$, $\mathcal{C}$. The prices of two kinds of corn are (as before) six and eight nummi per modius, and altogether one modius was bought for six and a half nummi; what were the two quantities?

Given thus $p_0^{(1)} = 6$, $p_0^{(2)} = 8$, $P = 6 + \frac{1}{2}$, $Q = 1$, required q_1, q_2. We have then a system of the form

$$\begin{cases} p_0^{(1)} \cdot q_1 + p_0^{(2)} \cdot q_2 = P \\ q_1 + q_2 = Q. \end{cases}$$

(α) *Preliminary remark.* Suppose we are given a system of the form

$$\begin{cases} a_1 \cdot q_1 + a_2 \cdot q_2 = k \\ q_1 + q_2 = l \end{cases}$$

with a_1, a_2, k, l given and positive.

Condition. If $a_1 < a_2$ then the system is solvable (that is, q_1, $q_2 > 0$) for $a_1 \cdot l < k < a_2 \cdot l$.
Indeed, from

$$a_1 \cdot q_1 + a_2 \cdot q_2 = a_1 \cdot q_1 + a_2\left(l - q_1\right) = a_2 \cdot l - \left(a_2 - a_1\right) q_1 = k,$$

$$a_1 \cdot q_1 + a_2 \cdot q_2 = a_1\left(l - q_2\right) + a_2 \cdot q_2 = a_1 \cdot l + \left(a_2 - a_1\right) q_2 = k,$$

we infer the formulae

$$q_1 = \frac{a_2 \cdot l - k}{a_2 - a_1}, \qquad q_2 = \frac{k - a_1 \cdot l}{a_2 - a_1}$$

or, in our case,

$$q_1 = \frac{p_0^{(2)} \cdot Q - P}{p_0^{(2)} - p_0^{(1)}}, \qquad q_2 = \frac{P - p_0^{(1)} \cdot Q}{p_0^{(2)} - p_0^{(1)}}.$$

This shows the condition mentioned in the text: the price paid P must lie between the two prices obtained if the whole quantity Q consists of one kind only, namely $p_0^{(1)} \cdot Q$ and $p_0^{(2)} \cdot Q$ (in this problem we have $Q = 1$, in the next $Q \ne 1$). We shall meet such systems again later (B.313–315, B.329), where the condition will be fully explained and, in addition, demonstrated.

(β) *Solution in the text.* The general form

$$\begin{cases} p_0^{(1)} \cdot q_1 + p_0^{(2)} \cdot q_2 = P \\ q_1 + q_2 = Q, \end{cases}$$

becomes, with the given quantities,

$$\begin{cases} 6 \cdot q_1 + 8 \cdot q_2 = 6 + \tfrac{1}{2} \\ q_1 + q_2 = 1. \end{cases}$$

Since $6 < 6 + \tfrac{1}{2} < 8$, that is, $p_0^{(1)} \cdot Q < P < p_0^{(2)} \cdot Q$, the condition is satisfied; the solutions as computed individually are then

$$q_1 = \frac{p_0^{(2)} \cdot Q - P}{p_0^{(2)} - p_0^{(1)}} = \frac{8 - (6 + \tfrac{1}{2})}{8 - 6} = \frac{1 + \tfrac{1}{2}}{2} = \frac{3}{4},$$

$$q_2 = \frac{P - p_0^{(1)} \cdot Q}{p_0^{(2)} - p_0^{(1)}} = \frac{(6 + \tfrac{1}{2}) - 6}{8 - 6} = \frac{\tfrac{1}{2}}{2} = \frac{1}{4}.$$

(B.47) MSS $\mathcal{A}$, $\mathcal{B}$, $\mathcal{C}$. The prices of barley and wheat are respectively six and ten nummi per modius, and altogether ten modii were bought for eighty-eight nummi; what were the two quantities?

Given thus $p_0^{(1)} = 6$ (barley), $p_0^{(2)} = 10$ (wheat), $Q = 10$, $P = 88$, required q_1, q_2. Same type of problem as before, but with $Q \ne 1$.

(α) *Preliminary remark.* We have therefore

$$\begin{cases} p_0^{(1)} \cdot q_1 + p_0^{(2)} \cdot q_2 = P \\ q_1 + q_2 = Q, \end{cases}$$

with in this case

$$\begin{cases} 6 \cdot q_1 + 10 \cdot q_2 = 88 \\ q_1 + q_2 = 10. \end{cases}$$

Here again the condition is that the price paid must lie between the prices obtained when the whole quantity is barley or wheat; that is, since $p_0^{(1)} <$

$p_0^{(2)}$, we must have $p_0^{(1)} \cdot Q < P < p_0^{(2)} \cdot Q$, as stated at the end of the problem —the limits being excluded if we are to buy from both.

(β) *Computation in the text.*

$$q_2 = \frac{P - p_0^{(1)} \cdot Q}{p_0^{(2)} - p_0^{(1)}} = \frac{88 - 6 \cdot 10}{10 - 6} = \frac{88 - 60}{4} = \frac{28}{4} = 7 \text{ (wheat)},$$

$$q_1 = \frac{p_0^{(2)} \cdot Q - P}{p_0^{(2)} - p_0^{(1)}} = \frac{10 \cdot 10 - 88}{10 - 6} = \frac{100 - 88}{4} = \frac{12}{4} = 3 \text{ (barley)}.$$

★ ★ ★

In the group B.48–50 we know the values of $p_0^{(1)}$, $p_0^{(2)}$, $p_0^{(3)}$ and we are told that the quantities q_i are equal. Furthermore, either P is given and Q required (B.48) or Q is given and P required (B.50; B.49 is particular).

(B.48) MSS $\mathcal{A}$, $\mathcal{B}$. The prices of three kinds of corn are respectively three, four, five nummi per sextarius, and for two nummi was bought an equal quantity of each; what was this quantity?

Given thus $p_0^{(1)} = 3$, $p_0^{(2)} = 4$, $p_0^{(3)} = 5$, $P = 2$, $q_1 = q_2 = q_3 = q$, required q (with $p_0^{(i)} \cdot q_i = p_i$ also computed for a verification). Similar to the (banal) problem B.44.

Computation. Since

$$p_0^{(1)} \cdot q_1 + p_0^{(2)} \cdot q_2 + p_0^{(3)} \cdot q_3 = \left(p_0^{(1)} + p_0^{(2)} + p_0^{(3)}\right) \cdot q = P,$$

then

$$q = \frac{P}{\sum p_0^{(i)}} = \frac{2}{3 + 4 + 5} = \frac{2}{12} = \frac{1}{6},$$

and therefore

$$p_1 = p_0^{(1)} \cdot q = \frac{1}{2}, \quad p_2 = \frac{2}{3}, \quad p_3 = \frac{5}{6},$$

so altogether $Q = \frac{1}{2}$ for $P = 2$.

Demonstration.

(α) *Preliminary remark.* If $p_0^{(i)}$ is the unit price of the ith kind, then

$$\frac{q}{1} = \frac{p_0^{(i)} \cdot q}{p_0^{(i)}} = \frac{p_i}{p_0^{(i)}}, \qquad (1)$$

and therefore

$$\sum p_i = P = \sum p_0^{(i)} \cdot q, \qquad (2)$$

so that

$$q = \frac{P}{\sum p_0^{(i)}}. \qquad (3)$$

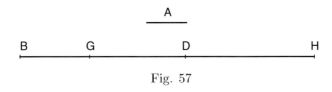

Fig. 57

(β) *Demonstration in the text.* Let (Fig. 57) $q_1 = q_2 = q_3 = A = q$, $p_1 = BG = p_0^{(1)} \cdot q$, $p_2 = GD = p_0^{(2)} \cdot q$, $p_3 = DH = p_0^{(3)} \cdot q$, so that $BH = P = 2$. Since

$$\frac{A}{1} = \frac{BG}{3} = \frac{GD}{4} = \frac{DH}{5}, \qquad (1)$$

then $3A = BG$, $4A = GD$, $5A = DH$, so (*Elements* II.1 = PE_1)

$$12A = BG + GD + DH = BH = P = 2 \qquad (2)$$

whence

$$q = A = \frac{2}{12} = \frac{P}{\sum p_0^{(i)}}. \qquad (3)$$

(B.49) MSS $\mathcal{A}$, $\mathcal{B}$. The prices of three kinds of corn are respectively three, four, five nummi per sextarius, and an equal quantity of each is bought, making altogether one sextarius; what is this quantity?

Given thus $p_0^{(1)} = 3$, $p_0^{(2)} = 4$, $p_0^{(3)} = 5$, $q_1 = q_2 = q_3 = q$, $Q = 1$ (P not required). This problem is said to be '*apertus*', that is, straightforward: one third of each kind is bought.

Remark. We may well qualify this problem as being absurd, for giving the values of the $p_0^{(i)}$ serves no purpose.

(B.50) MSS $\mathcal{A}$, $\mathcal{B}$. The prices of three kinds of corn are respectively three, four, five nummi per sextarius, and an equal quantity of each is bought, altogether one and a half sextarii; what is the price?

Given $p_0^{(1)} = 3$, $p_0^{(2)} = 4$, $p_0^{(3)} = 5$, $q_1 = q_2 = q_3 = q$, $Q = 1 + \frac{1}{2}$, and required is P. Hardly less banal than the problem before, but at least correctly formulated.

$$P = \sum p_0^{(i)} \cdot q_i = q \cdot \sum p_0^{(i)} = \frac{Q}{3} \cdot \sum p_0^{(i)} = \frac{1}{2} \cdot 12 = 6, \qquad q = \frac{1}{2}.$$

(B.51) MS $\mathcal{A}$ only. The prices of three kinds of corn are respectively three, seven, twelve nummi per sextarius, and altogether one sextarius was bought for ten nummi; what were the three quantities?

Given thus $p_0^{(1)} = 3$, $p_0^{(2)} = 7$, $p_0^{(3)} = 12$, $P = 10$, $Q = 1$, required q_i. Same type as B.46–47, but with three unknowns.

(α) *Preliminary remark.* We are to solve

$$\begin{cases} 3\,q_1 + 7\,q_2 + 12\,q_3 = P = 10 \\ q_1 + q_2 + q_3 = Q = 1. \end{cases}$$

This indeterminate problem, of a kind which will be fully treated later (B–XVIII), is reduced to a system with two unknowns, and thus rendered determinate, by an arbitrary assumption, namely setting q_1 and q_2 equal. The system then becomes

$$\begin{cases} 10\,q_1 + 12\,q_3 = 10 \\ 2\,q_1 + q_3 = 1, \end{cases}$$

or, taking as new unknown $2\,q_1 = q'$, and doubling the first equation (thus keeping the same coefficient of q_1),

$$\begin{cases} 10\,q' + 24\,q_3 = 20 \\ q' + q_3 = 1, \end{cases}$$

which is of the form seen in B.46. The fulfilment of the condition is readily seen from the first equation.

(β) *Computations in the text.* The initial computations correspond to the above transformations. The solution is then

$$q_1 = q_2 = \frac{1}{2} \cdot q' = \frac{1}{2} \cdot \frac{24 - 20}{24 - 10} = \frac{12 - 10}{14} = \frac{2}{14} = \frac{1}{7}, \quad \text{then} \quad q_3 = \frac{5}{7}.$$

As noted by the author at the end, such a problem will be impossible if the price given for the sextarius bought exceeds the highest price, here 12, or is less than the lowest, here 3. The extreme cases are excluded as well, for there must be corn of all three kinds, as stated in the question. So the condition is, as said in the text, $3 < P < 12$.

(B.52) MSS $\mathcal{A}$, $\mathcal{B}$. Ten sextarii are bought each at a different price, the first at three and the prices exceed one another by four; what are the prices of the last and of all?

We are given $Q = 10 = n$, $p_0^{(1)} = 3$, $p_0^{(k)} = p_0^{(1)} + (k-1) \cdot \delta$ ($k = 1, \ldots, n$, here $n = 10$, $\delta = 4$), required $p_0^{(n)}$, P.

(α) *Preliminary remark.* If $p_0^{(k)} = p_0^{(1)} + (k-1) \cdot \delta$, where $k = 1, \ldots, n$, their sum and the last term are determined respectively by

$$P = \sum_{i=1}^{n} p_0^{(i)} = \frac{n}{2} \left[2\,p_0^{(1)} + (n-1)\,\delta \right], \qquad p_0^{(n)} = p_0^{(1)} + (n-1)\,\delta.$$

(β) *Computations in the text.* The text calculates according to the above formulae:

$$P = \frac{10}{2} \cdot \left[2 \cdot 3 + (10-1) \cdot 4 \right] = 5\,[6 + 9 \cdot 4] = 5 \cdot 42 = 210,$$

$$p_0^{(10)} = 3 + (10-1) \cdot 4 = 39.$$

There is no justification whatsoever for these computations, nor any allusion to the similar problems A.268–270. For a demonstration, we are to await B–XII. This must be yet another example of a problem added later (by the author; see Part I, pp. lvi–lvii).

(B.53) MS $\mathcal{A}$ only. Twelve and a fourth sextarii are bought each at a different price, the first at three and the prices exceed one another by five; what is the total price?

Given thus $Q = 12 + \frac{1}{4}$, $p_0^{(1)} = 3$, $\delta = 5$, required P.

(α) *Preliminary remark.* The previous formula is only partly applicable. Indeed, twelve whole sextarii are bought at the respective prices $p_0^{(1)}$, $p_0^{(2)}$, ..., $p_0^{(12)}$, but the cost of the last quarter must be calculated for the unit price $p_0^{(13)}$. Whence the author's formula, where our n is the number of whole sextarii:

$$P = \frac{n}{2}\left[2p_0^{(1)} + (n-1)\cdot\delta\right] + \frac{1}{4}\left(p_0^{(1)} + n\cdot\delta\right).$$

(β) *Computations in the text.* Accordingly,

$$P = \frac{12}{2}\left[2\cdot 3 + (12-1)\cdot 5\right] + \frac{1}{4}(3 + 12\cdot 5)$$

$$= 366 + \frac{1}{4}\cdot 63 = 366 + 15 + \frac{3}{4} = 381 + \frac{3}{4}.$$

Chapter B–II:
Profit
Summary

Suppose a certain commodity is bought with two capitals c_1, c_2 and sold at v_1 and v_2, thus yielding the respective profits e_1, e_2. The two selling prices are then $v_1 = c_1 + e_1$, $v_2 = c_2 + e_2$, and these six magnitudes are linked by the relation
$$\frac{c_1}{c_2} = \frac{e_1}{e_2} = \frac{v_1}{v_2}.$$
The left-hand proportion is first stated in B.54; the third ratio is found by composition from $c_1 : e_1 = c_2 : e_2$.

1. The five fundamental problems.

As the author says, there are five types of mathematical problem which can be applied to profit. The required quantity may be c_2 or e_2, or both may be unknown but their sum, difference or product known. The author also reminds us that we have already encountered, in the chapter on buying and selling, these five basic types (B.1–2, B.15–17). It is with five such problems that the section opens (B.54–58). They are thus solved by the relations
$$e_2 = \frac{e_1 \cdot c_2}{c_1};$$
$$c_2 = \frac{c_1 \cdot e_2}{e_1};$$
$$c_2 = \frac{c_1(c_2 + e_2)}{c_1 + e_1}, \quad e_2 = \frac{e_1(c_2 + e_2)}{c_1 + e_1};$$
$$c_2 = \frac{c_1(c_2 - e_2)}{c_1 - e_1}, \quad e_2 = \frac{e_1(c_2 - e_2)}{c_1 - e_1};$$
$$c_2 = \sqrt{\frac{c_1(c_2 \cdot e_2)}{e_1}}, \quad e_2 = \sqrt{\frac{e_1(c_2 \cdot e_2)}{c_1}}.$$

The plain proportion is explained in B.54, and the formulae of the three cases are proved in the corresponding problems B.56–58 —this time fully (unlike in the first occurrence of this kind of problem in B.15–17).

2. The twenty kinds.

The above problems involve pure numbers, just like the corresponding ones in the previous chapter on buying and selling. Where the nature of the units is specified, however, a further sort of distinction must be made since capital and profit may be expressed in cash or in kind (namely both in cash, or both in kind, or one in cash and the other in kind). These four cases, combined with each of the five types mentioned above, give rise to twenty variants altogether, which as such will be the central subject of this chapter. As a matter of fact, there are twenty-four problems, all about bushels (*caficii* or *modii*), six for each of the four cases (B.59–82); for each group is introduced by a simple example, which first considers the case with a unit quantity (as usual: see B.4), then presents the five types of problem and mentions the possible use of other, larger units: the *almodi* to 12 caficii (cases *ii–iv* below) and the *solidus* to 12 nummi (cases *i–iii*). Anyway, as the author notes at the very beginning of this section, the theory he will present remains valid whatever the commodity or unit of measure.

Remark. In trading, corn was not weighed but measured by volume. Here the unit of capacity is a few times the *modius*, generally the *caficius* (Arabic *qafīz*) and once (B.89) the *sextarius*. The word *almodi* transcribes the Arabic *al-mudd*.

We shall then meet successively the cases:

(*i*) Capitals c_i in cash, profits e_i in cash (B.59–64). Solved by

$$\frac{c_1}{c_2} = \frac{e_1}{e_2}.$$

(*ii*) Capitals C_i in kind, profits e_i in cash (B.65–70). Solved by

$$\frac{C_1}{C_2} = \frac{e_1}{e_2}.$$

(*iii*) Capitals c_i in cash, profits E_i in kind (B.71–76). Solved by

$$\frac{c_1}{c_2} = \frac{E_1}{E_2}.$$

(*iv*) Capitals C_i in kind, profits E_i in kind (B.77–82). Solved by

$$\frac{C_1}{C_2} = \frac{E_1}{E_2}.$$

As we are told at the end of the introduction to this second section, before B.59, these four cases each belong to one of the two types mentioned at the beginning of the previous chapter: in (*i*) and (*ii*), when the profit in cash is required, the question is 'what is the price?', whereas in (*iii*) and (*iv*), when the profit in kind is required, the question is 'how much shall I have?'.

Note that in all these cases the pairs of analogous quantities are of the same type, either cash or kind. The situation is different in B.91 (termed 'foreign' by our author): the quantities involved are C_1, c_2, E_1, e_2 (but the proportion, namely $C_1 : c_2 = E_1 : e_2$ still holds).

3. Variation in capital.

In B.83–90 the required capital is different from that mentioned in the data, either because the capital actually invested turns out to be another or because we consider that the sale ends, or is interrupted, before the whole merchandise is sold, so that the profit is not what was expected.

A. *Different capital*.

We consider the sale for another capital, c'_2 or C'_2, which is a linear function of the initial (unknown) capital c_2 or C_2, for it differs from the latter by a given multiplicative factor (a fraction), sometimes also by a given additive amount, either of money r_2 or of bushels R_2; we are given the corresponding profit, e'_2 or E'_2. All these problems, the author says in B.88, the last problem of this type, are reducible to the twenty kinds seen before. Let us thus consider the same subdivisions as in the previous section.

(*i*) Capitals in cash, profits in cash. Since $c'_2 = \frac{k}{l} \cdot c_2 + r_2$ nummi, and $c'_2 : c_1 = e'_2 : e_1$, we have

$$c'_2 = \frac{k}{l} \cdot c_2 + r_2 = \frac{c_1 \cdot e'_2}{e_1}, \quad \text{whence} \quad c_2 = \frac{l}{k}\left(\frac{c_1 \cdot e'_2}{e_1} - r_2\right).$$

See the simple case, with $\frac{k}{l} = 1$, at the very beginning (B.83–83′), and the general case in B.88.

(*ii*) Capitals in kind, profits in cash. Since $C'_2 = \frac{k}{l} \cdot C_2 + R_2$ bushels, with $C'_2 : C_1 = e'_2 : e_1$, we have

$$C'_2 = \frac{k}{l} \cdot C_2 + R_2 = \frac{C_1 \cdot e'_2}{e_1}, \quad \text{whence} \quad C_2 = \frac{l}{k}\left(\frac{C_1 \cdot e'_2}{e_1} - R_2\right).$$

Examples in B.86–86′.

(*iii*) Capitals in cash, profits in kind. Since again $c'_2 = \frac{k}{l} \cdot c_2 + r_2$ nummi, with $c'_2 : c_1 = E'_2 : E_1$, we have

$$c'_2 = \frac{k}{l} \cdot c_2 + r_2 = \frac{c_1 \cdot E'_2}{E_1}, \quad \text{whence} \quad c_2 = \frac{l}{k}\left(\frac{c_1 \cdot E'_2}{E_1} - r_2\right).$$

See the examples in B.85–85″, B.87.

(*iv*) Capitals in kind, profits in kind. No example.

Condition. As stated in B.87, we must have in case *iii*

$$\frac{c_1 \cdot E'_2}{E_1} > r_2.$$

The same kind of condition must of course hold for the other cases when the supplementary amount is additive.

B. *Sale not completed*.

We consider the sale as coming to an end, or being interrupted, at an intermediate stage, without the whole merchandise being sold. Profit may then become loss, in the sense that the capital invested is not fully recovered.

Let c_1, v_1 (given) be the buying and selling prices of a given quantity C_1 of bushels; thus we can compute the buying price c_0 and the selling price v_0 per unit, namely

$$c_0 = \frac{c_1}{C_1}, \qquad v_0 = \frac{v_1}{C_1}.$$

Let next C_2 be another, unknown quantity of bushels, bought for $c_2 = C_2 \cdot c_0$ nummi and to be sold for $v_2 = C_2 \cdot v_0$ nummi. At an intermediate stage, suppose a fraction $\frac{k}{l}$ of the initial capital has been recovered, together with a quantity r_2 of nummi while R_2 bushels remain unsold. There are two ways of solving such a problem.

(α) The whole selling price v_2 (if all bushels sold) can be expressed as the sum of three quantities: the part recovered of the initial capital (bushels at the buying price), the bushels left (considered this time at their selling price), the quantity of nummi obtained in addition. We then have for v_2, in nummi,

$$v_2 = C_2 \cdot v_0 = \frac{k}{l} C_2 \cdot c_0 + R_2 \cdot v_0 + r_2$$

whence, expressed either in terms of v_0 or v_1,

$$C_2 = \frac{R_2 \cdot v_0 + r_2}{v_0 - \frac{k}{l} \cdot c_0} = \frac{R_2 \cdot v_1 + r_2 \cdot C_1}{v_1 - \frac{k}{l} \cdot c_1}$$

(and $c_2 = C_2 \cdot c_0$). In the group B.89 ($v_1 = v_0$), we find the simpler case $\frac{k}{l} = 1$ (the whole capital c_2 is recovered) with a few bushels left over. There the three possibilities are considered: $r_2 = 0$ (B.89, no profit in cash), $r_2 = 10$ (B.89′, small profit), $r_2 = -10$ (B.89″, small loss).

(β) Or else consider in a similar situation C_1, the first quantity of bushels, and determine what R_1, the number of bushels left, would be. Since

$$v_2 = \frac{k}{l} C_2 \cdot c_0 + R_2 \cdot v_0, \quad \text{then} \quad v_1 = C_1 \cdot v_0 = \frac{k}{l} C_1 \cdot c_0 + R_1 \cdot v_0,$$

and so (B.90)

$$R_1 = \frac{C_1 \cdot v_0 - \frac{k}{l} C_1 \cdot c_0}{v_0} = C_1 - \frac{\frac{k}{l} c_1}{v_0}.$$

Since we now know R_1 while, obviously, $C_1 : C_2 = R_1 : R_2 = c_1 : c_2$, we can determine C_2 and c_2 from

$$C_2 = \frac{C_1 \cdot R_2}{R_1}, \qquad c_2 = \frac{c_1 \cdot R_2}{R_1}.$$

This is valid if there is no supplement in nummi r_2 (B.84, B.90). If there is (B.84', B.90'; B.84'', B.90''), we shall incorporate it into R_2 by changing these r_2 nummi to $\dfrac{r_2}{v_0}$ caficii.

We have already commented on B.91 (p. 1396, bottom).

4. Other problems.

The last problems are of the type encountered at the end of the previous chapter. First, we have a problem with two kinds of corn, each bushel of one kind being sold at the buying price of the other (B.92); this leads to a quadratic equation. The rest is, with one exception, devoted to problems involving roots in the data, which belong to each of the four cases with data in cash and kind:

- The first group involves c_i and e_i (B.93–95, B.97, B.97'). We are always given c_1 and e_1, and, in addition, the quotient $e_2 : \sqrt{c_2}$ (B.93–94), or $\sqrt{c_2} \cdot \sqrt{e_2}$ (B.95), or $\sqrt{c_2} + \sqrt{e_2}$ (B.97), or finally $\sqrt{c_2} - \sqrt{e_2}$ (B.97'), thus all are problems with two unknowns.
- The second group involves C_i and e_i (B.96, B.102–103). We are given C_1, e_1 and $\sqrt{C_2} \cdot \sqrt{e_2}$ in B.96; in the other two, C_2, e_2 are now given while $e_1 = \alpha\sqrt{C_1} + \beta$, which leads to solving a quadratic equation in B.102 but not in B.103 ($\beta = 0$).
- The third group involves c_i, E_i (B.98, B.100). We know c_1 and E_1 and then $\sqrt{c_2} + \sqrt{E_2}$ in B.98, or $E_2 : \sqrt{c_2}$ in B.100.
- Finally B.99 involves C_i and E_i. We know C_1, E_1 and $\sqrt{C_2} - \sqrt{E_2}$.

An exception is B.101 since it does not include roots: c_1, v_0, c_2, E_2 are known and C_1 required; it is closely related to B.72.

---

1. The five fundamental problems.

A. *One unknown*.

(B.54) MSS $\mathcal{A}$, $\mathcal{B}$, $\mathcal{C}$. For a capital of five the gain is three; what will be the gain for eighty?

Given thus $c_1 = 5$, $e_1 = 3$, $c_2 = 80$, required e_2.

$$e_2 = \frac{e_1 \cdot c_2}{c_1} = \frac{3 \cdot 80}{5} = 48.$$

The two forms of the basic proportion ($e_1 : c_1 = e_2 : c_2$ and its inverted form $c_1 : e_1 = c_2 : e_2$) are mentioned and the second form is used to explain the formula. Here again, the author mentions the two banal 'other ways' of computing, namely

$$e_2 = \frac{c_2}{c_1} e_1, \quad e_2 = \frac{e_1}{c_1} c_2.$$

(B.55) MSS $\mathcal{A}$, $\mathcal{B}$, $\mathcal{C}$. For a capital of five the gain is three; what will be the capital for a gain of forty?

Given thus $c_1 = 5$, $e_1 = 3$, $e_2 = 40$, required c_2.

$$c_2 = \frac{c_1 \cdot e_2}{e_1} = \frac{5 \cdot 40}{3} = 66 + \frac{2}{3}.$$

B. *Two unknowns*.

In the next three problems B.56–58 we are given c_1, e_1 and successively $c_2 + e_2$, $c_2 - e_2$, $c_2 \cdot e_2$. The formulae for computing c_2 and e_2 are those mentioned in our summary to this chapter; two of them are demonstrated here.

(B.56) MSS $\mathcal{A}$, $\mathcal{B}$; $\mathcal{C}$ (computation only). For a capital of five the gain is three; what are capital and profit if their sum is a hundred?

Given thus $c_1 = 5$, $e_1 = 3$, $c_2 + e_2 = 100$, required c_2 and e_2.

$$c_2 = \frac{c_1\,(c_2 + e_2)}{c_1 + e_1} = \frac{5 \cdot 100}{5 + 3} = \frac{5 \cdot 100}{8} = 62 + \frac{1}{2},$$

$$e_2 = \frac{e_1\,(c_2 + e_2)}{c_1 + e_1} = \frac{3 \cdot 100}{8} = 37 + \frac{1}{2}.$$

Demonstration of the formulae.

(α) *Preliminary remark*. The demonstration relies on composition of the basic proportion and its inverted form. Since

$$\frac{c_1}{e_1} = \frac{c_2}{e_2}, \quad \text{so} \quad \frac{c_1 + e_1}{e_1} = \frac{c_2 + e_2}{e_2},$$

$$\frac{e_1}{c_1} = \frac{e_2}{c_2}, \quad \text{so} \quad \frac{c_1 + e_1}{c_1} = \frac{c_2 + e_2}{c_2},$$

whence e_2 and c_2.

```
A      B   G    D                  H            Z
|----------|   |------------------|------------|
```

Fig. 58

(β) *Demonstration in the text*. Let (Fig. 58) $AB = c_1 = 5$, $BG = e_1 = 3$, $DH = c_2$, $HZ = e_2$, thus $DZ = 100$. From the basic proportion

$$\frac{AB}{BG} = \frac{DH}{HZ}$$

we obtain, by composition,

$$\frac{AB + BG}{BG} = \frac{DH + HZ}{HZ}, \quad \text{or} \quad \frac{AG}{BG} = \frac{DZ}{HZ}$$

and so
$$HZ = e_2 = \frac{BG \cdot DZ}{AG} = \frac{e_1(c_2 + e_2)}{c_1 + e_1}.$$

Next, from the inverted proportion
$$\frac{BG}{AB} = \frac{HZ}{DH}$$

by composition again,
$$\frac{AB + BG}{AB} = \frac{DH + HZ}{DH}, \quad \text{or} \quad \frac{AG}{AB} = \frac{DZ}{DH}$$

whence
$$DH = c_2 = \frac{AB \cdot DZ}{AG} = \frac{c_1(c_2 + e_2)}{c_1 + e_1}.$$

Remark. This incidentally proves that
$$\frac{v_1}{e_1} = \frac{v_2}{e_2}, \qquad \frac{v_1}{c_1} = \frac{v_2}{c_2}.$$

(B.57) MSS $\mathcal{A}$, $\mathcal{B}$, $\mathcal{C}$. For a capital of six the gain is one and a half; what are capital and profit if subtracting profit from capital leaves ninety? Given thus $c_1 = 6$, $e_1 = 1 + \frac{1}{2}$, $c_2 - e_2 = 90$, required c_2 and e_2. Then:
$$c_2 = \frac{c_1(c_2 - e_2)}{c_1 - e_1} = \frac{6 \cdot 90}{6 - (1 + \frac{1}{2})} = \frac{6 \cdot 90}{4 + \frac{1}{2}} = 120,$$

$$e_2 = \frac{e_1(c_2 - e_2)}{c_1 - e_1} = \frac{(1 + \frac{1}{2}) 90}{4 + \frac{1}{2}} = 30.$$

As we are told, the proof is the same as before but with conversion instead of composition. Furthermore, the author says, the proportion has already been set in the chapter on buying and selling (B.16).

As usual, mention is made of the possibility of first dividing one of the factors.

Remark. In the previous figure, take $AG = c_1$, $BG = e_1$, $DZ = c_2$, $HZ = e_2$, thus $DH = DZ - HZ = 90$. From the proportion $AG : BG = DZ : HZ$, we indeed obtain by conversion
$$\frac{AG}{AG - BG} = \frac{DZ}{DZ - HZ} \quad \text{thus} \quad \frac{AG}{AB} = \frac{DZ}{DH} \quad \text{or} \quad \frac{c_1}{c_1 - e_1} = \frac{c_2}{c_2 - e_2},$$

and also, but this time by separation,
$$\frac{AG - BG}{BG} = \frac{DZ - HZ}{HZ} \quad \text{thus} \quad \frac{AB}{BG} = \frac{DH}{HZ} \quad \text{or} \quad \frac{c_1 - e_1}{e_1} = \frac{c_2 - e_2}{e_2}.$$

(B.58) MSS $\mathcal{A}$, $\mathcal{B}$; $\mathcal{C}$ (computation only). For a capital of five the gain is three; what are capital and profit if their product is sixty?

Given thus $c_1 = 5$, $e_1 = 3$, $c_2 \cdot e_2 = 60$, required c_2, e_2.
As in B.17, we are told, we shall compute

$$c_2 = \sqrt{\frac{c_1 (c_2 \cdot e_2)}{e_1}} = \sqrt{\frac{5 \cdot 60}{3}} = 10,$$

$$e_2 = \sqrt{\frac{e_1 (c_2 \cdot e_2)}{c_1}} = \sqrt{\frac{3 \cdot 60}{5}} = 6.$$

Unlike in the two previous problems, it is noted that we may also calculate either one and determine the other from the data, here the given product.

Demonstration of the formulae.

(α) *Preliminary remark.* Since

$$\frac{c_2}{e_2} = \frac{c_2^2}{c_2 \cdot e_2} = \frac{c_1}{e_1}, \qquad \frac{c_2}{e_2} = \frac{c_2 \cdot e_2}{e_2^2} = \frac{c_1}{e_1},$$

therefore

$$c_2 = \sqrt{\frac{c_1 (c_2 \cdot e_2)}{e_1}}, \qquad e_2 = \sqrt{\frac{e_1 (c_2 \cdot e_2)}{c_1}}.$$

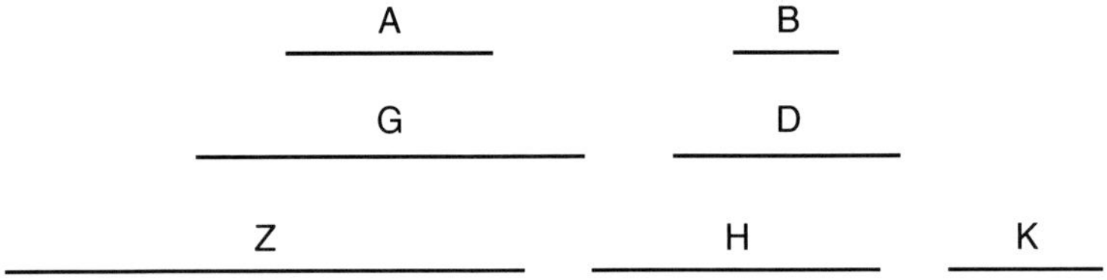

Fig. 59

(β) *Demonstration in the text.* Let (Fig. 59) $A = c_1 = 5$, $B = e_1 = 3$, $G = c_2$, $D = e_2$; then put $G \cdot D = H = c_2 \cdot e_2 = 60$. Let $G^2 = Z$, $D^2 = K$. Then, since (*Elements* VII.17–18)

$$\frac{Z}{H} = \frac{G^2}{G \cdot D} = \frac{G}{D} \quad \text{and} \quad \frac{H}{K} = \frac{G \cdot D}{D^2} = \frac{G}{D},$$

so $Z : H = G : D = H : K$. But since $G : D = A : B$, so

$$\frac{Z}{H} = \frac{A}{B} = \frac{H}{K},$$

and therefore

$$Z = \frac{A \cdot H}{B}, \qquad K = \frac{B \cdot H}{A}.$$

whence finally

$$G = c_2 = \sqrt{Z} = \sqrt{\frac{A \cdot H}{B}} = \sqrt{\frac{c_1 (c_2 \cdot e_2)}{e_1}},$$

$$D = e_2 = \sqrt{K} = \sqrt{\frac{B \cdot H}{A}} = \sqrt{\frac{e_1 (c_2 \cdot e_2)}{c_1}}.$$

2. The twenty kinds.

As said in our summary (p. 1395), a distinction is now made as to whether capital and profit are in cash or in kind. Although we shall be dealing with buying and selling bushels (*caficii* or *modii*), thus capacity measures, the subsequent theory will remain valid, as stated in the text at the very beginning of this second section, for any kind of commodity bought and sold.

(i) First case: Capital in cash, profit in cash (B.59–64)

The proportion used will therefore be

$$\frac{c_1}{c_2} = \frac{e_1}{e_2}.$$

(B.59) MSS $\mathcal{A}$, $\mathcal{B}$, $\mathcal{C}$. One modius bought for six is sold for seven and a half; how many nummi are gained from a hundred nummi?

Given thus $c_1 = 6 = c_0$ for one modius, $v_1 = 7 + \frac{1}{2} = v_0$ (thus $e_1 = 1 + \frac{1}{2} = e_0$), $c_2 = 100$, required e_2. As in B.54,

$$e_2 = \frac{e_1 \cdot c_2}{c_1} = \frac{\left(1 + \frac{1}{2}\right) \cdot 100}{6} = 25.$$

As he will do in all the introductory examples of the four cases, the author mentions first the possible use of other units and then, before going on to practical examples, reminds us of what the four other fundamental problems are. Thus:

— If c_2 is in solidi (1 solidus $= 12$ nummi), the result e_2 will be in solidi (and must be multiplied by 12 to be expressed in nummi).

— If e_2 is given and c_2 required, then (B.55)

$$c_2 = \frac{c_1 \cdot e_2}{e_1}.$$

— We may be given (besides c_1, e_1) $c_2 + e_2$, or $c_2 - e_2$, or $c_2 \cdot e_2$, with c_i and e_i either in nummi or in solidi. We shall solve these three cases as seen in B.56–58.

1404   PART THREE: MATHEMATICAL COMMENTARY, BIBLIOGRAPHY, INDEX

Remark. In the next five examples, one for each type, the numerical values adopted for the first pair c_1, e_1 are the same throughout. Thus, in B.60–64 we are initially told that 3 caficii are bought for 10 and 4 are sold for 20. Therefore, 3 are sold for 15, and to a capital in nummi $c_1 = 10$ corresponds a profit in nummi $e_1 = 5$. We shall omit this step in the computations and include the values $c_1 = 10$, $e_1 = 5$ in the data. Note that the required quantities often take simple values (in all five problems $e_2 = 50$, $c_2 = 100$). This is not surprising: as the author tells us, how to solve these problems is known from before and he will give examples just 'for the purpose of greater clarity'.

(B.60) MSS $\mathcal{A}$, $\mathcal{B}$. Caficii are bought three for ten nummi and sold four for twenty; how many nummi are gained from a hundred nummi?
Given thus $c_1 = 10$, $e_1 = 5$, $c_2 = 100$, required e_2. See B.54.

$$e_2 = \frac{e_1 \cdot c_2}{c_1} = 50.$$

(B.61) MSS $\mathcal{A}$, $\mathcal{B}$. Caficii are bought three for ten nummi and sold four for twenty; from how many nummi is there a profit of fifty nummi?
Given thus $c_1 = 10$, $e_1 = 5$, $e_2 = 50$, required c_2. See B.55.

$$c_2 = \frac{c_1 \cdot e_2}{e_1} = 100.$$

(B.62) MSS $\mathcal{A}$, $\mathcal{B}$. Caficii are bought three for ten nummi and sold four for twenty; what are profit in nummi and capital in nummi if their sum is a hundred and fifty?
Given thus $c_1 = 10$, $e_1 = 5$, $c_2 + e_2 = 150$, required c_2 and e_2. See B.56.

$$c_2 = \frac{c_1(c_2 + e_2)}{c_1 + e_1} = 100, \quad e_2 = \frac{e_1(c_2 + e_2)}{c_1 + e_1} = 50.$$

(B.63) MSS $\mathcal{A}$, $\mathcal{B}$. Caficii are bought three for ten nummi and sold four for twenty; what are capital in nummi and profit in nummi if subtracting the profit from the capital leaves fifty nummi?
Given thus $c_1 = 10$, $e_1 = 5$, $c_2 - e_2 = 50$, required c_2 and e_2. See B.57.

$$c_2 = \frac{c_1(c_2 - e_2)}{c_1 - e_1} = 100, \quad e_2 = \frac{e_1(c_2 - e_2)}{c_1 - e_1} = 50.$$

(B.64) MSS $\mathcal{A}$, $\mathcal{B}$. Caficii are bought three for ten nummi and sold four for twenty; what are capital in nummi and profit in nummi if their product is five thousand?
Given thus $c_1 = 10$, $e_1 = 5$, $c_2 \cdot e_2 = 5000$, required c_2 and e_2. See B.58.

$$c_2 = \sqrt{\frac{c_1(c_2 \cdot e_2)}{e_1}} = 100, \quad e_2 = \sqrt{\frac{e_1(c_2 \cdot e_2)}{c_1}} = 50.$$

(*ii*) Second case: Capital in kind, profit in cash (B.65–70)

Here the proportion is
$$\frac{C_1}{C_2} = \frac{e_1}{e_2}.$$

(B.65) MSS $\mathcal{A}$, $\mathcal{B}$, $\mathcal{C}$. One modius is bought for six nummi and sold for seven and a half nummi; how many nummi are gained from a hundred modii?

Given thus $c_1 = 6 = c_0$ for one modius, so $C_1 = 1$, $v_1 = 7 + \frac{1}{2} = v_0$ (whence $e_1 = 1 + \frac{1}{2} = e_0$), $C_2 = 100$, required e_2. Then
$$e_2 = \frac{e_1 \cdot C_2}{C_1} = 150.$$

— If $e_1 = e_0$ is the profit in nummi from one caficius and C_2 is expressed in almodis (1 almodi = 12 caficii), replace e_1 by $12e_1$, the profit in nummi from one almodi, and compute as above.

— If e_2 is given and C_2 required, then
$$C_2 = \frac{C_1 \cdot e_2}{e_1}.$$

— If e_1 is in nummi and e_2 in solidi, compute as above but multiply the result by 12 since C_2 needs to be 12 times larger to yield e_2 solidi.

— Finally, there are the three cases where we are given $e_2 + C_2$, or $e_2 - C_2$ (since $e_1 > C_1$), or also $e_2 \cdot C_2$.

Remark. In the subsequent problems B.66–70, we are again told at the outset that 3 caficii are bought for 10 and 4 sold for 20. Therefore, since 3 are bought for 10 and sold for 15, to a capital in caficii $C_1 = 3$ corresponds a profit in nummi $e_1 = 5$. We shall omit this initial step which recurs in all these problems. Note too that the last term in the data is systematically taken as 100; the same in the third and fourth case.

(B.66) MSS $\mathcal{A}$, $\mathcal{B}$. Caficii are bought three for ten nummi and sold four for twenty; how many nummi are gained from a hundred caficii?

Given thus $C_1 = 3$, $e_1 = 5$, $C_2 = 100$, required e_2.
$$e_2 = \frac{e_1 \cdot C_2}{C_1} = 166 + \frac{2}{3}.$$

(B.67) MSS $\mathcal{A}$, $\mathcal{B}$. Caficii are bought three for ten nummi and sold four for twenty; from how many caficii are gained a hundred nummi?

Given thus $C_1 = 3$, $e_1 = 5$, $e_2 = 100$, required C_2.
$$C_2 = \frac{C_1 \cdot e_2}{e_1} = 60.$$

There is a verification: $C_2 = 60$ corresponds to an initial capital in nummi $c_2 = 200$ (since 3 caficii are bought for 10 nummi) and a selling price $v_2 = 300$ (since 3 caficii are sold for 15 nummi); thus indeed $e_2 = 100$.

(B.68) MSS A, B. Caficii are bought three for ten nummi and sold four for twenty; what are profit in nummi and capital in caficii if their sum is a hundred?

Given thus $C_1 = 3$, $e_1 = 5$, $C_2 + e_2 = 100$, required C_2, e_2.

$$C_2 = \frac{C_1(C_2 + e_2)}{C_1 + e_1} = 37 + \frac{1}{2}, \quad e_2 = \frac{e_1(C_2 + e_2)}{C_1 + e_1} = 62 + \frac{1}{2}.$$

(B.69) MSS A, B. Caficii are bought three for ten nummi and sold four for twenty; what are profit in nummi and capital in caficii if subtracting the capital from the profit leaves a hundred?

Given thus $C_1 = 3$, $e_1 = 5$, $e_2 - C_2 = 100$, required C_2, e_2. As the author remarks, since in this problem $e_1 > C_1$, the given difference must be $e_2 - C_2$; otherwise the problem will not be solvable. Similarly, if we had $C_1 > e_1$, we should be given $C_2 - e_2$.

Then

$$C_2 = \frac{C_1(e_2 - C_2)}{e_1 - C_1} = 150, \quad e_2 = \frac{e_1(e_2 - C_2)}{e_1 - C_1} = 250.$$

(B.70) MSS A, B. Caficii are bought three for ten nummi and sold four for twenty; what are profit in nummi and capital in caficii if their product is a hundred?

Given thus $C_1 = 3$, $e_1 = 5$, $C_2 \cdot e_2 = 100$, required C_2, e_2.

$$C_2 = \sqrt{\frac{C_1(C_2 \cdot e_2)}{e_1}} = \sqrt{60}, \quad e_2 = \sqrt{\frac{e_1(C_2 \cdot e_2)}{C_1}} = \sqrt{166 + \frac{2}{3}}.$$

(*iii*) Third case: Capital in cash, profit in kind (B.71–76)

All these problems rely on the same proportion as before, now of the form

$$\frac{c_1}{c_2} = \frac{E_1}{E_2}.$$

(B.71) MSS A, B, C. One caficius is bought for six nummi and sold for seven and a half; how many caficii are gained with a hundred nummi?

Given thus $c_1 = 6 = c_0$, $v_1 = 7 + \frac{1}{2} = v_0$ (so $e_1 = 1 + \frac{1}{2} = e_0$), $c_2 = 100$, required E_2.

We know $e_1 = 1 + \frac{1}{2}$, in nummi, from which we must determine E_1, in caficii. Since $1 + \frac{1}{2}$ is gained on an investment of 6 and a sale of $7 + \frac{1}{2}$, by selling $\frac{4}{5}$ of a caficius we recover c_1, so a fifth of a caficius is gained on

each caficius sold. Thus, for $c_1 = 6$ (and $C_1 = 1$), we shall have $E_1 = \frac{1}{5}$. Then
$$E_2 = \frac{E_1 \cdot c_2}{c_1} = 3 + \frac{1}{3}.$$

— If $c_2 = 100$ solidi, the above result must be multiplied by 12 since the capital in nummi becomes twelve times larger; we may also change initially $c_1 = 6$ nummi to $\frac{1}{2}$ solidus and compute as before.

— If E_2 is given ($= 10$) and c_2 required, then
$$c_2 = \frac{c_1 \cdot E_2}{E_1} = \frac{c_1 \cdot 10}{\frac{1}{5}}$$

with either $c_1 = 6$ (result in nummi, to be divided by 12 to express c_2 in solidi) or $c_1 = \frac{1}{2}$ (result in solidi).‡

— The three remaining types propose $c_2 + E_2$, $c_2 - E_2$, $c_2 \cdot E_2$.

Remark. In the subsequent problems B.72–76, we are again told at the outset that 3 caficii are bought for 10 and 4 sold for 20. That is, two are sold for 10 and one caficius is gained. Therefore, a capital in nummi $c_1 = 10$ corresponds to a profit in caficii $E_1 = 1$. We shall include this directly in the enunciation of each problem.

(B.72) MSS $\mathcal{A}$, $\mathcal{B}$. Caficii are bought three for ten nummi and sold four for twenty; how many caficii are gained with a hundred nummi?

Given thus $c_1 = 10$, $E_1 = 1$, $c_2 = 100$, required E_2.
$$E_2 = \frac{E_1 \cdot c_2}{c_1} = 10.$$

(B.73) MSS $\mathcal{A}$, $\mathcal{B}$. Caficii are bought three for ten nummi and sold four for twenty; with how many nummi are gained a hundred caficii?

Given thus $c_1 = 10$, $E_1 = 1$, $E_2 = 100$, required c_2.
$$c_2 = \frac{c_1 \cdot E_2}{E_1} = 1000.$$

(B.74) MSS $\mathcal{A}$, $\mathcal{B}$. Caficii are bought three for ten nummi and sold four for twenty; what are profit in caficii and capital in nummi if their sum is a hundred?

Given thus $c_1 = 10$, $E_1 = 1$, $c_2 + E_2 = 100$, required c_2, E_2.
$$c_2 = \frac{c_1 (c_2 + E_2)}{c_1 + E_1} = 90 + \frac{10}{11}, \quad E_2 = \frac{E_1 (c_2 + E_2)}{c_1 + E_1} = 9 + \frac{1}{11}.$$

---

‡ We would expect here an allusion to E_2 expressed in almodis.

(B.75) MSS $\mathcal{A}$, $\mathcal{B}$. Caficii are bought three for ten nummi and sold four for twenty; what are profit in caficii and capital in nummi if subtracting the profit from the capital leaves a hundred?

Given thus $c_1 = 10$, $E_1 = 1$, $c_2 - E_2 = 100$, required c_2, E_2.

$$c_2 = \frac{c_1(c_2 - E_2)}{c_1 - E_1} = 111 + \frac{1}{9}, \qquad E_2 = \frac{E_1(c_2 - E_2)}{c_1 - E_1} = 11 + \frac{1}{9}.$$

(B.76) MSS $\mathcal{A}$, $\mathcal{B}$. Caficii are bought three for ten nummi and sold four for twenty; what are profit in caficii and capital in nummi if their product is a hundred?

Given thus $c_1 = 10$, $E_1 = 1$, $c_2 \cdot E_2 = 100$, required c_2, E_2.

$$c_2 = \sqrt{\frac{c_1(c_2 \cdot E_2)}{E_1}} = \sqrt{1000}, \qquad E_2 = \sqrt{\frac{E_1(c_2 \cdot E_2)}{c_1}} = \sqrt{10}.$$

(*iv*) Fourth case: Capital in kind, profit in kind (B.77–82)

The fundamental proportion becomes

$$\frac{C_1}{C_2} = \frac{E_1}{E_2}.$$

(B.77) MSS $\mathcal{A}$, $\mathcal{B}$, $\mathcal{C}$. One caficius is bought for six nummi and sold for seven and a half; how many caficii are gained from a hundred caficii?

Given thus $c_1 = 6 = c_0$, thus $C_1 = 1$, $v_1 = 7 + \frac{1}{2} = v_0$ (so $e_1 = 1 + \frac{1}{2} = e_0$), $C_2 = 100$, required E_2.

We have $e_1 = 1 + \frac{1}{2}$, in nummi, and, as seen before (B.71), this corresponds to one fifth of a caficius. Therefore $E_1 = \frac{1}{5}$ and

$$E_2 = \frac{E_1 \cdot C_2}{C_1} = 20.$$

— If $C_2 = 100$ is in almodis, so will the result $E_2 = 20$.
— If E_2 is given, either in caficii or in almodis, and C_2 required, then

$$C_2 = \frac{C_1 \cdot E_2}{E_1}$$

will accordingly be in caficii or in almodis.
— In the three remaining cases, $C_2 + E_2$, $C_2 - E_2$, $C_2 \cdot E_2$ are respectively given.

Remark. In the subsequent problems B.78–82, we are again told at the outset that 3 caficii are bought for 10 and 4 sold for 20, so 2 are sold for

10. Therefore, to a capital in caficii $C_1 = 3$ corresponds a profit in caficii $E_1 = 1$. We shall, as before, include this directly in the data.

(B.78) MSS $\mathcal{A}$, $\mathcal{B}$. Caficii are bought three for ten nummi and sold four for twenty; how many caficii are gained from a hundred caficii?
Given thus $C_1 = 3$, $E_1 = 1$, $C_2 = 100$, required E_2.
$$E_2 = \frac{E_1 \cdot C_2}{C_1} = 33 + \frac{1}{3}.$$

(B.79) MSS $\mathcal{A}$, $\mathcal{B}$. Caficii are bought three for ten nummi and sold four for twenty; from how many caficii are gained a hundred caficii?
Given thus $C_1 = 3$, $E_1 = 1$, $E_2 = 100$, required C_2.
$$C_2 = \frac{C_1 \cdot E_2}{E_1} = 300.$$

(B.80) MSS $\mathcal{A}$, $\mathcal{B}$. Caficii are bought three for ten nummi and sold four for twenty; what are capital in caficii and profit in caficii if their sum is a hundred?
Given thus $C_1 = 3$, $E_1 = 1$, $C_2 + E_2 = 100$, required C_2, E_2.
$$C_2 = \frac{C_1(C_2 + E_2)}{C_1 + E_1} = 75, \qquad E_2 = \frac{E_1(C_2 + E_2)}{C_1 + E_1} = 25.$$

(B.81) MSS $\mathcal{A}$, $\mathcal{B}$. Caficii are bought three for ten nummi and sold four for twenty; what are capital in caficii and profit in caficii if subtracting profit from capital leaves a hundred?
Given thus $C_1 = 3$, $E_1 = 1$, $C_2 - E_2 = 100$, required C_2, E_2.
$$C_2 = \frac{C_1(C_2 - E_2)}{C_1 - E_1} = 150, \qquad E_2 = \frac{E_1(C_2 - E_2)}{C_1 - E_1} = 50.$$

As the author notes, we may not suppose a (positive) value for $E_2 - C_2$ since $E_1 < C_1$.

(B.82) MSS $\mathcal{A}$, $\mathcal{B}$. Caficii are brought three for ten nummi and sold four for twenty; what are capital in caficii and profit in caficii if their product is a hundred?
Given thus $C_1 = 3$, $E_1 = 1$, $C_2 \cdot E_2 = 100$, required C_2, E_2.
$$C_2 = \sqrt{\frac{C_1(C_2 \cdot E_2)}{E_1}} = \sqrt{300}, \qquad E_2 = \sqrt{\frac{E_1(C_2 \cdot E_2)}{C_1}} = \sqrt{33 + \frac{1}{3}}.$$

---

3. Variation in capital.

As said in our summary, this is the result of two situations. We may consider the capital to be another, c_2' or C_2', which is a fraction of the original capital c_2 or C_2 plus a quantity of nummi r_2 or of bushels R_2, and the profit is in nummi or in bushels. Then

$$c'_2 = \frac{k}{l} \cdot c_2 + r_2 = \frac{c_1 \cdot e'_2}{e_1}, \quad \text{whence} \quad c_2 = \frac{l}{k}\left(\frac{c_1 \cdot e'_2}{e_1} - r_2\right). \quad \text{(B.83, B.88)}$$

$$c'_2 = \frac{k}{l} \cdot c_2 + r_2 = \frac{c_1 \cdot E'_2}{E_1}, \quad \text{whence} \quad c_2 = \frac{l}{k}\left(\frac{c_1 \cdot E'_2}{E_1} - r_2\right). \quad \text{(B.85, B.87)}$$

$$C'_2 = \frac{k}{l} \cdot C_2 + R_2 = \frac{C_1 \cdot e'_2}{e_1}, \quad \text{whence} \quad C_2 = \frac{l}{k}\left(\frac{C_1 \cdot e'_2}{e_1} - R_2\right). \quad \text{(B.86)}$$

The other situation is when part of the (unknown) initial capital, c_2 or C_2, has been recovered with a supplement of r_2 nummi while R_2 caficii are left over. Then, if c_1 and v_1 are known buying and selling prices,

$$C_2 = \frac{R_2 \cdot v_1 + r_2 \cdot C_1}{v_1 - \frac{k}{l} \cdot C_1}, \qquad c_2 = \frac{R_2 \cdot v_1 \cdot c_1 + r_2 \cdot C_1 \cdot c_1}{v_1 \cdot C_1 - \frac{k}{l} \cdot C_1 \cdot c_1}.$$

A. *Different capital*.

Remember that B.84 belongs to the second situation (sale not completed).

(B.83) MSS $\mathcal{A}$, $\mathcal{B}$, $\mathcal{C}$. A caficius is bought for six nummi and sold for seven and a half; if from a capital less ten nummi is gained sixty nummi, what is this capital?

Given thus $c_1 = 6 = c_0$, $v_1 = 7 + \frac{1}{2} = v_0$, thus $e_1 = 1 + \frac{1}{2} = e_0$, $c'_2 = c_2 - 10$, $e'_2 = 60$, required c_2. As in B.61,

$$c'_2 = \frac{c_1 \cdot e'_2}{e_1} = \frac{6 \cdot 60}{1 + \frac{1}{2}} = 240, \qquad c_2 = c'_2 + 10 = 250.$$

(B.83') Same, but $c'_2 = c_2 + 10$, so $c_2 = c'_2 - 10 = 230$.

* * *

(B.84) MSS $\mathcal{A}$, $\mathcal{B}$, $\mathcal{C}$. A caficius is bought for eight nummi and sold for ten; what was the capital in nummi such that three fourths of it are recovered from the sale and forty caficii remain?

Given thus $c_1 = 8 = c_0$, $v_1 = 10 = v_0$, $v_2 = \frac{3}{4} \cdot c_2 + 40$ caficii, required c_2.

Remark. If a remainder is in kind and we want to express it in cash, we shall do so by computing the value of the caficii at the selling price. If there are additional nummi and they are to be converted into bushels, this too will be at the selling price.

Computation. Let us designate with r_i the remainders in cash and R_i the remainders in kind (thus here $R_2 = 40$). For one caficius we have, by analogy with the data,

$$v_1 = 10 = \frac{3}{4} \cdot c_1 + r_1 = 6 + r_1,$$

so $r_1 = 4$ nummi, that is, in caficii (at the selling price), $R_1 = \frac{2}{5}$.

We can now apply the proportion $c_2 : c_1 = R_2 : R_1$ (or, as said in the justification, $c_1 : R_1 = c_2 : R_2$),

$$c_2 = \frac{c_1 \cdot R_2}{R_1} = \frac{8 \cdot 40}{\frac{2}{5}} = 800.$$

This formula will be proved in B.90.

Verification. Since $c_1 = 8$ is the buying price of one caficius, $c_2 = 800$ will buy 100 caficii. Now $\frac{3}{4} c_2 = 600$ will indeed be recovered by selling 60 caficii at $v_1 = 10$ each, and 40 caficii are left.

Remark. Since v_2 is expressed partly in nummi, partly in caficii, we may also convert the caficii into nummi. Taking for the caficii the selling price (since they are supposed to be sold), we shall have $v_2 = \frac{3}{4} \cdot c_2 + 400$ and, since $c_2 : c_1 = v_2 : v_1$,

$$c_2 = \frac{c_1 \cdot v_2}{v_1} = \frac{8 \left(\frac{3}{4} \cdot c_2 + 400 \right)}{10} = \frac{3}{5} \cdot c_2 + 320, \quad \text{whence } c_2.$$

(B.84′) Same, with a supplement in nummi to the part of the capital recovered: $v_2 = \frac{3}{4} \cdot c_2 + 30$ nummi $+ 40$ caficii (thus $r_2 = 30$, $R_2 = 40$). We can change that into the previous case by considering that the 30 nummi correspond (at the selling price since they result from the sale) to three caficii, so that $R_2 = 43$ caficii (whence $c_2 = 860$).

(B.84″) Same, with $v_2 = \frac{3}{4} \cdot c_2 - 30$ nummi $+ 40$ caficii. Then $R_2 = 37$ caficii (whence $c_2 = 740$).

* * *

With B.85–88, we return to the subject of partly invested capital (incomplete sales will be encountered again in B.89–90). Here too, capital and profit may be given in cash or in kind. In all these problems we are told that three caficii are bought for ten and four are sold for twenty, as in the group of problems about the twenty kinds. As we did in those cases, we shall omit the initial transformation.

(B.85) MSS $\mathcal{A}$, $\mathcal{B}$. Caficii are bought three for ten nummi and sold four for twenty; what is the capital in nummi such that, when caficii bought with two thirds of it are sold, this sum is recovered and ten caficii are gained?

Given thus $c_1 = 10$, $E_1 = 1$ (see our remark preceding B.72, p. 1407), $c'_2 = \frac{2}{3} \cdot c_2$, $E'_2 = 10$, required c_2.

Since $c'_2 : c_1 = E'_2 : E_1$, then (as in B.73)

$$c'_2 = \frac{c_1 \cdot E'_2}{E_1} = 100 = \frac{2}{3} c_2, \quad c_2 = 150.$$

(B.85′) Same, with $c'_2 = \frac{2}{3} \cdot c_2 - 2$ nummi. Then

$$\frac{2}{3} c_2 - 2 = 100, \qquad \frac{2}{3} \cdot c_2 = 102, \qquad c_2 = 153.$$

(B.85″) Same, with $c'_2 = \frac{2}{3} \cdot c_2 + 2$ nummi. Then

$$\frac{2}{3} c_2 + 2 = 100, \qquad \frac{2}{3} \cdot c_2 = 98, \qquad c_2 = 147.$$

(B.86) MSS $\mathcal{A}$, $\mathcal{B}$. Caficii are bought three for ten nummi and sold four for twenty; what was the capital in caficii such that, after trading with three fourths of it, ninety-three nummi are gained?

Given thus $C_1 = 3$, $e_1 = 5$ (see our remark preceding B. 66, p. 1405), $C'_2 = \frac{3}{4} \cdot C_2$, $e'_2 = 93$, required C_2.

According to what has been seen in B.67 (the relation there, $C_2 : C_1 = e_2 : e_1$, becomes here $C'_2 : C_1 = e'_2 : e_1$),

$$C'_2 = \frac{C_1 \cdot e'_2}{e_1} = \frac{3 \cdot 93}{5} = 55 + \frac{4}{5} = \frac{3}{4} \cdot C_2, \qquad C_2 = 74 + \frac{2}{5}.$$

(B.86′) Same, but with an additional quantity of caficii gained: $C'_2 = \frac{3}{4} \cdot C_2 + 5$ caficii.

$$C'_2 = 55 + \frac{4}{5} = \frac{3}{4} C_2 + 5 \quad \text{whence} \quad \frac{3}{4} \cdot C_2 \text{ and } C_2.$$

(B.87) MSS $\mathcal{A}$, $\mathcal{B}$. Caficii are bought three for ten nummi and sold four for twenty; what was the capital in nummi such that, after trading with three fourths of it plus ten nummi, one caficius is gained?

Given thus $c_1 = 10$, $E_1 = 1$ (see our remark preceding B.72, p. 1407), $c'_2 = \frac{3}{4} \cdot c_2 + 10$ nummi, $E'_2 = 1$. Same kind as B.85″, but here impossible. For the merchant trading with $\frac{3}{4} c_2 + 10$ gains one caficius while with just 10 he would also gain one caficius. For us, this would correspond to the solution $c_2 = 0$. For our author, this would be nonsensical since the existence of the capital c_2 was asserted in the formulation.

Generally, we are told, the number of caficii gained from c'_2, namely E'_2, must be larger than the number of caficii gained with r_2 alone, say E_{r_2}; we thus need $E'_2 > E_{r_2}$ in order that $c_2 > 0$. Now, just as E'_2, the profit in caficii from c'_2, is found by means of $c'_2 : E'_2 = c_1 : E_1$, we shall determine the profit in caficii from the amount r_2 by means of $r_2 : E_{r_2} = c_1 : E_1$.

(B.88) MSS $\mathcal{A}$, $\mathcal{B}$. Caficii are bought three for ten nummi and sold four for twenty; what was the capital in nummi such that, after two thirds of it plus ten nummi are traded, a hundred nummi are gained?

Given thus $c_1 = 10$, $e_1 = 5$ (see our remark preceding B.60, p. 1404), $c'_2 = \frac{2}{3} \cdot c_2 + 10$ nummi, $e'_2 = 100$.

By analogy with B.61, form

$$c_2' = \frac{c_1 \cdot e_2'}{e_1} = 200 = \frac{2}{3} \cdot c_2 + 10, \quad \frac{2}{3} \cdot c_2 = 190, \quad c_2 = 285.$$

Verification. With caficii being bought 3 for 10 and sold 4 for 20, $c_2' = 200$ buys 60, which are then sold for 300, thus the profit is $e_2' = 100$.

As said at the end, all such problems can be reduced to the twenty kinds. The proportion does indeed still apply whether the whole capital is used or not.

B. *Sale not completed*.

(B.89) MS *A* only. Sextarii* are bought each for three nummi and sold for five; how many were bought if, after the capital is recovered, twenty sextarii are left?

Given thus $c_1 = 3 = c_0$ (thus $C_1 = 1$), $v_1 = 5 = v_0$, $v_2 = c_2 + 20$ sextarii (thus $R_2 = 20$), required C_2.

The formula, not explained in the text, has been set out by us in the summary (p. 1398).

$$C_2 = \frac{R_2 \cdot v_0}{v_0 - c_0} = \frac{20 \cdot 5}{5 - 3} = \frac{100}{2} = 50.$$

Remark. We might also consider that, since $c_1 = 3$ and $v_1 = 5$, the capital c_1 is recovered when $\frac{3}{5}$ of a caficius is sold, and thus $R_1 = \frac{2}{5}$ is left from each caficius. Since $C_2 : C_1 = R_2 : R_1$,

$$C_2 = \frac{C_1 \cdot R_2}{R_1} = \frac{1 \cdot 20}{\frac{2}{5}} = 50.$$

We shall see that this is the formula used in B.90.

In the other parts of B.89 an amount of nummi, added or subtracted, occurs. Here again, this is solved without explanation.

(B.89′) Same, but c_2 plus 10 nummi is recovered.

$$C_2 = \frac{R_2 \cdot v_0 + r_2}{v_0 - c_0} = \frac{20 \cdot 5 + 10}{2} = 55.$$

(B.89″) Same, but c_2 less 10 nummi is recovered.

$$C_2 = \frac{R_2 \cdot v_0 - r_2}{v_0 - c_0} = \frac{20 \cdot 5 - 10}{2} = 45.$$

(B.90) MSS *A*, *B*. Caficii are bought three for ten nummi and sold four for twenty; what was the capital in nummi and the quantity of caficii if,

---

* Instead of caficii.

after three fourths of the capital are recovered, twenty caficii remain to be sold?

Given thus $c_1 = 10$, $C_1 = 3$, $v_1 = 15$, thus $v_0 = 5$, $v_2 = \frac{3}{4} \cdot c_2 + 20$ caficii, thus $R_2 = 20$, required C_2 and c_2. This problem is of the same sort as B.84.

Computation. Considering that $v_2 = \frac{3}{4} \cdot c_2 + R_2$, calculate first the value R_1 corresponding to v_1 and c_1. As seen in our summary (p. 1398), we have

$$R_1 = C_1 - \frac{\frac{k}{l} c_1}{v_0} = 3 - \frac{\frac{3}{4} \cdot 10}{5} = 3 - \frac{7 + \frac{1}{2}}{5} = 3 - \left(1 + \frac{1}{2}\right) = 1 + \frac{1}{2}.$$

Next, since $c_1 : c_2 = R_1 : R_2 = C_1 : C_2$,

$$c_2 = \frac{c_1 \cdot R_2}{R_1} = \frac{10 \cdot 20}{1 + \frac{1}{2}} = 133 + \frac{1}{3},$$

$$C_2 = \frac{C_1 \cdot R_2}{R_1} = \frac{3 \cdot 20}{1 + \frac{1}{2}} = 40.$$

Verification. With $c_2 = 133 + \frac{1}{3}$ are bought, at the buying price $c_0 = 3 + \frac{1}{3}$, 40 caficii; to recover $\frac{3}{4} \cdot c_2 = 100$, 20 must be sold at $v_0 = 5$ each and thus 20 are indeed left.

Demonstration.

(α) Preliminary remark.

• Formula for c_2. From the fundamental relation whereby we determine the buying price c_0,

$$\frac{C_1}{c_1} = \frac{C_2}{c_2} = \frac{1}{c_0}, \qquad (1)$$

we infer that

$$\frac{C_1}{\frac{3}{4} \cdot c_1} = \frac{C_2}{\frac{3}{4} \cdot c_2}. \qquad (2)$$

On the other hand, we know that, when the capital $\frac{3}{4} c_2$ is recovered, R_2 bushels are left of the initial quantity C_2. Thus by selling $C_2 - R_2$ caficii at v_0 we recover $\frac{3}{4} c_2$. If, under the same conditions, R_1 bushels are left from C_1, we shall have $(C_1 - R_1) v_0 = \frac{3}{4} \cdot c_1$, that is,

$$\frac{C_1 - R_1}{\frac{3}{4} \cdot c_1} = \frac{1}{v_0} = \frac{C_2 - R_2}{\frac{3}{4} \cdot c_2}. \qquad (3)$$

Subtracting (3) from (2), we are left with

$$\frac{R_1}{\frac{3}{4} \cdot c_1} = \frac{R_2}{\frac{3}{4} \cdot c_2}, \quad \text{or} \quad \frac{R_1}{c_1} = \frac{R_2}{c_2}, \qquad (4), (4')$$

whence
$$c_2 = \frac{c_1 \cdot R_2}{R_1}. \qquad (5)$$

• **Formula for C_2.** Inverting (1) and multiplying the result by the corresponding terms of (4'), we find
$$\frac{R_1}{C_1} = \frac{R_2}{C_2}, \qquad (6)$$

whence
$$C_2 = \frac{C_1 \cdot R_2}{R_1}. \qquad (7)$$

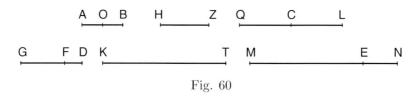

Fig. 60

(β) *Demonstration according to the text.*

• **Formula for c_2.** Let (Fig. 60) $AB = C_1 = 3$ caficii, bought for $GD = c_1 = 10$ nummi, $HZ = 4$ caficii, sold for $KT = 20$ nummi, $QL = C_2$ the second capital in caficii, $MN = c_2$ the second capital in nummi, $ME = \frac{3}{4} \cdot c_2 = \frac{3}{4} \cdot MN$, $CL = 20 = R_2$ the number of caficii left. Thus we have, since the buying price of one caficius is the same,
$$\frac{AB}{GD} = \frac{QL}{MN} \quad \left(= \frac{3}{10} = \frac{1}{c_0}\right) \qquad (1)$$

and further, considering the number of caficii actually sold ($QC = QL - R_2$) and their selling price,
$$\frac{HZ}{KT} = \frac{QC}{ME} \quad \left(= \frac{4}{20} = \frac{1}{v_0}\right). \qquad (1')$$

But, from (1)
$$\frac{AB}{\frac{3}{4} \cdot GD} = \frac{QL}{\frac{3}{4} \cdot MN},$$

that is, since $\frac{3}{4} \cdot MN = ME$ and putting $GF = \frac{3}{4} \cdot GD$,
$$\frac{AB}{GF} = \frac{QL}{ME}. \qquad (2)$$

Take AO such that°
$$\frac{HZ}{KT} = \frac{AO}{GF} \quad \left(= \frac{4}{20}\right).$$

° Thus AO is the caficii sold for $\frac{3}{4} \cdot c_1$, and therefore
$$AO = C_1 - R_1 = \frac{4}{20}\left(7 + \frac{1}{2}\right) = 1 + \frac{1}{2}.$$

So, using (1'):
$$\frac{AO}{GF} = \frac{QC}{ME}. \qquad (3)$$

Applying Premiss P_6 to (2) & (3), with $OB = AB - AO$,[†]
$$\frac{OB}{GF} = \frac{CL}{ME}. \qquad (4)$$

Since multiplying GF and ME by $\frac{4}{3}$ gives GD and MN, respectively,
$$\frac{OB}{GD} = \frac{CL}{MN}, \qquad (4')$$

whence
$$MN = c_2 = \frac{GD \cdot CL}{OB} = \frac{c_1 \cdot R_2}{R_1} = \frac{10 \cdot 20}{1 + \frac{1}{2}} \left(= 133 + \frac{1}{3}\right). \qquad (5)$$

• Formula for C_2.

As seen above, in (4') & (1),
$$\frac{OB}{GD} = \frac{CL}{MN} \quad \text{and} \quad \frac{GD}{AB} = \frac{MN}{QL},$$

whence, by multiplication of these ratios ('equal proportionality', *Elements* V.22),
$$\frac{OB}{AB} = \frac{CL}{QL} \qquad (6)$$

whence
$$QL = C_2 = \frac{AB \cdot CL}{OB} = \frac{C_1 \cdot R_2}{R_1} = \frac{3 \cdot 20}{1 + \frac{1}{2}} = 40. \qquad (7)$$

(**B.90'**) Same problem, but with a supplement in nummi. Given thus $c_1 = 10$, $C_1 = 3$, $v_1 = 15$ (thus $v_0 = 5$), $v_2 = \frac{3}{4} \cdot c_2 + 10$ nummi + 18 caficii $= \frac{3}{4} \cdot c_2 + r_2 + R_2$, required c_2 and C_2.

Since the 10 nummi correspond to 2 caficii according to the selling price, we have
$$v_2 = \frac{3}{4} \cdot c_2 + 20 \text{ caficii},$$

and this is just B.90.

(**B.90''**) Same problem, but with an amount in nummi subtracted. Given thus $c_1 = 10$, $C_1 = 3$, $v_1 = 15$ (thus $v_0 = 5$), $v_2 = \frac{3}{4} \cdot c_2 - 20$ nummi + 24 caficii.

[†] Thus $OB = R_1 = AB - AO = 1 + \frac{1}{2}$.

This is again B.90 since the 20 nummi are worth 4 caficii. Thus the answer is, as before, $C_2 = 40$ and $c_2 = 133 + \frac{1}{3}$. There is a verification of B.90" ($C_2 - R_2 = 16$ caficii sold gives 80 nummi, and this is indeed $\frac{3}{4} \cdot c_2 - 20$ nummi).

* * *

(B.91) MSS $\mathcal{A}$, $\mathcal{B}$. If three caficii are gained from twelve caficii, how many nummi will be gained from a hundred nummi?

This problem is said to be 'foreign'. In what we have seen so far, both capitals were of the same nature (cash or kind), and so were both profits; but here the first capital and the corresponding profit are in kind while the second capital and profit are in cash.

The statement is followed by an explanation, the purpose of which is not to say what is meant by 'foreign', but, as in other similar occurrences of a new type of problem, to clarify the formulation. Indeed, if 'a man gains three caficii from twelve caficii', this can mean either that the buying price of twelve caficii is recovered by selling nine of them or, possibly, that the buying price of fifteen caficii is recovered by selling twelve. The former applies here.

Therefore, we have here $C_1 = 12$, $E_1 = 3 = R_1$ (thus $C'_1 = C_1 - E_1 = 9$), $c_2 = 100$, and e_2 is required.

Computation. Despite the inhomogeneity of the data, the proportion remains valid, that is, $c_2 : e_2 = C'_1 : E_1$, and therefore

$$e_2 = \frac{E_1 \cdot c_2}{C'_1} = \frac{3 \cdot 100}{9} = 33 + \frac{1}{3}.$$

Demonstration of the formula.

(α) *Preliminary remark.* The ratio profit to capital remaining constant, we have

$$\frac{e_0}{c_0} = \frac{e_2}{c_2}. \qquad (1)$$

Next, by hypothesis, the capital in cash spent for buying C_1 is recovered by selling C'_1; then

$$c_0 \cdot C_1 = v_0 \cdot C'_1, \quad \text{so} \quad \frac{v_0}{c_0} = \frac{C_1}{C'_1}, \qquad (1')$$

whence, by separation,

$$\frac{v_0 - c_0}{c_0} = \frac{e_0}{c_0} = \frac{C_1 - C'_1}{C'_1} = \frac{E_1}{C'_1}. \qquad (2)$$

Thus, by (1) and (2),

$$\frac{e_2}{c_2} = \frac{e_0}{c_0} = \frac{E_1}{C'_1}, \qquad (3)$$

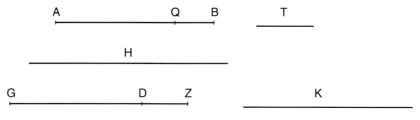

Fig. 61

so finally
$$e_2 = \frac{E_1 \cdot c_2}{C'_1}. \tag{4}$$

(β) *Demonstration according to the text.* Let (Fig. 61) $AB = C_1 = 12$ caficii, $GD = c_0$ (buying price per unit), and $H = AB \cdot GD = c_1$; then let $AQ = C'_1 = 9$, $GZ = v_0 = c_0 + e_0$ (selling price per unit), thus $DZ = e_0$; let finally $K = c_2$, $T = e_2$. Then we have the proportion (between homogeneous quantities)
$$\frac{DZ}{GD} = \frac{T}{K}. \tag{1}$$

But because the amount for buying twelve, $c_0 \cdot AB$, equals the amount for selling nine, $v_0 \cdot AQ$, that is, $GD \cdot AB = GZ \cdot AQ$, we find that
$$\frac{GZ}{GD} = \frac{AB}{AQ}; \tag{1'}$$

next, by separation,
$$\frac{DZ}{GD} = \frac{QB}{AQ}. \tag{2}$$

By (1) and (2),
$$\frac{T}{K} = \frac{DZ}{GD} = \frac{QB}{AQ}, \tag{3}$$

and therefore
$$T = e_2 = \frac{QB \cdot K}{AQ} = \frac{E_1 \cdot c_2}{C'_1}. \tag{4}$$

---

4. Other problems.

A. *A particular problem*.

(B.92) MSS $\mathcal{A}$, $\mathcal{B}$. Three caficii of wheat are bought for ten nummi and an unknown quantity of caficii of barley for twelve, and each is sold at the buying price of the other; what is the quantity of barley if the profit is four nummi?

Thus $q^{(w)} = 3$ caficii of wheat are bought for $c^{(w)} = 10$ and a (required) quantity $q^{(b)}$ of caficii of barley is bought for $c^{(b)} = 12$. Each caficius of wheat is sold at the buying price of each of barley, thus at $c_0^{(b)}$, and each caficius of barley at the buying price of each of wheat, thus at $c_0^{(w)}$, that is, $3 + \frac{1}{3}$, and the profit is $e = 4$ nummi. In other words, the amount earned from the sale is

$$q^{(w)} \cdot c_0^{(b)} + q^{(b)} \cdot c_0^{(w)} = v^{(w)} + v^{(b)} = \left(c^{(w)} + c^{(b)}\right) + 4 = 26.$$

Computation.

(α) *Preliminary remark.* We end in this problem with an equation of the type $x^2 + 120 = 26x$, which has the two positive solutions

$$x = \frac{26}{2} \pm \sqrt{\left(\frac{26}{2}\right)^2 - 120} = 20 \text{ and } 6.$$

(β) *Computation in the text.* The text computes successively the two possible solutions for $v^{(b)}$ (the other solution being then $v^{(w)}$):

$$v^{(b)} = \frac{1}{2}\left(c^{(w)} + c^{(b)} + e\right) \pm \sqrt{\left[\frac{1}{2}\left(c^{(w)} + c^{(b)} + e\right)\right]^2 - c^{(w)} \cdot c^{(b)}}$$

$$= \frac{1}{2}(10 + 12 + 4) \pm \sqrt{\left[\frac{1}{2}(10 + 12 + 4)\right]^2 - 10 \cdot 12}$$

$$= \frac{26}{2} \pm \sqrt{\left(\frac{26}{2}\right)^2 - 120} = 13 \pm \sqrt{169 - 120}$$

$$= 13 \pm \sqrt{49} = 13 \pm 7.$$

- Take first $v^{(b)} = 13 + 7 = 20$. Then, since

$$c_0^{(w)} = \frac{v^{(b)}}{q^{(b)}} = \frac{20}{q^{(b)}} \quad \text{and also} \quad c_0^{(w)} = \frac{c^{(w)}}{q^{(w)}} = \frac{10}{3}, \quad \text{so}$$

$$q^{(b)} = \frac{3}{10} \cdot 20 = 6 \quad \left(\text{and} \quad c_0^{(b)} = \frac{c^{(b)}}{q^{(b)}} = \frac{12}{6} = 2 = \frac{v^{(w)}}{q^{(w)}}\right).$$

- Take next $v^{(b)} = 13 - 7 = 6$. Then, similarly, since

$$c_0^{(w)} = \frac{v^{(b)}}{q^{(b)}} = \frac{6}{q^{(b)}} \quad \text{and also} \quad c_0^{(w)} = \frac{c^{(w)}}{q^{(w)}} = \frac{10}{3}, \quad \text{so}$$

$$q^{(b)} = \frac{3}{10} \cdot 6 = 1 + \frac{4}{5} \quad \left(\text{and} \quad c_0^{(b)} = \frac{c^{(b)}}{q^{(b)}} = \frac{12}{1 + \frac{4}{5}} = 6 + \frac{2}{3}\right).$$

Demonstration. The origin of the computations is totally obscure. The demonstration is meant to cope with this difficulty.

(α) *Preliminary remark.* If $q^{(w)}$ bushels of wheat are bought for $c^{(w)}$ nummi and sold for $v^{(w)}$, while an unknown quantity $q^{(b)}$ bushels of barley is bought for $c^{(b)}$ and sold for $v^{(b)}$, we shall have, for the imposed equality of buying price per bushel of one kind and selling price of the other,

$$\frac{q^{(w)}}{c^{(w)}} = \frac{q^{(b)}}{v^{(b)}} = \frac{1}{c_0^{(w)}}, \qquad (1)$$

$$\frac{q^{(b)}}{c^{(b)}} = \frac{q^{(w)}}{v^{(w)}} = \frac{1}{c_0^{(b)}}, \qquad (2)$$

from which we infer that

$$\frac{q^{(w)}}{q^{(b)}} = \frac{c^{(w)}}{v^{(b)}}, \qquad \frac{q^{(w)}}{q^{(b)}} = \frac{v^{(w)}}{c^{(b)}}, \qquad (3), (4)$$

and therefore $v^{(b)} \cdot v^{(w)} = c^{(b)} \cdot c^{(w)} = 10 \cdot 12 = 120$. On the other hand, the profit of 4 nummi means that $v^{(b)} + v^{(w)} - c^{(b)} - c^{(w)} = 4$ and therefore that $v^{(b)} + v^{(w)} = c^{(b)} + c^{(w)} + 4 = 26$. We obtain then, for determining the selling prices of the two quantities, the system

$$\begin{cases} v^{(b)} + v^{(w)} = 26 \\ v^{(b)} \cdot v^{(w)} = 120, \end{cases}$$

to be solved as usual:

$$\left(\frac{v^{(b)} - v^{(w)}}{2}\right)^2 = \left(\frac{v^{(b)} + v^{(w)}}{2}\right)^2 - v^{(b)} \cdot v^{(w)} = 169 - 120 = 49,$$

so that

$$\frac{|v^{(b)} - v^{(w)}|}{2} = 7.$$

There are then two possibilities:

• If $v^{(b)} > v^{(w)}$, then

$$v^{(b)} = \frac{v^{(b)} + v^{(w)}}{2} + \frac{v^{(b)} - v^{(w)}}{2} = 13 + 7 = 20 \quad \left(v^{(w)} = 26 - 20 = 6\right);$$

• If $v^{(w)} > v^{(b)}$, then

$$v^{(b)} = \frac{v^{(b)} + v^{(w)}}{2} - \frac{v^{(w)} - v^{(b)}}{2} = 13 - 7 = 6 \quad \left(v^{(w)} = 26 - 6 = 20\right).$$

There are accordingly two possibilities for the required quantity of barley and its buying price per unit, namely those seen before.

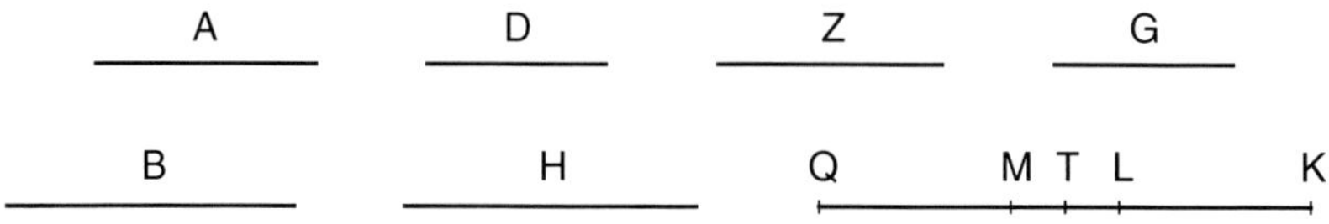

Fig. 62

(β) *Demonstration according to the text.* Let (Fig. 62; not found in the manuscripts) the quantity $A = 3 = q^{(w)}$ of wheat be bought at $B = 10 = c^{(w)}$, and the unknown quantity $D = q^{(b)}$ of barley be bought at $H = 12 = c^{(b)}$. Let next the quantity $Z = 3 \,(= A)$ of wheat be sold for $QT = v^{(w)}$ and the unknown quantity $G \,(= D)$ of barley be sold for $TK = v^{(b)}$.† The buying price of wheat being equal to the selling price of barley, we have first

$$\frac{A}{B} = \frac{G}{TK}. \qquad (1)$$

The buying price of barley being equal to the selling price of wheat, we have second

$$\frac{D}{H} = \frac{Z}{QT}. \qquad (2)$$

From the first relation, by alternation, and considering that Z and D are just other designations of the quantities A and G respectively,

$$\frac{A}{G} = \frac{B}{TK} = \frac{Z}{D}. \qquad (3)$$

Alternating the terms in (2) and using the result just found, we find

$$\frac{Z}{D} = \frac{QT}{H} = \frac{B}{TK}, \qquad (4)$$

whence $B \cdot H = QT \cdot TK$, that is, $QT \cdot TK = 120$.

On the other hand, the amount from the sale gives a profit of 4, so $QT + TK = B + H + 4 = 26$. Thus we are now to solve

$$\begin{cases} QT + TK = 26 \\ QT \cdot TK = 120. \end{cases}$$

This is done as usual. Bisecting the segment $QK = 26$, the section will fall either at L on TK (if $QT < TK$) or at M on QT (if $QT > TK$). Accordingly (by *Elements* II.5; this step, performed several times before, is omitted in the text):

$$QT \cdot TK + TL^2 = QL^2 \quad \text{or} \quad QT \cdot TK + MT^2 = QM^2$$

which gives $TL = 7$ or $MT = 7$.

- If $QT < TK$ (section at L), the amount of money resulting from selling the wheat is less than the amount from selling the barley ($v^{(w)} < v^{(b)}$); since $QL = LK = 13$, we shall have $TK = LK + TL = 20 = v^{(b)}$ and therefore $QT = 6 = v^{(w)}$, so $q^{(b)} = D = 6$ and the buying price per unit is $c_0^{(b)} = 2$.

---

† Z and G have been introduced as an aid to distinguishing between quantities bought and sold and thus buying and selling prices.

- If $QT > TK$ (section at M), $v^{(w)} > v^{(b)}$; since $QM = MK = 13$, we shall have $TK = MK - MT = 6 = v^{(b)}$ and $QT = 20 = v^{(w)}$, so $q^{(b)} = 1 + \frac{4}{5}$ and the buying price per unit is $c_0^{(b)} = 6 + \frac{2}{3}$.

B. *Data involving roots*.

The last group of problems uses the basic proportion between capital and profit for each of the four cases of cash and kind; for we need only to know that

$$\frac{c_1}{c_2} = \frac{e_1}{e_2}, \quad \frac{C_1}{C_2} = \frac{e_1}{e_2}, \quad \frac{c_1}{c_2} = \frac{E_1}{E_2}, \quad \frac{C_1}{C_2} = \frac{E_1}{E_2}.$$

(B.93) MS $\mathcal{A}$ only. In buying for five the gain is three; what is the capital if the corresponding profit is six roots of it?

Given thus $c_1 = 5$, $e_1 = 3$, $e_2 = 6\sqrt{c_2}$, required c_2. The text calculates

$$c_2 = \left(5 \cdot \frac{6}{3}\right)^2 = 100.$$

Remark. Using the basic relation, we obtain successively

$$c_2 = \frac{c_1 \cdot e_2}{e_1} = \frac{c_1 \cdot 6\sqrt{c_2}}{e_1}, \quad \sqrt{c_2} = \frac{c_1 \cdot 6}{e_1}, \quad c_2 = \left(c_1 \cdot \frac{6}{e_1}\right)^2.$$

This explains the author's computation, the origin of which can hardly be guessed from the calculations as presented.

(B.94) MS $\mathcal{A}$ only. In buying for five the gain is three; what is the capital if the profit is six roots of its half?

Given thus $c_1 = 5$, $e_1 = 3$, $e_2 = 6\sqrt{\frac{c_2}{2}}$, required c_2. The text calculates

$$c_2 = 2 \cdot \left(\frac{5 \cdot 6}{3 \cdot 2}\right)^2 = 50.$$

Remark. Here again, we have a sequence of unexplained computations. Doing as before, but with the factor $\sqrt{2}$ in the divisor, we find

$$\sqrt{c_2} = \frac{c_1 \cdot 6}{e_1 \cdot \sqrt{2}}, \quad c_2 = \frac{1}{2}\left(\frac{c_1 \cdot 6}{e_1}\right)^2 = 2\left(\frac{c_1 \cdot 6}{e_1 \cdot 2}\right)^2.$$

(B.95) MSS $\mathcal{A}$, $\mathcal{B}$. In buying for five the gain is three; what are capital and profit if the product of their roots is ten?

Given thus $c_1 = 5$, $e_1 = 3$, $\sqrt{c_2} \cdot \sqrt{e_2} = 10$, required c_2, e_2.

As said in the text, we shall just square the product of the roots, so that the problem becomes: Given $c_1 = 5$, $e_1 = 3$, $c_2 \cdot e_2 = 100$, required c_2 and e_2, a type of problem already solved (B.64 —or B.58).

(B.96) MSS $\mathcal{A}$, $\mathcal{B}$. Caficii are bought three for ten nummi and sold four for twenty; what are capital in caficii and profit in nummi if the product of their roots is twenty?

Given thus $C_1 = 3$, $e_1 = 5$ (see our remark preceding B.66, p. 1405), $\sqrt{C_2} \cdot \sqrt{e_2} = 20$, required C_2, e_2.
Since $C_2 \cdot e_2 = 400$, we shall treat this as seen before (B.70, p. 1406).

(B.97) MSS $\mathcal{A}$, $\mathcal{B}$. In buying for nine the gain is four; what are capital and profit if the sum of their roots is twenty?°

Given thus $c_1 = 9$, $e_1 = 4$ (thus c_1 and e_1 both squares), $\sqrt{c_2} + \sqrt{e_2} = 20$, required c_2, e_2.

As explained in the text, the solution is attained by operating on ratios (see B.56, pp. 1400–1401). Since

$$\frac{c_1}{e_1} = \frac{c_2}{e_2}, \quad \text{we have} \quad \frac{\sqrt{c_1}}{\sqrt{e_1}} = \frac{\sqrt{c_2}}{\sqrt{e_2}} \quad \text{and, inversely,} \quad \frac{\sqrt{e_1}}{\sqrt{c_1}} = \frac{\sqrt{e_2}}{\sqrt{c_2}}.$$

Therefore, by composition,

$$\frac{\sqrt{c_1}}{\sqrt{c_1} + \sqrt{e_1}} = \frac{\sqrt{c_2}}{\sqrt{c_2} + \sqrt{e_2}}, \quad \frac{\sqrt{e_1}}{\sqrt{c_1} + \sqrt{e_1}} = \frac{\sqrt{e_2}}{\sqrt{c_2} + \sqrt{e_2}},$$

whence

$$\sqrt{c_2} = \frac{\sqrt{c_1}\left(\sqrt{c_2} + \sqrt{e_2}\right)}{\sqrt{c_1} + \sqrt{e_1}} = \frac{3 \cdot 20}{5} = 12, \quad c_2 = 144;$$

$$\sqrt{e_2} = \frac{\sqrt{e_1}\left(\sqrt{c_2} + \sqrt{e_2}\right)}{\sqrt{c_1} + \sqrt{e_1}} = \frac{2 \cdot 20}{5} = 8, \quad e_2 = 64.$$

(B.97′) The case of subtraction $\sqrt{c_2} - \sqrt{e_2} = 4$, mentioned at the end, is indeed clear, and is solved by conversion of the ratio $\sqrt{c_1} : \sqrt{e_1} = \sqrt{c_2} : \sqrt{e_2}$ (and separation for obtaining $\sqrt{e_2}$).

(B.98) MSS $\mathcal{A}$, $\mathcal{B}$. Caficii are bought three for ten nummi and sold four for twenty; what are capital in nummi and profit in caficii if the sum of their roots is fifty?

Given thus $c_1 = 10$, $E_1 = 1$ (see our remark preceding B.72, p. 1407), $\sqrt{c_2} + \sqrt{E_2} = 50$, required c_2, E_2. See B.74 or B.97.
Since

$$\frac{\sqrt{E_1}}{\sqrt{c_1}} = \frac{\sqrt{E_2}}{\sqrt{c_2}}, \quad \frac{\sqrt{c_1}}{\sqrt{E_1}} = \frac{\sqrt{c_2}}{\sqrt{E_2}},$$

° B.97–97′, with capital and profit both in cash, should precede B.96 —or did the author prefer to present the simpler cases B.95–96 first?

then as before (we may just replace $\sqrt{e_i}$ by $\sqrt{E_i}$)

$$\frac{\sqrt{E_1}}{\sqrt{c_1}+\sqrt{E_1}} = \frac{\sqrt{E_2}}{\sqrt{c_2}+\sqrt{E_2}}, \qquad \frac{\sqrt{c_1}}{\sqrt{c_1}+\sqrt{E_1}} = \frac{\sqrt{c_2}}{\sqrt{c_2}+\sqrt{E_2}},$$

$$\sqrt{E_2} = \frac{\sqrt{E_1}\left(\sqrt{c_2}+\sqrt{E_2}\right)}{\sqrt{c_1}+\sqrt{E_1}} = \frac{1\cdot 50}{\sqrt{10}+1} = \frac{50}{\sqrt{10}+1};$$

$$\sqrt{c_2} = \frac{\sqrt{c_1}\left(\sqrt{c_2}+\sqrt{E_2}\right)}{\sqrt{c_1}+\sqrt{E_1}} = \frac{\sqrt{10}\cdot 50}{\sqrt{10}+1}.$$

(B.99) MSS $\mathcal{A}$**,** $\mathcal{B}$**.** Caficii are bought three for ten nummi and sold four for twenty; what are capital in caficii and profit in caficii if subtracting the root of the profit from the root of the capital leaves sixty?

Given thus $C_1 = 3$, $E_1 = 1$ (see our remark preceding B.78, pp. 1408–1409), $\sqrt{C_2} - \sqrt{E_2} = 60$, required C_2, E_2. See B.81, B.97′.

The formulae are analogous to the ones before, but we are to replace the addition by a subtraction (and consider the capitals to be in kind).

(B.100) MSS $\mathcal{A}$**,** $\mathcal{B}$**.** Caficii are bought three for ten nummi and sold four for twenty; what are capital in nummi and profit in caficii if the profit is the root of the capital?

Given thus $c_1 = 10$, $E_1 = 1$ (see our remark preceding B.72, p. 1407), $E_2 = \sqrt{c_2}$, required c_2, E_2. Since

$$\frac{c_1}{E_1} = \frac{10}{1} = \frac{c_2}{E_2} = \frac{c_2}{\sqrt{c_2}},$$

so $c_2 = 10\sqrt{c_2}$, $\sqrt{c_2} = 10 = E_2$, $c_2 = 100$.

(B.101) MSS $\mathcal{A}$**,** $\mathcal{B}$**.** A quantity of caficii bought for ten nummi are sold at four for twenty, while with a hundred nummi are gained ten caficii; what was the first quantity of caficii?

Given thus $c_1 = 10$, $v_0 = 5$, $c_2 = 100$, $E_2 = 10$, required this time C_1.

Since $v_0 = 5$, the initial capital is recovered by selling two caficii. Next, since $E_1 : c_1 = E_2 : c_2 = 10 : 100$ and $c_1 = 10$, so $E_1 = 1$. The two caficii sold and the caficius gained make $C_1 = 3$.

(B.102) MS $\mathcal{A}$**.** A quantity of caficii, bought for three times its root, is sold for twenty nummi, and three hundred and fifty nummi are gained with a hundred caficii; what was the initial quantity of caficii and the capital?

Given thus $c_1 = 3\sqrt{C_1}$, $v_1 = c_1 + e_1 = 20$ (thus $e_1 = 20 - 3\sqrt{C_1}$), $C_2 = 100$, $e_2 = 350$, required C_1 and c_1.

$$\frac{e_1}{C_1} = \frac{e_2}{C_2} \text{ gives } \frac{20 - 3\sqrt{C_1}}{C_1} = \frac{350}{100} = \frac{3+\frac{1}{2}}{1}, \text{ hence}$$

$$\left(3+\frac{1}{2}\right)C_1 + 3\sqrt{C_1} = 20, \quad \text{and} \quad C_1 + \frac{6}{7}\sqrt{C_1} = 5 + \frac{5}{7}.$$

Proceed then 'as has been shown in algebra, which is as follows':

$$\sqrt{C_1} = \sqrt{\left(\frac{3}{7}\right)^2 + 5 + \frac{5}{7}} - \frac{3}{7} = \sqrt{\frac{1}{7} + \frac{2}{7}\frac{1}{7} + 5 + \frac{5}{7}} - \frac{3}{7}$$

$$= \sqrt{5 + \frac{6}{7} + \frac{2}{7}\cdot\frac{1}{7}} - \frac{3}{7} = 2 + \frac{3}{7} - \frac{3}{7} = 2,$$

so $C_1 = 4$ and $c_1 = 6$.

Remark. Solving the equation $x^2 + bx = c$ has been taught before, and a direct computation by the formula, without any use of Euclid, has been seen in A.184 and A.186. But in neither case was the quadratic equation clearly stated before the computation of its solution.

(B.103) MS $\mathcal{A}$. A quantity of caficii is bought for three times its root, sold for five times its root, and a hundred and fifty nummi are gained with a hundred caficii; what was the initial quantity of caficii?

Given thus $c_1 = 3\sqrt{C_1}$, $v_1 = c_1 + e_1 = 5\sqrt{C_1}$ (thus $e_1 = 2\sqrt{C_1}$), $C_2 = 100$, $e_2 = 150$, required C_1.

$$\frac{e_1}{C_1} = \frac{e_2}{C_2} \quad \text{gives} \quad \frac{2\sqrt{C_1}}{C_1} = \frac{150}{100} = 1 + \frac{1}{2}, \quad \text{hence}$$

$$2\sqrt{C_1} = \left(1 + \frac{1}{2}\right)C_1, \quad \sqrt{C_1} = 1 + \frac{1}{3}, \quad C_1 = 1 + \frac{7}{9}.$$

Chapter B–III:
Profit in partnership

Summary

Suppose n men have each invested a capital c_i and the sum of all these capitals, $\sum c_i = c$, has produced a profit e; the part of the ith partner is then e_i, with $\sum e_i = e$. These quantities obey the relations (explained in B.104–105)

$$\frac{c_1}{e_1} = \frac{c_2}{e_2} = \ldots = \frac{c_n}{e_n} = \frac{\sum c_i}{\sum e_i} = \frac{c}{e}.$$

In all thirteen problems of this chapter we are given the individual capitals c_i, and we thus also know the total capital c (although mostly we do not need to).

1. Three partners.

The first group of problems, involving three partners, begins with two simple cases in which we are given, in addition to the capitals, the total profit e (B.104) or one individual profit, here e_2 (B.105); the formulae for the other profits are then

$$e_i = \frac{e \cdot c_i}{c}, \qquad e_i = \frac{e_2 \cdot c_i}{c_2}.$$

In the subsequent problems of this first group, banal operations on ratios, such as (considering that $c_1 : c_2 = e_1 : e_2$)

$$\frac{c_1 + c_2}{c_2} = \frac{e_1 + e_2}{e_2}, \qquad \frac{c_1^2}{c_1 \cdot c_2} = \frac{e_1^2}{e_1 \cdot e_2},$$

suffice to deduce the formulae used. These are the following:

— if (besides c_i) $e_1 + e_2$ is given (B.106)

$$\frac{e_i}{c_i} = \frac{e_1 + e_2}{c_1 + c_2}, \quad \text{whence} \quad e_i = \frac{e_1 + e_2}{c_1 + c_2} \cdot c_i\,;$$

— if $e_3 - e_1$, with $e_3 > e_1$, is given (B.107)

$$\frac{e_i}{c_i} = \frac{e_3 - e_1}{c_3 - c_1}, \quad \text{whence} \quad e_i = \frac{e_3 - e_1}{c_3 - c_1} \cdot c_i\,;$$

— if $e_3 - (e_1 + e_2)$, with $e_3 > e_1 + e_2$, is given (B.110)

$$\frac{e_i}{c_i} = \frac{e_3 - (e_1 + e_2)}{c_3 - (c_1 + c_2)}, \quad \text{whence} \quad e_i = \frac{e_3 - (e_1 + e_2)}{c_3 - (c_1 + c_2)} \cdot c_i\,;$$

— if $(e_1 + e_2) - e_3$, with $e_1 + e_2 > e_3$, is given (B.111)

$$\frac{e_i}{c_i} = \frac{(e_1 + e_2) - e_3}{(c_1 + c_2) - c_3}, \quad \text{whence} \quad e_i = \frac{(e_1 + e_2) - e_3}{(c_1 + c_2) - c_3} \cdot c_i \, ;$$

— if $e_1 \cdot e_2$ is given (B.108)

$$\frac{e_i^2}{c_i^2} = \frac{e_1 \cdot e_2}{c_1 \cdot c_2}, \quad \text{whence} \quad e_i = \sqrt{\frac{e_1 \cdot e_2}{c_1 \cdot c_2}} \cdot c_i \, ;$$

— if $(e_1 + e_2) e_3$ is given (B.112)

$$\frac{e_i^2}{c_i^2} = \frac{(e_1 + e_2) e_3}{(c_1 + c_2) c_3}, \quad \text{whence} \quad e_i = \sqrt{\frac{(e_1 + e_2) e_3}{(c_1 + c_2) c_3}} \cdot c_i \, .$$

B.115 is similar to B.112, but solved in another way.

2. Other cases.

Here too (as in B–I and B–II), the chapter ends with problems involving roots in the data, and then also with a particular problem.

A. *Data involving roots*.

In this group of problems (the first of which is misplaced) two partners are involved. In the first two problems, e_1 is equal to $\sqrt{e_2}$ (B.109) or to a multiple of it (B.113). We have then, if $e_1 = k \cdot \sqrt{e_2}$,

$$\frac{c_2}{c_1} = \frac{e_2}{e_1} = \frac{e_2}{k \cdot \sqrt{e_2}}, \quad \text{thus} \quad e_2 = \left(k \cdot \frac{c_2}{c_1} \right)^2.$$

In the last problem (B.114), the product $\sqrt{e_1} \cdot \sqrt{e_2}$ is known; this is reducible to B.108 since $e_1 \cdot e_2$ is known. This group reminds us of B.93–96.

B. *A particular problem*.

B.116 is a particular, but well-known problem on partnership. It usually takes the form of loaves shared with a passer-by, as in the example from the seventh section (*De regulis erraticis*) of the twelfth chapter in Leonardo Fibonacci's *Liber abaci*.[*] As we are told, two men, the first of whom has three loaves and the other, two, are about to eat; a traveller passing by, invited to partake of their meal, will afterwards give five bezants for his share.[†] Some incompetent people (*quidam imperiti*), notes Fibonacci, would attribute to each of the two men as many bezants as he had loaves, whereas the correct answer is 4 and 1.

---

[*] Edition of the Latin text, p. 283.
[†] The 'bezants' (*bizantii*), originally gold coins struck at Byzantium, correspond here to the nummi of the *Liber mahameleth*.

The tradition in Arabic writing has a similar problem, the solution of which is said to originate with Calif 'Alī: one man has five loaves, the other three, and the passer-by leaves eight dirhams; the first man claims five dirhams for himself and three for the other, whereas the second expects to receive half of the eight dirhams. They appealed to 'Alī, whose answer was seven dirhams for the first man and just one dirham for his partner; and he is said to have explained why.°

In our terms, suppose the first had q_1 and the second q_2. If they were to sell the *whole* to the passer-by for S, they would receive, respectively,

$$\frac{q_1}{q_1 + q_2} S, \qquad \frac{q_2}{q_1 + q_2} S,$$

just as we have seen in B.104 that

$$e_1 = \frac{c_1}{c_1 + c_2} e, \qquad e_2 = \frac{c_2}{c_1 + c_2} e.$$

Such was the answer of Fibonacci's *imperiti* and of the Arab with five loaves. But, in fact, since the first two men eat as well, each of the three men is to receive $\frac{1}{3}(q_1 + q_2)$; that is, the first two associates actually give up to the passer-by only what they each had in excess of $\frac{1}{3}(q_1 + q_2)$, that is, respectively,

$$q_1 - \frac{1}{3}(q_1 + q_2) = \frac{2}{3} q_1 - \frac{1}{3} q_2, \qquad q_2 - \frac{1}{3}(q_1 + q_2) = \frac{2}{3} q_2 - \frac{1}{3} q_1.$$

Since these are the parts handed over for the third's share $\frac{1}{3}(q_1 + q_2)$, the two partners are to receive from the sum S paid by the passer-by the amounts

$$\frac{\frac{2}{3} q_1 - \frac{1}{3} q_2}{\frac{1}{3}(q_1 + q_2)} \cdot S = \frac{2q_1 - q_2}{q_1 + q_2} \cdot S, \qquad \frac{\frac{2}{3} q_2 - \frac{1}{3} q_1}{\frac{1}{3}(q_1 + q_2)} \cdot S = \frac{2q_2 - q_1}{q_1 + q_2} \cdot S.$$

---

1. Three partners.

(B.104) MSS $\mathcal{A}$, $\mathcal{B}$, $\mathcal{C}$. Three partners have respectively invested eight, ten, fourteen *nummi*; what will their profits be if the total profit is twenty-two?

Given thus $c_1 = 8$, $c_2 = 10$, $c_3 = 14$ (thus $c = 32$), $e = 22$, required e_i. Then, according to the 'three ways' often seen before (from B.1 on),

$$e_i = \frac{e \cdot c_i}{c}, \qquad e_i = \frac{c_i}{c} \cdot e, \qquad e_i = \frac{e}{c} \cdot c_i.$$

° See Caetani's *Annali dell'Islām*, vol. X, pp. 442–443. To 'Alī is also attributed the problem of the camels (see summary to the next chapter).

Whence the ways of computing successively explained in the text:

$$e_i = \frac{22 \cdot c_i}{32},$$

$$e_1 = \frac{8}{32} \cdot 22 = \frac{1}{4} \cdot 22, \quad e_2 = \frac{10}{32} \cdot 22 = \left(\frac{2}{8} + \frac{1}{2}\frac{1}{8}\right) \cdot 22, \quad e_3 = \frac{14}{32} \cdot 22,$$

$$e_i = \frac{22}{32} \cdot c_i = \left(\frac{5}{8} + \frac{1}{2}\frac{1}{8}\right) \cdot c_i.$$

Since this is the first problem of this chapter, the justification explains the fundamental proportion, first seen in B.54, $c_i : e_i = c_k : e_k$, which becomes in the present case (since we know c, the total capital, and e, the total profit) $c_i : e_i = c : e$. The same is said to apply whatever the number of partners in such a problem: we shall begin by adding all the capitals and all the profits, and apply the previous relation.

(B.105) MSS $\mathcal{A}$, $\mathcal{B}$, $\mathcal{C}$. Three partners have respectively invested eight, ten, fourteen; if the profit of the investor of ten is four what will be the profits of the other two?

Given thus $c_1 = 8$, $c_2 = 10$, $c_3 = 14$, $e_2 = 4$, required e_1, e_3. Since we know e_2, we shall consider $e_i : c_i = e_2 : c_2$, whence

$$e_i = \frac{e_2 \cdot c_i}{c_2} = \frac{4 \cdot c_i}{10}, \quad e_i = \frac{e_2}{c_2} \cdot c_i = \frac{c_i}{c_2} \cdot e_2.$$

The origin of the formula is then explained (using the proportion $c_2 : e_2 = c_i : e_i$).

(B.106) MSS $\mathcal{A}$, $\mathcal{B}$, $\mathcal{C}$. In the same problem, we know this time not one profit but the sum of the profits of the first two men and we are asked about the individual profits.

Given thus $c_1 = 8$, $c_2 = 10$, $c_3 = 14$, $e_1 + e_2 = 12$, required e_i. As seen in our summary, and said in the text to be evident from a proportion already explained (the ratio profit to capital remains the same; see also B.56),

$$e_2 = \frac{(e_1 + e_2) c_2}{c_1 + c_2} = \frac{12 \cdot 10}{18},$$

and generally

$$e_i = \frac{12 \cdot c_i}{18} = \frac{12}{18} \cdot c_i = \frac{c_i}{18} \cdot 12.$$

(B.107) MSS $\mathcal{A}$, $\mathcal{B}$, $\mathcal{C}$. Three partners have again respectively invested eight, ten, fourteen; what will their profits be if subtracting that of the first from that of the last leaves four?

Given thus $c_1 = 8$, $c_2 = 10$, $c_3 = 14$, $e_3 - e_1 = 4$, required e_i.
Same as before, subtracting instead of adding:

$$e_i = \frac{(e_3 - e_1) c_i}{c_3 - c_1} = \frac{4 \cdot c_i}{14 - 8} = \frac{4 \cdot c_i}{6}.$$

(B.108) MSS A, B, C. The investments remaining the same, find the profits of the three partners if multiplying those of the first two produces forty-five.

Given thus $c_1 = 8$, $c_2 = 10$, $c_3 = 14$, $e_1 \cdot e_2 = 45$, required e_i.

The treatment of this is known from earlier analogous problems in the first two chapters, the text recalls (see indeed B.17, B.58).

$$e_1 = \sqrt{\frac{c_1 (e_1 \cdot e_2)}{c_2}} = \sqrt{\frac{8 \cdot 45}{10}} = 6,$$

$$e_2 = \sqrt{\frac{c_2 (e_1 \cdot e_2)}{c_1}} = \sqrt{\frac{10 \cdot 45}{8}} = 7 + \frac{1}{2} \quad \left(e_3 = \frac{e_1 \cdot c_3}{c_1}\right).$$

Remark. In our summary we have seen (p. 1427) the general formula

$$e_i = \sqrt{\frac{e_1 \cdot e_2}{c_1 \cdot c_2}} \cdot c_i = \sqrt{\frac{c_i^2 (e_1 \cdot e_2)}{c_1 \cdot c_2}},$$

which will indeed be employed in the similar problem B.114. Now this formula takes a reduced form for e_1 and e_2, which is used in the above computations. There is, however, no such reduced form for e_3, and this may explain why its computation is omitted.

★ ★ ★

The subsequent problems are in disorder. At least, their logical sequence should be: B.110, B.111, B.112, B.115, B.109, B.113, B.114, B.116. Of these, B.109–115 are found only in MS A. Some of their solutions are just unexplained computations.

(B.109) MS A. Two partners have respectively invested ten and twenty, and the profit of the first is the root of the profit of the second; what are these profits?

Given thus $c_1 = 10$, $c_2 = 20$, $e_1 = \sqrt{e_2}$, required e_i. This problem is related to B.113–114 and thus clearly misplaced.

$$\frac{e_1}{e_2} = \frac{c_1}{c_2} = \frac{1}{2}, \quad \text{thus} \quad \frac{1}{2} \cdot e_2 = e_1 = \sqrt{e_2},$$

then $e_2 = 4$, $e_1 = 2$.

(B.110) MS A. Three partners have respectively invested ten, thirty, fifty, and the profit of the third less the profit of the first two is three; what are their profits?

Given thus $c_1 = 10$, $c_2 = 30$, $c_3 = 50$, $e_3 - (e_1 + e_2) = 3$, required e_i. Then

$$e_i = \frac{e_3 - (e_1 + e_2)}{c_3 - (c_1 + c_2)} \cdot c_i = \frac{3}{50 - (10 + 30)} \cdot c_i = \frac{3}{10} \cdot c_i,$$

giving respectively 3, 9, 15.

(B.111) MS $\mathcal{A}$. Three partners have respectively invested twenty, fifty, thirty, and the profit of the first two less the profit of the third is four. Given thus $c_1 = 20$, $c_2 = 50$, $c_3 = 30$, $(e_1 + e_2) - e_3 = 4$, required e_i.

$$e_i = \frac{(e_1 + e_2) - e_3}{(c_1 + c_2) - c_3} \cdot c_i = \frac{4}{(20 + 50) - 30} \cdot c_i = \frac{4}{40} \cdot c_i = \frac{1}{10} \cdot c_i,$$

giving 2, 5, 3.

(B.112) MS $\mathcal{A}$. Three partners have respectively invested ten, twenty, forty, and multiplying the sum of the profits of the first two by that of the third produces forty-eight. Given thus $c_1 = 10$, $c_2 = 20$, $c_3 = 40$, $(e_1 + e_2) e_3 = 48$, required e_i.

$$\frac{e_i^2}{c_i^2} = \frac{(e_1 + e_2) e_3}{(c_1 + c_2) c_3} = \frac{48}{1200} = \frac{1}{25}, \quad \text{thus} \quad e_i = \frac{1}{5} \cdot c_i,$$

giving 2, 4, 8.

2. Other cases.

A. *Data involving roots*.

This applies to B.113–114 and B.109.

(B.113) MS $\mathcal{A}$. Two partners have invested ten and fifty, and the profit of the first is half the root of the profit of the second.

Given thus $c_1 = 10$, $c_2 = 50$, $e_1 = \frac{1}{2} \cdot \sqrt{e_2}$, required e_i. Similar to B.109, which should be in this section.

$$\frac{e_2}{e_1} = \frac{e_2}{\frac{1}{2} \cdot \sqrt{e_2}} = \frac{c_2}{c_1} = \frac{50}{10}, \quad \text{thus} \quad \sqrt{e_2} = \frac{1}{2} \cdot 5 = 2 + \frac{1}{2} \quad \text{and}$$

$$e_2 = 6 + \frac{1}{4}, \quad e_1 = \frac{1}{2} \cdot \sqrt{e_2} = 1 + \frac{1}{4}.$$

(B.114) MS $\mathcal{A}$. Two partners have invested eight and eighteen, and the product of the roots of their profits is six.

Given thus $c_1 = 8$, $c_2 = 18$, $\sqrt{e_1} \cdot \sqrt{e_2} = 6$, required e_i. Similar to B.108 since $e_1 \cdot e_2 = 36$. See B.95, p. 1422.

(*a*) As seen in our summary,

$$e_1 = \sqrt{\frac{e_1 \cdot e_2}{c_1 \cdot c_2}} \cdot c_1 = \sqrt{\frac{36}{144}} \cdot 8 = \frac{1}{2} \cdot 8 = 4,$$

$$e_2 = \sqrt{\frac{e_1 \cdot e_2}{c_1 \cdot c_2}} \cdot c_2 = \frac{1}{2} \cdot 18 = 9.$$

(b) Also computed, since $\dfrac{c_2}{c_1} = \dfrac{18}{8} = 2 + \dfrac{1}{4} = \left(1 + \dfrac{1}{2}\right)^2$, as

$$e_1 = \dfrac{\sqrt{e_1 \cdot e_2}}{\sqrt{\dfrac{c_2}{c_1}}} = \dfrac{6}{1 + \dfrac{1}{2}} = 4, \qquad e_2 = \dfrac{c_2}{c_1} \cdot e_1 = \left(2 + \dfrac{1}{4}\right) \cdot 4 = 9.$$

(B.115) MS A. Three partners have respectively invested ten, twenty, a hundred, and multiplying the sum of the profits of the first two by that of the third produces a hundred and twenty.

Given thus $c_1 = 10$, $c_2 = 20$, $c_3 = 100$, $(e_1 + e_2)\, e_3 = 120$, required e_i. Similar to B.112, but solved by analogy to a problem in B–II.

Since
$$\dfrac{e_1 + e_2}{c_1 + c_2} = \dfrac{e_3}{c_3}, \quad \text{so} \quad \dfrac{e_1 + e_2}{e_3} = \dfrac{c_1 + c_2}{c_3} = \dfrac{30}{100}.$$

The subsequent computation should be obvious:
$$\dfrac{(e_1 + e_2)\, e_3}{e_3^2} = \dfrac{30}{100}, \qquad e_3^2 = \dfrac{100 \cdot 120}{30} = 400,$$

whence $e_3 = 20$. Since we now know e_3 and c_3, the ratio of which is a fifth, e_1 and e_2 will equal a fifth of the corresponding capital.

Sometimes our author seems to prefer an intricate way. Considering in this case that

$$\dfrac{e_1 + e_2}{e_3} = \dfrac{30}{100}, \qquad (e_1 + e_2)\, e_3 = 120,$$

with thus $e_3 > e_1 + e_2$, he considers that e_3 could be a capital c' of which $e_1 + e_2$ were the profit e'. Since then

$$\dfrac{e'}{c'} = \dfrac{30}{100}, \qquad e' \cdot c' = 120,$$

we are to determine a capital and its profit when their ratio and product are known. Solving such a problem has been seen in the previous chapter (B.58). So we shall find

$$c'^2 = \dfrac{c'\,(e' \cdot c')}{e'} = \dfrac{100 \cdot 120}{30} = 400,$$

whence $c' = 20$ and $e' = 6$. Returning to the original problem, this means that $e_3 = 20$ and $e_1 + e_2 = 6$, which is the profit corresponding to $c_1 + c_2 = 30$. This is the situation of B.106. The remaining values e_1, e_2 are then found from

$$e_i = \dfrac{6}{30} \cdot c_i = \dfrac{1}{5} \cdot c_i,$$

whence $e_1 = 2$, $e_2 = 4$. We have indeed reached the answer, but in a way which will rather confuse the reader. For 6 was the profit of the capital 20 in the intermediate problem while the same 6 is the profit of the capital 30 in the original problem. Our author is so concerned with drawing parallels between one chapter and another that he also does it when it would be better not to.

B. *A particular problem.*

(B.116) MSS $\mathcal{A}, \mathcal{B}, \mathcal{C}$. Two men who own a hundred sheep, one sixty and the other forty, admit a third man into partnership, with each possessing then a third of the flock. If the third gives sixty nummi for his part, how will they share them?

As seen in our summary (pp. 1427–1428), owning three fifths and two fifths respectively does not entitle the two owners to receive such parts of the sixty nummi, for they do not sell the whole flock but each keeps a third of it. We must therefore, as the text clearly states, consider that what each man actually sold is the difference between his initial parts and a third of the flock. Thus, with $q_1 = 60$ and $q_2 = 40$, the first and the second sold, respectively,

$$q_1 - \frac{1}{3}(q_1 + q_2) = 60 - \left(33 + \frac{1}{3}\right) = 26 + \frac{2}{3}$$

$$q_2 - \frac{1}{3}(q_1 + q_2) = 40 - \left(33 + \frac{1}{3}\right) = 6 + \frac{2}{3},$$

and this will determine what they are to receive.

This problem is thereby reduced to the previous ones in the present chapter: given $c_1 = 26 + \frac{2}{3}$, $c_2 = 6 + \frac{2}{3}$, $e_1 + e_2 = 60$, required e_1, e_2. As in B.106, we shall have

$$e_1 = \frac{(e_1 + e_2)c_1}{c_1 + c_2} = \frac{60\left(26 + \frac{2}{3}\right)}{33 + \frac{1}{3}} = 48,$$

$$e_2 = \frac{(e_1 + e_2)c_2}{c_1 + c_2} = \frac{60\left(6 + \frac{2}{3}\right)}{33 + \frac{1}{3}} = 12.$$

Chapter B–IV:
Sharing out according to prescribed parts

Summary

There is one problem in the form of an anecdote frequently found in Arabic sources concerning a herd of seventeen camels to be shared out among three people, one of whom is to receive a half, the second a third and the last a ninth. The solution, which has been attributed to various sages, including Calif 'Alī, was to borrow a further camel and give half of the eighteen camels, thus nine, to the first, a third of them, thus six, to the second, and finally a ninth, thus two, to the third. When this was done, everyone had received his due share, and one camel was left which could be returned to its owner.

Another kind of sharing out, which does not this time belong to recreational mathematics but to what was called *'ilm al-farā'iḍ*, that is, the science of calculating legacies, became a major application of arithmetic and algebra in Islamic countries. Basically, Islamic succession law restricted the power of testamentary disposition to one third of the estate, the remaining two thirds being shared out among the legal heirs, that is, the relatives; each of whom was moreover entitled to receive a given fraction of the estate according to their degree of kinship with the deceased (Koran IV.11–12 & 176).* Suppose, for instance, that a woman dies, leaving her husband, two daughters and mother. Between them, they will share the estate, or at least two thirds of it. Now, by law, the husband is entitled to a quarter, the mother to a sixth and the daughters, since there is no son, to two thirds between them. Such problems of sharing, which can become arbitrarily complicated, are found from the earliest times of Islamic mathematics, as, for example, in the first Arabic treatise on algebra fully extant, Muḥammad al-Khwārizmī's (*fl.* 820), where they even make up the largest section, leaving us in no doubt as to importance of this particular application. By the time of Abū Kāmil (*fl.* 880), thus at the close of the same century, there were already separate works on the subject —which is understandable since those whose responsibility it was to share out estates did not, after all, need to master more than a few specific operations in elementary algebra and arithmetic.

Neither of the two above examples reduces, as we might expect at first sight, to a straightforward computation with fractions. Since both the

---

* Such restrictions are also known from antiquity. See D. E. Smith's *History of Mathematics*, II, pp. 544–545, or Tropfke (*et al.*)'s, (ed. 1980) p. 655.

parts and the number of persons may vary, so will the sum of the fractions. Thus, in the first example, that of the camels, adding the fractions gives $\frac{1}{2}+\frac{1}{3}+\frac{1}{9}=\frac{17}{18}<1$; in the second, that of the estate, the sum is $\frac{1}{4}+\frac{1}{6}+\frac{2}{3}=\frac{13}{12}>1$. These, then, are the characteristics and singularity of the topic 'sharing out according to prescribed parts': the sum of the fractions does not equal unity. In other words, we are to divide up some property S according to fractions

$$\frac{k_i}{l_i}, \quad \text{with} \quad \sum \frac{k_i}{l_i} = \frac{K}{L} \neq 1,$$

so that the sum of the theoretical parts does not equal the whole amount S.

Let us therefore see how the author of the *Liber mahameleth* proceeds in order to reach the result, keeping his cases and, apart from the mathematical form, his reasonings.

Consider first that the sum of the given fractions is less than unity; that is, $K < L$. Suppose we begin by distributing the shares as prescribed. We shall thus have shared out, from the whole sum S,

$$\sum \frac{k_i}{l_i} \cdot S = \frac{K}{L} \cdot S.$$

Since this sum is less than S, there will be a remainder, namely

$$R_1 = S - \frac{K}{L} \cdot S = \left(1 - \frac{K}{L}\right) S = \frac{L-K}{L} \cdot S = r \cdot S \quad \text{with} \quad r = \frac{L-K}{L}.$$

This remainder R_1 will then be shared out as before, each partner receiving his prescribed part of it. Again, there will remain something of R_1, namely

$$R_2 = R_1 - \frac{K}{L} \cdot R_1 = \left(1 - \frac{K}{L}\right) R_1 = r \cdot R_1 = r^2 \cdot S.$$

After repeating the procedure to the nth share-out, we shall be left with

$$R_n = R_{n-1} - \frac{K}{L} \cdot R_{n-1} = \left(1 - \frac{K}{L}\right) R_{n-1} = r^n \cdot S,$$

while the amount of S already distributed will be equal to

$$\frac{K}{L} \cdot S + \frac{K}{L} \cdot R_1 + \ldots + \frac{K}{L} \cdot R_{n-1}$$
$$= \frac{K}{L} \cdot S \left(1 + r + r^2 + \ldots + r^{n-1}\right) = \frac{K}{L} \cdot S \cdot \frac{1 - r^n}{1 - r}.$$

Suppose now we proceed in such a way to infinity. Since $r < 1$, $r^n \to 0$, and we shall ultimately have distributed

$$\frac{K}{L} \cdot S \cdot \frac{1}{1-r} = \frac{K}{L} \cdot S \cdot \frac{L}{L-(L-K)} = \frac{K}{L} \cdot S \cdot \frac{L}{K} = S.$$

This then accounts for the whole sum S, with the ith partner receiving the part of it

$$s_i = \frac{k_i}{l_i} \cdot S \cdot \frac{1}{1-r} = \frac{k_i}{l_i} S \cdot \frac{L}{K}.$$

That can be rewritten as

$$s_i = \frac{k_i}{l_i} \left(\frac{L}{K} S\right) = \frac{k_i}{l_i} S' \quad (\star)$$

with $S' > S$ a fictitious capital of which the shares s_i are the prescribed fractions.

Let us illustrate this with the example of the camels. We have $S = 17$ and $\frac{1}{2} + \frac{1}{3} + \frac{1}{9} = \frac{17}{18}$, so $K = 17$, $L = 18$, $S' = \frac{18}{17} S = 18$. The required parts are then $\frac{1}{2} \cdot 18 = 9$, $\frac{1}{3} \cdot 18 = 6$, $\frac{1}{9} \cdot 18 = 2$.

In the first such problem in our text, B.117, the fractions are $\frac{1}{2}$ and $\frac{1}{3}$, and $S = 30$. Then $\frac{1}{2} + \frac{1}{3} = \frac{5}{6}$, so $K = 5$, $L = 6$, hence $S' = \frac{6}{5} S = 36$. Thus the shares will be $\frac{1}{2} \cdot 36 = 18$, $\frac{1}{3} \cdot 36 = 12$. In the second problem (B.118), the fractions are the same, so again $K = 5$, $L = 6$, but $S = 10$. Then $S' = 12$ and the shares are $\frac{1}{2} \cdot 12 = 6$ and $\frac{1}{3} \cdot 12 = 4$.

In the above cases, we had $S' > S$. If the sum of the fractions is larger than unity, thus $K > L$, we shall take once again (see below)

$$s_i = \frac{k_i}{l_i} \left(\frac{L}{K} S\right) = \frac{k_i}{l_i} S' \quad (\star)$$

with, this time, $S' < S$.

This may be applied to the inheritance problem mentioned initially. Since $\frac{1}{4} + \frac{1}{6} + \frac{2}{3} = \frac{13}{12}$, thus $K = 13$ and $L = 12$, then $S' = \frac{12}{13} S$, and the parts will be $\frac{3}{13} S$, $\frac{2}{13} S$, $\frac{8}{13} S$.

It may also be applied to B.119 in our text. The parts are $\frac{2}{3}$, 1, $\frac{3}{2}$, the sum of which is $\frac{19}{6}$, thus $K = 19$, $L = 6$; since $S = 30$, so $S' = \frac{6}{19} \cdot 30$, and the partners' shares are $\frac{2}{3} \left(\frac{6}{19} \cdot 30\right) = 4 \cdot \frac{30}{19}$, $\frac{6}{19} \cdot 30$, $\frac{3}{2} \left(\frac{6}{19} \cdot 30\right) = 9 \cdot \frac{30}{19}$.

To obtain the formulae $(\star)$ we may also consider that S is a profit to be shared among partners having invested capitals $c_i = \frac{k_i}{l_i} C$. Then, since $e_i : c_i = e : c = S : \frac{K}{L} C$, we have

$$e_i = \frac{c_i \cdot S}{\frac{K}{L} C} = \frac{\frac{k_i}{l_i} C \cdot S}{\frac{K}{L} C} = \frac{k_i}{l_i} \frac{L}{K} S,$$

where C may thus be an optional number f (divisible by the l_i). This is essentially the analogy drawn by the author in B.117 to link, for mathematical treatment, these problems with those of the previous chapter. That is how B.118–119 are also solved.

In each of the above two cases, the recipients may be allocated certain amounts of money m_i, M altogether, along with the prescribed parts. Two situations are possible.

(*i*) <u>The amounts are received beforehand.</u>

What we have seen above is directly applicable, except that the actual capital to be shared out must first be reduced by the amounts in cash. Then $S^* = S - \sum m_i = S - M$. The shares from S^* will be, as determined before,

$$s_i^* = \frac{k_i}{l_i}\left(\frac{L}{K}S^*\right),$$

and the complete shares

$$s_i = \frac{k_i}{l_i}\left(\frac{L}{K}S^*\right) + m_i.$$

(*ii*) <u>The amounts are to be included in the parts.</u>

The fictitious capital to be shared out is then $S' = \frac{K}{L}S + M$, which is the sum of the individual parts $s_i' = \frac{k_i}{l_i}S + m_i$. Setting the proportion as before, we have

$$s_i = \frac{S \cdot s_i'}{S'} = S \cdot \frac{\frac{k_i}{l_i} \cdot S + m_i}{\frac{K}{L} \cdot S + M}.$$

The above two cases are treated successively in B.120, which closes this (short) chapter on sharing out according to prescribed parts.

(**B.117**) MSS $\mathcal{B}, \mathcal{C}$. An amount of thirty nummi must be divided between two men, one of whom is to receive a half and the other a third.

(*α*) *Preliminary remark.* The enunciation of each of the first two problems in this chapter is followed, appropriately enough, by an explanation, which can be expressed thus: Each participant is to receive first his prescribed fraction, then the same fraction of the remainder, and so on 'until nothing remains of the thirty'. Here, after the first share-out, a sixth of thirty remains, of which a half and a third are distributed. What then remains, a sixth of a sixth of thirty, is shared in the same way, and so on. Thus the two men will finally have

$$s_1 = \left(\frac{1}{2} + \frac{1}{2}\frac{1}{6} + \frac{1}{2}\frac{1}{6^2} + \cdots\right) \cdot 30 = \left(\frac{1}{2}\sum_0^\infty \frac{1}{6^m}\right) \cdot 30$$

$$= \left(\frac{1}{2} \cdot \frac{1}{1 - \frac{1}{6}}\right) \cdot 30 = \frac{3}{5} \cdot 30 = 18,$$

$$s_2 = \left(\frac{1}{3} + \frac{1}{3}\frac{1}{6} + \frac{1}{3}\frac{1}{6^2} + \cdots\right) \cdot 30 = \left(\frac{1}{3}\sum_0^\infty \frac{1}{6^m}\right) \cdot 30$$

$$= \left(\frac{1}{3} \cdot \frac{1}{1-\frac{1}{6}}\right) \cdot 30 = \frac{2}{5} \cdot 30 = 12.$$

This is just what we have seen in our introduction above: the parts are $\frac{1}{2}$ and $\frac{1}{3}$, the remainder is $\frac{1}{6}$, therefore $\frac{K}{L} = 1 - \frac{1}{6}$ and we shall take $K = 5$ and $L = 6$. Then

$$s_1 = \frac{k_1}{l_1} \frac{L}{K} \cdot S = \frac{1}{2} \frac{6}{5} \cdot 30 = \frac{3}{5} \cdot 30,$$

$$s_2 = \frac{k_2}{l_2} \frac{L}{K} \cdot S = \frac{1}{3} \frac{6}{5} \cdot 30 = \frac{2}{5} \cdot 30.$$

(β) *Treatment in the text.* The author does not go into these details, and he reduces the problem to one on profit. To clear the fractions he takes the common denominator 6, a half and a third of which are both integers; the parts, namely 3 and 2, are then considered as capitals invested by the recipients of a half and a third, respectively. The problem thus becomes: given $c_1 = 3$, $c_2 = 2$, $e_1 + e_2 = 30$, required e_1, e_2. Now this is B.106. We therefore consider

$$e_i = \frac{(e_1 + e_2) \cdot c_i}{c_1 + c_2}$$

whence $e_1 = 18 = s_1$, $e_2 = 12 = s_2$.

(**B.118**) MS *A*. Similar to the preceding problem, but this time the amount is ten and a third man is involved. His presence is in fact superfluous and only serves as a way to explain the method of sharing since he will just return each time the fractions of the remaining part. This problem may well have originally preceded the previous one. In any case the reader will better understand B.117 after studying B.118.

(***a***) According to the repeated distribution, the parts of the two partners will be

$$s_1 = \frac{1}{2}\left(1 + \frac{1}{6} + \frac{1}{6^2} + \ldots\right) \cdot 10,$$

$$s_2 = \frac{1}{3}\left(1 + \frac{1}{6} + \frac{1}{6^2} + \ldots\right) \cdot 10.$$

We have then

$$\frac{s_1}{s_2} = \frac{\frac{1}{2}\left(1 + \frac{1}{6} + \frac{1}{6^2} + \ldots\right) \cdot 10}{\frac{1}{3}\left(1 + \frac{1}{6} + \frac{1}{6^2} + \ldots\right) \cdot 10} = \frac{\frac{1}{2}}{\frac{1}{3}}.$$

By so doing, the author has eliminated the infinite series. This is just in keeping with what he usually does; for when there is a common factor in both terms of a division, it can be cancelled (rule *iii*, p. 1183). Reintroducing now any (more conventional) common factor f (this time in order to clear the fractions), we have

$$\frac{s_1}{s_2} = \frac{\frac{1}{2} \cdot f}{\frac{1}{3} \cdot f}.$$

With the simplest choice $f = 2 \cdot 3 = 6$, he obtains $s_1 : s_2 = 3 : 2$. Now $s_1 + s_2 = 10$. So, by composition,

$$\frac{s_1}{s_1 + s_2} = \frac{3}{5}, \quad s_1 = \frac{3}{5} \cdot (s_1 + s_2) = \frac{3 \cdot 10}{5} = 6,$$

$$\frac{s_2}{s_1 + s_2} = \frac{2}{5}, \quad s_2 = \frac{2}{5} \cdot (s_1 + s_2) = \frac{2 \cdot 10}{5} = 4.$$

The treatment is recapitulated at the end in the form of a formula, as is sometimes the case after a computation.[†] Taking an appropriate quantity f (divisible by the l_i), then

$$s_i = \frac{\frac{k_i}{l_i} \cdot f}{\frac{k_1}{l_1} \cdot f + \frac{k_2}{l_2} \cdot f} \cdot S.$$

(**b**) This and the following, other solution do not add anything new. Here the optional number is 1, that is, we have a direct computation without clearing the fractions. In c the author chooses $f = 10$.
Then ($f = 1$)

$$s_1 = \frac{\frac{1}{2}}{\frac{1}{2} + \frac{1}{3}} \cdot 10 = \frac{1}{2} \left(1 + \frac{1}{5}\right) \cdot 10 = \frac{1}{2} \cdot 12 = 6,$$

$$s_2 = \frac{\frac{1}{3}}{\frac{1}{2} + \frac{1}{3}} \cdot 10 = \frac{1}{3} \left(1 + \frac{1}{5}\right) \cdot 10 = \frac{1}{3} \cdot 12 = 4.$$

This way is then proved by calculating q such that $\left(\frac{1}{2} + \frac{1}{3}\right) q = 10$.[‡] Since $q : 10 = 1 : \left(\frac{1}{2} + \frac{1}{3}\right) = 1 : \frac{5}{6}$, so $q = \frac{6}{5} \cdot 10 = 12$. Then

$$s_2 : (s_1 + s_2) = \frac{1}{3} : \left(\frac{1}{2} + \frac{1}{3}\right) = \frac{1}{3} \cdot q : \frac{5}{6} \cdot q = \frac{1}{3} \cdot 12 : 10,$$

$$s_1 : (s_1 + s_2) = \frac{1}{2} : \left(\frac{1}{2} + \frac{1}{3}\right) = \frac{1}{2} \cdot q : \frac{5}{6} \cdot q = \frac{1}{2} \cdot 12 : 10,$$

whence, since $s_1 + s_2 = 10$, $s_2 = 4$, $s_1 = 6$.

(**c**) Same treatment as in a, but with a value $f \neq 6$.

$$s_1 = \frac{\frac{1}{2} \cdot f}{\left(\frac{1}{2} + \frac{1}{3}\right) \cdot f} \cdot 10, \quad s_2 = \frac{\frac{1}{3} \cdot f}{\left(\frac{1}{2} + \frac{1}{3}\right) \cdot f} \cdot 10.$$

Remark. The choice $f = 10$ is rather confusing since this is precisely the value of the amount to be shared.

[†] See B.160, B.163, B.164, B.236.
[‡] Thus q corresponds to our fictitious capital S' (see summary).

(d) The text concludes by generalizing to n parts $\frac{k_i}{l_i}$. Choose f divisible by all l_i. Then, if S is the sum to be divided, the shares will be

$$s_i = \frac{\frac{k_i}{l_i} \cdot f}{\sum \frac{k_i}{l_i} \cdot f} \cdot S.$$

(B.119) MSS $\mathcal{A}, \mathcal{B}, \mathcal{C}$. Thirty nummi are to be divided among three men, the first receiving two thirds, the second, the whole, the third, one and a half times.

The sum of the fractions is here greater than 1:

$$\frac{k_1}{l_1} = \frac{2}{3}, \quad \frac{k_2}{l_2} = 1, \quad \frac{k_3}{l_3} = \frac{3}{2}, \quad \text{with } S = 30.$$

The solution presented is explained as before. Taking $f = 2 \cdot 3 = 6$, thus the common denominator, the above fractions of f will be considered as representing invested capitals $c_1 = 4$, $c_2 = 6$, $c_3 = 9$, and we are then to determine the individual profits e_i for a total profit $e = 30$. Then, as seen in B.104,

$$e_i = \frac{e \cdot c_i}{c},$$

whence

$$e_1 = \frac{30 \cdot 4}{19} = 6 + \frac{6}{19}, \quad e_2 = \frac{30 \cdot 6}{19} = 9 + \frac{9}{19}, \quad e_3 = \frac{30 \cdot 9}{19} = 14 + \frac{4}{19}$$

which are the individual shares s_i.

(B.120) MSS $\mathcal{A}, \mathcal{B}, \mathcal{C}$. Thirty nummi are to be divided among three men, the first receiving two thirds plus three nummi, the second, the whole plus two nummi, the third, two and a fourth times plus one nummus.

Given thus the capital $S = 30$ and the fractions $\frac{2}{3}$, 1 and $2 + \frac{1}{4}$, with supplements $m_1 = 3$, $m_2 = 2$, $m_3 = 1$. Due to the ambiguous formulation, the additive amounts of nummi may be either given beforehand or included in the parts.

(i) Share-out of capital less the supplementary nummi, these being then added to the parts. So consider

$$\frac{k_1}{l_1} \cdot f = \frac{2}{3} \cdot f, \quad \frac{k_2}{l_2} \cdot f = f, \quad \frac{k_3}{l_3} \cdot f = \left(2 + \frac{1}{4}\right) f, \quad \text{and}$$

$$S^* = 30 - (3 + 2 + 1) = 24.$$

Since this is what has just been seen, the text does not compute further. To follow B.119, we may take $f = 12$, which gives, for the $\frac{k_i}{l_i} \cdot f$, 8, 12, 27, and for their sum, 47; the formula seen in B.118,

$$s_i^* = \frac{\frac{k_i}{l_i} \cdot f}{\sum \frac{k_i}{l_i} \cdot f} \cdot S^*,$$

then gives the shares (summing up to 24)

$$s_1^* = 4 + \frac{4}{47}, \quad s_2^* = 6 + \frac{6}{47}, \quad s_3^* = 13 + \frac{37}{47},$$

from which the required shares s_1, s_2, s_3 are found by adding the supplementary nummi.

(ii) The nummi being included in the parts, consider as the parts

$$s_i' = \frac{k_i}{l_i} \cdot S + m_i,$$

thus

$$s_1' = \frac{2}{3} \cdot S + 3, \quad s_2' = S + 2, \quad s_3' = \left(2 + \frac{1}{4}\right) S + 1,$$

which gives (with $S = 30$) $s_1' = 23$, $s_2' = 32$, $s_3' = 68 + \frac{1}{2}$ and $S' = \sum s_i' = 123 + \frac{1}{2}$. The actual shares s_i of 30 are then determined as in a problem on capital, the invested parts being $c_i = s_i'$ for a total capital $c = S' = 123 + \frac{1}{2}$, while the total profit to be shared is $e = S = 30$. Thus

$$e_i = \frac{c_i}{c} \cdot e \quad \text{becomes} \quad s_i = \frac{s_i'}{S'} \cdot S = \frac{s_i'}{123 + \frac{1}{2}} \cdot 30, \quad \text{whence}$$

$$s_1 = 5 + \frac{145}{247}, \quad s_2 = 7 + \frac{191}{247}, \quad s_3 = 16 + \frac{158}{247}.$$

Chapter B–V:
Masses

Summary

In the first section of this short chapter, a mass and smaller identical pieces made out of it are considered according to the proportion of their respective constituents; in the second, according to size and price; in the third and last, we are to determine the alloy in a mixture of gold and silver. As usual, Book B groups the problems by subject rather than by mathematical method.

1. Constituents of mixed masses.

Here (B.121) we are to infer from the given composition of a whole mixed mass what composition a known part of it will have, with the treatment being, we are told, just the same as in the chapter on partnership (B–III). Indeed, suppose that a mass weighing w_1 consists of given quantities $w_i^{(1)}$ of n metals and that required are the quantities $w_i^{(2)}$ of these metals in a similarly mixed mass weighing w_2. Comparing this with the problems of partnership involving a capital c, shares c_i, and profits e_i (above, B.104), we see that the relation there,

$$\frac{c_i}{c} = \frac{e_i}{e}, \quad \text{or} \quad \frac{c_i}{\sum c_i} = \frac{e_i}{\sum e_i},$$

may be adapted to the present case and so become

$$\frac{w_i^{(1)}}{w_1} = \frac{w_i^{(2)}}{w_2} \quad \text{or} \quad \frac{w_i^{(1)}}{\sum w_i^{(1)}} = \frac{w_i^{(2)}}{\sum w_i^{(2)}},$$

which is the relation solving B.121.

2. Making smaller pieces.

The second set of problems (B.122–123) is closely related to the ones on buying and selling (and also to those in the next two chapters). From a spherical mass of known diameter are made smaller, also spherical masses having all the same diameter; in one case, we seek the number q of smaller masses which can be made out of the larger one, in the other we ask for the price p_2 of the smaller when that of the larger, p_1, is given. In both cases, we shall make use of the fact that their volumes are proportional to the cube of the diameter $(\frac{4}{3}\pi \cdot r^3 = \frac{1}{6}\pi \cdot d^3)$; we find then, as applied in the two problems B.122–123,

$$q = \frac{V_1}{V_2} = \frac{d_1^3}{d_2^3} = \frac{p_1}{p_2}.$$

3. Alloy determination.

The last two problems, on determining the degree of purity of an alloy, are clearly related to the famous crown problem of Archimedes. Hieron, king of Sicily, is said to have entrusted a goldsmith with a mass of pure gold to be made into a crown.* When the work had been done, the weight was indeed equal to that of the initial quantity of gold, but Hieron suspected the goldsmith to have mixed in silver. He thereupon ordered Archimedes to detect and assess the extent of the fraud. Archimedes, we are told, discovered how to set about this whilst in the public bath: he realized that when a body is immersed in water as much water as its volume is displaced. This meant, on the one hand, that bodies equal in weight but not in volume were seen to displace accordingly different quantities of water, and, on the other hand, that a hard-to-measure volume (like that of the wreath) could thus be determined. This also explained the apparent loss of weight of bodies in water, that loss being equal to the weight of the water displaced.

To solve Hieron's problem, Archimedes ordered that two masses be made, one of pure gold and the other of pure silver, each equal in weight to the crown. Since the specific weight of gold is higher, it will when immersed in water displace a smaller volume than silver, while its apparent loss of weight in water will be less since it equals the weight of water dispersed. If we now immerse the crown and the quantity of water displaced is not equal to (and is in fact more than) that for the gold mass, the fraud will be evident. It remains now to find out what parts of gold and silver the crown contains. There are two ancient versions of Archimedes' procedure, according to whether *volume* or *weight* of the water displaced is considered.° These we shall recall first, before considering the method taught in our text, which is not based on equality of weight of the three bodies.

(1) <u>Weights equal: measuring volume or weight of the water displaced.</u>

(*a*) Let W be the common weight of each body, and their respective volumes V_C for the crown, V_G for the gold and V_S for the silver, with $V_S > V_G$. Each being in turn immersed in the same vessel completely filled with water, each will displace an amount of water equal to its volume; by measuring the volumes of these quantities of water, we shall therefore know the volumes of the solid bodies.

Now we know that $V_S > V_G$, and, if the crown is a mixture of gold and silver, $V_S > V_C > V_G$. The volume of the crown can then be expressed as $V_C = \alpha V_G + \beta V_S$, where α, β are fractions with $\alpha + \beta = 1$; the extreme cases are then $\alpha = 1$ (if the crown were pure gold) and $\alpha = 0$ (if pure silver). From

---

* Or a wreath dedicated to the gods. See Dijksterhuis, *Archimedes*, p. 19 and reference there.

° See e.g. Berthelot, *La chimie au moyen âge*, pp. 167–177; Cantor, *Vorlesungen*, I, pp. 310–312; Heath, *History of Greek Mathematics*, II, pp. 92–94.

$$\begin{cases} \alpha V_G + \beta V_S = V_C \\ \alpha + \beta = 1 \end{cases}$$

we infer that
$$\alpha = \frac{V_S - V_C}{V_S - V_G}, \qquad \beta = \frac{V_C - V_G}{V_S - V_G}.$$

These are the fractions of each metal in the crown. Since, for the weight of the crown, $W = w^{(G)} + w^{(S)}$, where $w^{(G)}$ and $w^{(S)}$ are the weights of pure gold and pure silver in the crown, we shall equate α and β to $\frac{w^{(G)}}{W}$ and $\frac{w^{(S)}}{W}$ and thus find

$$w^{(G)} = \frac{V_S - V_C}{V_S - V_G} \cdot W, \qquad w^{(S)} = \frac{V_C - V_G}{V_S - V_G} \cdot W.$$

Such is the method Archimedes used according to Book IX of Vitruvius' *De architectura* (IX, pr. 9–12).

Remark. We do not actually need to have a quantity of pure gold weighing as much as the crown: having one in pure silver will suffice, since the volume corresponding to the same weight in pure gold can be computed by finding what ratio the volumes of two smaller bodies, equal in weight and of pure silver and pure gold respectively, have to one another.

(b) Considering the same situation as before, let Δ_G, Δ_S, Δ_C be this time the three weights of water displaced. Then $\Delta_S > \Delta_C > \Delta_G$, according to the inequality of the volumes. So we know that when the bodies are immersed in water their initially equal weights drop to, respectively, $W - \Delta_S$, $W - \Delta_C$, $W - \Delta_G$. The loss of weight of the crown can then be expressed by

$$\begin{cases} \alpha \Delta_G + \beta \Delta_S = \Delta_C \\ \alpha + \beta = 1 \end{cases}$$

with $\alpha = 1$ if the crown is pure gold and α decreasing towards 0 with increasing quantity of silver mixed in, α and β being, as before, equal to $\frac{w^{(G)}}{W}$ and $\frac{w^{(S)}}{W}$ respectively. This brings us to

$$\alpha = \frac{\Delta_S - \Delta_C}{\Delta_S - \Delta_G}, \qquad \beta = \frac{\Delta_C - \Delta_G}{\Delta_S - \Delta_G}$$

and

$$w^{(G)} = \frac{\Delta_S - \Delta_C}{\Delta_S - \Delta_G} \cdot W, \qquad w^{(S)} = \frac{\Delta_C - \Delta_G}{\Delta_S - \Delta_G} \cdot W.$$

This is (more or less) the version of the anonymous *Carmen de ponderibus*, written about A.D. 500.[‡] It is essentially the same as before, the water volumes being replaced by their weights.

(2) <u>Volumes equal: measuring the weights of the three bodies.</u>

In the previous case the weight of the three bodies was supposed to be the same. We consider now that the three bodies have the same volume V

[‡] Lines 124–162 in Hultsch's edition, *Metrol. script. rel.*, II, pp. 95–97.

and the respective weights W_G, W_C, W_S; then $W_G > W_C > W_S$ since the specific weight of pure gold is the highest and assuming that the crown is mixed. For the weight of the crown, we have
$$\begin{cases} \alpha W_G + \beta W_S = W_C \\ \alpha + \beta = 1. \end{cases}$$
The closer W_C is to one of the weights, the more it contains of this metal and thus the closer to 1 the corresponding coefficient will be. By analogy to the previous cases, we shall find
$$\alpha = \frac{W_C - W_S}{W_G - W_S}, \qquad \beta = \frac{W_G - W_C}{W_G - W_S}$$
whence, since $w^{(G)} = \alpha \cdot W_G$ and $w^{(S)} = \beta \cdot W_S$, the two relations solving B.124–125:
$$w^{(G)} = \frac{W_C - W_S}{W_G - W_S} \cdot W_G, \qquad w^{(S)} = \frac{W_G - W_C}{W_G - W_S} \cdot W_S.$$

This method, besides being explained in the *Liber mahameleth*, also appears in the *Carmen de ponderibus*.

All these three methods involve considering two other pieces, one of pure gold and the other of pure silver, both equal either in weight or in volume to the given piece. An essential supposition was indeed to leave the wreath intact, as appropriate for a consecrated object. For otherwise it would have been simpler to cut off a piece, melt it down, and separate the two components by the then customary methods. At the beginning of B.123, our author explicitly says that we are not to use melting. He also excludes a supplementary weighing, as is required in the second aspect of the first method (weighing the water displaced in addition to ascertaining the equality in weight of the three bodies). Immersion will be used by him, though not for applying Archimedes' principle, but only as a mean to ascertain equality in volume of the bodies considered.

---

1. Constituents of mixed masses.

(B.121) MSS $\mathcal{A}, \mathcal{B}, \mathcal{C}$. From a mixed mass containing ten ounces of gold, fourteen of silver, twenty of brass a smaller mass is taken weighing twelve ounces; how much is there of each metal?

Given thus, in ounces, the quantities in the larger mass of the three metals gold, silver, brass, namely, respectively, $w_1^{(1)} = 10$, $w_2^{(1)} = 14$, $w_3^{(1)} = 20$, with $w_1 = w_1^{(1)} + w_2^{(1)} + w_3^{(1)}$ and knowing the weight of the second mass $w_2 = 12$, we are to find the quantities $w_i^{(2)}$ of each metal contained. As implied by the text, this is to be treated like problems on partnership: taking the $w_i^{(1)}$ as the invested capitals and w_2 as the total profit, while $w_i^{(2)}$ will be the required parts of it according to each capital. So, as seen in our summary, we have

$$\frac{w_i^{(1)}}{\sum w_i^{(1)}} = \frac{w_i^{(2)}}{\sum w_i^{(2)}},$$

whence

$$w_i^{(2)} = \frac{w_2}{w_1} \cdot w_i^{(1)} = \frac{12}{44} \cdot w_i^{(1)} = \frac{3}{11} \cdot w_i^{(1)}.$$

---

2. Making smaller pieces.

(B.122) MS $\mathcal{A}$. A spherical mass is ten palms in diameter and we want to make out of it smaller masses five palms in diameter.

Given thus $d_1 = 10$ and $d_2 = 5$, required the number q of smaller masses.

$$q = \frac{V_1}{V_2} = \frac{d_1^3}{d_2^3} = \frac{1000}{125} = 8.$$

This is obtained in the text first by computing d_1^3 and d_2^3 and dividing the first result by the second, then also by dividing d_1 by d_2 and cubing the quotient.

Remark. Since this is the first instance involving the volume of a sphere, we might have expected an allusion to *Elements* XII.18 ('Spheres are to one another in the triplicate ratio of their respective diameters').

(B.123) MS $\mathcal{A}$. A spherical mass ten palms in diameter costs a hundred nummi. What is the price of a smaller mass five palms in diameter?

Given thus $d_1 = 10$, $p_1 = 100$, $d_2 = 5$ and required p_2. Since

$$\frac{p_2}{p_1} = \frac{V_2}{V_1} = \frac{d_2^3}{d_1^3},$$

$$p_2 = \left(\frac{d_2}{d_1}\right)^3 \cdot p_1 = \left(\frac{1}{2}\right)^3 \cdot 100 = \frac{1}{8} \cdot 100 = 12 + \frac{1}{2};$$

or else, calculate

$$p_2 = \frac{d_2^3}{d_1^3} \cdot p_1 = \frac{125}{1000} \cdot 100 = \frac{1}{8} \cdot 100.$$

---

3. Alloy determination.

In B.124–125 (MS $\mathcal{A}$) we are to determine whether a mass is of pure gold or mixed with silver, without melting it down, or cutting a part of it off, and without weighing more than once.

(**a**) The first step is determining the ratio between the specific weights of gold and silver. For that purpose, we take any mass of pure gold and find a quantity of pure silver equal in volume; weighing them will give us the desired ratio. The equality in volume of two bodies which may be unequal in shape, with hard-to-measure irregularities or containing open cavities, is verified by immersing them in water: for equal volumes the rise in water level will be identical.

(**b**) Let now W_C be the weight of the given mass, and let us take a silver mass equal to it in volume. Suppose we find its weight to be W_S. From what has just been said, we may now calculate the weight W_G a gold mass of equal volume would have. According to our introduction to this chapter, we can then determine

$$w^{(G)} = \frac{W_C - W_S}{W_G - W_S} \cdot W_G, \qquad w^{(S)} = \frac{W_G - W_C}{W_G - W_S} \cdot W_S.$$

(**B.124**) The author takes $W_C = 500$, $W_S = 432$, and calculates with *his* ratio $5:4$ that $W_G = 540$. Thus he computes

$$w^{(G)} = \frac{W_C - W_S}{W_G - W_S} \cdot W_G = \frac{500 - 432}{540 - 432} \cdot 540 = \left(\frac{5}{9} + \frac{2}{3}\frac{1}{9}\right) \cdot 540 = 340,$$

$$w^{(S)} = \frac{W_G - W_C}{W_G - W_S} \cdot W_S = \frac{540 - 500}{540 - 432} \cdot 432 = \left(\frac{3}{9} + \frac{1}{3}\frac{1}{9}\right) \cdot 432 = 160.$$

Remark. Our author inferred W_G from W_S using the ratio $5:4$, which, he says, is the ratio between the (specific) weights of gold and silver. He tells us that he chose it not on the basis of experience but only for the sake of his example. This we may well believe. As noted in the Translation, this approximation is a poor one, with much better values already obtained in antiquity and in the Islamic world. But the purpose of such a choice may have been to facilitate the computations.

(**B.125**) The next example is about a craftsman supposed to produce vessels from a thousand ounces of gold, thus $W_C = 1000$. An equal volume of silver weighs $W_S = 864$, thus $W_G = 1080$. Since these are just twice the preceding values, so too, therefore, are the required quantities:

$$w^{(G)} = \frac{W_C - W_S}{W_G - W_S} \cdot W_G = \frac{136}{216} \cdot 1080 = \left(\frac{5}{9} + \frac{2}{3}\frac{1}{9}\right) \cdot 1080 = 680$$

$$w^{(S)} = \frac{W_G - W_C}{W_G - W_S} \cdot W_S = \frac{80}{216} \cdot 864 = \left(\frac{3}{9} + \frac{1}{3}\frac{1}{9}\right) \cdot 864 = 320.$$

Chapter B–VI:
Drapery

Summary

This chapter deals with cutting from a rectangular (or square) piece of cloth (*cortina*) a smaller one. In the first group of problems, whole and piece are entirely characterized by their dimensions and their weights (there is no question of price in the whole chapter); the treatment relies on the constancy of the ratio weight to size. (Since size depends on dimensions, we meet here for the first time a proportion with six terms.) In the second group, the cloth is of mixed material; then the constituents of the whole and the part will be found in the same proportions, and on this is based the treatment.

1. Cutting a piece.

If the whole is known by its length l_1, its width r_1 ($l_1 \geq r_1$) and its weight w_1, then we shall have for another piece of the same material, since the weight is proportional to the size,

$$\frac{l_1 \cdot r_1}{l_2 \cdot r_2} = \frac{w_1}{w_2}.$$

With this are solved B.126–132 and B.138 (the last perhaps misplaced). It is pointed out in B.127 that each dimension of the smaller (also rectangular) piece cannot exceed the corresponding one of the larger.

A. *One unknown*.

In the simplest cases, we are given all the magnitudes (weights and dimensions) but one, namely w_2 or r_2 (B.126–128). The two formulae are then

$$w_2 = \frac{l_2 \cdot r_2 \cdot w_1}{l_1 \cdot r_1}, \qquad r_2 = \frac{l_1 \cdot r_1 \cdot w_2}{l_2 \cdot w_1}.$$

B. *Two unknowns*.

In the next group, we are given all except two dimensions of which we know either the ratio, since the pieces are said to be similar (B.129, B.131–132), or the difference (B.130). We have already met problems of the latter kind before, but not of the first: this is because there are now six terms, not just four.

Suppose, in the first case, that we know of the whole the two dimensions and the weight (say l_1, r_1, w_1), and of the piece just the weight w_2; since this piece is said to be similar in shape to the former, we know its ratio length to width and we can calculate them individually, as shown in B.129, B.131–132. Indeed, since

$$\frac{l_1}{r_1} = \frac{l_2}{r_2} = \frac{l_2^2}{l_2 \cdot r_2} = \frac{l_2 \cdot r_2}{r_2^2} \quad \text{and} \quad l_2 \cdot r_2 = \frac{w_2}{w_1}(l_1 \cdot r_1),$$

we find

$$r_2^2 = \frac{r_1}{l_1} l_2 \cdot r_2 = \frac{r_1}{l_1} \cdot \frac{w_2}{w_1}(l_1 \cdot r_1), \qquad l_2^2 = \frac{l_1}{r_1} l_2 \cdot r_2 = \frac{l_1}{r_1} \cdot \frac{w_2}{w_1}(l_1 \cdot r_1),$$

so that finally

$$r_2 = \sqrt{\frac{r_1}{l_1} \cdot \frac{w_2}{w_1}(l_1 \cdot r_1)} = \sqrt{\frac{w_2}{w_1} \cdot r_1}, \qquad l_2 = \sqrt{\frac{l_1}{r_1} \cdot \frac{w_2}{w_1}(l_1 \cdot r_1)} = \sqrt{\frac{w_2}{w_1}} \cdot l_1,$$

the first (and less simple) forms of which are used in B.129.

Suppose now, in the second case, that we know of the whole the dimensions and the weight (say l_1, r_1, w_1), and of the piece just the weight w_2 and the difference between length and width, thus $l_2 - r_2$. Since

$$l_2 \cdot r_2 = \frac{w_2}{w_1}(l_1 \cdot r_1),$$

we also know the product $l_2 \cdot r_2$. Then, by the usual identity

$$\left(\frac{l_2 + r_2}{2}\right)^2 = \left(\frac{l_2 - r_2}{2}\right)^2 + l_2 \cdot r_2,$$

we can determine half the sum of l_2 and r_2, thus also l_2 and r_2 individually from

$$l_2, r_2 = \frac{l_2 + r_2}{2} \pm \frac{l_2 - r_2}{2}.$$

Whence the formula

$$r_2 = \sqrt{\left(\frac{l_2 - r_2}{2}\right)^2 + l_2 \cdot r_2} - \frac{l_2 - r_2}{2}$$

$$= \sqrt{\left(\frac{l_2 - r_2}{2}\right)^2 + \frac{w_2}{w_1}(l_1 \cdot r_1)} - \frac{l_2 - r_2}{2}$$

used in B.130 and (with a demonstration) B.138.

Remark. This corresponds to the (single) positive solution of the quadratic equation

$$r_2^2 + r_2(l_2 - r_2) = \frac{w_2}{w_1}(l_1 \cdot r_1).$$

2. Mixed material.

B.133–137 differ in that the weight of each, whole and piece cut, is the sum of the weights of its three constituents (silk, cotton, linen). In that case, we have, with $w_1 = \sum w_i^{(1)}$ and $w_2 = \sum w_i^{(2)}$,

$$\frac{w_1}{w_2} = \frac{\sum w_i^{(1)}}{\sum w_i^{(2)}} = \frac{l_1 \cdot r_1}{l_2 \cdot r_2} = \frac{w_i^{(1)}}{w_i^{(2)}}.$$

We are always given l_1, r_1 and the $w_i^{(1)}$ and, according to what else, must determine the remaining quantities, which may be the $w_i^{(2)}$ (B.133), the size $l_2 \cdot r_2$ of one piece (B.134), one of its dimensions (r_2 in B.135, l_2 in B.135′), or both if their ratio is known ($l_2 = r_2$ in B.136, $l_2 \neq r_2$ in B.137). The formulae used in B.133–135′ are all taken from the above relations; this is also true for the particular case of B.136, with

$$l_2 = r_2 = \sqrt{\frac{l_1 \cdot r_1 \cdot w_2}{\sum w_i^{(1)}}}.$$

---

1. Cutting a piece.

A. *One unknown*.

(B.126) MSS $\mathcal{A}, \mathcal{B}, \mathcal{C}$. From a cloth ten cubits long, eight wide, weighing fifty ounces, is cut a piece six cubits long, four wide; what is its weight?

Given thus $l_1 = 10$, $r_1 = 8$, $w_1 = 50$, $l_2 = 6$, $r_2 = 4$, required w_2.

We find (by the usual 'three ways')

$$w_2 = \frac{l_2 \cdot r_2 \cdot w_1}{l_1 \cdot r_1} = \frac{24 \cdot 50}{80} = 15, \qquad w_2 = \frac{l_2 \cdot r_2}{l_1 \cdot r_1} \cdot w_1 = \frac{w_1}{l_1 \cdot r_1} l_2 \cdot r_2.$$

The origin of the formula is then explained: for material of the same substance, the ratio size to weight will remain constant.

(B.127) MS $\mathcal{A}$. From a cloth ten cubits long, eight wide, weighing sixty ounces, is cut a piece fifteen cubits long, nine wide; what is its weight?

Given thus $l_1 = 10$, $r_1 = 8$, $w_1 = 60$, $l_2 = 15$, $r_2 = 9$, required w_2.

The answer to this problem can be calculated (its computation is like that of the previous one), but the result is inapplicable since the dimensions of the part are larger than those of the whole. (There will be, in the next chapter, further instances of computed results which are inapplicable in practice.)

(B.128) MSS $\mathcal{A}, \mathcal{B}$; $\mathcal{C}$ partly (the text of $\mathcal{C}$ breaks off before the end of the formulation). From a cloth ten cubits long, eight wide, weighing fifty ounces, is cut a piece weighing fifteen ounces and six cubits long; what is its width?

Given thus $l_1 = 10$, $r_1 = 8$, $w_1 = 50$, $l_2 = 6$, $w_2 = 15$, required $l_2 \cdot r_2$ and r_2. Same values as in B.126.

Since

$$l_2 \cdot r_2 = \frac{l_1 \cdot r_1}{w_1} \cdot w_2 = \frac{80}{50} \cdot 15 = 24, \quad \text{then} \quad r_2 = \frac{24}{6} = 4.$$

The origin of the formula is again explained by the proportion $l_1 \cdot r_1 : w_1 = l_2 \cdot r_2 : w_2$.

B. *Two unknowns*.

(B.129) MS $\mathcal{A}$. From a cloth weighing sixty ounces, is cut a piece five cubits long, four wide, weighing fifteen ounces, similar in shape to the whole; what are length and width of the original piece?

Given thus $w_1 = 60$, $l_2 = 5$, $r_2 = 4$, $w_2 = 15$, required l_1, r_1 with $l_1 : l_2 = r_1 : r_2$ (thus this time the dimensions of the whole are unknown). We find

$$r_1 = \sqrt{\frac{r_2}{l_2}\left(\frac{w_1}{w_2} l_2 \cdot r_2\right)} = \sqrt{\frac{4}{5}\left(\frac{60}{15} \cdot 20\right)} = \sqrt{\frac{4}{5} \cdot 80} = \sqrt{64} = 8$$

$$l_1 = \frac{l_2}{r_2} \cdot r_1, \quad \text{or} \quad l_1 = \sqrt{\frac{l_2}{r_2}\left(\frac{w_1}{w_2} l_2 \cdot r_2\right)} = 10.$$

Remark. There is no justification here, presumably because we have seen such problems earlier (B.17 & B.58). As noted in our summary (p. 1449), simpler formulae for r_1, l_1 would have been

$$r_1 = \sqrt{\frac{w_1}{w_2}} \cdot r_2, \quad l_1 = \sqrt{\frac{w_1}{w_2}} \cdot l_2,$$

giving directly $r_1 = 2 \cdot r_2$, $l_1 = 2 \cdot l_2$.

(B.130) MS $\mathcal{A}$. From a cloth sixty ounces in weight, of which length less width is two cubits, is cut a piece five cubits long and four wide weighing fifteen ounces; what are length and width of the original piece?

Given thus $l_1 - r_1 = 2$, $w_1 = 60$, $l_2 = 5$, $r_2 = 4$, $w_2 = 15$, required l_1 and r_1. Same values as before.
The text computes

$$r_1 = \sqrt{\left[\frac{1}{2}(l_1 - r_1)\right]^2 + \frac{w_1}{w_2}(l_2 \cdot r_2)} - \frac{1}{2}(l_1 - r_1)$$

$$= \sqrt{\left(\frac{1}{2} \cdot 2\right)^2 + \frac{60}{15} \cdot 20} - \frac{1}{2} \cdot 2 = 8, \quad l_1 = r_1 + 2 = 10.$$

Remark. There is no justification. We have indicated the origin of this formula in the summary.

(B.131) MSS $\mathcal{A}$, $\mathcal{B}$. From a cloth ten cubits long, eight wide, weighing fifty ounces, is cut a square piece weighing twenty-two and a half ounces; what are its dimensions?

Given thus $l_1 = 10$, $r_1 = 8$, $w_1 = 50$, $w_2 = 22 + \frac{1}{2}$, required $l_2 \cdot r_2$ and $l_2 = r_2$.
Finding the size of the part by means of

$$l_2 \cdot r_2 = \frac{w_2}{w_1}(l_1 \cdot r_1) = \frac{22 + \frac{1}{2}}{50} \cdot 80 = 36,$$

we have then
$$l_2 = r_2 = \sqrt{\frac{w_2}{w_1}(l_1 \cdot r_1)} = \sqrt{36} = 6.$$

If need be, the text adds, we shall find the approximate root as taught in Book A (A.275–283).

(**B.131′**)–(**B.131″**) Same with $w_2 = 20$ and, respectively, $l_2 = 2r_2$, $l_2 = 3r_2$.

If generally $l_2 = k \cdot r_2$, with k given, we calculate $l_2 \cdot r_2$ as above and multiply the value found by k or divide it by k; this gives l_2^2 and r_2^2, respectively. In the proposed examples,
$$l_2 \cdot r_2 = \frac{w_2}{w_1}(l_1 \cdot r_1) = \frac{2}{5} \cdot 80 = 32$$

gives
$$l_2^2 = k \cdot 32, \quad r_2^2 = \frac{1}{k} \cdot 32,$$

thus, if $k = 2$, $l_2 = 8$, $r_2 = 4$.

(**B.132**) MSS $\mathcal{A}$; $\mathcal{B}$ (omits b). From a cloth ten cubits long, eight wide, weighing sixty ounces, is cut a piece weighing fifteen ounces and similar in shape; what are its dimensions?

Given thus $l_1 = 10$, $r_1 = 8$, $w_1 = 60$, $w_2 = 15$, $l_2 : r_2 = l_1 : r_1$, required l_2, r_2. This is of the same type as B.129 (same values), but with dimensions of the smaller piece now required.

(***a***) The size of the piece being
$$l_2 \cdot r_2 = \frac{w_2}{w_1}(l_1 \cdot r_1) = 20,$$

we shall have, since $l_2^2 : r_2 = l_1 l_2 : r_1$ and $r_2^2 : l_2 = r_1 r_2 : l_1$,
$$l_2 = \sqrt{\frac{l_1(l_2 \cdot r_2)}{r_1}} = 5, \quad r_2 = \sqrt{\frac{r_1(l_2 \cdot r_2)}{l_1}} = 4.$$

Remark. The author makes here an analogy with similar problems encountered before: l_i is the number of bushels q_i, and r_i their price p_i (B.17), or l_i is the capital c_i and r_i the profit e_i (B.58). As usual, he is not analogizing about the nature of the quantities but merely drawing attention to the similarity of the formulae.°

(***b***) What follows in the extant text is odd. There is only a demonstration of the formulae
$$l_2 = \sqrt{\frac{l_2 \cdot r_2}{l_1 \cdot r_1}} \cdot l_1, \quad r_2 = \sqrt{\frac{l_2 \cdot r_2}{l_1 \cdot r_1}} \cdot r_1.$$

° See our General Introduction (Part I), p. xvii.

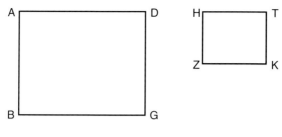

Fig. 63

We have attempted to supply what is missing, whereby the formulae would be

$$l_2 = \sqrt{\frac{w_2}{w_1}} \cdot l_1, \qquad r_2 = \sqrt{\frac{w_2}{w_1}} \cdot r_1$$

which is well in keeping with what precedes.

The demonstration is as follows. Let (Fig. 63) $l_1 \cdot r_1 = ABGD$ ($l_1 = BG$, $r_1 = AB$), $l_2 \cdot r_2 = HZKT$ ($l_2 = ZK$, $r_2 = HZ$). We know that

$$\frac{l_2 \cdot r_2}{l_1 \cdot r_1} = \frac{w_2}{w_1} = \frac{1}{4};$$

thus we shall have (since $l_1 : r_1 = l_2 : r_2$)

$$\frac{HZKT}{ABGD} = \frac{ZK \cdot HZ}{BG \cdot AB} = \frac{ZK}{BG} \cdot \frac{HZ}{AB} = \left(\frac{ZK}{BG}\right)^2 = \left(\frac{HZ}{AB}\right)^2 = \frac{1}{4},$$

$$\text{hence} \quad ZK = \frac{1}{2} \cdot BG = 5, \quad HZ = \frac{1}{2} \cdot AB = 4.$$

---

2. Mixed material.

This last group of problems (B.133–137) deals with mixed material. (B.138 seems, as said, to be misplaced, for it rather belongs to the previous group.) If the cloth is characterized by l_1, r_1 and the constituents $w_i^{(1)}$, and the piece by l_2, r_2 and the constituents $w_i^{(2)}$, then, as seen in our summary of this chapter,

$$\frac{l_1 \cdot r_1}{w_1} = \frac{l_2 \cdot r_2}{w_2} \quad \text{or} \quad \frac{l_1 \cdot r_1}{\sum w_i^{(1)}} = \frac{l_2 \cdot r_2}{\sum w_i^{(2)}}, \quad \text{and} \quad \frac{l_1 \cdot r_1}{w_i^{(1)}} = \frac{l_2 \cdot r_2}{w_i^{(2)}}.$$

(B.133) MSS $\mathcal{A}$, $\mathcal{B}$. From a cloth ten cubits long, eight wide, containing ten ounces of silk, fourteen of cotton, twenty of linen, is cut a piece six cubits long, four wide; how much does it contain of each material and what is its weight?

Given thus $l_1 = 10$, $r_1 = 8$, $w_1^{(1)} = 10$, $w_2^{(1)} = 14$, $w_3^{(1)} = 20$, $l_2 = 6$, $r_2 = 4$, required $w_i^{(2)}$, w_2.

$$w_i^{(2)} = \frac{l_2 \cdot r_2}{l_1 \cdot r_1} \cdot w_i^{(1)} = \frac{24}{80} \cdot w_i^{(1)},$$

$$w_2 = \sum w_i^{(2)} = \frac{l_2 \cdot r_2}{l_1 \cdot r_1} \cdot \sum w_i^{(1)}, \quad \text{or} \quad w_2 = \frac{l_2 \cdot r_2}{l_1 \cdot r_1} \cdot w_1.$$

The formulae are then explained, as well as their origin.

(B.134) MSS $\mathcal{A}$, $\mathcal{B}$. From a cloth ten cubits long and wide, containing thirty ounces of silk, forty of linen, fifty of cotton, is cut a piece weighing fourteen ounces; how much does it contain of each material and what is its size?

Given thus $l_1 = 10$, $r_1 = 10$, $w_1^{(1)} = 30$, $w_2^{(1)} = 40$, $w_3^{(1)} = 50$, $w_2 = 14$, required $w_i^{(2)}$, $l_2 \cdot r_2$.

$$l_2 \cdot r_2 = \frac{w_2}{w_1}(l_1 \cdot r_1) = \frac{14}{120} \cdot 100 = \frac{1400}{120} = 11 + \frac{2}{3}, \quad w_i^{(2)} = \frac{14}{120} \cdot w_i^{(1)}.$$

(B.135) MSS $\mathcal{A}$, $\mathcal{B}$. From a cloth ten cubits long, eight wide, containing thirty ounces of silk, forty of cotton, fifty of linen, is cut a piece five cubits long and weighing thirty ounces; what is its width and how much of each material does it contain?

Given thus $l_1 = 10$, $r_1 = 8$, $w_1^{(1)} = 30$, $w_2^{(1)} = 40$, $w_3^{(1)} = 50$, $w_2 = 30$, $l_2 = 5$, required r_2 and $w_i^{(2)}$.

(a) *Formulae.*

$$r_2 = \frac{(l_1 \cdot r_1) w_2}{l_2 \cdot \sum w_i^{(1)}} = \frac{80 \cdot 30}{5 \cdot 120} = \frac{2400}{600} = 4$$

and, as in B.134,

$$w_i^{(2)} = \frac{30}{120} \cdot w_i^{(1)}.$$

(b) The width r_2 is also computed as

$$r_2 = \frac{1}{l_2}\left(\frac{w_2}{w_1}(l_1 \cdot r_1)\right) = \frac{1}{5}\left(\frac{1}{4} \cdot 80\right) = \frac{1}{5} \cdot 20 = 4.$$

(B.135′) Same, but with the width $r_2 = 4$ given and the length l_2 required.

$$l_2 = \frac{(l_1 \cdot r_1) w_2}{r_2 \cdot w_1} = 5, \quad \text{or} \quad l_2 = \frac{1}{r_2}\left(\frac{w_2}{w_1}(l_1 \cdot r_1)\right).$$

(B.136) MSS $\mathcal{A}$, $\mathcal{B}$. From a cloth ten cubits long and wide, containing thirty ounces of silk, forty of cotton, fifty of linen, is cut a square piece weighing thirty ounces; what are its dimensions?

Given thus $l_1 = 10 = r_1$, $w_1^{(1)} = 30$, $w_2^{(1)} = 40$, $w_3^{(1)} = 50$, $w_2 = 30$, required $l_2 = r_2$.

$$l_2 = r_2 = \sqrt{\frac{w_2(l_1 \cdot r_1)}{w_1}} = \sqrt{\frac{30 \cdot 100}{120}} = \sqrt{25} = 5.$$

Remark. B.136–137 (like B.138) practically belong to the first group since the given weights of the constituents of the whole serve no other purpose than calculating its weight.

(B.137) MSS $\mathcal{A}$, $\mathcal{B}$. From a cloth ten cubits long, eight wide, containing ten ounces of silk, twenty of cotton, thirty of linen, is cut a piece similar in shape weighing fifteen ounces; what are its dimensions?

Given thus $l_1 = 10$, $r_1 = 8$, $w_1^{(1)} = 10$, $w_2^{(1)} = 20$, $w_3^{(1)} = 30$, $w_2 = 15$, $l_1 : r_1 = l_2 : r_2$ required l_2, r_2.

(*a*) *Formulae.*

$$l_2 \cdot r_2 = \frac{w_2 (l_1 \cdot r_1)}{w_1} = \frac{15 \cdot 80}{60} = \frac{1200}{60} = 20,$$

from which we infer (see B.132)

$$l_2 = \sqrt{\frac{l_1 (l_2 \cdot r_2)}{r_1}} = \sqrt{\frac{10 \cdot 20}{8}} = \sqrt{\frac{200}{8}} = \sqrt{25} = 5,$$

$$r_2 = \sqrt{\frac{r_1 (l_2 \cdot r_2)}{l_1}} = \sqrt{\frac{8 \cdot 20}{10}} = \sqrt{\frac{160}{10}} = \sqrt{16} = 4.$$

(*b*) Also computed by each of the following equivalent formulae (the last two seen in B.132):

$$l_2 = \frac{l_1}{\sqrt{\frac{l_1 \cdot r_1}{l_2 \cdot r_2}}}, \quad \frac{l_1}{\sqrt{\frac{w_1}{w_2}}}, \quad \sqrt{\frac{l_2 \cdot r_2}{l_1 \cdot r_1}} \cdot l_1, \quad \sqrt{\frac{w_2}{w_1}} \cdot l_1$$

$$r_2 = \frac{r_1}{\sqrt{\frac{l_1 \cdot r_1}{l_2 \cdot r_2}}}, \quad \frac{r_1}{\sqrt{\frac{w_1}{w_2}}}, \quad \sqrt{\frac{l_2 \cdot r_2}{l_1 \cdot r_1}} \cdot r_1, \quad \sqrt{\frac{w_2}{w_1}} \cdot r_1.$$

(B.138) MS $\mathcal{A}$. From a cloth ten cubits long, eight wide, weighing sixty ounces is cut a piece two cubits longer than wide and weighing eighteen ounces; what are its length and width?

Given thus $l_1 = 10$, $r_1 = 8$, $w_1 = 60$, $w_2 = 18$, $l_2 - r_2 = 2$, required l_2, r_2. Similar to B.130, where $l_1 - r_1$ was given. Therefore, as was the case in both B.132 and B.129, the rôles of whole and piece are merely interchanged. The equation obtained, of the type $x(x + b) = c$, thus $x^2 + bx = c$, is solved geometrically.

A   G D   B

Fig. 64

First (B.126)

$$\frac{l_2 \cdot r_2}{l_1 \cdot r_1} = \frac{w_2}{w_1} \quad \text{whence} \quad l_2 \cdot r_2 = \frac{18 \cdot 80}{60} = 24.$$

Therefore $l_2 \cdot r_2 = r_2\,(r_2 + 2) = 24$. Let then (Fig. 64) $l_2 = AB$, $r_2 = AG$. We are then to solve
$$\begin{cases} AB - AG = GB = 2 \\ AB \cdot AG = 24. \end{cases}$$
Let D be the mid-point of GB. By *Elements* II.6,
$$AB \cdot AG + GD^2 = AD^2, \quad AD^2 = 24 + 1 = 25.$$

Therefore $AD = 5$, so that $r_2 = AG = AD - GD = 4$, $l_2 = AB = AD + DB = 6$.

Remark. This is equivalent to computing

$$r_2 = \sqrt{\left(\frac{l_2 - r_2}{2}\right)^2 + l_2 \cdot r_2} - \frac{l_2 - r_2}{2}$$

$$\left(l_2 = r_2 + 2 = \sqrt{\left(\frac{l_2 - r_2}{2}\right)^2 + l_2 \cdot r_2} + \frac{l_2 - r_2}{2}\right).$$

Chapter B–VII:
Linen cloths
Summary

The problems on linen cloths (*lintei*, hereafter 'linens') differ from those on drapes, and not only on account of the material. First, and evidently, since we consider pure linen, no mixed material is involved. Second, instead of just one piece being taken from the whole, as in B–VI, here the whole is mostly to be cut into identical pieces. Therefore the question of feasibility, which arose only incidentally before (B.127), is here essential. If, as in the first part, both the whole piece and the smaller ones are rectangular, the dimensions of the former will have to be integral multiples of those of the latter if no material is to be left over (B.139). But with circular pieces, which is the subject of the second part, this condition cannot be satisfied.

1. Rectangular pieces.

The first set of problems is about cutting, from a rectangular piece of linen, l_1 in length, r_1 in width and costing p_1, smaller rectangular (or square) pieces, q in number, with individual price p_2 and dimensions l_2, r_2. The basic relation is then

$$q = \frac{l_1 \cdot r_1}{l_2 \cdot r_2} = \frac{p_1}{p_2}.$$

We may thus be asked for the number q of pieces cut (B.139), the fraction cut from the whole (B.140), one of the dimensions of one piece (B.141, B.144–145), or both when we know their ratio (B.146). We may also be asked for the price of a piece (B.142–143). The main formulae used are accordingly

$$q = \frac{l_1 \cdot r_1}{l_2 \cdot r_2}, \quad r_2 = \frac{l_1 \cdot r_1}{q \cdot l_2}, \quad r_2 = l_2 = \sqrt{\frac{l_1 \cdot r_1}{q}}, \quad p_2 = \frac{l_2 \cdot r_2}{l_1 \cdot r_1} \cdot p_1.$$

B.145′, with two unknowns, leads to a quadratic equation.

2. Circular pieces.

In the second part, the main piece and the smaller ones are generally both circular, with diameters d_1, d_2 and circumferences c_1, c_2 (B.147–153). The basic relation then becomes

$$q = \frac{d_1^2}{d_2^2} = \frac{c_1^2}{c_2^2} = \frac{p_1}{p_2}.$$

Suppose first that both original and smaller pieces are circular and we know three of four proportional quantities; we may then calculate the remaining one. Thus we can obtain one of the two prices when we know the circumferences (B.147) or the diameters (B.151), or conversely one circumference from the prices (B.148). With the other cases it is not so straightforward: the number of circular pieces the original piece contains can be calculated but this is of no practical use (B.149); if one, either whole or piece, is rectangular, the problem is not only of little practical use but can only be solved approximately since the ratio of a circular area to a rectangular area, thus of an approximate value to an exactly calculable value, is not exactly calculable (B.150, 153′); for this same reason determination of the price is also approximative (B.152).

B.153–154, which close this chapter, are mathematically analogous but with a change of subject: we are now paving floors with slabs.

Remark. Any cloth merchant would agree about the inconvenience of cutting circular pieces from another circular one. He might, however, take some persuading to accept the fact that no exact price can be calculated due to the impossibility of squaring the circle.

---

1. Rectangular pieces.

A. *One unknown*.

(B.139) MSS *A*, *B*. From a linen fifteen cubits long and eight wide are cut pieces four cubits long and three wide; how many are there?
Given thus $l_1 = 15$, $r_1 = 8$, $l_2 = 4$, $r_2 = 3$, required q.

$$q = \frac{l_1 \cdot r_1}{l_2 \cdot r_2} = \frac{120}{12} = 10.$$

Also computed as

$$q = \frac{l_1}{l_2} \cdot \frac{r_1}{r_2} = \left(3 + \frac{3}{4}\right)\left(2 + \frac{2}{3}\right) \quad \text{or} \quad q = \frac{l_1}{r_2} \cdot \frac{r_1}{l_2} = 5 \cdot 2.$$

Remark. Here, and for once, this latter 'other way' makes sense, for it shows how cutting is actually possible (both quotients must be integral).

(B.140) MS *A*. From a linen ten cubits long and eight wide is cut a piece six cubits long and five wide; what fraction of the whole is cut off?
Given thus $l_1 = 10$, $r_1 = 8$, $l_2 = 6$, $r_2 = 5$, required the fraction cut.

$$\frac{l_2 \cdot r_2}{l_1 \cdot r_1} = \frac{30}{80} = \frac{3}{8}.$$

(B.141) MSS *A*, *B*. From a linen fifteen cubits long and eight wide are cut ten pieces each four cubits long; how wide are they?

Given thus $l_1 = 15$, $r_1 = 8$, $q = 10$, $l_2 = 4$, required r_2. Same values as in B.139.
$$r_2 = \frac{1}{l_2}\left(\frac{l_1 \cdot r_1}{q}\right) = \frac{1}{4} \cdot 12 = 3.$$

Also computed as
$$r_2 = \frac{l_1 \cdot r_1}{l_2 \cdot q}.$$

For the equivalence of the second formula with the first, the reader is referred to Premiss P_4.

(B.141′) From a linen fifteen cubits long and eight wide are cut ten pieces three cubits wide; how long are they?

Same as before, but with width known and length required. Thus the rôles of r_2 and l_2 are just exchanged.

(B.141″) From a linen fifteen cubits long and eight wide are cut ten square pieces; what is their side?

If $r_2 = l_2$, then (corresponds to B.131 for the drapes, q replacing $\frac{w_1}{w_2}$)

$$r_2 = l_2 = \sqrt{\frac{l_1 \cdot r_1}{q}}.$$

Remark. This is not feasible.

(B.141‴) If $l_2 = kr_2$, do as in the corresponding problems on drapes (B.131′–131″).

(B.141ⁱᵛ) If $l_1 : r_1 = l_2 : r_2$, do as in the corresponding problem B.129. As a matter of fact, an example will follow (B.146).

(B.142) MSS $\mathcal{A}$, $\mathcal{B}$. A linen fifteen cubits long and eight wide costs twenty nummi; what is the price of a piece eight cubits long and four wide?

Given thus $l_1 = 15$, $r_1 = 8$, $p_1 = 20$, $l_2 = 8$, $r_2 = 4$, required p_2.

$$p_2 = \frac{(l_2 \cdot r_2)\,p_1}{l_1 \cdot r_1} = \frac{32 \cdot 20}{120} = 5 + \frac{1}{3}.$$

Since this is the first occurrence of the price, we are referred to the problems at the beginning of Book B, and the proportion (constancy of the ratio size to price) is explained.

(B.143) MSS $\mathcal{A}$, $\mathcal{B}$. A linen six cubits long and four wide costs ten nummi; what is the price of a piece fifteen cubits long and five wide?

Given thus $l_1 = 6$, $r_1 = 4$, $p_1 = 10$, $l_2 = 15$, $r_2 = 5$, required p_2. Similar to B.142, but this time the price of the larger piece is required.

$$p_2 = \frac{(l_2 \cdot r_2)\,p_1}{l_1 \cdot r_1} = \frac{75 \cdot 10}{24} = 31 + \frac{1}{4}.$$

(B.143′) So far the pieces have been rectangular. They now take other shapes. Thus, if the pieces are circular, find their areas by means of the following formulae (occurring here for the first time): if d is the diameter, c the circumference, s the area, then

$$s = \frac{1}{2}d \cdot \frac{1}{2}c \quad \text{or (approximately)} \quad s = d^2\left(1 - \frac{1}{7} - \frac{1}{2} \cdot \frac{1}{7}\right)$$

(as known from Archimedes' *Dimensio circuli*).

Remark. This formula is correctly stated in B.341. But here and in B.152, the manuscripts have 'sixths' instead of 'sevenths'.

If the pieces are of another shape, the reader, we are told, will calculate their size using a book on *taccir*; then the price will be computed as above by means of the relation size : price = size : price.

Remark. The Arabic *taksīr* is used to denote the practical textbooks on geometry which taught (often civil servants) how to calculate areas by means of handy formulae, exact or approximate. As a rule, in view of the intended readership, such works present numerous examples but avoid demonstrations. A typical example of such a book from Moslem Spain has recently been edited by A. Djebbar.*

(B.144) MSS 𝒜, ℬ. A linen twenty cubits long and eight wide costs twenty nummi; what is the length of a piece four cubits wide costing twelve nummi?

Given thus $l_1 = 20$, $r_1 = 8$, $p_1 = 20$, $r_2 = 4$, $p_2 = 12$, required l_2.

$$l_2 = \frac{(l_1 \cdot r_1) p_2}{r_2 \cdot p_1} = \frac{160 \cdot 12}{4 \cdot 20} = 24.$$

(B.145) MSS 𝒜; ℬ (omits B.145′). From a linen twenty cubits long and eight wide are cut six pieces each eight cubits long; how wide are they?

Given thus $l_1 = 20$, $r_1 = 8$, $q = 6$, $l_2 = 8$, required r_2; thus the length of the pieces equals the width of the linen. The genuineness of this problem is questionable for several reasons. First, it is of the same type as B.141 (except that there the required width is integral). Second, the place of this problem is inappropriate since here we are in the section on prices. Finally, and this is surely an interpolation, the formula, though seen in B.141, is again explained and Premiss P_4 is demonstrated.‖

---

* In this *Risāla fi'l-taksīr*, its author, Ibn ʿAbdūn, considers successively rectilinear figures (quadrilaterals, triangles, trapezia), solid figures (regular prisms and pyramids, also truncated), then the circle and its parts (Djebbar, p. 256).

‖ A similar situation, for Premiss P_5, occurs in B.277b (but this time only MS ℬ is concerned).

Formula.
$$r_2 = \frac{l_1 \cdot r_1}{q \cdot l_2} = \frac{160}{48} = 3 + \frac{1}{3}.$$

The formula is said to originate with
$$\left(\frac{l_1 \cdot r_1}{q}\right) \frac{1}{l_2}$$

by applying P_4, namely
$$\frac{1}{D}\left(\frac{A}{B}\right) = \frac{A}{D \cdot B}.$$

To prove this, let (Fig. 65) $\frac{A}{B} = G$ while $\frac{G}{D} = H$ and $D \cdot B = T$; then
$$\frac{A}{T} = \frac{A}{D \cdot B} = \frac{1}{D}\left(\frac{A}{B}\right) = H.$$

Indeed, since $G = D \cdot H$ and $T = D \cdot B$, then (*Elements* VII.17) $G : T = H : B$, and therefore
$$H = \frac{B \cdot G}{T} = \frac{A}{T}.$$

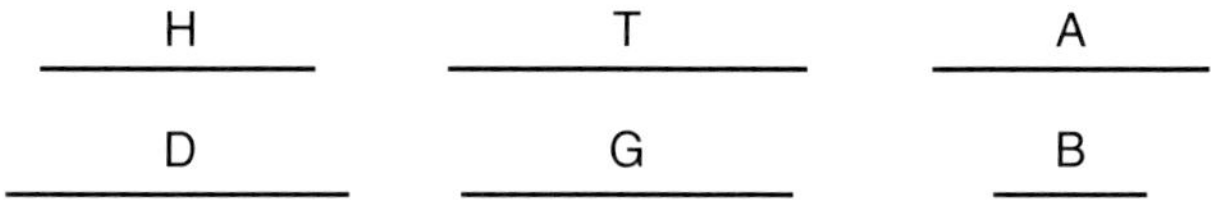

H	T	A
D	G	B

Fig. 65

B. *Two unknowns*.

(B.145′) From a linen (presumably fifteen cubits long and eight wide) are cut ten pieces the width of which is less than its length by three cubits; how long and wide are they?

Given thus l_1, r_1, then $q = 10$, and $l_2 - r_2 = 3$, required r_2. The enunciation makes it clear that this problem is related to the group B.141. Consider
$$r_2(r_2 + 3) = \frac{l_1 \cdot r_1}{q}$$

and proceed for the solution of this quadratic equation as 'explained in drapery' (B.130 has an algebraic solution & B.138 a geometrical one).

(B.146) MSS $\mathcal{A}$, $\mathcal{B}$. From a linen ten cubits long and eight wide are cut four pieces similar to it; how long and wide are they?

Given thus $l_1 = 10$, $r_1 = 8$, $q = 4$, $l_1 : r_1 = l_2 : r_2$, required l_2, r_2. See B.141[iv].

(a) *Formulae*:
$$l_2 = \sqrt{\frac{l_1\left(\frac{l_1 \cdot r_1}{q}\right)}{r_1}} = \sqrt{\frac{10 \cdot 20}{8}} = 5,$$

$$r_2 = \sqrt{\frac{r_1\left(\frac{l_1 \cdot r_1}{q}\right)}{l_1}} = \sqrt{\frac{8 \cdot 20}{10}} = 4.$$

Remark. These formulae are not explained. But how to obtain them has been seen in B.132; we consider that $l_1 : r_1 = l_2 : r_2 = (l_2 \cdot r_2) : r_2^2 = l_2^2 : (l_2 \cdot r_2)$, with $l_2 \cdot r_2 = \frac{1}{q}(l_1 \cdot r_1)$. All that is, however, a needless complication since the above formulae reduce to

$$l_2 = \frac{1}{\sqrt{q}} l_1 = \frac{1}{2} l_1, \qquad r_2 = \frac{1}{\sqrt{q}} r_1 = \frac{1}{2} r_1.$$

(b) Also computed as

$$l_2 = \sqrt{\frac{l_1^2\left(\frac{l_1 \cdot r_1}{q}\right)}{l_1 \cdot r_1}}, \qquad r_2 = \sqrt{\frac{r_1^2\left(\frac{l_1 \cdot r_1}{q}\right)}{l_1 \cdot r_1}}.$$

---

2. Circular pieces.

B.147–153 deal with circular pieces. The product of length and width is thus replaced in the proportion by the square of either the diameter d_i or the circumference c_i (*Elements* XII.2, see B.149). We have then

$$q = \frac{d_1^2}{d_2^2} = \frac{c_1^2}{c_2^2} = \frac{p_1}{p_2}.$$

(B.147) MSS $\mathcal{A}$, $\mathcal{B}$. A circular linen ten cubits in circumference costs sixty nummi; what is the price of a circular piece with a circumference of five cubits?

Given thus $c_1 = 10$, $p_1 = 60$, $c_2 = 5$, required p_2.

$$p_2 = \frac{c_2^2 \cdot p_1}{c_1^2} = \frac{25 \cdot 60}{100} = \frac{1500}{100} = 15.$$

Also computed as

$$p_2 = \frac{c_2^2}{c_1^2} \cdot p_1 = \frac{1}{4} \cdot p_1.$$

(B.148) MSS $\mathcal{A}$, $\mathcal{B}$. A circular linen ten cubits in circumference costs sixty nummi; what is the circumference of a circular piece priced at fifteen nummi?

Given thus $c_1 = 10$, $p_1 = 60$, $p_2 = 15$, required c_2. Same values as before.

$$c_2 = \sqrt{\frac{c_1^2 \cdot p_2}{p_1}} = \sqrt{\frac{100 \cdot 15}{60}} = \sqrt{\frac{1500}{60}} = \sqrt{25} = 5, \text{ or}$$

$$c_2 = \sqrt{\frac{p_2}{p_1}} \cdot c_1 = \frac{1}{2} \cdot c_1.$$

The author's remark, that 'these rules are generally applicable', means that these two problems are solvable both mathematically and practically, that is, that they make sense both in theory and application. This will not be the case in the next two problems where circular or rectangular pieces are to be taken from the whole of a circular linen.

(B.149) MS $\mathcal{A}$. From a circular linen ten cubits in diameter are cut pieces two cubits in diameter; how many are there?

Given thus $d_1 = 10$, $d_2 = 2$, required q.

Formula.
$$q = \frac{d_1^2}{d_2^2} = 25.$$

Proof. Areas of circles are to one another as the squares of their diameters (*Elements* XII.2). As the author notes, such problems are of theoretical interest only. See B.153.

(B.150) MS $\mathcal{A}$. From a circular linen ten cubits in diameter are cut pieces three cubits long and two wide; how many are there?

Given thus $d_1 = 10$, $l_2 = 3$, $r_2 = 2$, required q.

This problem is not solvable exactly in numbers, we are told, since the circle cannot be squared, the reason being that its circumference, like any arc segment, cannot be rectified.* Even if, the author goes on, Archimedes has calculated $3 + \frac{1}{7}$ as the value of the ratio circumference to diameter, this is still only an approximation.†

Remark. The author does not point out that a few pieces can actually be cut from the linen (but see B.153).

B.153′, probably misplaced, is the inverse case: cutting circular pieces from a rectangular linen.

(B.151) MS $\mathcal{A}$. A circular linen eight cubits in diameter costs twenty nummi; what is the price of a circular linen with a diameter of two cubits?

Given thus $d_1 = 8$, $p_1 = 20$, $d_2 = 2$, required p_2. Since $p_1 : p_2 = s_1^2 : s_2^2 = d_1^2 : d_2^2$ (with s_i areas), we shall have

$$p_2 = \frac{d_2^2 \cdot p_1}{d_1^2} = \frac{4 \cdot 20}{64} = 1 + \frac{1}{4}.$$

(B.152) MS $\mathcal{A}$. A circular linen ten cubits in diameter costs a hundred nummi; what is the price of a linen three cubits long and two wide?

---

* Since, as found by Archimedes, $s = \frac{1}{2} c \cdot \frac{1}{2} d$ and c cannot be computed exactly, neither can s.

† In his *Dimensio circuli*, Archimedes calculates that (our) π lies between $3 + \frac{10}{71}$ and $3 + \frac{10}{70}$.

Given thus $d_1 = 10$, $p_1 = 100$, $l_2 = 3$, $r_2 = 2$, required p_2. Since the ratio of the prices equals the ratio of the areas, one of which is circular and the other rectangular, we shall not obtain an exact result (B.150).

If c_1 is the circumference of the piece with diameter d_1 and area s_1, compute s_1 by one of the three relations:

$$\left(\frac{d_1}{2}\right)^2\left(3+\frac{1}{7}\right), \quad \frac{d_1}{2}\cdot\frac{1}{2}\left[d_1\left(3+\frac{1}{7}\right)\right] = \frac{d_1}{2}\cdot\frac{c_1}{2}, \quad d_1^2\left[1-\left(\frac{1}{7}+\frac{1}{2}\cdot\frac{1}{7}\right)\right].$$

This leads to the value $s_1 \cong 78 + \frac{4}{7}$. Then

$$p_2 = \frac{(l_2 \cdot r_2)p_1}{s_1} \cong \frac{6 \cdot 100}{78+\frac{4}{7}}.$$

Remark. Here again (see B.143′), the extant text has in the last of the three formulae 'sixths' instead of 'sevenths', which corresponds to the very bad approximation $\pi = 3$. That this error cannot have been in the original text is suggested by the correct occurrence elsewhere (B.341) and by the remark that all three ways indicated above will lead to the same (above) value for s_1.

An analogous problem: Paving floors.

(B.153) MS $\mathcal{A}$. A circular room is six cubits in diameter; how many circular slabs each three cubits in diameter does it hold?

Given thus for a room $d_1 = 6$ and for slabs $d_2 = 3$, required q. In theory (B.149),

$$q = \frac{d_1^2}{d_2^2} = 4,$$

but in practice only two slabs can be placed, thus the room cannot be fully paved, as is made clear by the geometric illustration (Fig. 66): the floor $ABGD$ can contain only two slabs like $HZKT$, say $AMQL$ and $NGCQ$, while the space left empty is equal in area to the two other.

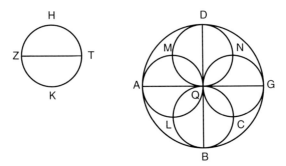

Fig. 66

(B.153′) From a linen thirty cubits long and ten wide are cut circular pieces with a diameter of three cubits; how many are there?

Here $l_1 = 30$, $r_1 = 10$, $d_1 = 3$. It would be more appropriate for this problem, which involves a rectangular linen and circular pieces, to follow B.150. The author does not attempt to solve it and says right away that it has no application.

Remark. As in the case of B.150, he does not mention the possibility that some pieces may be cut (actually, thirty).

(B.154) MSS $\mathcal{A}$, $\mathcal{B}$. A hall is twenty cubits long and eight wide; how many marble slabs each two cubits long and one and a half wide does it hold?

Given thus for a hall $l_1 = 20$, $r_1 = 8$, and for the slabs $l_2 = 2$, $r_2 = 1 + \frac{1}{2}$, required q. As in the first problem of this chapter (B.139), we shall compute

$$q = \frac{l_1 \cdot r_1}{l_2 \cdot r_2} = \frac{20 \cdot 8}{2\left(1 + \frac{1}{2}\right)} = \frac{160}{3} = 53 + \frac{1}{3}.$$

Remark. The author says nothing about the fractional result, nor about the impossibility of actually laying all the slabs without breaking one at the end of either each horizontal or each vertical row ($1 + \frac{1}{2}$ divides neither 8 nor 20).

Chapter B-VIII:
Grinding

Summary

A miller receives a certain quantity of bushels to grind, for which he will be paid in kind by receiving a fraction of one bushel for each bushel ground.* Two possibilities are considered, according to whether he is paid from the corn brought to be ground or from a separate quantity. This is the subject of the first two groups of problems (B.155–158 & B.164, B.159–163 & B.166–168). In the third group, a further condition occurs in each of these two cases since the corn whilst being ground increases in volume so that, even after the miller has been paid, a larger quantity of bushels may be taken away than was brought (B.165, B.170–172). Other problems about corn complete this chapter, either because three mills work together (B.169) or because further operations are involved (B.173–174).

Remark. The bushels are *caficii*, except in the last two problems (*sextarii*).

1. Payment from each bushel.

If q is the quantity of corn brought, q' the quantity taken away and $\frac{k}{l}$ the given fraction of q left as remuneration p, we shall have

$$q' = q - p = q - \frac{k}{l} \cdot q = \frac{l-k}{l} \cdot q,$$

$$p = \frac{k}{l} \cdot q = \frac{k}{l} \frac{l}{l-k} \cdot q' = \frac{k}{l-k} \cdot q'.$$

Thus, four magnitudes occur ($\frac{k}{l}$, q, q', p). The first, the fraction $\frac{k}{l}$, is always given. From the above relations, we see that we need to know just one of the three quantities q, q', p in order to determine the others. There are then four fundamental problems, the solution formulae of which are deduced from these relations.

(*i*) Given q, find p (B.155); solved by

$$p = \frac{k}{l} \cdot q \qquad \left(q' = q - p, \text{ thus } q' = \frac{l-k}{l} \cdot q\right).$$

(*ii*) Given q', find p (B.156); solved by

$$p = \frac{k}{l-k} \cdot q' \qquad (q = q' + p, \text{ thus } q \text{ as in } iii).$$

---

* Payment in kind was then customary in Seville. See Lévy-Provençal, *Séville musulmane*, pp. 118–119.

(*iii*) Given q', find q (B.157, B.164); solved by

$$q = \frac{l}{l-k} \cdot q' \qquad (p = q - q', \text{ thus } p \text{ as in } ii).$$

(*iv*) Given p, find q (B.158); solved by

$$q = \frac{l}{k} \cdot p \qquad \left(q' = q - p, \text{ thus } q' = \frac{l-k}{k} \cdot p\right).$$

2. Payment from another bushel.

A. *One unknown*.

As before, $\frac{k}{l}$ is given as well as one of q, q', p. Since now the miller is to receive the fraction $\frac{k}{l}$ of a bushel separately, a bushel brought will no longer be reduced to $1 - \frac{k}{l}$ of itself, but for one bushel taken away the quantity $1 + \frac{k}{l}$ must have been brought. Thus if q' is the quantity taken away and q the total quantity brought (including the payment p) we shall have

$$q = q' + p = q' + \frac{k}{l} q' = \frac{k+l}{l} \cdot q'$$

$$p = \frac{k}{l} \cdot q' = \frac{k}{l} \frac{l}{k+l} \cdot q = \frac{k}{k+l} \cdot q.$$

The four fundamental problems are then:

(*i′*) Given q', find p (B.159); solved by

$$p = \frac{k}{l} \cdot q' \qquad (q = q' + p, \text{ thus } q \text{ as in } iii').$$

(*ii′*) Given q, find p (B.160); solved by

$$p = \frac{k}{k+l} \cdot q \qquad \left(q' = q - p, \text{ thus } q' = \frac{l}{k+l} \cdot q\right).$$

(*iii′*) Given q', find q (B.161, B.163); solved by

$$q = \frac{k+l}{l} \cdot q' \qquad (p = q - q', \text{ thus } p \text{ as in } i').$$

(*iv′*) Given p, find q (B.162); solved by

$$q = \frac{k+l}{k} \cdot p \qquad \left(q' = q - p, \text{ thus } q' = \frac{l}{k} \cdot p\right).$$

B. *Two unknowns*.

The last three problems of this second section (B.166–168, misplaced) are of the kind 'involving unknowns' (*de ignoto*) seen before: instead of one

of q, q', p given, we know the sum, difference or product of two of them. Since, from above,
$$p = \frac{k}{k+l} \cdot q, \qquad q' = \frac{l}{k+l} \cdot q,$$
we infer that:

- if we are given $q \cdot p$, then
$$q \cdot p = \frac{k}{k+l} q^2, \quad \text{whence (B.166)} \quad q = \sqrt{\frac{k+l}{k}(q \cdot p)};$$

- if we are given $q' - p$, then
$$q' - p = \frac{l-k}{k+l} q, \quad \text{whence (B.167)} \quad q = \frac{k+l}{l-k}(q'-p);$$

- if we are given $q' \cdot p$, then
$$q' \cdot p = \frac{l}{k+l}\frac{k}{k+l} q^2, \quad \text{whence (B.168)} \quad q = \sqrt{\frac{(k+l)^2}{k \cdot l}(q' \cdot p)}.$$

Remark. Here we find in addition algebraic solutions (which, though, just repeat the treatment seen before).

3. Increment in volume.

B.165 (misplaced[†]) & 170–172 are concerned with the increment in volume, the multiplicative factor being $\frac{1}{r}$. This just means that the above q', thus the quantity ground which remains after payment is to be multiplied by $\left(1 + \frac{1}{r}\right)$. So we have for the two above modes of payment, respectively,

$$q' = \frac{r+1}{r} \cdot \frac{l-k}{l} q, \quad \text{so} \quad q = \frac{r}{r+1} \cdot \frac{l}{l-k} q',$$

$$q' = \frac{r+1}{r} \cdot \frac{l}{k+l} q, \quad \text{so} \quad q = \frac{r}{r+1} \cdot \frac{k+l}{l} q',$$

by means of which B.170, 165, 172 and 171 may be solved, respectively.

4. Other problems on corn.

There are several mills at work or various transformations of the corn. In B.169 (misplaced) we are asked how long it will take three mills of different productivity to grind, working simultaneously, a given amount of corn (analogous to the cistern problems B.331–333). In B.173–174, what is required is the cost of buying, grinding and baking a quantity q of bushels when we know how much each operation costs for a single bushel.

[†] The order of this and the subsequent problems should be: 166–168, 165, 170–172, 169, 173–174.

1. Payment from each bushel.

(B.155) MS $\mathcal{A}$ only. For grinding each caficius a fifth of it is due; how much is due if a hundred caficii are brought?

First problem of the first case (banal problem). Given $\frac{k}{l} = \frac{1}{5}$, $q = 100$, required p and q'. If $q_0 = 5$, then $p_0 = 1$ and $q'_0 = 4$. Since $p : q = p_0 : q_0$, we shall have
$$p = \frac{1}{5} \cdot 100 = 20, \quad \text{and} \quad q' = 80.$$

Remark. Here follows a problem (our B.155', in MS $\mathcal{B}$ only) treating the case $\frac{k}{l} = \frac{1}{6}$ which is clearly an interpolation.

(B.156) MSS $\mathcal{A}$, $\mathcal{B}$. For grinding each caficius two ninths of it is due; how much is due if a hundred caficii are taken away?

Second problem of the first case. Given $\frac{k}{l} = \frac{2}{9}$, $q' = 100$, required p (and q). If $q_0 = 9$, so $p_0 = 2$ and $q'_0 = 7$. Since $p : q' = p_0 : q'_0$, we shall have
$$p = \frac{2 \cdot 100}{7} = 28 + \frac{4}{7} \quad \left(\text{and} \quad q = 128 + \frac{4}{7}\right).$$

Or else, compute (usual 'other ways', also in the next problems) $\frac{2}{7} \cdot 100$, $\frac{100}{7} \cdot 2$.

(B.157) MSS $\mathcal{A}$, $\mathcal{B}$. For grinding each caficius two sevenths of it is due; how much is brought if eighty caficii are taken away?

Third problem of the first case. Given $\frac{k}{l} = \frac{2}{7}$, $q' = 80$, required q. If $q_0 = 7$, then $p_0 = 2$ and $q'_0 = 5$. Since $q : q' = q_0 : q'_0$, we shall have
$$q = \frac{7 \cdot 80}{5} = 112.$$

(B.158) MSS $\mathcal{A}$, $\mathcal{B}$. For grinding each caficius two ninths of it is due; how much is taken away if ten caficii are due?

Fourth problem of the first case. Given $\frac{k}{l} = \frac{2}{9}$, $p = 10$, required q and q'. If $q_0 = 9$ is brought, then $p_0 = 2$ is paid and $q'_0 = 7$ is taken away. Since $q : p = q_0 : p_0$ and $q' : p = q'_0 : p_0$, we shall have
$$q = \frac{9 \cdot 10}{2} = 45, \qquad q' = \frac{7 \cdot 10}{2} = 35.$$

---

2. Payment from another bushel.

A. *One unknown*.

(B.159) MS $\mathcal{A}$. For grinding each caficius a separate quantity equal to a fifth of it is due; how much is due if a hundred caficii are taken away?

First problem of the second case (banal). Given $\frac{k}{l} = \frac{1}{5}$, $q' = 100$, required p. If $q'_0 = 5$, so $p_0 = 1$. Since $p : q' = p_0 : q'_0$, we shall have

$$p = \frac{1 \cdot 100}{5} = 20.$$

(B.160) MSS $\mathcal{A}$; $\mathcal{B}$ (formulation only). For grinding each caficius a separate quantity equal to two sevenths of it is due; how much is due if a hundred caficii are brought?

Second problem of the second case. Given $\frac{k}{l} = \frac{2}{7}$, $q = 100$, required p (and q'). If $q'_0 = 7$, so $p_0 = 2$. Since $p : q' = p_0 : q'_0$ and, by composition, $p : q = p_0 : q_0$, with $q_0 = q'_0 + p_0 = 9$, we shall have

$$p = \frac{2 \cdot 100}{9} = 22 + \frac{2}{9} \quad \left(\text{and } q' = 77 + \frac{7}{9}\right).$$

(B.161) MSS $\mathcal{A}$, $\mathcal{B}$. For grinding each caficius a separate quantity equal to two ninths of it is due; how much was brought if a hundred is taken away?

Third problem of the second case. Given $\frac{k}{l} = \frac{2}{9}$, $q' = 100$, required q. If $q'_0 = 9$, then $p_0 = 2$ and $q_0 = 11$. Since $q : q' = q_0 : q'_0$, we shall have

$$q = \frac{11 \cdot 100}{9} = 122 + \frac{2}{9}.$$

(B.162) MSS $\mathcal{A}$, $\mathcal{B}$. For grinding each caficius a separate quantity equal to two ninths of it is due; how much was brought if ten caficii are due?

Fourth problem of the second case. Given $\frac{k}{l} = \frac{2}{9}$, $p = 10$, required q (and q'). If $q'_0 = 9$, then $p_0 = 2$ and $q_0 = 11$. Since $q : p = q_0 : p_0$, we shall have

$$q = \frac{11 \cdot 10}{2} = 55, \qquad q' = q - p = 45.$$

• **Further examples.**

(B.163) MS $\mathcal{A}$. For grinding each caficius a separate quantity equal to a fifth of it is due; how much was brought if a hundred is taken away?

Same problem as B.161, except for the value of the given fraction. Given thus $\frac{k}{l} = \frac{1}{5}$, $q' = 100$, required q. If $q'_0 = 5$ and $p_0 = 1$, so $q_0 = 6$. Then $q : q' = q_0 : q'_0$ and therefore

$$q = \frac{6 \cdot 100}{5} = 120, \quad \text{or} \quad q = \left(1 + \frac{1}{5}\right) 100.$$

(B.164) MS $\mathcal{A}$. For grinding each caficius a seventh and half a seventh of it is due; how much was brought if a hundred caficii are taken away?

Given thus $\frac{k}{l} = \frac{1}{7} + \frac{1}{2}\frac{1}{7}$, $q' = 100$, required q. Similar to B.157, but with a compound fraction. If $q_0 = 7$, then $p_0 = 1 + \frac{1}{2}$ and $q'_0 = 5 + \frac{1}{2}$, whence $q : q' = q_0 : q'_0$ and

$$q = \frac{7 \cdot 100}{5 + \frac{1}{2}} = 127 + \frac{3}{11}, \quad \text{or} \quad q = \frac{100}{1 - \left(\frac{1}{7} + \frac{1}{2}\frac{1}{7}\right)}.$$

(**B.165**) MS $\mathcal{A}$. For grinding each caficius a fifth of it is due, and in the process the quantity of corn increases by a tenth of itself; how much was brought if a hundred is taken away?

This problem is misplaced and should precede B.170. Given thus $\frac{k}{l} = \frac{1}{5}$, $\frac{1}{r} = \frac{1}{10}$, $q' = 100$, required q. First case (remuneration from each bushel), but involving an increment in volume.

(**a**) If $q_0 = 5$, then $p_0 = 1$, $q_0 - p_0 = 4$ which increases to $q'_0 = 4 + \frac{2}{5}$. Since $q : q' = q_0 : q'_0$, we shall have

$$q = \frac{5 \cdot 100}{4 + \frac{2}{5}} = \frac{25 \cdot 100}{22} = 113 + \frac{7}{11}.$$

(**b**) Also computed first considering the transformation for a single caficius: reduced by the remuneration, it becomes $\frac{4}{5}$; this quantity, when taken away, has increased to $\frac{4}{5} + \frac{2}{5}\frac{1}{5}$. Then $q : q' = 1 : \left(\frac{4}{5} + \frac{2}{5}\frac{1}{5}\right)$.

B. *Two unknowns*.

(**B.166**) MSS $\mathcal{A}$, $\mathcal{B}$. For grinding each caficius a separate quantity equal to a fifth of it is due, and the quantity brought multiplied by the quantity due produces one hundred and fifty; what was the quantity brought?

Given thus $\frac{k}{l} = \frac{1}{5}$, $q \cdot p = 150$, required q. Second case (remuneration separately).

(**a**) Since (as seen in B.163) if $p = \frac{1}{5}q'$, so $p = \frac{1}{6}q$, then

$$q \cdot p = \frac{1}{6}q^2 = 150, \quad q = \sqrt{900} = 30 \quad (\text{thus } q' = 25).$$

(**b**) Algebraic solution. Putting $q = x$, we are led in like manner to

$$q \cdot p = \frac{1}{6}x^2 = 150.$$

(**c**) Verification. Since $q = 30$, the remuneration is $p = \frac{1}{6}q = 5$, so indeed $p \cdot q = 150$.

(**B.167**) MSS $\mathcal{A}$, $\mathcal{B}$. For grinding each caficius a separate quantity equal to a fifth of it is due, and the quantity taken away diminished by the quantity due leaves twenty; what was the quantity brought?

Given thus $\frac{k}{l} = \frac{1}{5}$, $q' - p = 20$, required q. Second case.

(**a**) Since (B.163, B.166) $p = \frac{1}{6}q$, so $q' = \frac{5}{6}q$, whence

$$q' - p = \frac{4}{6}q = 20, \quad \text{thus} \quad q = 30.$$

(**b**) Algebraic solution. Same treatment, with x taking the place of our q.
(**c**) Verification. Since $q = 30$, so $p = 5$, then $q' = 25$ and indeed $q' - p = 20$.
Remark. We would expect here the problem where $q + p$ is given.

(**B.168**) MSS $\mathcal{A}$, $\mathcal{B}$. For grinding each caficius a separate quantity equal to a fifth of it is due, and the quantity taken away multiplied by the quantity due produces one hundred and twenty-five; what was the quantity taken away?

Given thus $\frac{k}{7} = \frac{1}{5}$, $q' \cdot p = 125$, required q'. Second case. This problem contains errors and omissions and its genuineness is questionable.

(**a**) As seen above, since $p = \frac{1}{6}q$, $q' = \frac{5}{6}q$, so

$$q' \cdot p = \frac{5}{6^2}q^2 = 125, \quad \text{so} \quad \left(\frac{5}{6}q\right)^2 = q'^2 = 625, \quad q' = 25 \ (q = 30).$$

Remark. In the text, the 'unknown number of caficii' designates at the beginning our q, at the end our q'.

(**b**) Same, by algebra (putting $x = q$ at the beginning, finding $x = q'$ at the end).

* * *

As in some other sections, we find, usually towards the end, problems which are mathematically completely different but share with the others the general theme, in this case grinding.

(**B.169**) MSS $\mathcal{A}$, $\mathcal{B}$. Three mills grind in the same time ('day and night') respectively twenty, thirty, forty caficii; how much will be put in each one for grinding ten caficii if all three are to work simultaneously?

Given thus the quantities ground simultaneously $(Q_i =)$ 20, 30 and 40 caficii, and the quantity $q = 10$ to be ground likewise, required the parts q_i of q each is to grind.

(**a**) As the author mentions at the outset, this is solved like a problem of partnership (B.104; see also B.121). Indeed, considering that the daily productions Q_i are the individual capitals, thus their sum Q the whole capital, and that $q = 10$ is the corresponding profit, we have a similar solving formula:

$$e_i = \frac{c_i \cdot e}{c} \quad \text{becomes} \quad q_i = \frac{Q_i \cdot q}{Q} \quad \text{with} \quad Q = \sum Q_i = 90.$$

The quantities in caficii are then, respectively,

$$q_1 = \frac{20 \cdot 10}{90} = 2 + \frac{2}{9}, \quad q_2 = \frac{30 \cdot 10}{90} = 3 + \frac{1}{3}, \quad q_3 = \frac{40 \cdot 10}{90} = 4 + \frac{4}{9}.$$

(*b*) By algebra. Since the quantities ground simultaneously are in the ratio $20 : 30 : 40$, if x is the part attributed to the third mill, $\frac{1}{2}x$ will be that of the first and $\frac{3}{4}x$ that of the second; hence

$$\left(2 + \frac{1}{4}\right)x = 10, \quad \text{and} \quad x = 4 + \frac{4}{9}.$$

3. Increment in volume.[†]

A. *Payment from each bushel.*

(B.170) MSS *A*, *B*. For grinding each caficius a sixth of it is due, and in the process the quantity of corn increases by a third of itself; how much is taken away if a hundred was brought?

Given thus $\frac{k}{l} = \frac{1}{6}$, $\frac{1}{r} = \frac{1}{3}$, $q = 100$, required q'. This problem with increment in volume was probably meant to come immediately after B.165, misplaced, and which like it belongs to the first case (remuneration from the same bushel). If $q_0 = 3 \cdot 6 = 18$ is brought, then $p_0 = 3$ and $q_0 - p_0 = 15$ is left, which increases to $q'_0 = 20$. Since $q : q' = q_0 : q'_0$, we shall have

$$q' = \frac{20 \cdot 100}{18} = 111 + \frac{1}{9}.$$

(B.170′) Same, but with $q' = 100$ and q required. Thus here $q : 100 = 18 : 20$, giving $q = 90$. Same case as B.165, except that here $q'_0 > q_0$.

B. *Payment from another bushel.*

Consider the fractional quantities q, q', p. Since in the two related problems B.171–172 we have $\frac{k}{l} = \frac{1}{6}$ and $r = \frac{1}{3}$, then, if q is brought, $p = \frac{1}{7}q$ must be due and the remainder of the quantity brought, namely $\frac{6}{7}q = q - p$, will increase whilst being ground to $q' = \frac{8}{7}q = \frac{4}{3}(q-p)$. This is the basis of the relations found below.

(B.171) MSS *A*, *B*. For grinding each caficius a separate quantity equal to a sixth of it is due, and in the process the quantity of corn increases by a third of itself; how much was brought if thirty is taken away?

Given thus $\frac{k}{l} = \frac{1}{6}$, $\frac{1}{r} = \frac{1}{3}$, $q' = 30$, required q.

(*a*) If $q_0 = 7$ is brought, $p_0 = 1$ and $q_0 - p_0 = 6$ will be taken away but, whilst being ground, will increase to $q'_0 = 8$; then $q : q' = q_0 : q'_0$. Justification: When $\frac{1}{7}q$ is paid, $\frac{6}{7}q$ remains, which becomes $\frac{8}{7}q$, thus $\frac{8}{7}q = 30$. Therefore

$$q = \frac{7 \cdot 30}{8} = 26 + \frac{1}{4}.$$

[†] See also B.165 above.

(b) Since $q' = \left[\left(1 - \frac{1}{7}\right)q\right]\left(1 + \frac{1}{3}\right) = \frac{8}{7}q$, then $q = \left(1 - \frac{1}{8}\right)q'$, and therefore

$$q = \left(1 - \frac{1}{8}\right)30 = 26 + \frac{1}{4}.$$

(c) First, $q' = \frac{8}{7}q = \left(1 + \frac{1}{3}\right)\frac{6}{7}q = \left(1 + \frac{1}{3}\right)(q - p)$, then $q - p = \left(1 - \frac{1}{4}\right)q'$; second, $q = \frac{7}{6}(q - p) = \left(1 + \frac{1}{6}\right)(q - p)$. Therefore

$$q = \left(1 + \frac{1}{6}\right)(q - p) = \left(1 + \frac{1}{6}\right)\left(1 - \frac{1}{4}\right)q',$$

that is,

$$q = \left(1 + \frac{1}{6}\right)\left(1 - \frac{1}{4}\right)30 = \left(1 + \frac{1}{6}\right)\left(22 + \frac{1}{2}\right) = 26 + \frac{1}{4}.$$

(d) Using algebra (like in part a, with x instead of q). Since q is first reduced to $\frac{6}{7}q$ and this then increases to $\frac{8}{7}q = q'$, so $\frac{8}{7}x = 30$ and therefore $x = 26 + \frac{1}{4}$.

(**B.172**) MSS $\mathcal{A}$, $\mathcal{B}$. For grinding each caficius a separate quantity equal to a sixth of it is due, and in the process the quantity of corn increases by a third of itself; how much is taken away if thirty was brought?

Given thus $\frac{k}{l} = \frac{1}{6}$, $\frac{1}{r} = \frac{1}{3}$, $q = 30$, required q'. Closely related to B.171, where q' was given and q required.

(a) Solved and justified by analogy to B.171a. If $q_0 = 7$ is brought, $q_0 - p_0 = 6$ will be ground and increase to $q'_0 = 8$, with $q_0 : q = q'_0 : q'$. Or else, q is brought, $\frac{6}{7}q$ is ground, $\frac{8}{7}q = q'$ is taken away. Then

$$q' = \frac{8 \cdot 30}{7} = 34 + \frac{2}{7}.$$

(b) By analogy to B.171c,

$$q' = \left(1 + \frac{1}{3}\right)(q - p) = \left(1 + \frac{1}{3}\right)\left(1 - \frac{1}{7}\right)30 = \left(1 + \frac{1}{3}\right)\left(25 + \frac{5}{7}\right).$$

---

4. Other problems on corn.*

In the last two problems what is asked for is the cost of buying, grinding and baking a quantity q of sextarii when we know the corresponding cost for one sextarius (if more, we are told at the end to reduce to the unit prices).

(**B.173**) MSS $\mathcal{A}$, $\mathcal{B}$. A sextarius is bought for three nummi, ground for a half and baked for one; what is the cost of ten sextarii bought, ground and baked?

---

* See also B.169 above.

Given for one sextarius, respectively, $p_1 = 3$, $p_2 = \frac{1}{2}$, $p_3 = 1$, required the total price p for $q = 10$.

$$p = q(p_1 + p_2 + p_3) = 10\left(4 + \frac{1}{2}\right) = 45.$$

(B.174) MSS $\mathcal{A}$, $\mathcal{B}$. A sextarius is bought for five nummi, ground for three and baked for one and a fourth; how many sextarii bought, ground and baked will there be for a hundred nummi?

Given $p_1 = 5$, $p_2 = 3$, $p_3 = 1 + \frac{1}{4}$, required q for which was paid $p = 100$.

$$q = \frac{p}{p_1 + p_2 + p_3} = \frac{100}{9 + \frac{1}{4}} = 10 + \frac{30}{37}.$$

Chapter B–IX:
Boiling must

Summary

This kind of problem has to do with the Islamic prohibition on alcoholic beverages. We have a clear account of its origin in 'Abd al-Qāhir ibn Ṭāhir al-Baghdādī's (d. 1037) *Takmila*.° According to the expert in law and holy tradition Abū Ḥanīfa (d. 767), founder of the school of Hanafites, a beverage obtained by boiling must (Arabic *'aṣīr*, here *mustum*) will be considered as legal (*ḥalāl*) when reduced by the process to a third of the original quantity. For that purpose, as al-Baghdādī goes on to explain, a flat-bottomed cylindrical vat is used, divided into three equal parts by two engraved circles and marked off for finer graduation. Now it may happen, during the boiling process, that part of the must overflows; we are then to pursue the reduction process until a new remainder is obtained —which will be neither the original third, since we now have less liquid than before, nor one third of the remaining must, since part of the reduction has already been effected. The subsequent problems, on the reduction of must, deal with this change in situation.

Suppose then the original quantity to be q, which is to be reduced to $\frac{1}{k}q$ ($k = 3$ according to Abū Ḥanīfa's stipulation), and v overflows just after d has evaporated. Without overflow, $q - d$ should continue boiling until the remainder is $\frac{1}{k}q$; after the overflow, the remaining quantity $q - d - v$ must continue boiling so as to attain the same concentration as the $q - d$ would have. Thus the new remainder r_2 will have the desired concentration if it bears to the quantity remaining just after the overflow the same ratio as r_1 to the quantity just before the overflow. The basic relation is therefore

$$\frac{r_2}{q - d - v} = \frac{r_1}{q - d}.$$

If $r_1 = \frac{1}{3}q$ is imposed, only the four quantities q, d, v, r_2 may vary. Thus, according to al-Baghdādī, four cases arise: we know q, d, v but not r_2; or q, d, r_2 but not v; or d, v, r_2 but not q; or, finally, q, v, r_2 but not d. Further cases are, however, to be considered if r_1 has not the fixed ratio 1 : 3 to q or if several overflows occur during the process; and this is also considered in the *Liber mahameleth*.

1. **One quantity unknown.**

A. *One overflow*.

The first two problems are on determining the remainders; thus, in B.175 (given q, r_1, d, v required r_2) and in B. 176 (given q, d, v, r_2 required

---

° VII.9, pp. 283–284 in Saidan's edition.

r_1), the formulae are, respectively,

$$r_2 = \frac{r_1 (q - d - v)}{q - d}, \qquad r_1 = \frac{r_2 (q - d)}{q - d - v}.$$

For the next three problems, considering that

$$q - d - v = \frac{r_2}{r_1} \cdot (q - d), \quad \text{thus} \quad q - d = \frac{r_1}{r_1 - r_2} \cdot v,$$

we infer the following three relations

$$d = q - \frac{r_1}{r_1 - r_2} \cdot v, \quad v = \frac{r_1 - r_2}{r_1} (q - d), \quad d = v = \frac{r_1 - r_2}{2r_1 - r_2} \cdot q,$$

with which B.177 (given q, r_1, v, r_2), B.178 (given q, r_1, d, r_2), B.179 (given q, r_1, r_2) may be solved respectively. The treatment in the text, however, does not use these direct formulae.

B. *Several overflows*.

In B.180–181, two overflows take place, and we are given q, r_1, d_1, v_1 and d_2, v_2, and we are to find r_2 and r_3. By analogy to what we have seen in B.175, we know that

$$r_2 = \frac{r_1 (q - d_1 - v_1)}{q - d_1}, \qquad r_3 = \frac{r_2 \Big[(q - d_1 - v_1) - d_2 - v_2\Big]}{(q - d_1 - v_1) - d_2},$$

and therefore

$$r_3 = \frac{r_1 (q - d_1 - v_1) \Big[(q - d_1 - v_1) - d_2 - v_2\Big]}{(q - d_1) \Big[(q - d_1 - v_1) - d_2\Big]}.$$

This is extended to three overflows in B.182, where the same quantities are given, with in addition d_3 and v_3. We have then

$$r_4 = \frac{r_3 \Big[\big([q - d_1 - v_1] - d_2 - v_2\big) - d_3 - v_3\Big]}{\big([q - d_1 - v_1] - d_2 - v_2\big) - d_3}.$$

C. *Particular cases of a single overflow*.

When fixing the quantity evaporated d, we must take $q - d \geq r_1$; if not, the problem, as in B.183, is not solvable because the process has gone too far and the desired concentration can no longer be obtained. Another particular case is considered in B.184, where the overflow occurs before any evaporation takes place (thus $d = 0$); then we shall just reformulate the problem with a new initial quantity $q - v$.

2. Several quantities unknown.

A. *Two quantities unknown*.

Cross-multiplication in the fundamental relation gives $qr_2 - dr_2 = qr_1 - dr_1 - vr_1$, thus

$$qr_1 + dr_2 = qr_2 + r_1(d+v).$$

If we suppose only q to be unknown and the other quantities given, the problem is banal. But suppose, as in B.185, q to be unknown as well as, therefore, the first remainder $r_1 = \frac{1}{k}q$, with k given ($k > 1$). We have then

$$\frac{1}{k}q^2 + dr_2 = q\left(r_2 + \frac{1}{k}d + \frac{1}{k}v\right), \quad \text{or} \quad q^2 + k \cdot dr_2 = q(kr_2 + d + v),$$

that is, a quadratic equation of the type $x^2 + c = bx$, thus with two positive solutions (if real). This is solved by our author in his usual way: since the equation can be written as

$$q(k \cdot r_2 + d + v - q) = k \cdot dr_2$$

while

$$q + (k \cdot r_2 + d + v - q) = k \cdot r_2 + d + v,$$

we are to find two quantities of which we know the sum and the product.

B. *Three quantities unknown*.

In B.186, we suppose that not only $r_1 = \frac{1}{k}q$ but also $d = \frac{1}{s}q$, with k and s (both > 1) given. Then, since $qr_1 - dr_1 = qr_2 + vr_1 - dr_2$,

$$q^2\left(\frac{1}{k} - \frac{1}{k}\frac{1}{s}\right) = q\left(r_2 + \frac{1}{k}v - \frac{1}{s}r_2\right), \quad \text{or}$$

$$q^2(s-1) = q(ks \cdot r_2 + s \cdot v - k \cdot r_2),$$

so that we end up with a linear equation.

C. *Particular cases*.

In the last group of problems B.187–189, where q is again unknown, v is a given fraction of $q - d$, say $v = \frac{m}{n}(q-d)$. Then, by the basic formula,

$$\frac{r_1}{r_2} = \frac{q-d}{q-d-v} = \frac{q-d}{(q-d)\left(1 - \frac{m}{n}\right)} = \frac{n}{n-m},$$

that is, the ratio of r_1 to r_2 is *de facto* imposed. Three cases are then considered in our text.

- If (B.187) $r_1 = \frac{1}{k}q$, $r_2 = \frac{1}{l}q$, with k and l given, we have

$$\frac{l}{k} = \frac{n}{n-m}$$

that is, the given values must satisfy this relation. If so, any q is a solution, otherwise the problem is impossible.

- If now (B.188) r_2 is a given numerical value (and $d = \frac{1}{s} q$, which plays no rôle), then
$$\frac{1}{k} q = \frac{n}{n-m} \cdot r_2.$$

- In B.189 the conditions are as in B.188 except that the numerical value of d is given. The solution will just remain the same; for here again $q - d$ disappears at the outset.

* * *

We have already mentioned in the General Introduction (I, pp. xviii–xix) that the *Liber mahameleth* left some definite traces in Italy. Indeed, some of the problems in this chapter, namely B.175, B.176, B.177, B.181, B.185, B.189 appear in a mid 15th-century treatise written by a student of Domenico d'Agostino Vaiaio (early 15th century).[||] Only for B.185 is the source explicitly indicated: as mentioned in the General Introduction, Domenico d'Agostino is said to draw on the work of Grazia de' Castellani (second half of the 14th century), who commented on the *Liber mahameleth*. The other problems clearly also originate in the *Liber mahameleth* because questions 'on boiling must' do not occur in Leonardo Fibonacci's works, the main source of inspiration for late mediaeval mathematicians. Such problems are also found in two contemporary Italian manuscripts: one has B.175, B.176, B.181, with the same indication as to their origin; the other has B.175, B.178, B.181, all three with all quantities multiplied by 10 (and, in B.178, the values $q = 100$, $r_1 = 33 + \frac{1}{3}$, $d = 20$, $r_2 = 25$).[**] Of these problems, the first (B.175) aroused the greatest interest: it is found in the 15th-century *Tractato d'abbacho* attributed to Pier Maria Calandri; then, doubtlessly transmitted from Italy, in the *Triparty* written in France in 1484 by Nicolas Chuquet; finally we find it, with all quantities multiplied by 10, in the *Trattato d'abacho* by Benedetto da Firenze, also from the fifteenth century.[†]

Since this preparation of must was an Islamic tradition unknown in the Christian world, the context of such problems came to be misrepresented. That is why Calandri (like Benedetto) thinks that the liquid is wine (*uno vuol quocere 10 barili di vino*) and that it is spilt through carelessness (*volendo spiccare la chaldaia ne versò 2 barili*). Grazia de' Castellani must

---

[||] See MS Firenze Palat. 577, fol. 157^v–158^v & 422^{r-v}.

[**] MS Vat. Ottob. lat. 3307, fol. 138^r–139^r (see General Introduction, pp. xviii–xix); MS Siena L.IV.21, fol. 120^{r-v}.

[†] See, respectively, Arrighi's edition of Calandri's text, pp. 116–117; the *Appendice au Triparty* by Marre, p. 454 (No 147), fol. 204^r of the manuscript; Arrighi, *Scritti scelti*, p. 352. Chuquet has a further problem from the *Liber mahameleth* (B.218).

have been more faithful to his source since in his 15th-century disciple's six problems we find *vino* instead of *mosto* only a few times.

---

1. One quantity unknown.

A. *One overflow*.

(B.175) MSS $\mathcal{A}$, $\mathcal{B}$. A quantity of ten measures of must is to be reduced by boiling to its third; but, after two measures have evaporated, two overflow. To what quantity does the rest have to be reduced?

Given thus $q = 10$, $r_1 = \frac{1}{3}q = 3 + \frac{1}{3}$, $d = 2$, $v = 2$, required r_2.

(*a*) *Formula*.

$$r_2 = \frac{(q - d - v) r_1}{q - d} = \frac{6 \cdot (3 + \frac{1}{3})}{8} = 2 + \frac{1}{2}.$$

The explanations set out the fundamental relation

$$\frac{q - d}{r_1} = \frac{q - d - v}{r_2}, \quad \text{or} \quad \frac{q - d}{q - d - v} = \frac{r_1}{r_2}.$$

For, we are told in substance, $q - d - v$ must continue boiling until reduced to r_2 as $q - d$ would have continued boiling until reduced to r_1.

Remark. The Italian sources have the same values and follow this first way (as said, Benedetto da Firenze takes values ten times those above).

(*b*) The subsequent treatments are just variations on the fundamental proportion. In the first, we calculate

$$r_2 = \frac{1}{3}\left[\frac{10}{8} \cdot 6\right] = \frac{1}{3}\left[\left(1 + \frac{1}{4}\right)6\right] = \frac{1}{3}\left(7 + \frac{1}{2}\right) = 2 + \frac{1}{2};$$

that is, we have calculated

$$r_2 = \frac{1}{3}\left[\frac{q}{q - d}(q - d - v)\right].$$

The justification is an immediate inference from the fundamental proportion:

$$\frac{q - d}{q - d - v} = \frac{r_1}{r_2} = \frac{3r_1}{3r_2} = \frac{q}{3r_2} \quad \text{gives} \quad 3r_2 = \frac{q}{q - d}(q - d - v).$$

In the text, this is deduced from $q : (q - d) = 3r_2 : (q - d - v)$ as well, but expressed as

$$\frac{d + (q - d)}{q - d} = \frac{[3r_2 - (q - d - v)] + (q - d - v)}{q - d - v}.$$

This form is alluded to in a later proof (B.185a).

(**c**) Next, since, from the data,

$$q - d - v = \left(1 - \frac{1}{4}\right)(q - d) \quad \text{while} \quad \frac{r_1}{r_2} = \frac{q - d}{q - d - v}, \quad \text{so}$$

$$r_2 = \left(1 - \frac{1}{4}\right) r_1 = r_1 - \frac{1}{4} r_1.$$

(**d**) Since

$$r_2 = \frac{(q - d - v) r_1}{q - d},$$

we may compute r_2 by the two 'other ways'

$$r_2 = \frac{r_1}{q - d}(q - d - v) = \left(\frac{2}{6} + \frac{1}{2} \cdot \frac{1}{6}\right) 6, \quad r_2 = \frac{q - d - v}{q - d} \cdot r_1.$$

The first appears here, the second was alluded to at the end of part *a*.

(**e**) Considering that $q = 10$ is reduced to $q - d = 8$, a fifth of q has evaporated. At which point $v = 2$ overflows, thus a fourth of $q - d$. Taking into account the quantity evaporated, we may regard $v = 2$ as originating from an initial quantity $2 + \frac{1}{2}$. We may then consider that the $2 + \frac{1}{2}$ were lost at the beginning and therefore that $7 + \frac{1}{2}$ had to be reduced to a third of itself, which is r_2. See B.184.

(**B.176**) MSS *A*, *B*. A quantity of ten measures of must is to be reduced by boiling to an unknown remainder; but, after two measures have evaporated, two overflow, and the rest is reduced to two and a half measures. What was the unknown remainder?

Given thus $q = 10$, $d = 2$, $v = 2$, $r_2 = 2 + \frac{1}{2}$, required r_1.

$$r_1 = \frac{(q - d) r_2}{q - d - v} = \frac{20}{6} = 3 + \frac{1}{3}.$$

Remark. If Abū Ḥanīfa's stipulations were strictly adhered to, this problem would be banal ($q = 3r_1$). But, as we shall see in the next problem, other reductions are also considered in the *Liber mahameleth*.

(**B.177**) MS *A*. A quantity of ten measures of must is to be reduced by boiling to its fourth; but, after an unknown quantity has evaporated, three measures overflow, and the rest is reduced to one and four and a half eighths measures. What is the quantity evaporated?

Given thus $q = 10$, $r_1 = \frac{1}{4} q = 2 + \frac{1}{2}$, $v = 3$, $r_2 = 1 + \frac{4}{8} + \frac{1}{2}\frac{1}{8}$ $\left(= 1 + \frac{9}{16}\right)$, required d.
Since

$$\frac{q - d}{r_1} = \frac{q - d - v}{r_2}, \quad \text{then} \quad q - d = \frac{r_1}{r_2}(q - d - v),$$

that is,
$$10 - d = \left(1 + \frac{3}{5}\right)(10 - d - 3),$$

which is solved using a figure. Putting (Fig. 67) $q = 10 = AB$, $d = GB$, $v = 3 = DG$, we have $10 - d = AG = \left(1 + \frac{3}{5}\right)AD$. Since $AG = AD + DG$, $DG = \frac{3}{5}AD$, so that, since $DG = 3$, we have $AD = 5$, whence $AG = 8$ and $GB = AB - AG = 2 = d$.

Remark. The geometric reasoning thus corresponds to the reduction
$$10 - d = \left(1 + \frac{3}{5}\right)(10 - d) - 3\left(1 + \frac{3}{5}\right), \quad \frac{3}{5}(10 - d) = 3\left(1 + \frac{3}{5}\right),$$

whence $10 - d$ and d. The Italian source solves it otherwise; since $r_1 : r_2 = (q - d) : (q - d - v)$, then

$$\frac{r_1}{r_1 - r_2} = \frac{q - d}{q - d - (q - d - v)} = \frac{q - d}{v}, \quad \text{so} \quad q - d = \frac{v \cdot r_1}{r_1 - r_2}.$$

(B.178) MS $\mathcal{A}$. A quantity of ten measures of must is to be reduced by boiling to its fourth; but, after two measures have evaporated, an unknown quantity overflows, and the rest is reduced to one and a half and half an eighth measures. What is the quantity overflowed?

Given now $q = 10$, $r_1 = \frac{1}{4}q = 2 + \frac{1}{2}$, $d = 2$, $r_2 = 1 + \frac{9}{16}$, required v.

$$\frac{q - d}{r_1} = \frac{q - d - v}{r_2}, \quad \text{thus} \quad \frac{8}{2 + \frac{1}{2}} = \frac{8 - v}{1 + \frac{9}{16}}, \quad 8 - v = 5, \quad v = 3.$$

(B.179) MS $\mathcal{A}$. A quantity of ten measures of must is to be reduced by boiling to its third; but, after an unknown quantity has evaporated, as much overflows, and the rest is reduced to two and a half measures. What is the quantity evaporated?

Given thus $q = 10$, $r_1 = \frac{1}{3}q = 3 + \frac{1}{3}$, $r_2 = 2 + \frac{1}{2}$, required $d = v$.

$$\frac{q - d}{q - 2d} = \frac{r_1}{r_2} \quad \text{so} \quad q - d = \frac{r_1}{r_2}(q - 2d), \quad \text{thus}$$

$$10 - d = \left(1 + \frac{1}{3}\right)(10 - 2d)$$

from which it is inferred that $10 - 2d = 6$, thus $d = v = 2$.

Remark. We may attain $10 - 2d = 6$ by transforming the ratio:
$$\frac{10 - d}{10 - 2d} = 1 + \frac{1}{3}, \quad \frac{20 - 2d}{10 - 2d} = 2 + \frac{2}{3}, \quad \frac{(20 - 2d) - (10 - 2d)}{10 - 2d} = 1 + \frac{2}{3},$$

$$\text{whence} \quad 10 - 2d = \frac{10}{1 + \frac{2}{3}} = 6.$$

B. *Several overflows*.

(B.180) MS $\mathcal{A}$. A quantity of sixty measures of must is to be reduced by boiling to its third; but, after ten measures have evaporated, five overflow; after nine measures have evaporated from the remainder, six overflow. To what does this last remainder have to be reduced?

Given thus $q = 60$, $r_1 = \frac{1}{3}q = 20$, $d_1 = 10$, $v_1 = 5$, from which r_2; $d_2 = 9$, $v_2 = 6$, required r_3.
In this case
$$r_2 = \frac{r_1(q - d_1 - v_1)}{q - d_1} = \frac{20 \cdot 45}{50} = 18,$$

$$r_3 = \frac{r_2\Big[(q - d_1 - v_1) - d_2 - v_2\Big]}{(q - d_1 - v_1) - d_2} = \frac{18 \cdot 30}{36} = 15.$$

(B.181) MSS $\mathcal{A}$, $\mathcal{B}$. A quantity of ten measures of must is to be reduced by boiling to its third; but, after two measures have evaporated, two overflow; after two measures have evaporated from the remainder, two overflow. To what does this last remainder have to be reduced?

Given thus $q = 10$, $r_1 = \frac{1}{3}q = 3 + \frac{1}{3}$, $d_1 = 2$, $v_1 = 2$, from which r_2; $d_2 = 2$, $v_2 = 2$, required r_3. Similar to the preceding, but this time solved with a single formula, namely

$$r_3 = \frac{r_1(q - d_1 - v_1)\Big[(q - d_1 - v_1) - d_2 - v_2\Big]}{(q - d_1)\Big[(q - d_1 - v_1) - d_2\Big]}$$

$$= \frac{(3 + \frac{1}{3}) \cdot 6 \cdot 2}{8 \cdot 4} = \frac{40}{32} = 1 + \frac{1}{4}.$$

(B.182) MS $\mathcal{A}$. A quantity of a hundred measures of must is to be reduced by boiling to its fifth; but, after ten measures have evaporated, nine overflow; after eight measures have evaporated from the remainder, seven overflow; after six measures evaporate from the remainder, five overflow. To what does this last remainder have to be reduced?

Given thus $q = 100$, $r_1 = \frac{1}{5}q = 20$, $d_1 = 10$, $v_1 = 9$, from which r_2; $d_2 = 8$, $v_2 = 7$, from which r_3; $d_3 = 6$, $v_3 = 5$, required r_4. Not treated numerically. We would find successively

$$r_2 = \frac{r_1(q - d_1 - v_1)}{q - d_1} = \frac{20 \cdot 81}{90} = 18$$

$$r_3 = \frac{18(81 - 15)}{73} = \frac{18 \cdot 66}{73} = 16 + \frac{20}{73}$$

$$r_4 = \frac{\left(16 + \frac{20}{73}\right)(66 - 11)}{60} = \frac{1089}{73} = 14 + \frac{67}{73}.$$

C. *Particular cases of a single overflow*.

Impossible problems, the author says, should be recognized as such before any attempt is made to solve them. Indeed, in the next example, we may well compute the answer by applying the formula, but the result obtained does not meet the conditions of the problem. That is why it is better to identify such a problem from the outset.

(B.183) MS $\mathcal{A}$. A quantity of ten measures of must is to be reduced by boiling to its third; but, after seven measures have evaporated, two overflow. To what does the remainder have to be reduced?

Given thus $q = 10$, $r_1 = \frac{1}{3}q = 3 + \frac{1}{3}$, $d = 7$, $v = 2$, required r_2.

Since the must has been boiling too long ($q - d < r_1$), the desired concentration can no longer be attained.

Remark. The formula would give $1 + \frac{1}{9}$.

(B.184) MS $\mathcal{A}$. A quantity of ten measures of must is to be reduced by boiling to its third; but two measures overflow. To what does the remainder have to be reduced?

Given thus $q = 10$, $r_1 = \frac{1}{3}q = 3 + \frac{1}{3}$, $v_1 = 2$, required r_2.

In this case $d = 0$, that is, the must boils over (or is spilt) before any evaporation, thus before the beginning of the reduction process. We shall just have to consider that the new initial quantity $q' = q - d$ must be reduced to a third of itself (see similar case in B.175e).

Remark. Considering the usual proportion would also lead to the right answer $2 + \frac{2}{3}$.

---

2. Several quantities unknown.

A. *Two quantities unknown*.

(B.185) MSS $\mathcal{A}$; $\mathcal{B}$, which omits c. An unknown quantity of must is to be reduced by boiling to its third; but, after two measures have evaporated, two overflow, and the rest has then been reduced to two and a half measures. What was the initial quantity of must?

Given thus $d = 2$, $v = 2$, $r_2 = 2 + \frac{1}{2}$, required q (and $r_1 = \frac{1}{k}q = \frac{1}{3}q$). As mentioned in the General Introduction, pp. xviii–xix, we know the name of the *Liber mahameleth*'s author from the Italian version of this problem.

(*a*) *Formula*.

$$q = \frac{k \cdot r_2 + d + v}{2} + \sqrt{\left(\frac{k \cdot r_2 + d + v}{2}\right)^2 - k \cdot r_2 \cdot d}$$

$$= \frac{3\left(2+\frac{1}{2}\right)+2+2}{2} + \sqrt{\left(\frac{3\left(2+\frac{1}{2}\right)+2+2}{2}\right)^2 - 3\left(2+\frac{1}{2}\right)\cdot 2}$$

$$= \frac{11+\frac{1}{2}}{2} + \sqrt{\left(\frac{11+\frac{1}{2}}{2}\right)^2 - 15} = 5 + \frac{3}{4} + \sqrt{\left(5+\frac{3}{4}\right)^2 - 15}$$

$$= 5 + \frac{3}{4} + \sqrt{33 + \frac{1}{2}\frac{1}{8} - 15} = 5 + \frac{3}{4} + \sqrt{18 + \frac{1}{16}}$$

$$= 5 + \frac{3}{4} + 4 + \frac{1}{4} = 10.$$

Remark. This is the larger positive solution of the equation

$$q^2 + k \cdot r_2 \cdot d = q\left(k \cdot r_2 + d + v\right).$$

The other, $1+\frac{1}{2}$, is incompatible with the conditions of the problem. Our author does not mention this, unlike one Italian text (see below).

Demonstration.

(α) *Preliminary remark.* In order to better understand the demonstration, consider the following algebraic development. We know, from B.175b, that (with $k=3$)

$$\frac{d+(q-d)}{q-d} = \frac{\left[k\cdot r_2 - (q-d-v)\right] + (q-d-v)}{q-d-v}. \qquad (1)$$

By separation, we obtain

$$\frac{d}{q-d} = \frac{k\cdot r_2 - (q-d-v)}{q-d-v},$$

and, by composition,

$$\frac{d}{q} = \frac{k\cdot r_2 - (q-d-v)}{k\cdot r_2}, \qquad (2)$$

whence

$$q\cdot\left[k\cdot r_2 - (q-d-v)\right] = k\cdot r_2 \cdot d \qquad (3)$$

with $k\cdot r_2 \cdot d$ a quantity which is known. But

$$q + \left[k\cdot r_2 - (q-d-v)\right] = k\cdot r_2 + d + v,$$

with $k\cdot r_2 + d + v$ another known quantity. Thus we are to solve

$$\begin{cases} q + \left[k\cdot r_2 - (q-d-v)\right] = k\cdot r_2 + d + v \\ q\cdot\left[k\cdot r_2 - (q-d-v)\right] = k\cdot r_2 \cdot d, \end{cases} \qquad (4)$$

that is, finding two quantities of which we know sum and product. This will be determined as usual (B.30), by means of the identity (corresponding to *Elements* II.5)
$$\left(\frac{a-b}{2}\right)^2 = \left(\frac{a+b}{2}\right)^2 - a \cdot b,$$
and the subsequent determination of a by
$$a = \frac{a+b}{2} + \frac{a-b}{2}.$$
In our case the two known terms are
$$\left(\frac{q + [k \cdot r_2 - (q - d - v)]}{2}\right)^2 = \left(\frac{k \cdot r_2 + d + v}{2}\right)^2,$$
$$q \cdot [k \cdot r_2 - (q - d - v)] = k \cdot r_2 \cdot d,$$
and the still unknown term will be determined by
$$\left(\frac{q - [k \cdot r_2 - (q - d - v)]}{2}\right)^2 = \left(\frac{k \cdot r_2 + d + v}{2}\right)^2 - k \cdot r_2 \cdot d, \qquad (5)$$
whence for the required quantity
$$q = \frac{q + [k \cdot r_2 - (q - d - v)]}{2} + \frac{q - [k \cdot r_2 - (q - d - v)]}{2}$$
$$= \frac{k \cdot r_2 + d + v}{2} + \sqrt{\left(\frac{k \cdot r_2 + d + v}{2}\right)^2 - k \cdot r_2 \cdot d}.$$

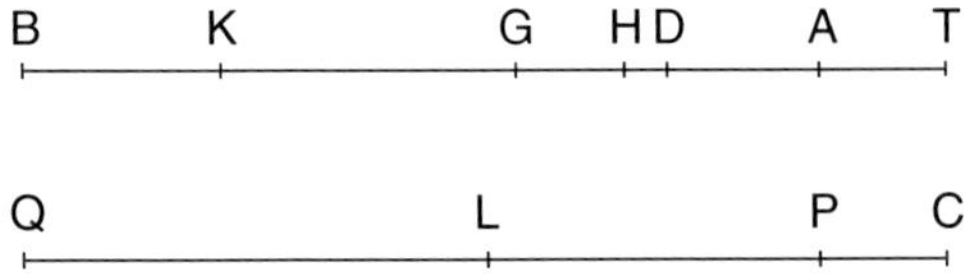

Fig. 68

(β) *Demonstration according to the text.* Let there be on $AB = q$ (Fig. 68) the segments $AD = d$, $DG = v$, $BK = r_2$, with $AB = q$ required; then $BG = q - d - v$.
As we know (B.175b),
$$\frac{d + (q - d)}{q - d} = \frac{[3r_2 - (q - d - v)] + (q - d - v)}{q - d - v}.$$
Put $BH = 3BK$ (thus $BH = 3r_2 = 7 + \frac{1}{2}$). Therefore, the previous relation becomes
$$\frac{AD + BD}{BD} = \frac{GH + BG}{BG}. \qquad (1)$$

By separation, we have
$$\frac{AD}{BD} = \frac{GH}{BG}$$
whence, by composition,
$$\frac{AD}{AB} = \frac{GH}{BH}. \qquad (2)$$

Since $AD = 2$ and $BH = 7 + \frac{1}{2}$, it means that
$$AB \cdot GH = 15. \qquad (3)$$

In order to subsequently apply *Elements* II.5, we annex $AT = GH$ to AB, so that now $AB \cdot AT = 15$. Furthermore, since $AT = GH$,
$$HT = GT - GH = GT - AT = GA,$$
whence (since $GA = 4$) $HT = 4$; therefore
$$BT = BH + HT = 7 + \frac{1}{2} + 4 = 11 + \frac{1}{2} = AB + AT.$$

Then we are to solve
$$\begin{cases} AB + AT = 11 + \frac{1}{2} \\ AB \cdot AT = 15. \end{cases} \qquad (4)$$

There is no need to add a second part to the figure here, as the author does, since all that remains now is to bisect TB. If L is the mid-point, then, by *Elements* II.5, $AB \cdot AT + AL^2 = TL^2$, so
$$AL^2 = TL^2 - AB \cdot AT = 33 + \frac{1}{16} - 15 = 18 + \frac{1}{16}, \qquad (5)$$
whence $AL = 4 + \frac{1}{4}$. Since $BL = 5 + \frac{3}{4}$, we find $AB = 10$.[k]

(**b**) Algebraic solution. Put $q = x$. Since, by the fundamental relation,
$$\frac{x-2}{\frac{1}{3}x} = \frac{x-4}{2+\frac{1}{2}}, \quad \text{then} \quad \left(2 + \frac{1}{2}\right)(x-2) = \frac{1}{3}x(x-4)$$

---

[k] In the second segment of straight line added in the text (where $QC = BT = 11 + \frac{1}{2}$, $PC = AT$, thus $PQ = AB$, with L as the mid-point of QC), we are to solve
$$\begin{cases} PQ + PC = 11 + \frac{1}{2} \\ PQ \cdot PC = 15. \end{cases}$$
Then $PQ \cdot PC + PL^2 = CL^2$, so that
$$PL^2 = CL^2 - PQ \cdot PC = 33 + \frac{1}{16} - 15 = 18 + \frac{1}{16},$$
and $PL = 4 + \frac{1}{4}$, $PQ = QL + PL = 5 + \frac{3}{4} + 4 + \frac{1}{4} = 10 = AB$.

which leads to the equation $x^2 + 15 = \left(11 + \frac{1}{2}\right)x$.

Remark. The final equation is not set in our treatise, which merely says that we are to proceed as 'taught above in algebra'. Domenico d'Agostino Vaiaio chose, among the *molti modi* presented by Grazia de' Castellani, this one. Setting the proportion as above, he arrives at the equation $x^2 + 15 = \left(11 + \frac{1}{2}\right)x$, of which he gives the two answers, namely $5 + \frac{3}{4} \pm \left(4 + \frac{1}{4}\right)$, retaining only the admissible one (*Adunque la chosa vale 10, onde e' barili furono 10. E chosì farai le simili. E nota che la chosa in questa ragione non puo valere $5\frac{3}{4}$ meno $4\frac{1}{4}$, che sarebbe inpossibile*). The question of the second solution does not arise in the *Liber mahameleth* since the quadratic equation is not solved with the direct formula.

(**c**) Another solution (using this time *Elements* II.6).

(α) *Preliminary remark.* What the author does (geometrically) corresponds to taking as the unknown $q - d - v = $ (say) y. Since

$$\frac{q-d}{3r_1} = \frac{q-d-v}{3r_2}, \quad \text{that is,} \quad \frac{y+v}{y+d+v} = \frac{y}{7+\frac{1}{2}}, \tag{1}$$

he obtains $y^2 = \left(7 + \frac{1}{2} - d - v\right)y + \left(7 + \frac{1}{2}\right)v$, or

$$y^2 = \left(3 + \frac{1}{2}\right)y + 15. \tag{2}$$

This being equivalent to

$$\begin{cases} y - \left[y - \left(3 + \frac{1}{2}\right)\right] = 3 + \frac{1}{2} \\ y \cdot \left[y - \left(3 + \frac{1}{2}\right)\right] = 15, \end{cases} \tag{3}$$

we are to find two quantities known by their difference and product. Since, by the usual identity,

$$\frac{y + \left[y - \left(3 + \frac{1}{2}\right)\right]}{2} = \sqrt{\left(\frac{y - \left[y - \left(3 + \frac{1}{2}\right)\right]}{2}\right)^2 + y\left[y - \left(3 + \frac{1}{2}\right)\right]},$$

we shall have

$$y = \frac{y + \left[y - \left(3 + \frac{1}{2}\right)\right]}{2} + \frac{y - \left[y - \left(3 + \frac{1}{2}\right)\right]}{2}$$

$$= \sqrt{\left(1 + \frac{3}{4}\right)^2 + 15} + 1 + \frac{3}{4} = \sqrt{18 + \frac{1}{16}} + 1 + \frac{3}{4}$$

$$= 4 + \frac{1}{4} + 1 + \frac{3}{4} = 6.$$

```
A     Q    T         D    G    B
|-----------|---------|----|----|
            H         K Z
            |---------|--|
```

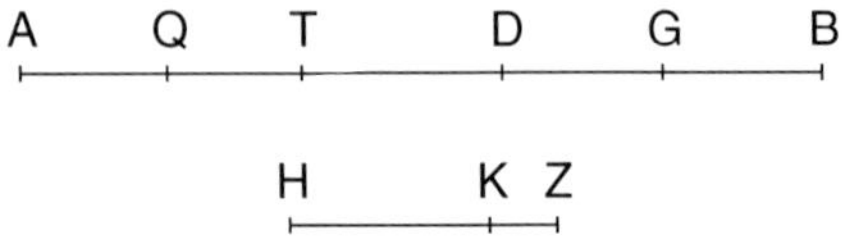

Fig. 69

(β) *Solution according to the text.* Let (Fig. 69) $d = GB$, $v = DG$, $r_2 = HK$, required $q = AB$ (or $r_1 = HZ$). Since

$$\frac{q-d}{r_1} = \frac{q-d-v}{r_2}, \quad \text{so} \quad \frac{AG}{HZ} = \frac{AD}{HK} \quad \text{and} \quad \frac{AG}{3HZ} = \frac{AD}{3HK}. \quad (1)$$

Then, since $3HZ = AB$, we have $3HK \cdot AG = AB \cdot AD$. But $3HK = 7 + \frac{1}{2}$ while $AG = AD + 2$ and $AB = AD + 4$; so we find

$$\left(7 + \frac{1}{2}\right)AD + 15 = AD^2 + 4AD, \quad \text{thus} \quad AD^2 = \left(3 + \frac{1}{2}\right)AD + 15. \quad (2)$$

Let $AT = 3 + \frac{1}{2}$. Then $AD^2 = AT \cdot AD + 15$ and, since $AD^2 = AD(AT + TD) = AD \cdot AT + AD \cdot TD$, we are left with $AD \cdot TD = 15$. Thus we are to solve

$$\begin{cases} AD - TD = 3 + \frac{1}{2} \\ AD \cdot TD = 15. \end{cases} \quad (3)$$

Let Q be the mid-point of AT, so that $AQ = QT = 1 + \frac{3}{4}$. Then, by *Elements* II.6,

$$AD \cdot TD + QT^2 = QD^2.$$

With $QT^2 = 3 + \frac{1}{16}$, we obtain $QD^2 = 18 + \frac{1}{16}$, thus $QD = 4 + \frac{1}{4}$, therefore $AD = AQ + QD = 6$ and finally $AB = AD + DG + GB = 10$.

B. **Three quantities unknown**.

Suppose we are given v and r_2 and, knowing that $r_1 = \frac{1}{k}q$, $d = \frac{1}{s}q$, we are to find q. The fundamental proportion

$$\frac{q-d}{r_1} = \frac{q-d-v}{r_2} \quad \text{gives} \quad \frac{\left(1-\frac{1}{s}\right)q}{\frac{1}{k}q} = \frac{\left(1-\frac{1}{s}\right)q - v}{r_2}$$

and thus leads to a linear equation for q, namely

$$\frac{k(s-1)}{s} r_2 = \frac{s-1}{s} q - v.$$

(**B.186**) MSS $\mathcal{A}$, $\mathcal{B}$. An unknown quantity of must is to be reduced by boiling to its third; but, after a fifth has evaporated, two measures overflow, and the rest is then reduced to two and a half measures. What was the initial quantity of must?

We are thus given $r_1 = \frac{1}{3}q$, $d = \frac{1}{5}q$, $v = 2$, $r_2 = 2 + \frac{1}{2}$.

(**a**) The fundamental proportion gives us

$$\frac{\frac{4}{5}q}{\frac{1}{3}q} = \frac{\frac{4}{5}q - 2}{2 + \frac{1}{2}}, \quad \text{so} \quad \frac{12}{5} \cdot \frac{5}{2} = 6 = \frac{4}{5}q - 2,$$

whence $q = 10$.

(b) Algebraic solution. The same proportion, but with $q = x$ (and no reduction), gives

$$\frac{4}{5}x \cdot \left(2 + \frac{1}{2}\right) = \frac{1}{3}x \cdot \left(\frac{4}{5}x - 2\right), \quad \text{so}$$

$$2x = \left(\frac{1}{5} + \frac{1}{3} \cdot \frac{1}{5}\right)x^2 - \frac{2}{3}x, \quad \text{thus} \quad \left(2 + \frac{2}{3}\right)x = \left(\frac{1}{5} + \frac{1}{3}\frac{1}{5}\right)x^2,$$

whence (multiplying by $3 + \frac{3}{5}$), $10x = x^2$, and $x = 10$.

Remark. After restoring, we would have cancelled x on both sides; not so in mediaeval algebra since $ax^2 = bx$ is one of the six final standard forms of linear and quadratic equations, considered as different from $ax = b$ (though the solution $x = 0$ was not considered).

C. *Particular cases*.

In the last group B.187–189, where q is again unknown, v is a fraction of $q - d$, say $v = \frac{m}{n}(q - d)$. As already seen in our summary, we then have

$$\frac{r_1}{r_2} = \frac{q - d}{q - d - v} = \frac{q - d}{(q - d)\left(1 - \frac{m}{n}\right)} = \frac{n}{n - m},$$

that is, the right-hand term is a fixed and known numerical quantity.

(B.187) MS $\mathcal{A}$. An unknown quantity of must is to be reduced by boiling to its third; but, after two measures have evaporated, a fourth of the liquid overflows, and the rest is then reduced to a fourth of the initial quantity. What was this initial quantity?

Given thus $d = 2$, required q and the other quantities depending on q, namely $r_1 = \frac{1}{3}q$, $v = \frac{1}{4}(q - d)$, $r_2 = \frac{1}{4}q$. Then

$$\frac{r_1}{r_2} = \frac{q - d}{q - d - v} \quad \text{gives}$$

$$\frac{\frac{1}{3}q}{\frac{1}{4}q} = \frac{q - 2}{(q - 2) - \frac{1}{4}(q - 2)} = \frac{q - 2}{\frac{3}{4}(q - 2)} = 1 + \frac{1}{3}.$$

Since q disappears, any value of it is acceptable here. This is verified in the text for $q = 30$. But should we, as proposed next in the text, put $r_2 = \frac{1}{5}q$ (leaving other values unchanged), then

$$\frac{\frac{1}{3}q}{\frac{1}{5}q} = 1 + \frac{2}{3}, \quad \text{thus} \quad \frac{5}{3} = \frac{4}{3},$$

and the problem is not solvable, as stated.

(B.188) MS $\mathcal{A}$. An unknown quantity of must is to be reduced by boiling to its third; but, after a fifth of the liquid has evaporated, a fourth overflows, and the rest is then reduced to two and a half measures. What was the initial quantity?

Given $r_2 = 2 + \frac{1}{2}$, and knowing that $r_1 = \frac{1}{3}q$, $d = \frac{1}{5}q$, $v = \frac{1}{4}(q-d)$, to find q. Since

$$\frac{q-d}{q-d-v} = \frac{r_1}{r_2} \quad \text{gives} \quad \frac{\frac{4}{5}q}{\frac{3}{4}(q-d)} = \frac{\frac{4}{5}q}{\frac{3}{4}\frac{4}{5}q} = 1 + \frac{1}{3} = \frac{\frac{1}{3}q}{2+\frac{1}{2}},$$

we have

$$\frac{1}{3}q = \left(2 + \frac{1}{2}\right)\left(1 + \frac{1}{3}\right) = 3 + \frac{1}{3}$$

and $q = 10$.

Remark. Fixing any value $d = \frac{r}{s}q$ would lead to the same result since this quantity drops out anyway.

(B.189) MS A. An unknown quantity of must is to be reduced by boiling to its third; but, after two measures have evaporated, a fourth of the liquid overflows, and the rest is then reduced to two and a half measures. What was the initial quantity?

Given $d = 2$, $r_2 = 2 + \frac{1}{2}$, and, knowing that $r_1 = \frac{1}{3}q$, $v = \frac{1}{4}(q-d)$, find q. As before,

$$\frac{q-d}{q-d-v} = \frac{r_1}{r_2} \quad \text{gives} \quad \frac{q-2}{\frac{3}{4}(q-2)} = 1 + \frac{1}{3} = \frac{\frac{1}{3}q}{2+\frac{1}{2}},$$

so here too we shall find $q = 10$.

This problem ends up being the same as the one before. In the text, the treatment is missing. We have added it but the author may originally have merely referred the reader to the previous problem. The Italian source (above, p. 1479) solves it by algebra, using the fundamental proportion without any reduction, *benché sieno molti modi*. (This statement does not imply that more was known of the *Liber mahameleth* than what we have at present.)

Chapter B–X:

Borrowing

Summary

Ancient and mediaeval trade was complicated by the fact that the many measure units in use varied from one locality to another (a situation which was in fact to last well into the nineteenth century), with identical names only adding to the confusion. Thus, when we are told about a distance expressed in cubits or in feet, we need to know *where* it was measured —and possibly also *when* since length of the measure unit might have been changed in the course of time as well. A characteristic illustration is the variety of ancient and mediaeval values of the Earth's circumference preserved: their differences are normally due more to conversions, or confusions between homonymous units, than to new measurements.

This chapter treats one particular aspect of this variety, involving two homonymous capacity measures. A certain quantity of bushels (*sextarii*) has been borrowed and is later returned, the quantity being also measured in sextarii; but the measuring-vessel is not the same. To determine the equivalence, the reference measure will be the number of bushels of each kind to one (common) *modius*.[†] Then, if a number q_1 of bushels of which there are m_1 to the modius has been borrowed and q_2 bushels of which there are m_2 to the modius are returned, we must have

$$\frac{q_1}{m_1} = \frac{q_2}{m_2}.$$

As the text asserts, how to treat this is well known from before: we have, as in the chapter on buying and selling, a simple proportion, the m_i's here representing bushels there and the q_i's, nummi. Because of this formal analogy, only the five fundamental problems will be presented: the two simple ones in which either q_2 or m_2 is required and the three where both are required but their sum, difference or product is known.

(B.190) MSS $\mathcal{A}$, $\mathcal{B}$. Six sextarii borrowed of which there are fourteen to the modius are returned with sextarii of which there are twenty to the modius. How many sextarii will be returned?

Given thus $q_1 = 6$, $m_1 = 14$, $m_2 = 20$, required q_2. Analogous, as stated, to similar earlier problems (B.1, B.3–6, B.11–12).

[†] In the present problems, there are 14 and 21 sextarii to the modius (14 and 20 in B.190).

(*a*) By the above proportion,
$$q_2 = \frac{m_2 \cdot q_1}{m_1} = \frac{20 \cdot 6}{14} = 8 + \frac{4}{7}.$$

Also computed as
$$\frac{q_1}{m_1} \cdot m_2, \qquad \frac{m_2}{m_1} \cdot q_1.$$

(*b*) Using algebra, put $q_2 = x$. Just as before: since $6 : 14 = x : 20$, we find $14x = 120$, whence x.

(**B.191**) MSS $\mathcal{A}$, $\mathcal{B}$. Six sextarii have been borrowed, of which there are fourteen to the modius, and the debt is discharged with nine sextarii. How many of these sextarii are there to the modius?

Given thus $q_1 = 6$, $m_1 = 14$, $q_2 = 9$, required m_2. Analogous, as stated, to earlier problems (B.2, B.7–10, B.13–14).

(*a*) By the known proportion,
$$m_2 = \frac{q_2 \cdot m_1}{q_1} = \frac{9 \cdot 14}{6} = 21.$$

Also computable as
$$\frac{m_1}{q_1} \cdot q_2.$$

(*b*) Using algebra, put $m_2 = x$, whence x from $6 : 14 = 9 : x$.

(**B.192**) MSS $\mathcal{A}$, $\mathcal{B}$. Six sextarii have been borrowed, of which there are fourteen to the modius, and the debt is discharged with a number of sextarii which, when multiplied by their number to the modius, produces a hundred and eighty-nine. How many sextarii are returned and how many are there to the modius?

Given thus $q_1 = 6$, $m_1 = 14$, $q_2 \cdot m_2 = 189$, required m_2, q_2. Analogous to B.17.

$$\frac{q_2}{q_1} = \frac{m_2}{m_1} \quad \text{gives} \quad \frac{q_2 \cdot m_2}{q_1 \cdot m_1} = \frac{q_2^2}{q_1^2} = \frac{m_2^2}{m_1^2}, \quad \text{whence}$$

$$m_2 = \sqrt{\frac{m_1 (q_2 \cdot m_2)}{q_1}} = \sqrt{\frac{14 \cdot 189}{6}} = 21 \quad \text{and}$$

$$q_2 = \frac{q_1}{m_1} \cdot m_2, \quad \text{or} \quad q_2 = \sqrt{\frac{q_1 (q_2 \cdot m_2)}{m_1}} = \sqrt{\frac{6 \cdot 189}{14}} = 9.$$

(**B.193**) MSS $\mathcal{A}$, $\mathcal{B}$. Six sextarii have been borrowed, of which there are fourteen to the modius, and the debt is discharged with a number of sextarii which, when added to their number to the modius, makes thirty. How many sextarii are returned and how many are there to the modius?

Given thus $q_1 = 6$, $m_1 = 14$, $q_2 + m_2 = 30$, required m_2, q_2. The reader is referred to B.15.

(**a**) As we have often seen before, we infer from $m_1 : q_1 = m_2 : q_2$ that

$$\frac{q_1 + m_1}{q_1} = \frac{q_2 + m_2}{q_2} \quad \text{and} \quad \frac{m_1}{q_1 + m_1} = \frac{m_2}{q_2 + m_2},$$

whence the relations found in the text:

$$m_2 = \frac{m_1(q_2 + m_2)}{q_1 + m_1} = \frac{14 \cdot 30}{20} = 21 \quad \text{and}$$

$$q_2 = 30 - m_2, \quad \text{or} \quad q_2 = \frac{q_1(q_2 + m_2)}{q_1 + m_1} = \frac{6 \cdot 30}{20} = 9.$$

(**b**) By algebra, put $m_2 = x$; then, without transforming the proportion,

$$q_2 + m_2 = 30 = \frac{q_1}{m_1} \cdot m_2 + m_2 = \frac{3}{7} \cdot x + x, \quad \text{whence} \quad x.$$

(**c**) Also computed as (this should occur before b)

$$m_2 = \frac{m_1}{q_1 + m_1}(q_2 + m_2), \qquad q_2 = \frac{q_1}{q_1 + m_1}(q_2 + m_2).$$

(**B.194**) MSS $\mathcal{A}$, $\mathcal{B}$. Six sextarii have been borrowed, of which there are fourteen to the modius, and the debt is discharged with a number of sextarii which, when subtracted from their number to the modius, leaves twelve. How many sextarii are returned and how many are there to the modius?

Given thus $q_1 = 6$, $m_1 = 14$, $m_2 - q_2 = 12$ (since $m_1 > q_1$), required m_2, q_2. Analogous to B.16.

Remark. The treatment is like that of the preceding B.193, but requires separation and conversion of the ratio. Indeed, we infer from $m_1 : q_1 = m_2 : q_2$ that

$$\frac{m_1 - q_1}{q_1} = \frac{m_2 - q_2}{q_2}, \qquad \frac{m_1}{m_1 - q_1} = \frac{m_2}{m_2 - q_2}.$$

Chapter B–XI:
Hiring workers

Summary

Calculating the wages of hired workers is —hardly surprisingly, that probably being the most common application— the main subject of Book B. The present chapter contains some fifty problems (B.195–243); but it does not exhaust the subject of hiring, for it is followed by three other chapters on the same topic. This chapter can be roughly divided into three parts, as mentioned by the author at the outset. In the first part (B.195–229), where a single worker is to do the work, the treatment is much the same as that of problems seen in the first chapter of this second book. The second part (B.230–239), where two workers are involved, deals with problems of the 'lazy workman': hired to do a certain task in a given number of days for a given wage, he stops working after a certain number of days, and a second worker is hired to finish the work at the first one's expense (though of course at a different monthly wage since otherwise we would be in the previous case); which means that with the wage of the second worker being higher for the same working time (exceptions are B.231, B.231'), the first worker may earn nothing at all, or even make a loss. The third part is about three workers completing the required time between them, with monthly wages either equal or different (B.240–243). Throughout this chapter the kind of work is not specified (but will be in B–XIII and B–XIV).

1. One worker.

A. *Simple problems*.

As in the problems on buying and selling, we have a simple proportion: if the wage for t_1 days is p_1 and the wage for t_2 days worked p_2, then

$$\frac{p_1}{p_2} = \frac{t_1}{t_2}.$$

This is applied in B.195–196 (one unknown, p_2 then t_2), while in B.197 the wage increases each day according to an arithmetical progression. We have already solved similar problems (B.1–14, B.52).

B. *Problems involving 'things'*.

In the group of problems B.198–204, B.207–217, B.220–222, B.224–226, we are given t_1, t_2, while p_1 and p_2 are linear functions of x, thus of the form $p_1 = a_1 x + b_1$, $p_2 = a_2 x + b_2$.‡ As in similar problems of the first chapter (B.21–29), two basic approaches are employed, one of which is to

---

‡ In B.223, $p_1 = a_1 x^2$. For the others, see below, C.

transform the proportion so as to obtain x equal to a number directly, while the other uses the unmodified proportion and, since x occurs on both sides, involves the two distinctive operations of algebraic reckoning. Since this is pure computation and no formula is applied, geometric demonstrations are not needed. Some are nonetheless found in the group B.220–221″.

— **Transformation of the proportion.**

This is applicable if $|a_1|=|a_2|$, or (as in B.202′–203) if the given coefficients of x can be reduced to that form.

• Case $a_1 = a_2$, thus $p_1 = ax + b_1$ and $p_2 = ax + b_2$. This is just as in B.21–25: the term in x will disappear on one of the sides by separation of the ratio. Indeed, since

$$\frac{p_1}{p_2} = \frac{t_1}{t_2}, \quad \text{then} \quad \frac{p_1 - p_2}{p_2} = \frac{t_1 - t_2}{t_2}, \quad \text{so}$$

$$p_2 = ax + b_2 = \frac{t_2}{t_1 - t_2}(p_1 - p_2), \quad \text{and} \quad ax = \frac{t_2}{t_1 - t_2}(b_1 - b_2) - b_2.$$

• Case $a_1 = -a_2$, thus $p_1 = ax + b_1$ and $p_2 = -ax + b_2$. As in B.27–29, the term in x will disappear from one side by composition of the ratio (B.202, B.211). Indeed, since

$$\frac{p_1}{p_2} = \frac{t_1}{t_2}, \quad \text{then} \quad \frac{p_1 + p_2}{p_2} = \frac{t_1 + t_2}{t_2}, \quad \text{so}$$

$$p_2 = -ax + b_2 = \frac{t_2}{t_1 + t_2}(p_1 + p_2), \quad \text{and} \quad ax = b_2 - \frac{t_2}{t_1 + t_2}(b_1 + b_2).$$

— **Using the unmodified proportion.**

Without these transformations, the proportion $p_1 : p_2 = t_1 : t_2$ will lead to the solution equation in two ways. We may either consider one of the p_i on one side, which gives

$$p_1 = a_1 x + b_1 = \frac{t_1}{t_2} p_2 = \frac{t_1}{t_2} a_2 x + \frac{t_1}{t_2} b_2,$$

$$p_2 = a_2 x + b_2 = \frac{t_2}{t_1} p_1 = \frac{t_2}{t_1} a_1 x + \frac{t_2}{t_1} b_1,$$

or we may multiply the opposite proportional terms, considering thus $t_1 \cdot p_2 = t_2 \cdot p_1$, which gives

$$t_1 \cdot a_2 x + t_1 \cdot b_2 = t_2 \cdot a_1 x + t_2 \cdot b_1.$$

As we see, using the unmodified proportion will lead to 'algebra'; for (unless a_1 or a_2 is zero) x will appear on both sides and thus the two algebraic operations (restoration and reduction), or one of them, will have to be applied. See examples from B.198b on.

C. *Solution formulae involving roots*.

In B.205–206, $p_1 = a_1 x + b_1$ and $p_2 = c_2 \sqrt{a_1 x + b_1} + d_2 x$, thus p_1 occurs under the radical sign in p_2, a situation encountered in the first chapter as well (B.39–40). We then have

$$p_1 = a_1 x + b_1 = \frac{t_1}{t_2} c_2 \sqrt{a_1 x + b_1} + \frac{t_1}{t_2} d_2 x$$

which leads to a second-degree equation for x (B.206), or to an equation of the first degree if $d_2 = 0$ since then

$$a_1 x + b_1 = \left(\frac{t_1}{t_2} \cdot c_2\right)^2$$

(but the author does not make use of this simplification in B.205b).

In B.218–219, we are given t_1, $t_2 \cdot p_2$, $(t_1 - t_2)(p_1 - p_2)$. To determine the unknown quantities, the following formulae occur:

$$p_1 = \frac{2\sqrt{t_2 \cdot p_2 \,(t_1 - t_2)(p_1 - p_2)} + t_2 \cdot p_2 + (t_1 - t_2)(p_1 - p_2)}{t_1}, \quad \text{(B.218)}$$

$$t_2 = \frac{\sqrt{t_2 \cdot p_2 \,(t_1 - t_2)(p_1 - p_2)} + t_2 \cdot p_2}{p_1}, \quad \text{(B.218)}$$

$$p_2 = \frac{\sqrt{t_2 \cdot p_2 \,(t_1 - t_2)(p_1 - p_2)} + t_2 \cdot p_2}{t_1} \quad \text{(B.219)}$$

equivalent to the previous one since $p_1 : t_1 = p_2 : t_2$, thus $p_1 \cdot t_2 = p_2 \cdot t_1$;

$$t_1 - t_2 = \frac{\sqrt{t_2 \cdot p_2 \,(t_1 - t_2)(p_1 - p_2)} + (t_1 - t_2)(p_1 - p_2)}{p_1}, \quad \text{(B.218)}$$

$$p_1 - p_2 = \frac{\sqrt{t_2 \cdot p_2 \,(t_1 - t_2)(p_1 - p_2)} + (t_1 - t_2)(p_1 - p_2)}{t_1} \quad \text{(B.219)}$$

equivalent to the previous one since, by separation of the ratios in the fundamental proportion, $(t_1 - t_2) p_1 = (p_1 - p_2) t_1$. In the formulae for t_2 and $t_1 - t_2$, p_1 is supposed to have been determined by the first formula.

Since these problems are related, but not identical, to the ones seen in B.35–36, all these formulae will be demonstrated geometrically.

Remark. B.218 is one of the problems found later in the Middle Ages.

D. *Wage in cash and kind*.

The first part of the present chapter ends with problems in which the monthly wage consists of cash and kind, namely a given sum of nummi s_1 and three bushels at different (unknown) prices c_i each one of which differs from the other by a given amount of d_i nummi. Thus, for t_1 the wage is

$$p_1 = c_1 + c_2 + c_3 + s_1, \quad \text{with} \quad c_1 = c_3 + d_1, \; c_2 = c_3 + d_2,$$

while for the given time t_2 the wage is said to be

$$p_2 = \frac{k_1}{l_1} \cdot c_1 + \frac{k_2}{l_2} \cdot c_2 + \frac{k_3}{l_3} \cdot c_3 + s_2,$$

where the fractions and s_2 are given.
Since $t_1 \cdot p_2 = t_2 \cdot p_1$, we have

$$t_1 \left(c_3 \sum \frac{k_i}{l_i} + \frac{k_1}{l_1} \cdot d_1 + \frac{k_2}{l_2} \cdot d_2 + s_2 \right) = t_2 \left(3c_3 + d_1 + d_2 + s_1 \right)$$

from which we find

$$c_3 = \frac{t_2 \left(d_1 + d_2 + s_1 \right) - t_1 \left(\frac{k_1}{l_1} \cdot d_1 + \frac{k_2}{l_2} \cdot d_2 + s_2 \right)}{t_1 \sum \frac{k_i}{l_i} - 3t_2},$$

whence the two other required quantities c_1, c_2.

The three problems presented (B.227–229) consider successively s_1 & $s_2 \neq 0$, $s_2 = 0$, finally $s_1 = 0 = s_2$. They are solved by algebra ($c_3 = x$).

2. Two workers (or single lazy worker).

In the first set of problems (B.230–233, B.235), a man is to do a certain task in t days for a certain wage. He stops after τ days and will be replaced, for the remaining $t - \tau$ days, by someone receiving a different wage (also initially fixed for t days). Since the difference in wage is charged to the first worker, he ends with a sum which may be positive, zero, or negative (loss, or penalty). In the second set (B.234, B.236–239), a single worker is again involved; but he is paid each day, and for each day not worked he will have to pay a certain sum to the hirer. The situation is the same as before: he may end with some earning, none, or a debt.

A. *Monthly wages*.

Let p be the monthly wage of the first worker, π that of the second. Working τ days at the wage p, the first will receive of p the fraction

$$p_\tau = \frac{\tau}{t} p.$$

Working $t - \tau$ days at the wage π the second is to receive

$$\pi_{t-\tau} = \frac{t - \tau}{t} \pi.$$

Now this can be written as

$$\pi_{t-\tau} = \frac{t - \tau}{t} \left(p + (\pi - p) \right) = \frac{t - \tau}{t} \cdot p + \frac{t - \tau}{t} (\pi - p),$$

so that the second worker will receive what the first worker would have earned for the $t - \tau$ days and, in addition, the amount

$$(\pi - p)_{t-\tau} = \frac{t - \tau}{t} (\pi - p)$$

to be taken from what the first worker has earned by working τ days. This first worker is therefore left with

$$\frac{\tau}{t} p - \frac{t-\tau}{t} (\pi - p) = S.$$

This is the fundamental relation, various forms of which are used and demonstrated in the problems. We may write it as

$$\tau \cdot p - (t - \tau)(\pi - p) = t \cdot S, \qquad (*)$$

which is the origin of three relations occurring in our text to determine p, τ, $t - \tau$. They are the following.

- Determination of p. Since

$$p = \frac{(t-\tau)(\pi - p) + t \cdot S}{\tau} = \frac{t-\tau}{\tau} (\pi - p) + \frac{t}{\tau} S,$$

we see that

$$p = \frac{t-\tau}{\tau} (\pi - p + S) + S, \qquad (B.235')$$

which, for $S = 0$, reduces to

$$p = \frac{t-\tau}{\tau} (\pi - p). \qquad (B.235)$$

- Determination of $t - \tau$. From the above relation again, we obtain

$$t - \tau = \frac{\tau \cdot p - t \cdot S}{\pi - p}, \qquad (B.235''')$$

which for $S = 0$ reduces to

$$t - \tau = \frac{\tau \cdot p}{\pi - p}. \qquad (B.235)$$

But $t - \tau$ can also be expressed in another way. Letting in relation $(*)$ the term $\tau \cdot p$ drop out, we are left with

$$t \cdot p - t \cdot \pi + \tau \cdot \pi = t \cdot S \qquad (**)$$

from which we see that

$$t - \tau = \frac{(p - S) t}{\pi},$$

and, if $S = 0$,

$$t - \tau = \frac{p \cdot t}{\pi}. \qquad (B.230)$$

- Determination of τ. Considering again $(**)$, we see that

$$\tau = \frac{(\pi - p + S) t}{\pi}, \qquad (B.232)$$

and, if $S = 0$,
$$\tau = \frac{(\pi - p) t}{\pi}. \qquad (B.230)$$

A second expression for τ with $\pi - p$ in the denominator and $\tau \cdot p$ given may be also deduced from the first relation (∗). From
$$t - \tau = \frac{\tau \cdot p - t \cdot S}{\pi - p}$$
we infer that
$$\tau = \frac{t(\pi - p) - \tau \cdot p + t \cdot S}{\pi - p} = \frac{(\pi - p + S) t - \tau \cdot p}{\pi - p}. \qquad (B.235')$$

In all these problems, the text pays particular attention to the value of S. It first considers the case $S = 0$ (B.230, B.231), with the proof of the first formula for τ in B.230. The case $S > 0$ is dealt with in B.231′ and B.232, and the first formula for τ is proved in B.232 (the last one in B.235′). Finally, the case $S < 0$ is treated in B.233. But such a problem may become impossible in some cases: thus if $\pi < p$, we must have $S \geq p - \pi$ since the first worker will at least be left with this difference (B.231–231′).

B. *Daily wages*.

In B.234, B.236–239 we no longer consider given monthly wages but daily ones (and thus t is not necessarily a month). The (single) worker receives p_0 as daily wage, but he will pay a penalty $\pi_0 - p_0$ for each day not worked. He will thus earn, for τ days worked, $p_0 \cdot \tau$, but will have to pay $(\pi_0 - p_0)(t - \tau)$. He will then leave with what he has earned less the whole penalty paid, thus
$$p_0 \cdot \tau - (\pi_0 - p_0)(t - \tau) = S, \quad \text{or} \quad \pi_0 \cdot \tau - (\pi_0 - p_0) t = S,$$
whence
$$\tau = \frac{(\pi_0 - p_0) t + S}{\pi_0},$$
and therefore
$$t - \tau = \frac{\pi_0 \cdot t - \pi_0 \cdot t + p_0 \cdot t - S}{\pi_0} = \frac{p_0 \cdot t - S}{\pi_0}.$$

These two relations become, for the case $S = 0$,
$$\tau = \frac{(\pi_0 - p_0) t}{\pi_0}, \quad t - \tau = \frac{p_0 \cdot t}{\pi_0}.$$

The text considers successively, as above, the cases $S = 0$ (B.234, B.236 & 239; in B.236 the formulae are proved), $S > 0$ (B.237, with proof),

$S < 0$ (B.238, with proof). The link with the problems on monthly wages is established in B.234 & B.239.

3. Three workers.

In the closing problems of this chapter, three workers are hired at respective monthly wages p_i. Between the three of them they complete the prescribed time t, each working τ_i days and receiving accordingly p_{τ_i}. Thus

$$t = \tau_1 + \tau_2 + \tau_3, \quad \text{and} \quad \frac{\tau_i}{t} = \frac{p_{\tau_i}}{p_i}.$$

Two cases are considered.

(i) The workers end up with an equal wage p' (B.240, B.241). Then, since

$$\frac{\tau_i}{t} = \frac{p'}{p_i},$$

we shall have

$$\frac{\sum \tau_k}{t} = 1 = p' \sum \frac{1}{p_k},$$

from which we infer that

$$p' = \frac{1}{\sum \frac{1}{p_k}}, \quad \tau_i = \frac{t}{p_i \sum \frac{1}{p_k}}.$$

(ii) The workers end up with wages p_{τ_i} differing by given amounts s_i (B.242). Suppose the wage p_{τ_3} is the lowest and $p_{\tau_1} = p_{\tau_3} + s_1$, $p_{\tau_2} = p_{\tau_3} + s_2$. Then, since

$$\frac{\tau_i}{t} = \frac{p_{\tau_i}}{p_i},$$

we shall have

$$\tau_i = \frac{t}{p_i} \cdot p_{\tau_i} = \frac{t}{p_i} \left(p_{\tau_3} + s_i \right) = \frac{t}{p_i} \cdot p_{\tau_3} + \frac{t}{p_i} \cdot s_i,$$

or, since $p_{\tau_3} \cdot t = p_3 \cdot \tau_3$,

$$\tau_i = \frac{p_3}{p_i} \cdot \tau_3 + \frac{t}{p_i} \cdot s_i.$$

The number of working days of the ith worker then consists of two parts: since $\frac{t}{p_i}$ is the number of days he takes to earn one nummus, the second term on the right-hand side represents the number of days worked to earn the extra s_i nummi, while the first term represents the number of days τ'_i he is to work to earn the common wage p_{τ_3}. Subtracting then from t the numbers of days corresponding to the two amounts s_i, we are left with the situation of three workers earning the same wage p_{τ_3} for working a different number of days; which is the situation of case i.

Remark. The author also considers that the common wage is not the lowest (B.242b). Some of the s_i then become subtractive, namely when the wage considered is lower than the reference wage.

The last problem, B.243, is of the same kind as B.240–241, except that the common wage $p_{T_i} = p'$ is given and the p_i's are multiples of a 'thing'.

---

1. One worker.

A. *Simple problems*.

(B.195) MSS $\mathcal{A}$ (without d), $\mathcal{B}$. The monthly wage is ten nummi; what is the wage for twelve days?

Given thus $t_1 = 30$ (month), $p_1 = 10$, $t_2 = 12$, required p_2. Analogous to B.1 (with t_i sextarii, p_i nummi).

(**a**) *Formula.*
$$p_2 = \frac{t_2 \cdot p_1}{t_1} = \frac{12 \cdot 10}{30} = 4.$$

(**b**) Also computed (first of the banal 'other ways') as
$$p_2 = \frac{t_2}{t_1} \cdot p_1 = \frac{12}{30} \cdot p_1 = \frac{2}{5} \cdot p_1.$$

(**c**) Considering $p_1 : t_1$, in the manner seen in B.171 & B.175,
$$p_1 = t_1 \left(1 - \frac{2}{3}\right), \text{ then } p_2 = t_2 \left(1 - \frac{2}{3}\right).$$

(**d**) Also computed (second 'other way', but here interpolated) as
$$p_2 = \frac{p_1}{t_1} \cdot t_2 = \frac{10}{30} \cdot t_2 = \frac{1}{3} \cdot t_2.$$

(**e**) Since, considering $t_2 : t_1$,
$$t_2 = t_1 \left(1 - \frac{3}{5}\right), \text{ then } p_2 = p_1 \left(1 - \frac{3}{5}\right).$$

(B.196) MSS $\mathcal{A}$, $\mathcal{B}$. The monthly wage is ten nummi; to how many working days do four nummi correspond?

Given thus $t_1 = 30$, $p_1 = 10$, $p_2 = 4$, required t_2 (same values as before). Analogous to B.2.

(**a**) *Formula.*
$$t_2 = \frac{p_2 \cdot t_1}{p_1} = \frac{4 \cdot 30}{10} = 12.$$

(**b**)—(**c**) Also computed as ('other ways')

$$t_2 = \frac{p_2}{p_1} \cdot t_1, \qquad t_2 = \frac{t_1}{p_1} \cdot p_2.$$

(**B.197**) MSS $\mathcal{A}$, $\mathcal{B}$. The first day's wage is one nummus and the wage increases each day by one nummus; what will the monthly wage be?

Given thus $p_1 = 1$, $\delta = 1$, required the wage for $t = 30$. The formula used, already encountered in B.52 and to be fully studied in the next chapter, is

$$p = \sum p_i = \frac{t}{2}\left((t-1)\delta + 2p_1\right).$$

In this case, where $t = 30$ and $\delta = 1 = p_1$, it reduces to the sum of the first thirty natural numbers, so p becomes equal to

$$\frac{t(t+1)}{2} = \frac{30 \cdot 31}{2} = 465.$$

B. *Problems involving 'things'*.

As said in our summary (p. 1496), we may use either the plain proportion or its transformed form, thus

$$\frac{p_1}{p_2} = \frac{t_1}{t_2} \quad \text{and} \quad \frac{p_1 \pm p_2}{p_2} = \frac{t_1 \pm t_2}{t_2}.$$

(**B.198**) MSS $\mathcal{A}$, $\mathcal{B}$. The monthly wage is ten nummi plus a thing and the wage for twelve days is one thing; what is the thing worth?

Given thus $t_1 = 30$, $p_1 = 10 + x$, $t_2 = 12$, $p_2 = x$, required x. See B.21, B.24.

(**a**) Transforming the proportion.

$$p_2 = x = \frac{t_2(p_1 - p_2)}{t_1 - t_2} = \frac{12 \cdot 10}{18} = 6 + \frac{2}{3}.$$

This is justified in two ways. First, since for one month $p_1 = x + 10$ and for twelve days $p_2 = x$, the wage $p_1 - p_2 = 10$ corresponds to $t_1 - t_2 = 18$ days worked; but the ratio wage to time for the remainder of the month is the same as for the days actually worked, and this proves the above relation. Alternatively, we may deduce the above formula by separation in the fundamental proportion.

(**b**) Without transforming the proportion. First, consider the wage for the month. Since

$$p_1 = \frac{t_1}{t_2} \cdot p_2,$$

we are led to the equation

$$x + 10 = \frac{30}{12} \cdot x = \left(2 + \frac{1}{2}\right)x, \quad \text{so} \quad \left(1 + \frac{1}{2}\right)x = 10.$$

The transformation of the partial wage p_2 to a month's wage is then explained and the general rule enunciated. It is pointed out that we shall end by applying 'algebra' in order to arrive at x equal to a number.

(**c**) Considering the wage for the days worked. (Here too we shall end with algebraic operations.)

$$\frac{t_2}{t_1} \cdot p_1 = p_2, \quad \text{thus} \quad \frac{2}{5}(x+10) = x, \quad \frac{3}{5}x = 4.$$

(**d**) By multiplication, which again leads to algebra. For $t_1 : p_1 = t_2 : p_2$ gives

$$t_1 \cdot p_2 = t_2 \cdot p_1, \quad \text{that is,} \quad 30\,x = 12\,(10+x).$$

(**e**) Verification. Since x is now known, we can compute the wage for the month as $p_1 = 10 + x = 16 + \frac{2}{3}$. He worked two fifths of the month to earn x, and two fifths of $16 + \frac{2}{3}$ indeed gives the value of x.

(**B.199**) MSS A, B. The monthly wage is ten nummi plus a thing and the wage for twelve days is a thing plus one nummus; what is the thing worth?

Given thus $t_1 = 30$, $p_1 = 10 + x$, $t_2 = 12$, $p_2 = x + 1$, required x. See B.24, B.25.

(**a**) Transforming the proportion.

$$p_2 = x + 1 = \frac{t_2\,(p_1 - p_2)}{t_1 - t_2} = \frac{12 \cdot 9}{18} = 6, \quad \text{thus} \quad x = 5.$$

Explained, as above, by considering the ratio wage to time applied to the remainder of the month, which must be equal to the ratio corresponding to the twelve days.

(**b**) Considering the wage for the month.

$$\frac{t_1}{t_2}(x+1) = \left(2 + \frac{1}{2}\right)(x+1),$$

and this is equal to $10 + x$, whence, after the algebraic operations, the solution.

(**c**) Considering the wage for the days worked.

$$\frac{t_2}{t_1}(10+x) = \frac{2}{5}(10+x) = 4 + \frac{2}{5}x,$$

and this is equal to $x + 1$, whence x.

(*d*) By multiplication, thus $t_1 \cdot p_2 = t_2 \cdot p_1$.

$$30(x+1) = 12(10+x), \quad 30x + 30 = 120 + 12x$$

whence x.

(**B.200**) MSS $\mathcal{A}$, $\mathcal{B}$. The monthly wage is ten nummi plus one thing and the wage for twelve days is a thing minus one nummus; what is the thing worth?

Given thus $t_1 = 30$, $p_1 = 10 + x$, $t_2 = 12$, $p_2 = x - 1$, required x. See B.23.

(*a*) Transforming the proportion.

$$p_2 = x - 1 = \frac{t_2(p_1 - p_2)}{t_1 - t_2} = \frac{12 \cdot 11}{18} = 7 + \frac{1}{3}, \quad \text{thus} \quad x = 8 + \frac{1}{3}.$$

(*b*) Considering the wage for the whole month.

$$p_1 = \frac{t_1}{t_2} \cdot p_2, \quad \text{thus} \quad 10 + x = \left(2 + \frac{1}{2}\right)(x-1)$$

whence x.

(*c*) Considering the wage for the days worked.

$$p_2 = \frac{t_2}{t_1} \cdot p_1, \quad \text{thus} \quad x - 1 = \frac{2}{5}(10+x) = 4 + \frac{2}{5}x.$$

(*d*) By multiplication.

$$t_1 \cdot p_2 = t_2 \cdot p_1, \quad \text{that is,} \quad 30(x-1) = 12(10+x),$$

whence x.

(**B.201**) MSS $\mathcal{A}$, $\mathcal{B}$. The monthly wage is ten nummi plus a thing and the wage for twelve days is ten nummi; what is the thing worth?

Given thus $t_1 = 30$, $p_1 = 10 + x$, $t_2 = 12$, $p_2 = 10$, required x. This problem, with just one term containing x, would be better put with B.217.

(*a*) Transforming the proportion.

$$p_1 - p_2 = x = \frac{(t_1 - t_2)p_2}{t_2} = \frac{18 \cdot 10}{12} = 15.$$

(*b*) Also computed as

$$p_1 - p_2 = x = \frac{t_1 - t_2}{t_2} \cdot p_2 = \left(1 + \frac{1}{2}\right)10.$$

(c) Considering the wage for the whole month.

$$p_1 = 10 + x = \frac{t_1}{t_2} \cdot p_2 = \left(2 + \frac{1}{2}\right) 10 = 25,$$

whence x.

Remark. Here an interpolated passage appears to contain elements of what may have been a verification, which normally would take the following form: since $x = 15$, so $x + 10 = 25$ is the monthly wage, two fifths of which is 10, the actual wage.

(d) Considering the wage for the days worked.

$$p_2 = 10 = \frac{t_2}{t_1} p_1 = \frac{2}{5}(10 + x) = 4 + \frac{2}{5}x,$$

whence x.

(e) By multiplication.

$$t_1 \cdot p_2 = t_2 \cdot p_1, \quad \text{that is,} \quad 30 \cdot 10 = 12(10 + x).$$

(B.202) MS $\mathcal{A}$. The monthly wage is ten nummi minus a thing and the wage for ten days is a thing plus two nummi; what is the thing worth? Given thus $t_1 = 30$, $p_1 = 10 - x$, $t_2 = 10$, $p_2 = x + 2$, required x.

$$p_2 = x + 2 = \frac{t_2(p_1 + p_2)}{t_1 + t_2} = \frac{10 \cdot 12}{40} = 3, \quad \text{so} \quad x = 1.$$

The justification explains the transformation of the ratio $p_1 : p_2 = t_1 : t_2$, now by composition.

(B.202′) The monthly wage is ten nummi minus a thing and the wage for twenty days is two things plus four nummi; what is the thing worth?

Given thus $t_1 = 30$, $p_1 = 10 - x$, $t_2 = 20$, $p_2 = 2x + 4$, required x. Reduced to B.202 since, for $t'_2 = 10$, the wage becomes $p'_2 = x + 2$.

(B.202″) The monthly wage is ten nummi minus a thing and the wage for five days is half a thing plus one nummus; what is the thing worth?

Given thus $t_1 = 30$, $p_1 = 10 - x$, $t_2 = 5$, $p_2 = \frac{1}{2} \cdot x + 1$, required x. Likewise reduced to B.202. Generally, the author concludes, whenever $p_2 = \frac{k}{l}x + m$ for t_2 days, consider $p'_2 = \frac{l}{k}p_2 = x + \frac{l \cdot m}{k}$, namely the wage for $t'_2 = \frac{l \cdot t_2}{k}$ days, which now contains just one thing.

(B.203) MS $\mathcal{A}$. The monthly wage is ten nummi plus a thing and the wage for three days is half a thing; what is the thing worth?

Given thus $t_1 = 30$, $p_1 = 10 + x$, $t_2 = 3$, $p_2 = \frac{1}{2}x$, required x. Take $t'_2 = 6$, $p'_2 = x$ (the new problem is similar to B.198, the first problem of this section involving things).

Remark. There is also the possibility of making appear in p_1 the fraction found in p_2, as the next group B.204 will incidentally illustrate.

(B.204) MSS $\mathcal{A}$, $\mathcal{B}$. The monthly wage is six nummi plus a thing and the wage for ten days is six nummi minus half a thing; what is the thing worth?

Given thus $t_1 = 30$, $p_1 = 6 + x$, $t_2 = 10$, $p_2 = 6 - \frac{1}{2}x$, required x. Taking $t'_1 = 15$, $p'_1 = 3 + \frac{1}{2}x$, we determine x from

$$x = \frac{(t_1 - t_2)\,6}{t'_1 + t_2} = \frac{20 \cdot 6}{25} = \frac{120}{25} = 4 + \frac{4}{5}.$$

Remark. The author, who is usually so willing to justify the most banal transformations, does not explain the basis for his computations here; their purpose is, however, clear: obtaining directly x equal to a number. Consider, generally, that we are given t_1, $p_1 = b + x$, t_2, $p_2 = b - \frac{k}{l}x$. As in the previous problems, we may change one pair and take $t'_1 = \frac{k}{l} \cdot t_1$, giving as the corresponding wage $p'_1 = \frac{k}{l} \cdot b + \frac{k}{l} \cdot x$. Turning now to the formula actually used in the text, it is found as follows. From

$$\frac{p_1}{p_2} = \frac{t_1}{t_2}, \qquad \frac{p'_1}{p_2} = \frac{t'_1}{t_2},$$

we infer respectively

$$\frac{p_1 - p_2}{p_2} = \frac{t_1 - t_2}{t_2}, \qquad \frac{p'_1 + p_2}{p_2} = \frac{t'_1 + t_2}{t_2};$$

therefore

$$\frac{p_2}{t_2} = \frac{p_1 - p_2}{t_1 - t_2} = \frac{p'_1 + p_2}{t'_1 + t_2},$$

whence

$$p_1 - p_2 = \frac{(t_1 - t_2)(p'_1 + p_2)}{t'_1 + t_2}, \quad \left(1 + \frac{k}{l}\right)x = \frac{(t_1 - t_2) \cdot b\left(1 + \frac{k}{l}\right)}{t'_1 + t_2},$$

and finally

$$x = \frac{(t_1 - t_2)\,b}{t'_1 + t_2}.$$

(B.204′) The monthly wage is again six nummi plus a thing and the wage for ten days is six nummi minus two thirds of a thing; what is the thing worth?

Given thus $t_1 = 30$, $p_1 = 6 + x$, $t_2 = 10$, but with $p_2 = 6 - \frac{2}{3}x$. Take then

$$t'_1 = \frac{2}{3} \cdot t_1 = 20, \quad p'_1 = \frac{2}{3} p_1 = 4 + \frac{2}{3}x,$$

and the same formula gives

$$x = \frac{20 \cdot 6}{30} = 4.$$

(**B.204″**) The monthly wage is again six nummi plus a thing and the wage for ten days is six nummi minus a thing; what is the thing worth?

Given thus $t_1 = 30$, $p_1 = 6+x$, $t_2 = 10$, but with $p_2 = 6-x$. Putting $t'_1 = t_1$ (thus $p'_1 = p_1$) gives

$$x = \frac{20 \cdot 6}{40} = 3.$$

Remark. This could have been solved by the simple proportion $p_1 : p_2 = t_1 : t_2$. But the above formula, once again, makes it possible to have x directly equal to a number, thus no 'algebra' is involved.

* * *

B.205–206 involve roots in the data. Their present place is questionable since they interrupt a sequence of closely related problems.

(**B.205**) MSS *A*, *B*. The monthly wage is three nummi plus a thing and the wage for ten days is three times the root of that wage; what is the thing worth?

Given here $t_1 = 30$, $p_1 = 3+x$, $t_2 = 10$, $p_2 = 3\sqrt{3+x}$, required x. B.205–206 are of the same kind as B.39–40.

(*a*) Considering the wage for the whole month.

$$p_1 = \frac{t_1}{t_2} \cdot p_2 \quad \text{thus} \quad 3+x = 3 \cdot 3\sqrt{3+x},$$

whence $\sqrt{3+x} = 9$, $3+x = 81$ and $x = 78$.

(*b*) By multiplication. The plain proportion $t_1 : p_1 = t_2 : p_2$ gives

$$\frac{30}{3+x} = \frac{10}{3\sqrt{3+x}} = \frac{10}{\sqrt{27+9x}},$$

Remark. Here is explained the transformation just performed of $3\sqrt{3+x}$ into $\sqrt{27+9x}$, and reference is made to Abū Kāmil's *Algebra*, where the operation $k\sqrt{a} = \sqrt{k^2 \cdot a}$ is indeed taught by means of particular examples and a geometrical proof.* As said in the Translation (p. 903, note 1362), this reference is odd since, first, the subject has been treated in the *Liber mahameleth* itself (A.285) and, second, in a similar situation (B.95) we are indeed referred to Book A.

---

* Arabic text, fol. 17^r, line 9 *seqq.*; Latin translation, line 799 *seqq.*

Considering now $30\left(\sqrt{27+9x}\right) = 10\left(3+x\right)$, we obtain successively

$$\sqrt{24\,300 + 8100x} = 30 + 10x,$$

$$24\,300 + 8100\,x = 900 + 600\,x + 100\,x^2,$$

$$100\,x^2 = 7500\,x + 23\,400,$$

whence $x^2 = 75\,x + 234$, with the positive solution

$$x = \sqrt{\left(\frac{75}{2}\right)^2 + 234} + \frac{75}{2} = \sqrt{1406 + \frac{1}{4} + 234} + 37 + \frac{1}{2}$$

$$= \sqrt{1640 + \frac{1}{4}} + 37 + \frac{1}{2} = 40 + \frac{1}{2} + 37 + \frac{1}{2} = 78.$$

Remark. This is one of the few instances in the *Liber mahameleth* where the quadratic equation is established and the solving formula presented quite explicitly. See also the next problem. But reducing this problem to a second-degree equation is quite unnecessary since from $10\left(3+x\right) = 90 \cdot \sqrt{3+x}$ we could have immediately inferred that, as in a, $\sqrt{3+x} = 9$.

(*c*) *Verification.* The wage for the whole month is $p_1 = 3 + x = 81$; thus for ten days it will be 27, which is indeed three times the root of the wage for the month.

(**B.206**) MSS $\mathcal{A}$, $\mathcal{B}$. The monthly wage is thirty nummi plus a thing and the wage for ten days is a thing plus the root of that wage; what is the thing worth?

Before, we had $p_2 = k\sqrt{p_1}$; here, $p_2 = x + \sqrt{p_1}$. Given indeed $t_1 = 30$, $p_1 = 30 + x$, $t_2 = 10$, $p_2 = x + \sqrt{30 + x}$, required x.

(*a*) Considering the wage for the days worked, we have

$$p_2 = \frac{t_2}{t_1} \cdot p_1 = \frac{1}{3}(30 + x) = 10 + \frac{1}{3}x,$$

which equals $x + \sqrt{30 + x}$. Thus, as computed successively,

$$10 - \frac{2}{3}x = \sqrt{30 + x},$$

$$\frac{4}{9}x^2 + 100 - \left(13 + \frac{1}{3}\right)x = 30 + x,$$

$$\frac{4}{9}x^2 + 70 = \left(14 + \frac{1}{3}\right)x,$$

$$x^2 + 157 + \frac{1}{2} = \left(32 + \frac{1}{4}\right)x,$$

with the solution

$$x = 16 + \frac{1}{8} - \sqrt{\left(16 + \frac{1}{8}\right)^2 - \left(157 + \frac{1}{2}\right)}$$

$$= 16 + \frac{1}{8} - \sqrt{260 + \frac{1}{8}\frac{1}{8}} - \left(157 + \frac{1}{2}\right) = 16 + \frac{1}{8} - \sqrt{102 + \frac{4}{8} + \frac{1}{8}\frac{1}{8}}$$

$$= 16 + \frac{1}{8} - \left(10 + \frac{1}{8}\right) = 6.$$

Remark. Since we have an equation of the form $x^2 + q = px$, there is a second positive solution, equal to $26 + \frac{1}{4}$. But this will not satisfy the conditions as it would make, as a third of $30 + x$, $x - \sqrt{30 + x}$.

(*b*) By multiplication. Since $t_1 \cdot p_2 = t_2 \cdot p_1$,

$$30\left(x + \sqrt{30 + x}\right) = 10\left(30 + x\right),$$

$$30\,x + \sqrt{27\,000 + 900\,x} = 300 + 10x,$$

$$\sqrt{27\,000 + 900\,x} = 300 - 20\,x,$$

$$27\,000 + 900\,x = 400x^2 + 90\,000 - 12\,000\,x,$$

$$400\,x^2 + 63\,000 = 12\,900\,x,$$

$$x^2 + 157 + \frac{1}{2} = \left(32 + \frac{1}{4}\right)x,$$

and continue as above.

(*c*) Considering the wage for the whole month.

$$p_1 = \frac{t_1}{t_2}\,p_2, \quad \text{thus} \quad 30 + x = 3\left(x + \sqrt{30 + x}\right),$$

and continuing as before.

(*d*) Verification. Since the wage for the month is $30 + x$, thus 36, the wage for the days worked will be 12. Now this is indeed $x + \sqrt{30 + x}$.

* * *

We now return (B.207–217) to a type of problem already encountered (B.198–204).

(**B.207**) MSS $\mathcal{A}$, $\mathcal{B}$. The monthly wage is ten nummi plus a thing and the wage for twelve days is ten nummi minus a thing; what is the thing worth?

Given thus $t_1 = 30$, $p_1 = 10 + x$, $t_2 = 12$, $p_2 = 10 - x$, required x. See B.202 and B.204 (similar problems).

(*a*) Equating the month's wages.

$$\frac{t_1}{t_2}\,p_2 = p_1, \quad \text{so} \quad \left(2 + \frac{1}{2}\right)(10 - x) = 10 + x,$$

$$25 - \left(2 + \frac{1}{2}\right)x = 10 + x, \quad 15 = \left(3 + \frac{1}{2}\right)x, \quad x = 4 + \frac{2}{7}.$$

• Verification. The wage for the month is $p_1 = 10 + x = 14 + \frac{2}{7}$, and two fifths of it, for 12 days, is $5 + \frac{5}{7}$. Now this is indeed $10 - x$.

(b) By multiplication.

$$t_1 \cdot p_2 = t_2 \cdot p_1, \quad \text{thus} \quad 30(10 - x) = 12(10 + x),$$

whence x.

(c) Equating the wages for the days worked.

$$\frac{t_2}{t_1} p_1 = p_2, \quad \text{thus} \quad \frac{2}{5}(10 + x) = 10 - x, \quad 4 + \frac{2}{5}x = 10 - x,$$

whence x.

(B.208) MSS $\mathcal{A}$, $\mathcal{B}$. The monthly wage is ten nummi plus a thing and the wage for twelve days is ten nummi minus a fourth of a thing; what is the thing worth?

Given thus $t_1 = 30$, $p_1 = 10 + x$, $t_2 = 12$, $p_2 = 10 - \frac{1}{4}x$, required x. Same as the preceding, but with a fraction of x subtracted in p_2. See B.204, B.204′.

(a) Equating the month's wages.

$$\frac{t_1}{t_2} p_2 = p_1, \quad \text{thus} \quad \left(2 + \frac{1}{2}\right)\left(10 - \frac{1}{4}x\right) = 10 + x, \quad \text{whence}$$

$$25 - \frac{5}{8} \cdot x = 10 + x, \quad 15 = \left(1 + \frac{5}{8}\right)x, \quad x = \frac{8}{13} \cdot 15 = 9 + \frac{3}{13}.$$

• Verification. Since $p_1 = 10 + x = 19 + \frac{3}{13}$, so $p_2 = \frac{2}{5} p_1 = 7 + \frac{9}{13}$, which is indeed $10 - \frac{1}{4}x$.

(b) Equating the wages for the days worked.

$$\frac{t_2}{t_1} p_1 = p_2, \quad \text{thus} \quad \frac{2}{5}(10 + x) = 10 - \frac{1}{4}x, \quad \text{whence}$$

$$4 + \frac{2}{5}x = 10 - \frac{1}{4}x, \quad \frac{13}{20}x = 6, \quad x = \left(1 + \frac{7}{13}\right)6 = 9 + \frac{3}{13}.$$

(c) By multiplication.

$$t_1 \cdot p_2 = t_2 \cdot p_1, \quad \text{thus} \quad 30\left(10 - \frac{1}{4} \cdot x\right) = 12(10 + x).$$

(B.209) MSS $\mathcal{A}$, $\mathcal{B}$. The monthly wage is ten nummi plus a thing and the wage for ten days is a thing minus a third of ten nummi; what is the thing worth?

Given thus $t_1 = 30$, $p_1 = 10 + x$, $t_2 = 10$, $p_2 = x - \frac{1}{3} \cdot 10$, required x.

(*a*) Equating the month's wages.

$$\frac{t_1}{t_2} p_2 = p_1, \quad \text{thus} \quad 3\left(x - \frac{1}{3}10\right) = 10 + x,$$

so $3x - 10 = 10 + x$, and $x = 10$.
- Verification. Since $p_1 = 10 + x = 20$, so $p_2 = \frac{1}{3}p_1 = 6 + \frac{2}{3}$, which is indeed $x - \frac{1}{3} \cdot 10$.

(*b*) By multiplication.

$$t_1 \cdot p_2 = t_2 \cdot p_1, \quad \text{thus} \quad 30\left[x - \left(3 + \frac{1}{3}\right)\right] = 10\left(10 + x\right).$$

(*c*) Equating the wages for the days worked.

$$\frac{t_2}{t_1} p_1 = p_2, \quad \text{thus} \quad 3 + \frac{1}{3} + \frac{1}{3}x = x - \left(3 + \frac{1}{3}\right).$$

(**B.210**) MSS $\mathcal{A}$, $\mathcal{B}$. The monthly wage is ten nummi plus a thing and the wage for twelve days is a thing minus ten nummi; what is the thing worth?

Given thus $t_1 = 30$, $p_1 = 10 + x$, $t_2 = 12$, $p_2 = x - 10$, required x.

(*a*) Forming the wage for the days not served, thus using separation in $p_1 : p_2 = t_1 : t_2$ ($p_1 > p_2$).

$$p_2 = x - 10 = \frac{t_2(p_1 - p_2)}{t_1 - t_2} = \frac{12 \cdot 20}{18} = \frac{240}{18} = 13 + \frac{1}{3}, \quad x = 23 + \frac{1}{3}.$$

The justification is of the same kind as in B.198*a*, first 'reason': considering the wage for the remainder of the month.

(*b*) Equating the month's wages.

$$\frac{t_1}{t_2} p_2 = p_1, \quad \text{thus} \quad \left(2 + \frac{1}{2}\right)(x - 10) = 10 + x,$$

whence x.

(*c*) Verification. Since $p_1 = 10 + x = 33 + \frac{1}{3}$, so $p_2 = \frac{2}{5}p_1 = 13 + \frac{1}{3}$, which is indeed $x - 10$.

(*d*) Considering the wages for the days worked.

$$\frac{t_2}{t_1} p_1 = p_2, \quad \text{thus} \quad \frac{2}{5}(10 + x) = x - 10, \quad \text{and} \quad \frac{3}{5}x = 14.$$

(*e*) By multiplication.

$$t_1 \cdot p_2 = t_2 \cdot p_1, \quad \text{thus} \quad 30\left(x - 10\right) = 12\left(10 + x\right).$$

(B.211) MS $\mathcal{A}$. The monthly wage is twenty nummi plus a thing and the wage for ten days is twelve nummi minus a thing; what is the thing worth?

Given thus $t_1 = 30$, $p_1 = 20 + x$, $t_2 = 10$, $p_2 = 12 - x$, required x. By composition in $p_1 : p_2 = t_1 : t_2$, we find

$$p_2 = 12 - x = \frac{t_2(p_1 + p_2)}{t_1 + t_2} = \frac{10 \cdot 32}{40} = 8, \quad x = 4.$$

The author adds that a problem with mx instead of x in one p_i will be treated in the same way (changing $p_i = k + mx$ to $p'_i = \frac{k}{m} + x$ and t accordingly, see B.202′–202″): the x's will disappear by composition.

(B.212) MSS $\mathcal{A}$, $\mathcal{B}$. The monthly wage is ten nummi minus a thing and the wage for twelve days is a thing minus two nummi; what is the thing worth?

Given thus $t_1 = 30$, $p_1 = 10 - x$, $t_2 = 12$, $p_2 = x - 2$, required x.

(*a*) Equating the month's wages.

$$\frac{t_1}{t_2} p_2 = p_1, \quad \text{thus} \quad \left(2 + \frac{1}{2}\right)(x - 2) = 10 - x, \quad \text{whence}$$

$$\left(2 + \frac{1}{2}\right)x - 5 = 10 - x, \quad \left(3 + \frac{1}{2}\right)x = 15, \quad x = 4 + \frac{2}{7}.$$

(*b*) Equating the wages for the days worked.

$$\frac{t_2}{t_1} p_1 = p_2, \quad \text{thus} \quad \frac{2}{5}(10 - x) = x - 2, \quad 4 - \frac{2}{5}x = x - 2.$$

(*c*) By multiplication.

$$t_1 \cdot p_2 = t_2 \cdot p_1, \quad \text{thus} \quad 30(x - 2) = 12(10 - x).$$

(B.213) MS $\mathcal{A}$. The monthly wage is ten nummi minus a thing and the wage for six days is four nummi minus a thing; what is the thing worth?

Given thus $t_1 = 30$, $p_1 = 10 - x$, $t_2 = 6$, $p_2 = 4 - x$, required x. Using separation in $p_1 : p_2 = t_1 : t_2$ ($p_1 > p_2$), we find

$$p_2 = 4 - x = \frac{t_2(p_1 - p_2)}{t_1 - t_2} = \frac{36}{24} = \frac{3}{2}, \quad \text{leading to} \quad x = 2 + \frac{1}{2}.$$

Remark. The aim in separating the ratio must have been to have x on one side only; for equating the two expressions for the month's wage, thus $5p_2 = p_1$, seems simpler.

(B.214) MSS $\mathcal{A}$, $\mathcal{B}$. The monthly wage is a thing and the wage for ten days is a thing minus ten nummi; what is the thing worth?

Given thus $t_1 = 30$, $p_1 = x$, $t_2 = 10$, $p_2 = x - 10$, required x.

(**a**) Equating the month's wages.

$$\frac{t_1}{t_2} p_2 = p_1, \quad \text{thus} \quad 3(x-10) = x, \quad x = 15.$$

(**b**) Equating the wages for the days worked.

$$\frac{t_2}{t_1} p_1 = p_2, \quad \text{thus} \quad \frac{1}{3}x = x - 10.$$

(**c**) By multiplication.

$$t_1 \cdot p_2 = t_2 \cdot p_1, \quad \text{thus} \quad 30(x-10) = 10x.$$

(**B.215**) MSS $\mathcal{A}$, $\mathcal{B}$. The monthly wage is a thing and the wage for ten days is a fourth of a thing plus two nummi; what is the thing worth?
Given thus $t_1 = 30$, $p_1 = x$, $t_2 = 10$, $p_2 = \frac{1}{4}x + 2$, required x.
(**a**) Considering the month's wage.

$$\frac{t_1}{t_2} p_2 = p_1, \quad \text{thus} \quad 3\left(\frac{1}{4}x + 2\right) = x, \quad x = 24.$$

(**b**) Considering the wage for the days worked.

$$\frac{t_2}{t_1} p_1 = p_2, \quad \text{thus} \quad \frac{1}{3}x = \frac{1}{4}x + 2, \quad \frac{1}{2}\frac{1}{6}x = 2.$$

(**c**) By multiplication.

$$t_1 \cdot p_2 = t_2 \cdot p_1, \quad \text{thus} \quad 30\left(\frac{1}{4}x + 2\right) = 10x.$$

(**B.216**) MS $\mathcal{B}$. The monthly wage is a thing minus ten nummi and the wage for twelve days is a fourth of a thing plus two nummi; what is the thing worth?
Given $t_1 = 30$, $p_1 = x - 10$, $t_2 = 12$, $p_2 = \frac{1}{4}x + 2$, required x.
(**a**) Considering the month's wage.

$$\frac{t_1}{t_2} p_2 = p_1, \quad \text{thus} \quad \left(2 + \frac{1}{2}\right)\left(\frac{1}{4}x + 2\right) = x - 10, \quad \frac{5}{8}x + 5 = x - 10,$$

so $x = 40$.
- Verification. Since $p_1 = x - 10 = 30$, so $p_2 = \frac{2}{5}p_1 = 12$, which is indeed $\frac{1}{4}x + 2$.

(**b**) Considering the wage for the days worked.

$$\frac{t_2}{t_1} p_1 = p_2, \quad \text{thus} \quad \frac{2}{5}(x - 10) = \frac{1}{4}x + 2,$$

$$\frac{3}{4}\frac{1}{5}x = 6, \quad x = \left(6 + \frac{2}{3}\right)6 = 40.$$

(*c*) By multiplication.

$$t_1 \cdot p_2 = t_2 \cdot p_1, \quad \text{thus} \quad 30\left(\frac{1}{4}x + 2\right) = 12\left(x - 10\right).$$

(**B.217**) MS *B*. The monthly wage is a thing minus ten nummi and the wage for twelve days is ten nummi; what is the thing worth?

Given thus $t_1 = 30$, $p_1 = x - 10$, $t_2 = 12$, $p_2 = 10$, required x. Similar to B.201 (where $p_1 = x + 10$).

(*a*) Considering the month's wage.

$$\frac{t_1}{t_2}p_2 = p_1, \quad \text{thus} \quad \left(2 + \frac{1}{2}\right)10 = x - 10, \quad x = 35.$$

(*b*) By multiplication.

$$t_1 \cdot p_2 = t_2 \cdot p_1, \quad \text{thus} \quad 30 \cdot 10 = 12\left(x - 10\right).$$

Remark. Forming $t_2 : (t_1 + t_2) = p_2 : (p_1 + p_2)$ would have given x directly.

C. *Solution formulae involving roots*.

The two problems B.218–219 lead to notably complicated formulae, involving roots, which are fully demonstrated in the text.

(**B.218**) MS *B*. The monthly wage is a thing, the days worked multiplied by the corresponding wage produce six, and the days not worked multiplied by the corresponding wage produce twenty-four; what are the unknown quantities?

Given thus $t_1 = 30$, $t_2 \cdot p_2 = 6$, $(t_1 - t_2)(p_1 - p_2) = 24$, required $p_1 = x$, $t_1 - t_2$, t_2. See B.35.

Formulae.

$$x = p_1 = \frac{2\sqrt{t_2 \cdot p_2\,(t_1 - t_2)(p_1 - p_2)} + t_2 \cdot p_2 + (t_1 - t_2)(p_1 - p_2)}{t_1}$$

$$= \frac{2\sqrt{6 \cdot 24} + 6 + 24}{30} = \frac{2\sqrt{144} + 30}{30} = \frac{24 + 30}{30} = \frac{54}{30} = 1 + \frac{4}{5}$$

while (calculating $t_1 - t_2$ and t_2 separately)

$$t_1 - t_2 = \frac{\sqrt{t_2 \cdot p_2\,(t_1 - t_2)(p_1 - p_2)} + (t_1 - t_2)(p_1 - p_2)}{p_1} = \frac{12 + 24}{1 + \frac{4}{5}} = 20,$$

$$t_2 = \frac{\sqrt{t_2 \cdot p_2\,(t_1 - t_2)(p_1 - p_2)} + t_2 \cdot p_2}{p_1} = \frac{12 + 6}{1 + \frac{4}{5}} = 10.$$

Remark. We have already noted the presence in Chuquet's work of a problem from the *Liber mahameleth* (B.175). The present problem (MS $\mathcal{B}$ only) is also found in the appendix to the *Triparty*, where the answer is given directly, without computation.*

Demonstration.

(α) *Preliminary remark.* By the constancy of the ratio time to wage we know that

$$\frac{t_2}{p_2} = \frac{t_1 - t_2}{p_1 - p_2}, \quad \text{or} \quad \frac{t_2}{t_1 - t_2} = \frac{p_2}{p_1 - p_2} \quad (1).$$

From the first form we infer that

$$t_2 \left(p_1 - p_2\right) = \left(t_1 - t_2\right) p_2 \quad (2)$$

whence

$$t_2 \cdot p_1 = \left(t_1 - t_2\right) p_2 + t_2 \cdot p_2. \quad (\star)$$

From the second form we infer that

$$\frac{t_2 \cdot p_2}{(t_1 - t_2) p_2} = \frac{(t_1 - t_2) p_2}{(t_1 - t_2)(p_1 - p_2)} \quad (3)$$

and therefore

$$\left[(t_1 - t_2) p_2\right]^2 = t_2 \cdot p_2 (t_1 - t_2)(p_1 - p_2), \quad \text{or}$$

$$(t_1 - t_2) p_2 = \sqrt{t_2 \cdot p_2 (t_1 - t_2)(p_1 - p_2)} \quad (4)$$

the left side of which is now determined since the quantity under the radical sign is known.

Let us next consider the two quantities $(t_1 - t_2) p_1$ and $t_1 \cdot p_1$ and express them in another form. The first quantity can be written as

$$(t_1 - t_2) p_1 = (t_1 - t_2) p_2 + (t_1 - t_2)(p_1 - p_2) \quad (\star\star)$$

---

* *Ung homme a loué ung serviteur par l'espace de 30 jours à certain pris, et l'a servy ledit serviteur tant de jours que multiplié l'argent qu'il doit avoir de ses jours qu'il a servy par lesdits jours monte 6, et multiplié l'argent qu'il devroit avoir des jours qu'il n'a pas servy par iceulx jours monte 24. Et noteras que le pris d'ung chascun jour est consideré à la rayson de ce qu'il gangne en 30 jours. L'on demande quantz jours il a servy et quantz non, et combien il gangne en 30 jours et en ung chascun jour. Responce. Procede selon la rigle de mediacion entre le plus et le moins, et trouveras que ledit serviteur a servy 10 jours, et 20 jours qu'il n'a pas servy. Item il a gangné $\frac{3}{5}$ par le temps qu'il a servy, et eusst gangné $1\frac{1}{5}$ par le temps qu'il n'a pas servy. Par ainsi il eust gangné $1\frac{4}{5}$ en 30 jours, qui est $\frac{3}{50}$ par jour.* See MS français 1346 (Bibliothèque Nationale de France), fol. 162^v (this problem in the margin; not in Marre's extract).

and the second one as

$$\begin{aligned} t_1 \cdot p_1 &= (t_1 - t_2)\, p_1 + t_1 \cdot p_2 \quad \text{(since } t_2 \cdot p_1 = t_1 \cdot p_2\text{)} \\ &= (t_1 - t_2)\, p_1 + (t_1 - t_2)\, p_2 + t_2 \cdot p_2 \\ &= (t_1 - t_2)(p_1 + p_2) + t_2 \cdot p_2 \\ &= (t_1 - t_2)(2 p_2 + p_1 - p_2) + t_2 \cdot p_2 \\ &= 2(t_1 - t_2)\, p_2 + (t_1 - t_2)(p_1 - p_2) + t_2 \cdot p_2. \end{aligned} \quad (5)$$

— From (5) and using (4), we find

$$\begin{aligned} p_1 &= \frac{2(t_1 - t_2)\, p_2 + (t_1 - t_2)(p_1 - p_2) + t_2 \cdot p_2}{t_1} \\ &= \frac{2\sqrt{t_2 \cdot p_2 (t_1 - t_2)(p_1 - p_2)} + (t_1 - t_2)(p_1 - p_2) + t_2 \cdot p_2}{t_1}, \end{aligned} \quad (6)$$

which was the formula by means of which x was computed.

— From ($\star\star$) and using (4), we find

$$\begin{aligned} t_1 - t_2 &= \frac{(t_1 - t_2)\, p_2 + (t_1 - t_2)(p_1 - p_2)}{p_1} \\ &= \frac{\sqrt{t_2 \cdot p_2 (t_1 - t_2)(p_1 - p_2)} + (t_1 - t_2)(p_1 - p_2)}{p_1}, \end{aligned} \quad (7)$$

which was the formula by means of which $t_1 - t_2$ was computed.

— From ($\star$) and using (4), we find

$$\begin{aligned} t_2 &= \frac{(t_1 - t_2)\, p_2 + t_2 \cdot p_2}{p_1} \\ &= \frac{\sqrt{t_2 \cdot p_2 (t_1 - t_2)(p_1 - p_2)} + t_2 \cdot p_2}{p_1}, \end{aligned} \quad (8)$$

which was used to compute t_2.

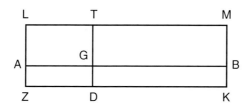

Fig. 70

(β) *Demonstration according to the text.* Let (Fig. 70) $t_1 = AB$, $t_2 = AG$, thus $t_1 - t_2 = GB$, $x = p_1 = DT$ and $p_2 = DG$ ($p_1 > p_2$). We know that $AG \cdot DG = 6 =$ area $AZDG$ and $GB \cdot GT = 24 =$ area $GBMT$. Complete the area $LZKM$. Since $t_2 : p_2 = (t_1 - t_2) : (p_1 - p_2)$,

$$\frac{AG}{DG} = \frac{GB}{GT} \quad \text{or} \quad \frac{AG}{GB} = \frac{DG}{GT}. \qquad (1)$$

This means (*Elements* VI.22) that the sides of the areas $AGTL$ and $GDKB$ are reciprocally proportional. Thus (*Elements* VI.14)

$$AGTL = GDKB. \qquad (2)$$

Now, by *Elements* VI.1 (rectangles with one side the same),

$$\frac{AZDG}{GDKB} = \frac{AG}{GB}.$$

For the same reason

$$\frac{GDKB}{GBMT} = \frac{DG}{GT}.$$

But, as seen

$$\frac{AG}{GB} = \frac{DG}{GT}, \quad \text{so} \quad \frac{AZDG}{GDKB} = \frac{GDKB}{GBMT}, \qquad (3)$$

whence

$$AZDG \cdot GBMT = (GDKB)^2, \quad \text{or}$$

$$GDKB = \sqrt{AZDG \cdot GBMT}. \qquad (4)$$

But we know the values of the right-hand terms ($AZDG = 6$ and $GBMT = 24$). So $GDKB = 12 = AGTL$. The four rectangles being now known, the whole area is

$$LZKM = GDKB + AGTL + AZDG + GBMT = 2 \cdot 12 + 6 + 24 = 54. \qquad (5)$$

We may therefore now compute p_1, $t_1 - t_2$, t_2, for

$$p_1 = DT = KM = \frac{LZKM}{ZK} = \frac{54}{30}, \qquad (6)$$

$$t_1 - t_2 = GB = DK = \frac{TDKM}{DT} = \frac{GDKB + GBMT}{DT} = \frac{12 + 24}{1 + \frac{4}{5}}, \qquad (7)$$

$$t_2 = AG = ZD = \frac{LZDT}{DT} = \frac{AGTL + AZDG}{DT} = \frac{12 + 6}{1 + \frac{4}{5}}. \qquad (8)$$

(B.219) MS $\mathcal{A}$. The monthly wage is a thing, the days served multiplied by the corresponding wage produce twenty-seven, and the days not served multiplied by the corresponding wage produce a hundred and forty-seven; what are the unknown quantities?

Given thus $t_1 = 30$, $t_2 \cdot p_2 = 27$, $(t_1 - t_2)(p_1 - p_2) = 147$, required $p_1 = x$, $t_1 - t_2$, t_2.

Remark. Same problem as the preceding, but this time p_2 and $p_1 - p_2$ are computed first. Since $p_2 \cdot t_1 = p_1 \cdot t_2$ and $p_1 (t_1 - t_2) = (p_1 - p_2) t_1$, the two formulae below can be inferred from the previous one. Note, however, that each of the two manuscripts has just one of these two problems.

Formulae.
$$p_2 = \frac{\sqrt{t_2 \cdot p_2 (t_1 - t_2)(p_1 - p_2)} + t_2 \cdot p_2}{t_1}$$
$$= \frac{\sqrt{27 \cdot 147} + 27}{30} = \frac{63 + 27}{30} = 3,$$

whence $\quad t_2 = \dfrac{t_2 \cdot p_2}{p_2} = \dfrac{27}{3} = 9\,;$

$$p_1 - p_2 = \frac{\sqrt{t_2 \cdot p_2 (t_1 - t_2)(p_1 - p_2)} + (t_1 - t_2)(p_1 - p_2)}{t_1}$$
$$= \frac{\sqrt{27 \cdot 147} + 147}{30} = \frac{63 + 147}{30} = 7,$$

whence $\quad t_1 - t_2 = \dfrac{(t_1 - t_2)(p_1 - p_2)}{p_1 - p_2} = \dfrac{147}{7} = 21\quad$ and

$$p_1 = x = p_2 + (p_1 - p_2) = 3 + 7 = 10.$$

Demonstration.

(α) *Preliminary remark.* As in the previous problem (formulae 2 and 4) we have
$$t_2 (p_1 - p_2) = (t_1 - t_2) p_2 \qquad (i)$$
$$(t_1 - t_2) p_2 = \sqrt{t_2 \cdot p_2 (t_1 - t_2)(p_1 - p_2)}. \qquad (ii)$$

Adding $t_2 \cdot p_2$ to both sides, we find
$$t_1 \cdot p_2 = \sqrt{t_2 \cdot p_2 (t_1 - t_2)(p_1 - p_2)} + t_2 \cdot p_2,$$

so we can calculate
$$p_2 = \frac{\sqrt{t_2 \cdot p_2 (t_1 - t_2)(p_1 - p_2)} + t_2 \cdot p_2}{t_1} \qquad (iii)$$

from which we determine
$$t_2 = \frac{t_2 \cdot p_2}{p_2}. \qquad (iv)$$

By the first two relations (i) and (ii), we have
$$t_2 (p_1 - p_2) = \sqrt{t_2 \cdot p_2 (t_1 - t_2)(p_1 - p_2)}\,;$$

adding in common $(t_1 - t_2)(p_1 - p_2)$, we obtain

$$t_1(p_1 - p_2) = \sqrt{t_2 \cdot p_2 (t_1 - t_2)(p_1 - p_2)} + (t_1 - t_2)(p_1 - p_2), \qquad (v)$$

so we can calculate

$$p_1 - p_2 = \frac{\sqrt{t_2 \cdot p_2 (t_1 - t_2)(p_1 - p_2)} + (t_1 - t_2)(p_1 - p_2)}{t_1} \qquad (vi)$$

from which we determine

$$t_1 - t_2 = \frac{(t_1 - t_2)(p_1 - p_2)}{p_1 - p_2}. \qquad (vii)$$

(β) *Demonstration according to the text.* Let (Fig. 71) $p_1 = x = AB$, $t_1 = 30 = DH$, $t_2 = DZ$, $p_2 = AG$, thus $t_1 - t_2 = ZH$ and $p_1 - p_2 = GB$. Let also $AG \cdot DZ = K$, $GB \cdot ZH = T$, with $K = 27$ and $T = 147$ by hypothesis. Since

$$\frac{AG}{DZ} = \frac{GB}{ZH}, \quad \text{so}$$

$$AG \cdot ZH = GB \cdot DZ. \qquad (i)$$

We now form the product $K \cdot T = (AG \cdot DZ)(GB \cdot ZH)$. Then, by Premiss P_3,

$$K \cdot T = (AG \cdot ZH)(GB \cdot DZ),$$

whence, using (i),

$$\sqrt{K \cdot T} = \sqrt{(AG \cdot ZH)(GB \cdot DZ)} = AG \cdot ZH = GB \cdot DZ. \qquad (ii)$$

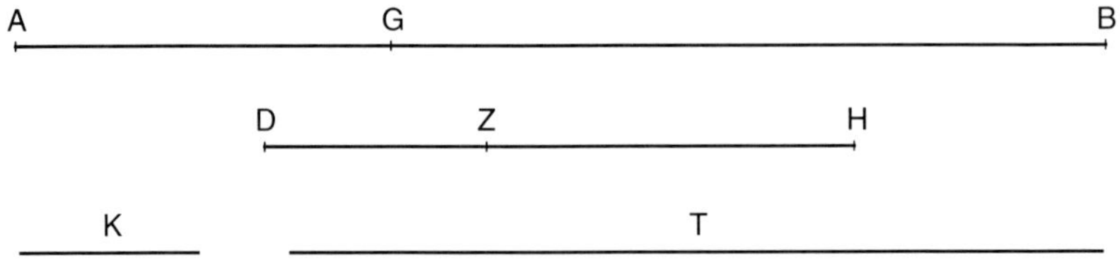

```
A               G                                       B
|_____|_____|
        D       Z               H
        |_____|_____|
  K                     T
  ___           _____
```

Fig. 71

• Determining AG and DZ. From $AG \cdot ZH = \sqrt{K \cdot T}$ and $AG \cdot DZ = K$, by addition, $AG \cdot DH = \sqrt{K \cdot T} + K$, so

$$AG = \frac{\sqrt{K \cdot T} + K}{DH} = \frac{63 + 27}{30} = \frac{90}{30} = 3 \qquad (iii)$$

whence

$$DZ = \frac{K}{AG} = \frac{27}{3} = 9. \qquad (iv)$$

- Determining GB and ZH. From $GB \cdot DZ = \sqrt{K \cdot T}$ and $GB \cdot ZH = T$, we obtain by addition

$$GB \cdot DH = \sqrt{K \cdot T} + T, \qquad (v)$$

whence

$$GB = \frac{\sqrt{K \cdot T} + T}{DH} = \frac{63 + 147}{30} = \frac{210}{30} = 7, \qquad (vi)$$

while, since $GB \cdot ZH = T$,

$$ZH = \frac{T}{GB} = \frac{147}{7} = 21. \qquad (vii)$$

* * *

Here we again have problems involving things, and thus B.220–226 would have been better grouped with B.198–204, 207–217.

(B.220) MSS $\mathcal{A}$, $\mathcal{B}$. The monthly wage is a thing minus ten nummi and the wage for twenty days is eight nummi; what is the thing worth?

Given thus $t_1 = 30$, $p_1 = x - 10$, $t_2 = 20$, $p_2 = 8$, required x. Similar to B.201 and hardly different from B.217 (not in MS $\mathcal{A}$).

(a) By composition, thus for the sum of the days,

$$p_1 + p_2 = x - 2 = \frac{(t_1 + t_2) p_2}{t_2} = \frac{50 \cdot 8}{20} = 20, \quad x = 22.$$

(b) Considering the wage for the month.

$$p_1 = \frac{t_1}{t_2} p_2, \quad \text{so} \quad x - 10 = \frac{30 \cdot 8}{20}.$$

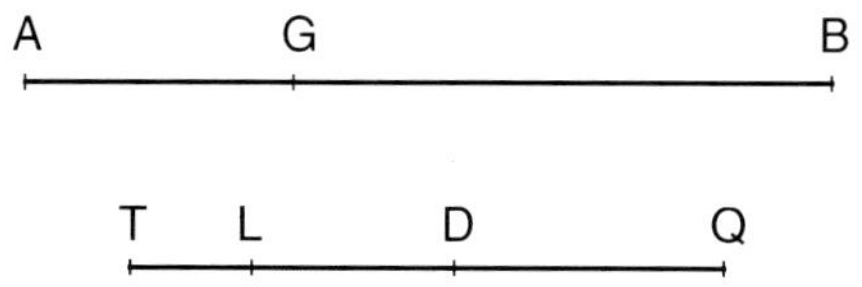

Fig. 72

Demonstration. Let (Fig. 72) $t_1 = 30 = AB$, $t_2 = 20 = GB$, $x = TQ$, $10 = DQ$, thus $DT = x - 10 = p_1$, $p_2 = 8 = DL$ ($< DT$). Then

$$\frac{AB}{GB} = \frac{DT}{DL}, \quad \text{so} \quad DT = \frac{AB \cdot DL}{GB} = 12, \quad DT + DQ = TQ = x.$$

Remark. Usually, in the *Liber mahameleth*, a demonstration explains a formula or performs a computation; it does not prove a simple algebraic computation. Still, the same is found in two of the coming problems.

(B.221) MSS $\mathcal{A}$; $\mathcal{B}$ (only 221′, 221″). The monthly wage is a thing minus ten nummi and the wage for forty days is one thing; what is the thing worth?

Given thus $t_1 = 30$, $p_1 = x - 10$, $t_2 = 40$, $p_2 = x$, required x.

Here both $p_2 > p_1$ and $t_2 > t_1$. Thus, as noted by the author, this problem is solvable (unlike B.226–226′, which should therefore precede it). From $p_1 : p_2 = t_1 : t_2$, we obtain

$$p_1 = x - 10 = \frac{t_1(p_2 - p_1)}{t_2 - t_1} = \frac{30 \cdot 10}{10} = 30, \quad x = 40.$$

(B.221′) The monthly wage is a thing minus ten nummi and the wage for twenty days is half a thing; what is the thing worth?

Given thus $t_1 = 30$, $p_1 = x - 10$, $t_2 = 20$, $p_2 = \frac{1}{2} \cdot x$. This problem turns out to be the same as the last one; nevertheless, it is solved independently.

Take $t'_2 = 2 \cdot t_2 = 40$, $p'_2 = 2 \cdot p_2 = x$ (this is B.221, in $\mathcal{A}$ only). Then

$$p_1 = x - 10 = \frac{t_1(p'_2 - p_1)}{t'_2 - t_1} = \frac{30 \cdot 10}{10} = 30, \quad x = 40.$$

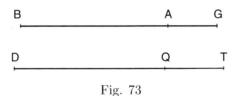

Fig. 73

Demonstration. Let (Fig. 73) $t_1 = AB = 30$, $t'_2 = BG = 40$, $p'_2 = x = DT$, $10 = QT$, thus $p_1 = x - 10 = DQ$. Since

$$\frac{t'_2}{t_1} = \frac{p'_2}{p_1}, \quad \text{or} \quad \frac{BG}{AB} = \frac{DT}{DQ},$$

we obtain, by separation,

$$\frac{t'_2 - t_1}{t_1} = \frac{p'_2 - p_1}{p_1}, \quad \text{or} \quad \frac{AG}{AB} = \frac{QT}{DQ},$$

whence

$$DQ = p_1 = \frac{AB \cdot QT}{AG} = \frac{t_1(p'_2 - p_1)}{t'_2 - t_1} = 30,$$

and therefore $DT = x = DQ + QT = 40$.

Remark. Formally, a demonstration should start with all the initially given quantities being set as segments of a straight line, and thus include t_2 and p_2. Same applies to B.221″.

(B.221″) The monthly wage is a thing minus ten nummi and the wage for twenty days is half a thing plus one nummus; what is the thing worth?

Given here again $t_1 = 30$, $p_1 = x - 10$, $t_2 = 20$, but $p_2 = \frac{1}{2} \cdot x + 1$. Similar to B.216, which MS $\mathcal{A}$ omits.

Take $t'_2 = 2 \cdot t_2 = 40$, $p'_2 = 2 \cdot p_2 = x + 2$. Then, as before,

$$p_1 = x - 10 = \frac{t_1(p'_2 - p_1)}{t'_2 - t_1} = \frac{30 \cdot 12}{10} = 36, \quad x = 46.$$

Fig. 74

Demonstration. Let (Fig. 74; comparing this figure with the previous one, the only new element is L) $t_1 = AB$, $t'_2 = BG$, $p'_2 = x+2 = DL$, $x = DT$, thus $TL = 2$, and put $QT = 10$, so $DQ = p_1 = x - 10$. Since

$$\frac{t'_2}{t_1} = \frac{p'_2}{p_1}, \quad \text{so} \quad \frac{BG}{AB} = \frac{DL}{DQ},$$

and, by separation,

$$\frac{t'_2 - t_1}{t_1} = \frac{p'_2 - p_1}{p_1} \quad \text{or} \quad \frac{AG}{AB} = \frac{QL}{DQ},$$

$$\text{whence} \quad DQ = p_1 = \frac{AB \cdot QL}{AG} = \frac{t_1(p'_2 - p_1)}{t'_2 - t_1} = 36,$$

and thus $DT = DQ + QT = 46$.

(B.222) MS $\mathcal{A}$. The monthly wage is a thing minus twenty nummi and the wage for fifty days is a thing plus ten nummi; what is the thing worth?

Given thus $t_1 = 30$, $p_1 = x - 20$, $t_2 = 50$, $p_2 = x + 10$, required x. From $p_1 : p_2 = t_1 : t_2$ ($p_2 > p_1$), by separation,

$$p_1 = x - 20 = \frac{t_1(p_2 - p_1)}{t_2 - t_1} = \frac{30 \cdot 30}{20} = 45, \quad x = 65.$$

(B.223) MS $\mathcal{A}$. The monthly wage is twenty squares and the wage for six days is eight times the root; what is the thing worth?

Given thus $t_1 = 30$, $p_1 = 20x^2$, $t_2 = 6$, $p_2 = 8x$, required x.

The way of solving this is said to be the same 'using algebra' (see B.198b) and 'using multiplication' (see B.198d). What should rather be said is that only these two ways are applicable here; indeed, transforming the proportion makes no sense since the powers of x are different. Then, 'using algebra',

$$p_1 = 20x^2 = \frac{t_1}{t_2} \cdot p_2 = 5 \cdot 8x = 40x,$$

whence $x = 2$, $x^2 = 4$.

(**B.224**) MS $\mathcal{A}$. The monthly wage is a thing minus twenty nummi and the wage for five days is a thing minus forty-five nummi; what is the thing worth?

Given thus $t_1 = 30$, $p_1 = x - 20$, $t_2 = 5$, $p_2 = x - 45$, required x. From $t_1 : t_2 = p_1 : p_2$, we obtain

$$\frac{t_1 - t_2}{t_2} = \frac{p_1 - p_2}{p_2}, \quad \frac{t_1 - t_2}{p_1 - p_2} = \frac{t_2}{p_2}, \quad \text{that is,} \quad \frac{25}{25} = \frac{5}{x - 45},$$

whence $x = 50$.

(**B.225**) MS $\mathcal{A}$. The monthly wage is a thing minus twenty nummi and the wage for three days is half a thing minus twenty nummi; what is the thing worth?

Given thus $t_1 = 30$, $p_1 = x - 20$, $t_2 = 3$, $p_2 = \frac{1}{2} \cdot x - 20$, required x. Take $t'_2 = 6$, $p'_2 = x - 40$, thus $x = 45$ found, as in the previous problem, by separation and alternation.

(**B.226**) MS $\mathcal{A}$. The monthly wage is a thing minus twenty nummi and the wage for ten days is a thing minus five nummi; what is the thing worth?

Given $t_1 = 30$, $p_1 = x - 20$, $t_2 = 10$, $p_2 = x - 5$, required x.

Absurd, because $p_1 < p_2$ whereas $t_1 > t_2$, so that more would be received, the text says, for part of the month than for the whole month.

(**B.226'**) The monthly wage is a thing minus ten nummi and the wage for five days is a thing, or a thing plus one nummus; what is the thing worth?

Given thus $t_1 = 30$, $p_1 = x - 10$, $t_2 = 5$, and $p_2 = x$ or $p_2 = x + 1$, required x. Both are unacceptable since again $p_1 < p_2$ but $t_1 > t_2$. See B.221.

D. *Wage in cash and kind*.

(**B.227**) MSS $\mathcal{A}$, $\mathcal{B}$. The monthly wage is ten nummi plus three sextarii, of which the price of the first exceeds that of the second by two nummi and the second the third by two nummi; for ten days is received half of the third, a third of the second, and a fourth of the first, but five nummi are returned; what is each sextarius worth?

Given thus $t_1 = 30$, $p_1 = c_1 + c_2 + c_3 + 10$, $t_2 = 10$, $p_2 = \frac{1}{4}c_1 + \frac{1}{3}c_2 + \frac{1}{2}c_3 - 5$, with $c_1 = c_2 + 2$, $c_2 = c_3 + 2$, required the prices c_i.

Putting $c_3 = x$, we have $c_2 = x + 2$, $c_1 = x + 4$, thus $p_1 = 3x + 16$; while, by hypothesis,

$$p_2 = \frac{1}{4}(x+4) + \frac{1}{3}(x+2) + \frac{1}{2}x - 5 = \left(1 + \frac{1}{12}\right)x - \left(3 + \frac{1}{3}\right).$$

Since $3t_2 = t_1$, so $3p_2 = p_1$, thus

$$\left(3 + \frac{1}{4}\right)x - 10 = 3x + 16, \quad \frac{1}{4}x = 26, \quad \text{so}$$

$x = 104 = c_3$, $c_2 = 106$, $c_1 = 108$.

• Verification. Since $p_1 = c_1 + c_2 + c_3 + 10 = 328$, then $p_2 = \frac{1}{3} \cdot p_1 = 109 + \frac{1}{3}$; now $\frac{1}{4}c_1 + \frac{1}{3}c_2 + \frac{1}{2}c_3 = 114 + \frac{1}{3}$, and by returning five nummi the worker will indeed have his due.

(B.228) MSS $\mathcal{A}$, $\mathcal{B}$. The monthly wage is ten nummi plus three sextarii, of which the price of the first exceeds that of the second by three nummi and the second the third by three nummi; the wage for ten days is half of the third, a third of the second, a fourth of the first; what is each sextarius worth?

Given thus $t_1 = 30$, $p_1 = c_1 + c_2 + c_3 + 10$, $t_2 = 10$, $p_2 = \frac{1}{4}c_1 + \frac{1}{3}c_2 + \frac{1}{2}c_3$, with $c_1 = c_2 + 3$, $c_2 = c_3 + 3$, required the prices c_i.

As before, put $c_3 = x$; then $c_2 = x + 3$, $c_1 = x + 6$, thus $p_1 = 3x + 19$; thus, by hypothesis,

$$p_2 = \frac{1}{4}(x+6) + \frac{1}{3}(x+3) + \frac{1}{2}x = \left(1 + \frac{1}{12}\right)x + 2 + \frac{1}{2}.$$

Since $3t_2 = t_1$, so $3p_2 = p_1$, thus

$$\left(3 + \frac{1}{4}\right)x + 7 + \frac{1}{2} = 3x + 19, \quad \frac{1}{4}x = 11 + \frac{1}{2},$$

so that $x = 46 = c_3$, $c_2 = 49$, $c_1 = 52$.

• Verification (as above). Since $p_1 = c_1 + c_2 + c_3 + 10 = 157$, then $p_2 = \frac{1}{3} \cdot p_1 = 52 + \frac{1}{3}$; now $p_2 = \frac{1}{4}c_1 + \frac{1}{3}c_2 + \frac{1}{2}c_3 = 52 + \frac{1}{3}$, indeed the same.

(B.229) MSS $\mathcal{A}$, $\mathcal{B}$. The monthly wage is three sextarii, of which the price of the first exceeds that of the second by three nummi and the second the third by two nummi; the wage for ten days is half of the third, a third of the second, a fourth of the first; what is each sextarius worth?

Given thus $t_1 = 30$, $p_1 = c_1 + c_2 + c_3$, $t_2 = 10$, $p_2 = \frac{1}{4}c_1 + \frac{1}{3}c_2 + \frac{1}{2}c_3$, with $c_1 = c_2 + 3$, $c_2 = c_3 + 2$, required c_i.

We have then $c_2 = c_3 + 2$ and $c_1 = c_2 + 3 = c_3 + 5$, therefore $c_1 + c_2 + c_3 = 3c_3 + 7$. In the text, however, the constants are interchanged at the beginning of the treatment, so that $c_2 = c_3 + 3$, $c_1 = c_2 + 2 = c_3 + 5$, and therefore $c_1 + c_2 + c_3 = 3c_3 + 8$. This will not change the value found for c_3, just those of c_1 and c_2, and in any event momentarily since after determining c_3 the text returns to the initial hypothesis and all three values turn out to be correct.

We thus find in the text successively, after putting $c_3 = x$,

$$p_1 = c_1 + c_2 + c_3 = 3x + 8. \qquad \left[\text{instead of } 3x + 7\right]$$

Calculating now the actual wage, we have

$$p_2 = \frac{1}{4}c_1 + \frac{1}{3}c_2 + \frac{1}{2}c_3 = \left(1 + \frac{1}{12}\right)x + 2 + \frac{1}{4} \qquad \left[\left(1 + \frac{1}{12}\right)x + 1 + \frac{11}{12}\right]$$

which corresponds for the month to

$$3p_2 = \left(3 + \frac{1}{4}\right)x + 6 + \frac{3}{4} \qquad \left[\left(3 + \frac{1}{4}\right)x + 5 + \frac{3}{4}\right]$$

whence the equation

$$\left(3 + \frac{1}{4}\right)x + 6 + \frac{3}{4} = 3x + 8 \qquad \left[\left(3 + \frac{1}{4}\right)x + 5 + \frac{3}{4} = 3x + 7\right]$$

and therefore, in both cases,

$$\frac{1}{4}x = 1 + \frac{1}{4}, \qquad x = 5.$$

So $x = 5 = c_3$, whence (according to the original hypothesis) $c_2 = c_3 + 2 = 7$, $c_1 = c_2 + 3 = 10$.

---

2. Two workers (or single lazy worker).

A worker is to do a certain task in t days for a wage p. He stops after τ days and will pay the hirer, for the remaining $t - \tau$ days, a penalty so that someone else may take his place at a different monthly wage π. The first worker thus ends up earning S instead of p, according to what he must pay. As noted in our summary, the formulae expressing p, τ, $t - \tau$ which are used in the text are the following (case $S \neq 0$ and particular case $S = 0$) :

$$p = \frac{(t-\tau)(\pi - p + S)}{\tau} + S, \qquad p = \frac{(t-\tau)(\pi - p)}{\tau},$$

$$\tau = \frac{(\pi - p + S)t}{\pi}, \qquad \tau = \frac{(\pi - p)t}{\pi},$$

$$\tau = \frac{(\pi - p + S)t - \tau \cdot p}{\pi - p}, \qquad \tau = \frac{(\pi - p)t - \tau \cdot p}{\pi - p},$$

$$t - \tau = \frac{(p - S)t}{\pi}, \qquad t - \tau = \frac{p \cdot t}{\pi},$$

$$t - \tau = \frac{\tau \cdot p - t \cdot S}{\pi - p}, \qquad t - \tau = \frac{\tau \cdot p}{\pi - p}.$$

The situation is the same if the worker hired receives p_0 for each day worked and has to pay a fine $\pi_0 - p_0$ for each day not worked.

A. *Monthly wages*.

(B.230) MSS $\mathcal{A}$, $\mathcal{B}$. A worker hired at the monthly wage of ten nummi will, if he does not work at all, pay the hirer a supplement of two nummi. After working part only he leaves with nothing. How many days did he work and how many not?

Given thus $t = 30$, $p = 10$, $\pi = 12$, $S = 0$, required τ, $t - \tau$.

(*a*) <u>Formulae</u>. See above, with $S = 0$.

$$\tau = \frac{(\pi - p)\, t}{\pi} = \frac{2 \cdot 30}{12} = 5$$

$$t - \tau = \frac{p \cdot t}{\pi} = \frac{10 \cdot 30}{12} = 25.$$

Since these two formulae are demonstrated in a similar way, the author will consider fully only the case of τ. Here he does so using first a reasoning, then a geometrical illustration. This reasoning can very well be rendered in algebraic symbols, which incidentally shows that our preliminary (algebraic) transcriptions of the author's purely geometric demonstrations are not at all out of place.

<u>Demonstration</u>.

(α) *Justification*. If p_τ is the wage of the first worker for τ days, then

$$\frac{\tau}{t} = \frac{p_\tau}{p}, \quad \text{whence} \quad \tau \cdot p = t \cdot p_\tau, \qquad (1)$$

and second, if $\pi_{t-\tau} - p_{t-\tau} = (\pi - p)_{t-\tau}$ designates the amount of $\pi - p$ corresponding to $t - \tau$ days, thus to be charged to the first worker, then

$$\frac{t - \tau}{t} = \frac{(\pi - p)_{t-\tau}}{\pi - p}, \quad \text{whence}$$

$$(t - \tau)(\pi - p) = t\,(\pi - p)_{t-\tau}. \qquad (2)$$

Now, by hypothesis, $S = 0$, that is, the amount given to the first for τ days, thus p_τ, equals the supplement he has to pay for the second one working $t - \tau$ days, thus $(\pi - p)_{t-\tau}$. After multiplying by t, we have

$$t \cdot p_\tau = t\,(\pi - p)_{t-\tau},$$

from which we infer, by (1) and (2), that:

$$\tau \cdot p = (t - \tau)(\pi - p). \qquad (*)$$

Therefore
$$\frac{\tau}{t-\tau} = \frac{\pi-p}{p} \qquad (3)$$

and, by composition,
$$\frac{\tau}{\tau+(t-\tau)} = \frac{\pi-p}{(\pi-p)+p}, \quad \text{thus} \quad \frac{\tau}{t} = \frac{\pi-p}{\pi},$$

that is,
$$\tau = \frac{(\pi-p)\,t}{\pi}.$$

As asserted in the text, the formula for $t-\tau$ is obtained in an analogous way: in (3), we shall use the other mode of composition, forming
$$\frac{(t-\tau)+\tau}{t-\tau} = \frac{p+(\pi-p)}{p}, \quad \text{which gives} \quad \frac{t}{t-\tau} = \frac{\pi}{p}.$$

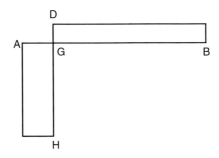

Fig. 75

(β) *Geometrical illustration* (of the latter part of this justification). Let (Fig. 75) $t = AB$, with $\tau = AG$ (thus $t-\tau = GB$), then, perpendicularly, $\pi - p = GD$, $p = GH$ (thus $\pi = DH$).

Since, as seen in (∗) above, $\tau \cdot p = (t-\tau)(\pi-p)$, we must have $AG \cdot GH = GB \cdot GD$. Therefore, the two rectangles AGH and DGB are equal in area and, by *Elements* VI.14, their sides are reciprocally proportional, that is,
$$\frac{AG}{GB} = \frac{GD}{GH}.$$

Then, by composition,
$$\frac{AG}{AB} = \frac{GD}{DH}, \quad \text{whence}$$

$$AG \cdot DH = GD \cdot AB \quad \text{and} \quad AG = \tau = \frac{GD \cdot AB}{DH} = \frac{(\pi-p)\,t}{\pi}.$$

Remark. This was equivalent to solving a system of the type

$$\begin{cases} AG + GB = AB \\ GH \cdot AG = GD \cdot GB, \end{cases}$$

that is (since AB, GH, GD are known), finding two quantities the sum and the ratio of which are given. See part *b*.

(**b**) *By algebra.* As we know,

$$\begin{cases} \tau + (t - \tau) = t \\ p \cdot \tau = (\pi - p)(t - \tau), \end{cases}$$

that is, numerically,

$$\begin{cases} \tau + (t - \tau) = 30 \\ 10 \cdot \tau = 2\,(t - \tau). \end{cases}$$

Put $\tau = x$; by the first equation, $t - \tau = 30 - x$, and so, by the second equation, $10\,x = 60 - 2\,x$, whence $12x = 60$ and $x = 5$.

(**c**) Put again $\tau = x$. We have first

$$\frac{t}{p} = \frac{\tau}{p_\tau}, \quad \text{or} \quad \frac{30}{10} = \frac{x}{p_\tau}, \quad \text{and therefore} \quad p_\tau = \frac{1}{3}x.$$

We have second

$$\frac{t}{\pi - p} = \frac{t - \tau}{(\pi - p)_{t-\tau}}, \quad \text{or} \quad \frac{30}{2} = \frac{30 - x}{(\pi - p)_{t-\tau}}, \quad \text{and therefore}$$

$$(\pi - p)_{t-\tau} = \frac{2\,(30 - x)}{30} = 2 - \frac{2}{3}\frac{1}{10}x.$$

But (case $S = 0$) $p_\tau = (\pi - p)_{t-\tau}$, so we obtain finally

$$p_\tau = \frac{1}{3}x = (\pi - p)_{t-\tau} = 2 - \frac{2}{3}\frac{1}{10}x, \quad \text{and} \quad \frac{2}{5}x = 2, \quad x = 5.$$

(**d**) Put again $\tau = x$. Since, as seen in *a*, relation (3),

$$\frac{\tau}{t-\tau} = \frac{\pi - p}{p}, \quad \text{then} \quad \frac{t - \tau}{\tau} = \frac{p}{\pi - p} = \frac{10}{2} = 5,$$

so $t - \tau$, or $30 - x$, equals $5\,x$, thus $6\,x = 30$, $x = 5$.

(**e**) *Verification that he leaves with nothing.* Working a sixth of the month, he will receive $\frac{1}{6} \cdot 10 = 1 + \frac{2}{3}$. Not working the remainder of the month, thus five sixths of it, he will pay $\frac{5}{6}$ of 2, which is just the same amount.

(**B.231**) MS *A*. A worker hired at the monthly wage of ten nummi will, if he does not work at all, pay the hirer eight nummi to hire someone else. After working part only he leaves with nothing. How many days did he work?

Given thus $t = 30$, $p = 10$, $\pi = 8$, $S = 0$, required τ.

This is impossible. Since $p > \pi$, the first worker will receive the difference $p - \pi$ anyway, whether he works or not, and, if he does at all, he will be left with a quantity $S > p - \pi$; in no case can he leave with $S = 0$.

(B.231′) So take here $S = 3$. Since the first worker leaves with 3 and receives $p - \pi = 2$ anyway, what matters in determining the days worked by him is $S - (p - \pi) = 1$ according to the monthly wage π, while the second worker will receive the remainder, thus $7 = \pi - [S - (p - \pi)] = p - S$. Therefore we have the proportions

$$\frac{\tau}{t} = \frac{S - (p - \pi)}{\pi} = \frac{\pi - (p - S)}{\pi} = \frac{1}{8} \quad \text{and} \quad \frac{t - \tau}{t} = \frac{p - S}{\pi} = \frac{7}{8},$$

so that

$$\tau = \frac{[\pi - (p - S)]\, t}{\pi} = \frac{1 \cdot 30}{8} = 3 + \frac{3}{4}$$

$$t - \tau = \frac{(p - S)\, t}{\pi} = \frac{7 \cdot 30}{8} = 26 + \frac{1}{4}.$$

Remark. That this pair of problems is misplaced and should follow B.232 is clear from the context —B.232 considers, as does B.230, the case $p < \pi$— as well as from a reference in B.232 to 'the problem preceding this one', which is B.230.

(B.232) MSS A, B. A worker hired at the monthly wage of ten nummi will, if he does not work at all, pay the hirer twelve nummi to hire someone else. After working part only he leaves with one nummus. How many days did he work?

Given thus $t = 30$, $p = 10$, $\pi = 12$, $S = 1$, required τ.

(a) Formula. As seen in our summary (and in the misplaced problem B.231′), and to be demonstrated below,

$$\tau = \frac{[\pi - (p - S)]\, t}{\pi} = \frac{(12 - 9)\, 30}{12} = 7 + \frac{1}{2}, \quad t - \tau = 22 + \frac{1}{2}.$$

Demonstration.

(α) *Preliminary remark.* As in B.230, the demonstration is part reasoning, part geometry; but here the second completes the first.

Suppose we are to determine two quantities x, y of which we know the sum or difference and the ratio:

$$\begin{cases} x \pm y = k \\ m x = n y \end{cases} \quad (n > m).$$

This is solved by operating with ratios. Since $x : y = n : m$, then

$$\frac{x \pm y}{y} = \frac{n \pm m}{m}, \qquad \frac{x}{x \pm y} = \frac{n}{n \pm m},$$

whence
$$x = \frac{n(x \pm y)}{n \pm m}, \qquad y = \frac{m(x \pm y)}{n \pm m}.$$

In the present problem, as in B.230, the sum is given. Thus here the pair of equations is
$$\begin{cases} \tau + (t - \tau) = t \\ (p - S)\tau = (\pi - p + S)(t - \tau) \end{cases}$$
where t, $p - S$ and $\pi - p + S$ are known.

(β) *Demonstration according to the text*. In B.230a, relation ($*$), we arrived at
$$\tau \cdot p = (t - \tau)(\pi - p)$$
and this was inferred from the equality of what the first one earned for working τ days and what he was to pay for $t - \tau$ days not worked, that is,
$$\frac{\tau}{t} p = \frac{t - \tau}{t}(\pi - p),$$
or, otherwise expressed, $p_\tau = (\pi - p)_{t-\tau}$. Since the first now leaves with $S > 0$ instead of $S = 0$, he earns more than he pays by S; thus $p_\tau - (\pi - p)_{t-\tau} = S$. Multiplying this by t, we find
$$t \cdot p_\tau - t \cdot (\pi - p)_{t-\tau} = t \cdot S.$$

Now, as we know from B.230, relations (1) & (2), $t \cdot p_\tau = \tau \cdot p$ and $t \cdot (\pi - p)_{t-\tau} = (t - \tau)(\pi - p)$; so we shall have $\tau \cdot p - (t - \tau)(\pi - p) = t \cdot S$, or, as expressed in the text,
$$\tau \cdot p = (t - \tau)(\pi - p) + t \cdot S.$$

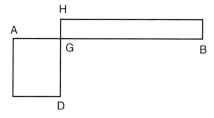

Fig. 76

The numerical solution is now continued geometrically (and is presented below in general terms). Let (Fig. 76) $t = AB$, $\tau = AG$ and $t - \tau = GB$. From what we have just seen,
$$AG \cdot p = GB(\pi - p) + AB \cdot S.$$

Since (*Elements* II.1 = Premiss PE_1) $AB \cdot S = AG \cdot S + GB \cdot S$, then

$$AG \cdot p = GB(\pi - p + S) + AG \cdot S,$$

that is,

$$AG(p - S) = GB(\pi - p + S).$$

We are therefore to solve

$$\begin{cases} AG + GB = AB \\ AG(p - S) = GB(\pi - p + S), \end{cases}$$

that is,

$$\begin{cases} \tau + (t - \tau) = t \\ \tau(p - S) = (t - \tau)(\pi - p + S). \end{cases}$$

Let us draw from G, perpendicularly, $GD = p - S = 9$ and $GH = \pi - p + S = 3$ (thus $HD = \pi$); then $AG \cdot GD$ and $GH \cdot GB$ are equal areas. Therefore

$$\frac{AG}{GB} = \frac{GH}{GD}, \quad \text{whence, by composition,} \quad \frac{AG}{AB} = \frac{GH}{HD}, \quad \text{thus}$$

$$AG = \tau = \frac{GH \cdot AB}{HD} = \frac{(\pi - p + S)t}{\pi}, \quad \text{and} \quad GB = t - \tau.$$

(b) Put $\tau = x$. From the above system,

$$9x = 3(30 - x) = 90 - 3x, \quad \text{and} \quad x = \frac{90}{12} = 7 + \frac{1}{2} = \tau.$$

(c) Also computed ('other way' in part *a*) as

$$\tau = \frac{\pi - (p - S)}{\pi} \cdot t = \frac{12 - 9}{12} \cdot 30 = \frac{1}{4} \cdot 30.$$

(d) Reducing this problem to the previous case (where $S = 0$). The sum the first was left with, $S =$ one nummus, corresponds to three days' work according to the remuneration $p = 10$ for the whole month. Consider thus the remaining number of days, $t' = 27$, and the corresponding wages $p' = p - 1 = p - \frac{3}{30}p$ and $\pi' = \pi - \frac{3}{30}\pi$. Thus doing, we are in the situation of B.230 since now $S' = 0$. Furthermore, since π and p in B.230 and B.232 have the same values and the multiplying factors changing π and p to π' and p' are the same, relation (3) seen in B.230 will retain the same value in both problems, so that

$$\frac{\tau}{t - \tau} = \frac{\pi - p}{p} = \frac{1}{5} \quad \text{gives} \quad \frac{\tau'}{t' - \tau'} = \frac{\pi' - p'}{p'} = \frac{1}{5}.$$

Putting now $\tau' = x$, we shall have accordingly $5x = 27 - x$ (or, as in the text, $x + 5x + 3 = 30$), whence

$$x = 4 + \frac{1}{2} = \tau', \quad \text{thus} \quad \tau = 7 + \frac{1}{2}.$$

(*e*) Verification. Since he worked a quarter of the month, he is to receive $\frac{1}{4}p = 2 + \frac{1}{2}$. Since he missed three quarters, he will pay $(\pi - p)_{t-\tau} = \frac{3}{4}(\pi - p) = 1 + \frac{1}{2}$. Subtracting the second quantity from the first does indeed leave $S = 1$.

(**B.233**) MSS $\mathcal{A}$, $\mathcal{B}$. A worker hired at the monthly wage of ten nummi will, if he does not work at all, pay the hirer twelve nummi to hire someone else. After working part only, he has to pay one nummus. How many days did he work?

Given thus $t = 30$, $p = 10$, $\pi = 12$, $S = -1$, required τ.

(*a*) Formula.

$$\tau = \frac{[\pi - (p - S)] \, t}{\pi} = \frac{(12 - 11) \, 30}{12} = 2 + \frac{1}{2}.$$

Justification.

The text is concise, and draws an analogy with B.232*a*. Before, we had $p_\tau > (\pi - p)_{t-\tau}$, and the worker hired first earned the difference, thus

$$p_\tau - (\pi - p)_{t-\tau} = \frac{\tau}{t} p - \frac{t - \tau}{t}(\pi - p) = S > 0;$$

but now we have $(\pi - p)_{t-\tau} > p_\tau$ with

$$(\pi - p)_{t-\tau} - p_\tau = \frac{t - \tau}{t}(\pi - p) - \frac{\tau}{t} p = |S| > 0.$$

Considering now the true (negative) value of S, this can be written as

$$\frac{t - \tau}{t}(\pi - p + S) = \frac{\tau}{t}(p - S),$$

whence the proportion

$$\frac{t - \tau}{\tau} = \frac{p - S}{\pi - p + S}$$

and the equality

$$\tau (p - S) = (t - \tau)(\pi - p + S) \qquad (*)$$

Thus, as stated, we have that t is divided into two parts τ and $t - \tau$ in a given ratio. Numerically, with $p - S = 11$ and $\pi - p + S = 1$,

$$\begin{cases} \tau + (t-\tau) = 30 \\ 11 \cdot \tau = 1 \cdot (t-\tau). \end{cases}$$

This, solved as before (B.232a), gives $\tau = 2 + \frac{1}{2}$, $t - \tau = 27 + \frac{1}{2}$.

(**b**) Putting, as in B.232b, $\tau = x$, we shall find, according to the two equations, $x \cdot 11 = 30 - x$.

(**c**) Verification. Computing the wage due p_τ and what is paid $(\pi - p)_{t-\tau}$ gives

$$p_\tau = \frac{\tau}{t} p = \frac{1}{2}\frac{1}{6} \cdot 10 = \frac{5}{6},$$

$$(\pi - p)_{t-\tau} = \frac{t-\tau}{t}(\pi - p) = \left(\frac{5}{6} + \frac{1}{2}\frac{1}{6}\right) \cdot 2 = 1 + \frac{5}{6},$$

and a difference of one nummus does indeed appear, which is what the first worker must pay.

(**d**) Also computed as ('other way' for part *a*)

$$\tau = \frac{(\pi - p) + S}{\pi} \cdot t = \frac{1}{2}\frac{1}{6} \cdot 30 = 2 + \frac{1}{2}.$$

Remark. The next two problems should be B.235 and B.236.

B. *Daily wages.*

We are now no longer given monthly wages but daily wages and the hiring time t may be less than a month. Thus, if p_0 is the daily wage of the first worker and π_0 the daily wage of the second, the first will earn, for τ days, $p_0 \cdot \tau$ and the second $\pi_0(t-\tau)$. Since the first will then leave with

$$p_0 \cdot \tau - (\pi_0 - p_0)(t-\tau) = S,$$

we have $\pi_0 \cdot \tau - \pi_0 \cdot t + p_0 \cdot t = S$, whence

$$\tau = \frac{(\pi_0 - p_0)t + S}{\pi_0}, \qquad t - \tau = \frac{p_0 \cdot t - S}{\pi_0}.$$

The text considers the cases $S = 0$ (B.234, B.236), $S > 0$ (B.237), $S < 0$ (B.238).

(**B.234**) MSS $\mathcal{A}$, $\mathcal{B}$. A worker hired for ten days at three nummi a day will pay five nummi for each day not worked. After working part of the ten days, he leaves with nothing. How many days did he work, and how many not?

Given thus $t = 10$, $p_0 = 3$, $\pi_0 - p_0 = 5$ (thus $\pi_0 = 8$), $S = 0$, required τ, $t - \tau$. This problem should follow B.236.

(**a**) *Formulae.*

$$\tau = \frac{(\pi_0 - p_0)t}{\pi_0} = \frac{5 \cdot 10}{8} = 6 + \frac{1}{4},$$

$$t - \tau = \frac{p_0 \cdot t}{\pi_0} = \frac{3 \cdot 10}{8} = 3 + \frac{3}{4}.$$

(**b**) Let us change the daily wages to wages for the whole time t. This brings us to the situation encountered in the previous group. Thus, since $t = 10$, we shall have $p = t \cdot p_0 = 30$, $\pi - p = t(\pi_0 - p_0) = 50$, and therefore $\pi = t \cdot \pi_0 = 80$ as the second worker's wage. Multiplying dividends and divisors of the expressions just seen by t, we find

$$\tau = \frac{(\pi - p)t}{\pi}, \qquad t - \tau = \frac{p \cdot t}{\pi},$$

which are the formulae seen in B.230.

(**B.235**) MSS $\mathcal{A}$, $\mathcal{B}$. A worker hired at the monthly wage of a thing will, if he does not work at all, pay the hirer a supplement of two nummi to hire someone else. After working part only he leaves with nothing, and the product of the days worked by the wage for the month is fifty. What are these two quantities?

Given thus $t = 30$, $p = x$, $\pi = x + 2$, $S = 0$, $\tau \cdot p = 50$, required τ and p. This problem is clearly misplaced and should follow B.233.

From relation (3) in B.230, we have

$$t - \tau = \frac{\tau \cdot p}{\pi - p} = \frac{50}{2} = 25, \qquad \tau = 5,$$

$$p = x = \frac{t - \tau}{\tau}(\pi - p) = \frac{25}{5} \cdot 2 = 10.$$

Remark. Here we would have expected the use of the direct formula for τ,

$$\tau = t - (t - \tau) = \frac{(\pi - p)t - \tau \cdot p}{\pi - p} = \frac{2 \cdot 30 - 50}{2} = 5,$$

which will occur in the next, related problem.

(**B.235**′) Same, with $S = 1$.

Formulae. In this case,

$$\tau = \frac{(\pi - p + S)t - \tau \cdot p}{\pi - p} = \frac{3 \cdot 30 - 50}{2} = 20, \qquad t - \tau = 10$$

$$p = \frac{t - \tau}{\tau}(\pi - p + S) + S = \frac{10}{20} \cdot 3 + 1 = 2 + \frac{1}{2} = x.$$

Demonstration.

(α) *Preliminary remark.* The steps to demonstrate the formulae for τ and p are the following.

- Formula for τ. As seen in B.233a, (∗), (and also in B.232a)

$$\tau(p - S) = (t - \tau)(\pi - p + S). \qquad (1)$$

Adding $\tau(\pi - p + S)$ in common gives

$$\tau \cdot \pi = t(\pi - p + S). \qquad (2)$$

Since, identically, $\tau \cdot \pi = \tau(\pi - p) + \tau \cdot p$, so

$$\tau(\pi - p) = t(\pi - p + S) - \tau \cdot p \qquad (3)$$

and finally

$$\tau = \frac{t(\pi - p + S) - \tau \cdot p}{\pi - p}. \qquad (4)$$

• **Formula for p.** From the same initial relation (1), we infer that

$$\frac{t - \tau}{\tau} = \frac{p - S}{\pi - p + S}, \qquad (5)$$

whence

$$p - S = \frac{t - \tau}{\tau}(\pi - p + S), \qquad (6)$$

and therefore

$$p = \frac{t - \tau}{\tau}(\pi - p + S) + S. \qquad (7)$$

Since we now know τ, and thus also $t - \tau$, the value of p can be calculated.

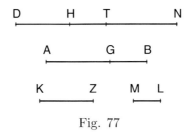

Fig. 77

(β) *Demonstration according to the text.* Let (Fig. 77) $p = x = DT$, $\pi - p = 2 = TN$ (thus $DN = \pi$), $t = 30 = AB$, with $\tau = AG$ and $t - \tau = GB$, $p_\tau = KZ$, $(\pi - p)_{t-\tau} = ML$, thus $KZ - ML = S = 1$, and put $HT = S = 1$, thus $HN = 3$.[†]

• **Formula for τ.** As seen, with here $DH = p - S$, $HN = \pi - p + S$,

$$AG \cdot DH = GB \cdot HN. \qquad (1)$$

Adding $AG \cdot HN$ in common gives

$$AG \cdot DN = AB \cdot HN = 90. \qquad (2)$$

[†] In what follows, KZ and ML no longer occur.

But since (PE_1) $AG \cdot DN = AG \cdot DT + AG \cdot TN$, with $AG \cdot DT\ (= \tau \cdot p) = 50$, so

$$AG \cdot TN = AG \cdot DN - AG \cdot DT = AB \cdot HN - AG \cdot DT = 40, \qquad (3)$$

thus

$$AG = \tau = \frac{AB \cdot HN - AG \cdot DT}{TN} = \frac{t(\pi - p + S) - \tau \cdot p}{\pi - p} = 20 \qquad (4)$$

and $GB = 10$.

- **Formula for p.** We start from the same initial relation as above,

$$\frac{GB}{AG} = \frac{DH}{HN}. \qquad (5)$$

Having just found GB and AG, we know their quotient. Then

$$DH = \frac{GB}{AG} \cdot HN = \frac{1}{2} \cdot HN = 1 + \frac{1}{2} \qquad (6)$$

(since $HN = \pi - p + S = 3$) and

$$DT = p = x = DH + HT = \frac{t - \tau}{\tau}(\pi - p + S) + S = 2 + \frac{1}{2}. \qquad (7)$$

(B.235″) Same, with $S = -1$. Not solvable: the above formula leads to

$$\tau = \frac{30 \cdot 1 - 50}{2} < 0$$

whereas we must have $t(\pi - p + S) > \tau \cdot p$.

(B.235‴) Same, also with $S = -1$, but choosing this time $\tau \cdot p = 25$. We could compute τ by the above formula:

$$\tau = \frac{t(\pi - p + S) - \tau \cdot p}{\pi - p} = \frac{30 \cdot 1 - 25}{2} = 2 + \frac{1}{2}.$$

The text computes (using the formula for τ) $t - \tau$:

$$t - \tau = \frac{t(\pi - p) - t(\pi - p + S) + \tau \cdot p}{\pi - p}$$

$$= \frac{\tau \cdot p - t \cdot S}{\pi - p} = \frac{25 + 30 \cdot 1}{2} = 27 + \frac{1}{2},$$

whence

$$\tau = 2 + \frac{1}{2}, \quad \text{and} \quad p = x = \frac{\tau \cdot p}{\tau} = \frac{25}{2 + \frac{1}{2}} = 10.$$

(B.236) MS $\mathcal{A}$. A worker hired for twenty days at three nummi a day will pay two nummi for each day not worked. After working part only he leaves with nothing. How many days did he work, and how many not?

Given thus $t = 20$, $p_0 = 3$, $p_0 - \pi_0 = 2$ (so $\pi_0 = 5$), $S = 0$, required τ, $t - \tau$. Since the signification of this problem is expounded, it must be the first in the group on daily wages, and thus B.234 should have followed it.

The computation, by means of the formulae, follow their demonstration in the text.
$$\tau = \frac{(\pi_0 - p_0)\,t}{\pi_0} = \frac{2 \cdot 20}{5} = 8, \qquad t - \tau = \frac{p_0 \cdot t}{\pi_0} = \frac{3 \cdot 20}{5} = 12.$$

<u>Demonstration</u>. It establishes the formulae already employed in the (misplaced) problem B.234.

(α) *Preliminary remark.* Since $S = 0$, what he earns equals what he pays, thus $p_0 \cdot \tau = (\pi_0 - p_0)(t - \tau)$. We are thus to solve
$$\begin{cases} \tau + (t - \tau) = t \\ p_0 \cdot \tau = (\pi_0 - p_0)(t - \tau) \end{cases}$$
We know how to solve this from B.232. The steps of the geometrical demonstration (which this time does not involve areas in the figure) are as follows. From the second equation, we infer that
$$\frac{p_0}{\pi_0 - p_0} = \frac{t - \tau}{\tau}. \qquad (1)$$

Then, by composition,
$$\frac{p_0 + (\pi_0 - p_0)}{\pi_0 - p_0} = \frac{(t - \tau) + \tau}{\tau},$$
thus
$$\frac{\pi_0}{\pi_0 - p_0} = \frac{t}{\tau} \qquad (2)$$
whence
$$\tau = \frac{(\pi_0 - p_0)\,t}{\pi_0}. \qquad (3)$$

From (2), by conversion,
$$\frac{\pi_0}{p_0} = \frac{t}{t - \tau}, \qquad (4)$$
whence
$$t - \tau = \frac{p_0 \cdot t}{\pi_0}. \qquad (5)$$

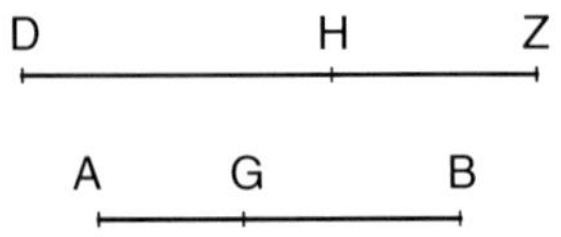

Fig. 78

(β) *Demonstration according to the text*. Let (Fig. 78) $t = 20 = AB$, $\tau = AG$, $t - \tau = GB$, $p_0 = 3 = DH$, $\pi_0 - p_0 = 2 = HZ$ (thus $\pi_0 = 5 = DZ$). By the conditions ($S = 0$), $DH \cdot AG = HZ \cdot GB$; then

$$\frac{DH}{HZ} = \frac{GB}{AG}, \qquad (1)$$

whence, by composition,

$$\frac{DZ}{HZ} = \frac{AB}{AG}, \qquad (2)$$

and thus

$$AG = \tau = \frac{HZ \cdot AB}{DZ}. \qquad (3)$$

From (2), by conversion,

$$\frac{DZ}{DH} = \frac{AB}{GB}, \qquad (4)$$

and thus

$$GB = t - \tau = \frac{DH \cdot AB}{DZ}. \qquad (5)$$

This proves the formulae.

(B.237) MS $\mathcal{A}$. A worker hired for thirty days at three nummi a day will pay two nummi for each day not worked. After working part only he leaves with ten nummi. How many days did he work, and how many not?

Given thus $t = 30$, $p_0 = 3$, $\pi_0 - p_0 = 2$ (thus $\pi_0 = 5$), $S = 10$, required τ, $t - \tau$.
Then:

$$\tau = \frac{(\pi_0 - p_0) t + S}{\pi_0} = \frac{2 \cdot 30 + 10}{5} = \frac{70}{5} = 14$$

$$t - \tau = \frac{p_0 \cdot t - S}{\pi_0} = \frac{3 \cdot 30 - 10}{5} = \frac{80}{5} = 16.$$

Demonstration.
(α) *Preliminary remark*. We know that (since $S > 0$) $p_0 \cdot \tau > (\pi_0 - p_0)(t - \tau)$, namely

$$p_0 \cdot \tau = (\pi_0 - p_0)(t - \tau) + S.$$

First, as stated in the text, we add $(\pi_0 - p_0)\tau$ in common; this gives

$$\pi_0 \cdot \tau = (\pi_0 - p_0) t + S \qquad (1)$$

whence

$$\tau = \frac{(\pi_0 - p_0) t + S}{\pi_0}. \qquad (2)$$

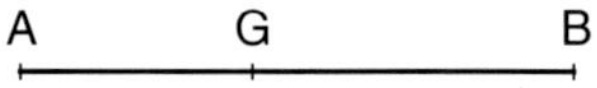

```
A         G            B
├─────────┼────────────┤
```

Fig. 79

Second, we add in common $p_0(t-\tau)$; this gives

$$p_0 \cdot t = \pi_0(t-\tau) + S, \qquad (3)$$

whence

$$t - \tau = \frac{p_0 \cdot t - S}{\pi_0}. \qquad (4)$$

(β) *Demonstration according to the text.* Let (Fig. 79) $t = AB$, $\tau = AG$, $t-\tau = GB$. Then $p_0 \cdot \tau = (\pi_0-p_0)(t-\tau)+S$, that is, $3 \cdot AG = 2 \cdot GB + 10$. Adding $2 \cdot AG$ in common gives

$$5 \cdot AG = 2 \cdot AB + 10, \qquad (1)$$

whence

$$AG = \tau = \frac{2 \cdot 30 + 10}{5} = \frac{(\pi_0 - p_0)t + S}{\pi_0}. \qquad (2)$$

Again, considering $3 \cdot AG = 2 \cdot GB + 10$, and adding this time $3 \cdot GB$ leads to

$$3 \cdot AB = 5 \cdot GB + 10, \qquad (3)$$

whence

$$GB = t - \tau = \frac{3 \cdot 30 - 10}{5} = \frac{p_0 \cdot t - S}{\pi_0}. \qquad (4)$$

(B.238) MS $\mathcal{A}$. A worker hired for a month at three nummi a day will pay two nummi for each day not worked. After working part only he has to pay ten nummi. How many days did he work, and how many not?

Given thus $t = 30$, $p_0 = 3$, $\pi_0 - p_0 = 2$ (thus $\pi_0 = 5$), $S = -10$, required τ, $t-\tau$.

Same as above, except that $|S|$ is considered and the sign changed. Since this is regarded as a new formula with $|S|$ subtracted, a new demonstration will be found below.

$$\tau = \frac{(\pi_0 - p_0)t - |S|}{\pi_0} = \frac{2 \cdot 30 - 10}{5} = \frac{50}{5} = 10$$

$$t - \tau = \frac{p_0 \cdot t + |S|}{\pi_0} = \frac{3 \cdot 30 + 10}{5} = \frac{100}{5} = 20.$$

Demonstration.

(α) *Preliminary remark.* According to the hypothesis, we shall consider in this case

$$(\pi_0 - p_0)(t - \tau) = p_0 \cdot \tau + |S|,$$

and proceed just as before. Adding in common $(\pi_0 - p_0)\tau$ gives

$$(\pi_0 - p_0)t = \pi_0 \cdot \tau + |S|, \qquad (1)$$

whence

$$\tau = \frac{(\pi_0 - p_0)t - |S|}{\pi_0}. \qquad (2)$$

Adding next $p_0(t - \tau)$ gives

$$\pi_0(t - \tau) = p_0 \cdot t + |S|, \qquad (3)$$

whence

$$t - \tau = \frac{p_0 \cdot t + |S|}{\pi_0}. \qquad (4)$$

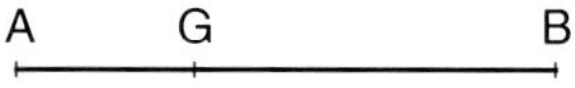

```
A      G              B
|──────┼──────────────|
```

Fig. 80

(β) *Demonstration according to the text.* Let (Fig. 80) $t = AB$, $\tau = AG$, $t - \tau = GB$. Then

$$2 \cdot GB = 3 \cdot AG + 10.$$

Adding in common $2 \cdot AG$ gives

$$2 \cdot AB = 5 \cdot AG + 10, \qquad (1)$$

whence

$$AG = \tau = \frac{2 \cdot 30 - 10}{5} = \frac{(\pi_0 - p_0)t - |S|}{\pi_0}. \qquad (2)$$

Adding now $3 \cdot GB$ gives

$$5 \cdot GB = 3 \cdot AB + 10, \qquad (3)$$

whence

$$GB = t - \tau = \frac{3 \cdot 30 + 10}{5} = \frac{p_0 \cdot t + |S|}{\pi_0}. \qquad (4)$$

(B.239) MS $\mathcal{A}$. A worker hired for twenty days at two nummi a day will pay one and a half for each day not worked. After working part only he leaves with nothing. How many days did he work, and how many not?

We have thus $t = 20$, $p_0 = 2$, $\pi_0 - p_0 = 1 + \frac{1}{2}$ (thus $\pi_0 = 3 + \frac{1}{2}$), $S = 0$. It is, the text explains, as if a second worker were hired to complete the work and were to receive for each day worked the two which were initially given to the first plus the one and a half the first worker pays, thus altogether three and a half. Now this corresponds for the twenty days to a total wage $p = 2 \cdot 20 = 40$ for the first and $\pi = (3 + \frac{1}{2}) 20 = 70$ for the second. We are then in the situation of B.230 and may apply the formulae seen there, namely

$$\tau = \frac{(\pi - p)\, t}{\pi} = \frac{30 \cdot 20}{70} = 8 + \frac{4}{7},$$

$$t - \tau = \frac{p \cdot t}{\pi} = \frac{40 \cdot 20}{70} = 11 + \frac{3}{7}.$$

Remark. This transformation from daily wages to wages for the whole month has been seen in B.234b. If $S \neq 0$, a situation alluded to at the end, we can effect the same transformation by multiplying numerator and denominator in the formulae of B.237 by t and putting as before $p = t \cdot p_0$ and $\pi = t \cdot \pi_0$, thus obtaining

$$\tau = \frac{(\pi - p)\, t + S \cdot t}{\pi}, \qquad t - \tau = \frac{p \cdot t - S \cdot t}{\pi}.$$

---

3. Three workers.

In the problems which follow three workers are hired at respective monthly wages p_i. Between the three of them they complete the prescribed time t, each working τ_i days (thus $\tau_1 + \tau_2 + \tau_3 = t$). They are said to receive wages p'_i either equal (B.240–241) or differing by given amounts of nummi (B.242–243).

(B.240) MSS $\mathcal{A}$, $\mathcal{B}$. Three workers, hired for a month at the respective wages of three, five and six nummi, complete the month between them and receive equal wages. How many days did each of them work?

Given thus $t = 30$, $p_1 = 3$, $p_2 = 5$, $p_3 = 6$, $p_{\tau_i} = p'$, required τ_i, with $\tau_1 + \tau_2 + \tau_3 = t$.

(*a*) Dividing p_i by t will give the amount earned each day. Because of the equality of the earnings and the completion of the month between them, we are then to solve

$$\begin{cases} \frac{p_1}{t} \cdot \tau_1 = \frac{p_2}{t} \cdot \tau_2 = \frac{p_3}{t} \cdot \tau_3 \\ \tau_1 + \tau_2 + \tau_3 = t \end{cases}$$

or, numerically,

$$\begin{cases} \frac{1}{10} \cdot \tau_1 = \frac{1}{6} \cdot \tau_2 = \frac{1}{5} \cdot \tau_3 \\ \tau_1 + \tau_2 + \tau_3 = 30. \end{cases}$$

This is said to be analogous to a problem of partnership, the invested capitals c_i being $\frac{t}{p_i}$ (thus $c_1 = 10$, $c_2 = 6$, $c_3 = 5$, and $c = \sum c_k = 21$), the whole profit e being $t = 30$ and the required parts e_i being the days worked τ_i (proof in B.241). The solution of the partnership problem being, as seen in B.104,

$$e_i = \frac{c_i \cdot e}{c},$$

the analogous solution of the present problem will be

$$\tau_i = \frac{\frac{t}{p_i} \cdot t}{\sum \frac{t}{p_k}}.$$

This gives, since $\sum \frac{1}{p_k} = \frac{21}{30}$,

$$\tau_1 = \frac{10}{\frac{21}{30}} = 14 + \frac{2}{7}, \quad \tau_2 = \frac{6}{\frac{21}{30}} = 8 + \frac{4}{7}, \quad \tau_3 = \frac{5}{\frac{21}{30}} = 7 + \frac{1}{7}.$$

• Verification. Knowing the τ_i and the p_i, we can calculate what is due to each:

$$\frac{\tau_1 \cdot p_1}{t} = \frac{(14 + \frac{2}{7}) \, 3}{30}, \quad \frac{\tau_2 \cdot p_2}{t} = \frac{(8 + \frac{4}{7}) \, 5}{30}, \quad \frac{\tau_3 \cdot p_3}{t} = \frac{(7 + \frac{1}{7}) \, 6}{30},$$

all of which do indeed give the same wage $p' = 1 + \frac{3}{7}$.

(**b**) We have seen in part *a* that

$$\tau_i = \frac{\frac{t}{p_i} \cdot t}{\sum \frac{t}{p_k}}.$$

Since t in the fractions disappears anyway, we may choose any value f (which will, for convenience, be divisible by the p_i) and consider

$$\tau_i = \frac{\frac{f}{p_i} \cdot t}{\sum \frac{f}{p_k}} = \frac{f_i \cdot t}{\sum f_k} \quad \left(\text{writing } f_i = \frac{f}{p_i}\right).$$

Take $f = 60$ (instead of 30 above, thus the f_i are just twice the previous quantities). We obtain $f_i = \{20, 12, 10\}$, to be treated again as would be a problem on partnership, with $c = f = \sum f_i = 42$, $c_i = f_i$, and $e = t = 30$.

(**c**) By algebra. Put $\tau_1 = x$; then, since $\tau_1 \cdot p_1 = \tau_2 \cdot p_2 = \tau_3 \cdot p_3$,

$$\tau_2 = \frac{p_1}{p_2} \cdot \tau_1 = \frac{3}{5} x, \quad \tau_3 = \frac{p_1}{p_3} \cdot \tau_1 = \frac{1}{2} x;$$

hence, since they complete the month,

$$\tau_1 + \tau_2 + \tau_3 = \left(2 + \frac{1}{10}\right) x = t = 30, \quad \text{so} \quad x = 14 + \frac{2}{7} = \tau_1,$$

whence τ_2 and τ_3. The equal wage is then

$$p_{\tau_i} = p' = \frac{14 + \frac{2}{7}}{30} \cdot 3 = \frac{14 + \frac{2}{7}}{10} = 1 + \frac{4}{10} + \frac{2}{7} \cdot \frac{1}{10} = 1 + \frac{3}{7}.$$

(d) Take now the common wage to be the unknown, thus $p' = x$. Since $t : p_i = \tau_i : x$, we find successively $\tau_1 = 10\,x$, $\tau_2 = 6\,x$, $\tau_3 = 5\,x$, thus

$$\tau_1 + \tau_2 + \tau_3 = t = 30 = 21x,$$

whence x and the τ_i.

Remark. In the text, the τ_i are not computed from their expressions in terms of x, but, less conveniently, from the initial proportion $t : p_i = \tau_i : x$. The same in B.242d.

(B.241) MS A. Three workers, hired for a month at the respective wages of four, six and twelve nummi, complete the month between them and receive equal wages. How many days did each of them work, and what is the common wage?

Similar problem; but here the formulae for τ_i and p' are demonstrated. Given $t = 30$, $p_1 = 4$, $p_2 = 6$, $p_3 = 12$, $p\tau_i = p'$, required τ_i with $\tau_1 + \tau_2 + \tau_3 = t$.

Treated as in B.240b, but with $f = 12$; thus $f_1 = 3$, $f_2 = 2$, $f_3 = 1$, so that $\sum f_i = 6$. Then, since

$$\tau_i = \frac{f_i \cdot t}{\sum f_k},$$

$$\tau_1 = \frac{3 \cdot 30}{6} = 15, \qquad \tau_2 = \frac{2 \cdot 30}{6} = 10, \qquad \tau_3 = \frac{1 \cdot 30}{6} = 5.$$

Next, since

$$p' : p_i = \tau_i : t = f_i : \sum f_k = \frac{f}{p_i} : \sum f_k,$$

$$p' = \frac{f}{\sum f_k} = \frac{12}{6} = 2.$$

Demonstration.

(α) *Preliminary remark.* The two demonstrations, of the formula for the τ_i and of that for the wage p', correspond to what follows.

• Demonstration of the formula for the τ_i. Since, according to the hypothesis,

$$\frac{\tau_i}{t} = \frac{p\tau_i}{p_i} = \frac{p'}{p_i},$$

we have $\tau_i \cdot p_i = t \cdot p'$. This means that all $\tau_i \cdot p_i$ have the same value, thus

$$\tau_1 \cdot p_1 = \tau_2 \cdot p_2 = \tau_3 \cdot p_3 = t \cdot p'. \tag{1}$$

From this, we infer that

$$\frac{p_1}{p_2} = \frac{\tau_2}{\tau_1}, \qquad \frac{p_2}{p_3} = \frac{\tau_3}{\tau_2}. \tag{2}$$

To determine the τ_i, find first three numbers f_i such that
$$f_1 \cdot p_1 = f_2 \cdot p_2 = f_3 \cdot p_3, \qquad (3)$$
which is attained by taking any number f and putting
$$f_1 = \frac{f}{p_1}, \qquad f_2 = \frac{f}{p_2}, \qquad f_3 = \frac{f}{p_3};$$
we then obtain, by (2) and (3),
$$\frac{\tau_1}{\tau_2} = \frac{p_2}{p_1} = \frac{f_1}{f_2}, \qquad \frac{\tau_2}{\tau_3} = \frac{p_3}{p_2} = \frac{f_2}{f_3}. \qquad (4), (4')$$
(Thus we now have direct proportionality between each τ_i and the corresponding f_i, a transformation procedure already used in B.45a, p. 1387, there too in order to make the analogy with problems on partnership.) Considering (4), we find, by composition,
$$\frac{\tau_1}{\tau_1 + \tau_2} = \frac{f_1}{f_1 + f_2} \qquad (5)$$
while, considering now both (4) and (4'), by 'equal proportionality',
$$\frac{\tau_1}{\tau_3}\left(= \frac{\tau_1}{\tau_2} \cdot \frac{\tau_2}{\tau_3} = \frac{f_1}{f_2} \cdot \frac{f_2}{f_3}\right) = \frac{f_1}{f_3}. \qquad (6)$$
Next, by inversion of (5) and (6),
$$\frac{\tau_1 + \tau_2}{\tau_1} = \frac{f_1 + f_2}{f_1}, \qquad \frac{\tau_3}{\tau_1} = \frac{f_3}{f_1},$$
whence, by P'_7,
$$\frac{\tau_1 + \tau_2 + \tau_3}{\tau_1} = \frac{f_1 + f_2 + f_3}{f_1}, \qquad \text{or} \qquad \frac{t}{\tau_1} = \frac{f}{f_1} = \frac{\sum f_k}{f_1},$$
and thus
$$\tau_1 = \frac{f_1 \cdot t}{\sum f_k}. \qquad (7)$$
We shall find further that
$$\tau_2 = \frac{f_2 \cdot t}{\sum f_k}, \qquad \tau_3 = \frac{f_3 \cdot t}{\sum f_k},$$
as is directly inferred from (7) using (4) and (4'), namely $\tau_2 : \tau_1 = f_2 : f_1$, $\tau_3 : \tau_2 = f_3 : f_2$.*

---

* The text seems to allude to a complete analogous computation. This would mean inferring these results from (4) and (4'), respectively from (4') and (6):
$$\frac{\tau_1 + \tau_2}{\tau_2} = \frac{f_1 + f_2}{f_2}, \quad \frac{\tau_3}{\tau_2} = \frac{f_3}{f_2} \quad \text{so} \quad \frac{\tau_1 + \tau_2 + \tau_3}{\tau_2} = \frac{f_1 + f_2 + f_3}{f_2},$$
$$\frac{\tau_2 + \tau_3}{\tau_3} = \frac{f_2 + f_3}{f_3}, \quad \frac{\tau_1}{\tau_3} = \frac{f_1}{f_3} \quad \text{so} \quad \frac{\tau_1 + \tau_2 + \tau_3}{\tau_3} = \frac{f_1 + f_2 + f_3}{f_3}.$$

The relations we have thus found, namely

$$\tau_i = \frac{f_i \cdot t}{\sum f_k},$$

make the analogy with the problems on partnership clear: the latter were solved by

$$e_i = \frac{c_i \cdot e}{c}, \quad c = \sum c_i, \quad e = \sum e_i,$$

and here $t = 30$ is like a total profit, which is the sum of the shares τ_i, while the total capital is f, which is the sum of the individual capitals f_i.

- Demonstration of the formula for p'. Since, as seen above,

$$\frac{f_2}{\sum f_k} = \frac{\tau_2}{t} \quad \text{and} \quad \frac{\tau_2}{t} = \frac{p'}{p_2}, \quad \text{then}$$

$$\frac{p'}{p_2} = \frac{f_2}{\sum f_k} = \frac{\frac{f}{p_2}}{\sum f_k}, \quad \text{so} \quad p' = \frac{f}{\sum f_k}.$$

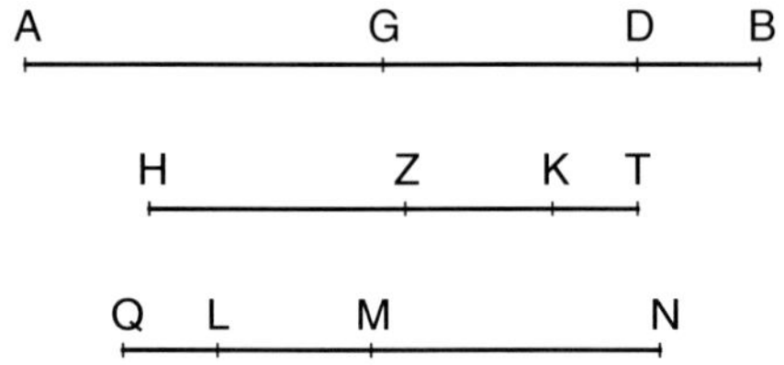

Fig. 81

(β) *Demonstrations according to the text.* Let (Fig. 81) $t = AB$, with $\tau_1 = AG$, $\tau_2 = GD$, $\tau_3 = DB$, then $p_1 = QL$, $p_2 = LM$, $p_3 = MN$. From

$$\frac{AG}{AB} = \frac{p'}{QL}, \quad \frac{GD}{AB} = \frac{p'}{LM}, \quad \frac{DB}{AB} = \frac{p'}{MN},$$

we infer that

$$p' = \frac{AG \cdot QL}{AB} = \frac{GD \cdot LM}{AB} = \frac{DB \cdot MN}{AB},$$

and therefore

$$p' \cdot AB = AG \cdot QL = GD \cdot LM = DB \cdot MN \qquad (1)$$

whence

$$\frac{QL}{LM} = \frac{GD}{AG}, \quad \frac{LM}{MN} = \frac{DB}{GD}. \qquad (2)$$

We now determine three numbers f_i such that $f_1 \cdot QL = f_2 \cdot LM = f_3 \cdot MN$. To that purpose, take any number f and let

$$f_1 = HZ = \frac{f}{QL}, \quad f_2 = ZK = \frac{f}{LM}, \quad f_3 = KT = \frac{f}{MN}. \quad \text{So}$$

$$f = HZ \cdot QL = ZK \cdot LM = KT \cdot MN. \quad (3)$$

Therefore, we have, using (3) then (2),

$$\frac{HZ}{ZK} = \frac{LM}{QL} = \frac{AG}{GD}, \quad \frac{ZK}{KT} = \frac{MN}{LM} = \frac{GD}{DB},$$

and so

$$\frac{AG}{GD} = \frac{HZ}{ZK}, \quad \frac{GD}{DB} = \frac{ZK}{KT}. \quad (4), (4')$$

Remark. At this point the extant text notes that we may adopt another value f; for instance, $f = 60$ (as in B.240b) instead of $f = 12$ will give $f_1 = HZ = 15$, $f_2 = ZK = 10$, $f_3 = KT = 5$.

Then from (4), by composition,

$$\frac{AG}{AG + GD} = \frac{HZ}{HZ + ZK}, \quad \text{so} \quad \frac{AG}{AD} = \frac{HZ}{HK}. \quad (5)$$

Next, from (4) and (4'), by equal proportionality,

$$\frac{AG}{GD} \cdot \frac{GD}{DB} = \frac{HZ}{ZK} \cdot \frac{ZK}{KT}, \quad \text{or} \quad \frac{AG}{DB} = \frac{HZ}{KT}. \quad (6)$$

Inverting the two results (5) and (6), we have

$$\frac{AD}{AG} = \frac{HK}{HZ}, \quad \frac{DB}{AG} = \frac{KT}{HZ},$$

whence, using P'_7,

$$\frac{AD + DB}{AG} = \frac{HK + KT}{HZ}, \quad \text{or} \quad \frac{AB}{AG} = \frac{HT}{HZ},$$

whence

$$AG = \frac{HZ \cdot AB}{HT} = \tau_1 = 15. \quad (7)$$

We shall likewise find, the text says,

$$\tau_2 = GD = \frac{ZK \cdot AB}{HT} = 10, \quad \tau_3 = DB = \frac{KT \cdot AB}{HT} = 5.$$

- To demonstrate the formula for p', the author considers $f_2 = ZK$. Then, as seen just above and, respectively, by (1),

$$\frac{ZK}{HT} = \frac{GD}{AB} \quad \text{and} \quad \frac{GD}{AB} = \frac{p'}{LM}.$$

Therefore

$$\frac{p'}{LM} = \frac{ZK}{HT}, \quad \text{and} \quad p' = \frac{ZK \cdot LM}{HT} = \frac{f}{\sum f_k}.$$

(**B.242**) MSS $\mathcal{A}$, $\mathcal{B}$. Three workers, hired for a month at the respective wages of six, five and three nummi, complete the month between them and receive unequal wages, the first receiving one nummus more than the second and the second, one nummus more than the third. How many days did each of them work?

Given thus $t = 30$, $p_1 = 6$, $p_2 = 5$, $p_3 = 3$, $p_{\tau_1} = p_{\tau_2}+1$, $p_{\tau_2} = p_{\tau_3}+1$, required τ_i with $\tau_1 + \tau_2 + \tau_3 = t$.

Let us consider the wages actually received to be $p_{\tau_i} = p_{\tau_k} + s_i$, p_{τ_k} being one of the wages serving as reference (thus s_i may be additive or subtractive). Since $p_{\tau_i} : p_i = \tau_i : t$, we have (as already seen in our summary)

$$\tau_i = \frac{t}{p_i} \cdot p_{\tau_i} = \frac{t}{p_i} \cdot p_{\tau_k} + \frac{t}{p_i} \cdot s_i = \frac{t}{p_i} \cdot \frac{\tau_k}{t} \cdot p_k + \frac{t}{p_i} \cdot s_i$$

whence

$$\tau_i = \frac{p_k}{p_i} \cdot \tau_k + \frac{t}{p_i} \cdot s_i. \qquad (*)$$

Since $\frac{t}{p_i}$ is the number of days the ith worker takes to earn one nummus, the second term is the number of days worked corresponding to the difference of s_i nummi while the first term represents the number of working days τ'_i to earn the amount p_{τ_k}. Subtracting then from t the total number of days corresponding to the s_i's, we have the situation of the two previous problems, with three men hired at different monthly wages but earning the same wage p_{τ_k} and working t' days altogether.

(**a**) Let first the common wage be the lowest. Then $p_{\tau_1} = p_{\tau_3} + 2$, $p_{\tau_2} = p_{\tau_3} + 1$, that is, $s_1 = 2$ and $s_2 = 1$. The author explains what we have just seen, but mixes computations in with his explanation. The first worker receives one nummus for working $\frac{t}{p_1}$ days, thus for s_1 nummi he is to work $\frac{t}{p_1} \cdot s_1$ days, that is, 10 days. For the second, it will be $\frac{t}{p_2} \cdot s_2 = 6$ days. Therefore subtracting these two quantities from t will leave the time t' corresponding to equal wages, namely 14 days. This being the situation of the preceding problems, the formula seen there (B.240a) applies, which will give us the respective numbers τ'_i of days worked to earn the reference wage; adding to each the individual numbers of days corresponding to the supplements leads to the τ_i. The formula in the present case is therefore

$$\tau_i = \tau'_i + \frac{t}{p_i} \cdot s_i = \frac{\frac{t}{p_i} t'}{\sum \frac{t}{p_k}} + \frac{t}{p_i} \cdot s_i = \frac{\frac{t}{p_i}\left(t - \frac{t}{p_1} s_1 - \frac{t}{p_2} s_2\right)}{\sum \frac{t}{p_k}} + \frac{t}{p_i} \cdot s_i.$$

Thus, as computed in the text,

$$\tau_1 = \frac{\frac{30}{6}\left(30 - \frac{30}{6}\cdot 2 - \frac{30}{5}\cdot 1\right)}{\frac{30}{6} + \frac{30}{5} + \frac{30}{3}} + \frac{30}{6}\cdot 2$$

$$= \frac{5\cdot 14}{21} + 10 = 3 + \frac{1}{3} + 10 = 13 + \frac{1}{3}$$

$$\tau_2 = \frac{6\cdot 14}{21} + \frac{30}{5}\cdot 1 = 4 + 6 = 10$$

$$\tau_3 = \frac{10\cdot 14}{21} = 6 + \frac{2}{3}.$$

(**b**) Taking p_{τ_1} as the reference wage (thus $s_1 = 0$), the actual wages can be expressed as, respectively, p_{τ_1}, $p_{\tau_2} = p_{\tau_1} - s_2$, $p_{\tau_3} = p_{\tau_1} - s_3$, with $s_2 = 1$, $s_3 = 2$. We shall in this case write τ_2 and τ_3 in terms of τ_1 as follows:

$$\tau_2 = \frac{p_1}{p_2}\cdot \tau_1 - \frac{t}{p_2}\cdot s_2$$

$$\tau_3 = \frac{p_1}{p_3}\cdot \tau_1 - \frac{t}{p_3}\cdot s_3.$$

To this corresponds the second treatment, using algebra. Let $\tau_1 = x$. Since $\tau_1 + \tau_2 + \tau_3 = t = 30$ and with the given values for the p_i's and s_i's, we find

$$x + \left(\frac{6}{5}\cdot x - \frac{30}{5}\cdot 1\right) + \left(\frac{6}{3}\cdot x - \frac{30}{3}\cdot 2\right) = x + \left(1 + \frac{1}{5}\right)x - 6 + 2x - 20$$

$$= \left(4 + \frac{1}{5}\right)x - 26 = 30,$$

whence

$$x = \tau_1 = 13 + \frac{1}{3}, \quad \tau_2 = \left(1 + \frac{1}{5}\right)x - 6 = 10, \quad \tau_3 = 2x - 20 = 6 + \frac{2}{3}.$$

We can now compute the p_{τ_i}:

$$p_{\tau_1} = \frac{\tau_1}{t}\cdot p_1 = 2 + \frac{2}{3}, \quad p_{\tau_2} = p_{\tau_1} - 1 = 1 + \frac{2}{3}, \quad p_{\tau_3} = p_{\tau_1} - 2 = \frac{2}{3}.$$

(**c**) Put now $\tau_1 = \frac{1}{2}\cdot x$ to give in like manner

$$\tau_1 + \tau_2 + \tau_3 = \frac{1}{2}\cdot x + \left(\frac{3}{5}\cdot x - 6\right) + (x - 20) = \left(2 + \frac{1}{10}\right)x - 26 = 30,$$

whence $x = 26 + \frac{2}{3}$, from which the τ_i's are computed.

(**d**) Put now, as the unknown, one of the wages, say $p_{\tau_3} = x$; then $p_{\tau_2} = x + 1$, $p_{\tau_1} = x + 2$; since $p_{\tau_i} : p_i = \tau_i : t$, we shall have

$$\tau_1 = \frac{30(x+2)}{6} = 5x + 10, \quad \tau_2 = \frac{30(x+1)}{5} = 6x + 6, \quad \tau_3 = \frac{30\cdot x}{3} = 10x,$$

and therefore
$$\tau_1 + \tau_2 + \tau_3 = 21x + 16 = 30,$$
whence
$$p_{\tau_3} = x = \frac{14}{21} = \frac{2}{3}, \quad p_{\tau_2} = 1 + \frac{2}{3}, \quad p_{\tau_1} = 2 + \frac{2}{3}.$$

Remark. The number of days each served is computed not by inserting x into the relations found before but, as in B.240d, by means of the initial proportion $\tau_i : t = p_{\tau_i} : p_i$, three terms of which are now known.

(B.243) MS $\mathcal{A}$. Three workers, hired for a month at the respective wages of a thing, half a thing, a third of a thing, complete the month between them and receive an equal wage of two nummi. What is the thing worth and how many days did each of them serve?

As is frequently the case, we end with problems involving 'things' in the data. Given $t = 30$, $p_1 = x$, $p_2 = \frac{1}{2}x$, $p_3 = \frac{1}{3}x$, $p_{\tau_i} = p' = 2$, required x and τ_i with $\tau_1 + \tau_2 + \tau_3 = t$.

According to what was seen in B.241 a,
$$p' = \frac{f}{\sum f_k} = \frac{f}{\sum \frac{f}{p_k}}, \quad \text{thus} \quad p' \cdot \sum \frac{f}{p_k} = f.$$

Taking $f = x$, we obtain first $\sum \frac{f}{p_k} = 1 + 2 + 3 = 6$. Therefore $6\,p' = f = x$, with $p' = 2$, so
$$x = 12 = p_1, \quad p_2 = 6, \quad p_3 = 4, \quad \text{while, since } \tau_i = \frac{t \cdot p'}{p_i} = \frac{60}{p_i},$$
$$\tau_1 = 5, \quad \tau_2 = 10, \quad \tau_3 = 15.$$

Chapter B–XII:
Wages in arithmetical progression

Summary

In the previous chapter, we studied the subject of hiring workers, first a single one then two or three with different wages, who had to do a job within a certain period of time. We are now to consider workers hired at individual wages in arithmetical progression (no specified working time). Mathematically, the whole chapter is about summing arithmetical progressions. Note that we had already encountered such problems before (A.268–270, B.52); but the treatment will now be systematic, and the formulae used explained and demonstrated.

A number n of workers are each paid a different wage p_i, with common difference δ. Let thus p_1 be the lowest wage; the others are then $p_i = p_1 + (i-1)\delta$, so that $p_n = p_1 + (n-1)\delta$ is the highest wage. The sum of all wages is then

$$S = \sum_{i=1}^{n} p_i = (p_1 + p_n)\frac{n}{2} = \left[2p_1 + (n-1)\delta\right]\frac{n}{2}.$$

Of the five quantities involved, n, p_1, p_n, S, δ, three may be given and the other two (or one of them) required. Here, we are given successively: n, p_1, p_n (B.244); n, p_1, δ (B.245); n, S, δ (B.246); n, p_n, δ (B.247); n, p_1, S (B.248); n, p_n, S (B.249); then (with this time n unknown) p_1, p_n, δ (B.250); p_1, S, δ (B.251); p_n, S, δ (B.252); p_1, p_n, S (B.253). Therefore every possible combination is considered.

The chapter thus falls into two sections, according to whether the number of workers is known or required; the latter case may lead to second-degree problems. From the above formula are inferred all the relations used and demonstrated in the text. Note finally that in each problem the quantities involved, either given or required, have the same value ($n = 10$, $p_1 = 3$, $p_n = 21$, $S = 120$, $\delta = 2$).

1. Number of workers known.

The two fundamental formulae, used and then demonstrated in B.244, are

$$S = (p_1 + p_n)\frac{n}{2}, \qquad \delta = \frac{p_n - p_1}{n-1}.$$

Since they link together n, p_1, p_n, S and n, p_1, p_n, δ respectively, they are directly applicable to the cases where one quantity of either set is unknown and the other three given, which is the case in B.244–245 and B.247–249. In B.246, where n, S, δ are given, p_1 (whence p_n) is found from

$$2p_1 = \frac{S}{\frac{n}{2}} - (n-1)\delta.$$

2. Number of workers required.

The same two fundamental formulae are used to determine n in B.250 (p_1, p_n, δ given) and B.253 (p_1, p_n, S given).* The other two problems are less simple and bring us to quadratic equations. From the two above formulae, we know that $2S = n \cdot p_1 + n \cdot p_n$ and $p_n = p_1 + (n-1)\delta$. We infer that

$$2S = n \cdot p_1 + n \left(p_1 + (n-1)\delta\right), \quad 2S = n\left(p_n - (n-1)\delta\right) + n \cdot p_n,$$

that is, respectively,

$$S = n \cdot p_1 + n(n-1)\frac{\delta}{2}, \quad S = n \cdot p_n - n(n-1)\frac{\delta}{2}.$$

In B.251 (p_1, S, δ given), n is found from the left-hand equality:

$$n^2 \cdot \frac{\delta}{2} + n\left(p_1 - \frac{\delta}{2}\right) = S.$$

In B.252 (p_n, S, δ given), n is found from the right-hand equality:

$$n^2 \cdot \frac{\delta}{2} + S = n\left(p_n + \frac{\delta}{2}\right).$$

These two quadratic equations are solved as usual. In the first case, we have

$$n\left(n \cdot \frac{\delta}{2} + p_1 - \frac{\delta}{2}\right) = S$$

and we are to determine two quantities of which we know the product and (in the particular situation with $\delta = 2$) the difference $|p_1 - \frac{\delta}{2}|$.† In the second case, where

$$n\left(p_n + \frac{\delta}{2} - n \cdot \frac{\delta}{2}\right) = S,$$

we know their product and, again since $\delta = 2$, their sum $p_n + \frac{\delta}{2}$. The unknown is determined in both cases by the usual identities, corresponding to *Elements* II.5 (sum given) and II.6 (difference given).

---

1. Number of workers known.

(B.244) MSS $\mathcal{A}$ (incomplete), $\mathcal{B}$. Ten workers are hired at wages in arithmetical progression; the wage of the first is three nummi, that of the last twenty-one. What are the common difference and the sum of all the wages? Given thus $n = 10$, $p_1 = 3$, $p_n = 21$, required S and δ.

---

* B.253, put by us at the end, is found only in MS $\mathcal{D}$.

† If $\delta \neq 2$ we shall just multiply both sides by $\frac{\delta}{2}$.

Formulae.

$$S = (p_1 + p_n)\frac{n}{2} = (3+21)\frac{10}{2} = 24 \cdot 5 = 120,$$

$$\delta = \frac{p_n - p_1}{n-1} = \frac{21-3}{9} = \frac{18}{9} = 2.$$

Demonstration. The demonstration for each formula consists of two parts: first the reasoning from which we are to deduce it and then a geometrical illustration. Both are unsatisfactory. Furthermore, the text of the geometrical parts (found only in MS $\mathcal{B}$) is incomplete.

(α) *Justification*.
- Formula for S.

In the case of an arithmetical progression, we shall have, with n terms,

$$p_1 + p_n = p_2 + p_{n-1} = p_3 + p_{n-2} = \ldots,$$

so that the sum S of the terms will be $\frac{n}{2}$ times the amount of such a pair of opposite terms, that is,

$$S = (p_1 + p_n)\frac{n}{2}.$$

Remark. The first part has already been proved, namely in Premiss P_1, but there is no allusion to that here. Furthermore, since the text speaks only about pairs, it takes for granted that there is an even number of terms.
- Formula for δ.

We know that $p_2 - p_1 = \delta$, $p_3 - p_1 = 2\delta$, $p_4 - p_1 = 3\delta$, ..., $p_n - p_1 = (n-1)\delta$. In this last relation all the quantities except δ are known. So we shall calculate

$$\delta = \frac{p_n - p_1}{n-1}.$$

(β) *Geometrical illustration*.
- Formula for S. It consists in successively transferring δ from one term to another:

$$p_1 + p_n = p_1 + \delta + p_n - \delta = p_2 + p_{n-1} = \ldots .$$

Let (Fig. 82) $n = AB$, with AT, TZ, ZL, ..., DF, FB representing each of the n workers, and the perpendicular lines TG ($= p_1$), ZQ ($= p_1 + \delta$), LN, ..., DH, FO, BK the wages with the common difference $\delta = PQ = XN = Y\!SO = \ldots = SO = CK$. The extant text begins by telling us to draw parallels to AB through G, Q, N, SO, ..., H, O, meeting the perpendicular to BA at B in I, R, E, ..., C, respectively; then we shall

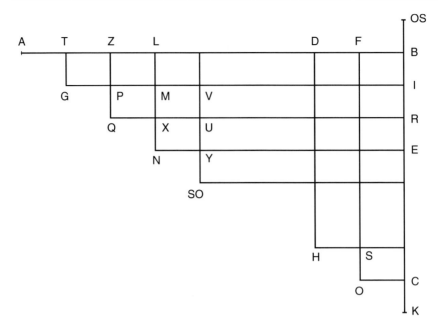

Fig. 82

have (by *Elements* I.34) rectangles the widths of which are segments on IC of uniform length δ. Then

$$TG + BK = p_1 + p_n$$
$$= ZP + PQ + BK - CK = ZQ + BC = ZQ + FO = p_2 + p_{n-1}$$
$$= LX + XN + FO - SO = LN + FS = LN + DH = p_3 + p_{n-2}$$
$$= \ldots .$$

Thus the required sum S does indeed consist of $\frac{n}{2}$ times $TG + BK$, that is, the sum of the first and the last wages.

• Formula for δ. We consider in the same figure that $p_1 = TG$, $p_2 - p_1 = PQ = \delta$, $p_3 - p_1 = MN = XN + MX = 2\,PQ = 2\,\delta$, then $p_4 - p_1 = V\!,\!SO = Y\!,\!SO + UY + VU = 3\,PQ = 3\,\delta$ and so on to $p_n - p_1 = IK = (n-1) \cdot PQ = (n-1)\,\delta$, whence

$$\delta = \frac{p_n - p_1}{n-1}.$$

(B.245) MSS $\mathcal{A}$, $\mathcal{B}$. Ten workers are hired at wages in arithmetical progression; the wage of the first is three nummi and the common difference is two. What are the highest wage and the sum of all the wages?

Given thus $n = 10$, $p_1 = 3$, $\delta = 2$, required p_n and S.

(*a*) The text computes

$$p_n = p_1 + (n-1)\,\delta = 3 + 9 \cdot 2 = 3 + 18 = 21,$$

$$S = (p_1 + p_n)\frac{n}{2} = 24 \cdot 5.$$

The second relation is known from before. As to the formula for p_n, we are told, it originates from that seen in B.244, namely

$$\delta = \frac{p_n - p_1}{n-1}.$$

(**b**) Using algebra, put $p_n = x$; since $p_n - p_1 = (n-1)\delta$, so $x - 3 = 2 \cdot 9 = 18$ and $x = 21$, from which S may be determined (B.244).

(**B.246**) MSS $\mathcal{A}$ (incomplete), $\mathcal{B}$. Ten workers are hired at wages in arithmetical progression; the sum of the wages is a hundred and twenty nummi and the common difference, two. What are the lowest and the highest wages?

Given thus $n = 10$, $S = 120$, $\delta = 2$, required p_1 and p_n.

(**a**) <u>Formula</u>.

$$2p_1 = \frac{S}{\frac{n}{2}} - (n-1)\delta = \frac{120}{5} - 9 \cdot 2 = 24 - 18 = 6,$$

whence $p_1 = 3$, and $p_{10} = p_1 + 9 \cdot \delta = 3 + 9 \cdot 2 = 21$.

<u>Demonstration</u>. Here again, there is a reasoning and a geometrical illustration.

(α) *Justification*. From B.244, we know that

$$S = (p_1 + p_n)\frac{n}{2}, \qquad \delta = \frac{p_n - p_1}{n-1},$$

from which we infer, respectively, that

$$p_1 + p_n = \frac{S}{\frac{n}{2}}, \qquad p_n - p_1 = (n-1)\delta,$$

so that

$$2p_1 = (p_1 + p_n) - (p_n - p_1) = \frac{S}{\frac{n}{2}} - (n-1)\delta. \qquad (*)$$

(β) *Geometrical illustration*. (MS $\mathcal{B}$ only, Fig. 82 again; illustrates the left-hand part of (*)). Add to $BK = p_n$ the segment $B{,}OS = TG$. Then $OS{,}K = p_1 + p_n$. But $IK = (n-1)\delta$, as already seen. So $OS{,}K - IK = OS{,}I = 2TG$. Therefore, since $IK = p_n - p_1$, $2p_1$ is the difference between $p_1 + p_n$ and $p_n - p_1$.

(**b**) Using algebra, put $p_1 = x$; then

$$S = \frac{n}{2}\Big(p_1 + \big[p_1 + (n-1)\delta\big]\Big) = 5(2x + 9 \cdot 2) = 10x + 90 = 120,$$

whence $x = 3$.

(B.247) MSS $\mathcal{A}$, $\mathcal{B}$. Ten workers are hired at wages in arithmetical progression; the highest wage is twenty-one nummi and the common difference, two. What are the lowest wage and the sum of all the wages?

Given thus $n = 10$, $p_n = 21$, $\delta = 2$, required p_1 and S.

$$p_1 = p_n - (n-1)\delta = 21 - 9 \cdot 2 = 21 - 18 = 3, \qquad S = (p_1 + p_n)\frac{n}{2}.$$

(B.248) MSS $\mathcal{A}$, $\mathcal{B}$. Ten workers are hired at wages in arithmetical progression; the lowest wage is three nummi and the sum of all the wages, a hundred and twenty. What are the highest wage and the common difference?

Given thus $n = 10$, $p_1 = 3$, $S = 120$, required p_n and δ.

(**a**) *Formulae.*

$$p_n = \frac{S}{\frac{n}{2}} - p_1 = \frac{120}{5} - 3 = 24 - 3 = 21, \qquad \delta = \frac{p_n - p_1}{n-1} = \frac{21-3}{9} = 2.$$

Since the formula for p_n occurs in this form for the first time, it is justified; as we are told, it comes from the relation (seen in B.244)

$$S = (p_1 + p_n)\frac{n}{2}.$$

The second relation has been seen several times.

(**b**) Using algebra, put $p_n = x$; then

$$\frac{n}{2}(p_n + p_1) = S \quad \text{gives} \quad \frac{10}{2}(x+3) = 120,$$

whence $5x + 15 = 120$, and $x = 21$.

(B.249) MSS $\mathcal{A}$, $\mathcal{B}$. Ten workers are hired at wages in arithmetical progression; the highest wage is twenty-one nummi and the sum of all the wages, a hundred and twenty. What is the lowest wage?

Given thus $n = 10$, $p_n = 21$, $S = 120$, required p_1. By the formula just seen

$$p_1 = \frac{S}{\frac{n}{2}} - p_n = \frac{120}{5} - 21.$$

---

2. Number of workers required.

(B.250) MSS $\mathcal{A}$, $\mathcal{B}$. An unknown number of workers are hired at wages in arithmetical progression; the lowest wage is three nummi, the highest twenty-one, and the common difference is two. What is the number of workers?

Given thus $p_1 = 3$, $p_n = 21$, $\delta = 2$, required n.

(**a**) *Formula.*
$$\frac{p_n - p_1}{\delta} + 1 = n.$$

The origin is the formula seen in B.244 (*seqq.*) for known n, namely
$$\frac{p_n - p_1}{n - 1} = \delta, \quad \text{from which} \quad \frac{p_n - p_1}{\delta} = n - 1.$$

(**b**) Using algebra, put $n = x$; then
$$(x - 1)\delta + p_1 = p_n, \quad \text{so} \quad 2(x - 1) + 3 = 21,$$
thus $2x + 1 = 21$ and $x = 10$.

(**B.251**) MSS $\mathcal{A}$; $\mathcal{B}$, but without a. An unknown number of workers are hired at wages in arithmetical progression; the sum of all the wages is a hundred and twenty nummi, the lowest is three, and the common difference is two. What are the highest wage and the number of workers?

Given thus $p_1 = 3$, $S = 120$, $\delta = 2$, required n and p_n.

(**a**) *Formulae.*

• Determination of p_n. This will be determined immediately, once we know n, by the formula seen from B.244 on, namely
$$p_n = (n - 1)\delta + p_1.$$

• Determination of n.

(α) *Preliminary remark.* The subsequent solving of the text corresponds to what follows. From before (B.244), we know or infer the relations
$$p_1 + p_n = (n - 1)\delta + 2p_1 \qquad (1)$$
$$(p_1 + p_n)\frac{n}{2} = S. \qquad (2)$$

Combining (1) and (2), we obtain
$$\left[(n - 1)\delta + 2p_1\right]\frac{n}{2} = S,$$
that is,
$$n^2\frac{\delta}{2} + n\left(p_1 - \frac{\delta}{2}\right) = S, \qquad (3)$$
which takes the form
$$n\left(n\frac{\delta}{2} + p_1 - \frac{\delta}{2}\right) = S.$$

In our text, $\delta = 2$. This simplifies the computations, but the way of solving remains unchanged. For, if $\frac{\delta}{2} \neq 1$, multiply the whole by $\frac{\delta}{2}$ to produce

$$n\frac{\delta}{2}\left(n\frac{\delta}{2} + p_1 - \frac{\delta}{2}\right) = \frac{\delta}{2} S.$$

Whatever the sign of $p_1 - \frac{\delta}{2}$, we are then to solve the system

$$\begin{cases} \left|\left(n\frac{\delta}{2} + p_1 - \frac{\delta}{2}\right) - n\frac{\delta}{2}\right| = \left|p_1 - \frac{\delta}{2}\right| \\ n\frac{\delta}{2} \cdot \left(n\frac{\delta}{2} + p_1 - \frac{\delta}{2}\right) = \frac{\delta}{2} S, \end{cases}$$

that is, finding two quantities of which we know difference and product. This will be treated in the usual way, by *Elements* II.6.

(β) *Solution in the text.* The quadratic equation

$$n^2 \frac{\delta}{2} + n\left(p_1 - \frac{\delta}{2}\right) = S$$

is established and solved geometrically. Because this is primarily a geometrical solution and not a demonstration as such, the segments are often replaced by their values.

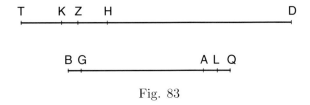

Fig. 83

Let (Fig. 83) two segments of straight line be drawn with, on the second, $n = AB$, $1 = GB$; thus $AG = n - 1$, and therefore $\delta \cdot AG + p_1 = 2AG + 3 = p_n$.

We now take on the other segment $DZ = n \cdot \delta = AB \cdot \delta = 2 \cdot AB$, $HZ = \delta = 2$; thus $DH = (n-1)\delta = AG \cdot HZ = 2AB - 2$.

Then we take $p_1 = HK$; so $DK = DH + HK = (n-1)\delta + p_1 = 2AB + 1$, which equals p_n. Put finally $KT = HK = p_1 = 3$;[†] so

$$DT = DK + KT = p_n + p_1 = (n-1)\delta + 2p_1 = 2 \cdot AB + 4. \quad (1)$$

Therefore we must have

$$S = (p_1 + p_n)\frac{n}{2} = DT \cdot \frac{AB}{2} \quad \text{or} \quad 120 = (2 \cdot AB + 4)\frac{AB}{2}. \quad (2)$$

[†] By the data, $HK = p_1 > HZ = \delta$, so *a fortiori* $HT = 2p_1 > HZ = \delta$. See the author's final remark.

On the other hand, $DT = DZ + ZT$. Since $DZ = n \cdot \delta = AB \cdot \delta$ while $ZT = TH - HZ = 2\,p_1 - \delta$, so

$$S = DT \cdot \frac{AB}{2} = (DZ + ZT)\frac{AB}{2} = (AB \cdot \delta + 2\,p_1 - \delta)\frac{AB}{2}$$

$$= AB^2 \cdot \frac{\delta}{2} + AB\left(p_1 - \frac{\delta}{2}\right), \qquad (3)$$

or also

$$S = AB\left(AB \cdot \frac{\delta}{2} + \left(p_1 - \frac{\delta}{2}\right)\right),$$

which, with the given values, becomes $AB\,(AB + 2) = 120$. This is the required equation. The author then introduces the segment $QA = 2$, and the previous equation becomes

$$AB\,(AB + QA) = AB \cdot QB = S,$$

with S given. We are then to solve

$$\begin{cases} QB - AB = 2 \\ QB \cdot AB = 120. \end{cases}$$

The subsequent procedure is not included by the author, for it is known from before (e.g. B.31). Let L be the mid-point of QA. Then, by *Elements* II.6,

$$AB \cdot QB + LA^2 = LB^2.$$

Since we know $AB \cdot QB = S = 120$ and $LA = 1$, we now know $LB = 11$, so also $LB - LA = AB = 10$.

The author concludes his solution by remarking that this 'proof' (or, rather, way of solving) is applicable whatever the relative sizes of p_1 and δ. For we see that if $2\,p_1 < \delta$ (with T, K, Z, H replaced in the figure by Z, T, K, H respectively), the coefficient of AB in equation (3) will change its sign, and we shall end up with an equation not of the type $x^2 + p\,x = q$, but of the type $x^2 = p\,x + q$ (p, $q > 0$) —possibly also $x^2 = q$ if $p_1 = \frac{\delta}{2}$. Thus the above general equation, instead of reducing to

$$AB \cdot \frac{\delta}{2}\left(AB \cdot \frac{\delta}{2} + \left(p_1 - \frac{\delta}{2}\right)\right) = S \cdot \frac{\delta}{2},$$

will reduce to

$$AB \cdot \frac{\delta}{2}\left(AB \cdot \frac{\delta}{2} - \left(p_1 - \frac{\delta}{2}\right)\right) = S \cdot \frac{\delta}{2}.$$

In both cases, though, with product and difference of the required segments being known, positive quantities, we shall indeed proceed in the same way, namely by applying *Elements* II.6.

(b) *Using algebra, put* $n = x$. *What follows is just the same as above. That is, considering that*

$$p_n + p_1 = \left[(n-1)\delta + p_1\right] + p_1 = \left[(2x-2) + 3\right] + 3 = 2x + 4,$$

$$\frac{n}{2}(p_1 + p_n) = \frac{x}{2}(2x + 4) = x^2 + 2x = S = 120$$

whence, as 'shown in algebra' (B.38, B.102), $x = 10$.

(B.252) MS A. An unknown number of workers are hired at wages in arithmetical progression; the highest wage is twenty-one nummi, the common difference, two, and the sum of all the wages is a hundred and twenty. What are the lowest wage and the number of workers?

Given thus $p_n = 21$, $S = 120$, $\delta = 2$, required p_1, n.

(α) *Preliminary remark.* By means of the formulae

$$p_1 = p_n - (n-1)\delta, \quad S = (p_1 + p_n)\frac{n}{2}, \quad \text{thus} \quad S = \left(2p_n - (n-1)\delta\right)\frac{n}{2},$$

is established the equation

$$\frac{n^2}{2} \cdot \delta + S = n\left(p_n + \frac{\delta}{2}\right),$$

thus $n^2 + 120 = 22n$, with the two positive solutions $n = 10$ and $n = 12$, the second of which is unacceptable since it would make p_1 negative. In any event, the subsequent geometrical consideration leads to a single solution. What the author does corresponds to the following. Starting from

$$\frac{n}{2}(p_1 + p_n) = S,$$

we have

$$n(p_1 + p_n) = 2S. \qquad (1)$$

Introducing

$$p_1 + p_n = (p_n - (n-1)\delta) + p_n = 2p_n + \delta - n \cdot \delta \qquad (2)$$

into (1) gives

$$n(2p_n + \delta - n \cdot \delta) = 2S. \qquad (3)$$

Considering now

$$n \cdot \delta (2p_n + \delta - n \cdot \delta) = 2S \cdot \delta, \qquad (3')$$

we are to solve

$$\begin{cases} n \cdot \delta + (2p_n + \delta - n \cdot \delta) = 2p_n + \delta \\ n \cdot \delta (2p_n + \delta - n \cdot \delta) = 2S \cdot \delta, \end{cases}$$

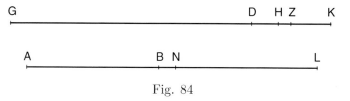

Fig. 84

that is, finding two quantities of which we know sum and product. We may then determine the half-difference by *Elements* II.5 and then each individually as seen several times before.

Remark. In order to obtain a known value for the sum of the two quantities, we have changed equation (3) to (3'), whereby the unknown term (our $n \cdot \delta$) is eliminated. In the text, (3) takes the form

$$n \left(\frac{2\, p_n + \delta}{\delta} - n \right) = \frac{2\, S}{\delta} \qquad (*)$$

which has the very same purpose and thus corresponds to our modified form.

(β) *Geometrical solution in the text.* We again find, as in the previous problem, the abstract representation of quantities as segments mixed in with their numerical values. Let us represent (Fig. 84), on two segments of straight line, $n = AB$, $(n-1)\,\delta = GD = 2 \cdot AB - 2$, $\delta = DH = 2$ and thus $GD + DH = GH = \delta \cdot n = DH \cdot AB = 2 \cdot AB$, $p_1 = DZ$ and thus $GD + DZ = GZ = p_n = 21$, and let us append to GZ the segment $ZK = DZ = p_1$.*

Since then $GK = GZ + ZK = p_1 + p_n$, so

$$\frac{AB}{2} \cdot GK = \frac{n}{2} (p_1 + p_n) = S \quad (= 120), \quad \text{whence}$$

$$AB \cdot GK = 2S \quad (= 240). \qquad (1)$$

Let us now express GK in another way, considering its various constituents.
Since $GH = 2 \cdot AB$ and $GZ = 21$, so $HZ = 21 - 2 \cdot AB$.
Since $DH = 2$, so $DZ = p_1 = DH + HZ = 23 - 2 \cdot AB$.
Since $ZK = p_1$ also, so $DK = DZ + ZK = 2\,p_1 = 46 - 4 \cdot AB$.
Since $GD = GH - DH = 2 \cdot AB - 2$, so

$$GK = GD + DK = 44 - 2 \cdot AB. \qquad (2)$$

Therefore, (1) becomes:

$$AB \cdot GK = AB\,(44 - 2 \cdot AB) = 240,$$

$$\left(\text{that is,} \quad AB \cdot GK = AB \left([2 \cdot GZ + DH] - DH \cdot AB \right) = 2\,S \right)$$

---

* DH being within DZ in the figure, this means that $p_1 = DZ > \delta = DH$. See our final remark.

$$AB \cdot \frac{GK}{2} = AB\,(22 - AB) = 120,$$

$$\left(\text{that is,}\quad AB \cdot \frac{GK}{2} = AB\left(\frac{2 \cdot GZ + DH}{DH} - AB\right) = \frac{2\,S}{DH} \quad (*)\right)$$

and, by putting $22 = AL\;(> AB)$,

$$\left(\text{that is,}\quad \frac{2 \cdot GZ + DH}{DH} = AL, \qquad AL = \frac{2p_n + \delta}{\delta}\right)$$

$$AB\,(AL - AB) = AB \cdot BL = 120. \qquad (2)$$

Thus
$$\begin{cases} AB + BL = AL = 22 \\ AB \cdot BL = 120. \end{cases}$$

Halve the known quantity AL at N; so $AN = 11 = NL$. Then, by *Elements* II.5,
$$AB \cdot BL + NB^2 = AN^2 \;(= 121)$$
whence $NB^2 = AN^2 - AB \cdot BL = 1$, then $AB = AN - NB = 10$.

Remark. Here an early reader noted that we must have $DZ = p_1 > DH = \delta$, which is the situation in the figure. But even if $p_1 \le \delta$ (thus Z between D and H, or on H if $p_1 = \delta$), we shall arrive at the same equation. For HZ becomes $GH - GZ = 2AB - 21$ (instead of $GZ - GH = 21 - 2AB$), and DZ becomes $DH - HZ = 2 - (2AB - 21) = 23 - 2AB$ (instead of $DH + HZ = 2 + (21 - 2AB) = 23 - 2AB$); whence the same result.

(B.253) MS $\mathcal{D}$ only, of which it is the first problem taken from the *Liber mahameleth*. An unknown number of workers are hired at wages in arithmetical progression; the total sum of the wages is a hundred and twenty nummi, the lowest wage, three, the highest, twenty-one. What is the number of workers?

Given thus $p_1 = 3$, $p_n = 21$, $S = 120$, required n.

(a) *Formula.*
$$n = 2 \cdot \frac{S}{p_1 + p_n}$$
which is, the text explains, inferred from the known relation
$$(p_1 + p_n)\frac{n}{2} = S.$$

(b) By algebra, put $n = x$. Then
$$(p_1 + p_n)\frac{x}{2} = 24\,\frac{x}{2} = 12\,x,$$
whence $12x = S = 120$, and $x = 10$.

Chapter B–XIII:
Hiring a carrier

Summary

In this and the following chapters the kind of work is mostly specified. Here a single worker is hired to carry q_1 bushels (*sextarii*) over a distance d_1 miles for a wage of p_1 nummi, but he actually carries q_2 bushels d_2 miles for a wage of p_2 nummi. The fundamental relation (proved in B.255) is

$$\frac{p_1}{p_2} = \frac{q_1 \cdot d_1}{q_2 \cdot d_2}.$$

This is the first real occurrence of the compound rule of three, which will also be the subject of the next four chapters.[||] For the case of carrying we may have one quantity required (B.254–256), or two when their relation is known (B.257–261). These are the 'five fundamental problems' first seen in the chapter on buying and selling and then several times thereafter, but with a further development: since now six instead of four quantities are involved, the ratio of the two required quantities may also be given. The chapter concludes with two problems containing 'things' in the data (B.262–263).

1. One unknown.

The required quantity is p_2 in B.254–255 and d_2 in B.256; the case of q_2 required does not occur, presumably because it would be solved in the very same way as for d_2. In the treatment, the author reduces the problem to simple proportions involving four terms by making two terms of the same kind equal.

(α) Let p_2 be the required quantity while p_1, q_1, d_1, q_2, d_2 are known. Suppose first q_2 bushels are carried the *same* distance d_1. The wage, being proportional to the quantity carried, will be p'_1, determined by $p'_1 : p_1 = q_2 : q_1$, thus

$$p'_1 = \frac{q_2}{q_1} \cdot p_1.$$

Suppose now these same q_2 bushels are carried a different distance d_2. The wage will become p''_1, with $p''_1 : p'_1 = d_2 : d_1$, so that

$$p''_1 = \frac{d_2}{d_1} \cdot p'_1.$$

[||] More than four quantities were involved when we were dealing with textiles (B–V & B–VI).

We now have q_2 bushels carried d_2 miles, as required; that is, the wage p_1'' is our p_2, and therefore

$$p_2 = \frac{d_2}{d_1} \cdot p_1' = \frac{q_2 \, d_2}{q_1 \, d_1} \cdot p_1.$$

This is the essence of the justifications found below, in which the given set q_1, d_1, p_1 is progressively transformed, as above, into q_2, d_2, p_2.

Remark. Another approach (α') consists in reducing initially the first set to unit quantities $q_0 = 1 = d_0$, giving a wage p_0; p_2 is then simply equal to $q_2 \cdot d_2 \cdot p_0$.

(β) Suppose second d_2 to be required. Carrying q_2 instead of q_1 for the same wage, the distance will change to d_1' with $d_1' : d_1 = q_1 : q_2$ (inverse proportionality: the greater the distance, the smaller the quantity carried). Hence

$$d_1' = \frac{q_1}{q_2} \cdot d_1.$$

But if now p varies, so will d_1 according to $d_1'' : d_1' = p_2 : p_1$ (direct proportionality: the quantity remaining the same, the distance increases with the wage). We find then the final result

$$d_1'' = d_2 = \frac{p_2}{p_1} \cdot d_1' = \frac{q_1 \, p_2}{q_2 \, p_1} \cdot d_1.$$

The same will be done if q_2 is required.

These successive transformations may be represented as follows —and this corresponds to the numerical transformations found in the text as justifications.

(α) *Direct proportionality.* Required the wage. The greater the quantity and the distance, the higher the wage. See B.254.

q (bushels)	d (miles)	p (nummi)
q_1	d_1	p_1
q_2	d_1	$\dfrac{q_2}{q_1} \cdot p_1$
q_2	d_2	$\dfrac{q_2 \, d_2}{q_1 \, d_1} \cdot p_1.$

(α') Unit quantities.

q (bushels)	d (miles)	p (nummi)
q_1	d_1	p_1
q_0	d_0	$\dfrac{q_0 \, d_0}{q_1 \, d_1} \cdot p_1$
q_2	d_2	$q_2 \cdot d_2 \cdot \dfrac{q_0 \, d_0}{q_1 \, d_1} \cdot p_1.$

Same formula as before since $q_0 = 1 = d_0$.

(β) *Involving an inverse proportionality.* Required the distance. See B.256.

q (bushels)	d (miles)	p (nummi)
q_1	d_1	p_1
q_2	$\dfrac{q_1}{q_2} \cdot d_1$	p_1
q_2	$\dfrac{q_1}{q_2} \dfrac{p_2}{p_1} \cdot d_1$	$p_2.$

The justifications in the next four chapters will be performed in the same way.

2. Two unknowns.

We are now familiar with this kind of problem. But, as said before, since six quantities are involved instead of four, the two required quantities of the second set may be known by not only their product, sum or difference but also their quotient (B.257–261).

(i) q_2 and product $d_2 \cdot p_2$ known. Since

$$\frac{p_2}{p_1} = \frac{q_2 \cdot d_2}{q_1 \cdot d_1} \quad \text{and} \quad \frac{d_2}{d_1} = \frac{q_1 \cdot p_2}{q_2 \cdot p_1},$$

we infer that

$$\frac{p_2^2}{p_1} = \frac{q_2 \left(d_2 \cdot p_2\right)}{q_1 \cdot d_1}, \quad \frac{d_2^2}{d_1} = \frac{q_1 \left(d_2 \cdot p_2\right)}{q_2 \cdot p_1},$$

and thus obtain the formulae (B.257)

$$p_2 = \sqrt{\frac{q_2 \cdot p_1 \left(d_2 \cdot p_2\right)}{q_1 \cdot d_1}}, \quad d_2 = \sqrt{\frac{q_1 \cdot d_1 \left(d_2 \cdot p_2\right)}{q_2 \cdot p_1}}.$$

(ii) q_2 and sum or difference $d_2 \pm p_2$ known. Since

$$\frac{p_1}{p_2} = \frac{q_1 \cdot d_1}{q_2 \cdot d_2}, \quad \text{and therefore} \quad \frac{d_2}{p_2} = \frac{q_1 \cdot d_1}{q_2 \cdot p_1},$$

we infer from the latter by composition and separation or conversion that

$$\frac{d_2 \pm p_2}{p_2} = \frac{q_1 \cdot d_1 \pm q_2 \cdot p_1}{q_2 \cdot p_1} \quad \text{and} \quad \frac{d_2}{d_2 \pm p_2} = \frac{q_1 \cdot d_1}{q_1 \cdot d_1 \pm q_2 \cdot p_1},$$

whence the formulae (B.258–259)

$$p_2 = \frac{q_2 \cdot p_1 \left(d_2 \pm p_2\right)}{q_1 \cdot d_1 \pm q_2 \cdot p_1}, \quad d_2 = \frac{q_1 \cdot d_1 \left(d_2 \pm p_2\right)}{q_1 \cdot d_1 \pm q_2 \cdot p_1}.$$

(iii) p_2 and quotient $q_2 : d_2 = k$ known. Since

$$\frac{p_1}{p_2} = \frac{q_1 \cdot d_1}{q_2 \cdot d_2} = \frac{q_2}{d_2} \cdot \frac{q_1 \cdot d_1}{q_2^2} = \frac{k \cdot q_1 \cdot d_1}{q_2^2} \quad \text{and}$$

$$\frac{p_1}{p_2} = \frac{q_1 \cdot d_1}{q_2 \cdot d_2} = \frac{d_2}{q_2} \cdot \frac{q_1 \cdot d_1}{d_2^2} = \frac{q_1 \cdot d_1}{k \cdot d_2^2},$$

we have the formulae (B.260–261)

$$q_2 = \sqrt{\frac{k \cdot q_1 \cdot d_1 \cdot p_2}{p_1}}, \qquad d_2 = \sqrt{\frac{q_1 \cdot d_1 \cdot p_2}{k \cdot p_1}}.$$

3. Problems involving 'things'.

The last two problems involve, as said, an unknown x. As a matter of fact, here we also have the occurrence of x^2 (B.262) and, for the first time, of x^3 (B.263). In one case, $p_1 = x^2 + bx$ and $p_2 = ax$, and in the other $p_1 = x^3 + bx^2$ and $p_2 = ax^2$; thus in both cases

$$\frac{p_1}{p_2} = \frac{1}{a}x + \frac{b}{a} = \frac{q_1 \cdot d_1}{q_2 \cdot d_2}, \quad \text{and} \quad x = \frac{q_1 \cdot d_1}{q_2 \cdot d_2} \cdot a - b,$$

and so x is known since a, b, q_i, d_i are given.

---

1. One unknown.

(B.254) MSS $\mathcal{A}$, $\mathcal{B}$. Sixty nummi is the wage for carrying twelve sextarii thirty miles; what is due for carrying three sextarii ten miles?

Given thus $q_1 = 12$, $d_1 = 30$, $p_1 = 60$, $q_2 = 3$, $d_2 = 10$, required p_2.

(a) Computation.

$$p_2 = \frac{q_2 \cdot d_2 \cdot p_1}{q_1 \cdot d_1} = \frac{3 \cdot 10 \cdot 60}{12 \cdot 30} = 5.$$

Justification. The origin of this formula appears from the following successive transformations, of the first set of data and then of the second, in order to bring q_1 and q_2 to the same value (here 1) and set the proportion between the remaining terms.

q (sextarii)	d (miles)	p (nummi)
$q_1 = 12$	$d_1 = 30$	$p_1 = 60$,
$q_1' = q_0 = 1$	$d_1' = q_1 \cdot d_1 = 360$	$p_1 = 60$;†
$q_2 = 3$	$d_2 = 10$	p_2,
$q_2' = q_0 = 1$	$d_2' = q_2 \cdot d_2 = 30$	p_2.

Considering the cases where $q_0 = 1$, we shall have $d_1' : p_1 = d_2' : p_2$ (direct proportionality), whence

$$p_2 = \frac{d_2'}{d_1'} \cdot p_1 = \frac{q_2 \cdot d_2}{q_1 \cdot d_1} \cdot p_1,$$

---

† For the same wage, the fewer the sextarii, the greater the distance (inverse proportionality: $q_1' : q_1 = d_1 : d_1'$).

which explains the formula for p_2.

(**b**) Although the other treatments of these problems mostly consist in different ways of associating the factors, they are justified as well. Here, we have
$$p_2 = \frac{d_2}{d_1}\left(\frac{q_2}{q_1} \cdot p_1\right) = \frac{10}{30}\left(\frac{3}{12} \cdot 60\right).$$

<u>Justification</u>. Transformation of the first set.

q (sextarii)	d (miles)	p (nummi)
$q_1 = 12$	$d_1 = 30$	$p_1 = 60;$
$q'_1 = 3 = q_2$	$d_1 = 30$	$p'_1 = \frac{q_2}{q_1} \cdot p_1 = 15,$
$q_2 = 3$	$d'_1 = 10 = d_2$	$p''_1 = \frac{d_2}{d_1} \cdot p'_1 = 5,$

whence
$$p_2 = p''_1 = \frac{d_2}{d_1} \cdot p'_1 = \frac{d_2}{d_1}\left(\frac{q_2}{q_1} \cdot p_1\right).$$

(**c**) Another justification, which may not be genuine.

q (sextarii)	d (miles)	p (nummi)
$q_1 = 12$	$d_1 = 30$	$p_1 = 60,$
$q'_1 = q_0 = 1$	$d_1 = 30$	$p'_1 = \frac{q_0}{q_1} \cdot p_1 = \frac{p_1}{q_1} = 5,$
$q''_1 = 3 = q_2$	$d'_1 = 10 = d_2$	$p_2.$

Thus $p_2 = p'_1 = 5$ because of $q_0 \cdot d_1 = q_2 \cdot d_2$ (what we have here being the particular situation $d_1 = q_2 \cdot d_2$).

(**d**) Also computed as ('other ways')
$$p_2 = \frac{q_2}{q_1} \cdot \left(\frac{d_2}{d_1} \cdot p_1\right), \quad \left(\frac{q_2}{q_1} \cdot \frac{d_2}{d_1}\right) \cdot p_1,$$
the latter being justified by applying Premiss P'_3 to the formula seen in a:
$$\frac{q_2 \cdot d_2}{q_1 \cdot d_1} \cdot p_1 = \frac{q_2}{q_1} \cdot \frac{d_2}{d_1} \cdot p_1.$$

(**B.255**) MS $\mathcal{A}$. A hundred nummi is the wage for carrying five sextarii ten miles; what is due for carrying three sextarii four miles?

Similar problem. Given thus $q_1 = 5$, $d_1 = 10$, $p_1 = 100$, $q_2 = 3$, $d_2 = 4$, required p_2.

(**a**) <u>Formula</u>. As known from the previous problem,
$$\frac{p_1}{p_2} = \frac{q_1 \cdot d_1}{q_2 \cdot d_2}, \quad \text{so}$$
$$p_2 = \frac{q_2 \cdot d_2}{q_1 \cdot d_1} \cdot p_1 = \frac{3 \cdot 4}{5 \cdot 10} \cdot 100 = \frac{12}{50} \cdot 100 = 24.$$

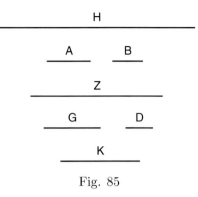

Fig. 85

Demonstration of the general formula. Let (Fig. 85) $q_1 = A = 5$, $d_1 = G = 10$, $q_2 = B = 3$, $d_2 = D = 4$, $p_1 = H = 100$. Consider first the relation between wages and distances, thus keeping q_1 the same. Evidently, to that situation corresponds a wage p'_1 determined by

$$\frac{p'_1}{p_1} = \frac{d_2}{d_1}.$$

Let $p'_1 = Z$; then $Z : H = D : G$ and

$$Z = p'_1 = \frac{D}{G} \cdot H = \frac{d_2}{d_1} \cdot p_1 = 40;$$

thus $Z = p'_1$ is a fraction of the (full) wage, due for carrying the same number of sextarii but over the new distance.

Consider next the relation between wages and number of sextarii carried, the distance remaining the same. Since $p'_1 = Z$ was the wage for q_1, the wage p_2 for carrying the quantity q_2 will be determined by the relation

$$\frac{p_2}{p'_1} = \frac{q_2}{q_1}.$$

Let $p_2 = K$; then $K : Z = B : A$ and

$$K = p_2 = \frac{B}{A} \cdot Z = \frac{q_2}{q_1} \cdot p'_1 = 24.$$

Thus we have found that $K : Z = B : A$ while $Z : H = D : G$, and therefore (applying P'_3)

$$\frac{K}{H} = \frac{K}{Z} \cdot \frac{Z}{H} = \frac{B}{A} \cdot \frac{D}{G} = \frac{B \cdot D}{A \cdot G},$$

whence

$$\frac{A \cdot G}{B \cdot D} = \frac{q_1 \cdot d_1}{q_2 \cdot d_2} = \frac{H}{K} = \frac{p_1}{p_2}.$$

(**b**) Also computed (as in B.254d)

$$p_2 = \frac{q_2}{q_1} \cdot \frac{d_2}{d_1} \cdot p_1.$$

(**B.256**) MSS $\mathcal{A}$, $\mathcal{B}$. Sixty nummi is the wage for carrying twelve sextarii thirty miles; how many miles will three sextarii be carried for five nummi?

Given thus $q_1 = 12$, $d_1 = 30$, $p_1 = 60$, $q_2 = 3$, $p_2 = 5$, required d_2. Same values as in B.254.

(**a**) Computation.

$$d_2 = \frac{q_1 \cdot d_1 \cdot p_2}{q_2 \cdot p_1} = \frac{12 \cdot 30 \cdot 5}{3 \cdot 60} = \frac{1800}{180} = 10.$$

Justification. The above formula is inferred from the proportion seen in B.254a and established in B.255a, namely

$$\frac{q_1 \cdot d_1}{p_1} = \frac{q_2 \cdot d_2}{p_2}, \quad \text{here} \quad \frac{12 \cdot 30}{60} = \frac{3 \cdot d_2}{5}.$$

(**b**) Using algebra, put $d_2 = x$. From what we have just seen, $360 : 60 = 3x : 5$, whence $180\, x = 1800$.

Remark. Use of algebra generally facilitates treatment, or at least makes it shorter. In this example, it merely reproduces, with the usual naming of the unknown, the previous computations. We have seen instances of such a pointless treatment earlier (B.166–168). See B.258c.

(**c**) Also computed as

$$d_2 = \frac{q_1}{q_2}\left(\frac{p_2}{p_1}\cdot d_1\right).$$

Justification. Changing first p_1 to p_2.

q (sextarii)	d (miles)	p (nummi)
$q_1 = 12$	$d_1 = 30$	$p_1 = 60,$
$q_1 = 12$	$d'_1 = \dfrac{p_2}{p_1}\cdot d_1 = 2 + \dfrac{1}{2}$	$p'_1 = 5 = p_2;$
$q_2 = 3$	$d''_1 = \dfrac{q_1}{q_2}\cdot d'_1\ *$	$p_2 = 5,$

whence

$$d_2 = d''_1 = \frac{q_1}{q_2}\cdot d'_1 = \frac{q_1}{q_2}\left(\frac{p_2}{p_1}\cdot d_1\right).$$

(**d**) Also computed as

$$d_2 = \frac{p_2}{\frac{q_2}{q_1}\cdot p_1}\cdot d_1.$$

---

* Since $q_2 : q_1 = d'_1 : d''_1$ (inverse proportion: the shorter the distance, the greater the quantity for the same wage).

Justification. Changing first q_1 to q_2 in the first set.

$$\begin{array}{lll} q \text{ (sextarii)} & d \text{ (miles)} & p \text{ (nummi)} \\ q_1 = 12 & d_1 = 30 & p_1 = 60, \\ q_1' = 3 = q_2 & d_1' = 30 & p_1' = \dfrac{q_2}{q_1} \cdot p_1 = 15; \\ q_2 = 3 & d_2 & p_2 = 5, \end{array}$$

whence, since (same q) $d_1 : p_1' = d_2 : p_2$,

$$d_2 = \frac{p_2}{p_1'} \cdot d_1 = \frac{p_2}{\frac{q_2}{q_1} \cdot p_1} \cdot d_1.$$

Remark. As noted in the summary (p. 1563), the case with q_2 required is not considered, presumably because it is solved just like this one.

---

2. Two unknowns.

We know product, sum, difference or ratio of the two unknowns.

(B.257) MSS A, B. Sixty nummi is the wage for carrying twelve sextarii thirty miles; what are wage and number of miles for carrying three sextarii if their product is fifty?

Given in the present problem $q_1 = 12$, $d_1 = 30$, $p_1 = 60$, $q_2 = 3$, $d_2 \cdot p_2 = 50$, required d_2 and p_2. Same values as in B.254.

(a) Formulae.

$$d_2 = \sqrt{\frac{q_1 \cdot d_1 \, (d_2 \cdot p_2)}{q_2 \cdot p_1}} = \sqrt{\frac{12 \cdot 30 \cdot 50}{3 \cdot 60}} = \sqrt{100} = 10,$$

$$p_2 = \frac{50}{d_2} = 5, \quad \text{or} \quad p_2 = \sqrt{\frac{q_2 \cdot p_1 \, (d_2 \cdot p_2)}{q_1 \cdot d_1}} = \sqrt{\frac{3 \cdot 60 \cdot 50}{12 \cdot 30}}.$$

Justification. As known from before,

$$\frac{q_1 \cdot d_1}{p_1} = \frac{q_2 \cdot d_2}{p_2},$$

so $q_1 \cdot d_1 \cdot p_2 = q_2 \cdot d_2 \cdot p_1$. Multiplying this by d_2 and by p_2, we find respectively

$$q_1 \cdot d_1 \, (d_2 \cdot p_2) = q_2 \cdot p_1 \cdot d_2^2, \quad q_1 \cdot d_1 \cdot p_2^2 = q_2 \cdot p_1 \, (d_2 \cdot p_2),$$

from which the above formulae are inferred. (The text has the numerical values and, as usual, lengthy transformations involving commutativity and associativity.)

(b) Using algebra, put $d_2 = x$; then $p_2 = \dfrac{50}{x}$ and $q_2 \cdot d_2 = 3\,x$. Therefore, since

$$\frac{q_1 \cdot d_1}{p_1} = \frac{q_2 \cdot d_2}{p_2}, \quad \text{or} \quad \frac{360}{60} = \frac{3\,x}{\frac{50}{x}},$$

we have
$$360 \cdot \frac{50}{x} = 60 \cdot 3\,x, \quad \text{or} \quad \frac{18\,000}{x} = 180\,x,$$

so $180 x^2 = 18\,000$ and $x^2 = 100$ by 'reduction', whence $x = d_2 = 10$ and $p_2 = 5$.

(*c*) Same, again with $d_2 = x$ and $p_2 = \frac{50}{x}$, but merely with the proportion inverted, thus
$$\frac{60}{360} = \frac{\frac{50}{x}}{3\,x},$$

whence, successively,
$$\frac{50}{x} = \frac{1}{6} \cdot 3\,x = \frac{1}{2}\,x, \quad \frac{1}{2} x^2 = 50,$$

and so $x^2 = 100$.

(**B.258**) MSS $\mathcal{A}$, $\mathcal{B}$; $\mathcal{D}$, which omits the demonstration and *b–d*. Sixty nummi is the wage for carrying twelve sextarii thirty miles; what are, for carrying three sextarii, wage and number of miles if their sum is fifteen?

Given thus $q_1 = 12$, $d_1 = 30$, $p_1 = 60$, $q_2 = 3$, $d_2 + p_2 = 15$, required d_2 and p_2. Same numerical values as before.

(*a*) <u>Computation</u>.
$$d_2 = \frac{q_1 \cdot d_1\,(d_2 + p_2)}{q_1 \cdot d_1 + q_2 \cdot p_1} = \frac{12 \cdot 30 \cdot 15}{12 \cdot 30 + 3 \cdot 60} = \frac{360 \cdot 15}{540} = 10$$

$$p_2 = \frac{q_2 \cdot p_1\,(d_2 + p_2)}{q_1 \cdot d_1 + q_2 \cdot p_1} = \frac{3 \cdot 60 \cdot 15}{12 \cdot 30 + 3 \cdot 60} = \frac{180 \cdot 15}{540} = 5.$$

<u>Demonstration</u>.

(α) *Preliminary remark*. It will lead to the usual demonstration for a linear system of the type
$$\begin{cases} x + y = k \\ m\,x = n\,y, \end{cases}$$
with k, m, n given (B.232). Geometrically, a segment of straight line with length k is divided into two parts in a given ratio. For the computation, since
$$\frac{x}{y} = \frac{n}{m} \quad \text{and} \quad \frac{y}{x} = \frac{m}{n},$$

then
$$\frac{x}{x+y} = \frac{n}{m+n}, \qquad \frac{y}{x+y} = \frac{m}{m+n}.$$

whence
$$x = \frac{n\,(x+y)}{m+n} = \frac{n \cdot k}{m+n}, \qquad y = \frac{m\,(x+y)}{m+n} = \frac{m \cdot k}{m+n}.$$

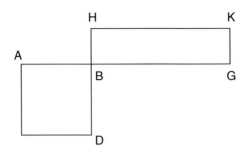

Fig. 86

In the present case, $x = p_2$, $y = d_2$, $m = q_1 \cdot d_1$, $n = q_2 \cdot p_1$.

(β) *Demonstration according to the text.* Let (Fig. 86) $d_2 + p_2 = AG = 15$, with $AB = p_2$, $BG = d_2$. Since $q_1 \cdot d_1 : p_1 = q_2 \cdot d_2 : p_2$, thus $q_1 \cdot d_1 \cdot p_2 = q_2 \cdot p_1 \cdot d_2$, then $360 \cdot AB = 180 \cdot BG$. Therefore we are to solve
$$\begin{cases} AB + BG = 15 \\ 360 \cdot AB = 180 \cdot BG. \end{cases}$$

Let $BH = 180 = q_2 \cdot p_1$ and $DB = 360 = q_1 \cdot d_1$ be perpendicular to AG at B in opposite directions. Since the areas thus formed $AB \cdot DB$ and $BG \cdot BH$ are rectangular and equal, but not identical, their sides must be reciprocally proportional (*Elements* VI.14); thus

$$\frac{DB}{BH} = \frac{q_1 \cdot d_1}{q_2 \cdot p_1} = \frac{BG}{AB} = \frac{d_2}{p_2},$$

whence, by composition,

$$\frac{DB}{DB + BH} = \frac{DB}{DH} = \frac{q_1 \cdot d_1}{q_1 \cdot d_1 + q_2 \cdot p_1} = \frac{BG}{BG + AB} = \frac{BG}{AG} = \frac{d_2}{d_2 + p_2}$$

and therefore $BG = d_2 = \dfrac{DB \cdot AG}{DH} = \dfrac{q_1 \cdot d_1 (d_2 + p_2)}{q_1 \cdot d_1 + q_2 \cdot p_1}.$

Similarly, taking the inverted form,

$$\frac{BH}{DB} = \frac{q_2 \cdot p_1}{q_1 \cdot d_1} = \frac{AB}{BG} = \frac{p_2}{d_2},$$

so, again by composition,

$$\frac{BH}{DH} = \frac{q_2 \cdot p_1}{q_1 \cdot d_1 + q_2 \cdot p_1} = \frac{AB}{AG} = \frac{p_2}{d_2 + p_2},$$

and therefore $AB = p_2 = \dfrac{BH \cdot AG}{DH} = \dfrac{q_2 \cdot p_1 (d_2 + p_2)}{q_1 \cdot d_1 + q_2 \cdot p_1}.$

(*b*) Also computed as

$$p_2 = \frac{q_2 \cdot p_1}{q_1 \cdot d_1 + q_2 \cdot p_1} \cdot (d_2 + p_2) = \frac{180}{540} \cdot 15 = \frac{1}{3} \cdot 15 = 5$$

$$d_2 = \frac{q_1 \cdot d_1}{q_1 \cdot d_1 + q_2 \cdot p_1} \cdot (d_2 + p_2) = \frac{360}{540} \cdot 15 = \frac{2}{3} \cdot 15 = 10.$$

(*c*) By algebra, which here makes sense (unlike in B.256*b*) since it avoids transforming the proportion. Put $d_2 = x$, so $p_2 = 15 - x$. Therefore the basic relation $q_1 \cdot d_1 : p_1 = q_2 \cdot d_2 : p_2$ gives

$$360 : 60 = 3\,x : (15-x), \quad 5400 - 360\,x = 180\,x,$$

whence, after 'completing' and dividing, $x = 10$.

(*d*) Same, this time for the wage. Put $p_2 = x$, so $d_2 = 15 - x$; the same proportion gives

$$360 : 60 = 3\,(15-x) : x, \quad 360\,x = 2700 - 180\,x,$$

whence $x = 5$.

(**B.259**) MSS $\mathcal{A}$, $\mathcal{B}$. A hundred nummi is the wage for carrying ten sextarii fifty miles; what are wage and number of miles for carrying three sextarii if the number of miles less the wage is four?

Given thus $q_1 = 10$, $d_1 = 50$, $p_1 = 100$, $q_2 = 3$, $d_2 - p_2 = 4$, required d_2 and p_2.

(*a*) Computation.

$$d_2 = \frac{q_1 \cdot d_1 \,(d_2 - p_2)}{q_1 \cdot d_1 - q_2 \cdot p_1} = \frac{10 \cdot 50 \cdot 4}{10 \cdot 50 - 3 \cdot 100} = \frac{500 \cdot 4}{200} = 10$$

$$p_2 = \frac{q_2 \cdot p_1 \,(d_2 - p_2)}{q_1 \cdot d_1 - q_2 \cdot p_1} = \frac{3 \cdot 100 \cdot 4}{10 \cdot 50 - 3 \cdot 100} = \frac{300 \cdot 4}{200} = 6.$$

Demonstration.

(α) *Preliminary remark.* We have this time a relation of the type
$$\begin{cases} x - y = k \\ m\,x = n\,y, \end{cases}$$
with k, m, n given ($n > m$). Solved in the same way as before. Indeed, by separation,

$$\frac{y}{x} = \frac{m}{n} \quad \text{gives} \quad \frac{y}{x-y} = \frac{m}{n-m}, \qquad (1)$$

and, inverting (1) and by composition,

$$\frac{x-y}{y} = \frac{n-m}{m} \quad \text{gives} \quad \frac{x-y}{x} = \frac{n-m}{n}, \qquad (2)$$

and we thus find

$$x = \frac{n\,(x-y)}{n-m} = \frac{n \cdot k}{n-m}, \quad y = \frac{m\,(x-y)}{n-m} = \frac{m \cdot k}{n-m}.$$

In this case, $x = d_2$, $y = p_2$, $n = q_1 \cdot d_1$, $m = q_2 \cdot p_1$.

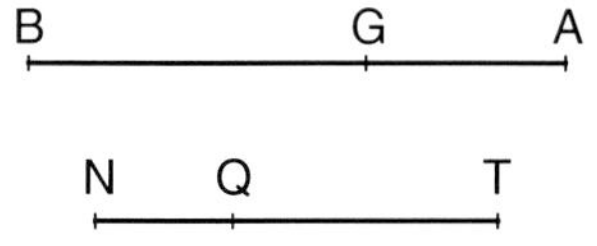

B   G   A

N Q   T

Fig. 87

(β) *Demonstration according to the text.* Let (Fig. 87) $d_2 = AB$, $p_2 = GB$, so $AG = 4$. Since $q_1 \cdot d_1 : p_1 = q_2 \cdot d_2 : p_2$, so

$$500 \cdot GB = (q_1 \cdot d_1) \, p_2 = 300 \cdot AB = (q_2 \cdot p_1) \, d_2,$$

so that we are to solve

$$\begin{cases} AB - GB = 4 \\ 300 \cdot AB = 500 \cdot GB. \end{cases}$$

Then, drawing $TN = 500$, and (on TN) $TQ = 300$, we shall have, according to the conditions, $TQ \cdot AB = TN \cdot GB$ and thus

$$\frac{TN}{TQ} = \frac{q_1 \cdot d_1}{q_2 \cdot p_1} = \frac{AB}{GB} = \frac{d_2}{p_2};$$

so, by inversion and by separation, successively,

$$\frac{TQ}{TN} = \frac{GB}{AB}, \qquad \frac{TQ}{TN-TQ} = \frac{GB}{AB-GB}, \qquad (1)$$

and therefore

$$\frac{TQ}{QN} = \frac{q_2 \cdot p_1}{q_1 \cdot d_1 - q_2 \cdot p_1} = \frac{GB}{AG} = \frac{p_2}{d_2 - p_2}, \qquad (1')$$

whence, first,

$$GB = p_2 = \frac{TQ \cdot AG}{QN} = \frac{q_2 \cdot p_1 \, (d_2 - p_2)}{q_1 \cdot d_1 - q_2 \cdot p_1},$$

and, secondly, by inversion of the above relation (1') and then by composition,

$$\frac{QN}{TQ} = \frac{AG}{GB}, \quad \frac{QN}{QN+TQ} = \frac{QN}{TN} = \frac{AG}{AG+GB} = \frac{AG}{AB}, \qquad (2)$$

and therefore

$$AB = d_2 = \frac{TN \cdot AG}{QN} = \frac{q_1 \cdot d_1 \, (d_2 - p_2)}{q_1 \cdot d_1 - q_2 \cdot p_1}.$$

(**b**) *Also computed as*

$$d_2 = \frac{q_1 \cdot d_1}{q_1 \cdot d_1 - q_2 \cdot p_1} \cdot (d_2 - p_2) = \frac{500}{200} \cdot 4 = \left(2 + \frac{1}{2}\right) 4 = 10$$

$$p_2 = \frac{q_2 \cdot p_1}{q_1 \cdot d_1 - q_2 \cdot p_1} \cdot (d_2 - p_2) = \frac{300}{200} \cdot 4 = \frac{3}{2} \cdot 4.$$

Remark. In the latter case, the text computes

$$\frac{q_1 \cdot d_1 - q_2 \cdot p_1}{q_2 \cdot p_1} \cdot p_2 = d_2 - p_2, \quad \text{whence} \quad \frac{2}{3} \cdot p_2 = 4.$$

(**c**) By algebra, put $d_2 = x$, so $p_2 = x - 4$; thus $q_1 \cdot d_1 \cdot p_2 = q_2 \cdot d_2 \cdot p_1$ gives

$$500\,(x - 4) = 300\,x, \quad 500\,x - 2000 = 300\,x,$$

whence $x = d_2$.

(**d**) Or put $p_2 = x$, so $d_2 = x + 4$, and

$$500\,x = 300\,(x + 4) = 300\,x + 1200,$$

whence $x = p_2$.

(**B.260**) MSS $\mathcal{A}$, $\mathcal{B}$. Sixty nummi is the wage for carrying twelve sextarii thirty miles; six nummi being given, what are number of sextarii and number of miles if they are equal?

In this and the subsequent problems we know the ratio (here $k = 1$) between the two unknowns. Given thus $q_1 = 12$, $d_1 = 30$, $p_1 = 60$, $p_2 = 6$, required $q_2 = d_2$.

(**a**) Computation.

$$q_2 = d_2 = \sqrt{\frac{q_1 \cdot d_1 \cdot p_2}{p_1}} = \sqrt{\frac{12 \cdot 30 \cdot 6}{60}} = \sqrt{\frac{2160}{60}} = \sqrt{36} = 6.$$

Justification. From

$$\frac{q_1 \cdot d_1}{p_1} = \frac{q_2 \cdot d_2}{p_2}, \quad \text{we infer} \quad q_2 \cdot d_2 = q_2^2 = d_2^2 = \frac{q_1 \cdot d_1 \cdot p_2}{p_1}.$$

(**b**) Also computed ('other way') as

$$\frac{q_1 \cdot d_1}{p_1} \cdot p_2 = \frac{360}{60} \cdot 6 = q_2^2 = d_2^2,$$

inferred from the previous relation by Premiss P_5.

(**c**) Also computed ('other ways') as

$$q_2^2 = d_2^2 = \left(\frac{p_2}{p_1} \cdot d_1\right) q_1 \quad \text{or} \quad \left(\frac{p_2}{p_1} \cdot q_1\right) d_1,$$

justified by P_2.

(**d**) By algebra, putting $q_2 = d_2 = x$; so, from the fundamental relation directly,

$$q_1 \cdot d_1 \cdot p_2 = p_1 \cdot q_2 \cdot d_2 = p_1 \cdot x^2,$$

whence x.

(B.261) MSS $\mathcal{A}$, $\mathcal{B}$. A hundred nummi is the wage for carrying twelve sextarii thirty miles, and ten nummi are given for carrying a number of sextarii which is three fourths the number of miles; what are they?

Given thus $q_1 = 12$, $d_1 = 30$, $p_1 = 100$, $p_2 = 10$, $q_2 = \frac{3}{4} d_2$, required q_2 and d_2.

(*a*) <u>Computation</u>.

$$d_2 = \sqrt{\frac{q_1 \cdot d_1 \cdot p_2}{\frac{3}{4} p_1}} = \sqrt{\frac{12 \cdot 30 \cdot 10}{75}} = \sqrt{\frac{3600}{75}} = \sqrt{48} \cong 7 - \frac{1}{14},$$

whence q_2.

<u>Remark</u>. The author employs here the approximation formula by subtraction already applied in A.277; for this is one case where the more common root approximation by addition, applied in A.275, is of no use (see p. 1304).

<u>Justification</u>. From the fundamental relation used before, namely here in the form

$$q_2 \cdot d_2 = \frac{3}{4} \cdot d_2^2 = \frac{q_1 \cdot d_1 \cdot p_2}{p_1}.$$

(*b*) Also computed as

$$d_2^2 = \frac{q_1 \cdot d_1 \cdot p_2}{p_1} \cdot \frac{4}{3} = \frac{3600}{100}\left(1 + \frac{1}{3}\right).$$

(*c*) Or as

$$d_2^2 = \left[\left(\frac{p_2}{p_1} \cdot d_1\right) q_1\right]\left(1 + \frac{1}{3}\right), \qquad d_2^2 = \left[\left(\frac{p_2}{p_1} \cdot q_1\right) d_1\right]\left(1 + \frac{1}{3}\right).$$

(*d*) By algebra, putting $d_2 = x$, so $q_2 = \frac{3}{4} x$; hence

$$q_1 \cdot d_1 \cdot p_2 = p_1 \cdot \frac{3}{4} x \cdot x, \qquad 3600 = 75 \, x^2,$$

whence $x^2 = 48$.

---

3. Problems involving 'things'.

As often, a section with 'things' in the data is found at the end. Here x^2 and even x^3 occur. MS $\mathcal{A}$ only.

(B.262) MS $\mathcal{A}$. A square plus its root is the wage for carrying fifteen sextarii forty miles, and the root is given for carrying ten sextarii ten miles; what is the square?

Given then $q_1 = 15$, $d_1 = 40$, $p_1 = x^2 + x$, $q_2 = 10$, $d_2 = 10$, $p_2 = x$, required x^2. According to the text, we shall compute

$$x = \frac{15 \cdot 40}{10 \cdot 10} - 1 = \frac{600}{100} - 1 = 5, \text{ so } x^2 = 25.$$

Remark. The text notes that if, generally, $p_1 = x^2 + bx$, $p_2 = ax$, then

$$x = \frac{q_1 \cdot d_1}{q_2 \cdot d_2} \cdot a - b.$$

As often with such problems found at the end of chapters, there is no explanation of the formula. This one originates in the fundamental proportion, since

$$\frac{p_1}{p_2} = \frac{x^2 + bx}{ax} = \frac{x + b}{a} = \frac{q_1 \cdot d_1}{q_2 \cdot d_2}.$$

(B.263) MS $\mathcal{A}$. A cube plus the corresponding square is the wage for carrying six sextarii ten miles, and the square is given for carrying four sextarii five miles; what are the square and the cube?

Given thus $q_1 = 6$, $d_1 = 10$, $p_1 = x^3 + x^2$, $q_2 = 4$, $d_2 = 5$, $p_2 = x^2$, required x^3 and x^2.

The text computes

$$x = \frac{6 \cdot 10}{4 \cdot 5} - 1 = 2, \text{ so } x^2 = 4, \ x^3 = 8.$$

Remark. It is again noted that if, generally, $p_1 = x^3 + bx^2$, $p_2 = ax^2$, then

$$x = \frac{q_1 \cdot d_1}{q_2 \cdot d_2} \cdot a - b,$$

which is just the same as before. Indeed, here too the origin of the formula is

$$\frac{p_1}{p_2} = \frac{x^3 + bx^2}{ax^2} = \frac{x + b}{a} = \frac{q_1 \cdot d_1}{q_2 \cdot d_2}.$$

As said in our summary, we have here the first occurrence of the cubic power of the unknown.

Chapter B-XIV:
Hiring stone-cutters
Summary

In the previous chapter the kind of work was specified and involved one worker at a time. There are now several workers (B.264–270), except at the end (B.271–275). But the chapter heading is misleading: only the first two problems are concerned with stone-cutting (B.264–265); in the others, the kind of work is not specified (B.266–270) or is different (B.271–275). As before, there are two sets of either three or four quantities, with those of the first always given while, in the second, one or two are unknown; in the latter case we know their product, sum or difference, as often seen before.

1. Cutting stones.

If n_1 workers cut q_1 stones in t_1 days for p_1 nummi, then n_2 men working under the same conditions would cut q_2 stones in t_2 days for p_2 nummi. Two cases are considered:

(i) A total wage is paid according to how many stones are cut and days worked, irrespective of the number of workers in each group. We have then the relation
$$\frac{p_1}{p_2} = \frac{q_1 \cdot t_1}{q_2 \cdot t_2}.$$
Two different groups of men will thus receive the same sum for working the same number of days and cutting the same number of stones. The workers of the larger group are then to receive individually a correspondingly lower wage. Thus, in the first part of B.264, we are to compute
$$p_2 = \frac{q_2 \cdot t_2 \cdot p_1}{q_1 \cdot t_1},$$
to be shared among the n_2 workers.

Case i being just like the problems seen in the previous chapter, the author gives only one example, as he did once before in a similar situation (B.121).

(ii) The wage is also proportional to the number of workers. Then four types of quantity occur, and we have
$$\frac{p_1}{p_2} = \frac{n_1 \cdot q_1 \cdot t_1}{n_2 \cdot q_2 \cdot t_2}.$$
In this case the workers of two different groups will *each* receive the same wage if the two groups do the same work in the same time. In the two problems dealing with this case (B.264ii–265), the formulae used are
$$p_2 = \frac{n_2 \cdot q_2 \cdot t_2 \cdot p_1}{n_1 \cdot q_1 \cdot t_1}, \qquad t_2 = \frac{n_1 \cdot q_1 \cdot t_1 \cdot p_2}{n_2 \cdot q_2 \cdot p_1}.$$

Only the formula for case ii is proved (in B.264). That for case i is not, being similar to the formula seen before (and proved in B.255).

Remark. In either case, the situation may become absurd, at least for the employer, because of the wage depending on the number of days worked and also, in the second case, the number of workers employed. Thus, in the first case, the less time taken by a worker or a group of workers to cut the same number of stones, the less will be earned; moreover, it will cost the hirer the same if a normal worker cuts q stones in t days, a lazy one cuts one stone in $q \cdot t$ days, or a zealous one cuts $q \cdot t$ stones in one day. The situation becomes even odder in the second case where, as said, the wage is also proportional to the number of workers. For, to have the same number of stones cut in the same time will cost the employer n times more if n men do it instead of just one; furthermore, n men cutting a single stone in $q \cdot t$ days costs the employer as much as a single worker cutting $n \cdot q \cdot t$ stones in a single day. These absurdities cannot have escaped our author; he nonetheless does not hesitate to resort to such (mathematically equivalent) transformations whenever he needs to justify, in the same manner as in the problems on carriers, a formula. His purpose is to make sense mathematically, not suit the labor market.

2. Kind of work not specified.

Since in the group B.266–270 the kind of work is not mentioned, the q_i disappear while the n_i remain: a number n of men are to work a number t of days for a wage p. Payment being proportional to the time spent and the number of workers, we shall have

$$\frac{p_1}{p_2} = \frac{n_1 \cdot t_1}{n_2 \cdot t_2}.$$

The kind and quantity of work not being specified, we avoid the absurdities seen above. B.266 and B.267 are the simpler case (p_2 or t_2 required). In B.268–270, t_2 and p_2 are required individually when we know their product, their sum or their difference. Evidently, all these problems are similar to those on carrying: the sextarii become the workers and the distances are the days. This analogy is alluded to by the author in all problems but one (B.268). The formulae used are accordingly

$$t_2 = \frac{n_1 \cdot t_1 \cdot p_2}{n_2 \cdot p_1}, \qquad p_2 = \frac{n_2 \cdot t_2 \cdot p_1}{n_1 \cdot t_1}.$$

$$t_2 = \sqrt{\frac{n_1 \cdot t_1 (t_2 \cdot p_2)}{n_2 \cdot p_1}}, \qquad p_2 = \sqrt{\frac{n_2 \cdot p_1 (t_2 \cdot p_2)}{n_1 \cdot t_1}}.$$

$$t_2 = \frac{n_1 \cdot t_1 (t_2 \pm p_2)}{n_1 \cdot t_1 \pm n_2 \cdot p_1}, \qquad p_2 = \frac{n_2 \cdot p_1 (t_2 \pm p_2)}{n_1 \cdot t_1 \pm n_2 \cdot p_1}.$$

3. Other kinds of work.

The last problems in this chapter are of the same type mathematically, but the kind of work changes. First (B.271–271′), a shepherd tends q_i sheep for a certain time t_i at a wage p_i. Since, for the mathematical treatment, the sheep just replace the workers, or the bushels, the relation will be

$$\frac{p_1}{p_2} = \frac{q_1 \cdot t_1}{q_2 \cdot t_2}$$

giving the formulae actually used

$$p_2 = \frac{p_1 \cdot q_2 \cdot t_2}{q_1 \cdot t_1}, \qquad t_2 = \frac{p_2 \cdot q_1 \cdot t_1}{p_1 \cdot q_2}.$$

In the following problems, about digging holes (B.272–274) and building chests (B.275), the dimensions replace quantity and time. That is, if l_i, r_i, h_i are length, width and depth or height (and V_i the volumes), the proportion will be

$$\frac{p_1}{p_2} = \frac{V_1}{V_2} = \frac{l_1 \cdot r_1 \cdot h_1}{l_2 \cdot r_2 \cdot h_2}.$$

The formulae occurring in B.272–274 are

$$p_2 = \frac{l_2 \cdot r_2 \cdot h_2 \cdot p_1}{l_1 \cdot r_1 \cdot h_1}, \qquad h_2 = \frac{l_1 \cdot r_1 \cdot h_1 \cdot p_2}{l_2 \cdot r_2 \cdot p_1}, \qquad l_2 = r_2 = \sqrt{\frac{l_1 \cdot r_1 \cdot h_1 \cdot p_2}{h_2 \cdot p_1}}.$$

For chests the situation is slightly different since only the sizes of the six sides, and not the volume of the chest, matters (B.275). In this case we shall use

$$\frac{p_1}{p_2} = \frac{2(l_1 \cdot r_1 + l_1 \cdot h_1 + r_1 \cdot h_1)}{2(l_2 \cdot r_2 + l_2 \cdot h_2 + r_2 \cdot h_2)}.$$

---

1. Cutting stones.

(B.264) MSS $\mathcal{A}$; $\mathcal{B}$ (omits the demonstration of the general formula), $\mathcal{D}$ (omits the demonstration and what follows). Three workers cut four stones in thirty days for sixty nummi; how much is due to two workers cutting two stones in ten days?

As mentioned above, the case where the wage is independent of the number of workers, is only briefly mentioned in this problem.

Given then $n_1 = 3$, $q_1 = 4$, $t_1 = 30$, $p_1 = 60$, $n_2 = 2$, $q_2 = 2$, $t_2 = 10$, required p_2, the wage of the two workers.

• Case *i*. Analogous to B.254–255 on carrying: we need only replace the stones cut by sextarii and the days worked by miles; since the number of workers is irrelevant, it is as if a single man were involved. Thus in this case

$$p_2 = \frac{q_2 \cdot t_2 \cdot p_1}{q_1 \cdot t_1} = \frac{2 \cdot 10 \cdot 60}{4 \cdot 30} = 10$$

and all 'other' ways seen in these two problems on carrying are applicable.

• Case *ii* (considering the number of workers).

(*a*) Computation.

$$p_2 = \frac{n_2 \cdot q_2 \cdot t_2 \cdot p_1}{n_1 \cdot q_1 \cdot t_1} = \frac{2 \cdot 2 \cdot 10 \cdot 60}{3 \cdot 4 \cdot 30} = \frac{2400}{360} = 6 + \frac{2}{3}.$$

This is established in two manners: first, by transforming the sets of quantities as seen before, the wage remaining the same and all other factors but one (the days) being reduced to 1; second, and more formally, by a geometrical demonstration.

Justification. Both sets are transformed in the same way into unit quantities n_0, q_0:

n (workers)	q (stones)	t (days)	p (nummi)
$n_1 = 3$	$q_1 = 4$	$t_1 = 30$	$p_1 = 60$
$n'_1 = 1 = n_0$	$q'_1 = n_1 \cdot q_1 = 12$	$t_1 = 30$	$p_1 = 60$
$n'_1 = 1 = n_0$	$q''_1 = 1 = q_0$	$t'_1 = q'_1 \cdot t_1 = 360$	$p_1 = 60;$
$n_2 = 2$	$q_2 = 2$	$t_2 = 10$	p_2
$n'_2 = 1 = n_0$	$q'_2 = 1 = q_0$	$t'_2 = n_2 \cdot q_2 \cdot t_2 = 40$	$p_2.$

The reduction of each of the two sets has thus led to a simple proportion, as in the elementary problems on buying and selling: since in the two transformed sets now $n_0 = q_0 = 1$, we must have $t'_2 : p_2 = t'_1 : p_1$, thus

$$p_2 = \frac{t'_2 \cdot p_1}{t'_1} = \frac{n_2 \cdot q_2 \cdot t_2 \cdot p_1}{q'_1 \cdot t_1} = \frac{n_2 \cdot q_2 \cdot t_2 \cdot p_1}{n_1 \cdot q_1 \cdot t_1}.$$

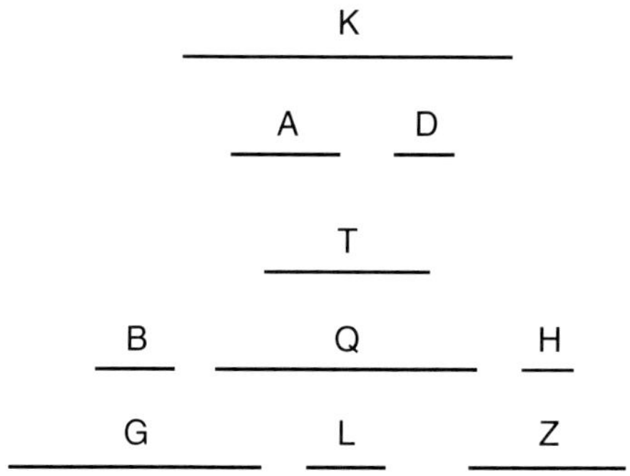

```
                    K
               ─────────────
              A         D
              ───       ───
                    T
                   ───
          B           Q           H
          ───      ──────        ───
          G           L           Z
          ───      ──────        ───
```

Fig. 88

Demonstration of the general formula

$$\frac{p_1}{p_2} = \frac{n_1 \cdot q_1 \cdot t_1}{n_2 \cdot q_2 \cdot t_2}.$$

(Analogous demonstration for three quantities in B.255*a*.)

Let (Fig. 88) $n_1 = A = 3$, $q_1 = B = 4$, $t_1 = G = 30$, $n_2 = D = 2$, $q_2 = H = 2$, $t_2 = Z = 10$, $p_1 = K = 60$.

If $A = 3$ workers cut 4 stones in 30 days for $K = 60$ nummi, then $D = 2$ workers cut 4 stones in 30 days for $p'_1 = \frac{2}{3} \cdot 60 = 40$ nummi. Let $T = p'_1 = 40$. This means that

$$\frac{p'_1}{p_1} = \frac{n_2}{n_1}, \quad \text{or} \quad \frac{T}{K} = \frac{D}{A}.$$

Next, given that 2 workers cut $B = 4$ stones in 30 days for $T = 40$ nummi, then 2 workers cutting $H = 2$ stones in 30 days will receive $p''_1 = \frac{1}{2} \cdot 40 = 20$ nummi. Let $Q = p''_1 = 20$. This means that

$$\frac{p''_1}{p'_1} = \frac{q_2}{q_1}, \quad \text{or} \quad \frac{Q}{T} = \frac{H}{B}.$$

Finally, given that 2 workers cut 2 stones in $G = 30$ days for $Q = 20$ nummi, then 2 workers will cut 2 stones in $Z = 10$ days for $\frac{1}{3} \cdot 20 = 6 + \frac{2}{3}$ nummi. Let $L = 6 + \frac{2}{3} = p_2$. This means that

$$\frac{p_2}{p''_1} = \frac{t_2}{t_1}, \quad \text{or} \quad \frac{L}{Q} = \frac{Z}{G}.$$

Thus we have found

$$\frac{L}{Q} = \frac{Z}{G}, \quad \frac{Q}{T} = \frac{H}{B}, \quad \frac{T}{K} = \frac{D}{A}$$

and therefore, since we have continued proportionals and by P'_3 (extended to three factors),

$$\frac{L}{K} = \frac{L}{Q} \cdot \frac{Q}{T} \cdot \frac{T}{K} = \frac{Z}{G} \cdot \frac{H}{B} \cdot \frac{D}{A} = \frac{Z \cdot H \cdot D}{G \cdot B \cdot A}$$

which proves that

$$\frac{K}{L} = \frac{A \cdot B \cdot G}{D \cdot H \cdot Z}, \quad \text{or} \quad \frac{p_1}{p_2} = \frac{n_1 \cdot q_1 \cdot t_1}{n_2 \cdot q_2 \cdot t_2}.$$

(b) Also computed as

$$p_2 = \frac{p_1}{n_1 \cdot q_1 \cdot t_1} \cdot n_2 \cdot q_2 \cdot t_2.$$

(c) Also computed by taking successively fractions of p_1:

$$p_2 = \frac{n_2}{n_1} \left[\frac{t_2}{t_1} \left(\frac{q_2}{q_1} \cdot p_1 \right) \right].$$

<u>Justification</u>. Transforming the first set into the second.

n (workers) $\quad\quad q$ (stones) $\quad\quad t$ (days) $\quad\quad\quad\quad p$ (nummi)

$n_1 = 3$	$q_1 = 4$	$t_1 = 30$	$p_1 = 60$
$n_1 = 3$	$q_1' = 2 = q_2$	$t_1 = 30$	$p_1' = \dfrac{q_2}{q_1} \cdot p_1 = 30$
$n_1 = 3$	$q_2 = 2$	$t_1' = 10 = t_2$	$p_1'' = \dfrac{t_2}{t_1} \cdot p_1' = 10$
$n_1' = 2 = n_2$	q_2	t_2	$p_2 = \dfrac{n_2}{n_1} \cdot p_1'' = 6 + \dfrac{2}{3}.$

This gives the above formula.

(**d**) Also computed by taking successively fractions of p_1 in another order:

$$p_2 = \frac{t_2}{t_1}\left[\frac{q_2}{q_1}\left(\frac{n_2}{n_1}\cdot p_1\right)\right].$$

Justification. In the same way as before,

n (workers)	q (stones)	t (days)	p (nummi)
$n_1 = 3$	$q_1 = 4$	$t_1 = 30$	$p_1 = 60$
$n_1' = 2 = n_2$	$q_1 = 4$	$t_1 = 30$	$p_1' = \dfrac{n_2}{n_1} \cdot p_1 = 40,$[†]
$n_2 = 2$	$q_1' = 2 = q_2$	$t_1 = 30$	$p_1'' = \dfrac{q_2}{q_1} \cdot p_1' = 20,$
n_2	q_2	$t_1' = 10 = t_2$	$p_2 = \dfrac{t_2}{t_1} \cdot p_1'' = 6 + \dfrac{2}{3}.$

This justifies the above formula.

(**e**) Also computed as

$$p_2 = \left(\frac{t_2}{t_1}\cdot\frac{q_2}{q_1}\cdot\frac{n_2}{n_1}\right)p_1,$$

which does not need to be justified since that has already been done, the text says (in B.254d, but with two fractions).

(**B.265**) MSS $\mathcal{A}$, $\mathcal{B}$. Three workers cut four stones in thirty days at sixty nummi; how many days will two of them work for cutting two stones at six nummi and two thirds?

Given thus $n_1 = 3$, $q_1 = 4$, $t_1 = 30$, $p_1 = 60$, $n_2 = 2$, $q_2 = 2$, $p_2 = 6 + \frac{2}{3}$, required t_2. Same values as before.

(**a**) Computation.

$$t_2 = \frac{q_1 \cdot n_1 \cdot t_1 \cdot p_2}{q_2 \cdot n_2 \cdot p_1} = \frac{4\cdot 3\cdot 30\cdot (6+\frac{2}{3})}{2\cdot 2\cdot 60} = \frac{2400}{240} = 10.$$

[†] Because (the author says) 'the work of two workers is two thirds of the work of three' —though what he means is 'wage' rather than 'work', since the same amount of work is done by the two groups.

Justification. The origin is (as seen and proved in B.264), the constancy of the ratio stones times workers times days to wage:

$$\frac{q_1 \cdot n_1 \cdot t_1}{p_1} = \frac{q_2 \cdot n_2 \cdot t_2}{p_2}, \quad \text{thus} \quad \frac{4 \cdot 3 \cdot 30}{60} = \frac{2 \cdot 2 \cdot t_2}{6 + \frac{2}{3}},$$

whence the previous formula.

(**b**) Also computed as

$$t_2 = \frac{1}{q_2 \cdot n_2} \left(\frac{p_2}{p_1} \cdot q_1 \cdot n_1 \cdot t_1 \right).$$

Justification. As in B.264a (both sets transformed to unitary quantities n_0, q_0).

n (workers)	q (stones)	t (days)	p (nummi)
$n_1 = 3$	$q_1 = 4$	$t_1 = 30$	$p_1 = 60,$
$n_1' = 1 = n_0$	$q_1' = 1 = q_0$	$t_1' = q_1 \cdot n_1 \cdot t_1 = 360$	$p_1 = 60.$‖
$n_2 = 2$	$q_2 = 2$	t_2	$p_2 = 6 + \frac{2}{3},$
$n_2' = 1 = n_0$	$q_2' = q_2 \cdot n_2 = 4$	t_2	$p_2 = 6 + \frac{2}{3},$*
$n_0 = 1$	$q_2'' = 1 = q_0$	$t_2' = q_2' \cdot t_2 = 4\, t_2$	$p_2 = 6 + \frac{2}{3}.$‡

Since now $p_2 : p_1 = t_2' : t_1'$, then

$$t_2' = \frac{p_2}{p_1} \cdot t_1' = \frac{p_2}{p_1} \cdot q_1 \cdot n_1 \cdot t_1;$$

but $t_2' = q_2' \cdot t_2$, so we obtain finally

$$t_2 = \frac{1}{q_2 \cdot n_2} \left(\frac{p_2}{p_1} \cdot q_1 \cdot n_1 \cdot t_1 \right) = \frac{1}{4} \left(\frac{6 + \frac{2}{3}}{60} \cdot 360 \right) = \frac{1}{4} \left(\frac{2}{3} \frac{1}{6} \cdot 360 \right) = 10.$$

Remark. The text seems to have been somewhat altered here. This may again be due to reworking by some early reader on account of the odd fact that a stone-cutter receives the same wage while producing four times less in a number of days four times greater.

(**c**) Also computed as

$$t_2 = \frac{n_1 \cdot q_1}{n_2 \cdot q_2} \left(\frac{p_2}{p_1} \cdot t_1 \right) = \frac{12}{4} \left(\frac{6 + \frac{2}{3}}{60} \cdot 30 \right) = 3 \left(\frac{2}{3} \frac{1}{6} \cdot 30 \right) = 10.$$

‖ Known from B.264ii.a.

* Since 'two workers cutting two stones is the same as one worker cutting four stones'.

‡ This last step is not found in the text (see remark below).

Justification. From, respectively, B.264a and b above, we know the equivalences leading to $n'_1 = 1 = n'_2$ (first two steps and last two steps below); we shall just pursue the transformation to obtain the same value for q, namely $q''_1 (= q_1) = q'_2$.

$n_1 = 3$	$q_1 = 4$	$t_1 = 30$	$p_1 = 60$
$n'_1 = 1 = n_0$	$q'_1 = n_1 \cdot q_1 = 12$	$t_1 = 30$	$p_1 = 60$,
$n'_1 = 1$	$q'_1 = 12$	$t'_1 = \dfrac{p_2}{p_1} \cdot t_1 = 3 + \dfrac{1}{3}$	$p_2 = 6 + \dfrac{2}{3}$,[†]
$n'_1 = 1$	$q''_1 = 4$	$t''_1 = \dfrac{q'_1}{q''_1} \cdot t'_1 = 10$	$p_2 = 6 + \dfrac{2}{3}$;[∘]
$n_2 = 2$	$q_2 = 2$	t_2	$p_2 = 6 + \dfrac{2}{3}$,
$n'_2 = 1 = n_0$	$q'_2 = n_2 \cdot q_2 = 4$	t_2	$p_2 = 6 + \dfrac{2}{3}$.

The quantities in the fourth and sixth steps being the same, we must have

$$t_2 = t''_1 = \frac{q'_1}{q''_1} \cdot t'_1 = \frac{q'_1}{q'_2} \cdot t'_1 = \frac{q'_1}{q'_2} \cdot \frac{p_2}{p_1} \cdot t_1 = \frac{n_1 \cdot q_1}{n_2 \cdot q_2} \left(\frac{p_2}{p_1} \cdot t_1\right).$$

(**d**) Also computed as

$$t_2 = \frac{n_1}{n_2}\left[\frac{q_1}{q_2}\left(\frac{p_2}{p_1} \cdot t_1\right)\right],$$

with reference to part c, the terms just being associated differently.

---

2. Kind of work not specified.

In the next group of problems the kind of work is not specified: a number n of men are to work a number t of days for a wage p. These problems are similar to those on carrying: the sextarii and the distances become the workers and the days.

A. *One unknown*.

(**B.266**) MSS $\mathcal{A}$, $\mathcal{B}$. Five workers serve thirty days at sixty nummi; what is due to two workers for ten days?

Given thus $n_1 = 5$, $t_1 = 30$, $p_1 = 60$, $n_2 = 2$, $t_2 = 10$, required p_2 (same values in B.267–270). Similar to B.254–255, as stated in the text.

(**a**) Computation.

$$p_2 = \frac{n_2 \cdot t_2 \cdot p_1}{n_1 \cdot t_1} = \frac{2 \cdot 10 \cdot 60}{5 \cdot 30} = \frac{1200}{150} = 8.$$

[†] Since $p_2 : p_1 = t'_1 : t_1$ (wage proportional to working time).
[∘] For $q'_1 : q''_1 = t''_1 : t'_1$ (time inversely proportional to number of stones cut (!)).

Justification. Again obtained by transformation of the two sets (n becoming the same).

n (workers)	t (days)	p (nummi)
$n_1 = 5$	$t_1 = 30$	$p_1 = 60$,
$n_1' = 1 = n_0$	$t_1' = n_1 \cdot t_1 = 150$	$p_1 = 60$;
$n_2 = 2$	$t_2 = 10$	p_2,
$n_2' = 1 = n_0$	$t_2' = n_2 \cdot t_2 = 20$	p_2.

Thus, with n_0 the same and t_1', p_1, t_2' known, we have the simple proportion $t_1' : p_1 = t_2' : p_2$, whence

$$p_2 = \frac{t_2' \cdot p_1}{t_1'} = \frac{n_2 \cdot t_2 \cdot p_1}{n_1 \cdot t_1}.$$

(b) Also computed as (see B.254b)

$$p_2 = \frac{t_2}{t_1}\left(\frac{n_2}{n_1} \cdot p_1\right).$$

(c) Also computed as (see B.254d)

$$p_2 = \frac{n_2}{n_1}\left(\frac{t_2}{t_1} \cdot p_1\right).$$

(d) Also computed as (see B.254d)

$$p_2 = \left(\frac{t_2}{t_1} \cdot \frac{n_2}{n_1}\right) p_1.$$

(B.267) MSS $\mathcal{A}$, $\mathcal{B}$. Five workers serve thirty days at sixty nummi; how many days will two of them serve for eight nummi?

Given thus $n_1 = 5$, $t_1 = 30$, $p_1 = 60$, $n_2 = 2$, $p_2 = 8$, required t_2. Analogous to B.256, as stated.

(a) Same computation as in B.256a.

$$t_2 = \frac{n_1 \cdot t_1 \cdot p_2}{n_2 \cdot p_1} = \frac{5 \cdot 30 \cdot 8}{2 \cdot 60} = 10.$$

(b) Also computed as (see B.256d)

$$t_2 = \frac{p_2}{\frac{n_2}{n_1} \cdot p_1} \cdot t_1.$$

(c) Also computed as (see B.256c)

$$t_2 = \frac{n_1}{n_2}\left(\frac{p_2}{p_1} \cdot t_1\right) = \left(2 + \frac{1}{2}\right)\left[\left(\frac{1}{10} + \frac{1}{3}\frac{1}{10}\right)30\right].$$

Remark. B.256 has in addition an algebraic treatment.

B. *Two unknowns*.

(B.268) MSS $\mathcal{A}$, $\mathcal{B}$. Five workers serve thirty days at sixty nummi, and two of them serve a number of days which when multiplied by the wage produces eighty; what are number of days and wage?

Given thus $n_1 = 5$, $t_1 = 30$, $p_1 = 60$, $n_2 = 2$, $t_2 \cdot p_2 = 80$, required t_2, p_2. See B.257.

Computation.

$$t_2 = \sqrt{\frac{n_1 \cdot t_1 \, (t_2 \cdot p_2)}{n_2 \cdot p_1}} = \sqrt{\frac{5 \cdot 30 \cdot 80}{2 \cdot 60}} = 10, \quad \text{and} \quad p_2 = \frac{80}{t_2} = 8 \quad \text{or}$$

$$p_2 = \sqrt{\frac{n_2 \cdot p_1 \, (t_2 \cdot p_2)}{n_1 \cdot t_1}}.$$

Justification. The formula for t_2 is obtained in the same way as for d_2 in B.257. Since

$$\frac{n_1 \cdot t_1}{p_1} = \frac{n_2 \cdot t_2}{p_2},$$

so $n_1 \cdot t_1 \cdot p_2 = p_1 \cdot n_2 \cdot t_2$, which we shall multiply by t_2 (or by p_2 to obtain the formula for p_2), whence

$$n_1 \cdot t_1 \, (t_2 \cdot p_2) = p_1 \cdot n_2 \cdot t_2^2 \quad \left(\text{and} \ \ n_1 \cdot t_1 \cdot p_2^2 = p_1 \cdot n_2 \, (t_2 \cdot p_2)\right),$$

after wordy transformations involving associativity and commutativity in the text.

(B.269) MSS $\mathcal{A}$, $\mathcal{B}$. Five workers serve thirty days at sixty nummi, and two of them serve a number of days which when added to the wage makes eighteen; what are number of days and wage?

Given thus $n_1 = 5$, $t_1 = 30$, $p_1 = 60$, $n_2 = 2$, $t_2 + p_2 = 18$, required t_2, p_2. The formulae are known from B.258, the statement of which is repeated for comparison (the sextarii become workers, the miles days).

(*a*) *Formulae.*

$$t_2 = \frac{n_1 \cdot t_1 \, (t_2 + p_2)}{n_1 \cdot t_1 + n_2 \cdot p_1} = \frac{5 \cdot 30 \cdot 18}{5 \cdot 30 + 2 \cdot 60} = \frac{150 \cdot 18}{270} = 10$$

$$p_2 = \frac{p_1 \cdot n_2 \, (t_2 + p_2)}{n_1 \cdot t_1 + n_2 \cdot p_1} = \frac{60 \cdot 2 \cdot 18}{5 \cdot 30 + 2 \cdot 60} = \frac{120 \cdot 18}{270} = 8.$$

(*b*) By algebra, which avoids transforming the proportion (see B.258c). Put $t_2 = x$, so $p_2 = 18 - x$; then

$$\frac{n_1 \cdot t_1}{p_1} = \frac{n_2 \cdot t_2}{p_2} \quad \text{or} \quad \frac{150}{60} = \frac{2 \cdot t_2}{p_2},$$

gives
$$120\,x = 150\,(18-x) = 2700 - 150\,x,$$
whence x.

(*c*) Or put $p_2 = x$, thus $t_2 = 18 - x$, which gives
$$150\,x = 120\,(18-x) = 2160 - 120\,x,$$
whence x. See B.258*d*.

(**B.270**) MSS *A*, *B*, *D*. Five workers serve thirty days at sixty nummi, and two of them serve a number of days which when diminished by the wage leaves two; what are number of days and wage?

Given thus $n_1 = 5$, $t_1 = 30$, $p_1 = 60$, $n_2 = 2$, $t_2 - p_2 = 2$. To be treated, the text says, like B.259 (four ways).

---

3. Other kinds of work.

The last problems in this chapter are, mathematically, of the same kind but the type of work done is specified. First (B.271–271$'$), a shepherd tends q_i sheep for a certain time t_i at a wage p_i. As before, the relation will be
$$\frac{p_1}{p_2} = \frac{q_1 \cdot t_1}{q_2 \cdot t_2},$$
since the wage depends on the number of sheep and tending time. In the subsequent problems, about digging a hole (B.272–274) and making a chest (B.275), the dimensions replace quantity and time.

(**B.271**) MSS *A*, *B*; *D* (*a* only, without the justification). A shepherd tends a hundred sheep for thirty days at ten nummi; what is due to him for tending sixty for twenty days?

Given thus $q_1 = 100$, $t_1 = 30$, $p_1 = 10$, $q_2 = 60$, $t_2 = 20$, required p_2.

(*a*) Computation.
$$p_2 = \frac{p_1 \cdot q_2 \cdot t_2}{q_1 \cdot t_1} = \frac{10 \cdot 60 \cdot 20}{100 \cdot 30} = 4.$$

Justification. The origin of the formula lies in the proportion, similar to that seen before,
$$\frac{q_1 \cdot t_1}{p_1} = \frac{q_2 \cdot t_2}{p_2}, \quad \text{thus} \quad \frac{100 \cdot 30}{10} = \frac{60 \cdot 20}{p_2}.$$

(*b*) Also computed as
$$p_2 = \frac{t_2}{t_1}\left(\frac{q_2}{q_1} \cdot p_1\right).$$

Justification. Transform the first set into the second.

q (sheep)	t (days)	p (nummi)
$q_1 = 100$	$t_1 = 30$	$p_1 = 10,$
$q'_1 = 60 = q_2$	$t_1 = 30$	$p'_1 = \dfrac{q_2}{q_1} \cdot p_1 = \dfrac{3}{5} \cdot 10 = 6;$
q_2	$t'_1 = 20 = t_2$	$p''_1 = \dfrac{t_2}{t_1} \cdot p'_1 = \dfrac{2}{3} \cdot 6.$

Since this last step satisfies the requirement, we must have
$$p_2 = p''_1 = \frac{t_2}{t_1} \cdot p'_1 = \frac{t_2}{t_1}\left(\frac{q_2}{q_1} \cdot p_1\right).$$

(*c*) Also computed as
$$p_2 = \frac{q_2}{q_1}\left(\frac{t_2}{t_1} \cdot p_1\right).$$

(**B.271′**) Same problem, but with $p_2 = 4$ and t_2 required.
(*a*) Formula.
$$t_2 = \frac{p_2 \cdot q_1 \cdot t_1}{p_1 \cdot q_2} = \frac{4 \cdot 100 \cdot 30}{10 \cdot 60} = 20.$$

The author refers to the similar problem B.256 (the distance there being replaced here by the time). B.267 is of the same type.
(*b*) Also computed as (see B.256*d*, B.267*b*)
$$t_2 = \frac{p_2}{\frac{q_2}{q_1} \cdot p_1} \cdot t_1.$$

(*c*) Also computed as (see B.256*c*, B.267*c*)
$$t_2 = \frac{q_1}{q_2}\left(\frac{p_2}{p_1} \cdot t_1\right).$$

(**B.272**) MSS $\mathcal{A}, \mathcal{B}; \mathcal{D}$ (*a* only, without the justification). Eighty nummi are given for digging a hole ten cubits long, eight wide, six deep; what is due for digging a hole four cubits long, three wide and five deep?

Given here $l_1 = 10$, $r_1 = 8$, $h_1 = 6$, $p_1 = 80$, $l_2 = 4$, $r_2 = 3$, $h_2 = 5$, required p_2. When holes are dug, the wage is proportional to their size. First will be considered parallelipipedal holes, then (B.272′) cylindrical ones.
(*a*) Computation.
$$p_2 = \frac{l_2 \cdot r_2 \cdot h_2 \cdot p_1}{l_1 \cdot r_1 \cdot h_1} = \frac{4 \cdot 3 \cdot 5 \cdot 80}{10 \cdot 8 \cdot 6} = \frac{4800}{480} = 10.$$

Justification. The wage being proportional to the volume, we have
$$\frac{V_1}{p_1} = \frac{V_2}{p_2}, \text{ then } \frac{p_1}{p_2} = \frac{V_1}{V_2} = \frac{l_1 \cdot r_1 \cdot h_1}{l_2 \cdot r_2 \cdot h_2} = \frac{480}{60}.$$

(*b*) Also computed as

$$p_2 = \frac{V_2}{V_1} \cdot p_1, \quad \text{or} \quad p_2 = \frac{p_1}{V_1} \cdot V_2.$$

(**B.272′**) For cylindrical holes, proceed as in *b* (or *a*), but calculating the volume as $s \cdot h$, with the area of the circular cross section $s = \frac{1}{2} c \cdot \frac{1}{2} d$ (*c* circumference, *d* diameter; see B.150) and *h* the depth.

(**B.273**) MSS 𝒜, ℬ. A hundred nummi are given for digging a hole ten cubits long, six wide, five deep; how deep will be a hole three long and two wide if the wage is ten nummi?

   Given thus $l_1 = 10$, $r_1 = 6$, $h_1 = 5$, $p_1 = 100$, $l_2 = 3$, $r_2 = 2$, $p_2 = 10$, required h_2.

Formula. Here the establishment of the formula precedes the computation. Since, as seen before,

$$\frac{p_1}{p_2} = \frac{V_1}{V_2} = \frac{l_1 \cdot r_1 \cdot h_1}{l_2 \cdot r_2 \cdot h_2}, \quad \text{or} \quad \frac{100}{10} = \frac{10 \cdot 6 \cdot 5}{3 \cdot 2 \cdot h_2} = \frac{300}{6 \cdot h_2}, \quad \text{then}$$

$$h_2 = \frac{l_1 \cdot r_1 \cdot h_1 \cdot p_2}{l_2 \cdot r_2 \cdot p_1} = \frac{10 \cdot 6 \cdot 5 \cdot 10}{3 \cdot 2 \cdot 100} = \frac{300 \cdot 10}{6 \cdot 100} = 5.$$

(**B.274**) MSS 𝒜, ℬ. Sixty nummi are given for digging a hole three cubits long, two wide and five deep, and a hole having length and width equal but one and a fourth cubits deep is dug for ten nummi; what are its length and width?

   Given thus $l_1 = 3$, $r_1 = 2$, $h_1 = 5$, $p_1 = 60$, $l_2 = r_2 = l$, $h_2 = 1 + \frac{1}{4}$, $p_2 = 10$.

Formula.

$$\frac{p_1}{p_2} = \frac{V_1}{V_2} = \frac{l_1 \cdot r_1 \cdot h_1}{l^2 \cdot h_2}, \quad \text{thus} \quad l = \sqrt{\frac{l_1 \cdot r_1 \cdot h_1 \cdot p_2}{h_2 \cdot p_1}}.$$

Remark. The text computes successively

$$V_2 = \frac{V_1 \cdot p_2}{p_1} = \frac{30 \cdot 10}{60} = 5, \quad \text{then} \quad l^2 = \frac{V_2}{h_2} = \frac{5}{1 + \frac{1}{4}} = 4, \quad l = 2.$$

(**B.275**) MSS 𝒜, ℬ, 𝒟. A hundred and seventy nummi are given for building a chest ten cubits long, five wide, eight high; how much is due for a chest two cubits long, three wide, four high?

   As remarked by the author, we are not to form, as 'many' think, the ratio of the volumes as we did in the problems on holes, for only the sides count; thus what will be compared is the sum of the areas of the six sides;

or, rather, the sum of twice the areas of three sides meeting at one corner since opposite sides are equal.

Given then the first chest's dimensions $l_1 = 10$, $r_1 = 5$, $h_1 = 8$ and its price $p_1 = 170$, next the second chest's dimensions $l_2 = 2$, $r_2 = 3$, $h_2 = 4$, required p_2.

Formula

$$\frac{p_1}{p_2} = \frac{2(l_1 \cdot r_1 + l_1 \cdot h_1 + r_1 \cdot h_1)}{2(l_2 \cdot r_2 + l_2 \cdot h_2 + r_2 \cdot h_2)} = \frac{100 + 160 + 80}{12 + 16 + 24} = \frac{340}{52},$$

thus

$$p_2 = \frac{52 \cdot 170}{340} = 26.$$

Chapter B–XV:
Consumption of lamp-oil

Summary

In this and the next two chapters, we see the same type of problem as before, except that the cost is no longer a wage but a quantity consumed: first of lamp-oil, then of food for animals and finally of loaves. In the present chapter the consumption of lamp-oil is measured by the *arrova* (the Arabic *ar-rub'*, *ar-ruba'* in Spain) —which we shall encounter again in the consumption of loaves (B–XVII). But an arrova being larger than the quantity of oil burned by one lamp in a night, this nightly consumption will always be a fraction of an arrova, namely, in all problems, some multiple of an eighth of an arrova. This is not fortuitous: the eighth (*thumn*) is a customary subdivision of the arrova.

Thus, if a number n_1 of lamps consumes in t_1 nights q_1 arrovae of oil, a number n_2 of identical lamps will consume in t_2 nights q_2 arrovae of oil. In the first part (B.276–280) we have $n_1 = 1$, with the consumption of a single lamp being given at the outset (a standard reference, as mentioned in B.4). The problems in the second part, B.281–285, are similar, but with $n_1 > 1$ as the given number of lamps. Since the oil consumption is proportional to the number of lamps and the number of nights, we shall have

$$\frac{q_1}{q_2} = \frac{n_1 \cdot t_1}{n_2 \cdot t_2}.$$

This is the relation on which all the following problems rest. For the only relations we employ are

$$q_2 = \frac{n_2 \cdot t_2 \cdot q_1}{n_1 \cdot t_1}, \quad n_2 = \frac{q_2 \cdot n_1 \cdot t_1}{q_1 \cdot t_2}, \quad t_2 = \frac{q_2 \cdot n_1 \cdot t_1}{q_1 \cdot n_2},$$

in accordance with the three types of problems mentioned by the author in his brief introduction. Indeed, the first set of quantities n_1, t_1, q_1 is always given and so are two quantities of the second set, namely, respectively: n_2 and t_2, thus q_2 is required (B.276, then, with $n_1 > 1$, B.281–283); t_2 and q_2, thus n_2 is required (B.277–278, then B.284); n_2 and q_2, thus t_2 is required (B.279–280, then B.285). Again as said in the introduction, these problems vary according to the form of the given quantities, which may be an integer (larger than 1), or 1, or a (proper) fraction.

In his solutions, the author first applies the formula, then shows how to obtain this formula by transforming the triads as in the justifications seen before, the cases with $n_1 > 1$ being also often reduced to the case

$n_0 = 1$. As usual, the other methods presented merely associate the factors differently, the relations thus obtained being also often justified.

1. Consumption of one lamp known.

(B.276) MSS $\mathcal{A}$, $\mathcal{B}$. One lamp consumes in one night a fourth of an eighth of an arrova; how much will three hundred lamps consume in twenty nights?

Given thus $n_1 = 1$, $t_1 = 1$, $q_1 = \frac{1}{4}\frac{1}{8}$, $n_2 = 300$, $t_2 = 20$, required q_2.

(*a*) Computation.

$$q_2 \left(= \frac{n_2 \cdot t_2 \cdot q_1}{n_1 \cdot t_1} \right) = n_2 \cdot t_2 \cdot q_1 = \frac{300 \cdot 20}{4 \cdot 8} = \frac{6000}{32} = 187 + \frac{1}{2}.$$

Justification. Transforming the first set into the second (simple multiplication since n_1, t_1 are unity).

n (lamps)	t (nights)	q (arrovae)
$n_1 = 1 = n_0$	$t_1 = 1 = t_0$	$q_1 = \dfrac{1}{32}$,
$n'_1 = 32$	$t'_1 = 1$	$q'_1 = n'_1 \cdot q_1 = 1,^\star$
$n''_1 = 300 = n_2$	$t'_1 = 20 = t_2$	$q_2 = n_2 \cdot t_2 \cdot q_1 = \dfrac{300 \cdot 20}{32}.$

(*b*) Also computed as

$$q_2 = (n_2 \cdot q_1) \, t_2 = \frac{300}{32} \cdot 20 = \frac{6000}{32}, \quad \text{or} \quad \left(9 + \frac{3}{8} \right) 20.$$

Justification (precedes the computation). Making first the quantity of lamps the same.

n (lamps)	t (nights)	q (arrovae)
$n_1 = 1 = n_0$	$t_1 = 1 = t_0$	$q_1 = \dfrac{1}{32}$,
$n'_1 = 300 = n_2$	$t'_1 = 1$	$q'_1 = n_2 \cdot q_1 = \dfrac{300}{32}$,
n_2	$t'_1 = 20 = t_2$	$q_2 = q'_1 \cdot t_2 = \dfrac{6000}{32}$,

and so

$$q_2 = q'_1 \cdot t_2 = (n_2 \cdot q_1) \, t_2.$$

(*c*) Also computed as

$$(t_2 \cdot q_1) \, n_2 = \frac{20}{32} \cdot 300 = \frac{5}{8} \cdot 300.$$

---

* This step could have been omitted. Same in B.277*a*.

Justification (within the computations). Again, transforming the first set, but making first the number of nights the same.

n (lamps)	t (nights)	q (arrovae)
$n_1 = 1 = n_0$	$t_1 = 1 = t_0$	$q_1 = \dfrac{1}{32}$,
$n_1 = 1$	$t_1' = 20 = t_2$	$q_1' = t_2 \cdot q_1 = \dfrac{20}{32} = \dfrac{5}{8}$,
$n_1' = 300 = n_2$	t_2	$q_2 = q_1' \cdot n_2 = \dfrac{5}{8} \cdot 300$.

Thus we find
$$q_2 = q_1' \cdot n_2 = (t_2 \cdot q_1)\, n_2.$$

(B.277) MSS $\mathcal{A}$, $\mathcal{B}$. One lamp consumes in one night half an eighth of an arrova; how many lamps will consume a hundred arrovae in thirty nights? Given thus $n_1 = 1$, $t_1 = 1$, $q_1 = \tfrac{1}{2}\tfrac{1}{8}$, $t_2 = 30$, $q_2 = 100$, required n_2.

(a) <u>Computation.</u>
$$n_2 \left(= \frac{q_2 \cdot n_1 \cdot t_1}{q_1 \cdot t_2}\right) = \frac{q_2}{q_1 \cdot t_2} = \frac{16 \cdot 100}{30} = \frac{1600}{30} = 53 + \frac{1}{3}.$$

Remark. Note the fractional number of lamps. The same in B.284.

<u>Justification.</u> Transforming the first step into the second.

n (lamps)	t (nights)	q (arrovae)
$n_1 = 1 = n_0$	$t_1 = 1 = t_0$	$q_1 = \dfrac{1}{16}$,
$n_1' = \dfrac{1}{q_1} = 16$	$t_1 = 1$	$q_1' = 1 = q_0$,
$n_1'' = q_2 \cdot n_1' = 1600$	$t_1 = 1$	$q_1'' = 100 = q_2$,
$n_1''' = \dfrac{n_1''}{t_2} = \dfrac{1600}{30}$	$t_1' = 30 = t_2$	q_2,

n_1'' being in the last step divided by t_2 since the more lamps there are the shorter the burning time for the same quantity of oil. Evidently, $n_1''' = n_2$, so that
$$n_2 = \frac{n_1''}{t_2} = \frac{q_2 \cdot n_1'}{t_2} = \frac{q_2}{q_1 \cdot t_2}.$$

(b) Also computed as
$$n_2 = \left(\frac{q_2}{t_2}\right) \frac{1}{q_1} = \frac{100}{30} \cdot 16,$$
which is in the text immediately inferred from the previous treatment by P_5.

Remark. According to Premiss P_5, for any three quantities a, b, c, we have
$$\frac{a}{b} \cdot c = \frac{a \cdot c}{b}.$$

Now the justification in the (genuine) text infers that the form $\frac{16 \cdot 100}{30}$ is equivalent to $\frac{100}{30} \cdot 16$, thus that for the three quantities a, b, c we have

$$\frac{a \cdot c}{b} = \frac{c}{b} \cdot a.$$

This banal inversion of terms has prompted an interpolator to append a demonstration of it (in MS $\mathcal{B}$ only, and with no figure).[†]

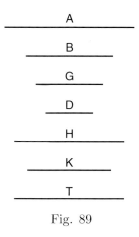

Fig. 89

This demonstration, relying as the original one on *Elements* VII.17–19, is as follows. Let (Fig. 89) $G \cdot B = A$, $\frac{A}{D} = H$, $\frac{B}{D} = K$, $K \cdot G = T$. To prove that

$$H = \frac{A}{D} = \frac{G \cdot B}{D} \quad \text{is equal to} \quad T = K \cdot G = \frac{B}{D} \cdot G.$$

First, $G \cdot B = A$ and $D \cdot H = A$, so $G \cdot B = D \cdot H$.
Second, $K \cdot D = B$ and $K \cdot G = T$, so $D : G = B : T$ and $G \cdot B = T \cdot D$.
Comparing the two results, $D \cdot H = T \cdot D$ and therefore $H = T$.

(*c*) Also computed as

$$n_2 = \frac{1}{t_2 \cdot q_1} \cdot q_2 = \frac{100}{1 + \frac{7}{8}}.$$

Justification. Making the number of nights the same and setting the proportion between the other terms.

n (lamps)	t (nights)	q (arrovae)
$n_1 = 1 = n_0$	$t_1 = 1 = t_0$	$q_1 = \frac{1}{16}$,
$n_1 = 1$	$t'_1 = 30 = t_2$	$q'_1 = t_2 \cdot q_1 = 1 + \frac{7}{8}$;

[†] A similar addition was found in B.145.

n_2 t_2 $q_2 = 100$.

So, for t the same, $n_2 : q_2 = n_1 : q'_1$, that is,

$$n_2 = \frac{n_1}{q'_1} \cdot q_2 = \frac{1}{t_2 \cdot q_1} \cdot q_2.$$

(B.278) MSS $\mathcal{A}$, $\mathcal{B}$. One lamp consumes in one night two ninths of an eighth of an arrova; how many lamps will consume a hundred arrovae in thirty nights?

Given thus $n_1 = 1$, $t_1 = 1$, $q_1 = \frac{2}{9}\frac{1}{8}$, $t_2 = 30$, $q_2 = 100$, required n_2. Similar to B.277, but with a numerator $\neq 1$ in q_1.

(a) Computation.

$$n_2 \left(= \frac{q_2 \cdot n_1 \cdot t_1}{q_1 \cdot t_2} \right) = \frac{q_2}{q_1 \cdot t_2} = \frac{1}{2} \cdot \frac{72 \cdot 100}{30} = 120.$$

Justification. Take $q_u = \frac{1}{9}\frac{1}{8}$, thus a unitary fraction for the nightly consumption of one lamp; then, solving as in B.277 a,

n (lamps)	t (nights)	q (arrovae)
$n_1 = 1 = n_0$	$t_1 = 1 = t_0$	$q_u = \frac{1}{72}$,
n'_2	$t_2 = 30$	$q_2 = 100$,

with

$$n'_2 = \frac{q_2}{q_u \cdot t_2} = \frac{72 \cdot 100}{30} = 240.$$

But since the actual nightly consumption is twice q_u, n'_2 must be halved to give n_2.

Likewise, we are told, if $q_1 = \frac{3}{4}\frac{1}{8}$, the result found by taking $q_u = \frac{1}{4}\frac{1}{8}$ will be divided by 3. Generally, it will be divided at the end by k if $q_1 = \frac{k}{l}\frac{1}{m}$ and we consider for the computation $q_u = \frac{1}{l}\frac{1}{m}$.

(b) Also computed as

$$n_2 = \frac{1}{t_2}\left(\frac{1}{q_1} \cdot q_2\right) = \frac{1}{30}\left(\frac{72}{2} \cdot 100\right).$$

(c) Also computed as

$$n_2 = \frac{q_2}{\frac{1}{2}q_1 \cdot 2 t_2} = \frac{72 \cdot 100}{60},$$

which amounts to now considering in the data $q_u = \frac{1}{9}\frac{1}{8}$ and $t'_2 = 60$, thus again a unitary fraction for the consumption but with a correspondingly longer time; we then find ourselves in the situation of B.277, but this time without any correction at the end. This is quite in keeping with our author's use of transferring a factor from one multiplicative term to another, as seen in the above justifications and earlier (B–XIV).

(**d**) Also computed as
$$n_2 = \left(\frac{1}{t_2 \cdot q_1}\right) q_2.$$

Justification. Making first the number of nights the same.

n (lamps)	t (nights)	q (arrovae)
$n_1 = 1 = n_0$	$t_1 = 1 = t_0$	$q_1 = \frac{2}{9}\frac{1}{8}$,
$n_1 = 1$	$t'_1 = 30 = t_2$	$q'_1 = t_2 \cdot q_1 = \frac{60}{72} = \frac{5}{6}$;
n_2	t_2	$q_2 = 100$.

This justification is only partly described in the text since it is the same as in B.277c, making the number of nights the same and considering then $n_2 : q_2 = n_1 : q'_1$.

Remark. We find here the only explicit reference to the *Liber mahameleth* being divided into two books. The authorship of this passage (see Translation, p. 982, note 1677) is in this respect irrelevant.

(**B.279**) MSS $\mathcal{A}$, $\mathcal{B}$. One lamp consumes in one night a ninth of an eighth of an arrova; in how many nights will forty lamps consume twenty arrovae? Given thus $n_1 = 1$, $t_1 = 1$, $q_1 = \frac{1}{9}\frac{1}{8}$, $n_2 = 40$, $q_2 = 20$, required t_2.

(**a**) Computation.
$$t_2 \left(= \frac{q_2 \cdot n_1 \cdot t_1}{q_1 \cdot n_2}\right) = \frac{q_2}{q_1 \cdot n_2} = \frac{72 \cdot 20}{40} = 36.$$

Justification. Making first the quantity consumed the same.

n (lamps)	t (nights)	q (arrovae)
$n_1 = 1 = n_0$	$t_1 = 1 = t_0$	$q_1 = \frac{1}{72}$,
$n'_1 = \frac{1}{q_1} = 72$	$t_1 = 1$	$q'_1 = 1 = q_0$,
$n''_1 = q_2 \cdot n'_1 = 72 \cdot 20 = 1440$	$t_1 = 1$	$q''_1 = 20 = q_2$;
$n_2 = 40$	t_2	q_2.

The q's now being the same, we shall have $t_2 : t_1 = n''_1 : n_2$ (with the time decreasing as the number of lamps increases since the quantity consumed remains unchanged), whence the above formula for t_2. Making first the number of lamps the same leads to a direct proportion; see hereafter.

(**b**) Also computed as
$$t_2 = \frac{1}{q_1 \cdot n_2} \cdot q_2 = \frac{20}{\frac{40}{72}} = \frac{20}{\frac{5}{9}}.$$

Justification. Making first the number of lamps the same.

n (lamps)	t (nights)	q (arrovae)
$n_1 = 1 = n_0$	$t_1 = 1 = t_0$	$q_1 = \dfrac{1}{72},$
$n_1' = 40 = n_2$	$t_1 = 1$	$q_1' = q_1 \cdot n_2 = \dfrac{40}{72} = \dfrac{5}{9};$
n_2	t_2	$q_2.$

The n's being now the same, we shall determine t_2 from the direct proportion $t_2 : q_2 = t_1 : q_1'$.

(**c**) Or else put $t_2 = x$. Thus, as just seen, $x : 20 = 1 : \frac{5}{9}$, whence $x = 36$.

Remark. This algebraic way adds nothing since x merely replaces t_2 in the above proportion. See above, p. 1569, remark.

(**B.280**) MSS $\mathcal{A}$, $\mathcal{B}$. One lamp consumes in one night two ninths of an eighth of an arrova; in how many nights will forty lamps consume twenty arrovae?

Given thus $n_1 = 1$, $t_1 = 1$, $q_1 = \frac{2}{9}\frac{1}{8}$, $n_2 = 40$, $q_2 = 20$, required t_2. Same problem as the preceding B.279, but with a numerator $\neq 1$ in q_1. (B.279–280 are analogous to the pair B.277–278.)

(**a**) Computation.

$$t_2 \left(= \frac{q_2 \cdot n_1 \cdot t_1}{q_1 \cdot n_2} \right) = \frac{q_2}{q_1 \cdot n_2} = \frac{1}{2} \cdot \frac{72 \cdot 20}{40} = \frac{1}{2} \cdot \frac{1440}{40} = 18.$$

Justification. The text is confused. What we would expect is, by analogy to B.278a:

n (lamps)	t (nights)	q (arrovae)
$n_1 = 1 = n_0$	$t_1 = 1 = t_0$	$q_1 = \dfrac{2}{72} = 2\,q_u,$
$n_1 = 1$	$t_1 = 1$	$q_u = \dfrac{1}{72},$
$n_1' = \dfrac{1}{q_u} = 72$	$t_1 = 1$	$q_u' = 1,$
$n_1'' = n_1' \cdot q_2 = 72 \cdot 20 = 1440$	$t_1 = 1$	$q_u'' = 20 = q_2;$
$n_2 = 40$	t_2'	$q_2.$

The q's being the same, we have (the greater the number of lamps, the fewer the nights) $t_2' : t_1 = n_1'' : n_2$, with $t_1 = 1$, so that

$$t_2' = \frac{n_1''}{n_2} = \frac{n_1' \cdot q_2}{n_2} = \frac{q_2}{q_u \cdot n_2},$$

with $t_2 = \frac{1}{2} \cdot t_2'$ since the actual consumption of each lamp is twice q_u.

(**b**) Also computed as

$$t_2 = \frac{1}{n_2} \left(\frac{1}{q_1} \cdot q_2 \right) = \frac{1}{40} \left(\frac{72}{2} \cdot 20 \right).$$

(c) Also computed as (see B.279b)

$$t_2 = \frac{1}{q_1 \cdot n_2} \cdot q_2 = \frac{20}{\frac{80}{72}} = \frac{20}{1 + \frac{1}{9}}.$$

<u>Justification</u> (within the solution).

n (lamps)	t (nights)	q (arrovae)
$n_1 = 1 = n_0$	$t_1 = 1 = t_0$	$q_1 = \frac{2}{72}$,
$n'_1 = 40 = n_2$	$t_1 = 1$	$q'_1 = q_1 \cdot n_2 = \frac{80}{72} = 1 + \frac{1}{9}$;
n_2	t_2	$q_2 = 20$.

The n's being the same, $t_2 : t_1 = q_2 : q'_1$ (direct proportionality), whence the above formula.

2. Consumption of several lamps known.

(B.281) MSS $\mathcal{A}$, $\mathcal{B}$. Six lamps consume in one night three eighths of an arrova; how many arrovae will they consume in thirty nights?

Given thus $n_1 = 6$, $t_1 = 1$, $q_1 = \frac{3}{8}$, $n_2 = 6$, $t_2 = 30$ (one month), required q_2. Since $n_1 = n_2$ and $t_1 = 1$, we shall just multiply the nightly consumption by the number of nights:

$$q_2 = q_1 \cdot t_2 = \frac{3}{8} \cdot 30 = 11 + \frac{1}{4}.$$

Remark. The sheer banality of this problem suggests interpolation. Indeed, it may result from an early reader's computation added to the justification in B.284a. See B.364.

(B.282) MSS $\mathcal{A}$, $\mathcal{B}$. Three lamps consume in one night an eighth of an arrova; how many arrovae will ten lamps consume in thirty nights?

Given thus $n_1 = 3$, $t_1 = 1$, $q_1 = \frac{1}{8}$, $n_2 = 10$, $t_2 = 30$, required q_2.

(*a*) <u>Computation</u>.

$$q_2 \left(= \frac{n_2 \cdot t_2 \cdot q_1}{n_1 \cdot t_1} \right) = \frac{n_2 \cdot t_2 \cdot q_1}{n_1} = \frac{10 \cdot 30}{3 \cdot 8} = 12 + \frac{1}{2}.$$

<u>Justification</u>. Take the consumption q_1 to hold for a single lamp (thus $n_u = 1$), whereby this problem is reduced to B.276; the final result is then divided by n_1.

n (lamps)	t (nights)	q (arrovae)
$n_1 = 3$	$t_1 = 1 = t_0$	$q_1 = \frac{1}{8}$;
$n_u = 1$	$t_1 = 1$	$q_1 = \frac{1}{8}$,
$n'_u = 10 = n_2$	$t'_1 = 30 = t_2$	$q'_1 = n_2 \cdot t_2 \cdot q_1$.

whence
$$q'_1 = n_2 \cdot t_2 \cdot q_1 = \frac{10 \cdot 30}{8} = 37 + \frac{1}{2}.$$

But, for $n_1 = 3$ lamps with the same conditions, divide the quantity obtained by 3 since the consumption of a single lamp was taken to be three times larger than it actually is. Therefore
$$q_2 = \frac{1}{n_1} \cdot q'_1 = \frac{1}{3}\left(37 + \frac{1}{2}\right) = \frac{n_2 \cdot t_2 \cdot q_1}{n_1}.$$

(b) Reduction to the consumption of a single lamp, which again leads to B.276, but this time without any correction at the end since we consider the effective consumption. This corresponds to computing
$$q_2 = n_2 \cdot t_2 \cdot \frac{q_1}{n_1}.$$

<u>Justification.</u>

n (lamps)	t (nights)	q (arrovae)
$n_1 = 3$	$t_1 = 1 = t_0$	$q_1 = \frac{1}{8}$,
$n'_1 = 1 = n_0$	$t_1 = 1$	$q'_1 = \frac{q_1}{n_1} = \frac{1}{3}\frac{1}{8}$;

proceeding as in B.276, we shall have
$$q_2 = n_2 \cdot t_2 \cdot q'_1 = n_2 \cdot t_2 \cdot \frac{q_1}{n_1} = \frac{10 \cdot 30}{3 \cdot 8}.$$

Then, we are told, we may also apply the other ways seen in B.276 to this reduced problem.

(c) Also computed as
$$q_2 = \left(\frac{n_2}{n_1} \cdot q_1\right) t_2.$$

<u>Justification.</u> Transforming the first set into the second:

n (lamps)	t (nights)	q (arrovae)
$n_1 = 3$	$t_1 = 1 = t_0$	$q_1 = \frac{1}{8}$,
$n'_1 = 10 = n_2$	$t_1 = 1$	$q'_1 = \frac{n_2}{n_1} \cdot q_1 = \left(3 + \frac{1}{3}\right)\frac{1}{8}$,
n_2	$t'_1 = 30 = t_2$	$q_2 = q'_1 \cdot t_2 = 12 + \frac{1}{2}$,

whence the formula for q_2.

(B.283) MSS $\mathcal{A}$, $\mathcal{B}$. Six lamps consume in one night three eighths of an arrova; how many arrovae will ten lamps consume in thirty nights?

Given $n_1 = 6$, $t_1 = 1$, $q_1 = \frac{3}{8}$, $n_2 = 10$, $t_2 = 30$, required q_2. Similar to B.282, except that q_1 is not a unitary fraction.

(*a*) Computation.
$$q_2 \left(= \frac{n_2 \cdot t_2 \cdot q_1}{n_1 \cdot t_1}\right) = \frac{n_2 \cdot t_2 \cdot q_1}{n_1} = \frac{10 \cdot 30 \cdot 3}{6 \cdot 8} = \frac{900}{48} = 18 + \frac{3}{4}.$$

Justification. Assume here again, as in B.282*a*, the given consumption q_1 to hold for a single lamp ($n_u = 1$), thus reducing the problem to B.276.

	n (lamps)	t (nights)	q (arrovae)
	$n_1 = 6$	$t_1 = 1 = t_0$	$q_1 = \frac{3}{8}$;
	$n_u = 1$	$t_1 = 1$	$q_1 = \frac{3}{8}$,
	$n'_u = 10 = n_2$	$t'_1 = 30 = t_2$	$q'_1 = n_2 \cdot t_2 \cdot q_1$,

whence
$$q'_1 = n_2 \cdot t_2 \cdot q_1 = 10 \cdot 30 \cdot \frac{3}{8} = \frac{900}{8} = 112 + \frac{1}{2}.$$

But the consumption of six lamps was made that of a single one. So the actual consumption must be six times less, thus take

$$q_2 = \frac{1}{6} q'_1 = \frac{1}{n_1} q'_1 = \frac{n_2 \cdot t_2 \cdot q_1}{n_1}.$$

(*b*) Reduction to the consumption of a single lamp.

	n (lamps)	t (nights)	q (arrovae)
	$n_1 = 6$	$t_1 = 1$	$q_1 = \frac{3}{8}$,
	$n'_1 = 1 = n_0$	$t_1 = 1$	$q'_1 = \frac{q_1}{n_1} = \frac{1}{2}\frac{1}{8}$.

We may then, the text says, apply the various ways seen in B.276.

(**B.284**) MSS *A*, *B*. Six lamps consume in one night three eighths of an arrova; how many lamps will consume twenty arrovae in thirty nights?

Given thus $n_1 = 6$, $t_1 = 1$, $q_1 = \frac{3}{8}$, $t_2 = 30$, $q_2 = 20$, required n_2.

(*a*) Computation.
$$n_2 \left(= \frac{q_2 \cdot n_1 \cdot t_1}{q_1 \cdot t_2}\right) = \frac{q_2 \cdot n_1}{q_1 \cdot t_2} = \frac{8 \cdot 20 \cdot 6}{3 \cdot 30} = \frac{960}{90} = 10 + \frac{2}{3}.$$

Remark. The result is thus 'ten lamps and two thirds of one lamp'. See above, p. 1594.

Justification. Making the number of nights the same.

	n (lamps)	t (nights)	q (arrovae)
	$n_1 = 6$	$t_1 = 1 = t_0$	$q_1 = \frac{3}{8}$,
	$n_1 = 6$	$t'_1 = 30 = t_2$	$q'_1 = q_1 \cdot t_2 = \frac{90}{8} \left(= 11 + \frac{1}{4}\right);$

$$n_2 \qquad t_2 = 30 \qquad q_2 = 20 = \frac{160}{8}.$$

The middle term being the same, we have $n_2 : q_2 = n_1 : q'_1$, thus

$$n_2 = \frac{q_2 \cdot n_1}{q'_1} = \frac{q_2 \cdot n_1}{q_1 \cdot t_2} = \frac{\frac{160}{8} \cdot 6}{\frac{3}{8} \cdot 30}.$$

This is then also computed as

$$n_2 = \frac{q_2}{q_1 \cdot t_2} \cdot n_1 = \frac{\frac{160}{8}}{\frac{90}{8}} \cdot 6 = \left(1 + \frac{7}{9}\right) 6.$$

(**b**) For both computation and justification, not significantly different from part *a* —merely using the term bracketed by us in the second step (thus, computation with arrovae instead of eighths). The above proportion then becomes $n_2 : 20 = 6 : \left(11 + \frac{1}{4}\right)$ and

$$n_2 = \frac{20 \cdot 6}{11 + \frac{1}{4}}.$$

(**c**) Reduction to the consumption of a single lamp, as in B.282*b* & B.283*b*.

n (lamps)	t (nights)	q (arrovae)
$n_1 = 6$	$t_1 = 1 = t_0$	$q_1 = \dfrac{3}{8},$
$n'_1 = 1 = n_0$	$t_1 = 1$	$q'_1 = \dfrac{1}{2}\dfrac{1}{8};$
n_2	$t_2 = 30$	$q_2 = 20.$

We are then in the situation of B.277 for calculating n_2.

(**B.285**) MSS $\mathcal{A}$, $\mathcal{B}$. Six lamps consume in one night three eighths of an arrova; in how many nights will ten lamps consume twenty arrovae? Given $n_1 = 6$, $t_1 = 1$, $q_1 = \frac{3}{8}$, $n_2 = 10$, $q_2 = 20$, required t_2.

(**a**) *Formula.*

$$t_2 \left(= \frac{q_2 \cdot n_1 \cdot t_1}{q_1 \cdot n_2}\right) = \frac{q_2 \cdot n_1}{q_1 \cdot n_2} = \frac{8 \cdot 20 \cdot 6}{3 \cdot 10} = 32.$$

(**b**) Reduction to the consumption of a single lamp.

n (lamps)	t (nights)	q (arrovae)
$n_1 = 6$	$t_1 = 1 = t_0$	$q_1 = \dfrac{3}{8},$
$n'_1 = 1 = n_0$	$t_1 = 1$	$q'_1 = \dfrac{1}{2}\dfrac{1}{8}.$

We are then in the situation of B.279.

Chapter B–XVI:
Consumption by animals

Summary

A number n_1 of animals eat in t_1 days (or nights) q_1 bushels and a number n_2 of animals eat in t_2 days q_2 bushels. Then

$$\frac{q_1}{q_2} = \frac{n_1 \cdot t_1}{n_2 \cdot t_2}.$$

The measuring-vessel is the *caficius* (the *sextarius* in B.297), as used in Toledo; in an appendix to this chapter we are given the conversion to the *modius* used in Segovia.

1. One quantity unknown.

These problems, our author says, are just like those about lamps. They are indeed presented in the same way: first with $n_1 = 1$, $t_1 = 1$, q_1 given and only one quantity required, namely, successively, q_2 (B.286–287, q_1 once aliquot fraction once not), n_2 (B.288), t_2 (B.289–290, again differentiated by the form of q_1); next, with n_1, t_1, $q_1 > 1$ (all integers), q_2, n_2, t_2 are successively required (B.291–293). The formulae are just the same as in the previous chapter, namely

$$q_2 = \frac{n_2 \cdot t_2 \cdot q_1}{n_1 \cdot t_1}, \quad n_2 = \frac{q_2 \cdot n_1 \cdot t_1}{q_1 \cdot t_2}, \quad t_2 = \frac{q_2 \cdot n_1 \cdot t_1}{q_1 \cdot n_2},$$

with their simplified form in the initial problems.

2. Two quantities unknown.

A difference occurs, again as indicated by our author, in the subsequent problems B.294–296: the number of animals n_1 is unknown and q_2 (sometimes q_1 as well) depends on it. In all these problems, the fundamental relation is thus

$$q_2 \cdot n_1 = \frac{q_1 \cdot n_2 \cdot t_2}{t_1}.$$

Successively imposed are, with a given fraction k, $q_2 = k \cdot n_1$ in B.294–295 (with, in addition, $q_1 = l \cdot n_1$, l given, in B.294) and $q_2 = \sqrt{n_1}$ in B.296 (in addition $q_1 = l \cdot n_1$), leading respectively to

$$q_2 = \frac{l \cdot n_2 \cdot t_2}{t_1}, \quad n_1 = \sqrt{\frac{q_1 \cdot n_2 \cdot t_2}{k \cdot t_1}}, \quad \sqrt{n_1} = \frac{l \cdot n_2 \cdot t_2}{t_1}.$$

We are told after B.296 that similar problems could have been proposed for lamps as well; in other words, the difference in question is not exclusive to the present topic.

3. Consumption in arithmetical progression.

In the last problem, B.297, we are given an initial number of geese, n, and their consumption in $t = 30$ days (a month), namely q bushels (*sextarii*). Dividing q by $t \cdot n$, we can then calculate the individual consumption for one day, which is also the quantity eaten by one goose in one meal since there is just one meal a day. The problem is then to determine how much will remain of the q bushels supposing that one goose is killed each day (after being fed, we are told). Since this remainder is equal to the monthly consumption less the daily consumption of the geese still alive, we shall have

$$R = q - \frac{q}{t \cdot n}[n + (n-1) + \ldots + 2 + 1] = q - \frac{q}{t \cdot n} \cdot \frac{n(n+1)}{2}.$$

---

1. One quantity unknown.
A. *Consumption of one animal known*.

(B.286) MSS $\mathcal{A}$ (formulation only), $\mathcal{B}$. One animal consumes a fourth of a caficius in one night; how many caficii will twenty animals consume in thirty nights?

Given thus $n_1 = 1$, $t_1 = 1$, $q_1 = \frac{1}{4}$, $n_2 = 20$, $t_2 = 30$ (a month), required q_2.

(a) Computation.

$$q_2 \left(= \frac{n_2 \cdot t_2 \cdot q_1}{n_1 \cdot t_1} \right) = n_2 \cdot t_2 \cdot q_1 = \frac{20 \cdot 30}{4} = \frac{600}{4} = 150.$$

Justification. Transforming the first set into the second.

n (animals)	t (nights)	q (caficii)
$n_1 = 1 = n_0$	$t_1 = 1 = t_0$	$q_1 = \frac{1}{4}$,
$n'_1 = 20 = n_2$	$t'_1 = 1$	$q'_1 = n_2 \cdot q_1 = \frac{20}{4} = 5$,
n_2	$t''_1 = 30 = t_2$	$q''_1 = t_2 \cdot q'_1 = \frac{600}{4} = 150$,

from which we infer that

$$q_2 = q''_1 = t_2 \cdot q'_1 = n_2 \cdot t_2 \cdot q_1.$$

(b) Also computed as

$$n_2(t_2 \cdot q_1) = 20 \cdot \frac{30}{4} = 20 \left(7 + \frac{1}{2}\right).$$

Justification (not separated from the computation). Transforming the first set into the second, as before, but in a different order.

n (animals)	t (nights)	q (caficii)
$n_1 = 1 = n_0$	$t_1 = 1 = t_0$	$q_1 = \dfrac{1}{4}$,
$n_1 = 1$	$t'_1 = 30 = t_2$	$q'_1 = t_2 \cdot q_1 = \dfrac{30}{4} = 7 + \dfrac{1}{2}$,
$n'_1 = 20 = n_2$	t_2	$q''_1 = n_2 \cdot q'_1 = 150$,

so that
$$q_2 = q''_1 = n_2 \cdot q'_1 = n_2 \left(t_2 \cdot q_1\right).$$

(**c**) *Remark about another unit.* We have already met (p. 1405 *seqq.*) the almodi (Arabic *al-mudd*), which has a fixed conversion rate to the caficius. One almodi being 12 caficii, a consumption of q caficii in one month means a consumption of $12\,q$ caficii in a year, thus of q almodis. For instance, if one animal eats in a month $7 + \tfrac{1}{2}$ caficii, this animal will eat in a year $7 + \tfrac{1}{2}$ almodis.

Remark. Since the solidus also has a fixed conversion rate —to the nummus (12 as well)— we shall meet analogous rules later on (B.309*b*, p. 1636).

(**d**) Again computed as
$$n_2 \left(t_2 \cdot q_1\right) = 20 \cdot \frac{30}{4},$$
but includes the (unnecessary) intermediate step $q'_1 = 1$ caficius (as in B.276*a* & B.277*a*).

n (animals)	t (nights)	q (caficii)
$n_1 = 1 = n_0$	$t_1 = 1 = t_0$	$q_1 = \dfrac{1}{4}$,
$n_1 = 1$	$t'_1 = \dfrac{1}{q_1} = 4$	$q'_1 = 1 = q_0$,
$n_1 = 1$	$t''_1 = 30 = t_2$	$q''_1 = \dfrac{t_2}{t'_1} \cdot q'_1 = \dfrac{30}{4} = 7 + \dfrac{1}{2}$,
$n'_1 = 20 = n_2$	t_2	$q'''_1 = n_2 \cdot q''_1 = 20 \left(7 + \dfrac{1}{2}\right)$,

whence
$$q_2 = q'''_1 = n_2 \cdot q''_1 = n_2 \left(\frac{t_2}{t'_1} \cdot q'_1\right) = n_2 \left(t_2 \cdot q_1\right).$$

(**B.287**) MS *B*. One animal consumes two fifths of a caficius in one night; how many caficii will twenty animals consume in thirty nights?

Given thus $n_1 = 1$, $t_1 = 1$, $q_1 = \tfrac{2}{5}$, $n_2 = 20$, $t_2 = 30$, required q_2. Variant of B.286 with a numerator $\neq 1$ in q_1. We have met analogous situations for lamps (B.277–278, B.279–280).

$$q_2 \left(= \frac{n_2 \cdot t_2 \cdot q_1}{n_1 \cdot t_1}\right) = n_2 \cdot t_2 \cdot q_1 \quad \text{thus} \quad q_2 = 2 \cdot \frac{20 \cdot 30}{5} = 2 \cdot \frac{600}{5} = 240.$$

Justification. The author's intention is to explain why the quotient is doubled. The justification should in that case take the following, usual form, considering first the daily consumption to be a unitary fraction q_u and continuing as in B.286:

n (animals)	t (nights)	q (caficii)
$n_1 = 1 = n_0$	$t_1 = 1 = t_0$	$q_1 = \dfrac{2}{5},$
$n_1 = 1$	$t_1 = 1$	$q_u = \dfrac{1}{5},$
$n_2 = 20$	$t_2 = 30$	$q'_2 = n_2 \cdot t_2 \cdot q_u.$

But with each animal eating twice as much ($q_1 = 2 \cdot q_u$), the actual consumption is doubled:

$$q_2 = 2 \cdot q'_2 = 2\, n_2 \cdot t_2 \cdot q_u = n_2 \cdot t_2 \cdot q_1.$$

However, the text just computes $q_2 = q_1 \cdot n_2 \cdot t_2$ and transforms it into $q_2 = 2\,(q_u \cdot n_2 \cdot t_2)$. A similar justification seen above was also unclear (B.280a, p. 1598).

(B.288) MS $\mathcal{B}$. One animal consumes a third of a caficius in one night; how many animals will consume forty caficii in thirty nights?

Given thus $n_1 = 1$, $t_1 = 1$, $q_1 = \frac{1}{3}$, $t_2 = 30$, $q_2 = 40$, required n_2.

(a) Computation.

$$n_2 \left(= \frac{q_2 \cdot n_1 \cdot t_1}{q_1 \cdot t_2}\right) = \frac{q_2}{q_1 \cdot t_2} = \frac{3 \cdot 40}{30} = \frac{120}{30} = 4.$$

Justification. Partial transformation of the first set ($q''_1 = q_2$).

n (animals)	t (nights)	q (caficii)
$n_1 = 1$	$t_1 = 1$	$q_1 = \dfrac{1}{3},$
$n'_1 = \dfrac{1}{q_1} = 3$	$t_1 = 1$	$q'_1 = 1,$‖
$n''_1 = q_2 \cdot n'_1 = 120$	$t_1 = 1$	$q''_1 = 40 = q_2;$
n_2	$t_2 = 30$	$q_2;$

whence, since $n_2 : n''_1 = t_1 : t_2$ (the greater the number of animals, the shorter the time taken to consume the same quantity), so

$$n_2 = \frac{n''_1}{t_2} = \frac{q_2 \cdot n'_1}{t_2} = \frac{q_2}{q_1 \cdot t_2}.$$

(b) Also computable as

$$n_2 = \frac{1}{q_1 \cdot t_2} \cdot q_2 = \frac{3}{30} \cdot 40 = \frac{40}{10}.$$

Justification. Partial transformation of the first step ($t'_1 = t_2$).

‖ Added by us.

n (animals)	t (nights)	q (caficii)
$n_1 = 1 = n_0$	$t_1 = 1 = t_0$	$q_1 = \dfrac{1}{3},$
$n_1 = 1$	$t'_1 = 30 = t_2$	$q'_1 = q_1 \cdot t_2 = 10;$
n_2	t_2	$q_2 = 40.$

The middle term now being the same, $n_2 : q_2 = n_1 : q'_1$ and

$$n_2 = \frac{n_1}{q'_1} \cdot q_2 = \frac{n_1}{q_1 \cdot t_2} \cdot q_2 = \frac{1}{q_1 \cdot t_2} \cdot q_2.$$

(B.289) MS $\mathcal{B}$. One animal consumes a fourth of a caficius in one night; in how many nights will fifty animals consume a hundred caficii?
Given thus $n_1 = 1$, $t_1 = 1$, $q_1 = \frac{1}{4}$, $n_2 = 50$, $q_2 = 100$, required t_2.
(*a*) <u>Computation.</u>

$$t_2 \left(= \frac{q_2 \cdot n_1 \cdot t_1}{q_1 \cdot n_2}\right) = \frac{q_2}{q_1 \cdot n_2} = \frac{4 \cdot 100}{50} = \frac{400}{50} = 8.$$

<u>Justification.</u> Partial transformation of the first set ($q''_1 = q_2$).

n (animals)	t (nights)	q (caficii)
$n_1 = 1 = n_0$	$t_1 = 1 = t_0$	$q_1 = \dfrac{1}{4},$
$n_1 = 1$	$t'_1 = \dfrac{1}{q_1} = 4$	$q'_1 = 1,$[†]
$n_1 = 1$	$t''_1 = q_2 \cdot t'_1 = 400$	$q''_1 = 100 = q_2;$
$n_2 = 50$	t_2	$q_2.$

To find t_2, we shall have to divide t''_1 by the actual number of animals, for $t_2 : t''_1 = n_1 : n_2 = 1 : n_2$ (as in B.288*a*: the more animals there are, the shorter the feeding time with a fixed quantity). Thus

$$t_2 = \frac{t''_1}{n_2} = \frac{q_2 \cdot t'_1}{n_2} = \frac{q_2}{q_1 \cdot n_2}.$$

(*b*) Also computable as

$$t_2 = \frac{1}{q_1 \cdot n_2} \cdot q_2 = \frac{100}{12 + \frac{1}{2}}.$$

<u>Justification</u> (not separated from the computations). Other partial transformation of the first set ($n'_1 = n_2$).

n (animals)	t (nights)	q (caficii)
$n_1 = 1 = n_0$	$t_1 = 1 = t_0$	$q_1 = \dfrac{1}{4},$

[†] Step added by us.

$$n'_1 = 50 = n_2 \qquad t_1 = 1 \qquad q'_1 = q_1 \cdot n_2 = \frac{50}{4} = 12 + \frac{1}{2};$$
$$n_2 \qquad\qquad t_2 \qquad\qquad q_2 = 100.$$

For a fixed number of animals, dividing the whole quantity consumed q_2 by their nightly consumption q'_1 will give the number of nights ($t_2 : t_1 = q_2 : q'_1$, with $t_1 = 1$). Thus

$$t_2 = \frac{q_2}{q'_1} = \frac{1}{q_1 \cdot n_2} \cdot q_2.$$

(B.290) MS *B***.** One animal consumes three eighths of a caficius in one night; in how many nights will twenty animals consume a hundred caficii?

Given thus $n_1 = 1$, $t_1 = 1$, $q_1 = \frac{3}{8}$, $n_2 = 20$, $q_2 = 100$, required t_2. Similar to B.289 but, by analogy to B.287–288, with a numerator $\neq 1$ in q_1.

$$t_2 \left(= \frac{q_2 \cdot n_1 \cdot t_1}{q_1 \cdot n_2}\right) = \frac{q_2}{q_1 \cdot n_2} = \frac{1}{3} \cdot \frac{8 \cdot 100}{20} = 13 + \frac{1}{3}.$$

Justification. Assume to begin with that the unit fraction $q_u = \frac{1}{8}$ is the nightly consumption of one animal (cf. B.278, B.280). So

n (animals)	t (nights)	q (caficii)
$n_1 = 1 = n_0$	$t_1 = 1 = t_0$	$q_1 = \frac{3}{8}$,
$n_1 = 1$	$t_1 = 1$	$q_u = \frac{1}{8}$,
$n_1 = 1$	$t'_1 = \frac{1}{q_u} = 8$	$q'_u = 1$,[‡]
$n_1 = 1$	$t''_1 = q_2 \cdot t'_1 = 800$	$q''_u = 100 = q_2$;
$n_2 = 20$	t_2	q_2.

Now, since each animal in fact eats $3\, q_u$, the actual number of nights must be three times less than t''_1, whence

$$t'''_1 = \frac{1}{3} t''_1 = \frac{800}{3} = 266 + \frac{2}{3}.$$

Then, for q the same, $t_2 : t'''_1 = n_1 : n_2$ (the longer the time, the fewer the animals for a fixed quantity of fodder), thus

$$t_2 = \frac{n_1 \cdot t'''_1}{n_2} = \frac{t'''_1}{n_2} = \frac{1}{3} \cdot \frac{t''_1}{n_2} = \frac{1}{3} \cdot \frac{q_2 \cdot t'_1}{n_2} = \frac{1}{3} \cdot \frac{q_2}{q_u \cdot n_2} = \frac{q_2}{q_1 \cdot n_2}.$$

Other ways of solving this problem are, the author says, known from before (B.279–280 rather than B.289).

B. *Consumption of several animals known*.

As in the later problems on lamps, we now start with a given number of animals $n_1 > 1$.

[‡] Added by us.

(**B.291**) MS $\mathcal{B}$. Five animals consume eight caficii in six nights; how many caficii will twenty animals consume in thirty nights?
Given thus $n_1 = 5$, $t_1 = 6$, $q_1 = 8$, $n_2 = 20$, $t_2 = 30$, required q_2.

(*a*) Computation.
$$q_2 = \frac{n_2 \cdot t_2 \cdot q_1}{n_1 \cdot t_1} = \frac{20 \cdot 30 \cdot 8}{5 \cdot 6} = 160.$$

Justification. Since now $n_1, t_1 \neq 1$, we are shown in detail how the proportion is established.

n (animals)	t (nights)	q (caficii)
$n_1 = 5$	$t_1 = 6$	$q_1 = 8,$
$n'_1 = 20 = n_2$	$t_1 = 6$	$q'_1;$
n_2	$t_2 = 30$	$q_2.$

Since in the first two steps t_1 is the same, we have $n_2 : q'_1 = n_1 : q_1$, thus
$$q'_1 = \frac{n_2 \cdot q_1}{n_1} = \frac{20 \cdot 8}{5} = 32.$$

Since in the last two steps n_2 is the same, we have $t_2 : q_2 = t_1 : q'_1$, and thus
$$q_2 = \frac{t_2 \cdot q'_1}{t_1} = \frac{30 \cdot 32}{6}, \quad \text{so} \quad q_2 = \frac{n_2 \cdot t_2 \cdot q_1}{n_1 \cdot t_1}.$$

(*b*) Also computed as
$$q_2 = \left(\frac{t_2}{t_1} \cdot q_1\right) \frac{n_2}{n_1} = \left(\frac{30}{6} \cdot 8\right) \frac{20}{5} = \frac{40 \cdot 20}{5}.$$

Justification. Changing t first instead of n in the first set.

n (animals)	t (nights)	q (caficii)
$n_1 = 5$	$t_1 = 6$	$q_1 = 8,$
$n_1 = 5$	$t'_1 = 30 = t_2$	$q'_1 = \frac{t_2}{t_1} \cdot q_1 = 40;$
$n_2 = 20$	t_2	$q_2.$

Since now the middle term is the same, we have $q_2 : n_2 = q'_1 : n_1$, therefore
$$q_2 = q'_1 \cdot \frac{n_2}{n_1} = \left(\frac{t_2}{t_1} \cdot q_1\right) \frac{n_2}{n_1}.$$

This is then also computed as
$$q_2 = \frac{q'_1}{n_1} \cdot n_2.$$

(*c*) Reduction to B.286–287, by first determining the consumption of a single animal in a single night.

n (animals)	t (nights)	q (caficii)
$n_1 = 5$	$t_1 = 6$	$q_1 = 8,$
$n_1' = 1$	$t_1' = 1$	$q_1' = \dfrac{q_1}{n_1 \cdot t_1} = \dfrac{1}{5}\dfrac{1}{6} \cdot 8 = \dfrac{1}{5} + \dfrac{1}{3}\dfrac{1}{5}.$

This is the situation of B.286–287. So we infer that, for $n_2 = 20$ and $t_2 = 30$,

$$q_2 = n_2 \cdot t_2 \cdot q_1' = \dfrac{n_2 \cdot t_2 \cdot q_1}{n_1 \cdot t_1} = 20 \cdot 30 \left(\dfrac{1}{5} + \dfrac{1}{3}\dfrac{1}{5}\right) = \dfrac{20 \cdot 30}{5}\left(1 + \dfrac{1}{3}\right).$$

(B.292) MS $\mathcal{B}$. Five animals consume six caficii in eight nights; how many animals will consume fifty caficii in thirty nights?

Given thus $n_1 = 5$, $t_1 = 8$, $q_1 = 6$, $t_2 = 30$, $q_2 = 50$, required n_2.

(**a**) Computation.

$$n_2 = \dfrac{q_2 \cdot n_1 \cdot t_1}{q_1 \cdot t_2} = \dfrac{50 \cdot 5 \cdot 8}{6 \cdot 30} = 11 + \dfrac{1}{9}.$$

Remark. The fractional result is expressed here as: 'eleven animals and a ninth of the consumption of one animal'.

Justification. Partial transformation of the second set $(t_2' = t_1)$

n (animals)	t (nights)	q (caficii)
$n_1 = 5$	$t_1 = 8$	$q_1 = 6,$
n_2	$t_2 = 30$	$q_2 = 50;$
n_2'	$t_2' = 8 = t_1$	$q_2.$

Since in the first and third step the middle term is the same, so $n_2' : q_2 = n_1 : q_1$, and

$$n_2' = \dfrac{q_2 \cdot n_1}{q_1} = \dfrac{50 \cdot 5}{6} = 41 + \dfrac{2}{3}.$$

Since in the last two steps q is the same, we have $n_2 : n_2' = t_2' : t_2$, with inversion of the second ratio because 'the number of animals eating fifty caficii in eight nights is actually greater than the number of animals eating fifty caficii in thirty nights'; then

$$n_2 = \dfrac{n_2' \cdot t_2'}{t_2} = \dfrac{q_2 \cdot n_1 \cdot t_1}{q_1 \cdot t_2} = \dfrac{(41 + \tfrac{2}{3})8}{30} = \dfrac{50 \cdot 5 \cdot 8}{6 \cdot 30}.$$

(**b**) Transformation of the first set with intermediate reduction to $t_1' = 1$.

n (animals)	t (nights)	q (caficii)
$n_1 = 5$	$t_1 = 8$	$q_1 = 6,$
n_1	$t_1' = 1 = t_0$	$q_1' = \dfrac{q_1}{t_1} = \dfrac{3}{4},$
n_1	$t_1'' = 30 = t_2$	$q_1'';^{\|}$

$^{\|}$ Step not in the text.

| | n_2 | t_2 | $q_2 = 50.$ |

From the second and third steps (same n), we infer, since $q_1'' : t_1'' = q_1' : t_1'$, that
$$q_1'' = \frac{q_1' \cdot t_1''}{t_1'} = \frac{q_1 \cdot t_2}{t_1},$$

and, from the last two (same t), since $n_2 : n_1 = q_2 : q_1''$, that
$$n_2 = \frac{q_2 \cdot n_1}{q_1''} = \frac{q_2 \cdot n_1 \cdot t_1}{q_1 \cdot t_2}.$$

In the text, which has not the third step, we are just told to calculate
$$n_2 = \frac{4 \cdot 50 \cdot 5}{3 \cdot 30} = \frac{q_2 \cdot n_1}{q_1' \cdot t_2}.$$

(**c**) Further reduction of the first set to $n_1' = 1$.

n (animals)	t (nights)	q (caficii)
$n_1 = 5$	$t_1' = 1 = t_0$	$q_1' = \dfrac{q_1}{t_1} = \dfrac{3}{4},$
$n_1' = 1 = n_0$	$t_1' = 1$	$q_1''' = \dfrac{q_1'}{n_1} = \dfrac{3}{4}\dfrac{1}{5};$
n_2	$t_2 = 30$	$q_2 = 50.$

By the last two steps, the problem is reduced to B.288, and we can use the relation seen there
$$n_2 = \frac{q_2}{q_1''' \cdot t_2} = \frac{50 \cdot 4 \cdot 5}{3 \cdot 30}, \quad \text{or} \quad n_2 = \frac{1}{3} \cdot \frac{50\,(4 \cdot 5)}{30},$$

according to whether we consider q_1' or replace it, as in B.290, provisionally by $q_u = \frac{1}{3} q_1'$.

(**B.293**) MS $\mathcal{B}$. Six animals consume eight caficii in ten nights; in how many nights will forty animals consume a hundred caficii?

Given thus $n_1 = 6$, $t_1 = 10$, $q_1 = 8$, $n_2 = 40$, $q_2 = 100$, required t_2.

(**a**) *Formula.* (Justification in part *b*.)
$$t_2 = \frac{q_2 \cdot n_1 \cdot t_1}{q_1 \cdot n_2} = \frac{100 \cdot 6 \cdot 10}{8 \cdot 40} = 18 + \frac{3}{4}.$$

(**b**) Reduction of the first set to $t_1' = 1$.

n (animals)	t (nights)	q (caficii)
$n_1 = 6$	$t_1 = 10$	$q_1 = 8,$
n_1	$t_1' = 1 = t_0$	$q_1' = \dfrac{q_1}{t_1} = \dfrac{8}{10},$

n_1 t_1'' $q_1'' = 100 = q_2;^\dagger$
$n_2 = 40$ t_2 $q_2.$

The required quantity t_2 is then found by means of two proportions, as in B.292b: $t_1'' : t_1' = q_1'' : q_1'$ (n the same), $n_2 : n_1 = t_1'' : t_2$ (q the same), so

$$t_2 = \frac{t_1'' \cdot n_1}{n_2} = \frac{q_1'' \cdot t_1' \cdot n_1}{q_1' \cdot n_2} = \frac{q_1'' \cdot n_1}{q_1' \cdot n_2} = \frac{q_2 \cdot n_1 \cdot t_1}{q_1 \cdot n_2}.$$

(*c*) Further reduction to $n_1' = 1$.

n (animals)	t (nights)	q (caficii)
$n_1 = 6$	$t_1' = 1 = t_0$	$q_1' = \dfrac{q_1}{t_1} = \dfrac{8}{10}$,
$n_1' = 1 = n_0$	$t_1' = 1$	$q_1''' = \dfrac{q_1'}{n_1} = \dfrac{1}{6}\dfrac{8}{10} = \dfrac{1}{10} + \dfrac{1}{3}\dfrac{1}{10};$
$n_2 = 40$	t_2	$q_2 = 100.$

The problem is thus reduced to B.289. Then (and introducing the common factor 30 in order to turn the divisor into an integer)

$$t_2 = \frac{q_2}{q_1''' \cdot n_2} = \frac{30 \cdot 100}{30 \left(\frac{1}{10} + \frac{1}{3}\frac{1}{10}\right) \cdot 40} = \frac{30 \cdot 100}{4 \cdot 40} = \frac{3000}{160} = 18 + \frac{3}{4}.$$

2. Two quantities unknown.

Here begins the section 'on unknown numbers of animals' mentioned by the author at the beginning of B–XVI. In fact, B.288 and B.292 also involved an unknown number of animals, but here the purpose is different: one number of animals is unknown and another quantity, or a pair, depends on it; thus, as in other problems *de ignoto*, there are now two unknowns (q_2 and n_1 in all three instances). As said by the author, that is (apart from the subject) the main difference between this and the preceding chapter on lamps. But this mathematical difference does not have to do with the subject since, as he points out at the end of B.296, such problems could have been proposed in the case of lamps as well.

Remark. In this section —possibly added by the author later— the animals will change their habits and eat during the day.

(**B.294**) MSS *A*, *B*. A certain number of animals consume in a month five times their number, and six of them consume in ten days a quantity equal to a fourth of this same number; how many are they?

Given thus $t_1 = 30$, $q_1 = 5n_1$, $n_2 = 6$, $t_2 = 10$ (nights, then days), with $q_2 = \frac{1}{4}n_1$, required q_2 or n_1.

(*a*) Reasoning. Transforming the first set.

---

† Step omitted in the text; this transformation is known from before (B.292b).

n (animals)	t (days)	q (caficii)
n_1	$t_1 = 30$	$q_1 = 5\,n_1,$
$n'_1 = 6 = n_2$	$t_1 = 30$	$q'_1 = 5\,n'_1 = 30,$
n_2	$t'_1 = 10 = t_2$	$q''_1 = \dfrac{t'_1}{t_1} \cdot q'_1 = \dfrac{1}{3} q'_1$

(with consumption reduced proportionally to time). Thus $q''_1 = q_2 = 10$, whence $n_1 = 4 q_2 = 40$.

(**b**) Using algebra, put $n_1 = x$, thus $q_1 = 5\,x$, $q_2 = \frac{1}{4} x$. Here the second set is transformed so as to obtain the same t.

n (animals)	t (days)	q (caficii)
$n_2 = 6$	$t_2 = 10$	$q_2 = \dfrac{1}{4} n_1 = \dfrac{1}{4} x,$
n_2	$t'_2 = 30 = t_1$	$q'_2 = \dfrac{t'_2}{t_2} \cdot q_2 = \dfrac{3}{4} \cdot x;$
$n_1 = x$	t_1	$q_1 = 5\,n_1 = 5\,x.$

Since t is now the same in the last two steps, $n_1 : q_1 = n_2 : q'_2$, thus

$$x : 5\,x = 6 : \frac{3}{4} x, \quad \text{so} \quad \frac{3}{4} x^2 = 30\,x,$$

whence $x^2 = 40\,x$ and $x = 40 = n_1$.

Remark. As usual, there is no reduction of the common term x in the proportion (see B.296b, remark). Note too that the difference between this and the previous solution lies not so much in the use of algebra as in the set considered for transformation. The same is seen in the two other problems of this section.

(**B.295**) MSS $\mathcal{A}$, $\mathcal{B}$. A certain number of animals consume in a month sixty caficii, and six of them consume in five days a quantity equal to three fifths of this same number; how many are they?

Given thus $t_1 = 30$, $q_1 = 60$, $n_2 = 6$, $t_2 = 5$, with $q_2 = \frac{3}{5} n_1$, required n_1.

(**a**) Transforming the first set.

n (animals)	t (days)	q (caficii)
n_1	$t_1 = 30$	$q_1 = 60,$
n_1	$t'_1 = 5 = t_2$	$q'_1 = \dfrac{t_2}{t_1} \cdot q_1 = 10;$
$n_2 = 6$	t_2	$q_2 = \dfrac{3}{5} n_1.$

Since the middle term is the same, we shall have $n_1 : q'_1 = n_2 : q_2$, thus

$$n_1 : 10 = 6 : \frac{3}{5} n_1, \quad \text{whence} \quad \frac{3}{5} n_1^2 = 60$$

and $n_1^2 = 100$, $n_1 = 10$.

The 'short solution' presented subsequently is the formula, expressed more generally. From the proportion $n_1 : q_1' = n_2 : q_2$ seen above,

$$n_1 = \frac{q_1' \cdot n_2}{q_2} = \frac{t_2}{t_1} \cdot \frac{q_1 \cdot n_2}{q_2} = \frac{t_2}{t_1} \cdot \frac{q_1 \cdot n_2}{\frac{3}{5} n_1}, \quad \text{whence}$$

$$n_1 = \sqrt{\frac{t_2}{t_1} \cdot \frac{q_1 \cdot n_2}{\frac{3}{5}}} = \sqrt{\frac{5}{30} \cdot \frac{60 \cdot 6}{\frac{3}{5}}} = \sqrt{\frac{\frac{1}{6} 60 \cdot 6}{\frac{3}{5}}}.$$

(**b**) Solved as in B.294*b*: Put $n_1 = x$ and determine the quantity consumed for $t_2' = 30$ days $(= t_1)$; that is, transform the second set.

n (animals)	t (days)	q (caficii)
$n_2 = 6$	$t_2 = 5$	$q_2 = \dfrac{3}{5} x,$
n_2	$t_2' = 30 = t_1$	$q_2' = \dfrac{t_1}{t_2} \cdot q_2 = \left(3 + \dfrac{3}{5}\right) x;$
$n_1 = x$	t_1	$q_1 = 60.$

Since t is now the same, we shall have $n_1 : q_1 = n_2 : q_2'$, that is

$$x : 60 = 6 : \left(3 + \frac{3}{5}\right)x, \quad \text{or} \quad \left(3 + \frac{3}{5}\right)x^2 = 360,$$

whence $x^2 = 100$, $x = 10 = n_1$.

(**B.296**) MSS 𝒜, ℬ. A certain number of animals consume in a month ten times their number, and five of them consume in six days a quantity equal to the root of this same number; how many are they?

Given thus $t_1 = 30$, $q_1 = 10\,n_1$, $n_2 = 5$, $t_2 = 6$, $q_2 = \sqrt{n_1}$, required n_1.

(**a**) Transform the first set into the second (using the condition $q = 10n$).

n (animals)	t (days)	q (caficii)
n_1	$t_1 = 30$	$q_1 = 10\,n_1,$
$n_1' = 5 = n_2$	$t_1 = 30$	$q_1' = 10\,n_1' = 50,$
n_2	$t_1'' = 6 = t_2$	$q_1'' = \dfrac{t_2}{t_1} \cdot q_1' = \dfrac{1}{5} \cdot q_1' = 10;$
n_2	t_2	$q_2 = \sqrt{n_1}.$

Identifying the q's in the last two steps, we find that $\sqrt{n_1} = 10$, and therefore $n_1 = 100$.

(**b**) Using algebra, put $n_1 = x^2$ (whereby $q_2 = x$), and consider the consumption for t_1 days, thus for a month, by transforming the second set.

n (animals)	t (days)	q (caficii)
$n_2 = 5$	$t_2 = 6$	$q_2 = x,$
$n_2 = 5$	$t_2' = 30 = t_1$	$q_2' = \dfrac{t_1}{t_2} \cdot q_2 = 5\,x;$

$$n_1 = x^2 \qquad t_1 \qquad q_1 = 10\,x^2.$$

So, since t is the same in the last two steps, $n_1 : q_1 = n_2 : q'_2$,

$$x^2 : 10\,x^2 = 5 : 5\,x, \quad \text{then} \quad 5\,x^3 = 50\,x^2,$$

whence $5x = 50$, $x = 10$, $n_1 = x^2 = 100$.

Remark. This is the second, and last, occurrence of the third power of the unknown. Its use here is in any event somewhat contrived since we could have cancelled the common x^2 right away. The other occurrence was in B.263.

---

3. Consumption in arithmetical progression.

(B.297) MSS $\mathcal{A}$ (lacunary), $\mathcal{B}$. Thirty geese consume six sextarii in a month, and one is killed each day after eating; what will remain of the six sextarii at the end of the month?

The only point this problem has in common with the others in this section is the subject (animal feeding).

Computation. With $n = 30$, $t = 30$, $q = 6$, the required remainder is

$$R = q - \frac{1}{\frac{n}{q} \cdot t} \cdot \frac{n(n+1)}{2} = 6 - \frac{1}{\frac{30}{6} \cdot 30} \cdot \frac{30(30+1)}{2} \qquad (*)$$

$$= 6 - \frac{1}{150} \cdot 465 = 6 - \left(3 + \frac{1}{10}\right) = 3 - \frac{1}{10}.$$

Justification. Since each of the two factors forming the subtractive term in (∗) needs to be explained, the justification is divided into two parts.

In the first part, the author explains the origin of the first factor, which is what is consumed daily —that is, in one meal— by each goose. Instead of simply dividing q by $n \cdot t$, he considers, here too, a transformation:

n (geese)	t (days)	q (sextarii)
$n = 30$	$t = 30$	$q = 6,$
$n' = \dfrac{n}{q} = 5$	$t = 30$	$q' = 1,$ °
$n'' = 1$	$t' = 1$	$q_0 = \dfrac{q'}{n' \cdot t} = \dfrac{1}{150},$

so that the daily consumption of one goose will be

$$q_0 = \frac{q'}{n' \cdot t} = \frac{1}{\frac{n}{q} \cdot t}.$$

To compute now the number of meals, we consider that the goose killed first will have only one meal, the second two, and so on to the last, which

---

° Thus n' is the number of geese eating one sextarius in a month. As in other instances, this intermediate step could have been left out.

receives thirty during the month ($t = 30$). The number of meals altogether will thus be

$$1 + 2 + \ldots + n = \frac{n(n+1)}{2}.$$

The whole consumption in sextarii will then be

$$q_0 (1 + 2 + \ldots + n) = q_0 \cdot \frac{n(n+1)}{2}.$$

In order to explain this relation, and in particular the summation formula, the author makes a parallel with the problem on wages B.245. Reckoning the quantity eaten in a month with a number of geese decreasing in arithmetical progression comes to about the same as summing a quantity of meals increasing in arithmetical progression; which is like finding the sum of the wages when each worker's pay exceeds the other's by a constant amount (the increment being in the present case equal to the first wage). Putting thus, in the formula seen in B.245, $\delta = p_1$ with n the number of workers, the sum of the wages will be

$$S = \frac{n}{2}(p_n + p_1) = \frac{n}{2}\Big([\delta(n-1) + \delta] + \delta\Big) \quad \Big(= \delta \cdot \frac{n(n+1)}{2}\Big),$$

which, with $\delta = \frac{1}{150} = q_0$ and $n = 30$, is the above subtractive term.

Remark. Being given the same value for n and t must have confused the reader.

Appendix: Bushels from different regions
Summary

Such a topic is introduced at this point because we have been dealing with capacity measures. Let q_M be the number of Segovian modii measuring a certain quantity and q_C be the number of Toledan caficii measuring the *same* quantity. We are told that $1 + \frac{1}{4}$ modii is equivalent to $1 + \frac{2}{3}$ caficii; thus the Segovian modius is larger and, accordingly, for a same total quantity measured, q_M will be smaller than q_C. Indeed, we shall have

$$q_M : q_C = \Big(1 + \frac{1}{4}\Big) : \Big(1 + \frac{2}{3}\Big), \quad \text{that is,} \quad q_M : q_C = 3 : 4.$$

We can thus find the number of modii from the number of caficii (B.298) or conversely (B.299). Only these two problems are presented for, as the author notes, they are similar to the initial problems on buying and selling which were solved by a simple proportion (B.1–14).

(B.298) MSS $\mathcal{A}$, $\mathcal{B}$, $\mathcal{D}$. One and a fourth modii in Segovia equal one and two thirds caficii in Toledo; to how many modii correspond twenty caficii?

Given thus $q_C = 20$, required q_M. Since solving this involves a proportion with four quantities, three of which are known, the present problem is conveniently transformed into a simple case of buying and selling: one and a fourth modii being given for one and two thirds nummi, how much is obtained for twenty nummi? Thus, the solution is found in the same ways as in B.9–10, B.13–14.

(*a*) As detailed in the proof,

$$\frac{1+\frac{1}{4}}{1+\frac{2}{3}} = \frac{q_M}{q_C}, \quad \text{so} \quad q_M = \frac{\left(1+\frac{1}{4}\right)q_C}{1+\frac{2}{3}} = \frac{\left(1+\frac{1}{4}\right)20}{1+\frac{2}{3}} = \frac{25}{1+\frac{2}{3}} = 15.$$

(*b*) Change the fractions to integers by multiplying them by the common denominator $3 \cdot 4 = 12$. Then

$$\frac{q_M}{q_C} = \frac{\left(1+\frac{1}{4}\right)12}{\left(1+\frac{2}{3}\right)12} = \frac{15}{20},$$

from which q_M is (banally) inferred.

(**B.299**) MSS $\mathcal{A}$, $\mathcal{B}$. One and a fourth modii in Segovia being equal to one and two thirds caficii in Toledo, to how many caficii correspond fifteen modii?

Given now $q_M = 15$, required q_C. The solution (and the result) being well known, the author omits it.

Chapter B–XVII:
Consumption of bread

Summary

Suppose that a quantity of loaves, made from q_1 bushels (*arrovæ*) of corn, is eaten by n_1 men in t_1 days, and another quantity, made from q_2 bushels, by n_2 men in t_2 days. (It is assumed that the men each eat the very same quantity every day.) Just as for lamp-oil and animals, the consumption is proportional to the number of consumers and the time, so that the basic relation is
$$\frac{q_1}{q_2} = \frac{n_1 \cdot t_1}{n_2 \cdot t_2}.$$

Since we are concerned with the consumption of loaves, we may as well consider the quantities of (identical) loaves b_i made from these q_i bushels, and thus take
$$\frac{b_1}{b_2} = \frac{n_1 \cdot t_1}{n_2 \cdot t_2}.$$

Thus, as said by the author in the introduction to this chapter, there are four kinds of quantity occurring in these problems: the number q of arrovae, the corresponding number b of loaves, the number n of men eating, and the number t of days in which they do so.

The above two formulae can be combined, that is, we may be given the quantity of loaves consumed by the n_1 men in t_1 days, namely b_1 loaves, while knowing that from one arrova β loaves are made. We then have
$$\frac{b_1}{q_2 \cdot \beta} = \frac{n_1 \cdot t_1}{n_2 \cdot t_2}.$$

This last relation, established in its complete form in B.302, is used in all six problems of this chapter. In the first three, we are to find q_2, either in the simple case where $n_1 = 1$ and $t_1 = 1$ (B.300–301), or when $n_1 \neq 1$ and $t_1 \neq 1$ (B.302). The formulae are then, respectively,
$$q_2 = \frac{n_2 \cdot t_2 \cdot b_1}{\beta}, \qquad q_2 = \frac{n_2 \cdot t_2 \cdot b_1}{n_1 \cdot t_1 \cdot \beta},$$

or, if we choose to calculate with loaves,
$$b_2 = n_2 \cdot t_2 \cdot b_1, \qquad b_2 = \frac{n_2 \cdot t_2 \cdot b_1}{n_1 \cdot t_1}.$$

In the other three problems, t_2 is required (B.303–304), next n_2 (B.305); therefore the relations used are, respectively,
$$t_2 = \frac{n_1 \cdot t_1 \cdot q_2 \cdot \beta}{n_2 \cdot b_1} = \frac{n_1 \cdot t_1 \cdot b_2}{n_2 \cdot b_1}, \qquad n_2 = \frac{n_1 \cdot t_1 \cdot q_2 \cdot \beta}{t_2 \cdot b_1}.$$

Note too that we are sometimes given $b_1 = 1$ (B.300, B.304, B.305) and sometimes $b_1 \neq 1$ (B.301, B.302, B.303, B.304', B.305').

In all these problems the unit of capacity is the same, either the arrova or (B.302–303, MS $\mathcal{A}$) the caficius. Appended to this chapter is a problem involving two together, namely the arrova of Toledo and the emina of Segovia.

(**B.300**) MSS $\mathcal{A}$, $\mathcal{B}$. From one arrova twenty loaves are made and one man eats one of them each day; how many arrovae will forty men consume in thirty days?

Given thus $\beta = 20$, $n_1 = 1$, $t_1 = 1$, $b_1 = 1$, $n_2 = 40$, $t_2 = 30$, required q_2.

(**a**) Computation.
$$q_2 \left(= \frac{n_2 \cdot t_2 \cdot b_1}{n_1 \cdot t_1 \cdot \beta} \right) = \frac{n_2 \cdot t_2}{\beta} = \frac{40 \cdot 30}{20} = 60.$$

Justification. Transformation of the first set.

n (men)	t (days)	b (loaves)	q (arrovae)
$n_1 = 1 = n_0$	$t_1 = 1 = t_0$	$b_1 = 1 = b_0$	$q_1 = \dfrac{1}{\beta} = \dfrac{1}{20}$,
$n'_1 = 40 = n_2$	$t_1 = 1$	$b'_1 = n_2 \cdot b_1 = 40$,	
n_2	$t'_1 = 30 = t_2$	$b_2 = t_2 \cdot b'_1 = 1200$.	

This is the required situation and we are just to convert b_2 into arrovae:
$$q_2 = \frac{b_2}{\beta} = \frac{t_2 \cdot b'_1}{\beta} = \frac{n_2 \cdot t_2 \cdot b_1}{\beta} = \frac{n_2 \cdot t_2}{\beta}.$$

(**b**) As in a, but with the conversion of the 40 loaves into arrovae in the second step; thus we arrive at the result directly in arrovae.

n (men)	t (days)	b (loaves)	q (arrovae)
$n_1 = 1 = n_0$	$t_1 = 1 = t_0$	$b_1 = 1 = b_0$	$q_1 = \dfrac{1}{\beta} = \dfrac{1}{20}$,
$n'_1 = 40 = n_2$	$t_1 = 1$	$b'_1 = n_2 \cdot b_1 = 40$	$q'_1 = n_2 \cdot q_1 = 2$,
n_2	$t'_1 = 30 = t_2$		$q_2 = t_2 \cdot q'_1 = 60$.

(**c**) Also computed as
$$q_2 = \frac{t_2}{\beta} \cdot n_2.$$

Justification. Changing t first instead of n.

n (men)	t (days)	b (loaves)	q (arrovae)
$n_1 = 1 = n_0$	$t_1 = 1 = t_0$	$b_1 = 1 = b_0$	$q_1 = \dfrac{1}{\beta} = \dfrac{1}{20}$,

$n_1 = 1$ $t'_1 = 30 = t_2$ $b'_1 = b_1 \cdot t_2 = 30$ $q'_1 = q_1 \cdot t_2 = 1 + \dfrac{1}{2}$,

$n'_1 = 40 = n_2$ t_2 $q_2 = q'_1 \cdot n_2 = 60$,

whence
$$q_2 = q'_1 \cdot n_2 = (q_1 \cdot t_2) n_2 = \dfrac{t_2}{\beta} \cdot n_2.$$

(B.301) MSS $\mathcal{A}$, $\mathcal{B}$. From one arrova forty loaves are made and one man eats two of them each day; how many arrovae will twenty men consume in thirty days?

Given thus $\beta = 40$, $n_1 = 1$, $t_1 = 1$, $b_1 = 2$, $n_2 = 20$, $t_2 = 30$, required q_2. Similar to B.300, but with $b_1 \neq 1$.

(a) Computation.
$$q_2 \left(= \dfrac{n_2 \cdot t_2 \cdot b_1}{n_1 \cdot t_1 \cdot \beta} \right) = \dfrac{n_2 \cdot t_2 \cdot b_1}{\beta} = \dfrac{20 \cdot 30 \cdot 2}{40} = 2 \cdot \dfrac{20 \cdot 30}{40} = 30.$$

Justification. Transforming the first set.

n (men)	t (days)	b (loaves)	q (arrovae)
$n_1 = 1 = n_0$	$t_1 = 1 = t_0$	$b_1 = 2$	$q_1 = \dfrac{b_1}{\beta} = \dfrac{2}{40}$,
$n'_1 = 20 = n_2$	$t_1 = 1$	$b'_1 = n_2 \cdot b_1 = 40$,	
n_2	$t'_1 = 30 = t_2$	$b_2 = b'_1 \cdot t_2 = 1200$,	

so that
$$q_2 = \dfrac{b_2}{\beta} = \dfrac{b'_1 \cdot t_2}{\beta} = \dfrac{n_2 \cdot t_2 \cdot b_1}{\beta} = b_1 \dfrac{n_2 \cdot t_2}{\beta}.$$

The aim of the justification is supposedly to explain the presence of the factor 2. We would therefore expect to see the introduction of $q_u = \dfrac{1}{40}$ (one loaf a day) in the transformation, then a reduction to B.300 and doubling the result since the consumption is doubled (see B.287, subject to the same criticism).

(b) As in a, but, as seen in B.300b, changing the loaves to arrovae.

n (men)	t (days)	b (loaves)	q (arrovae)
$n_1 = 1 = n_0$	$t_1 = 1 = t_0$	$b_1 = 2$,	
$n'_1 = 20 = n_2$	$t_1 = 1$	$b'_1 = n_2 \cdot b_1 = 40$	$q'_1 = \dfrac{b'_1}{\beta} = 1 = q_0$,
n_2	$t'_1 = 30 = t_2$		$q_2 = q'_1 \cdot t_2 = 30$,

which gives the result.

(c) The time is changed first instead of the number of men.

n (men)	t (days)	b (loaves)	q (arrovae)
$n_1 = 1 = n_0$	$t_1 = 1 = t_0$	$b_1 = 2$	$q_1 = \dfrac{b_1}{\beta} = \dfrac{2}{40}$,
$n_1 = 1$	$t'_1 = 30 = t_2$	$b'_1 = b_1 \cdot t_2 = 60$	$q'_1 = q_1 \cdot t_2 = 1 + \dfrac{1}{2}$,

$n'_1 = 20 = n_2$ t_2 $q_2 = q'_1 \cdot n_2 = 30$,

which gives the result.

(B.302) MS $\mathcal{A}$. From one caficius forty loaves are made and two men eat ten of them in three nights; how many caficii will twenty men consume in forty-five nights?

We now consider $n_1 \neq 1$, $t_1 \neq 1$. Given $\beta = 40$, $n_1 = 2$, $t_1 = 3$, $b_1 = 10$, $n_2 = 20$, $t_2 = 45$, required q_2 (caficii). As noted in the Translation, the men now (B.302–303) eat during the night and the measure is the caficius. These problems (in $\mathcal{A}$ only) must have been inserted by the author later on.

(a) Analogous for the treatment, we are told, to B.254–255, the men, nights and loaves being replaced by sextarii carried, miles and nummi, respectively. Therefore the fundamental relation will be

$$\frac{b_1}{b_2} = \frac{n_1 \cdot t_1}{n_2 \cdot t_2},$$

whereby we can determine the number b_2 of loaves, to be divided by β in order to obtain the number of bushels. This enables us to calculate

$$q_2 = \frac{n_2 \cdot t_2 \cdot b_1}{n_1 \cdot t_1} \cdot \frac{1}{\beta} = \frac{20 \cdot 45 \cdot 10}{2 \cdot 3} \cdot \frac{1}{40} = \frac{9000}{6} \cdot \frac{1}{40} = 37 + \frac{1}{2}.$$

The problem is solved in two steps, b_2 and q_2 being successively calculated. This further establishes the second form of the general formula, namely

$$q_2 = \frac{n_2 \cdot t_2 \cdot b_1}{n_1 \cdot t_1 \cdot \beta},$$

which is stated in the text at the end.

(b) Reducing first to unit consumption ($n'_1 = 1$, $t'_1 = 1$), which is the situation of B.301.

n (men)	t (nights)	b (loaves)
$n_1 = 2$	$t_1 = 3$	$b_1 = 10$,
$n'_1 = 1 = n_0$	$t_1 = 3$	$b'_1 = \dfrac{b_1}{n_1} = 5$,
$n'_1 = 1$	$t'_1 = 1 = t_0$	$b''_1 = \dfrac{b'_1}{t_1} = 1 + \dfrac{2}{3}$;
$n''_1 = 20 = n_2$	$t'_1 = 1$	$b'''_1 = b''_1 \cdot n_2 = 33 + \dfrac{1}{3}$,
n_2	$t''_1 = 45 = t_2$	$b^{iv}_1 = b'''_1 \cdot t_2 = 1500$,

so that

$$q_2 = b^{iv}_1 \cdot \frac{1}{\beta} = b'''_1 \cdot t_2 \cdot \frac{1}{\beta} = (b''_1 \cdot n_2) t_2 \cdot \frac{1}{\beta}$$

$$= \left(\frac{b'_1}{t_1} \cdot n_2\right) t_2 \cdot \frac{1}{\beta} = \left(\frac{b_1}{n_1 \cdot t_1} \cdot n_2\right) t_2 \cdot \frac{1}{\beta}.$$

(B.303) MS $\mathcal{A}$. From one caficius thirty loaves are made and two men eat four of them in one night; in how many nights will forty men consume fifty caficii?

Given thus $\beta = 30$, $n_1 = 2$, $t_1 = 1$, $b_1 = 4$, $n_2 = 40$, $q_2 = 50$, required t_2.

(a) Converting first the caficii into loaves. Thus $b_2 = q_2 \cdot \beta = 1500$. Then
$$t_2 = \frac{n_1 \cdot t_1 \cdot b_2}{n_2 \cdot b_1} = \frac{2 \cdot 1500}{40 \cdot 4} = 18 + \frac{3}{4}.$$

The author explains the formula with reference to the proportion established in B.302 a:
$$\frac{n_1 \cdot t_1}{n_2 \cdot t_2} = \frac{b_1}{b_2} = \frac{b_1}{q_2 \cdot \beta}, \quad \text{that is} \quad \frac{2 \cdot 1}{40\, t_2} = \frac{4}{1500}, \quad \text{thus}$$
$$t_2 = \frac{n_1 \cdot t_1 \cdot q_2 \cdot \beta}{n_2 \cdot b_1}.$$

(b) Reduction to the consumption of one man.

n (men)	t (nights)	b (loaves)	q (caficii)
$n_1 = 2$	$t_1 = 1 = t_0$	$b_1 = 4,$	
$n'_1 = 1 = n_0$	$t_1 = 1$	$b'_1 = \dfrac{b_1}{n_1} = 2,$	
$n''_1 = 40 = n_2$	$t_1 = 1$	$b''_1 = b'_1 \cdot n_2 = 80;$	$q''_1 = \dfrac{b''_1}{\beta} = 2 + \dfrac{2}{3};$
n_2	t_2		$q_2 = 50.$

The term n being the same in the last two steps, we shall have $t_2 : q_2 = t_1 : q''_1$, whence
$$t_2 = \frac{q_2 \cdot t_1}{q''_1} = \frac{q_2}{q''_1} = \frac{50}{2 + \frac{2}{3}}.$$

(B.304) MSS $\mathcal{A}$, $\mathcal{B}$. From one arrova forty loaves are made and one man eats one of them in a day; in how many days will twenty men consume sixty arrovae?

Given thus $\beta = 40$, $n_1 = 1$, $t_1 = 1$, $b_1 = 1$, $n_2 = 20$, $q_2 = 60$, required t_2 (days). B.304 and B.305 (not B.304' and B.305') return to unitary data ($n_1 = t_1 = b_1 = 1$); remember that B.302–303 seem to have been added by the author later.

(a) <u>Computation</u>.
$$t_2 \left(= \frac{n_1 \cdot t_1 \cdot q_2 \cdot \beta}{n_2 \cdot b_1}\right) = \frac{q_2 \cdot \beta}{n_2} = \frac{60 \cdot 40}{20} = 120.$$

<u>Justification</u>. Transformation of the first set.

n (men)	t (days)	b (loaves)	q (arrovae)
$n_1 = 1 = n_0$	$t_1 = 1 = t_0$	$b_1 = 1 = b_0$	$q_1 = \dfrac{b_1}{\beta} = \dfrac{1}{40},$

$$n_1 = 1 \qquad t'_1 = t_1 \cdot \beta = \beta = 40 \qquad b'_1 = b_1 \cdot \beta = 40 \qquad q'_1 = q_1 \cdot \beta = 1,^{\|}$$
$$n_1 = 1 \qquad t''_1 = q_2 \cdot t'_1 = 2400 \qquad b''_1 = q_2 \cdot b'_1 = 2400 \qquad q''_1 = 60 = q_2;$$
$$n_2 = 20 \qquad t_2 \qquad\qquad\qquad b_2 \qquad\qquad\qquad q_2.$$

The more men there are, the fewer the days for a fixed quantity of 2400 loaves or 60 arrovae ($t_2 : t''_1 = n_1 : n_2$), thus

$$t_2 = \frac{n_1 \cdot t''_1}{n_2} = \frac{t''_1}{n_2} = \frac{q_2 \cdot t'_1}{n_2} = \frac{q_2 \cdot t_1 \cdot \beta}{n_2} = \frac{q_2 \cdot \beta}{n_2}.$$

(**b**) Changing first n_1 to n_2, with a direct conversion of loaves into arrovae:

n (men)	t (days)	b (loaves)	q (arrovae)
$n_1 = 1 = n_0$	$t_1 = 1 = t_0$	$b_1 = 1 = b_0$	$q_1 = \dfrac{b_1}{\beta} = \dfrac{1}{40}$,
$n'_1 = 20 = n_2$	$t_1 = 1$	$b'_1 = 20$	$q'_1 = n_2 \cdot q_1 = \dfrac{1}{2}$;
n_2	t_2		$q_2 = 60.$

Since n is the same, $t_2 : t_1 = q_2 : q'_1$, and

$$t_2 = \frac{q_2 \cdot t_1}{q'_1} = \frac{q_2 \cdot t_1}{n_2 \cdot q_1} = \frac{q_2 \cdot t_1 \cdot \beta}{n_2 \cdot b_1} \left(= \frac{q_2 \cdot \beta}{n_2}\right).$$

(**B.304'**) Same, with $b_1 \neq 1$, namely $b_1 = 2$.

(**a**) Proceed as above, but assuming initially a unitary consumption $b_u = 1$ and then halving the result: since the quantity consumed is doubled, the number of days must be halved. If generally $b_1 = k$, calculate again with $b_u = 1$ and divide by k at the end. A similar situation was seen in B.278a.

Justification.

n (men)	t (days)	b (loaves)	q (arrovae)
$n_1 = 1 = n_0$	$t_1 = 1 = t_0$	$b_1 = 2$	$q_1 = \dfrac{b_1}{\beta} = \dfrac{2}{40}$;
$n_1 = 1$	$t_1 = 1$	$b_u = 1$	$q'_1 = \dfrac{b_u}{\beta} = \dfrac{1}{40}$;
$n_2 = 20$	t'_2	$b_2 = \beta \cdot q_2 = 2400$	$q_2 = 60.$

Since this is the situation of the previous problem, we must have

$$t'_2 = \frac{q_2 \cdot t_1 \cdot \beta}{n_2 \cdot b_u} = \frac{q_2 \cdot \beta}{n_2 \cdot b_u} = 120.$$

Then, according to what has been said,

$$t_2 = \frac{1}{2} t'_2 = \frac{1}{2} \frac{q_2 \cdot \beta}{n_2 \cdot b_u} = \frac{q_2 \cdot \beta}{n_2 \cdot b_1} = 60.$$

$^{\|}$ This step is left out in the text.

(*b*) Considering arrovae directly this time:

n (men)	t (days)	b (loaves)	q (arrovae)
$n_1 = 1 = n_0$	$t_1 = 1 = t_0$	$b_1 = 2$	$q_1 = \dfrac{b_1}{\beta} = \dfrac{2}{40}$,
$n_1' = 20 = n_2$	$t_1 = 1$	$b_1' = n_2 \cdot b_1 = 40$	$q_1' = n_2 \cdot q_1 = 1$,
n_2	$t_1' = q_2 \cdot t_1 = 60$		$q_1'' = q_2 \cdot q_1' = 60$,

then $t_2 = t_1' = q_2$ (in this particular case $t_1' : t_1 = q_1'' : q_1'$ becomes $t_1' = q_1'' = q_2$).

(**B.305**) MSS *A*, *B*. From one arrova twenty loaves are made and one man eats one of them in a day; how many men will consume forty arrovae in thirty days?

Given thus $\beta = 20$, $n_1 = 1$, $t_1 = 1$, $b_1 = 1$, $t_2 = 30$, $q_2 = 40$, required n_2.

(*a*) Computation.

$$n_2 \left(= \frac{n_1 \cdot t_1 \cdot q_2 \cdot \beta}{t_2 \cdot b_1} \right) = \frac{q_2 \cdot \beta}{t_2} = \frac{40 \cdot 20}{30} = \frac{800}{30} = 26 + \frac{2}{3}.$$

Justification. Transforming the first set.

n (men)	t (days)	b (loaves)	q (arrovae)
$n_1 = 1 = n_0$	$t_1 = 1 = t_0$	$b_1 = 1 = b_0$	$q_1 = \dfrac{b_1}{\beta} = \dfrac{1}{20}$,
$n_1' = q_2 \cdot \beta = 800$	$t_1 = 1$	$b_1' = q_2 \cdot \beta \cdot b_1 = 800$	$q_1' = \dfrac{b_1'}{\beta} = 40$;
n_2	$t_2 = 30$	$b_2 = 800$	$q_2 = 40$.

In the second step we have found that $n_1' = 800$ men will consume in one day $b_1' = 800$ loaves. But since the consumption of these 800 loaves must take place not in one day but in thirty, n_1' must be divided by the given number of days t_2 to give the actual number of men ($n_2 : n_1' = t_1 : t_2$):

$$n_2 = \frac{n_1'}{t_2} = \frac{q_2 \cdot \beta}{t_2}.$$

Remark. The text once gives $26 + \frac{2}{3}$ as 'the number of men' but then, as in B.292*a*, speaks of 26 men 'and two thirds of one man's consumption'.

(*b*) Making t the same and reducing to arrovae.

n (men)	t (days)	b (loaves)	q (arrovae)
$n_1 = 1 = n_0$	$t_1 = 1 = t_0$	$b_1 = 1 = b_0$	$q_1 = \dfrac{b_1}{\beta} = \dfrac{1}{20}$,
$n_1 = 1$	$t_1' = 30 = t_2$	$b_1' = t_2 \cdot b_1 = 30$	$q_1' = t_2 \cdot q_1 = 1 + \dfrac{1}{2}$;
n_2	$t_2 = 30$		$q_2 = 40$.

Since t is the same and the proportionality is direct, so $n_2 : q_2 = n_1 : q_1'$ and
$$n_2 = \frac{n_1 \cdot q_2}{q_1'} = \frac{n_1 \cdot q_2}{t_2 \cdot q_1} = \frac{n_1 \cdot q_2}{t_2 \cdot \frac{b_1}{\beta}} = \frac{q_2}{t_2 \cdot \frac{1}{\beta}} = \frac{40}{\frac{30}{20}} = \frac{40}{1 + \frac{1}{2}}.$$

(**B.305′**) Same with $b_1 \neq 1$ ($b_1 = 2$).

(*a*) Proceed as above in *a* and divide the result by b_1 at the end. That is, assume initially the consumption to be a unitary one ($b_u = 1$).

n (men)	t (days)	b (loaves)	q (arrovae)
$n_1 = 1$	$t_1 = 1$	$b_1 = 2$	$q_1 = \dfrac{b_1}{\beta} = \dfrac{2}{20}$;
$n_1 = 1$	$t_1 = 1$	$b_u = 1$	$q_u = \dfrac{1}{20}$.

This being the situation of B.305, we conclude that for $t_2 = 30$, $q_2 = 40$ we shall have, for the number of men n_2',
$$n_2' = \frac{q_2 \cdot \beta}{t_2} = 26 + \frac{2}{3}.$$

But since the actual consumption is twice as much ($b_1 = 2\,b_u$), the number of men eating the same quantity for the same time must be halved, thus
$$n_2 = \frac{1}{2} n_2' = \frac{1}{2} \frac{q_2 \cdot \beta}{t_2} \left(= \frac{1}{2} \frac{q_2 \cdot \beta}{t_2 \cdot b_u} = \frac{q_2 \cdot \beta}{t_2 \cdot b_1} \right) = 13 + \frac{1}{3},$$

thus 'thirteen men and a third'.

(*b*) As in B.305*b*, making t the same, and also considering arrovae instead of loaves.

n (men)	t (days)	b (loaves)	q (arrovae)
$n_1 = 1 = n_0$	$t_1 = 1 = t_0$	$b_1 = 2$	$q_1 = \dfrac{b_1}{\beta} = \dfrac{2}{20}$,
$n_1 = 1$	$t_1' = 30 = t_2$	$b_1' = t_2 \cdot b_1 = 60$	$q_1' = t_2 \cdot q_1 = 3$;
n_2	t_2		$q_2 = 40$.

For t the same we must have $n_2 : q_2 = n_1 : q_1'$, and
$$n_2 = \frac{n_1 \cdot q_2}{q_1'} = \frac{n_1 \cdot q_2}{t_2 \cdot q_1} = \frac{q_2}{t_2 \cdot q_1} \left(= \frac{q_2 \cdot \beta}{t_2 \cdot b_1} \right).$$

Appendix: Other units of capacity

Summary

We have already encountered the case of different kinds of measuring vessel in the chapter on borrowing (B.190–194) and in the appendix to the one on consumption by animals (B.298–299). Here we have a further such instance, with two comparable hollow measures also in use locally: that of Toledo being the arrova, that of Segovia the larger emina (the Graeco-Roman *hemina*). The ratio between them is, we are told,

$$\text{arrova} : \text{emina} = \left(1 + \frac{1}{4}\right) : \left(1 + \frac{1}{2}\right) = \frac{5}{4} : \frac{3}{2} = 5 : 6.$$

Thus, if the *same* quantity of corn measured first with the smaller arrova is q_A and then with the larger emina becomes q_E, we shall have $q_A > q_E$ according to the conversion

$$q_A : q_E = \left(1 + \frac{1}{2}\right) : \left(1 + \frac{1}{4}\right) = 6 : 5,$$

whence the formulae

$$q_A = \frac{6}{5} \cdot q_E, \qquad q_E = \frac{5}{6} \cdot q_A.$$

Suppose now, pursuing the subject of the previous chapter, that n_i designates a number of persons, t_i the number of days in which they consume, $b_1^{(A)}$ the consumption of loaves by n_1 in t_1, and $\beta^{(A)}$ the number of loaves to the arrova. The number of arrovae eaten by n_2 men in t_2 days is therefore calculated, as seen in B.302, by

$$q_A = \frac{n_2 \cdot t_2 \cdot b_1^{(A)}}{n_1 \cdot t_1 \cdot \beta^{(A)}}.$$

From what we have just seen, the same quantity of corn in eminae will be

$$q_E = \frac{5}{6} \cdot q_A = \frac{5}{6} \cdot \frac{n_2 \cdot t_2 \cdot b_1^{(A)}}{n_1 \cdot t_1 \cdot \beta^{(A)}}. \qquad (*)$$

This formula may be transformed into one involving eminae only in two ways, each giving the same result but based on a different assumption. They correspond, mathematically, to the inclusion of the numerical factor either in $\beta^{(A)}$ or in $b_1^{(A)}$.

(*i*) The loaves remain identical in size, irrespective of the unit of capacity. Consequently, the global consumption of loaves by n_1 men in t_1 days will remain the same, thus $b_1^{(A)} = b_1^{(E)} = b_1$, but each emina, being larger, will produce more loaves, in the proportion $\beta^{(E)} : \beta^{(A)} = 6 : 5$. Formula $(*)$ becomes

$$q_E = \frac{n_2 \cdot t_2 \cdot b_1}{n_1 \cdot t_1 \cdot \beta^{(E)}}.$$

(*ii*) The number of loaves to each kind of bushel remains the same. That is, each emina will produce as many loaves as the arrova, thus $\beta^{(A)} = \beta^{(E)} = \beta$, but, the loaves now being larger, the number of them consumed by n_1 men in t_1 days will drop from $b_1^{(A)}$ to $b_1^{(E)}$, with $b_1^{(E)} : b_1^{(A)} = 5 : 6$. Formula (*) becomes

$$q_E = \frac{n_2 \cdot t_2 \cdot b_1^{(E)}}{n_1 \cdot t_1 \cdot \beta}.$$

These two aspects are considered below (although not very clearly distinguished in the text).

(B.306) MSS $\mathcal{A}$, $\mathcal{B}$. From one arrova are made twenty loaves, one of which is eaten by one man in a day; how many eminae will forty men eat in thirty days, knowing that one and a half arrovae in Toledo equal one and a fourth eminae in Segovia?

Given thus $\beta^{(A)} = 20$ (from 1 arrova), $n_1 = 1$, $t_1 = 1$, $b_1^{(A)} = 1$, $n_2 = 40$, $t_2 = 30$, required q_E.

(*i*) Hypothesis of identical loaves.

(*a*) First, according to B.300–302 (and what has been seen in the above introduction), the quantity of arrovae will be

$$q_A \left(= \frac{n_2 \cdot t_2 \cdot b_1^{(A)}}{n_1 \cdot t_1 \cdot \beta^{(A)}} \right) = \frac{n_2 \cdot t_2}{\beta^{(A)}} = \frac{40 \cdot 30}{20} = 60.$$

We can now transform it into eminae:

$$q_E = \frac{q_A \left(1 + \frac{1}{4}\right)}{1 + \frac{1}{2}} = \frac{60 \left(1 + \frac{1}{4}\right)}{1 + \frac{1}{2}} = 50.$$

(*b*) Otherwise, if $\beta^{(A)} = 20$ (from 1 arrova), the number of identical loaves made from the larger emina $\beta^{(E)}$ will be

$$\beta^{(E)} = \frac{6}{5} \cdot \beta^{(A)} = 24.$$

We are then in the situation of B.300, and therefore

$$q_E = \frac{n_2 \cdot t_2}{\beta^{(E)}} = \frac{40 \cdot 30}{24} = 50.$$

(*ii*) Hypothesis of different loaves.

(*c*) Since, for an identical number of loaves to the bushel, a loaf in Toledo is $\frac{5}{6}$ of a loaf in Segovia, we shall have $b_1^{(E)} = \frac{5}{6} \cdot b_1^{(A)} = \frac{5}{6}$, while $\beta^{(E)} = \beta = 20$, so we shall compute, as in B.301,

$$q_E = \frac{n_2 \cdot t_2 \cdot b_1^{(E)}}{\beta^{(E)}} = \frac{40 \cdot 30 \cdot \frac{5}{6}}{20}.$$

(**d**) Transforming the first set, with direct conversion into eminae. A man who ate one loaf of Toledo will have eaten $\frac{5}{6}$ of a loaf of Segovia.

$\quad$ n (men) $\qquad$ t (days) $\qquad\qquad$ b (loaves) $\qquad\qquad$ q (arrovae)

$n_1 = 1 = n_0 \qquad t_1 = 1 = t_0 \qquad b_1^{(E)} = \dfrac{5}{6} \qquad\qquad \dfrac{b_1^{(E)}}{\beta},$

$n_1 = 1 \qquad t_1' = 30 = t_2 \quad b_1'^{(E)} = t_2 \cdot b_1^{(E)} = 25 \quad \dfrac{b_1'^{(E)}}{\beta} = 1 + \dfrac{1}{4};$

$n_1' = 40 = n_2 \qquad\quad t_2 \qquad\qquad\qquad\qquad\qquad q_E = n_2 \cdot \dfrac{b_1'^{(E)}}{\beta},$

whence $q_E = 40\left(1 + \frac{1}{4}\right) = 50.$

Chapter B–XVIII:
Exchanging moneys

Summary

We have already met in A.60–71 the gold coin used in mediaeval Spain, the *morabitinus*. We shall now encounter it again, but this time in connection with problems of exchange: a certain number of morabitini are to be converted either into nummi or into solidi —which is only formally different since the solidus is equivalent to a fixed amount of twelve nummi.[†] Since, however, the coinage was local, the relative values of morabitini and nummi were local as well, but the wide circulation of coins made it likely that an exchange would involve morabitini and nummi of unequal values. Thus, a problem of exchange between these two types of coins in its most general form would involve a quantity m of morabitini of l different kinds to be converted into a quantity q of nummi (or solidi) of n different types, or inversely.

Let us thus designate by a_{ij} the quantity of nummi of the jth type received for one morabitinus of the ith kind. Consider, then, that we are to exchange m_1 morabitini of the first kind into various nummi. Of these m_1 morabitini, let m_{11} be converted into nummi of the first type; we shall then receive from the change $a_{11} \cdot m_{11}$ nummi. Likewise, m_{12} morabitini of this same first kind changed at the rate a_{12} will be worth $a_{12} \cdot m_{12}$ nummi of the second type. And so on for the remaining morabitini of the first kind; thus their total number, namely

$$m_1 = \sum_{j=1}^{n} m_{1j}$$

will result in a total quantity of

$$\sum_{j=1}^{n} a_{1j} \cdot m_{1j}$$

nummi of the various types. The same will apply to each of the other kinds of the m morabitini. That is, the exchange of

$$m_i = \sum_{j=1}^{n} m_{ij} \qquad (i = 1, \ldots, l)$$

[†] The *Liber augmenti et diminutionis*, which has two problems of exchange, speaks in the same context of 'aurei', 'dragme' and 'solidi' (Libri, I, pp. 357–359 & 363–365). In his *Liber algorismi*, Johannes Hispalensis uses morabitini and aurei indiscriminately.

morabitini of the ith kind will produce

$$\sum_{j=1}^{n} a_{ij} \cdot m_{ij}$$

nummi belonging to the various types.

To find now the total number of nummi of one and the same type obtained from the exchange, we shall consider the quantity received from the l different m_i with their respective exchange rates. Thus we shall have

$$q_1 = a_{11} \cdot m_{11} + a_{21} \cdot m_{21} + \ldots + a_{l1} \cdot m_{l1}$$

nummi of the first type, and generally

$$q_j = \sum_{i=1}^{l} a_{ij} \cdot m_{ij} \qquad (j = 1, \ldots, n)$$

for the jth type of nummus.

The fundamental relations of the exchange problem are then the pair of equations

$$\begin{cases} \sum_{i=1}^{l} m_i = \sum_{i=1}^{l} \sum_{j=1}^{n} m_{ij} = m \\ \sum_{j=1}^{n} q_j = \sum_{j=1}^{n} \sum_{i=1}^{l} a_{ij} \cdot m_{ij} = q \end{cases}$$

with, as said, m the total number of morabitini and q the total number of nummi.

The *Liber mahameleth* considers the simpler case in which we have a single kind of money of either sort, that is, either one kind of morabitinus (mostly a single morabitinus) to be converted into various types of nummus, or (in a few problems only) several kinds of morabitinus to be converted into just one type of nummus or solidus. Thus one of the indices drops out and the pair of equations becomes

$$\begin{cases} \sum m_k = m \\ \sum q_k = \sum a_k \cdot m_k = q. \end{cases}$$

In the case with one kind of morabitinus, m_k represents the (integral or fractional) quantity among the m identical morabitini which is changed into the kth type of nummus, of which there will be the quantity q_k. In the case with one type of nummus, q_k represents the quantity among the q identical nummi obtained from the kth kind of morabitinus, of which there are m_k.

Let us consider various possibilities, according to the quantity of different moneys involved.

1. One kind of money exchanged for one kind of money.

If m morabitini are converted into q nummi at exchange rate a, we have simply
$$q = a\,m.$$
This banal case does not occur as such. What we find in B.307–308 is the conversion of an amount of morabitini of one and the same kind first into one type of nummus and then into another. Thus m morabitini, each worth a_1 nummi of one type and a_2 nummi of another, will yield respective quantities q_1 and q_2 of nummi, with
$$q_1 : q_2 = a_1\,m : a_2\,m = a_1 : a_2,$$
one of the q_i or a_i being required. B.309–310 involve, instead of two types of nummus, nummi and solidi: given the exchange rate a into nummi, find the number of morabitini m obtained for a given quantity of solidi q_s, or inversely. Since the solidus is always worth twelve nummi, then $a_s = 12a$, so that $a \cdot m$ nummi divided by 12 will give the q_s solidi. This is therefore just a particular case of what precedes.

2. Two kinds of money exchanged for one kind of money.

We have either one kind of morabitinus exchanged for an amount comprising two types of nummus (B.311, B.313–315) or two kinds of morabitinus exchanged for one type of solidus (B.329). In both cases we are to solve
$$\begin{cases} m_1 + m_2 = m \\ q_1 + q_2 = a_1\,m_1 + a_2\,m_2 = q, \end{cases}$$
where the exchange of m_i morabitini at the exchange rate a_i gives q_i coins. Since, from the first equation, $m_1 = m - m_2$, we then obtain for the second $a_1\,(m - m_2) + a_2\,m_2 = q$, so
$$m_2 = \frac{q - a_1\,m}{a_2 - a_1}$$
$$m_1 = m - m_2 = \frac{a_2\,m - q}{a_2 - a_1}$$
whence
$$q_2 = a_2\,m_2 = \frac{a_2\,q - a_1\,a_2 \cdot m}{a_2 - a_1}$$
$$q_1 = a_1\,m_1 = \frac{a_1\,a_2 \cdot m - a_1\,q}{a_2 - a_1}.$$

With $m = 1$, the formulae for q_1 and q_2 are demonstrated in B.313 and the one for m_1 (from which m_2 is inferred) in B.314 and B.315 (with a_1 subtractive); with $m \neq 1$, in B.329. (Each with its own demonstration on account of the absence of symbolism.)

The condition for a (strictly) positive solution is that q lies between $a_1\, m$ and $a_2\, m$, thus that

$$a_1\, m < q < a_2\, m$$

if a_2 is assumed to be the larger of a_1 and a_2, the second type of nummus being then of lesser value. This condition is stated in B.313 and B.329, and proved in B.313.

As mentioned above, most of the problems consider that a single morabitinus is exchanged, and the system takes the form

$$\begin{cases} m_1 + m_2 = m = 1 \\ a_1\, m_1 + a_2\, m_2 = q \end{cases}$$

with the condition $a_1 < q < a_2$. In B.329b, with $m \neq 1$ given, the author changes his unknowns to $m'_1 = \frac{1}{m} \cdot m_1$ and $m'_2 = \frac{1}{m} \cdot m_2$. Then the system can be written as

$$\begin{cases} m'_1 + m'_2 = 1 \\ a'_1\, m'_1 + a'_2\, m'_2 = q \end{cases}$$

with $a'_i = m\, a_i$. The system proposed thus takes the same form as the previous one, and the solving condition is accordingly, as before, $a'_1 < q < a'_2$. This is more practical: a glance at the three numerical quantities in the second equation tells us at once whether or not the problem is solvable.

Remark. There is no allusion to the similar systems of two equations already seen (B.46–47).

3. More than two kinds of money exchanged for one kind of money.

We have then, as already seen,

$$\begin{cases} \sum_{k=1}^{n} m_k = m \quad (n > 2) \\ \sum_{k=1}^{n} q_k = \sum_{k=1}^{n} a_k\, m_k = q. \end{cases}$$

Since such a problem is indeterminate, we need further information; this is provided in B.312. But if, as in B.316–318, only the a_k and q are known (with $m = 1$), the problem will be reduced to the preceding case with two unknowns by two different assumptions (two kinds of 'determination'): either supposing the equality of $n - 1$ unknowns or choosing the values of $n - 2$ unknowns.

— **Equating all unknowns but one.**

Consider thus that the quantity of morabitini exchanged is the same for $n - 1$ of the m_k's, say m_0. This m_0 is then to be determined, as well as the only quantity not put equal, m_j. The system

$$\begin{cases} \sum_{k \neq j} m_k + m_j = m \\ \sum_{k \neq j} a_k\, m_k + a_j\, m_j = q \end{cases}$$

then becomes

$$\begin{cases} (n-1)\, m_0 + m_j = m \\ \left(\sum_{k \neq j} a_k\right) m_0 + a_j\, m_j = q. \end{cases}$$

To have it in a form similar to that of the previous system with two unknowns, namely

$$\begin{cases} m_1 + m_2 = m \\ a_1\, m_1 + a_2\, m_2 = q, \end{cases}$$

where the first equation is just the sum of two unknowns, the author takes $(n-1)\, m_0$ as unknown instead of m_0 and multiplies the second equation by $n-1$ in order to introduce this new unknown; we then have

$$\begin{cases} (n-1)\, m_0 + m_j = m \\ \left(\sum_{k \neq j} a_k\right) \cdot (n-1)\, m_0 + (n-1)\, a_j \cdot m_j = (n-1)\, q. \end{cases}$$

By analogy to the former solutions,

$$m_1 = \frac{a_2 \cdot m - q}{a_2 - a_1}, \qquad m_2 = \frac{q - a_1 \cdot m}{a_2 - a_1},$$

we shall now have

$$(n-1)\, m_0 = \frac{(n-1)\, a_j \cdot m - (n-1)\, q}{(n-1)\, a_j - \sum_{k \neq j} a_k},$$

whence

$$m_0 = \frac{a_j \cdot m - q}{(n-1)\, a_j - \sum_{k \neq j} a_k},$$

and

$$m_j = \frac{(n-1)\, q - \left(\sum_{k \neq j} a_k\right) \cdot m}{(n-1)\, a_j - \sum_{k \neq j} a_k},$$

with the condition that, for a positive solution,

$$\left(\sum_{k \neq j} a_k\right) \cdot m \; \gtrless \; (n-1)\, q \; \gtrless \; (n-1)\, a_j \cdot m.$$

The formula for m_0 is demonstrated by means of a geometrical figure in B.316c (case $n = 3$). The transformation of the second equation is also demonstrated for the case $n = 3$ (B.316a) and the case $n = 4$ (B.317a).

It is observed in B.316–318 that if the condition for a positive solution is not fulfilled, the original problem is not solvable with the chosen 'determination' (*determinatio*), that is, assuming the equality of the $n - 1$ unknowns m_k as selected. We are then to try the same with some other set of $n - 1$ unknowns. We may also, as is done in B.317d, equate half of the unknowns m_k, each taking a same value m_0 to be determined, and do the same with the other half $m_{k'}$, each being equal to m'_0. Generally (this is not done by the author), if $n = n_1 + n_2$, we shall set each of n_1 quantities m_k equal to a quantity to be determined m_0, and each of the remaining n_2 quantities $m_{k'}$ to some other, also to be determined, quantity m'_0, and multiply the second equation by $n_1 \cdot n_2$. The system then becomes

$$\begin{cases} n_1 \cdot m_0 + n_2 \cdot m'_0 = m \\ \left(n_2 \cdot \sum a_k\right)(n_1 \cdot m_0) + \left(n_1 \cdot \sum a_{k'}\right)(n_2 \cdot m'_0) = n_1 n_2 \cdot q, \end{cases}$$

with the pair of unknowns $n_1 \cdot m_0$ and $n_2 \cdot m'_0$ and the condition

$$n_2 \sum a_k \cdot m \gtrless n_1 n_2 \cdot q \gtrless n_1 \sum a_{k'} \cdot m.$$

Although the author does not consider this case (except for the particular situation $n_1 = n_2$), he surely has it in mind when referring in B.317 to an 'innumerable' number of possibilities (literally, an 'infinite' number, a term which merely means more than he can calculate).

Remark. Here again, as with the case of two unknowns, there is no allusion to the earlier presence of a similar problem (B.51).

— **Choosing the values of all unknowns but two.**

Instead of equating $n - 1$ of the m_k's to some unknown m_0 to be determined, we may directly adopt numerical values for $n - 2$ of the m_k's and solve the problem left for the two remaining unknowns. Thus the original system

$$\begin{cases} \sum_{k=1}^{n} m_k = m \\ \sum_{k=1}^{n} q_k = \sum_{k=1}^{n} a_k m_k = q \end{cases}$$

becomes

$$\begin{cases} m_l + m_j = m - \sum_{k \neq l, j} m_k = m' \\ a_l \cdot m_l + a_j \cdot m_j = q - \sum_{k \neq l, j} a_k m_k = q - \sum_{k \neq l, j} q_k = q' \end{cases}$$

with m_l, m_j unknowns and m' and q' known quantities. The resulting problem has thus once again been reduced to the form seen in Section 2, p. 1631, with the condition that q' must be included between $a_l \cdot m'$ and $a_j \cdot m'$. If the problem thus derived is solvable, so is the original problem with the 'determination' chosen, that is, with the $n-2$ values assumed (and the two then calculated). If not, we shall take another set of $n-2$ values.

4. Particular problems.

In the problems closing this chapter we have either $n = 2$ or $n = 3$, but in the latter case the problems are rendered determinate with further imposed conditions for the m_k a_k. Indeed, unlike before, we are now primarily interested in the quantities of nummi obtained and not in the fraction of the morabitinus attributed to the individual moneys. For instance, both quantities are equal (B.319–320) or, in the case $n = 3$, all three are equal (B.321), or just two and the third is given (B.323), or the three stand in a given relation to one another (B.322). After a problem involving a root in the data and leading to a second-degree equation (B.324) and two problems of the kind seen just before, we proceed, in B.327–329, with problems involving more than one morabitinus (of which the only one of mathematical interest, B.329, has already been mentioned in our summary of Section 2, pp. 1631–1632). This section, and thus this rather long chapter on exchange, ends with a somewhat peculiar problem, to do with the use of the word nummus (or, rather, of its Arabic equivalent *dirham*) to mean both the coin and a unit of weight.

---

1. One kind of money exchanged for one kind of money.

A. *One morabitinus for two types of nummus*.

In B.307–308, one morabitinus is exchanged for a_1 coins of one type and a_2 of another and we receive q_1 and q_2 coins respectively.

(B.307) MS $\mathcal{A}$. One morabitinus is exchanged for thirty nummi of one money and forty of another. What are ten nummi of the first money worth in nummi of the second?

Given thus $a_1 = 30$, $a_2 = 40$, $q_1 = 10$, required q_2 equivalent to q_1.

Since $q_1 = 10$ is a third of $a_1 = 30$, q_2 will be a third of $a_2 = 40$. Indeed, $q_1 : a_1 = q_2 : a_2$, so

$$q_2 = \frac{q_1 \cdot a_2}{a_1} = \frac{10 \cdot 40}{30} = 13 + \frac{1}{3}.$$

(B.308) MS $\mathcal{A}$. One morabitinus is exchanged for twenty-five nummi of one money and for an unknown number of nummi of another, and ten nummi of the first are worth fifteen of the second. What is the exchange rate of the second money?

Given thus $a_1 = 25$, $q_1 = 10$, $q_2 = 15$, required a_2.

As above, from $q_1 : a_1 = q_2 : a_2$ follows

$$a_2 = \frac{a_1 \cdot q_2}{q_1} = \frac{25 \cdot 15}{10} = 37 + \frac{1}{2}.$$

B. *Morabitini for nummi and solidi*.

B.309–310 involve nummi, solidi and a quantity m of morabitini, with a fixed solidus-nummus exchange rate since 1 solidus equals 12 nummi, as known from before (B.65, B.71); with a_s the morabitinus-solidus exchange rate and a that for nummi, q the number of nummi and q_s that of solidi, we have $12 a_s = a$, $q = a \cdot m$, $q_s = a_s \cdot m$ and thus the situation of the preceding problems, except that a_1 and a_2 are in a constant ratio.

(B.309) MSS 𝒜, ℬ. A hundred morabitini are exchanged each for fourteen nummi. To how many solidi do the hundred morabitini correspond?

Given thus $m = 100$, $a = 14$, required q_s (solidi).

(a) Since $q = a \cdot m$ and $q = 12 q_s$, so

$$q_s = \frac{q}{12} = \frac{a \cdot m}{12} = \frac{14 \cdot 100}{12} = \frac{1400}{12} = 116 + \frac{2}{3}.$$

(b) Also computed as

$$q_s = \frac{m}{12} \cdot a = \frac{100}{12} \cdot 14 = \left(8 + \frac{1}{3}\right) 14.$$

Before that, we are taught conversion rules. Since a nummi are worth one morabitinus, so $\frac{a}{12}$ solidi are worth one morabitinus, or else a solidi are worth 12 morabitini. In other words, the exchange rate in nummi for one morabitinus will correspond to 12 morabitini when applied to solidi.‡ Thus, if 14 nummi are one morabitinus, 14 solidi will be worth 12 morabitini, or if 5 solidi, thus 60 nummi, are worth one morabitinus, 60 solidi will be equivalent to 12 morabitini.

This is used to explain the second method of solving: a solidi being 12 morabitini, we shall divide m by 12 before multiplying it by a.

(c) Also computed as

$$q_s = \frac{a}{12} \cdot m = \frac{14}{12} \cdot 100 = \left(1 + \frac{1}{6}\right) 100.$$

Although this is again just one 'other' way, it is justified by the previous rules as well: a nummi being one morabitinus, thus a solidi 12 morabitini,

---

‡ Analogous rules (for capacity measures) seen in B.286c (p. 1605).

divide a by 12 to find the solidi in one morabitinus and then multiply the result by m. That is, consider that $12 : a = m : q_s$.

(**d**) Assuming arbitrarily (false position) that one morabitinus is worth f solidi (here $f = 2$), m morabitini are worth mf solidi. But it should be $\frac{a}{12} m$ solidi (or am nummi). Therefore there is a supplement of

$$mf - \frac{a}{12} m = \frac{(12f - a) m}{12}$$

to be subtracted from mf. Whence the required result is

$$q_s = mf - \frac{(12f - a) m}{12} = 200 - \frac{(24 - 14) 100}{12} = 200 - \left(83 + \frac{1}{3}\right).$$

This is a good illustration of how to unduly complicate matters, as often happens when the method of the false position is used.

(**B.310**) MSS $\mathcal{A}$, $\mathcal{B}$. One morabitinus is exchanged for fifteen nummi; how many morabitini will there be for five hundred solidi?

Given thus $a = 15$ (nummi), $q_s = 500$ (solidi), required m.

(**a**) Since, as seen above, $q = a \cdot m = 12 \cdot q_s$,

$$m = \frac{12 \cdot q_s}{a} = \frac{12 \cdot 500}{15} = 400.$$

(**b**) Also computed as

$$m = \frac{1}{a} \cdot 12\, q_s = \left(\frac{1}{3} \cdot \frac{1}{5}\right) 12 \cdot 500.$$

(**c**) Also computed as

$$m = \frac{q_s}{a} \cdot 12 = \frac{500}{15} \cdot 12 = \left(33 + \frac{1}{3}\right) 12 = 400.$$

This is justified in two ways. The first uses the rules seen in B.309b (a being the exchange rate into nummi, a solidi are 12 morabitini; so we divide q_s by a and multiply the result by 12). The second infers, by means of P_5 (or the rules at the beginning of Book B), the present computation from that in part a.

(**d**) Take first any number f of solidi (say $f = 10$, or $f = 12$), calculate the corresponding number of morabitini, and then transform it into the given quantity. Thus

$$m = \frac{12 f}{a} \cdot \frac{q_s}{f},$$

where, as seen in part a, the first factor is the number of morabitini corresponding to f solidi.

(*e*) Also computed as

$$m = \frac{12}{a} \cdot q_s = \frac{4}{5} \cdot 500,$$

which is justified, since a solidi are 12 morabitini, by the proportion $12 : a = m : q_s$.

(*f*) Assuming that one morabitinus is f ($f = 2$, as in B.309*d*) solidi, we shall calculate

$$m = \frac{q_s}{f} + \frac{(12f - a)\, q_s}{a \cdot f} = 250 + \frac{9 \cdot 250}{15}.$$

The first term thus corresponds to 250 morabitini. Since the 500 solidi are worth $12 \cdot 500 = 24 \cdot 250$ nummi, while the 250 morabitini are worth only $15 \cdot 250$ nummi, to each of the ($q_s : f =$) 250 morabitini we must add $(12f - a) = 9$ nummi more, thus 2250 nummi altogether. This result, divided by a for the conversion into morabitini, is the second term.

(*g*) Also computed in this last case as

$$m = \frac{q_s}{f} + \frac{12f - a}{a} \cdot \frac{q_s}{f} = 250 + \frac{3}{5} \cdot 250.$$

(**B.310′**) The different ways seen above remain the same if a contains a fraction; for example we might take, in the previous problem, $a = 14 + \frac{3}{4}$.

---

2. Two kinds of money exchanged for one kind of money.

A. *Final number of nummi not specified*.

In the next two problems, one morabitinus is exchanged for a_i nummi of the ith money and we know, in addition to the a_i's, how many coins $a_i\, m_i$ are received from all but one of the moneys; thus we know all but one of the fractions m_i of the morabitinus converted into the ith money while we have $m_1 + m_2 = 1$ and $m_1 + m_2 + m_3 = 1$ respectively.

(**B.311**) MSS $\mathcal{A}, \mathcal{B}$. One morabitinus is exchanged for ten nummi of one money or for twenty of another, and two nummi are received from the first; how many will be received from the second?

Given thus $m = 1 = m_1 + m_2$, $a_1 = 10$, $a_2 = 20$, $a_1 m_1 = 2$, required $a_2 m_2$.

(*a*) Computation.

$$a_2\, m_2 = \frac{(a_1 - a_1 m_1)\, a_2}{a_1} = \frac{(10 - 2)\, 20}{10} = \frac{160}{10} = 16.$$

Justification. Since we know a_1 and $a_1 m_1$, thus m_1, we can compute

$$m_2 = 1 - m_1 = 1 - \frac{a_1 m_1}{a_1} = \frac{a_1 - a_1 m_1}{a_1} = \frac{8}{10}.$$

whence, by multiplying both sides by a_2, the above formula.

(*b*) Also computed as
$$\frac{a_1 - a_1 m_1}{a_1} \cdot a_2 = \frac{4}{5} \cdot 20.$$

(**B.312**) MSS $\mathcal{A}$; $\mathcal{B}$, but *a* & *b* only. One morabitinus is exchanged for ten nummi of one money, for twenty of another or for thirty of a third, and two nummi are received from the first and four from the second; how many will be received from the third?

Given thus $a_1 = 10$, $a_2 = 20$, $a_3 = 30$, $m_1 + m_2 + m_3 = 1$, $a_1 m_1 = 2$, $a_2 m_2 = 4$, required $a_3 m_3$.

(*a*) Computation.
$$a_3 m_3 = a_3 - \left(\frac{a_1 m_1 \cdot a_3}{a_1} + \frac{a_2 m_2 \cdot a_3}{a_2}\right) = 30 - \left(\frac{2 \cdot 30}{10} + \frac{4 \cdot 30}{20}\right) = 18.$$

Justification. We know a_1, $a_1 m_1$, a_2, $a_2 m_2$, thus we can first compute
$$m_3 = 1 - m_1 - m_2 = 1 - \frac{a_1 m_1}{a_1} - \frac{a_2 m_2}{a_2} = 1 - \frac{1}{5} - \frac{1}{5} = \frac{3}{5}$$
and then multiply this by a_3 to obtain the formula. The justification indeed corresponds to these two steps.

Remark. This amounts to converting the first two moneys into the third (B.307) and subtracting the quantities found from a_3.

(*b*) Direct computation:
$$m_1 = m_2 = \frac{1}{5}, \quad \text{so} \quad a_3 m_3 = \frac{3}{5} \cdot 30.$$

(*c*) Converting the two nummi received from the money a_1 into nummi of the money a_2 as taught in the previous section (B.307) reduces this problem to B.311. The two nummi of the first money being worth four of the second, it is as if we had 8 nummi from the money at 20 and an unknown number from a second money at 30.

Restrictions. We must have, if we convert the first money into the second,
$$\frac{a_1 m_1 \cdot a_2}{a_1} + a_2 m_2 < a_2,$$
otherwise the problem is unsolvable (*falsus*). In case of equality, nothing will remain for the third money ($m_3 = 0$).

Furthermore, if just (besides a_1, a_2, a_3) one value $a_1 m_1$ is given, the problem will be indeterminate for $n \geq 3$. This will be treated in detail later on (B.316–318).

B. Final number of nummi specified.

As said in our summary of this chapter, B.313–315 are about the conversion of one morabitinus, thus $m = 1$, into two moneys and correspond to the linear system

$$\begin{cases} m_1 + m_2 = 1 \\ a_1 m_1 + a_2 m_2 = q, \end{cases}$$

with q, a_1, a_2 given.

(B.313) MSS $\mathcal{A}$; $\mathcal{B}$, but b only. One morabitinus is exchanged for twenty nummi of a first money or for thirty of a second, and a man changing one morabitinus receives twenty-four nummi; how many does he receive from each?

Given thus $a_1 = 20$, $a_2 = 30$, $q = 24$, $m_1 + m_2 = 1$, required $q_1 = a_1 m_1$, $q_2 = a_2 m_2$.

(a) Computation.

$$q_2 = a_2 m_2 = \frac{(q - a_1) a_2}{a_2 - a_1} = \frac{(24 - 20) 30}{30 - 20} = \frac{2}{5} \cdot 30 = 12$$

$$q_1 = a_1 m_1 = \frac{(a_2 - q) a_1}{a_2 - a_1} = \frac{(30 - 24) 20}{30 - 20} = \frac{3}{5} \cdot 20 = 12$$

(thus, in this particular case, $q_1 = q_2$). In order for the solution to be positive, we must have (since $a_2 > a_1$) $a_1 < q < a_2$.

Demonstration.

(α) *Preliminary remark.* Since $a_1 m_1 + a_2 m_2 = q$ and $m_1 = 1 - m_2$, $a_1 + (a_2 - a_1) m_2 = q$, $(a_2 - a_1) m_2 = q - a_1$, and thus, multiplying by a_2,

$$a_2 m_2 = \frac{(q - a_1) a_2}{a_2 - a_1}.$$

Similarly, $a_1 m_1 + a_2 (1 - m_1) = q$ gives $(a_1 - a_2) m_1 + a_2 = q$, and thus

$$a_1 m_1 = \frac{(a_2 - q) a_1}{a_2 - a_1}.$$

The subsequent demonstration corresponds exactly to this deduction.

A———————G————B

Fig. 90

(β) *Demonstration in the text.* Let (Fig. 90) $1 = AB$ (the morabitinus), $m_1 = AG$, $m_2 = GB$; thus $a_1 \cdot AG + a_2 \cdot GB = q$.

• Demonstration of the condition. Evidently, $a_1 \neq a_2$, so assume that $a_2 > a_1$. For the problem to be solvable, we must have $a_2 > q > a_1$. Suppose this not to be the case. Consider first $q > a_2 > a_1$. Then

$$q = a_1 \cdot AG + a_2 \cdot GB > a_2 = a_2 \cdot AB = a_2 \cdot AG + a_2 \cdot GB,$$

thus we would have $a_1 \cdot AG > a_2 \cdot AG$, which is impossible since $a_1 < a_2$.

So consider next $a_2 > a_1 > q$. Then

$$q = a_1 \cdot AG + a_2 \cdot GB \; < \; a_1 = a_1 \cdot AB = a_1 \cdot AG + a_1 \cdot GB,$$

thus we would have $a_2 \cdot GB < a_1 \cdot GB$, which is impossible since $a_2 > a_1$.

Therefore the only possibility left is $a_2 > q > a_1$, which is the condition. We cannot have either $q = a_1$ or $q = a_2$ if the problem states that conversion of the single morabitinus must involve the two moneys.

Remark. The proof of the condition in the general case of m morabitini would be similar; for AB may be taken as m, and then we shall find $a_2 \cdot m > q > a_1 \cdot m$. See B.329.

- *Demonstration of the formulae.* Since

$$q = a_1 \cdot AG + a_2 \cdot GB = a_1 \cdot AG + a_1 \cdot GB + (a_2 - a_1) GB$$

$$= a_1 \cdot AB + (a_2 - a_1) GB = a_1 + (a_2 - a_1) GB,$$

we can find $GB = m_2$, and $a_2 m_2$ (by the proportion $m_2 : 1 = a_2 m_2 : a_2$ in the text).

Next, adding $a_2 \cdot AG$ to both sides of $a_1 \cdot AG + a_2 \cdot GB = q$, we shall have

$$a_1 \cdot AG + a_2 \cdot GB + a_2 \cdot AG = q + a_2 \cdot AG,$$

thus

$$a_1 \cdot AG + a_2 = q + a_2 \cdot AG \quad \text{and} \quad AG(a_2 - a_1) = a_2 - q.$$

So we can find $AG = m_1$ and then (using $m_1 : 1 = a_1 m_1 : a_1$ in the text) $a_1 \cdot m_1$.

(b) Algebraically, put $m_1 = x$. Then $m_2 = 1 - x$ and therefore

$$20\,x + 30\,(1 - x) = 24, \quad 30 - 10x = 24,$$

whence $x = \frac{3}{5}$.

(B.314) MSS $\mathcal{A}$ (omits the computation), $\mathcal{B}$. One morabitinus is exchanged for ten nummi of a first money or for twenty of a second, and a man changing one morabitinus receives fifteen nummi; which fraction of the morabitinus does he receive from each?

We again have a problem of the form
$$\begin{cases} m_1 + m_2 = 1 \\ a_1 m_1 + a_2 m_2 = q, \end{cases}$$
but this time we are to find the m_i instead of the $a_i \cdot m_i$, and this will be the standard form in the coming group (the $a_i \cdot m_i$ being immediately obtainable since the a_i are always given). Given thus $a_1 = 10$, $a_2 = 20$, $q = 15$, required m_1, m_2.

Computation.

$$m_1 = \frac{a_2 - q}{a_2 - a_1} = \frac{20 - 15}{20 - 10} = \frac{1}{2}, \quad \text{whence} \quad m_2 = \frac{1}{2}.$$

Demonstration.

(α) *Preliminary remark.*
$$\begin{cases} m_1 + m_2 = 1 \\ a_1 m_1 + a_2 m_2 = q \end{cases} \quad (a_2 > a_1).$$

We have deduced in the previous problem the formulae for $a_1 m_1$ and $a_2 m_2$. Consider now the geometrical demonstration proving the above formula for m_1. In Fig. 91, $AB = m = 1$, $m_1 = AG$, $m_2 = GB$, $a_1 = GD$, $a_2 = GK$. Then $AD = a_1 m_1$ and $BK = a_2 m_2$, the sum of which is q. But the completed rectangle, AH, has the area $a_2 m$. Therefore the area of the rectangle KQ equals on the one hand $(a_2 - a_1) m_1$ and on the other $a_2 m - q$. This gives the formula for m_1.

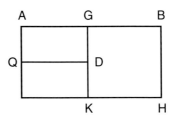

Fig. 91

(β) *Demonstration in the text.* Let (Fig. 91) $1 = AB$ (the morabitinus), $m_1 = AG$, $m_2 = GB$, and erect, perpendicular to AB at G, $GD = a_1$, $GK = a_2$. For the areas of the two rectangles thus constructed we have

$$AGDQ + GBHK = q.$$

Completing the outer rectangle, we have for its area $ABH = AB \cdot BH = AB \cdot a_2 = a_2$. Therefore

$$\text{rect. } QDK = a_2 - q, \quad \text{but also}$$

$$\text{rect. } QDK = QD \cdot DK = AG \,(GK - GD) = m_1 \,(a_2 - a_1)$$

whence the formula for m_1.

(B.315) MSS $\mathcal{A}$, $\mathcal{B}$. One morabitinus is exchanged for ten nummi of a first money or for thirty of a second, and changing one morabitinus gives twenty nummi more from the second money than from the first; what fraction of the morabitinus is received from each?

In this case, the proposed system takes the form

$$\begin{cases} m_1 + m_2 = 1 \\ a_2 m_2 - a_1 m_1 = q. \end{cases}$$

Given then $a_1 = 10$, $a_2 = 30$, $q = 20$, required m_1, m_2.

(**a**) Computation.

$$m_1 = \frac{a_2 - q}{a_1 + a_2} = \frac{30 - 20}{10 + 30} = \frac{1}{4}, \quad \text{whence} \quad m_2 = \frac{3}{4}.$$

Algebraically, this just means changing the sign of a_1 in B.314. In the *Liber mahameleth*, this is considered as a new problem, and the demonstration is indeed different.

Demonstration.

(α) *Preliminary remark.*

$$\begin{cases} m_1 + m_2 = m \\ a_2 m_2 - a_1 m_1 = q. \end{cases}$$

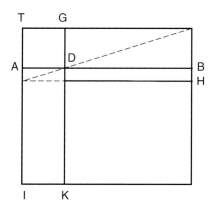

Fig. 92

Let (Fig. 92) $AB = m$, with $AD = m_1$, $DB = m_2$, $DK = a_2$, $DG = a_1$. Then $AG = a_1 m_1$, $KB = a_2 m_2$. Construct on DB an area DH equal to AG (*Elements* I.43). Now $KH = KB - DH = KB - AG = q$, while $IB = AB \cdot DK = a_2 m$. Therefore $IB - KH = a_2 m - q$. On the other hand, $IB - KH = DI + DH = DI + AG = (a_1 + a_2) m_1$. This gives the formula

$$m_1 = \frac{a_2 m - q}{a_1 + a_2}.$$

(β) *Demonstration in the text.* Let (Fig. 93) $1 = AB$ (the morabitinus), $m_1 = AD$, $m_2 = DB$, and erect, perpendicular to AB at D, $DG = a_1$, $DK = a_2$. Then rect. $ADGT = a_1 m_1$ and rect. $KDB = a_2 m_2$.

We now construct within KDB the rectangle DH having an area equal to that of $ADGT$. So the six-sided figure KHB equals $a_2 m_2 - a_1 m_1 = q$.*

---

* Fig. 93 is that of MS $\mathcal{A}$ ($\mathcal{B}$ has a blank space); the rectangle AG has just been reproduced within BK since $m_1 < m_2$ and $a_1 < a_2$.

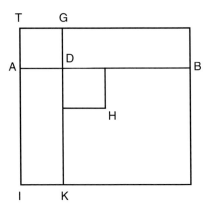

Fig. 93

Let us complete the rectangle IAB; then $IAB = AB \cdot DK = DK = a_2$. We then have

$$a_2 - q = \text{rect. } IAB - \text{fig. } KHB = IKDA + \text{rect. } DH$$

$$= IKDA + ADGT = IKGT = AD \cdot GK = m_1 (a_1 + a_2)$$

and this demonstrates the formula.

(**b**) Algebraically, put $m_1 = x$. Thus $m_2 = 1 - x$ and therefore

$$30(1-x) - 10x = 20,$$

which gives $40x = 10$.

3. More than two kinds of money exchanged for one kind of money.

We find here an exhaustive treatment of indeterminate linear systems of two equations. We have already seen an example of such a problem (B.51) but now the procedures to make the system determinate are quite clearly explained.

(**B.316**) MSS $\mathcal{A}$, without demonstration in c; $\mathcal{B}$, but only the formulation (in a different version) and b–c. One morabitinus is exchanged for ten nummi of a first money, twenty of a second and thirty of a third, and exchanging one morabitinus gives twenty-five nummi from all three; what fraction of the morabitinus is received from each?

Consider now the system

$$\begin{cases} m_1 + m_2 + m_3 = 1 \\ a_1 m_1 + a_2 m_2 + a_3 m_3 = q, \end{cases}$$

where we are given $a_1 = 10$, $a_2 = 20$, $a_3 = 30$, $q = 25$, with the m_i required. This problem is therefore indeterminate. Its treatment is very long, for in it are presented five possibilities to make it determinate; but just two principles are applied: either equating two of the fractional parts, as in a–d, or choosing two values, as in e–g. In addition, a related problem is inserted (B.316′).

(**a**) <u>First determination</u>. Put $m_1 = m_2 = m_0$. Thus
$$\begin{cases} 2m_0 + m_3 = 1. \\ (a_1 + a_2)\, m_0 + a_3\, m_3 = q. \end{cases}$$
In order to reduce this system to the form seen in B.313–314, we reformulate it in terms of $2m_0$ and m_3, and thus multiply the second equation by 2:
$$\begin{cases} (2m_0) + m_3 = 1 \\ (a_1 + a_2)(2m_0) + 2a_3\, m_3 = 2q \end{cases}$$
which is then, numerically,
$$\begin{cases} (2m_0) + m_3 = 1 \\ 30\,(2m_0) + 60 \cdot m_3 = 50, \end{cases}$$
of the form
$$\begin{cases} m' + m_2 = 1 \\ a'_1\, m' + a'_2\, m_2 = q', \qquad (a'_1 < a'_2) \end{cases}$$
known from B.314, which is solvable for $a'_1 < q' < a'_2$. Since $30 < 50 < 60$, this is solvable. We then find $2m_0 = \frac{1}{3}$, thus $m_0 = \frac{1}{6} = m_1 = m_2$ and $m_3 = \frac{2}{3}$.

<u>Remark</u>. Here two of the quantities of nummi received are fractional $\left(1 + \frac{2}{3},\ 3 + \frac{1}{3},\ 20\right)$.

<u>Demonstration</u> (to illustrate the modification made in the equations). Let (Fig. 94) $1 = AB$ (the morabitinus), $m_1 = AD$, $m_2 = DG$, $m_3 = GB$. Then $a_1 \cdot AD + a_2 \cdot DG + a_3 \cdot GB = q$, which becomes, with $AD = DG$, $(a_1 + a_2)\, AD + a_3 \cdot GB = q$, or also $(a_1 + a_2)\, 2\, AD + 2\, a_3 \cdot GB = 2\, q$, whence finally
$$(a_1 + a_2) AG + 2\, a_3 \cdot GB = 2\, q.$$

```
A   D   G                    B
|---+---+--------------------|
```

Fig. 94

(**b**) Algebraically, putting $m_1 = x = m_2$, we have $m_3 = 1 - 2x$, and thus the second equation becomes
$$30x + 30\,(1 - 2x) = 30 - 30x = 25, \quad \text{whence} \quad x = \frac{1}{6}.$$

<u>Remark</u>. The algebraic way thus avoids the above transformations.

(**c**) <u>Rule and generalization</u>. Let the given system be

$$\begin{cases} \sum_{i=1}^{n} m_i = 1 \\ \sum_{i=1}^{n} a_i m_i = q. \end{cases}$$

Put $m_1 = m_2 = \ldots = m_{n-1} = m_0$; thus, as seen in our introduction, and as described in the text, we shall calculate

$$m_0 = \frac{a_n - q}{(n-1)a_n - \sum_{i=1}^{n-1} a_i},$$

from which the other unknown m_1 may be inferred.

The demonstration is analogous to that of B.314, but here extended to the case $n = 3$, thus for the system

$$\begin{cases} m_1 + m_2 + m_3 = 1 \\ a_1 m_1 + a_2 m_2 + a_3 m_3 = q. \end{cases}$$

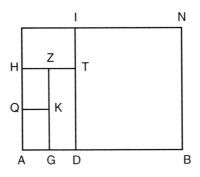

Fig. 95

<u>Demonstration</u>. Let (Fig. 95) $1 = AB$ (the morabitinus), $m_1 = AG$, $m_2 = GD$, $m_3 = DB$. Erect, perpendicular to AB at G, $GK = a_1 = 10$, $GZ = a_2 = 20$, and, at D, $DI = a_3 = 30$. Then

$$AGKQ = GK \cdot AG = a_1 m_1,$$
$$GDTZ = DT \cdot GD = a_2 m_2,$$
$$DBNI = DI \cdot DB = a_3 m_3.$$

Assume $AG = GD = m_0$. So $AGKQ + GDTZ + DBNI = (a_1 + a_2)m_0 + a_3 \cdot m_3 = 25 = q$. Completing the rectangle ABN, we have for its area

$$ABN = AB \cdot BN = BN = 30 = a_3,$$

which leaves for the two complementary areas

$$QKZH + HTI = 5 = a_3 - q.$$

Now this is $KZ \cdot ZH + TI \cdot TH$. But
$$TI \cdot TH = TI\left(TZ + ZH\right) = TI \cdot TZ + TI \cdot ZH,$$
and this, since $TZ = ZH$, equals
$$2\,TI \cdot ZH = TI \cdot 2\,ZH = (a_3 - a_2)\,2\,m_0.$$
Therefore we have for the complementary areas
$$QKZH + HTI = KZ \cdot ZH + 2\,TI \cdot ZH = ZH\left(KZ + 2\,TI\right).$$
At this point we have demonstrated, since $ZH = AG = m_0$, that
$$m_0 = \frac{QKZH + HTI}{KZ + 2\,TI} = \frac{a_3 - q}{(a_2 - a_1) + 2\,(a_3 - a_2)} = \frac{a_3 - q}{2\,a_3 - (a_1 + a_2)}$$
which is our previous formula, but here for the case $n = 3$. The text, however, ends the demonstration with the particular data of this problem, where $TI = KZ$, thus $a_3 - a_2 = a_2 - a_1$. The last two steps are then
$$QKZH + HTI = ZH\left(KZ + 2\,TI\right) = ZH \cdot 3\,KZ,$$
whence
$$m_0 = ZH = \frac{QKZH + HTI}{3\,KZ} = \frac{a_3 - q}{3\,(a_2 - a_1)} = \frac{5}{30} = \frac{1}{6}.$$

(**d**) <u>Second determination</u>. Put $m_1 = m_3 = m_0$. Then $10\,m_1 + 20\,m_2 + 30\,m_3 = 25$ becomes $40\,m_0 + 20\,m_2 = 25$ and we are to solve
$$\begin{cases} (2m_0) + m_2 = 1 \\ 40\,(2m_0) + 40 \cdot m_2 = 50. \end{cases}$$
Since the condition seen in B.313 is not satisfied, this is not solvable nor, therefore, is the original problem with the chosen determination $m_1 = m_3$. (The two equations are evidently inconsistent.)

Remark. If we were to assume $m_2 = m_3 = m_0$ in B.316, which is the remaining possibility not considered by the author, we would arrive at
$$\begin{cases} m_1 + (2m_0) = 1 \\ 10\,m_1 + 50\,m_0 = 25 \end{cases}$$
or
$$\begin{cases} m_1 + (2m_0) = 1 \\ 20\,m_1 + 50\,(2m_0) = 50, \end{cases}$$
solvable only for $m_1 = 0$. The author must have deliberately avoided this case.

(**B.316′**) Consider now, modifying the values of a_2 and q,
$$\begin{cases} m_1 + m_2 + m_3 = 1 \\ 10\,m_1 + 24\,m_2 + 30\,m_3 = 23. \end{cases}$$
This will be solvable for both above determinations in parts *a* and *d*. Indeed, the first ($m_1 = m_2 = m_0$) gives

$$\begin{cases} (2m_0) + m_3 = 1 \\ 34\,(2m_0) + 60\,m_3 = 46 \end{cases}$$
while the second ($m_1 = m_3 = m_0$) gives
$$\begin{cases} (2m_0) + m_2 = 1 \\ 40\,(2m_0) + 48\,m_2 = 46. \end{cases}$$

The last possible determination of this kind, it is noted, namely $m_2 = m_3 = m_0$, would be acceptable for this new problem as well, since we shall have
$$\begin{cases} (2m_0) + m_1 = 1 \\ 20\,m_1 + 54\,(2m_0) = 46 \end{cases}$$
which again meets the condition of solvability.

(**e**) <u>Third determination</u>. Let us now return to B.316 and apply the second type of assumption, that is, setting the value of just one (since $n = 3$) of the m_i. Thus, if in
$$\begin{cases} m_1 + m_2 + m_3 = 1 \\ 10\,m_1 + 20\,m_2 + 30\,m_3 = 25 \end{cases}$$
we take $m_1 = \frac{1}{5}$, thus $a_1 \cdot m_1 = 2$, we shall obtain
$$\begin{cases} m_2 + m_3 = \frac{4}{5} \\ 20\,m_2 + 30\,m_3 = 23. \end{cases}$$
Then, putting $m_2 = \frac{4}{5} m'_2$, $m_3 = \frac{4}{5} m'_3$, we attain the usual form
$$\begin{cases} m'_2 + m'_3 = 1 \\ 16\,m'_2 + 24\,m'_3 = 23. \end{cases}$$
Solving this as before, we find $m'_2 = \frac{1}{8}$, $m'_3 = \frac{7}{8}$, thus for the original problem $m_2 = \frac{1}{10}$, $m_3 = \frac{7}{10}$. The quantities of coins $a_i \cdot m_i$ are therefore 2, 2 and 21 respectively.

Remark. The system
$$\begin{cases} m_2 + m_3 = \frac{4}{5} \\ 20\,m_2 + 30\,m_3 = 23 \end{cases}$$
could be solved with the general formula, seen in B.314; but the transformation with m'_i enables us to verify the criterion for solvability right away.

(**f**) <u>Fourth determination</u>. Assuming now $m_3 = \frac{1}{5}$, thus $m_3 \cdot a_3 = 6$, gives
$$\begin{cases} m_1 + m_2 = \frac{4}{5} \\ 10\,m_1 + 20\,m_2 = 19 \end{cases}$$
or, with $m_1 = \frac{4}{5} m'_1$, $m_2 = \frac{4}{5} m'_2$,
$$\begin{cases} m'_1 + m'_2 = 1 \\ 8\,m'_1 + 16\,m'_2 = 19 \end{cases}$$
which is clearly unsolvable.

(**g**) <u>Fifth determination</u>. Taking $m_2 = \frac{1}{4}$ is acceptable. Indeed (the text does not pursue the treatment), we see that since $a_2 \cdot m_2 = 5$, the original system becomes

$$\begin{cases} m_1 + m_3 = \frac{3}{4} \\ 10\,m_1 + 30\,m_3 = 20 \end{cases}$$

or, with $m_i = \frac{3}{4}\,m'_i$,

$$\begin{cases} m'_1 + m'_3 = 1 \\ \left(7 + \frac{1}{2}\right) m'_1 + \left(22 + \frac{1}{2}\right) m'_3 = 20. \end{cases}$$

Despite the great number of possible determinations, the text concludes, the guiding principles for solving such problems remain the two seen above: either assuming $n-1$ of the m_i's to be equal or imposing $n-2$ of their values. We have just seen this for the case $n=3$ and shall now consider examples for the cases $n=4$ (B.317) and $n=5$ (B.318).

(B.317) MS $\mathcal{A}$. One morabitinus is exchanged for ten nummi of a first money, twenty of a second, thirty of a third and forty of a fourth, and a man exchanging one morabitinus receives twenty-five nummi from all four; what fraction of the morabitinus does he receive from each?

To solve the system

$$\begin{cases} m_1 + m_2 + m_3 + m_4 = 1 \\ 10\,m_1 + 20\,m_2 + 30\,m_3 + 40\,m_4 = 25. \end{cases}$$

(*a*) <u>First determination</u>. $m_1 = m_2 = m_3 = m_0$, thus $m_1 + m_2 + m_3 = 3m_0$. The system becomes

$$\begin{cases} (3m_0) + m_4 = 1 \\ 60\,(3m_0) + 120\,m_4 = 75. \end{cases}$$

This is solvable; we find then $3m_0 = \frac{3}{4}$, thus $m_1 = m_2 = m_3 = \frac{1}{4} = m_4$.

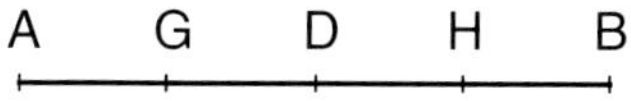

A   G   D   H   B

Fig. 96

<u>Demonstration</u> (explains, as in B.316a, the transformation of the equation, and solves it). Let (Fig. 96) $1 = AB$ (the morabitinus), $m_1 = AG$, $m_2 = GD$, $m_3 = DH$, $m_4 = HB$. By the above hypothesis, $10 \cdot AG + 20 \cdot GD + 30 \cdot DH + 40 \cdot HB = 60 \cdot AG + 40 \cdot HB = 25$, so $60\,(3 \cdot AG) + 120 \cdot HB = 75$, and we are to solve

$$\begin{cases} AH + HB = 1 \\ 60 \cdot AH + 120 \cdot HB = 75. \end{cases}$$

We find then (substituting $HB = AB - AH$ gives $60 \cdot AH + 120\,(AB - AH) = 75$, whence $60 \cdot AH = 120 - 75 = 45$)

$$AH = \frac{3}{4}, \quad \text{so that} \quad AG = GD = DH = \frac{1}{4} = HB.$$

(*b*) <u>Second determination</u>. Putting $m_1 = m_2 = m_4$, we arrive at a solvable problem, the text asserts. Indeed, for the second equation we are led to

$$70\,(3m_0) + 90\,m_3 = 75.$$

(c) Third determination. Putting $m_2 = m_3 = m_4$, we arrive at a solvable problem, the text again asserts. Indeed, for the second equation we are led to
$$30\, m_1 + 90\, (3 m_0) = 75.$$

(d) Fourth determination. Putting $m_1 = m_2 = m_0$ and $m_3 = m_4 = m'_0$, we shall have the system
$$\begin{cases} 2\, m_0 + 2\, m'_0 = 1 \\ 30\, m_0 + 70\, m'_0 = 25, \end{cases}$$
and we only need to multiply the second equation by the quantities of the unknowns made equal, namely 2. This produces the system
$$\begin{cases} (2\, m_0) + (2\, m'_0) = 1 \\ 30\, (2\, m_0) + 70\, (2\, m'_0) = 50, \end{cases}$$
which is solvable. We find $2\, m_0 = \frac{1}{2} = 2\, m'_0$, thus $m_1 = m_2 = m_3 = m_4 = \frac{1}{4}$.

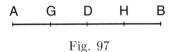

Fig. 97

Demonstration (illustrating the transformation of the equation). Let (Fig. 97) $1 = AB$ (the morabitinus), $m_1 = AG$, $m_2 = GD$, $m_3 = DH$, $m_4 = HB$. Putting $AG = GD$ and $DH = HB$, $10 \cdot AG + 20 \cdot GD + 30 \cdot DH + 40 \cdot HB$ becomes $AG \cdot 30 + HB \cdot 70 = 25$; doubling this last equation, we are left with solving
$$\begin{cases} AD + DB = 1 \\ AD \cdot 30 + DB \cdot 70 = 50, \end{cases}$$
leading to
$$AD = \frac{1}{2}, \quad DB = \frac{1}{2}, \quad \text{thus} \quad AG = GD = DH = HB = \frac{1}{4}.$$

(e) Fifth determination. Considering again
$$10\, m_1 + 20\, m_2 + 30\, m_3 + 40\, m_4 = 25,$$
we now put $m_1 = \frac{3}{10}$, $m_2 = \frac{1}{10}$, so that $a_1 \cdot m_1 = 3$, $a_2 \cdot m_2 = 2$. We then have to solve
$$\begin{cases} m_3 + m_4 = \frac{3}{5} \\ 30\, m_3 + 40\, m_4 = 20 \end{cases}$$
or, with $m_i = \frac{3}{5}\, m'_i$,
$$\begin{cases} m'_3 + m'_4 = 1 \\ 18\, m'_3 + 24\, m'_4 = 20. \end{cases}$$
We find then
$$m'_3 = \frac{2}{3}, \quad \text{so} \quad m_3 = \frac{2}{3}\frac{3}{5} = \frac{2}{5}, \quad m'_4 = \frac{1}{3}, \quad \text{so} \quad m_4 = \frac{1}{3}\frac{3}{5} = \frac{1}{5},$$

the respective numbers of nummi being: $a_1 \cdot m_1 = 3$, $a_2 \cdot m_2 = 2$, $a_3 \cdot m_3 = 12$, $a_4 \cdot m_4 = 8$.

Our author concludes by repeating the two ways of making such problems determinate, which will again be illustrated in the following problem.

(B.318) MS $\mathcal{A}$. One morabitinus is exchanged for eight nummi of a first money, twelve of a second, fifteen of a third, eighteen of a fourth and twenty of a fifth, and a man exchanging one morabitinus receives sixteen nummi from all five; what fraction of the morabitinus does he receive from each?

$$\begin{cases} m_1 + m_2 + m_3 + m_4 + m_5 = 1 \\ 8\,m_1 + 12\,m_2 + 15\,m_3 + 18\,m_4 + 20\,m_5 = 16. \end{cases}$$

(a) <u>First determination</u>. Put $m_1 = m_2 = m_4 = m_5 = m_0$. The system is unsolvable, since we shall have

$$\begin{cases} 4m_0 + m_3 = 1 \\ 58\,(4m_0) + 60\,m_3 = 64. \end{cases}$$

(b) <u>Second determination</u>. Put $m_1 = m_2 = m_3 = m_4 = m_0$. Solvable, since we obtain

$$\begin{cases} (4m_0) + m_5 = 1 \\ 53\,(4m_0) + 80\,m_5 = 64. \end{cases}$$

We find

$$4m_0 = \frac{5}{9} + \frac{1}{3} \cdot \frac{1}{9} = \frac{16}{27}, \quad \text{so}$$

$$\frac{1}{9} + \frac{1}{3}\frac{1}{9} = \frac{4}{27} = m_1 = m_2 = m_3 = m_4, \quad m_5 = \frac{3}{9} + \frac{2}{3} \cdot \frac{1}{9} = \frac{11}{27},$$

corresponding to the respective (fractional) amounts in nummi

$$1 + \frac{5}{27}, \quad 1 + \frac{7}{9}, \quad 2 + \frac{2}{9}, \quad 2 + \frac{2}{3}, \quad 8 + \frac{4}{27}.$$

(c) <u>Third determination</u>. Putting $m_2 = m_3 = m_4 = m_5 = m_0$ gives

$$\begin{cases} m_1 + (4m_0) = 1 \\ 32\,m_1 + 65\,(4m_0) = 64 \end{cases}$$

which is solvable. (We would find $m_1 = \frac{1}{33}$, $4m_0 = \frac{32}{33}$, so $m_0 = \frac{8}{33}$.)

(d) <u>Fourth determination</u> (second type). Assume now $n - 2$ values, for instance

$$m_1 = \frac{1}{8}, \quad m_3 = \frac{1}{4}, \quad m_4 = \frac{1}{8},$$

with $a_1 \cdot m_1 = 1$, $a_3 \cdot m_3 = 3 + \frac{3}{4}$, $a_4 \cdot m_4 = 2 + \frac{1}{4}$. Since this makes up half of the morabitinus and 7 nummi, we are left with solving

$$\begin{cases} m_2 + m_5 = \frac{1}{2} \\ 12\,m_2 + 20\,m_5 = 9, \end{cases}$$

which becomes, with $m_i = \frac{1}{2} m'_i$,

$$\begin{cases} m'_2 + m'_5 = 1 \\ 6\,m'_2 + 10\,m'_5 = 9, \end{cases}$$

solved by $m_2' = \frac{1}{4}$, thus $m_2 = \frac{1}{8}$, and $m_5' = \frac{3}{4}$, thus $m_5 = \frac{3}{8}$; the respective quantities of nummi $a_i \cdot m_i$ are then

$$1, \quad 1 + \frac{1}{2}, \quad 3 + \frac{3}{4}, \quad 2 + \frac{1}{4}, \quad 7 + \frac{1}{2}.$$

4. Particular problems.

The closing problems are, in themselves, mostly of little mathematical interest. However, as often in the *Liber mahameleth*, this banality is offset by the intricate way of solving.

A. *One morabitinus exchanged*.

(B.319) MS $\mathcal{B}$. One morabitinus is exchanged for ten nummi of a first money or for twenty of a second, and a man exchanging one morabitinus receives as many nummi from each; how many does he receive from each?

Given thus $a_1 = 10$, $a_2 = 20$, $m_1 + m_2 = 1$, required $a_1 m_1 = a_2 m_2$ (direct formula).

(a) Computation.

$$a_1 m_1 = \frac{a_2}{\frac{a_2}{a_1} + \frac{a_2}{a_2}} = \frac{20}{\frac{20}{10} + \frac{20}{20}} = \frac{20}{3} = 6 + \frac{2}{3} = a_2 m_2.$$

Remark. A similar particular form of the divisor is seen in B.45b (p. 1388).

Justification (a demonstration will be given in B.320a). This justification is somewhat awkward (note that only MS $\mathcal{B}$ has this problem). It amounts to the following. Since $a_1 m_1 = a_2 m_2$, so

$$m_1 = \frac{a_2}{a_1} m_2 \quad \text{and} \quad m_2 = \frac{a_1}{a_2} m_1.$$

Since $m_1 + m_2 = 1$, so

$$\frac{m_1}{1} = \frac{m_1}{m_1 + m_2} = \frac{\frac{a_2}{a_1} m_2}{\frac{a_2}{a_1} m_2 + \frac{a_2}{a_2} m_2} = \frac{\frac{a_2}{a_1}}{\frac{a_2}{a_1} + \frac{a_2}{a_2}},$$

and therefore

$$\frac{a_1 m_1}{a_1} = \frac{\frac{a_2}{a_1}}{\frac{a_2}{a_1} + \frac{a_2}{a_2}} \quad \text{whence} \quad a_1 m_1 = \frac{a_2}{\frac{a_2}{a_1} + \frac{a_2}{a_2}}.$$

Similarly,

$$\frac{m_2}{1} = \frac{m_2}{m_2 + m_1} = \frac{\frac{a_1}{a_2} m_1}{\frac{a_1}{a_2} m_1 + \frac{a_1}{a_1} m_1} = \frac{\frac{a_1}{a_2}}{\frac{a_1}{a_2} + \frac{a_1}{a_1}} = \frac{a_2 m_2}{a_2},$$

whence $\quad a_2 m_2 = \dfrac{a_1}{\frac{a_1}{a_1} + \frac{a_1}{a_2}}.$

(**b**) Also computed as
$$a_1\, m_1 = \frac{\frac{a_2}{a_1}}{\frac{a_2}{a_1}+\frac{a_2}{a_2}}\cdot a_1, \qquad a_2\, m_2 = \frac{\frac{a_1}{a_2}}{\frac{a_1}{a_1}+\frac{a_1}{a_2}}\cdot a_2.$$

(**c**) Take any f, preferably divisible by a_1 and a_2, such as here 40, and form the sum of the quotients of f divided by a_1 and a_2. Then
$$a_1\, m_1 = a_2\, m_2 = \frac{f}{\frac{f}{a_1}+\frac{f}{a_2}}.$$

Remark. The solution in part a corresponds to $f = a_2$ and $f = a_1$.

(**d**) This also amounts to solving
$$\begin{cases} m_1 + m_2 = 1 \\ a_1\, m_1 = a_2\, m_2. \end{cases}$$
The rest (algebraic solution as in B.320*b*, or solving as known from B.258*a*) is missing.

(**B.320**) MSS $\mathcal{A}$; $\mathcal{B}$ (algebraic solutions only). One morabitinus is exchanged for twenty nummi of a first money or for thirty of a second, and a man exchanging one morabitinus receives as many nummi from each; what fraction of the morabitinus does he receive from each?

Similar problem, with $a_1\, m_1 = a_2\, m_2$, $a_1 = 20$, $a_2 = 30$, required m_1, m_2 (but also $a_1\, m_1$, $a_2\, m_2$).

(**a**) <u>Computation.</u> Take any f, say $f = 30$; then
$$a_1\, m_1 = a_2\, m_2 = \frac{f}{\frac{f}{a_1}+\frac{f}{a_2}} = \frac{30}{\left(1+\frac{1}{2}\right)+1} = \frac{30}{2+\frac{1}{2}} = 12, \quad \text{so}$$
$$m_1 = \frac{12}{20} = \frac{3}{5}, \qquad m_2 = \frac{12}{30} = \frac{2}{5}.$$

This treatment has already been seen (B.319*c*). The purpose of the subsequent demonstration is just to show that f is indeed arbitrary.

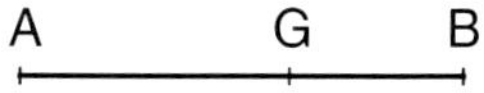

Fig. 98

<u>Demonstration.</u> Let (Fig. 98) $1 = AB$, $m_1 = AG$, $m_2 = GB$; since $a_1\, m_1 = a_2\, m_2$,
$$\frac{GB}{AG} = \frac{a_1}{a_2}.$$
Take two numbers n_1, n_2 such that $n_1 : n_2 = a_2 : a_1$, thus $a_1\, n_1 = a_2\, n_2$. This is done (B.45*a*, B.241) by choosing any f and forming
$$n_1 = \frac{f}{a_1}, \qquad n_2 = \frac{f}{a_2}$$

(which, with the author choosing again $f = 30$ —so $f = a_2$, is somewhat confusing for the reader). Then

$$\frac{n_2}{n_1} = \frac{a_1}{a_2} = \frac{m_2}{m_1} = \frac{GB}{AG},$$

whence, by composition, the two equalities

$$\frac{GB}{AG+GB} = \frac{GB}{AB} = \frac{n_2}{n_1+n_2}, \quad \frac{GB}{AB} = \frac{m_2}{m_1+m_2} = \frac{m_2}{1} = \frac{a_2\, m_2}{a_2};$$

therefore $n_2\, a_2 = a_2\, m_2\, (n_1 + n_2)$ and, since $n_2\, a_2 = f$,

$$a_2\, m_2 = \frac{f}{n_1 + n_2} = \frac{f}{\frac{f}{a_1} + \frac{f}{a_2}} = a_1\, m_1.$$

We are told at the end that treatment and proof remain the same for $n > 2$; we shall indeed see the case $n = 3$ in B.321–322.

(**b**) Using algebra, put $m_1 = x$; then $m_2 = 1 - x$. Thus, in our case,

$$20\, x = 30\, (1 - x), \quad x = \frac{3}{5} = m_1, \quad m_2 = \frac{2}{5},$$

and so $a_1\, m_1 = 12 = a_2\, m_2$.

(**c**) Using again algebra, put this time $a_1\, m_1 = x = a_2\, m_2$ (in order to find $a_i\, m_i$ directly); then, in our case,

$$m_1 + m_2 = \frac{x}{20} + \frac{x}{30} = 1 = \frac{5}{60}\, x,$$

and $x = 12 = a_1\, m_1 = a_2\, m_2$.

(**B.321**) MSS $\mathcal{A}$, $\mathcal{B}$. One morabitinus is exchanged for ten nummi of a first money, for twenty of a second, or for thirty of a third, and a man exchanging one morabitinus receives as many nummi from each; how many does he receive from each?

Extension to $n = 3$ of what we have seen in B.319–320. Given then $a_1 = 10$, $a_2 = 20$, $a_3 = 30$, with $m_1 + m_2 + m_3 = 1$ and $a_1\, m_1 = a_2\, m_2 = a_3\, m_3$, required $a_i\, m_i$.

(**a**) Computation.

$$a_i\, m_i = \frac{a_3}{\frac{a_3}{a_1} + \frac{a_3}{a_2} + \frac{a_3}{a_3}} = \frac{30}{3 + 1 + \frac{1}{2} + 1} = \frac{30}{5 + \frac{1}{2}} = 5 + \frac{5}{11}.$$

For the origin of the formula we are referred to B.319a.[‡]

[‡] Since $m_1 + m_2 + m_3 = 1$, we have $a_1\, a_2\, a_3 \cdot m_1 + a_1\, a_2\, a_3 \cdot m_2 + a_1\, a_2\, a_3 \cdot m_3 = a_1\, a_2\, a_3$; but, the $a_i\, m_i$'s being equal, this can be written as $a_3\, m_3\, (a_2\, a_3 + a_1\, a_3 + a_1\, a_2) = a_1\, a_2\, a_3$, whence

$$a_3\, m_3 = a_i\, m_i = \frac{a_3}{\frac{a_2\, a_3}{a_1\, a_2} + \frac{a_1\, a_3}{a_1\, a_2} + \frac{a_1\, a_2}{a_1\, a_2}},$$

which gives the formula.

(b) Also computed as (see B.319b)

$$a_i m_i = \frac{\frac{a_3}{a_i}}{\frac{a_3}{a_1} + \frac{a_3}{a_2} + \frac{a_3}{a_3}} \cdot a_i.$$

(c) Or, with an optional number f divisible by each of the a_i (see B.319c),

$$a_i m_i = \frac{f}{\frac{f}{a_1} + \frac{f}{a_2} + \frac{f}{a_3}}.$$

This is applied with $f = 60$, so

$$a_i m_i = \frac{60}{6 + 3 + 2} = \frac{60}{11} = 5 + \frac{5}{11}.$$

Remark. In the formula seen in b, a_3 may be cancelled, thus also any number replacing it.

(d) Let $a_i m_i = x$ (see B.320c); then, since

$$m_1 + m_2 + m_3 = \frac{x}{a_1} + \frac{x}{a_2} + \frac{x}{a_3} = 1 = \left(\frac{1}{10} + \frac{1}{20} + \frac{1}{30}\right) x = \frac{11}{60} x,$$

the equal number of nummi received from each money will be

$$x = \frac{60}{11} = 5 + \frac{5}{11}.$$

(e) Verification. Summing the parts, namely

$$m_1 = \frac{1}{10} \frac{60}{11} = \frac{6}{11}, \quad m_2 = \frac{1}{20} \frac{60}{11} = \frac{3}{11}, \quad m_3 = \frac{1}{30} \frac{60}{11} = \frac{2}{11},$$

we indeed find 1.

(**B.322**) MS $\mathcal{A}$. One morabitinus is exchanged for ten nummi of a first money, for twenty of a second, or for thirty of a third, and a man exchanging one morabitinus from all three receives from the second twice as many nummi as from the first, and from the third thrice as many as from the second; what fraction of the morabitinus does he receive of each?

We have this time $a_2 m_2 = k \cdot a_1 m_1$, $a_3 m_3 = l \cdot a_2 m_2$, $m_1 + m_2 + m_3 = 1$ with $a_1 = 10$, $a_2 = 20$, $a_3 = 30$, $k = 2$, $l = 3$.

Remark. With the values a_i, k, l given ($20 m_2 = 20 m_1$, $30 m_3 = 60 m_2$), the problem is mathematically banal: we see at once that $m_1 = m_2 = \frac{1}{2} m_3$. Since $a_2 m_2 = k \cdot a_1 m_1$ and $a_3 m_3 = l \cdot a_2 m_2 = k \cdot l \cdot a_1 m_1$, put

$$a'_2 = \frac{a_2}{k}, \quad a'_3 = \frac{a_3}{k \cdot l},$$

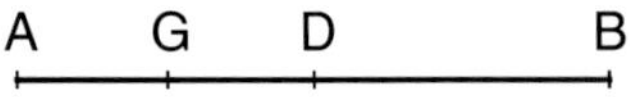

```
A    G    D           B
|────┼────┼───────────|
```

Fig. 99

whereby $a_1 m_1 = a'_2 m_2 = a'_3 m_3$, as in B.321. Since, in our case, $a'_2 = 10$, $a'_3 = 5$, we are to solve $10 m_1 = 10 m_2 = 5 m_3$ with $m_1 + m_2 + m_3 = 1$. Applying the formula seen in B.321a,

$$a_1 m_1 = \frac{a'_3}{\frac{a'_3}{a_1} + \frac{a'_3}{a'_2} + \frac{a'_3}{a'_3}} = \frac{5}{\frac{1}{2} + \frac{1}{2} + 1} = 2 + \frac{1}{2},$$

whence $a_2 m_2 = 2 a_1 m_1 = 5$, $a_3 m_3 = 3 a_2 m_2 = 15$. Thus finally:

$$m_1 = \frac{1}{4} = m_2, \qquad m_3 = \frac{1}{2}.$$

<u>Demonstration</u> (illustration of the transformation). Let (Fig. 99) $1 = AB$ (the morabitinus), $m_1 = AG$, $m_2 = GD$, $m_3 = DB$, with $20 \cdot GD = 2 \cdot 10 \, AG$ and $30 \cdot DB = 6 \cdot 10 \, AG$. Thus $10 \cdot AG = 10 \cdot GD = 5 \cdot DB$, which is the above $a_1 m_1 = a'_2 m_2 = a'_3 m_3$.

(B.323) MS $\mathcal{A}$. One morabitinus is exchanged for ten nummi of a first money, for twenty of a second, or for thirty of a third, and a man exchanging one morabitinus from all three receives two nummi from the first and an equal number from each of the two other; how many does he receive from these two moneys?

Given thus $a_1 = 10$, $a_2 = 20$, $a_3 = 30$, $m_1 + m_2 + m_3 = 1$ with $a_1 m_1 = 2$ and $a_2 m_2 = a_3 m_3$ (required).
Since $m_1 = \frac{1}{5}$, we are to solve

$$20 m_2 = 30 m_3 \quad \text{with} \quad m_2 + m_3 = \frac{4}{5}.$$

As known from B.316e, we may put

$$m_i = \frac{4}{5} m'_i, \quad \text{whereby} \quad 16 m'_2 = 24 m'_3, \quad m'_2 + m'_3 = 1.$$

Solving this as known from B.320 (not done in the text), we shall obtain

$$m'_2 = \frac{3}{5}, \quad m'_3 = \frac{2}{5}, \quad m_2 = \frac{12}{25}, \quad m_3 = \frac{8}{25}, \quad \left(m_1 = \frac{5}{25},\right)$$

with $a_2 m_2 = a_3 m_3 = 9 + \frac{3}{5}$.

(B.324) MS $\mathcal{A}$. One morabitinus is exchanged for fifteen nummi of a first money or for sixty of a second, and a man exchanging one morabitinus from both receives from the first the root of what he receives from the second; how many nummi does he receive from each money?

Given thus $a_1 = 15$, $a_2 = 60$, $m_1 + m_2 = 1$ with $a_1 m_1 = \sqrt{a_2 m_2}$.

<u>Geometrical treatment</u>. We end by solving a quadratic equation, namely $15 m_1^2 + 4 m_1 = 4$.

(α) *Preliminary remark*. Since $a_1 m_1 = \sqrt{a_2 m_2} = \sqrt{a_2 (1 - m_1)}$, we have $a_1^2 m_1^2 = a_2 (1 - m_1) = a_2 - a_2 m_1$, whence

$$m_1^2 + \frac{a_2}{a_1^2} m_1 = \frac{a_2}{a_1^2}, \quad \text{or} \quad m_1 \left(m_1 + \frac{a_2}{a_1^2} \right) = \frac{a_2}{a_1^2},$$

that is,

$$\begin{cases} \left(m_1 + \dfrac{a_2}{a_1^2} \right) - m_1 = \dfrac{a_2}{a_1^2} \\ m_1 \left(m_1 + \dfrac{a_2}{a_1^2} \right) = \dfrac{a_2}{a_1^2}, \end{cases}$$

to be solved as usual, using *Elements* II.6.

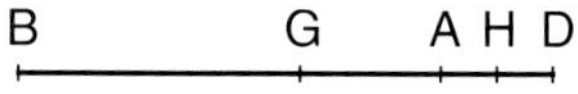

```
B           G      A H D
├───────────┼──────┼─┼─┤
```

Fig. 100

(β) *Treatment in the text*. Let (Fig. 100) $1 = AB$, $m_1 = AG$, $m_2 = GB$. Since $a_1 m_1 = \sqrt{a_2 m_2}$, so

$$(a_1 m_1)^2 = (15 \cdot AG)^2 = 225 \cdot AG^2 = a_2 m_2 = 60 \cdot GB.$$

But, since $m_2 = 1 - m_1$,

$$60 \cdot GB = 60 \cdot AB - 60 \cdot AG = 60 - 60 \cdot AG;$$

therefore

$$225 \cdot AG^2 + 60 \cdot AG = 60, \quad \text{or} \quad AG^2 + \frac{4}{15} AG = \frac{4}{15}.$$

Extend AB by DA, equal to the coefficient of AG, thus $DA = \frac{4}{15}$; the above equation takes the form

$$AG^2 + DA \cdot AG = AG \cdot DG = \frac{4}{15}.$$

But $DG - AG = DA$, so we are to solve

$$\begin{cases} DG - AG = \dfrac{4}{15} \\ DG \cdot AG = \dfrac{4}{15}. \end{cases}$$

Let H be the mid-point of DA; thus $AH = \frac{2}{15}$. By *Elements* II.6:

$$DG \cdot AG + AH^2 = HG^2 = \frac{4}{15} + \frac{4}{225} = \frac{64}{225}, \quad \text{so} \quad HG = \frac{8}{15},$$

whence

$$AG = HG - AH = \frac{8}{15} - \frac{2}{15} = \frac{2}{5} = m_1, \qquad GB = \frac{3}{5} = m_2,$$

and therefore $a_1 m_1 = 15 m_1 = 6$, $a_2 m_2 = 60 m_2 = 36$, and indeed $a_1 m_1 = \sqrt{a_2 m_2}$, as required.

(B.325) MSS $\mathcal{A}$, $\mathcal{B}$. One morabitinus is exchanged for ten nummi of a first money or for an unknown number of a second, and a man exchanging one morabitinus from both receives two nummi from the first and twenty from the second; what is the exchange rate of the second?

Given thus $a_1 = 10$, $a_1 m_1 = 2$, $a_2 m_2 = 20$, $m_1 + m_2 = 1$, required a_2.

Remark. Mathematically banal, since we have at once $m_1 = \frac{1}{5}$, thus $m_2 = \frac{4}{5}$ and $a_2 = 25$. The purpose seems rather to be a reduction to the initial problems of this chapter.

(a) Since

$$m_1 = \frac{a_1 m_1}{a_1} = \frac{1}{5}, \qquad m_2 = 1 - m_1 = \frac{4}{5},$$

we have received 2 nummi from the money at 10, and it is as if 8 nummi of the same money were to remain for the other money. Now this is the situation seen in B.307–308 on the conversion of one type of nummus into another ($q_1 : a_1 = q_2 : a_2$). This means in our case that

$$\frac{a_1 - a_1 m_1}{a_1} = \frac{a_2 m_2}{a_2}, \quad \text{and} \quad a_2 = \frac{a_1 (a_2 m_2)}{a_1 - a_1 m_1} = \frac{10 \cdot 20}{10 - 2}.$$

Also computed as

$$a_2 = \frac{a_1}{a_1 - a_1 m_1} \cdot a_2 m_2 = \left(1 + \frac{1}{4}\right) 20.$$

(b) Using algebra, put $a_2 = x$; then, after determining m_1 and m_2 as above,

$$a_2 m_2 = \frac{4}{5} x = 20,$$

whence x.

Remark. This algebraic treatment does not add anything. See also next problem.

(B.326) MSS $\mathcal{A}$, $\mathcal{B}$. One morabitinus is exchanged for ten nummi of a first money, twenty of a second or for an unknown number of a third, and a man exchanging one morabitinus from all three receives two nummi from the first, four from the second and thirty from the third; what is the exchange rate of the third?

Given thus $a_1 = 10$, $a_2 = 20$, $a_1 m_1 = 2$, $a_2 m_2 = 4$, $a_3 m_3 = 30$, $m_1 + m_2 + m_3 = 1$, required a_3. Similar to the preceding, but here for $n = 3$.

Remark. Again mathematically banal: m_1 and m_2 are immediately determinable, so also therefore m_3. Moreover, it appears from the data that $m_1 = m_2$, which lacks generality.

(*a*) Convert the nummi received at a_1 into nummi at a_2 (B.307):

$$a_2 m_1 = \frac{20 \cdot 2}{10} = 4;$$

then $a_2 m_1 + a_2 m_2 = a_2 m' = 8$ (with $m' = m_1 + m_2$), so that we are now to solve $20 m' = 8$, $a_3 m_3 = 30$, $m' + m_3 = 1$, with a_3 required, for which we know the way of solving from B.325:

$$a_3 = \frac{a_2 (a_3 m_3)}{a_2 - a_2 m'} = \frac{20 \cdot 30}{20 - 8} = \frac{600}{12} = 50.$$

Also computed, after the justification, as

$$a_3 = \frac{20}{12} \cdot 30 = \left(1 + \frac{2}{3}\right) 30.$$

(*b*) Direct computation (see our remark above). Since, from the data, $m_1 = m_2 = \frac{1}{5}$, so

$$m_3 = \frac{3}{5}, \quad \text{thus} \quad \frac{3}{5} a_3 = 30.$$

(*c*) The algebraic way here is exactly as in *b*, merely with x taking the place of a_3.

B. *Various kinds of morabitinus exchanged*.

Let there be m morabitini, with $m > 1$, to be exchanged for various types of solidi. If a_i is the exchange rate for one morabitinus of the ith type, the change of the quantity m_i of such morabitini will result in $a_i m_i$ solidi, and we shall obtain altogether q solidi from the $m = \sum m_i$ morabitini.

(**B.327**) MS *A*. Equal quantities of morabitini of three different types are exchanged for sixty solidi at the respective rates of three, four and five solidi; what is the equal quantity?

Given thus $a_1 = 3$, $a_2 = 4$, $a_3 = 5$, $q = a_1 m_1 + a_2 m_2 + a_3 m_3 = 60$ with $m_1 = m_2 = m_3 = m_0$. We just have to solve

$$3 m_1 + 4 m_2 + 5 m_3 = 12 m_0 = 60.$$

whence $m_0 = 5 = m_1 = m_2 = m_3$.

A   B

Fig. 101

Demonstration. Put (Fig. 101) $m_1 = m_2 = m_3 = AB$, whence $3 \cdot AB + 4 \cdot AB + 5 \cdot AB = (3 + 4 + 5) AB$ by Premiss $PE_1 = $ *Elements* II.1, whence $12 AB = 60$. The proof in B.48 (p. 1392) is similar.

Remark. This is the only place we are referred to A–II for one of the theorems of *Elements* II.1–10 (the usual reference being to the *Elements* itself).

(B.328) MS $\mathcal{A}$. Unequal quantities of morabitini of three different types are exchanged for a hundred solidi at the respective rates of three, four and five solidi, and the second quantity has four more morabitini than the first and the third five more than the second; what are these three quantities?

Given thus $a_1 = 3$, $a_2 = 4$, $a_3 = 5$, $m_2 = m_1 + 4$, $m_3 = m_2 + 5$, with $q = a_1 m_1 + a_2 m_2 + a_3 m_3 = 100$, required m_i.

Since $m_2 = m_1 + 4$ and $m_3 = m_2 + 5 = m_1 + 9$, we must have $3 m_1 + 4 (m_1 + 4) + 5 (m_1 + 9) = 100$, so

$$12 m_1 = 100 - 4 \cdot 4 - 9 \cdot 5 = 39,$$

which is then like the preceding problem since the three quantities are now the same. (We would find $m_1 = 3 + \frac{1}{4}$, $m_2 = m_1 + 4 = 7 + \frac{1}{4}$, $m_3 = m_1 + 9 = 12 + \frac{1}{4}$.)

The next problem involves morabitini from ancient Baetica, thus the region of Seville, that is, the part of Spain from which our (presumed) author was a native or one-time resident.

(B.329) MSS $\mathcal{A}$; $\mathcal{D}$, but without the demonstration and b. A hundred morabitini, comprising morabitini from Baeza (*baetes*) at ten solidi each and morabitini from Malaga (*melequini*) at fifteen solidi each, are exchanged for one thousand two hundred solidi; how many morabitini of each kind are there?

Given thus $a_1 = 10$, $a_2 = 15$, $q = 1200$, $m = 100$, required m_1 and m_2. Analogous to B.313, but with a quantity $m \neq 1$ of exchanged morabitini. We are to solve

$$\begin{cases} m_1 + m_2 = m \\ a_1 m_1 + a_2 m_2 = q \end{cases} \quad \text{or} \quad \begin{cases} m_1 + m_2 = 100 \\ 10 m_1 + 15 m_2 = 1200. \end{cases}$$

Condition. The problem is solvable if, for $a_2 > a_1$, $a_1 m < q < a_2 m$ (as is the case here since $1000 < 1200 < 1500$). As the text says, this condition will appear from the demonstration.

(a) Formulae.

$$m_1 = \frac{a_2 m - q}{a_2 - a_1} = \frac{1500 - 1200}{15 - 10} = 60 \text{ (baetes)}$$

$$m_2 = \frac{q - a_1 m}{a_2 - a_1} = \frac{1200 - 1000}{15 - 10} = 40 \text{ (melequini)}.$$

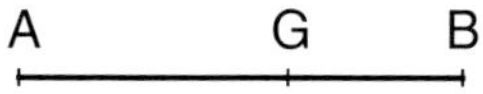

```
        A       G    B
        |-----------|-----|
```

Fig. 102

Remark. As in the similar problem B.313 the quantities $a_i\, m_i$ are equal.
Demonstration. Let (Fig. 102) $m = AB$, $m_1 = AG$, $m_2 = GB$. So

$$a_1 \cdot AG + a_2 \cdot GB = q.$$

First, since $a_2 \cdot GB = a_1 \cdot GB + (a_2 - a_1)\, GB$, the above equation becomes

$$a_1\, (AG + GB) + (a_2 - a_1)\, GB = a_1 \cdot m + (a_2 - a_1)\, GB = q,$$

which gives the formula for $GB = m_2$ and, incidentally, shows that $a_1 \cdot m < q$ (since $a_2 > a_1$).
Second, adding $a_2 \cdot AG$ on both sides of the initial equation results in

$$a_1 \cdot AG + a_2\, (AG + GB) = a_1 \cdot AG + a_2\, m = q + a_2 \cdot AG,$$

whence $AG\, (a_2 - a_1) = a_2\, m - q$, which gives the formula for $AG = m_1$ and, incidentally, shows that $a_2\, m > q$ (again since $a_2 > a_1$).
(*b*) Putting, in the manner seen in B.316e, $m_i = 100 \cdot m'_i$, we shall have to solve

$$\begin{cases} m'_1 + m'_2 = 1 \\ 1000 \cdot m'_1 + 1500 \cdot m'_2 = 1200, \end{cases}$$

the treatment of which is known from B.314 (the formulae found there give $m'_2 = \frac{2}{5}$, $m'_1 = \frac{3}{5}$, to be multiplied by 100 to find m_2, m_1). This shows that the treatment of the general case $m \neq 1$ can be inferred from the previous problems with $m = 1$.

If more than two quantities m_1, m_2 are involved, the author notes at the end, we shall introduce determinations just as we did with the m_i earlier (assuming, for n quantities m_i, $n - 2$ values m_i or equating $n - 1$ of the m_i).

Remark. According to M. Crusafont y Sabater (personal communication), the *melequini* and the *baetes* were in use in the mid-twelfth century.[†]

C. *Another kind of problem.*

(**B.330**) MSS *A*, *B*. Of a morabitinus, exchanged for fourteen nummi, a piece is cut of which the value plus the weight is four nummi. What is its weight?
Remark. This problem makes sense only when it known that the Arabic equivalent of the *nummus*, the *dirham*, is both a weight and a silver coin.[°]

[†] See our *Le Liber mahameleth*, p. 95. See also Gautier Dalché's study, p. 62.

[°] The Latin *dragma* would have conveyed this better.

(**a**) Let the morabitinus have, in nummi, the value P and the weight W, and the piece cut the value p and the weight w. Clearly, $w : p = W : P$, so

$$\frac{w}{p+w} = \frac{W}{P+W}, \quad \text{thus} \quad w = \frac{W}{P+W}(p+w).$$

As we know, $p + w = 4$ and $P = 14$; supposing $W = 2$, we shall obtain

$$w = \frac{2}{16} \cdot 4 = \frac{1}{2}.$$

Also computed as

$$w = \frac{W(p+w)}{P+W} = \frac{2 \cdot 4}{16}.$$

(**b**) Using algebra, which avoids transforming the proportion. Put $w = x$, thus $p = 4 - x$ and

$$\frac{w}{p} = \frac{x}{4-x} = \frac{W}{P} = \frac{2}{14},$$

thus $14\,x = 8 - 2\,x$, whence $16\,x = 8$ and $x = \frac{1}{2}$.

(**c**) Verification. We found that the part is a fourth of the morabitinus, so its value is $3 + \frac{1}{2}$. Since its weight is $\frac{1}{2}$, we have indeed $p + w = 4$.

Chapter B–XIX:
Cisterns
Summary

The problems in this chapter can be divided into three groups, all dealing with cisterns but from a different aspect each time. (As we know, in the *Liber mahameleth* the chapters are organized mostly by subject and not according to the mathematical methods or treatments addressed.)

1. Pipes filling a cistern.

In B.331–333 we have the classical cistern problem: water is running into a cistern from a certain number of pipes each of which would fill it separately in t_i days. Thus, in one day the ith pipe fills $\frac{1}{t_i}$ of the cistern. So all of them will fill the multiple $\sum \frac{1}{t_i}$ of one cistern in one day, that is, they will fill a single cistern completely in

$$\frac{1}{\sum \frac{1}{t_i}} \text{ days.}$$

If instead of the kth pipe there is an outlet, the term $\frac{1}{t_k}$ must be subtracted (B.332).

Remark. We have encountered a problem of this kind before (B.169, with three mills).

2. Stone thrown into a cistern.

Into a cistern of given volume filled with a given quantity of water a stone is thrown; knowing the volume of the stone we can determine the quantity of water overflowing (B.334, B.337) or, conversely, we may determine the volume, or one unknown dimension, of the stone when the quantity of water overflowing is known (B.335–336, B.338); furthermore, knowing the height of the cistern, its content and the quantity overflowing enables us also to calculate the drop in water level when the stone is removed (B.339–340).

Indeed, if v_1 is the volume of the cistern, v_2 the volume of the stone, with l_i, r_i, h_i the respective dimensions, q the quantity of water contained in the full cistern, q' the quantity overflowing and h'_1 the drop in water level, then

$$\frac{q'}{q} = \frac{v_2}{v_1} = \frac{l_2 \cdot r_2 \cdot h_2}{l_1 \cdot r_1 \cdot h_1} = \frac{h'_1}{h_1}.$$

The relation between a quantity of water and its volume, or the volume of its containing vessel, is explained in the first problem, B.334. (We have proportionality and not equality because each is measured with its own unit and we do not know the conversion factor between one 'measure' (*mensura*) of water and a (cubic) 'cubit'.) The proportion involving the respective heights of the water level is demonstrated in B.339 (since the water surface before and after removal of the stone remains the same, only the height of the water level is modified).

3. Content of a cask.

The last problem (B.341) is about the wine content of a cylindrical cask of which we know diameter of the base and height. Calculating the volume of the cask is already known from a previous problem (B.272′); what is new is its conversion into 'measures' of liquid (two filling a cube with side one cubit in this particular instance).

---

1. Pipes filling a cistern.

(B.331) MSS $\mathcal{A}$, $\mathcal{B}$, $\mathcal{D}$. A pipe fills a cistern in one day, a second in half a day and a third in a third of a day; how much time will it take for all three running together to fill the cistern?

Given thus $t_1 = 1$, $t_2 = \frac{1}{2}$, $t_3 = \frac{1}{3}$. Since the pipes fill in one day

$$\sum \frac{1}{t_i} = 1 + 2 + 3 = 6$$

such cisterns, they will fill a single one in $\frac{1}{6}$ day.

(B.332) MSS $\mathcal{A}$, $\mathcal{B}$, $\mathcal{D}$. A pipe fills a cistern in one day, a second in half a day and a third in a third of a day, while an outlet empties it in a third of a day; how long will it take to fill the cistern?

Given thus, as before, $t_1 = 1$, $t_2 = \frac{1}{2}$, $t_3 = \frac{1}{3}$, but $t_4 = -\frac{1}{3}$ (outlet). Since

$$\sum \frac{1}{t_i} = \frac{1}{t_1} + \frac{1}{t_2} + \frac{1}{t_3} - \frac{1}{t_4} = 1 + 2 + 3 - 3 = 3,$$

$\frac{1}{3}$ day is the required time.

Remark. This reduces to a problem with two pipes since the action of the third is cancelled out by the outlet.

(B.333) MSS $\mathcal{A}$, $\mathcal{B}$. A pipe fills a cistern in two days, a second in three days and a third in four days; how much time will it take for all three running together to fill the cistern?

Given thus $t_1 = 2$, $t_2 = 3$, $t_3 = 4$. Then

$$\sum \frac{1}{t_i} = \frac{1}{2} + \frac{1}{3} + \frac{1}{4} = \frac{13}{12}$$

and the required time is $\frac{12}{13}$ of a day.

---

2. Stone thrown into a cistern.

(B.334) MSS $\mathcal{A}$, $\mathcal{B}$. Into a cistern ten cubits long, eight wide, six deep, containing one thousand measures of water, a stone four cubits long, three wide, five thick is thrown; how much water will overflow?

Given thus $l_1 = 10$, $r_1 = 8$, $h_1 = 6$, $q = 1000$, $l_2 = 4$, $r_2 = 3$, $h_2 = 5$, required q'. (Each dimension of the stone being half one of the dimensions of the cistern, the two bodies are similar in shape.)

(*a*) <u>Computation</u>.

$$q' = \frac{v_2 \cdot q}{v_1} = \frac{4 \cdot 3 \cdot 5 \cdot 1000}{10 \cdot 8 \cdot 6} = \frac{60 \cdot 1000}{480} = \frac{60\,000}{480} = 125.$$

<u>Justification</u>. The ratio volume to water contained (or, in the case of the stone, overflowing) is constant. This means that, the volume of the cistern being $v_1 = 10 \cdot 8 \cdot 6 = 480$ and that of the stone $v_2 = 4 \cdot 3 \cdot 5 = 60$, then $v_1 : q = v_2 : q'$, whence the formula.

(*b*) Also computed as

$$q' = \frac{v_2}{v_1} \cdot q = \frac{1}{8} \cdot 1000.$$

(*c*) Also computed as

$$q' = \frac{q}{v_1} \cdot v_2 = \left(2 + \frac{1}{2} \cdot \frac{1}{6}\right) 60.$$

(**B.335**) MSS $\mathcal{A}$, $\mathcal{B}$. Into a cistern ten cubits long, eight wide, six deep, containing one thousand measures of water, a stone four cubits long, three wide is thrown and one hundred and twenty-five measures overflow; how thick is the stone?

Given thus $l_1 = 10$, $r_1 = 8$, $h_1 = 6$, $q = 1000$, $l_2 = 4$, $r_2 = 3$, $q' = 125$, required h_2. Same numerical values as before.

(*a*) <u>Computation</u>.

$$h_2 = \frac{v_1 \cdot q'}{l_2 \cdot r_2 \cdot q} = \frac{480 \cdot 125}{4 \cdot 3 \cdot 1000} = \frac{480 \cdot 125}{12\,000} = 5.$$

<u>Justification</u>. As we know from before,

$$\frac{v_1}{q} = \frac{v_2}{q'} = \frac{l_2 \cdot r_2 \cdot h_2}{q'}, \quad \text{thus} \quad \frac{480}{1000} = \frac{12 \cdot h_2}{125},$$

whence the formula.

(*b*) Also computed as

$$h_2 = \left(\frac{q'}{q} \cdot v_1\right) \frac{1}{l_2 \cdot r_2} = \left(\frac{1}{8} \cdot 480\right) \frac{1}{12} = \frac{60}{12}.$$

(*c*) Also computed as

$$h_2 = \left(\frac{v_1}{q} \cdot q'\right) \frac{1}{l_2 \cdot r_2} = \left(\left[\frac{2}{5} + \frac{2}{5} \cdot \frac{1}{5}\right] \cdot 125\right) \frac{1}{12} = \frac{60}{12}.$$

(B.336) MSS $\mathcal{A}$, $\mathcal{B}$. Into a cistern ten cubits long, eight wide, six deep, containing one thousand measures of water, a stone similar in shape and three cubits thick is thrown and one hundred and twenty-five measures overflow; how long and wide is the stone?

Given thus $l_1 = 10$, $r_1 = 8$, $h_1 = 6$, $q = 1000$, $l_2 : r_2 = l_1 : r_1$, $h_2 = 3$, $q' = 125$, required l_2 and r_2 (same numerical values as before).

Remark. The text specifies that one (rectangular) side of the cistern and the corresponding side of the stone are similar figures. If the *whole* stone were said to be similar in shape to the cistern (as it in fact is), then we would immediately have the answer: the stone's dimensions, being to those of the cistern in the ratio $h_2 : h_1$, are just half the corresponding ones of the cistern.

(a) Let s_2 be the aforesaid stone's surface area; then

$$s_2 = l_2 \cdot r_2 = \frac{v_1 \cdot q'}{q \cdot h_2} = \frac{480 \cdot 125}{1000 \cdot 3} = \frac{60\,000}{3000} = 20,$$

whence (since $l_1 : r_1 = l_2 : r_2 = l_2^2 : s_2 = s_2 : r_2^2$)

$$l_2^2 = \frac{l_1 \cdot s_2}{r_1} = \frac{10 \cdot 20}{8} = 25, \qquad r_2^2 = \frac{r_1 \cdot s_2}{l_1} = \frac{8 \cdot 20}{10} = 16,$$

and $l_2 = 5$, $r_2 = 4$.

(b) Also computed as

$$\left(\frac{q'}{q} \cdot v_1\right) \frac{1}{h_2} = \left(\frac{125}{1000} \cdot 480\right) \frac{1}{3} = \left(\frac{1}{8} \cdot 480\right) \frac{1}{3} = \frac{60}{3} \left(= \frac{v_2}{h_2}\right) = 20 = s_2,$$

with

$$s_2 = l_2 \cdot r_2 = l_2 \left(\frac{r_1}{l_1} \cdot l_2\right) = l_2 \cdot \frac{4}{5} l_2,$$

and consequently

$$\frac{4}{5} l_2^2 = 20, \qquad l_2^2 = 25, \qquad r_2 = \frac{4}{5} l_2.$$

There is no justification; for both ways, we are referred to a similar situation in the chapter on drapery (B.129 & B.132, involving two similar, rectangular areas).

(B.337) MSS $\mathcal{A}$, $\mathcal{B}$. Into a cistern ten cubits long, wide and deep, containing a hundred measures of water, a stone four cubits long, wide and thick is thrown; how much water will overflow?

Given thus $l_1 = r_1 = h_1 = 10$, $q = 100$, $l_2 = r_2 = h_2 = 4$, required q'.

(a) Since $q : q' = v_1 : v_2$ (B.334),

$$q' = \frac{l_2^3 \cdot q}{l_1^3} = \frac{64 \cdot 100}{1000} = 6 + \frac{2}{5}.$$

(**b**) Also computed as

$$q' = \frac{q}{l_1^3} \cdot l_2^3 = \frac{100}{1000} \cdot 64 = \frac{1}{10} \cdot 64.$$

(**B.338**) MSS $\mathcal{A}$, $\mathcal{B}$. Into a cistern ten cubits long, wide and deep, containing two hundred measures of water, a stone having length and width equal and four cubits thick is thrown and twenty measures overflow; how long and wide is the stone?

Given thus $l_1 = r_1 = h_1 = 10$, $q = 200$, $l_2 = r_2$, $h_2 = 4$, $q' = 20$, required $l_2 = r_2$. Related to B.336, which this might have followed.

(**a**) We have

$$l_2^2 = \frac{l_1 \cdot r_1 \cdot h_1 \cdot q'}{q \cdot h_2} = \frac{v_1 \cdot q'}{q \cdot h_2} = \frac{1000 \cdot 20}{200 \cdot 4} = 25,$$

thus $l_2 = r_2 = 5$.

(**b**) Also computed as

$$l_2^2 = \left(\frac{v_1}{q} \cdot q'\right) \frac{1}{h_2}.$$

(**B.339**) MS $\mathcal{A}$. Into a cistern ten cubits long, eight wide, five deep, containing one thousand measures of water, a stone is thrown and one hundred measures overflow; by how much will the water level drop when the stone is removed?

Given thus $l_1 = 10$, $r_1 = 8$, $h_1 = 5$, $q = 1000$, $q' = 100$, required this time h'_1, by which the water level drops.

Remark. Giving l_1 and r_1 is superfluous; for $h'_1 : h_1 = q' : q$. The demonstration will prove precisely that.

Computation.

$$h'_1 = \frac{q' \cdot h_1}{q} = \frac{100 \cdot 5}{1000} \left(= \frac{1}{2}\right).$$

Demonstration. The water level after overflow of q' separates the cistern of volume v_1 into two solids of volumes v'_1 (empty part) and $v''_1 = v_1 - v'_1$, namely I and II in our Fig. 103. The water level also divides each pair of vertical opposite walls into two rectangular areas; thus, on the front side, $s_1 = ABFE$ is divided into $s'_1 = ABDC$, which is the empty part, and $s''_1 = CDFE$, with respective heights $h'_1 = BD$ and $h''_1 = DF$, summing up to the cistern's height h_1. To prove that $q' : q = h'_1 : h_1$.

Indeed, first, v'_1 and v''_1 have a common separating area; therefore, by *Elements* XI.32, $v'_1 : v''_1 = s'_1 : s''_1$. Second, s'_1 and s''_1 have one side common (CD in the figure); therefore, by *Elements* VI.1, $s'_1 : s''_1 = h'_1 : h''_1$. Consequently, $v'_1 : v''_1 = h'_1 : h''_1$. Now, by composition,

$$\frac{v'_1}{v'_1 + v''_1} = \frac{v'_1}{v_1} = \frac{h'_1}{h'_1 + h''_1} = \frac{h'_1}{h_1}.$$

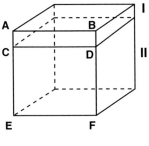

Fig. 103

Since the quantities of water may replace the volumes ($v'_1 : v_1 = q' : q$), the formula is proved.

(B.340) MSS $\mathcal{A}$, $\mathcal{D}$. Into a cistern ten cubits long, eight wide, five deep, containing one thousand measures of water, a stone three cubits long, two wide, one thick is thrown; by how much will the water level drop when the stone is removed?

Given thus $l_1 = 10$, $r_1 = 8$, $h_1 = 5$, $q = 1000$, $l_2 = 3$, $r_2 = 2$, $h_2 = 1$, required again h'_1.

Since q' is unknown, we compute it as seen in B.334:

$$q' = \frac{v_2 \cdot q}{v_1} = \frac{6 \cdot 1000}{400} = 15,$$

and we are thus in the situation of B.339.

Remark. Here again, giving q is superfluous since we could compute h'_1 as

$$h'_1 = \frac{v_2 \cdot h_1}{v_1} = \frac{3}{40}.$$

---

3. Content of a cask.

(B.341) MSS $\mathcal{A}$, $\mathcal{B}$, $\mathcal{D}$. A cylindrical cask has a diameter of ten cubits and a depth of eight; how much wine does it contain?

Given thus diameter $d = 10$, height $h = 8$, and required the number q of measures of wine. Let s be the area of its base. Since two measures fill a cube with side one cubit,‖ then, since the volume is $s \cdot h$ (see B.272'),

$$q = 2\,s \cdot h,$$

with s computed by either of the two ways:

$$s = d^2 \left[1 - \left(\frac{1}{7} + \frac{1}{2} \cdot \frac{1}{7} \right) \right] \quad \left(= \frac{d^2}{4} \left(3 + \frac{1}{7} \right) \right)$$

‖ This did not hold for the water measures mentioned in the previous problems.

$$s = \frac{c}{2} \cdot \frac{d}{2} = \left[d \left(3 + \frac{1}{7} \right) \cdot \frac{1}{2} \right] \frac{d}{2},$$

where c is the circumference of the circle; for, we are told, 'the Ancients' have proved that the circumference of a circle is equal to the diameter multiplied by $3+\frac{1}{7}$. (We find the same calculation for the area in B.143' and B.152; in B.150, our author mentions that the value $3+\frac{1}{7}$ is approximative and that it goes back to Archimedes.)

But if the cask has a parallelepipedal form, the text notes, its wine content will be equal to $2 \cdot v$, with the volume v determined as we have been taught previously (B.272, B.334).

Chapter B–XX:
Ladders

Summary

Here again (as in B–XIV) the heading is misleading: only the initial problems are on ladders. What these and most of the others do have in common is reliance on the Pythagorean theorem. After the examples with a ladder leaning against a wall (B.342–349), there will be the broken tree (B.350–353) and the two towers of unequal height (B.359–361). Those not thus related are the problems on falling trees (B.354-356) and three problems not found in the two main manuscripts (B.357–358, B.362).

It is here that we first encounter problems of a more recreational nature. Beginning with that of the ladder is, in fact, appropriate: known since Mesopotamian times, it is the oldest.[†] Its twofold aspect, computational and recreational, becomes evident in Arabic times, for it is then found both in algebra text books and in collections of mathematical recreations used for social entertainment. The 11th-century Persian scholar al-Bīrūnī mentions these two aspects explicitly.[°] He also speaks (elsewhere) about the occurrence of the broken tree in algebra treatises.[||] Both these types are found in contemporary Chinese and Indian texts.[‡]

1. The ladder.

A. *One unknown*.

Let (Fig. 104) $l = h$ be both the length of the ladder and the height of the wall, g the distance the upper end of the ladder drops from the top of the wall after the lower end is drawn away and d the resulting distance between this lower end and the bottom of the wall. Thus three quantities occur, which are linked by the relation $(l - g)^2 + d^2 = l^2$, with $l > g$. Accordingly, each can be determined from the other two. First (l, d given),

$$g = l - \sqrt{l^2 - d^2},$$

[†] See, e.g., Vogel's *Vorgriechische Mathematik II*, p. 67

[°] *Mas'ala al-sullam al-mudawwana fī ḥisābāt al-muṭāraḥa wa-furū' al-jabr wa'l-muqābala*, in his treatise *On Shadows*; see Kennedy's translation p. 119 and commentary p. 57, Arabic text p. 76. The word *muṭāraḥāt* designates recreational problems of the kind found in the *Anthologia græca*; see the occurrence of this word in Wiedemann's *Beiträge*, XIV.1, p. 28, XV, p. 125; and also Dozy's or Kazimirski's dictionaries.

[||] In his *Istikhrāj al-awtār*, pp. 81–82, pp. 44–45 in Suter's translation (pp. 313–314 in his *Beiträge*).

[‡] References to the individual problems in the 1987 edition of B–XX (in our *Survivance médiévale*).

which is demonstrated in B.342; next (l, g given),

$$d = \sqrt{l^2 - (l-g)^2},$$

which is demonstrated in B.343; finally (d, g given), since, from the relation just seen, $d = \sqrt{2gl - g^2}$, thus $g^2 + d^2 = 2gl$,

$$l = \frac{1}{g}\frac{g^2 + d^2}{2}, \quad \text{or, also,} \quad l = g + \frac{1}{g}\frac{d^2 - g^2}{2},$$

as it appears and is demonstrated in B.344 ($d > g$ proved in B.346).

Fig. 104

B. *Two unknowns*.

— In the group B.345–347, l is known while the sum, difference or quotient of d and g is known. The treatment of the first two, which corresponds for us to solving a second-degree problem, proceeds in our text in the usual way, by means of identities.

In B.345, where l and $d + g$ are given, d and g are individually determined by

$$d = \sqrt{\left[\frac{l - (d+g)}{2}\right]^2 + \frac{l^2 - [l - (d+g)]^2}{2}} - \frac{l - (d+g)}{2},$$

$$g = l - \left(\sqrt{\left[\frac{l - (d+g)}{2}\right]^2 + \frac{l^2 - [l - (d+g)]^2}{2}} + \frac{l - (d+g)}{2}\right),$$

demonstrated there.

In B.346, l and $d - g$ are given, and the quantities d and g required. The first is determined by

$$d = l - \left(\sqrt{\left[\frac{l - (d-g)}{2}\right]^2 - \frac{(d-g)^2}{2}} + \frac{l - (d-g)}{2}\right)$$

(demonstrated there), whereby we infer $g = d - (d - g)$.

In B.347, $d = k \cdot g$. Then $g^2 + d^2 = g^2(k^2 + 1)$ while, by the relation mentioned above, $g^2 + d^2 = 2gl$; therefore

$$g = \frac{2l}{k^2 + 1},$$

which is demonstrated there.

— In the last two problems, l is required and we know $d+g$ and $d \cdot g$ (B.348), $d - g$ and $d \cdot g$ (B.349). We shall first determine d and g individually by means of the formulae (although in frequent use before, again demonstrated here)

$$d, g = \frac{d+g}{2} \pm \sqrt{\left(\frac{d+g}{2}\right)^2 - d \cdot g},$$

$$d, g = \sqrt{\left(\frac{d-g}{2}\right)^2 + d \cdot g} \pm \frac{d-g}{2},$$

from which l may be calculated.

2. The broken tree.

A tree has broken (or bends to the ground) at a certain height; the upper part remains attached, but the top touches the ground. Let (Fig. 105) l be the original height of the tree, d the horizontal distance between top and root after the trunk breaks, and g the vertical distance between root and break (thus $l - g$ is the length of the bending part). Here again three quantities are involved, and we are successively given l and g (B.350), l and d (B.351; this is the problem described by al-Bīrūnī), g and d (B.352), $g : l = k$ and d (B.353). All are solved by means of the relation

$$d^2 + g^2 = (l-g)^2, \quad \text{or} \quad d^2 = l^2 - 2l \cdot g,$$

which underlies the four formulae used and demonstrated in these problems:

$$d = \sqrt{(l-g)^2 - g^2} \qquad \text{(B.350)}$$

$$g = \frac{1}{l}\left(\frac{l^2 - d^2}{2}\right) \qquad \text{(B.351)}$$

$$l = \sqrt{d^2 + g^2} + g \qquad \text{(B.352)}$$

$$l = \frac{d}{\sqrt{1 - 2k}} \quad \left(k < \frac{1}{2}\right) \qquad \text{(B.353).}$$

Fig. 105

3. The falling tree.

Two cases are considered here.

A. *Forward motion*.

An initially vertical tree falls towards the ground at a constant daily angular velocity (B.354, B.356). If l is the height of the tree, its top will describe an arc of length $d = l \cdot \varphi$, which becomes in our case the arc of a quadrant; dividing d by a, the arc-length covered daily, will give the number of days before the tree lies on the ground. Thus we have for the number of days (B.354) and the height of the tree (B.356)

$$t = \frac{l \cdot \frac{\pi}{2}}{a} \cong \frac{\frac{1}{2}l \cdot (3 + \frac{1}{7})}{a}, \qquad l \cong 2\frac{a \cdot t}{3 + \frac{1}{7}},$$

according to the approximation of π already encountered (B.341).

B. *Forward and retrograde motion*.

The downward movement is partly compensated by an upward one (B.355; see also B.370–374). Let d be the distance which must be covered by the top, a the arc-length covered daily, u the daily retrogression partly compensating the groundward movement ($u < a$). The number of days to reach the ground, assuming that once there the tree will not rise again and that both movements take place at the same uniform angular velocity, is, according to the text,

$$t = \left[\frac{d-u}{a-u}\right] + \frac{u+r}{a+u},$$

where the brackets indicate that the integral part of the quotient must be taken and r is the remainder.

The author does not explain the origin of this formula, which can be reconstructed as follows. Knowing the values of a and u, we divide the day into $a+u$ equal parts since the total displacement, irrespective of direction, is $a + u$ and the velocity is uniform. Downward motion is then supposed to take place during part $\frac{a}{a+u}$ of the day and upward motion during the remaining part $\frac{u}{a+u}$. Consider now the time taken to cover part $d-a$ of the quadrant. Since this distance is covered only by the daily mean progression, namely $a - u$, it will require a duration of

$$\frac{d-a}{a-u} = q + \frac{r}{a-u} \quad \text{days}$$

where q is the integral part of the quotient and $r < a - u$ the remainder.

- Case $r = 0$. The remaining distance to be covered by the top of the tree is a, namely the distance we have subtracted from d. Since a is the purely forward daily motion, the ground will be reached at the end of this motion during the day $q+1$, and the duration is then expressed by

$$t = \frac{d-a}{a-u} + \frac{a}{a+u} \quad \text{or else} \quad t = \frac{d-u}{a-u} - \frac{u}{a+u}.$$

• Case $r \neq 0$. At the beginning of the day $q+1$, the top of the tree will be distant from the ground by an arc of length $a+r$. It will thus first descend by a, then return by u. At the beginning of the day $q+2$, the distance $u+r$ will remain to be covered. Since $r < a - u$, thus $u + r < a$, this distance will be fully covered during the time of downward motion. The tree will thus reach the ground after the fraction $\frac{u+r}{a+u}$ of the $(q+2)$th day has elapsed. So in this case the whole duration of the movement will be

$$t = \left[\frac{d-a}{a-u}\right] + 1 + \frac{u+r}{a+u} = \left[\frac{d-u}{a-u}\right] + \frac{u+r}{a+u}.$$

This is the formula seen initially, which is thus valid only if $r \neq 0$.

Remark. It was certainly obtained by the rule of three: if a is the daily forward motion, how long will it take to cover $u + r$? Answer: $\frac{u+r}{a}$ of the time of forward motion, or $\frac{u+r}{a+u}$ of the whole $(q+2)$th day.

4. Planting trees.

B.357–358, found only in MS $\mathcal{D}$, are concerned with trees planted at regular intervals on a plot. If l, r are the length and width of the plot and d the distance between any two adjacent trees, the number q of trees is given by

$$q = \left(\frac{l}{d}+1\right)\left(\frac{r}{d}+1\right) = q_l \cdot q_r,$$

where q_l is the number of trees lined up in length and q_r those in width. They are integers, for l and r are chosen in the text as multiples of d. (The addition of 1 takes into account the tree planted on the border.)

5. The two towers.

Let (Fig. 106) l_1, l_2 be the respective heights of the two towers ($l_1 \neq l_2$), d the distance between their bases and e the distance between their tops. These four magnitudes thus obey the relation

$$(l_1 - l_2)^2 + d^2 = e^2.$$

Three of these quantities being known, we must calculate the fourth. Thus we are successively given l_1, l_2, d (B.359), l_1, l_2, e (B.360), and l_1 or l_2, e, d (B.361). The formulae, each time demonstrated, are accordingly

$$e = \sqrt{(l_1-l_2)^2 + d^2}, \quad d = \sqrt{e^2 - (l_1-l_2)^2}, \quad l_1 - l_2 = \sqrt{e^2 - d^2}.$$

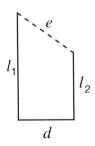

Fig. 106

Finally B.362, of a practical nature, teaches how to determine the height of a tower (or any other object the top of which cannot be reached). Only MS $\mathcal{D}$ contains this. It may very well have been in the *Liber mahameleth* originally, for there is another instance of a chapter ending with a practical measuring method (B.124–125, pp. 1446–1447).

1. The ladder.
A. *One unknown*.
(**B.342**) MSS $\mathcal{A}$, $\mathcal{B}$; $\mathcal{D}$, but only partly (formulation and computation in *a*). A ladder ten cubits long standing against a wall of equal height is withdrawn from the bottom of the wall by six cubits; by how much will it come down from the top?

Given thus $l = h = 10$, $d = 6$, required g. Throughout this section, the quantities involved, whether given or required, have the same values.

(***a***) Computation.

$$g = l - \sqrt{l^2 - d^2} = 10 - \sqrt{10^2 - 6^2} = 10 - 8 = 2.$$

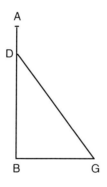

Fig. 107

Demonstration. Let (Fig. 107) $h = AB = l = DG$, $d = BG$. Since

$$DB^2 + BG^2 = DG^2 \quad (\textit{Elements I.47}),$$

$$DB^2 = DG^2 - BG^2 \, (= 64), \quad \text{whence} \quad DB = \sqrt{DG^2 - BG^2},$$

thus DB is known, while AB is given, so $AD = AB - DB$ can be calculated.

(***b***) Algebraic solution. Put $AD = x$; then $DB = 10 - x$, thus $(10-x)^2 + 36 = DG^2 = 100$, or $x^2 + 136 - 20x = 100$, therefore

$$x^2 + 36 = 20x, \quad x = AD = \frac{20}{2} - \sqrt{\left(\frac{20}{2}\right)^2 - 36} = 2,$$

which is the smaller (and only acceptable) positive solution.

Remark. Here the solution formula of the quadratic equation is quite explicit.

(B.343) MSS $\mathcal{A}, \mathcal{B}; \mathcal{D}$, but only partly. A ladder ten cubits long standing against a wall of equal height comes down, when withdrawn, by two cubits from the top of the wall; by how much is it withdrawn from the bottom of the wall?

Given thus $l = h = 10$, $g = 2$, required d.

(*a*) <u>Computation.</u>

$$d = \sqrt{l^2 - (l-g)^2} = \sqrt{10^2 - (10-2)^2} = \sqrt{36} = 6.$$

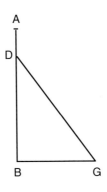

Fig. 108

<u>Demonstration.</u> Let (Fig. 108) $h = AB = l = DG$, $g = AD$. Since

$$BG^2 = DG^2 - DB^2 = DG^2 - (AB - AD)^2,$$

so $BG = \sqrt{DG^2 - (AB - AD)^2}$ can be calculated.

(*b*) Algebraic solution. Put $BG = x$; then

$$BG^2 + DB^2 = x^2 + 64 = 100,$$

whence $x = 6$.

(B.344) MSS $\mathcal{A}, \mathcal{B}; \mathcal{D}$, but only partly. A ladder of unknown length standing against a wall of the same height, when withdrawn by six cubits from the base of the wall, comes down two cubits from the top of the wall; what is its length?

Given thus $d = 6$, $g = 2$, required $l = h$.

(*a*) <u>Computation.</u>

$$l = g + \frac{1}{g}\frac{d^2 - g^2}{2} = 2 + \frac{1}{2}\frac{6^2 - 2^2}{2} = 10.$$

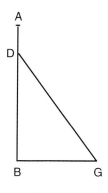

Fig. 109

Demonstration. Let (Fig. 109) $h = AB = l = DG$, $d = BG$, $g = AD$. Evidently (*Elements* II.4 = PE_4)
$$AB^2 = (AD + DB)^2 = AD^2 + 2AD \cdot DB + DB^2,$$
thus also ($AB = DG$)
$$DG^2 = AD^2 + 2\,AD \cdot DB + DB^2.$$
On the other hand, $DG^2 = BG^2 + DB^2$. Therefore
$$AD^2 + 2AD \cdot DB = BG^2,$$
whence
$$DB = \frac{1}{AD}\frac{BG^2 - AD^2}{2}, \quad AB = AD + DB = AD + \frac{1}{AD}\frac{BG^2 - AD^2}{2},$$
and this is the formula used above.

(*b*) Algebraic solution. Put $AB = x$; then $DB = x - 2$. Thus
$$(x - 2)^2 + 36 = x^2$$
which reduces to $4x = 40$, whence $x = 10$.

B. *Two unknowns*.

(**B.345**) MS $\mathcal{A}$. A ladder ten cubits long standing against a wall of the same height is withdrawn from the base of the wall by a distance which, when added to that by which the ladder comes down from the top of the wall, makes eight; by how much is it withdrawn and by how much does it come down?

Given thus $l = h = 10$, $d + g = 8$, required d and g.

Computation.

$$g = l - \left(\sqrt{\left[\frac{l - (d + g)}{2}\right]^2 + \frac{l^2 - [l - (d + g)]^2}{2}} + \frac{l - (d + g)}{2} \right)$$

$$= 10 - \left(\sqrt{\left[\frac{10-8}{2}\right]^2 + \frac{10^2 - (10-8)^2}{2}} + \frac{10-8}{2}\right)$$

$$= 10 - (\sqrt{49} + 1) = 2; \text{ then } d = 8 - g = 6, \text{ or else}$$

$$d = \sqrt{\left[\frac{l - (d+g)}{2}\right]^2 + \frac{l^2 - [l - (d+g)]^2}{2}} - \frac{l - (d+g)}{2} = \sqrt{49} - 1.$$

Demonstration.

(α) *Preliminary remark.* The demonstration below corresponds in our symbolism to what follows; basically, it amounts to using the two usual identities

$$\left(\frac{\alpha + \beta}{2}\right)^2 - \alpha \cdot \beta = \left(\frac{\alpha - \beta}{2}\right)^2, \quad \alpha, \beta = \frac{\alpha + \beta}{2} \pm \frac{\alpha - \beta}{2}.$$

As we know,
$$l^2 = (l - g)^2 + d^2. \tag{1}$$

On the other hand, $[(l - g) - d]^2 = (l - g)^2 + d^2 - 2d(l - g)$, so that

$$(l - g)^2 + d^2 = 2d(l - g) + [(l - g) - d]^2,$$

whence, with (1) and since $(l - g) - d = l - (d + g)$,

$$l^2 = 2d(l - g) + [l - (d + g)]^2, \tag{2}$$

from which we infer that

$$d(l - g) = \frac{l^2 - [l - (d + g)]^2}{2}, \tag{3}$$

the right side of which is known. But $(l - g) - d = l - (d + g)$ is also known. Thus we are to solve

$$\begin{cases} (l - g) - d = l - (d + g) \\ (l - g)d = \dfrac{l^2 - [l - (d + g)]^2}{2}. \end{cases}$$

By the first identity above,

$$\left[\frac{(l - g) + d}{2}\right]^2 = \left[\frac{(l - g) - d}{2}\right]^2 + (l - g)d, \tag{4}$$

whence

$$\frac{(l - g) + d}{2} = \sqrt{\left[\frac{(l - g) - d}{2}\right]^2 + (l - g)d}$$

$$= \sqrt{\left[\frac{l - (d + g)}{2}\right]^2 + \frac{l^2 - [l - (d + g)]^2}{2}}.$$

Then, using the second identity,

$$l - g = \frac{(l-g)+d}{2} + \frac{(l-g)-d}{2}$$

$$= \sqrt{\left[\frac{l-(d+g)}{2}\right]^2 + \frac{l^2 - [l-(d+g)]^2}{2}} + \frac{l-(d+g)}{2},$$

whence the above formula for g; next,

$$d = \frac{(l-g)+d}{2} - \frac{(l-g)-d}{2}$$

$$= \sqrt{\left[\frac{l-(d+g)}{2}\right]^2 + \frac{l^2 - [l-(d+g)]^2}{2}} - \frac{l-(d+g)}{2},$$

which is the above formula for d.

Remark. Calculating d corresponds to solving the quadratic equation

$$d^2 + d[l-(d+g)] = \frac{l^2 - [l-(d+g)]^2}{2},$$

and the formula just seen represents its (single) positive root. Likewise, $l - g$ is the positive solution of

$$(l-g)^2 = (l-g)\left[l-(d+g)\right] + \frac{l^2 - [l-(d+g)]^2}{2}.$$

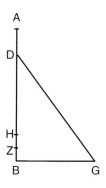

Fig. 110

(β) *Demonstration according to the text.* We know (Fig. 110) that $AB = DG = l = 10$ and that $BG + AD = d + g = 8$. Let us put $DH = d = BG$;[†]

[†] H within DB according to the data ($l > d + g$). The demonstration would remain the same for $HB = (d+g) - l = DH - DB$, thus with H on the other side of B.

then $AH\ (= d+g = 8)$ and $HB = AB - AH\ (= l - (d+g) = 2)$ are known. Since
$$DG^2 = DB^2 + BG^2 = DB^2 + DH^2 \quad (=100) \qquad (1)$$
while (*Elements* II.7 = PE_7)
$$DG^2 = DH^2 + DB^2 = 2\,DB \cdot DH + HB^2, \qquad (2)$$
we shall have
$$DB \cdot DH = \frac{DG^2 - HB^2}{2}, \quad \text{known } (=48). \qquad (3)$$
Since $HB = DB - DH$ is known, we are to solve
$$\begin{cases} DB - DH = 2 \\ DB \cdot DH = 48. \end{cases}$$
Let Z be the mid-point of HB (thus $HZ = 1 = ZB$). Then (*Elements* II.6 = PE_6)
$$DB \cdot DH + HZ^2 = DZ^2. \qquad (4)$$
Since the terms on the left are known, DZ is determined $(=7)$. We know thus $DZ \pm ZB$, therefore first $DZ + ZB = DB$ and then $g = AD = AB - DB$, second $DZ - ZB = DZ - HZ = DH = BG = d$.

Putting together the intermediate results, we see that, for the case of d, we have
$$d = DZ - HZ = \sqrt{HZ^2 + DB \cdot DH} - HZ$$
$$= \sqrt{\left(\frac{HB}{2}\right)^2 + \frac{DG^2 - HB^2}{2}} - \frac{HB}{2}$$
with $HB = l - (d+g)$ and $DG = l$. Thus the formula seen before is demonstrated.

(B.346) MSS $\mathcal{A}$; $\mathcal{B}$, end part only. A ladder ten cubits long standing against a wall of the same height is withdrawn from the base of the wall by a distance which, when diminished by the distance by which the ladder comes down from the top, leaves four; by how much is it withdrawn and by how much does it come down?

Given thus $l = 10$, $d - g = 4$, required d and g.

Condition. The author first shows that we could not attribute a positive value to $g - d$ since always $d > g$.

Indeed, let (Fig. 111) $AB = DG = l$, $BG = d$, $AD = g$; to prove that $BG > AD$. Since $AB^2 = DG^2 = DB^2 + BG^2$ while (*Elements* II.4 = PE_4) $AB^2 = AD^2 + 2AD \cdot DB + DB^2$, it follows that $BG^2 - AD^2 = 2AD \cdot DB$, thus that $BG > AD$.

Remark. The relation $BG^2 - AD^2 = 2AD \cdot DB$, or $d^2 - g^2 = 2\,g\,(l-g)$, had in fact already been established for solving B.344.

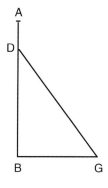

Fig. 111

Computation. The distance d is expressed in function of the given quantities l and $d-g$ by the relation

$$d = l - \left(\sqrt{\left[\frac{l-(d-g)}{2}\right]^2 - \frac{(d-g)^2}{2}} + \frac{l-(d-g)}{2}\right)$$

$$= 10 - \left(\sqrt{\left[\frac{10-4}{2}\right]^2 - \frac{4^2}{2}} + \frac{10-4}{2}\right) = 10 - (1+3) = 6$$

whereby we infer that $g = d - (d-g) = 6 - 4 = 2$.

Demonstration.

(α) *Preliminary remark.* As just seen in the demonstration of the condition,

$$d^2 = g^2 + 2g(l-g). \qquad (1)$$

On the other hand,

$$d^2 = [g + (d-g)]^2 = g^2 + 2g(d-g) + (d-g)^2.$$

Comparing these two results,

$$2g(l-g) = 2g(d-g) + (d-g)^2, \qquad (2)$$

whence

$$2g(l-d) = (d-g)^2. \qquad (3)$$

Since $d-g$ is given, $g(l-d)$ is known. But $l-(d-g) = (l-d) + g$, so we are to solve

$$\begin{cases} (l-d) + g = l - (d-g) \\ (l-d)\,g = \dfrac{(d-g)^2}{2}, \end{cases}$$

the right sides of which are known. By the identity (corresponding to *Elements* II.5)

$$\left[\frac{(l-d)-g}{2}\right]^2 = \left[\frac{(l-d)+g}{2}\right]^2 - (l-d)g$$

we can determine

$$\frac{(l-d)-g}{2} = \sqrt{\left[\frac{l-(d-g)}{2}\right]^2 - \frac{(d-g)^2}{2}}.$$

Therefore

$$l - d = \frac{(l-d)+g}{2} + \frac{(l-d)-g}{2}$$

$$= \frac{l-(d-g)}{2} + \sqrt{\left[\frac{l-(d-g)}{2}\right]^2 - \frac{(d-g)^2}{2}} \quad \text{and}$$

$$d = l - \left(\sqrt{\left[\frac{l-(d-g)}{2}\right]^2 - \frac{(d-g)^2}{2}} + \frac{l-(d-g)}{2}\right)$$

which is the formula used above and demonstrated below.

Remark. The expression for $l - d$ is obtained by solving the quadratic equation

$$(l-d)^2 + \frac{(d-g)^2}{2} = (l-d)\bigl[l-(d-g)\bigr],$$

with the two positive solutions

$$l - d = \frac{l-(d-g)}{2} \pm \sqrt{\left[\frac{l-(d-g)}{2}\right]^2 - \frac{(d-g)^2}{2}},$$

only the larger of which is considered, namely $l - d = 4$, whence $d = 6$, $g = 2$. But the second solution, $l - d = 2$, would also have led to an acceptable solution, namely $d = 8$, $g = 4$.

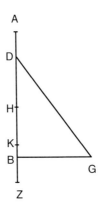

Fig. 112

(β) *Demonstration according to the text.* Let (Fig. 112) $AB = DG = l$, $BG = d$, $AD = g$. We know that $BG - AD = 4$. Let us put $AH = BG = d$; thus $AH - AD = DH = d - g$ is known $(= 4)$. Since

$$BG^2 = AD^2 + 2AD \cdot DB, \qquad (1)$$

as established at the beginning of the problem, and because $BG^2 = AH^2 = AD^2 + 2AD \cdot DH + DH^2$, we shall have

$$2AD \cdot DB = 2AD \cdot DH + DH^2. \qquad (2)$$

But $2AD \cdot DB = 2AD \cdot DH + 2AD \cdot HB$; there then remains

$$2AD \cdot HB = DH^2 \quad (= 16), \qquad (3)$$

so that $AD \cdot HB$ is known $(= 8)$.

Let us now make $BZ = AD$; thus $BZ \cdot HB = 8$. Since $DZ = DB + BZ = DB + AD = AB = 10$, so $HZ = DZ - DH = 6 = HB + BZ$. Then we are to solve

$$\begin{cases} HB + BZ = 6 \\ HB \cdot BZ = 8. \end{cases}$$

The segment HZ is already divided into two unequal parts at B (unequal since, by the above equations, $HB \neq BZ$). Taking its mid-point K (thus $KZ = 3$) it will be divided into two equal parts as well, and we shall have according to *Elements* II.5 = PE_5

$$HB \cdot BZ + KB^2 = KZ^2,$$

whence the value of KB $(= 1)$. So $AD = BZ = KZ - KB$ and therefore $BG = AH = AD + DH$ are determined.

We have thus found that $d = BG = AH = AD + DH = BZ + DH = DZ - HB = DZ - (KB + HK)$, with

$DZ = AB = l$,

$KB = \sqrt{KZ^2 - HB \cdot BZ} = \sqrt{KZ^2 - \frac{1}{2} DH^2}$,

$KZ = HK = \dfrac{HZ}{2} = \dfrac{DZ - DH}{2} = \dfrac{AB - DH}{2} = \dfrac{l - (d-g)}{2}$;

which demonstrates the above formula.

As already remarked, the author does not consider the second solution. He could have demonstrated its existence in the same way, the only difference being that, with the mid-point K of segment HZ falling in this latter case between B and Z, we would have $AD = BZ = KZ + KB = 4$.

(B.347) MSS $\mathcal{A}$, $\mathcal{B}$. A ladder ten cubits long standing against a wall of the same height is withdrawn from the base of the wall by thrice the distance by which the ladder comes down from the top; by how much is it withdrawn and by how much does it come down?

Given thus $l = 10$, $d = k \cdot g$ ($k = 3$), required d and g.
Computation (included in the geometrical demonstration).

$$g = \frac{2l}{k^2 + 1} = \frac{2 \cdot 10}{3^2 + 1} = 2, \qquad d = k \cdot g = 6.$$

Demonstration. Since $l^2 = (l - g)^2 + d^2$ while $d = k \cdot g$, with k given, we have $2lg = g^2 + d^2 = g^2(k^2 + 1)$, so

$$g = \frac{2l}{k^2 + 1}, \qquad d = k \cdot g.$$

This corresponds to the demonstration. Let (Fig. 113) $l = AB = DG$, $d = BG$, $g = AD$. Since (B.346) $AD^2 + 2AD \cdot DB = BG^2 = k^2 \cdot AD^2$, we obtain by adding AD^2 on both sides

$$2AD(AD + DB) = 2AD \cdot AB = (k^2 + 1)AD^2; \quad \text{hence}$$

$$AD = \frac{2AB}{k^2 + 1}, \quad \text{and} \quad BG = k \cdot AD.$$

Restriction. At the end of the problem it is pointed out that, according to what has been demonstrated in B.346 (that $d > g$), we must have $k > 1$.

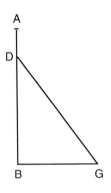

Fig. 113

(**B.347′**) MSS $\mathcal{A}$, $\mathcal{B}$. What follows in the text is a repetition, with the same data but differently formulated, of B.344 (solution and demonstration). It may have been an early gloss recapitulating B.344 (referred to in the next two problems).

(**B.348**) MSS $\mathcal{A}$, $\mathcal{B}$. A ladder of unknown length standing against a wall of the same height is withdrawn from the base of the wall by a distance which, when added to the distance by which the ladder comes down from the top, makes eight, while multiplying the two distances produces twelve; what is its length?

Given thus $d + g = 8$ and $d \cdot g = 12$, required l.

Geometrical solution.

(α) *Preliminary remark*. The subsequent geometrical solution amounts, by means of the identity

$$\left(\frac{d-g}{2}\right)^2 = \left(\frac{d+g}{2}\right)^2 - d \cdot g,$$

to calculating the left-side term, and then determining

$$d, g = \frac{d+g}{2} \pm \frac{d-g}{2} = \frac{d+g}{2} \pm \sqrt{\left(\frac{d+g}{2}\right)^2 - d \cdot g}.$$

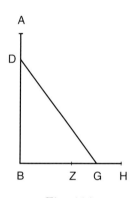

Fig. 114

(β) *Solution in the text*. Let (Fig. 114) $g = AD$, $d = BG$, with $BG + AD = 8$ and $BG \cdot AD = 12$; required $l = AB = DG$. Draw $GH = AD$ as an extension of BG; then $BH = 8$ and $BG \cdot GH = 12$. Thus we are to solve

$$\begin{cases} BG + GH = 8 \\ BG \cdot GH = 12. \end{cases}$$

BH being divided into unequal parts at G ($BG > GH = AD$), let now Z be the mid-point of BH (thus $BZ = ZH = 4$). By *Elements* II.5 = PE_5,

$$BG \cdot GH + ZG^2 = BZ^2;$$

thus ZG is known ($= 2$), therefore also $BZ + ZG = BG = 6$ and $BZ - ZG = ZH - ZG = GH = AD = 2$.

We are left with calculating the length of the ladder knowing the two displacements; this has been done in B.344 (& B.347'), with the same values.

(B.349) MSS A, B. A ladder of unknown length standing against a wall of the same height is withdrawn from the base of the wall by a distance which, when diminished by the distance by which the ladder comes down from the top, leaves four, while multiplying the two distances produces twelve; what is its length?

Given thus $d - g = 4$ and $d \cdot g = 12$, required l.

Geometrical solution.

(α) *Preliminary remark*. The subsequent geometrical solution relies on the same identities as before, giving here

$$d = \frac{d+g}{2} + \frac{d-g}{2} = \sqrt{\left(\frac{d-g}{2}\right)^2 + d \cdot g} + \frac{d-g}{2}.$$

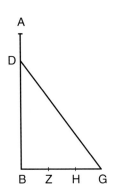

Fig. 115

(β) *Solution in the text*. Let (Fig. 115) $BG = d$, $AD = g$ with $BG - AD = 4$ and $BG \cdot AD = 12$; required $AB = DG = l$. Let us put $GH = AD$ (with H perforce within GB). Thus we are to solve

$$\begin{cases} BG - GH = 4 \\ BG \cdot GH = 12. \end{cases}$$

So let Z be the mid-point of BH (thus, since $BH = 4$, $BZ = ZH = 2$). By *Elements* II.6 = PE_6

$$BG \cdot GH + ZH^2 = ZG^2;$$

so ZG is known $(= 4)$, therefore also $ZG + BZ = BG = d = 6$ and $ZG - BZ = ZG - ZH = GH = AD = g = 2$.

We are left, for determining the length of the ladder, with the computations already performed in B.344 (& B.347′).

2. The broken tree.

If l is the initial height of the tree, g the vertical distance between root and break, and d the horizontal distance between top and root (thus $l - g$ is the length of the broken part), we have

$$(l - g)^2 = d^2 + g^2.$$

In the four problems B.350–353, d, g and l (twice) are required successively.

(B.350) MSS $\mathcal{A}$, $\mathcal{B}$; $\mathcal{D}$ partly. A tree (initially) thirty cubits high bends to the earth at a height of ten cubits; how far from the root does the top touch the ground?

Given thus $l = 30$ and $g = 10$ (so we know $l - g = 20$), required d.

(*a*) <u>Computation</u>.

$$d = \sqrt{(l-g)^2 - g^2} = \sqrt{20^2 - 10^2} = \sqrt{300}.$$

Applying the approximation formula taught for square roots (rule i p. 1316 above, with $N = 300$, $a = 17$, $N - a^2 = 11$), we find $d \cong 17 + \frac{11}{34}$.

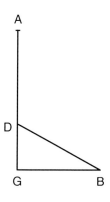

Fig. 116

<u>Demonstration</u>. Let (Fig. 116) $AG = l$, $DG = g$, $GB = d$. Thus

$$GB = \sqrt{DB^2 - DG^2},$$

where DG and $DB = AG - DG$ are known.

(*b*) Algebraic solution. We put $GB = x$ and proceed as above (or as seen in B.343*b*).

(B.351) MSS $\mathcal{A}$, $\mathcal{B}$. A tree thirty cubits high bends to the earth and its top touches the ground at a distance of ten cubits from its root; how high is its bending point?

Given thus $l = 30$ and $d = 10$, required g.

(*a*) <u>Computation</u>.

$$g = \frac{1}{l}\left(\frac{l^2 - d^2}{2}\right) = \frac{1}{30}\left(\frac{30^2 - 10^2}{2}\right) = \frac{1}{30}\frac{800}{2} = 13 + \frac{1}{3}.$$

<u>Demonstration</u>.

(α) *Preliminary remark.* The subsequent demonstration corresponds to the following. Since
$$l^2 = \left[(l-g)+g\right]^2 = (l-g)^2 + 2g(l-g) + g^2$$
while $(l-g)^2 = d^2 + g^2$, so
$$l^2 = 2g^2 + d^2 + 2g(l-g), \qquad (1) \qquad \text{thus}$$
$$g^2 + g(l-g) = \frac{l^2 - d^2}{2}, \qquad (2)$$
that is, $g \cdot l = \dfrac{l^2 - d^2}{2}$, whence $g = \dfrac{1}{l}\left(\dfrac{l^2 - d^2}{2}\right)$.

Fig. 117

(β) *Demonstration according to the text.* Let (Fig. 117) $l = AG$, $d = GB$, required $g = DG$. Since $AG^2 = AD^2 + 2AD \cdot DG + DG^2$, and thus (since $AD^2 = DB^2$) $AG^2 = DB^2 + 2AD \cdot DG + DG^2$, while $DB^2 = DG^2 + GB^2$, then
$$AG^2 = 2 \cdot DG^2 + GB^2 + 2AD \cdot DG \qquad (1)$$
and
$$DG^2 + AD \cdot DG = \frac{AG^2 - GB^2}{2} \quad (=400). \qquad (2)$$
On the other hand,
$$DG^2 + AD \cdot DG = DG(DG + AD) = DG \cdot AG, \quad \text{so}$$
$$DG = \frac{1}{AG} \frac{AG^2 - GB^2}{2}.$$
Therefore DG, being the quotient of two known quantities, is known.

Remark. A reader noted in the margin of MS $\mathcal{A}$ that the main point of this demonstration is that from $DG \cdot AG = 400$ and $AG = 30$ is inferred the required quantity DG.

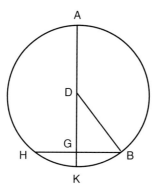

Fig. 118

(**b**) Other solution. The formula differs only in aspect from the previous one.

$$g = \frac{1}{2}\left(l + \frac{d^2}{l}\right) - \frac{d^2}{l} = \frac{1}{2}\left(30 + 3 + \frac{1}{3}\right) - \left(3 + \frac{1}{3}\right).$$

Demonstration. Let again (Fig. 118) $l = AG$, $d = GB$, $g = DG$. Since $AD = DB$, draw, with centre D, a circle passing through A and B. Let it meet at K the extension of AG and at H the extension of BG. By Elements III.35, we know that

$$AG \cdot GK = GB \cdot GH;$$

but AK and BH being perpendicular to one another, then (by Elements III.3) $GB = GH$, and therefore

$$GK = \frac{GB^2}{AG}.$$

Now GB^2 ($= 100$) and AG ($= 30$) are known, so GK is determined ($= 3 + \frac{1}{3}$). We may therefore compute $AK = AG + GK$ ($= 33 + \frac{1}{3}$), so also $AD = DK$ ($= 16 + \frac{2}{3}$), then $DG = DK - GK$ ($= 13 + \frac{1}{3}$). We have thus indeed found that

$$DG = DK - GK = \frac{1}{2}AK - GK$$

$$= \frac{1}{2}(AG + GK) - GK = \frac{1}{2}\left(AG + \frac{GB^2}{AG}\right) - \frac{GB^2}{AG}.$$

(**c**) Algebraic solution. Putting $DG = g = x$, we have $AD = 30 - x = DB$, whence the equation

$$(30 - x)^2 = x^2 + 10^2,$$

which reduces to the linear equation $60x = 800$.

(B.352) MSS 𝒜, ℬ; 𝒟 partly. A tree of unknown height bends beyond six cubits and its top touches the ground at a distance of eight cubits from its root; what is its height?

Given thus $g = 6$ and $d = 8$, required l.

Computation.
$$l = \sqrt{d^2 + g^2} + g = \sqrt{8^2 + 6^2} + 6 = 16.$$

Demonstration. Let (Fig. 119) $AB = l$, $BD = g$, $BG = d$. Then
$$AB = AD + BD = DG + BD = \sqrt{BG^2 + BD^2} + BD.$$

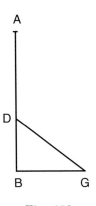

Fig. 119

(B.353) MSS 𝒜, ℬ. A tree of unknown height bends beyond its three eighths and its top touches the ground eight cubits away from its root; what is its height?

Given thus $g = \frac{3}{8} l$, $d = 8$, required l (and g).

(α) *Preliminary remark.* Since $d^2 + g^2 = (l - g)^2$, we have, with $g = k \cdot l$ (k given), $d^2 = (l - g)^2 - g^2 = l^2(1 - k)^2 - k^2 l^2 = l^2(1 - 2k)$, thus
$$l = \frac{d}{\sqrt{1 - 2k}}.$$

If in addition $d = r \cdot l$, with r given, we must impose $r = \sqrt{1 - 2k}$ and then any l is solution (B.353′, see below).

Condition. As the author says, when we suppose $g = k \cdot l$, k given, we must have $k < \frac{1}{2}$ (otherwise the part originally above the break will remain hanging, thus $d = 0$).

(β) *Solution in the text.* Let (Fig. 120) $l = AB$, $g = BD$, $d = BG$. Since $BD = k \cdot AB$, we have $AD = (1 - k)AB = DG$, whence
$$(1 - k)^2 AB^2 = DG^2 = BD^2 + BG^2 = k^2 \cdot AB^2 + BG^2$$

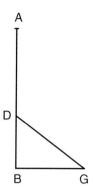

Fig. 120

$$\left(\text{numerically,} \quad \left(\frac{3}{8} + \frac{1}{8}\frac{1}{8}\right) AB^2 = \left(\frac{1}{8} + \frac{1}{8}\frac{1}{8}\right) AB^2 + BG^2\right)$$

where BG and k are known, thus

$$(1 - 2k) AB^2 = BG^2, \quad \text{or} \quad \frac{1}{4} AB^2 = BG^2,$$

that is,

$$AB = \frac{BG}{\sqrt{1-2k}} = \frac{8}{\sqrt{\frac{1}{4}}} = 16.$$

(B.353′) If we suppose that $g = \frac{3}{8}l$ (thus $l - g = \frac{5}{8}l$, as above) and, in addition, that $d = \frac{1}{2}l$, the problem will be indefinite (*multiplex*), meaning here that any l can solve it. For we are led by the Pythagorean theorem to the identity

$$\frac{25}{64}l^2 = \frac{9}{64}l^2 + \frac{16}{64}l^2.$$

3. The falling tree.

A. *Forward motion*.

(B.354) MSS $\mathcal{A}$, $\mathcal{B}$. The top of a tree thirty cubits high descends each day by one cubit; after how many days will it reach the ground?

Given thus the height of the tree $l = 30$ and the length $a = 1$ of the arc described daily by its top, required the number of days t during which the tree falls.

<u>Computation.</u> If d is the distance covered by the top, then

$$d = \left(l \cdot \frac{\pi}{2} \cong\right) \frac{1}{2} l \cdot \left(3 + \frac{1}{7}\right), \quad \text{so}$$

$$t = \frac{d}{a} \cong \frac{\frac{1}{2}l \cdot \left(3 + \frac{1}{7}\right)}{a} = 47 + \frac{1}{7} \text{ days.}$$

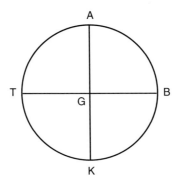

Fig. 121

<u>Demonstration</u> (of the formula for d). After noting the rather obvious fact that the trajectory of the tree-top will be the arc of a quadrant, the author tells us (Fig. 121, missing in the manuscripts) that the length of the whole circumference is obtained by multiplying the diameter $AK = 2\,AG$ by $3 + \frac{1}{7}$ (B.341). Then arc AB is a quarter of it. Therefore

$$\widehat{AB} = \frac{AK\left(3+\frac{1}{7}\right)}{4} = \frac{AG\left(3+\frac{1}{7}\right)}{2} = \frac{1}{2}AG \cdot \left(3+\frac{1}{7}\right).$$

This is then to be divided by the daily progression of the top.

B. *Forward and retrograde motion*.

(Note that B.356 is again about forward motion.)

(B.355) MS *A*. The top of a tree seventy cubits high descends each day by three cubits and rises again each day by one cubit; after how many days will it reach the ground?

Given thus the height of the tree, $l = 70$, the arc-length covered daily by its top, $a = 3$, and that of the daily retrogression, which partly compensates the downward movement, $u = 1$.

According to what we have just seen, a quarter of the circumference, thus the trajectory of the top, is $d = \frac{1}{2}l \cdot (3+\frac{1}{7}) = 110$ cubits. We are then told to compute (with the brackets meaning that the integral part is considered and r being the remainder)

$$t = \left[\frac{d-u}{a-u}\right] + \frac{u+r}{a+u} = \left[\frac{110-1}{3-1}\right] + \frac{1+1}{3+1} = \left[\frac{109}{2}\right] + \frac{1}{2} = 54 + \frac{1}{2}.$$

The origin of this formula is explained in our summary.

(B.356) MS *A*. The top of a tree of unknown height descends each day by two cubits and reaches the ground after forty-four days; what is its height?

Given thus $a = 2$, $t = 44$, required l. As in B.354, there is no retrogression, so this problem would better fit before B.355.

The distance covered by the top being $d = a \cdot t$, with d (approximately) equal to $\frac{1}{2} l \cdot (3 + \frac{1}{7})$, we have $a \cdot t = \frac{1}{2} l \cdot (3 + \frac{1}{7})$, so the required length is

$$l = 2 \cdot \frac{a \cdot t}{3 + \frac{1}{7}} = 2 \cdot \frac{2 \cdot 44}{3 + \frac{1}{7}} = 2 \cdot 28 = 56.$$

4. Planting trees.

B.357–358 deal with trees planted at regular intervals on a plot. As said in our summary, if l, r are the length and width of the plot while d is the distance between any two adjacent trees, the number q of trees is given by

$$q = \left(\frac{l}{d} + 1\right)\left(\frac{r}{d} + 1\right) = q_l \cdot q_r.$$

Such problems are known from Roman sources.[‡] In one of them, the number of trees is required and calculated as in B.357.

(B.357) MS $\mathcal{D}$ only. Trees are planted with a distance between them of two cubits, and the plot is twenty cubits long and ten wide; how many trees are there?

Given thus $l = 20$, $r = 10$, $d = 2$, required q. The number of trees is then

$$q = \left(\frac{20}{2} + 1\right)\left(\frac{10}{2} + 1\right) = 11 \cdot 6 = 66.$$

(B.358) MS $\mathcal{D}$ only. Eleven trees are planted on the length of a plot and six on its width, and the distance between them is two cubits; what are the length and width of the plot?

Given thus $q_l = 11$, $q_r = 6$ (same values as before), $d = 2$, required l and r. Since $q = q_l \cdot q_r$ with

$$q_l = \frac{l}{d} + 1, \qquad q_r = \frac{r}{d} + 1,$$

then $l = d(q_l - 1) = 2(11 - 1) = 20$, $r = d(q_r - 1) = 2(6 - 1) = 10$.

5. The two towers.

As said in our summary, if l_1, l_2 are the respective heights of two towers on flat ground ($l_1 \neq l_2$), d the distance between their bases, e the distance between their tops, we have the relation

$$(l_1 - l_2)^2 + d^2 = e^2,$$

[‡] See Cantor, *Agrimensoren*, pp. 208–209 & 119; Bubnov, *Gerberti Opera*, pp. 344–345 & 523–525.

by which are solved the three problems B.359–361, where each of the four magnitudes is required in turn.

(B.359) MSS $\mathcal{A}$, $\mathcal{B}$; $\mathcal{D}$ without the demonstration. The bases of two towers, one thirty cubits high and the other twenty, are eight cubits apart; what is the distance between their tops?

Given thus $l_1 = 30$, $l_2 = 20$, $d = 8$, required e.

Computation. From $(l_1 - l_2)^2 + d^2 = e^2$, we infer that

$$e = \sqrt{(l_1 - l_2)^2 + d^2} = \sqrt{10^2 + 8^2} = \sqrt{164} \cong 12 + \frac{20}{24} = 12 + \frac{5}{6},$$

using the usual root approximation formula (above, B.350 and rule i, p. 1316 above, with the positive sign).

Remark. In that case, with 164 being closer to 169 than to 144, we would apply the rule with the subtractive sign and thus find

$$e \cong 13 - \frac{5}{26} = 12 + \frac{21}{26},$$

a better approximation.

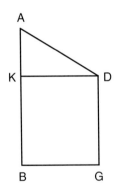

Fig. 122

Demonstration. Let (Fig. 122) $l_1 = AB$, $l_2 = DG$, $d = BG$, $e = AD$. Let KD be drawn parallel to BG. Since we know AB, $KB = DG$, $KD = BG$, then

$$AD = \sqrt{(AB - KB)^2 + KD^2} = \sqrt{AK^2 + KD^2}$$

is known.

(B.360) MSS $\mathcal{A}$, $\mathcal{B}$; $\mathcal{D}$ without the demonstration. The tops of two towers, one thirty cubits high and the other twenty, are twelve and five sixths cubits apart; what is the distance between their bases?

Given thus $l_1 = 30$, $l_2 = 20$, $e = 12 + \frac{5}{6}$, required d.

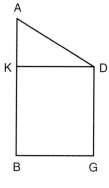

Fig. 123

Computation. From $(l_1 - l_2)^2 + d^2 = e^2$, we infer that

$$d = \sqrt{e^2 - (l_1 - l_2)^2}.$$

The author does not calculate d. Indeed, he would not obtain $d = 8$, as desired since the numerical values are supposed to be the same as in the previous problem, but an approximation.

Demonstration. Let (Fig. 123) $l_1 = AB$, $l_2 = DG$, $e = AD$, $d = BG$ (required) and, as before, draw KD parallel to BG so that $KB = DG$. Thus

$$BG = KD = \sqrt{AD^2 - (AB - KB)^2} = \sqrt{AD^2 - AK^2}.$$

(**B.361**) MSS $\mathcal{A}$, $\mathcal{B}$; $\mathcal{D}$ (omits the demonstration of *i* and all of *ii*, but has *iii*). Two towers, one eighteen cubits high and the other of unknown height, have their tops ten cubits apart and their bases, six; what is the unknown height?

Given thus the height of one of the two towers, $l = 18$, then $e = 10$ and $d = 6$, required the height of the other tower.

(*i*) First case: the tower of unknown height is the taller one; so $l_2 = 18$.

Computation (the numerical results are found within the demonstration). From $(l_1 - l_2)^2 + d^2 = e^2$, we infer that

$$l_1 = l_2 + \sqrt{e^2 - d^2} = 26.$$

Demonstration. Let (Fig. 124) $l_2 = AB$, $e = AD$, $d = BG$, required $l_1 = DG$. Drawing AH parallel to BG, we have $HG = AB$. Then

$$DG = HG + \sqrt{AD^2 - AH^2}.$$

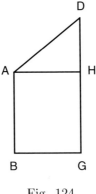

Fig. 124

(*ii*) Second case: the tower of unknown height is the lower one; so $l_1 = 18$. Computation (numerical results within the demonstration). From $\overline{(l_1 - l_2)^2 + d^2 = e^2}$, we infer that

$$l_2 = l_1 - \sqrt{e^2 - d^2} = 10.$$

Demonstration. Let (Fig. 125) $l_1 = DG$, $e = AD$, $d = BG$, required $l_2 = AB$. Let AH be drawn parallel to BG. Then

$$AB = HG = DG - DH = DG - \sqrt{AD^2 - AH^2}.$$

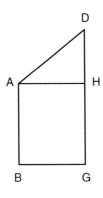

Fig. 125

(*iii*) *Restriction*. We must always have, the author adds, $e > d$; $e = d$ is excluded since the towers are supposed to be of unequal height.

* * *

Measurement of an inaccessible height

(**B.362**) MS $\mathcal{D}$ only. Measurement of the height of an object, the base of which is accessible.

Two sticks of different, known lengths are planted upright in the same vertical plane as the object and in such a way that all three upper ends lie in a straight line. Let the lengths of the two sticks be l_1 and l_2 ($l_1 > l_2$), the distance between their bases, δ, the height of the object, h, and the distance between the base of the shorter stick and that of the object, d. Then

$$h = d \cdot \frac{l_1 - l_2}{\delta} + l_2$$

which is the formula found in the text.

Demonstration (not in the text). Let (Fig. 126) $l_1 = DF$, $l_2 = GH$, $\delta = FH$, $h = AC$ and $d = CH$. Since the two triangles ABG and DEG are similar, we have

$$\frac{AB}{BG} = \frac{DE}{EG}, \quad \text{thus} \quad \frac{h - l_2}{d} = \frac{l_1 - l_2}{\delta},$$

whence the above formula.

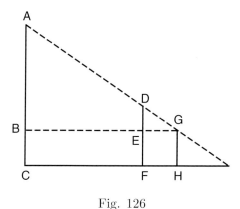

Fig. 126

Remark. We have not only added the figure but also introduced the designations. If this method was indeed originally in the *Liber mahameleth*, so surely was the demonstration we have appended. MS $\mathcal{D}$, which has transmitted the explanation of the method, would have left it out anyway, as it does all geometric demonstrations.

Chapter B–XXI:
Bundles
Summary

A rope of length l_1 surrounds a bundle of (identical) rods of which we know either the quantity q_1 or the price p_1, and another rope of length l_2 surrounds another set of rods, q_2 in quantity or with price p_2. The relations between these magnitudes are

$$\frac{q_1}{q_2} = \frac{p_1}{p_2} = \frac{l_1^2}{l_2^2}.$$

We infer the origin of this from B.363 and a gloss, or a misplaced sentence, in B.365: two circular areas being in the same ratio as the squares of their diameters (*Elements* XII.2) while the diameters are in the same ratio as their circumferences (B.354), the circular areas will be in the same ratio as the squares of their circumferences. Consequently, for identical rods, the numbers of them in the two bundles, and the corresponding prices, will be in the same ratio as the squares of the rope lengths.

In four of the five problems, just one term of the proportion is unknown. Thus, in the first, the number of rods q_2 is required (in the following problem both terms are unknown and only their ratio is sought); in the third, the price p_2 is required; finally, in the last two problems, l_2 is unknown, which leads us to square roots. The formulae are accordingly

$$q_2 = \frac{l_2^2}{l_1^2} \cdot q_1, \qquad \frac{q_2}{q_1} = \frac{l_2^2}{l_1^2}, \qquad \text{(B.363, B.364)}$$

$$p_2 = \frac{l_2^2}{l_1^2} \cdot p_1, \qquad \text{(B.365)}$$

$$l_2 = \sqrt{\frac{q_2}{q_1}} \cdot l_1, \qquad l_2 = \sqrt{\frac{p_2}{p_1}} \cdot l_1. \qquad \text{(B.366, B.367)}$$

Remark. The length unit is the cubit, as in B–XX, and once the palm.

(**B.363**) MSS $\mathcal{A}$, $\mathcal{B}$, $\mathcal{D}$. A rope four cubits long surrounds a bundle of one hundred rods; how many rods will be surrounded by a rope ten cubits long?

Given thus $l_1 = 4$, $q_1 = 100$, $l_2 = 10$, required q_2.

(*a*) As seen above,

$$q_2 = \frac{l_2^2 \cdot q_1}{l_1^2} = \frac{10^2 \cdot 100}{4^2} = \frac{10\,000}{16} = 625.$$

(*b*) Also computed as

$$q_2 = \frac{l_2^2}{l_1^2} \cdot q_1 = \left(6 + \frac{1}{4}\right)100.$$

The justification asserts that the two quantities of rods are in the same ratio as the squares of the ropes' lengths, which does not justify much. Referring the reader to B.149 (p. 1463 above) would also have been appropriate.

(**B.364**) MSS $\mathcal{A}$, $\mathcal{B}$, $\mathcal{D}$. Two ropes, one four cubits long and the other twelve, surround two bundles; how many times is the smaller contained in the larger?

Given thus $l_1 = 4$, $l_2 = 12$, required $q_2 : q_1$.

$$\frac{q_2}{q_1} = \frac{l_2^2}{l_1^2} = \frac{144}{16} = 9.$$

Remark. This problem is closely related to the next, where the computation of this ratio forms a preliminary step.

(**B.365**) MSS $\mathcal{A}$, $\mathcal{B}$, $\mathcal{D}$. A rope four palms long surrounds a bundle of harvest costing half a nummus; what will be the price of a bundle surrounded by a rope twelve palms long?

Given thus $l_1 = 4$, $p_1 = \frac{1}{2}$, $l_2 = 12$ required p_2.

$$p_2 = \frac{l_2^2}{l_1^2} \cdot p_1 = \frac{144}{16} \cdot \frac{1}{2} = 9 \cdot \frac{1}{2} = 4 + \frac{1}{2}.$$

(**B.366**) MSS $\mathcal{A}$, $\mathcal{B}$, $\mathcal{D}$. A rope ten cubits long surrounds one thousand rods; what is the length of a rope surrounding two hundred and fifty rods?

Given thus $l_1 = 10$, $q_1 = 1000$, $q_2 = 250$, required l_2.

$$l_2^2 = \frac{q_2 \cdot l_1^2}{q_1} = \frac{250 \cdot 100}{1000} = 25, \qquad l_2 = 5.$$

(**B.367**) MSS $\mathcal{A}$, $\mathcal{B}$, $\mathcal{D}$. A rope three cubits long surrounds a bundle costing eighteen nummi; what will be the length of a rope surrounding a bundle costing two nummi?

Given thus $l_1 = 3$, $p_1 = 18$, $p_2 = 2$, required l_2.

$$l_2^2 = \frac{p_2 \cdot l_1^2}{p_1} = \frac{2 \cdot 9}{18} = 1.$$

Chapter B–XXII:
Messengers

Summary

As in the case of the falling tree (B–XX, Section 3), the motion is considered to be uniform throughout.

1. Forward motion.

A first courier advances by a_1 miles daily and a second, advancing daily by $a_2 > a_1$, is sent after τ days. As seen in B.368d, the number of days travelled by each until they meet is obtained by setting equal the distance d covered by each during the number t_i of days he has been on his way:

$$t_1 \cdot a_1 = t_2 \cdot a_2 = d, \quad \text{with} \quad t_1 = t_2 + \tau.$$

Then $(t_2 + \tau)\, a_1 = t_2 \cdot a_2$ and $(t_2 + \tau)\, a_1 = d$ give, respectively,

$$t_2 = \frac{\tau \cdot a_1}{a_2 - a_1}, \qquad t_2 = \frac{d - \tau \cdot a_1}{a_1}.$$

The first relation is used and demonstrated in B.368, where we are given a_1, a_2, τ and asked for t_2, the second is found in B.369, where we are given a_1, τ, d and asked for a_2 (once t_2 is known, then $a_2 \cdot t_2 = d$).

2. Forward and retrograde motion.

We have already established the formulae used in such problems (see our summary of B–XX, pp. 1673–1674). The general formula for the number of days (a daily progression, u daily retrogression), namely

$$\frac{d - a}{a - u} = q + \frac{r}{a - u},$$

took two forms, according to whether the remainder of the division was $r = 0$ or $r \neq 0$, namely, respectively,

$$t = \frac{d - u}{a - u} - \frac{u}{a + u}, \qquad t = \left[\frac{d - u}{a - u}\right] + \frac{u + r}{a + u}.$$

The first problem in this section (B.370) concerns a boat covering a given distance, the other four (B.371–374) a snake coming out of a hole the same length as itself. The second formula is correctly applied in the case of the boat, but the treatment of the other problems is faulty. The snake problems B.371, B.372, B.374 are found almost word for word, thus faulty as well, in an Arabic source, the *Arithmetic* of the Egyptian 'Alī the Reckoner.°

° '*Alī al-ḥāsib.* See our *Recueil du XIIIe siècle*, pp. 120 & 131–132.

1. Forward motion.

(**B.368**) MSS $\mathcal{A}$, $\mathcal{B}$; $\mathcal{D}$, but only the computation in part a. A messenger advances daily by twenty miles, and another, sent five days later, advances daily by thirty miles; when will they meet?

Given thus $a_1 = 20$, $\tau = 5$, $a_2 = 30$, required t_2 (and t_1).

(**a**) Computing first t_2.
$$t_2 = \frac{\tau \cdot a_1}{a_2 - a_1} = \frac{5 \cdot 20}{10} = 10,$$
so that $t_1 = t_2 + \tau = 15$.

<u>Demonstration</u>. Let (Fig. 127) $t_1 = AB$, $\tau = GB = 5$, thus $AG = t_2$. From $a_2 \cdot t_2 = a_1 \cdot t_1$, or $30 \cdot AG = 20 \cdot AB$, we infer that
$$30 \cdot AG = 20 \cdot AB = 20\,(AG + GB) = 20 \cdot AG + 100,$$
whence
$$AG = t_2 = \frac{100}{30 - 20} = \frac{\tau \cdot a_1}{a_2 - a_1} = 10.$$

A   G B

Fig. 127

(**b**) Computing first t_1. Since $30 \cdot AG = 20 \cdot AB$, so
$$AG = \frac{2}{3} AB, \quad \text{therefore} \quad GB = \frac{1}{3} AB,$$
whence $AB = 3 \cdot GB = 15$, $AG = 10$, that is, $t_1 = 3 \cdot \tau$ and $t_2 = 2 \cdot \tau$.

(**c**) Algebraic solution. Put $t_2 = x$, then $30\,x = (x+5)\,20 = 20\,x + 100$, whence $x = 10$.

<u>Remark</u>. This is the solution presented by Abū Kāmil for the very same problem.*

(**d**) Verification. Checking that the number of miles covered is indeed the same in both cases. Since the first messenger covers 20 miles daily and walks for 15 days, he covers 300 miles. Since the second messenger covers 30 miles daily and walks for 10 days, he covers 300 miles, and overtakes the first one.

(**B.369**) MSS $\mathcal{A}$, $\mathcal{D}$ (from here on $\mathcal{D}$ has all the problems while $\mathcal{B}$ has none since the end leaves are missing). A messenger, advancing daily by twenty miles, is to reach a town four hundred miles away, and another, sent fifteen days later, meets him at the entry of this town; how many miles had he to cover daily?

---

 * *Algebra*, fol. 103^r, p. 205 of the facsimile edition.

Given thus $a_1 = 20$, $\tau = 15$, $d = 400$, required a_2.
The text computes successively

$$t_2 \left(= t_1 - \tau = \frac{d}{a_1} - \tau\right) = \frac{d - \tau \cdot a_1}{a_1} = \frac{400 - 15 \cdot 20}{20} = 5,$$

$$a_2 = \frac{d}{t_2} = \frac{400}{5} = 80.$$

---

2. Forward and retrograde motion.

(B.370) MSS $\mathcal{A}$, $\mathcal{D}$. A ship, which is to cover a distance of three hundred miles, moves forward daily twenty miles but is driven back by the wind five miles daily; in how many days will it reach its destination?

Given thus $d = 300$, with $a = 20$, $u = 5$, required t.

Dividing $d - a$ by $a - u$ gives 18 and a remainder $r = 10$, thus the second formula seen in our summary applies. Accordingly, the text computes

$$t = \left[\frac{d-u}{a-u}\right] + \frac{u+r}{a+u} = \left[\frac{300-5}{20-5}\right] + \frac{5+10}{20+5} = \left[\frac{295}{15}\right] + \frac{15}{25} = 19 + \frac{3}{5}.$$

(B.371) MSS $\mathcal{A}$, $\mathcal{D}$. A snake comes out of a hole daily by a third of its length and returns daily by a fourth of its length; in how many days will it be out?

Given thus $a = \frac{1}{3}d$ and $u = \frac{1}{4}d$ where d is the length of the snake, and t is required.

In these problems with a snake, where $d - a$ is divisible by $a - u$ and so $r = 0$, the author computes as before, and therefore uses the wrong formula

$$t' = \frac{d-u}{a-u} + \frac{u}{a+u} \quad \text{instead of} \quad t = \frac{d-u}{a-u} - \frac{u}{a+u};$$

that is, he adds the fraction which should be subtracted. What he finds in the present case, instead of $8 + \frac{4}{7}$, is then (by introducing the factor $f = 3 \cdot 4 = 12$ to cancel the denominators)

$$t' = \frac{(1-\frac{1}{4})\,12}{(\frac{1}{3}-\frac{1}{4})\,12} + \frac{\frac{1}{4}\cdot 12}{(\frac{1}{3}+\frac{1}{4})\,12} = \frac{12-3}{4-3} + \frac{3}{4+3} = 9 + \frac{3}{7}.$$

(B.372) MSS $\mathcal{A}$, $\mathcal{D}$. A snake seven cubits long comes out of a hole daily by one cubit and returns daily by a third of a cubit; in how many days will it be out?

Given thus, in cubits, $d = 7$, with $a = 1$, $u = \frac{1}{3}$. Proceeding as before, the author finds, instead of $t = 9 + \frac{3}{4}$,

$$t' = \frac{7 - \frac{1}{3}}{1 - \frac{1}{3}} + \frac{\frac{1}{3}}{1 + \frac{1}{3}} = 10 + \frac{1}{4}.$$

(**B.373**) MSS $\mathcal{A}$, $\mathcal{D}$. A snake comes out of a hole daily by a third of its length, returns daily by an unknown length, and is completely out after nine days and three sevenths; what is the unknown fraction by which it returns?

Given thus $a = \frac{1}{3}d$, with d the length of the snake, $t' = 9 + \frac{3}{7}$, required u. This problem being related to B.371 for the data (including the wrong answer), the author had recourse, in order to find the same u, to a meaningless formula, involving both the correct time t and the erroneous t':

$$u = \frac{a \cdot t' - d}{t} = \frac{\frac{1}{3}\left(9 + \frac{3}{7}\right) - 1}{9 - \frac{3}{7}} = \frac{2 + \frac{1}{7}}{8 + \frac{4}{7}} = \frac{1}{4} \quad \left(\text{that is, } \frac{1}{4}d\right).$$

(**B.374**) MSS $\mathcal{A}$, $\mathcal{D}$. A snake seven cubits long, coming out of a hole daily by one cubit and returning daily by an unknown fraction, is completely out after ten days and a fourth; what is the unknown fraction?

Given thus $d = 7$, $a = 1$, $t' = 10 + \frac{1}{4}$, required u. Related to B.372; here too, the wrong formula leads to the right answer:

$$u = \frac{a \cdot t' - d}{t} = \frac{1\left(10 + \frac{1}{4}\right) - 7}{10 - \frac{1}{4}} = \frac{3 + \frac{1}{4}}{9 + \frac{3}{4}} = \frac{1}{3}.$$

Chapter B–XXIII:
Mutual lending

Summary

The two types of problem closing the *Liber mahameleth*, involving a group of men, are commonly found in mathematical treatises of the Middle Ages. In the first, each partner, taken successively, doubles the amount of each of the others (B.375–378); in the second, each partner wishes to buy a horse but, not having enough money, borrows either from his immediate neighbour (B.379–380) or from all the other partners (B.381).‖

1. Successive doubling.

In a group of n partners each one doubles successively the amount of the others, and the final amounts are equal. The rule used, the origin of which is not explained, was no doubt found by reckoning backwards, starting from the final, equal amounts.

Indeed, suppose that $S_1, S_2, \cdots, S_n$, are the initial amounts (with $\sum S_i = S$), C_i the amount owned by the ith partner just before doubling those of the others, while C, with $n \cdot C = S$, represents the equal amounts at the end. Before distribution by the nth partner, the respective amounts must have been

$$\frac{1}{2}C, \ \frac{1}{2}C, \cdots, \ \frac{1}{2}C, \ C_n,$$

and therefore the last partner had

$$C_n = \frac{n-1}{2}C + C = \frac{n+1}{2}C.$$

Before distribution by the $(n-1)$th partner, the amounts were

$$\frac{1}{4}C, \ \frac{1}{4}C, \cdots, \ \frac{1}{4}C, \ C_{n-1}, \ \frac{n+1}{4}C,$$

and therefore

$$C_{n-1} = \frac{n-2}{4}C + \frac{1}{2}C + \frac{n+1}{4}C = \frac{2n+1}{4}C.$$

Before distribution by the $(n-2)$th partner, the amounts were

$$\frac{1}{8}C, \ \frac{1}{8}C, \cdots, \ \frac{1}{8}C, \ C_{n-2}, \ \frac{2n+1}{8}C, \ \frac{n+1}{8}C,$$

‖ For examples of such occurrences see Tropfke's *Geschichte* (ed. 1980), pp. 608–611.

and therefore

$$C_{n-2} = \frac{n-3}{8}C + \frac{1}{4}C + \frac{2n+1}{8}C + \frac{n+1}{8}C = \frac{4n+1}{8}C.$$

From this we may already infer that the initial amount of the ith partner must have been

$$S_i = \frac{2^{n-i} \cdot n + 1}{2^n} C.$$

Since the problem is indeterminate, we may set $C = 2^n$. This gives the integral solution

$$S_i = 2^{n-i} \cdot n + 1,$$

or, as the text expresses it (B.375–376),

$$S_n = n+1, \quad S_{n-1} = 2\,S_n - 1, \quad S_{n-2} = 2\,S_{n-1} - 1, \quad \ldots \, .$$

2. Buying a horse by borrowing from partners.

Each of n partners wishes to buy a horse but has not enough money to pay the price. He will, however, be able to buy it by borrowing from his partners a certain fraction of what they possess. In B.379–380 each partner borrows from the next one, in B.381 from all the others.

A. *Each borrows from his neighbour*.

The formula used to solve the first two problems, not explained in the text, can be established as follows. Let generally y be the price of the horse, x_i the amount of the ith partner, and $\frac{1}{q_i}$ the corresponding given fraction. The system is then

$$x_i + \frac{1}{q_i}x_{i+1} = y. \qquad (i = 1, \ldots, n, \text{ cyclically})$$

Therefore

$$\begin{aligned}
x_{i+1} &= q_i \cdot y - q_i \cdot x_i = q_i \cdot y - q_i\big[q_{i-1} \cdot y - q_{i-1} \cdot x_{i-1}\big] \\
&= \big[q_i - q_i q_{i-1}\big]y + q_i q_{i-1} \cdot x_{i-1} \\
&= \big[q_i - q_i q_{i-1} + q_i q_{i-1} q_{i-2}\big]y - q_i q_{i-1} q_{i-2} \cdot x_{i-2} \\
&= \ldots \\
&= \big[q_i - q_i q_{i-1} \pm \ldots + (-1)^{n+1} q_i q_{i-1} \ldots q_{i+1}\big]y + (-1)^n q_i q_{i-1} \ldots q_{i+1} x_{i+1}
\end{aligned}$$

and we may thus write

$$x_{i+1}\left((-1)^{n+1}\prod_{k=1}^{n} q_k + 1\right) = y\left[\sum_{j=0}^{n-1}(-1)^j \prod_{l=0}^{j} q_{i-l}\right]$$

or

$$x_{i+1}\left(\prod_{k=1}^{n} q_k + (-1)^{n+1}\right) = y\left[\sum_{j=0}^{n-1}(-1)^{n+(j+1)} \prod_{l=0}^{j} q_{i-l}\right].$$

Identifying the factors, we obtain

— if n is even

$$y = \prod_{k=1}^{n} q_k - 1 = q_1 \cdot q_2 \cdot \ldots \cdot q_n - 1$$

$$x_{i+1} = \sum_{j=0}^{n-1} (-1)^{j+1} \prod_{l=0}^{j} q_{i-l}$$

$$= q_i \cdot q_{i-1} \cdot \ldots \cdot q_{i+1} - q_i \cdot q_{i-1} \cdot \ldots \cdot q_{i+2} \pm \ldots + q_i \cdot q_{i-1} - q_i$$

$$= q_i \left\{ q_{i-1} \left[q_{i-2} \left(\ldots \left[q_{i+1} - 1 \right] \ldots \right) + 1 \right] - 1 \right\};$$

— if n is odd

$$y = \prod_{k=1}^{n} q_k + 1 = q_1 \cdot q_2 \cdot \ldots \cdot q_n + 1$$

$$x_{i+1} = \sum_{j=0}^{n-1} (-1)^{j} \prod_{l=0}^{j} q_{i-l}$$

$$= q_i \cdot q_{i-1} \ldots q_{i+1} - q_i \cdot q_{i-1} \cdot \ldots \cdot q_{i+2} \pm \ldots - q_i \cdot q_{i-1} + q_i$$

$$= q_i \left\{ q_{i-1} \left[q_{i-2} \left(\ldots \left[q_{i+1} - 1 \right] \ldots \right) - 1 \right] + 1 \right\}.$$

One value x_{i+1} and y being thus determined (to a multiplicative factor if the value of y is not imposed), we can deduce the remaining unknowns more simply, by means of the equations of the system, for we shall compute successively

$$x_{i+2} = q_{i+1}(y - x_{i+1})$$

$$x_{i+3} = q_{i+2}(y - x_{i+2})$$

and so on.

Such are, in a general form, the formulae used in the text and applied to $n = 3$ and $n = 4$; the author points out the link between the parity of n and the sign preceding 1 in the formula for y.

B. *Each borrows from all partners*.

In the second situation, where each partner borrows from all the others (B.381), the equations are

$$x_i + \frac{1}{q_i} \sum_{k \neq i} x_k = y. \qquad (i = 1, \ldots, n \text{ cyclically}; \ q_i > 1)$$

Such a problem is often solved in mediaeval mathematics by a formula obtained with a reasoning which corresponds, in modern symbolism, to the following. (The *Liber mahameleth* follows another way.)

Adding to both sides $\sum_{k \neq i} x_k$ and putting $S = \sum_l x_l$, we find

$$S = \left(1 - \frac{1}{q_i}\right) \sum_{k \neq i} x_k + y,$$

whence

$$\sum_{k \neq i} x_k = \frac{q_i}{q_i - 1}(S - y). \qquad (*)$$

Adding now all these n equations, in each of which one of the x_i's is lacking, we shall obtain

$$(n-1)S = (S-y) \sum_l \frac{q_l}{q_l - 1}, \quad \text{so that}$$

$$S = \frac{S-y}{n-1} \sum_l \frac{q_l}{q_l - 1}. \qquad (**)$$

Using the two relations $(*)$ and $(**)$, we obtain finally

$$x_i = S - \sum_{k \neq i} x_k = (S - y)\left[\frac{1}{n-1} \sum_l \frac{q_l}{q_l - 1} - \frac{q_i}{q_i - 1}\right].$$

For a solution, we must certainly have $S > y$, otherwise the problem makes no sense (the whole amount being either insufficient to buy the horse or just equal to its price). Still, it may happen that the expression in brackets, depending on the values chosen for the q_i's, takes a negative value. As a matter of fact, such problems came to be closely connected with the acceptance of negative numbers in the fifteenth century.[†]

1. Successive doubling.

(B.375) MSS $\mathcal{A}$, $\mathcal{D}$. Each of three men doubled successively the sums of money of the others, and they were found in the end to possess equal amounts; how much did each possess at the outset?

Given thus $n = 3$. Then, if S_i is the initial amount of the ith partner, we shall compute successively

$S_3 = n + 1 = 4$,
$S_2 = 2 S_3 - 1 = 7$,
$S_1 = 2 S_2 - 1 = 13$.

This rule is general, the author concludes. It will be illustrated by the next problem, while from B.377 it will appear that any multiple of a solution is also a solution.

[†] See our study on the appearance of negative solutions (or our *Introduction to the History of Algebra*, pp. 103–117).

(B.376) MSS $\mathcal{A}$, $\mathcal{D}$. Each of four men doubled successively the sums of money of the others, and they were found in the end to possess equal amounts; how much did each possess at the outset?

Given now $n = 4$. We find then

$S_4 = n + 1 = 5$,
$S_3 = 2\, S_4 - 1 = 9$,
$S_2 = 2\, S_3 - 1 = 17$,
$S_1 = 2\, S_2 - 1 = 33$.

(B.377) MSS $\mathcal{A}$, $\mathcal{D}$. Each of three men, having altogether seventy-two nummi, doubled successively the sums of money of the others, and they were found in the end to possess equal amounts; how much did each possess at the outset?

Here again $n = 3$, but, unlike in B.375, $S = 72$ is imposed. Since we have found in B.375 $S = \sum S_i = 24$ instead of the required $S = 72$, and any multiple of a solution is a solution, we need only to multiply the result found before by $72 : 24 = 3$. Thus the required quantities are $S_3 = 12$, $S_2 = 21$, $S_1 = 39$.

(B.378) MSS $\mathcal{A}$, $\mathcal{D}$. Each of three men doubled successively the sums of money of the others, and the first was found in the end to possess as much as the second plus two nummi, and the second as much as the third plus one nummus; how much did each possess at the outset?

Here the final amounts are no longer equal and their differences are given. In this case, the author mentions explicitly that the initial amounts will be found by computing backwards (*secundum almencuz*, the latter transcribing the Arabic *al-mankūs*). In other words, we start from the final situation and return step by step to the initial one.‖

Let C_i be the final amounts. According to the hypotheses, we have $C_2 = C_3 + 1$, $C_1 = C_2 + 2 = C_3 + 3$. Now, since the problem is indeterminate, one of the C's may be chosen. The author takes $C_3 = 5$, whence $C_2 = 6$, $C_1 = 8$.

Before the third and last distribution, by the third man, both the first and the second had half of what they had later and the other halves belonged to the third; thus the amounts were

$$C_1''' = \frac{C_1}{2} = 4, \quad C_2''' = \frac{C_2}{2} = 3, \quad C_3''' = \frac{C_1}{2} + \frac{C_2}{2} + C_3 = 12.$$

‖ Instances of such a way of computing are indeed found in Arabic mathematical literature. See al-Karajī's *Badīʿ* (ed. Anbouba, fol. 114^r–115^v, pp. 329-331 & 363-364 in our commented translation) or Tropfke (ed. 1980), p. 643. This *regula versa* (as opposed to the *regula recta*, the 'direct' way) is used several times in the *Liber abaci* by Leonardo Fibonacci (Latin text, pp. 203–204, 260, 263, 265). The *Liber augmenti et diminutionis* calls it *regula infusa* (Libri, I, pp. 312 & 320). Such a way, but with a geometrical computation, has been used in A.213.

Before distribution by the second the amounts were

$$C_1'' = \frac{C_1''''}{2} = 2, \quad C_2'' = \frac{C_1''''}{2} + C_2'''' + \frac{C_3''''}{2} = 11, \quad C_3'' = \frac{C_3''''}{2} = 6.$$

Finally, before the first distribution, the partners had

$$C_1' = C_1'' + \frac{C_2'''}{2} + \frac{C_3'''}{2} = 10 + \frac{1}{2}, \quad C_2' = \frac{C_2'''}{2} = 5 + \frac{1}{2}, \quad C_3' = \frac{C_3'''}{2} = 3,$$

and such were the initial amounts.

---

2. Buying a horse by borrowing from partners.

A. *Each borrows from his neighbour*.

(B.379) MSS $\mathcal{A}$, $\mathcal{D}$. Of three men each wanted to buy a horse but could attain its price only by borrowing a given fraction of his neighbour's property: the first by borrowing half of the second's money, the second by borrowing a third of the third's, the third by borrowing a fourth of the first's; how much did each have and what was the price of the horse?

We are thus to solve, with $n = 3$, the indeterminate system:

$$\begin{cases} x_1 + \dfrac{1}{q_1} x_2 = y \\ x_2 + \dfrac{1}{q_2} x_3 = y \\ x_3 + \dfrac{1}{q_3} x_1 = y \end{cases}$$

with $q_1 = 2$, $q_2 = 3$, $q_3 = 4$. According to the computations of the text, we shall form the product $q_1 \cdot q_2 \cdot q_3$. If n is odd, the price of the horse y will be found, we are told, by adding 1 to this product, whereas 1 will be subtracted if n is even. Thus, in our case, form

$$y = q_1 \cdot q_2 \cdot q_3 + 1 = 2 \cdot 3 \cdot 4 + 1 = 24 + 1 = 25.$$

The respective parts are then computed as

$$x_1 = q_3 \left(q_2 \left[q_1 - 1\right] + 1\right) = 4 \left(3 \left[2 - 1\right] + 1\right) = 4 \left(3 + 1\right) = 16,$$
$$x_2 = q_1 \left(y - x_1\right) = 2 \left(25 - 16\right) = 18,$$
$$x_3 = q_2 \left(y - x_2\right) = 3 \left(25 - 18\right) = 21.$$

(B.380) MSS $\mathcal{A}$, $\mathcal{D}$. Of four men each wanted to buy a horse but could attain its price only by borrowing a given fraction of his neighbour's property: the first by borrowing half of the second's money, the second by borrowing a third of the third's, the third by borrowing a fourth of the fourth's, the fourth by borrowing a fifth of the first's; how much did each have and what was the price of the horse?

The proposed indeterminate system is now

$$\begin{cases} x_1 + \dfrac{1}{q_1} x_2 = y \\ x_2 + \dfrac{1}{q_2} x_3 = y \\ x_3 + \dfrac{1}{q_3} x_4 = y \\ x_4 + \dfrac{1}{q_4} x_1 = y, \end{cases}$$

thus with $n = 4$, even, and $q_1 = 2$, $q_2 = 3$, $q_3 = 4$, $q_4 = 5$. We find by analogy to the previous case, except that 1 is subtracted,

$$y = q_1 \cdot q_2 \cdot q_3 \cdot q_4 - 1 = 2 \cdot 3 \cdot 4 \cdot 5 - 1 = 120 - 1 = 119,$$

and for the parts

$$x_1 = q_4 \left\{ q_3 \left(q_2 \left[q_1 - 1 \right] + 1 \right) - 1 \right\} = 5 \left\{ 4 \left(3 \left[2 - 1 \right] + 1 \right) - 1 \right\} = 75,$$
$$x_2 = q_1 \left(y - x_1 \right) = 2 \left(119 - 75 \right) = 88,$$
$$x_3 = q_2 \left(y - x_2 \right) = 3 \left(119 - 88 \right) = 93,$$
$$x_4 = q_3 \left(y - x_3 \right) = 4 \left(119 - 93 \right) = 104.$$

B. *Each borrows from all partners*.

(B.381) MSS $\mathcal{A}$, $\mathcal{D}$. Of four men each wanted to buy a horse but could attain its price only by borrowing from his partners a given fraction of their property: the first by borrowing half of the others' money, the second by borrowing a third, the third by borrowing a fourth, the fourth by borrowing a fifth; how much did each have and what was the price of the horse?

This example is found, we are told, in Abū Kāmil's *Algebra*; indeed, there it opens the part devoted to problems of a more recreational nature.°
Since $n = 4$ with $q_1 = 2$, $q_2 = 3$, $q_3 = 4$, $q_4 = 5$, we are thus to solve

$$\begin{cases} x_1 + \dfrac{1}{2}(x_2 + x_3 + x_4) = y & (1) \\ x_2 + \dfrac{1}{3}(x_3 + x_4 + x_1) = y & (2) \\ x_3 + \dfrac{1}{4}(x_4 + x_1 + x_2) = y & (3) \\ x_4 + \dfrac{1}{5}(x_1 + x_2 + x_3) = y. & (4) \end{cases}$$

° *Algebra*, fol. 95^r–97^r (pp. 189–193 of the facsimile edition).

The solution found in the *Liber mahameleth*, said to be algebraic though different from Abū Kāmil's, proceeds as follows. Since the problem is indeterminate, we may assume $x_1 = 1$; putting $x_2 + x_3 + x_4 = x$, we have then, if S represents the sum of the amounts,

$$S = \sum x_j = x + 1.$$

Then the price of the horse and the individual amounts are computed in the text as follows.

$$y = \frac{1}{2}x + 1,$$

$$x_2 = \frac{1}{2}(3 \cdot y - S) = \frac{1}{2}\left(\frac{3}{2}x + 3 - x - 1\right) = \frac{1}{2}\left(\frac{1}{2}x + 2\right) = \frac{1}{4}x + 1,$$

$$x_3 = \frac{1}{3}(4 \cdot y - S) = \frac{1}{3}\left(2x + 4 - x - 1\right) = \frac{1}{3}\left(x + 3\right) = \frac{1}{3}x + 1,$$

$$x_4 = \frac{1}{4}(5 \cdot y - S) = \frac{1}{4}\left(\frac{5}{2}x + 5 - x - 1\right) = \frac{1}{4}\left(\frac{3}{2}x + 4\right) = \frac{3}{8}x + 1.$$

Remark. There is no justification whatsoever for these computations. The expression for y is an obvious inference from equation (1) since $x_1 = 1$ and $x_2 + x_3 + x_4 = x$. To determine the other individual amounts, the remaining equations (2) to (4) are transformed by multiplying them by the denominators of their fractions. This gives for example, for equation (2),

$$3x_2 + (x_3 + x_4 + x_1) = 3y;$$

since the left side equals $2x_2 + S$, thus $2x_2 + x + 1$, we have

$$2x_2 + x + 1 = 3y = \frac{3}{2}x + 3,$$

whence the above, final expression for x_2. The expressions for x_3 and x_4 are similarly found by equating $3x_3 + S$ to $4y$ and $4x_4 + S$ to $5y$.

The text then proceeds with the addition of the expressions found for x_2, x_3, x_4, thus finding

$$x = x_2 + x_3 + x_4 = \left(\frac{5}{6} + \frac{3}{4}\frac{1}{6}\right)x + 3 = \frac{23}{24}x + 3, \quad \text{or} \quad \frac{1}{24}x = 3,$$

whence $x = 24 \cdot 3 = 72$, so that finally, substituting the value of x in the expressions found, $y = 37$, $(x_1 = 1,)$ $x_2 = 19$, $x_3 = 25$, $x_4 = 28$.

Commentary. Since we are told that the algebraic solution of this problem is different from Abū Kāmil's, we may compare them. As a matter of fact, Abū Kāmil presents two ways of solving.
Let us again consider the system

$$\begin{cases} x_1 + \dfrac{1}{2}(x_2 + x_3 + x_4) = y & (1) \\ x_2 + \dfrac{1}{3}(x_3 + x_4 + x_1) = y & (2) \\ x_3 + \dfrac{1}{4}(x_4 + x_1 + x_2) = y & (3) \\ x_4 + \dfrac{1}{5}(x_1 + x_2 + x_3) = y. & (4) \end{cases}$$

(α) *Abū Kāmil's first treatment.* This is, he says, the way 'mathematicians' usually proceed. Equating the first two equations, we find

$$x_2 = \frac{4}{3}x_1 + \frac{1}{3}x_3 + \frac{1}{3}x_4 \qquad (5)$$

which we introduce into the second equation, whence

$$y = \frac{5}{3}x_1 + \frac{2}{3}x_3 + \frac{2}{3}x_4. \qquad (6)$$

Equating (3) and (6), we obtain

$$x_3 = \frac{13}{5}x_1 + \frac{4}{5}x_4.$$

This may now be introduced into equations (5) and (6); whence, for x_2 and y,

$$x_2 = \frac{11}{5}x_1 + \frac{3}{5}x_4$$
$$y = \frac{17}{5}x_1 + \frac{6}{5}x_4.$$

Introducing finally into (4), the remaining equation, the relations just found for x_3, x_2, y, we shall have

$$x_4 = 28x_1.$$

Putting, for instance, $x_1 = 1$ gives $x_2 = 19$, $x_3 = 25$, $x_4 = 28$, $y = 37$.

In order to perform these computations, each unknown must have its own designation. Thus, after *shay'* for x_1, the designations for x_2, x_3 and x_4 are, respectively, *dīnār*, *dirham* and *fals*, which are names of coins, as is commonly the case in Arabic mathematical literature.* Were there more unknowns, we would have to introduce further names. Thus Abū Kāmil concludes:† *This treatment is that acknowledged among the mathematicians. It is a correct procedure though intricate for the user, particularly with a growing number (of partners).*

(β) *Abū Kāmil's second treatment.* This treatment, which he attributes to himself, is said to be simpler and more appropriate for a greater number of unknowns.

Put $x_1 + x_2 + x_3 + x_4 = S$. Since $x_2 + x_3 + x_4 = S - x_1$, equation (1) becomes

$$y = \frac{1}{2}x_1 + \frac{1}{2}S. \qquad (*)$$

Similarly, since $x_3 + x_4 + x_1 = S - x_2$, equation (2) becomes

$$y = \frac{2}{3}x_2 + \frac{1}{3}S.$$

---

* See, e.g., our *Introduction to the History of Algebra*, p. 76.
† Arabic text, fol. 96^v, 1–3 (p. 192 of the facsimile).

Equating these two expressions gives

$$\frac{2}{3}x_2 = \frac{1}{2}x_1 + \frac{1}{6}S, \quad \text{and thus} \quad x_2 = \frac{3}{4}x_1 + \frac{1}{4}S,$$

and x_2 is now expressed, as was y, by means of x_1 and S.
Since $x_4 + x_1 + x_2 = S - x_3$, equation (3) becomes

$$y = \frac{3}{4}x_3 + \frac{1}{4}S,$$

which, being equated to $(*)$, leads to

$$\frac{3}{4}x_3 = \frac{1}{2}x_1 + \frac{1}{4}S, \quad \text{and thus} \quad x_3 = \frac{2}{3}x_1 + \frac{1}{3}S,$$

and x_3 is expressed by means of x_1 and S.
Therefore, we have

$$x_4 = S - (x_1 + x_2 + x_3) = S - \left[x_1 + \left(\frac{3}{4}x_1 + \frac{1}{4}S\right) + \left(\frac{2}{3}x_1 + \frac{1}{3}S\right)\right]$$

$$= S - \left[\frac{7}{12}S + \frac{29}{12}x_1\right] = \frac{5}{12}S - \frac{29}{12}x_1.$$

With these last computations we may write equation (4) as

$$\frac{5}{12}S - \frac{29}{12}x_1 + \frac{1}{5}\left[\frac{7}{12}S + \frac{29}{12}x_1\right] = y,$$

or

$$\frac{32}{60}S - \frac{116}{60}x_1 = y.$$

Thus, by $(*)$,

$$\frac{32}{60}S - \frac{116}{60}x_1 = \frac{1}{2}x_1 + \frac{1}{2}S,$$

whence $S = 73x_1$. Putting $x_1 = 1$, we obtain the above results.

As we see, in Abū Kāmil's second treatment one unknown disappears at each step, and during the sequence of computations there are no more than three unknowns, namely S, x_1 and the unknown to be eliminated. Compared with the former treatment, this is, at least from an algebraic point of view, a futile difference. But in the context of verbal algebra, it has a practical advantage: one name used for an unknown may be employed repeatedly, once the quantity represented has been expressed in function of two specific unknowns. That is why Abū Kāmil concludes by saying:[°] *You may use this treatment for any problem you might encounter of such a kind, even if the number (of partners) grows; indeed, you do not need to use more than three designations of unknowns.*

[°] Arabic text, fol. 97^r, 12–13 (p. 193 of the facsimile).

(γ) *On the solution in the 'Liber mahameleth'.* This way of solving is said to be different from that (or those) of Abū Kāmil, though also algebraic. In our symbolism, consider generally that we are to solve

$$x_i + \frac{1}{q_i} \sum_{k \neq i} x_k = y \quad (i = 1, \ldots, n, \quad \text{cyclically}).$$

In the first equation,

$$x_1 + \frac{1}{q_1} \sum_{k \neq 1} x_k = y,$$

we take $\sum_{k \neq 1} x_k = x$ as unknown, so that

$$y = \frac{1}{q_1} x + x_1. \quad (1)$$

Transforming now the remaining equations, we obtain successively, starting from the initial equations and writing for convenience $\sum_{i=1}^{n} x_i = S = x + x_1$,

$$q_i \cdot x_i + \sum_{k \neq i} x_k = q_i \cdot y, \quad \text{then} \quad (q_i - 1) x_i + S = q_i \cdot y,$$

whence

$$x_i = \frac{1}{q_i - 1} (q_i \cdot y - S) = \frac{1}{q_i - 1} \left(\frac{q_i}{q_1} x + q_i \cdot x_1 - (x + x_1) \right)$$

$$= \frac{1}{q_i - 1} \left(\frac{q_i - q_1}{q_1} x + (q_i - 1) x_1 \right),$$

so finally

$$x_i = \frac{q_i - q_1}{q_1 (q_i - 1)} x + x_1. \quad (2)$$

First, we may choose x_1; the x_i's are then expressed in function of x by (2). Then x is determined by summing the expressions for the x_i's, $i \geq 2$, which gives the solving equation. Next y can be calculated from (1) and the individual x_i's from (2). As in Abū Kāmil's text —and in any algebraic treatment— the $n - 2$ unknowns are successively expressed in function of two chosen unknowns, and summing these expressions leads us to the final equation. But the treatment in the *Liber mahameleth* is more general than those presented by Abū Kāmil and is easily represented in formulae. Our author is thus again proving himself to be a worthy disciple of Abū Kāmil, if not his superior. Which is also the impression we are left with by the *Liber mahameleth* as a whole.

Bibliography

Abū Kāmil: *The Book of Algebra, Kitāb al-Jabr wa l-muqābala* [= Publications of the Institute for the History of Arabic-Islamic Science, C.24]. Frankfurt am Main 1986.

—— *See also* Lorch; Sesiano.

Abū'l-Wafā': *See* Saidan.

Anaritius [al-Nairīzī]: *Anaritii in decem libros priores Elementorum Euclidis commentarii* [= *Euclidis Opera omnia, Supplementum*], ed. M. Curtze. Leipzig 1899.

Anbouba, A.: *L'algèbre al-Badī' d'al-Karagī*, ed. with an introd. Beirut 1964.

Arrighi, G.: *La matematica dell'Età di Mezzo. Scritti scelti*, ed. F. Barbieri, R. Franci, L. Toti Rigatelli. Pisa 2004.

—— *See also* Calandri.

Avicenna: *Al-Shifā'*, ed. I. Madkūr et al. (10 vols). Cairo 1952–1963.

—— *Avicenne perhypatetici philosophi ac medicorum facile primi opera in lucem redacta: ac nuper quantum ars niti potuit per canonicos emendata. Logyca. Sufficientia. De celo et mundo. De anima. De animalibus. De intelligentiis. Alpharabius de intelligentiis. Philosophia prima.* Venice 1508.

—— *See also* Marmura.

al-Baghdādī: *See* Saidan.

Baur, L.: *See* Gundissalinus.

de Beauvais [Bellovacensis]: *See* Vincent de Beauvais.

Bernardinello, S.: *Catalogo dei codici della Biblioteca capitolare di Padova* (2 vols). Padua 2007.

Berthelot, M.: *La chimie au moyen âge* (3 vols). Paris 1893.

Birkenmajer, A.: "Bibljoteka Ryszarda de Fournival", *Rozprawy wydział filologiczny*, LX, 4 (1922).

al-Bīrūnī: *Risāla fī istikhrāj al-awtār fi'l-dā'ira li-khawāṣṣ al-khaṭṭ al-munhanī 'l-wāqi' fīhā* [= *Rasā'ilu 'l-Bīrūnī*, 1]. Hyderabad 1948.

—— *Risāla ifrād al-maqāl fī amr al-ẓilāl* [= *Rasā'ilu 'l-Bīrūnī*, 2]. Hyderabad 1948.

—— *See also* Kennedy; Suter.

Bombelli, R. *L'algebra, opera di Rafael Bombelli da Bologna, divisa in tre libri.* Bologna 1572. Complete edition (includes Books IV & V) by E. Bertolotti, Milan 1966.

Boncompagni, B.: *Trattati d'aritmetica* [I: *Algoritmi de numero Indorum* (other editions: *see* Folkerts, Vogel); II: *Iohannis Hispalensis liber algorismi de pratica arismetrice*]. Rome 1857.

——— *Scritti di Leonardo Pisano* (2 vols). Rome 1857–1862.

——— *See also* Fibonacci.

Bubnov, N.: *Gerberti postea Silvestri II papae Opera mathematica (972-1003)*. Berlin 1899.

Busard, H.: "L'algèbre au moyen âge: Le 'Liber mensurationum' d'Abû Bekr", *Journal des Savants*, 1968, pp. 65–124.

——— "Die Vermessungstraktate *Liber Saydi Abuothmi* und *Liber Aderamati*", *Janus*, 56 (1969), pp. 161–174.

Caetani, L., Duca di Sermoneta: *Annali dell'Islām*, vol. X. Milano 1926.

Calandri, P.M.: *Tractato d'abbacho*, ed. G. Arrighi. Pisa 1974.

Camerarius, G.: *See* al-Fārābī.

Campanus, J.: *Preclarissimus liber elementorum Euclidis perspicacissimi in artem Geometrie* (or: *Opus elementorum Euclidis Megarensis in geometriam artem, in id quoque Campani perspicacissimi Commentationes*). Venice 1482.

du Cange, Ch. du Fresne: *Glossarium mediæ et infimæ latinitatis* (10 vols). Niort 1883-1887.

Cantor, M.: *Die römischen Agrimensoren und ihre Stellung in der Geschichte der Feldmesskunst*. Leipzig 1875.

——— *Vorlesungen über Geschichte der Mathematik* (4 vols). Leipzig & Berlin 1880–1908.

Catalogus codicum manuscriptorum Bibliothecae Regiae (4 vols). Paris 1739–1744.

Chuquet, N.: *See* Marre.

Delisle, L.: *Inventaire des manuscrits latins conservés à la Bibliothèque Nationale sous les numéros 8823–18613*. Paris 1863–1871.

——— "Etat des manuscrits latins de la Bibliothèque Nationale au 1er août 1871", *Bibliothèque de l'Ecole des chartes*, 32 (1871), pp. 20–62.

Dictionary of medieval Latin from British sources. Oxford 1975– .

Dijksterhuis, E.: *Archimedes*. Copenhagen 1956.

Diophantos: *See* Tannery, Sesiano.

Djebbar, A.: "Ibn 'Abdūn's Epistle on surface measuring: a witness to the pre-algebraical tradition", *Suhayl*, 5 (2005), pp. 7–68 of the Arabic part; 6 (2006), pp. 255–257 & pp. 81–86 of the Arabic part.

Dozy, R.: *Supplément aux dictionnaires arabes* (2 vols). Leiden 1881.

Euclid: *Euclidis Opera omnia*, edd. J. Heiberg & H. Menge (9 vol.). Leipzig 1883–1916.

——— *Elementa* [= *Euclidis Opera omnia*, vols 1–4].

——— *See also* Heath; Vitrac.

al-Fārābī: *Alpharabii (...) opera omnia, quae, latina lingua conscripta, reperiri potuerunt*, ed. G. Camerarius. Paris 1638.

Fibonacci, L.: *Il Liber abbaci di Leonardo Pisano*, ed. B. Boncompagni [= *Scritti di Leonardo Pisano*, I]. Rome 1857.

Folkerts, M. "Die älteste lateinische Schrift über das indische Rechnen nach al-Ḥwārizmī", *Abhandlungen der Bayerischen Akademie der Wissenschaften, philos.-histor. Klasse*, N.F., 113 (1997).

Gaeta, F.: "Barozzi, Pietro", in *Dizionario biografico degli Italiani*, 6 (Rome 1964), pp. 510–512.

Gautier Dalché, J.: "L'histoire monétaire de l'Espagne septentrionale et centrale du IXe au XIIe siècles", *Anuario de estudios medievales*, 6 (1969), pp. 43–95.

Govi, E.: *Patavinae cathedralis ecclesiae capitularis bibliotheca, Librorum XV saec. impressorum index. Appendix: Petri Barocii bibliothecae inventarium*. Padua 1958.

Grand, P.: "Le Quodlibet XIV de Gérard d'Abbeville", *Archives d'histoire doctrinale et littéraire du moyen âge*, 39 (1964 [1965]), pp. 207–269.

Gundissalinus [Gundisalvus], D.: *De divisione philosophiae*, hrsg. u. philosophiegeschichtlich unters. v. L. Baur [= *Beiträge zur Geschichte der Philosophie des Mittelalters*, IV, 2–3]. Münster 1903.

al-Ḥaṣṣār: *See* Suter.

Heath, T.: *A History of Greek mathematics* (2 vols). Oxford 1921.

—— *The Thirteen Books of Euclid's Elements* (2nd ed., 3 vols). Cambridge 1926.

Hochheim, A.: *Kâfî fîl Hisâb (Genügendes über Arithmetik) des (...) Alkarkhî* (3 vols). Halle 1878–1880.

Hughes, B.: "Gerard of Cremona's translation of al-Khwārizmī's al-Jabr", *Mediaeval studies*, 48 (1986), pp. 211–263.

Hultsch, Fr.: *Metrologicorum scriptorum reliquiae* (2 vols). Leipzig 1864–1866.

Ibn 'Abdūn: *See* Djebbar.

Ibn Sīnā: *See* Avicenna.

Junge, G.: "Das Fragment der lateinischen Übersetzung des Pappuskommentars zum X. Buche Euklids, *Quellen und Studien zur Geschichte der Mathematik, Astronomie und Physik*, Abt. B, 3 (1936), pp. 1–17.

al-Karajī: *See* Anbouba; Hochheim; Woepcke; Sesiano.

Kazimirski, A. de Biberstein: *Dictionnaire arabe-français* (2 vols). Paris 1860.

Kennedy, E.: *The exhaustive treatise on shadows by (...) al-Bīrūnī*, transl. and comm. (2 vols). Aleppo 1976.

al-Khwārizmī: *See* Boncompagni (*Trattati*, I); Folkerts; Vogel.

Klamroth, M.: "Über den arabischen Euklid", *Zeitschrift der Deutschen morgenländischen Gesellschaft*, 35 (1881), pp. 270–326, 788.

Lévi-Provençal, E.: *Séville musulmane au début du XIIe siècle. Le traité d'Ibn 'Abdun*. Paris 1947.

Libri, G.: *Histoire des sciences mathématiques en Italie* (4 vols). Paris 1838.

Lorch, R.: "Abū Kāmil on the Pentagon and Decagon", in *Vestigia Mathematica* (Amsterdam 1993), pp. 215–252.

Marmura, M.: "Avicenna on the Division of the Sciences in the *Isagoge* of his *Shifā'*", *Journal for the history of Arabic sciences*, 4 (1980), pp. 239–251.

Marre, A.: "Notice sur Nicolas Chuquet et son *Triparty en la science des nombres*", *Bulletino di bibliografia e di storia delle scienze matematiche e fisiche*, XIII (1880), pp. 555–659, 693–814.

——— "Appendice au Triparty en la science des nombres de Nicolas Chuquet parisien", *Bullettino di bibliografia e di storia delle scienze matematiche e fisiche*, XIV (1881), pp. 413–460.

Mittellateinisches Wörterbuch bis zum ausgehenden 13. Jahrhundert. Munich 1967– .

Murdoch, J.: "Euclid: Transmission of the Elements", in *Dictionary of scientific biography*, IV (New York 1971), pp. 437–459.

Nagl, A.: "Der arithmetische Traktat des Radulf von Laon", *Zeitschrift für Mathematik und Physik* (hist.-lit. Abt., Suppl.), 34 (1890), pp. 87–133.

al-Nairīzī: *See* Anaritius.

Niermeyer, J.: *Mediae latinitatis lexicon minus*. Leiden 1976.

Ouy, G. & Gerz-von Büren, V.: *Catalogue de la Bibliothèque de l'Abbaye de Saint-Victor de Paris de Claude de Grandrue, 1514*. Paris 1983.

Rebstock, U.: *Rechnen im islamischen Orient*. Darmstadt 1992.

——— "Der Mu'āmalāt-Traktat des Ibn al-Haiṯam", *Zeitschrift für Geschichte der arabisch-islamischen Wissenschaften*, 10 (1995/96), pp. 61–121.

Rouse, R.: "Manuscripts belonging to Richard de Fournival", *Revue d'histoire des textes*, 3 (1973), pp. 253–259.

Saidan, A. S.: *The Arithmetic of Abu al-Wafa' al-Buzajani*, edited and annotated. Amman 1971.

——— *Al-Takmila fi'l-Hisab (The Completion of Arithmetic)* with a tract on Mensuration by (...) *Al-Baghdadi*, edited and annotated. Kuwait 1985.

Sesiano, J.: "Le Traitement des équations indéterminées dans le Badī' (...) d'Abū Bakr al-Karajī", *Archive for History of Exact Sciences*, 17 (1977), pp. 297–379.

——— *Books IV to VII of Diophantus 'Arithmetica' in the Arabic Translation attributed to Qusṭā ibn Lūqā*. New York 1982.

—— "The appearance of negative solutions in mediaeval mathematics", *Archive for History of Exact Sciences*, 32 (1985), pp. 105–150.

—— "Survivance médiévale en Hispanie d'un problème né en Mésopotamie", *Centaurus*, 30 (1987), pp. 18–61.

—— "Le Liber mahameleth, un traité mathématique latin composé au XIIe siècle en Espagne", in *Histoire des mathématiques arabes, Actes du Colloque* (Alger 1986 [1988]), pp. 69–98.

—— "La version latine médiévale de l'Algèbre d'Abū Kāmil", in *Vestigia Mathematica* (Amsterdam 1993), pp. 315–452.

—— "Un recueil du XIIIe siècle de problèmes mathématiques", *SCIAMVS*, 1 (2000), pp. 71–132.

—— *An Introduction to the History of Algebra*. Providence 2009.

Smith, D.E.: *History of Mathematics* (2 vols). Boston 1923–1925.

Suter, H.: *Beiträge zur Geschichte der Mathematik und Astronomie im Islam* (2 vols). Frankfurt am Main 1986.

—— "Das Rechenbuch des Abû Zakarîjâ el-Ḥaṣṣâr", *Bibliotheca mathematica*, 3. F., 1 (1901), pp. 12–40 [= *Beiträge*, II, pp. 115–143].

—— "Über den Kommentar des Muhammed ben 'Abdelbâqî zum zehnten Buche des Euklides", *Bibliotheca mathematica*, 3. F., 7 (1906–1907), pp. 234–251 [= *Beiträge*, II, pp. 198–215].

—— "Das Buch der Auffindung der Sehnen im Kreise von (...) el-Bīrūnī", *Bibliotheca mathematica*, 3. F., 11 (1910), pp. 11–78 [= *Beiträge*, II, pp. 280–347].

Tannery, P.: *Diophanti Alexandrini Opera omnia cum graecis commentariis*, ed. & lat. vert. P. Tannery (2 vols). Leipzig 1893–1895.

Thesaurus linguae latinae. Leipzig 1900– .

Thorndike, L.: "John of Seville", *Speculum*, 34 (1959), pp. 20–38.

Tropfke, J.: *Geschichte der Elementar-Mathematik in systematischer Darstellung*. Berlin/Leipzig 1921–1924 (7 vols, 2nd ed.), 1930–1940 (vols 1–4, 3rd ed.).

Tropfke, J. et al.: *Geschichte der Elementarmathematik*. Berlin 1980.

Vincent de Beauvais [Vincentius Bellovacensis]: *Bibliotheca mundi. Vincentii Bellovacensis Speculum quadruplex: naturale, doctrinale, morale, historiale* (4 vols). Douai 1624.

Vitrac, B.: *Les Eléments* (4 vols). Paris 1990–2001.

Vlasschaert, A.-M.: *Le Liber mahameleth, édition critique et commentaires*. Stuttgart 2010.

Vogel, K.: *Vorgriechische Mathematik II: Die Mathematik der Babylonier*. Hannover/Paderborn 1959.

—— *Mohammed ibn Musa Alchwarizmi's Algorismus*. Aalen 1963.

——— *Ein byzantinisches Rechenbuch des frühen 14. Jahrhunderts.* Vienna 1968.

Wiedemann, E.: *Aufsätze zur arabischen Wissenschaftsgeschichte* (2 vols). Hildesheim 1970.

——— "Über *al Fârâbîs* Aufzählung der Wissenschaften (De Scientiis)" (= Beiträge zur Geschichte der Naturwissenschaften, XI), *Sitzungsberichte der Physikalisch-medizinischen Societät in Erlangen*, 39 (1907), pp. 74–101 [= *Aufsätze*, I, pp. 323–350].

——— "Über die Geometrie und Arithmetik nach den *Mafâtîḥ al 'Ulûm*" (= Beiträge zur Geschichte der Naturwissenschaften, XIV.1), *Sitzungsberichte der Physikalisch-medizinischen Societät in Erlangen*, 40 (1908), pp. 1–29 [= *Aufsätze*, I, pp. 400–428].

Woepcke, F.: *Etudes sur les mathématiques arabo-islamiques* (2 vols). Frankfurt am Main 1986.

——— *Extrait du Fakhrî*. Paris 1853. [= *Etudes*, I, pp. 269–426].

Index

d'Abbeville, G. (*c.* 1250): **I** xxxiii.

Abū Bakr (*c.* 1100): **I** xxi.

Abū Ḥanīfa (*c.* 740): **III** 1476, 1481.

Abū Kāmil (*c.* 880): **I** xviii, xxi, xxiv, xxv, lvi, lxii–lxv **II** 592_{51}, 595_{64}, 663_{362}, 675_{402}, 687_{438}, 724_{584}, 1075, 1081, 1103 **III** 1144, 1147–1149, 1208, 1218, 1303, 1315–1316, 1324, $1362n$, 1374, 1383, 1434, 1508, 1701, 1710–1714 ◊ (name mentioned in the *Liber mahameleth*) **II** 590, 591 (2), 748, 903, 1073 ◊ (knowledge of his *Algebra* required) **I** lxii–lxiv **III** 1303.

Abū'l-Wafā' (*c.* 975): **III** $1362n$.

Abū 'Uthmān al-Dimashqī (*c.* 900): **I** xxi.

addition: (of integers) **I** lxiv **II** 626 & 626_{210}, 630–631, 707–710 **III** 1159, 1236–1238, 1251–1253 ◊ (written arrangement) **III** 1172 ◊ (of fractions) **II** 694–703 **III** 1162–1163, 1234–1235, 1238–1245, 1295–1297 ◊ (of roots) **II** 753–754, 763–765 **III** 1305–1306, 1308–1309, 1320–1322, 1329–1331.

al-Baghdādī: *See* Baghdādī.

Alcuin (*c.* 780): **I** xiii **II** 863_{1186}.

algebra: (in the *Liber mahameleth*) **I** lxv, lxvii–lxxvii **II** 582_9, 748_{685}, 793_{917}, 794_{921}, 809_{1000}, 876_{1247}, 876_{1250} **III** (e.g.) 1235–1236, 1247, 1345, 1347–1349, 1360, 1362–1369, 1383–1386, 1425, 1468, 1471–1474, 1487–1488, 1490, 1493–1494, 1495–1497, 1503–1515, 1521–1526, 1529, 1532–1533, 1543–1544, 1549–1550, 1555–1557, 1560, 1562, 1569–1571, 1573, 1575–1577, 1587–1588, 1598, 1613–1615, 1641, 1644–1645, 1654, 1658–1659, 1662, 1675–1677, 1687, 1689, 1701, 1711–1714 ◊ (name mentioned in the *Liber mahameleth*) **II** 748, 773, 796, 809, 811 (2), 841, 888, 890, 893, 897 (3), 898, 899, 900, 901, 903, 907, 908 (2), 909, 910, 911 (2), 912 (2), 913, 914, 919 (2), 921, 925, 929, 941 (3), 946, 948, 950, 961 (2), 964, 974, 975, 1016, 1029, 1047, 1048, 1056, 1059, 1069, 1073, 1077–1078 ◊ (algebraic operations) **I** lxv, lxviii–lxx **II** 704, 723_{580}, 794_{921}, 795_{930}, 890_{1309}, 897_{1341}, 904_{1369}, 904_{1371}, 1074_{2020} **III** 1275, 1345, 1347–1349, 1363, 1496.
See also equations; verbal mathematics.

'Alī (calif, *c.* 650): **III** 1428, 1434.

'Alī the Reckoner (Egypt): **III** 1700.

alloy determination: **II** 853–855 **III** 1443–1445, 1446–1447.

almodi (metrology; *see* **III** 1396): **I** lxxviii **II** 823 (4), 828 (5), 988 (5), 1078.

Almoravides: **III** 1185.

'ambiguity' (*see* **III** 1200): **II** 687–688, 689–692, 701, 718–719 **III** 1227–1229, 1229–1231, 1244 –1245, 1265–1266.

'amounts' (*see* **III** 1234–1236, 1256): **II** 702–707, 719–725, 731_{610} **III** 1245–1251, 1266–1274.

analogy of formulae (*see* **I** xvii): **II** 812, 846, 848, 850, 851, 852, 859, 878, 895–896, 935–936, 967,

974, 997 (2), 1000   **III** 1388, 1432–1433, 1438, 1440, 1441, 1445–1446, 1452, 1472, 1502, 1542–1543, 1580, 1587, 1616, 1617, 1621.

Anaritius: *See* al-Nairīzī.

Anbouba, A.: **III** 1708*n*.

the 'Ancients': **II** 1045.

Anthologia graeca: **III** 1670*n*.

Apollonios (*c.* 220 B.C.): **I** xiii.

apotome (*see* **III** 1306): **II** 757–759, 759₇₅₀, 766, 767–769, 772–775, 1124   **III** 1310–1315, 1326, 1327, 1332, 1333–1335, 1339–1342, 1376.

the 'Arabs': **I** xviii   **II** 605   **III** 1159.

Archimedes (*c.* 250 B.C.): **I** xiii, xvi, xxxii, lxiv   **II** 868₁₂₁₄, 1045₁₉₃₈, 1080, 1081   **III** 1443–1445, 1460, 1463, 1669 ◊ (name mentioned in the *Liber mahameleth*) **II** 868.

area formulae: **I** lxxviii   **II** 856 ff., 865 ff.   **III** 1460.

Aristotle (*c.* 340 B.C.): **III** 1139.

arithmetic: (theoretical) **II** 581 **III** 1141 ◊ (practical) **II** 581–582, 582₆, 582₇, 587₃₂, 588, 604, 605, 609   **III** 1141.

arithmetic progressions: *See* progressions.

Arrighi, G.: **I** xviii*n*, lxi*n*   **III** 1479*n*.

arrova (metrology): **I** lxxviii–lxxix   **II** 979–987 (liquid measure), 999–1005 (dry measure).

'articles' (*see* **II** 584): **II** 582, 584–586, 605, 607, 609, 613–615, 618–626, 630, 648–650, 654.

'associate' (terms, in a proportion): **II** 777 (4), 1128   **III** 1343, 1345.

associativity: **II** 589₃₆   **III** 1146, 1570, 1587.

Avicenna (*c.* 1010): **III** 1139.

Baeza: **III** 1660.

al-Baghdādī, 'Abd al-Qāhir ibn Ṭāhir (*c.* 1000): **III** 1476.

al-Baghdādī, Muḥammad ibn 'Abd al-Bāqī (*c.* 1100): **III** 1313*n*.

Barozzi, P. (*c.* 1480): **I** xxvii, xxix–xxxiii.

Baur, L.: **I** lx*n*   **III** 1140*n*.

de Beauvais, V. (*c.* 1250): **I** lx.

Beldomandi, P. de' (*c.* 1410): **I** lxi*n*.

Benedetto da Firenze (*c.* 1460): **I** lxi   **III** 1479, 1480.

Bernardinello, S.: **I** xxvii*n*.

Berthelot, M.: **III** 1443*n*.

Bianchini, G. (*c.* 1450): **I** lxi*n*.

bimedial: **II** 770 (3), 774 (3), 1081   **III** 1313, 1336, 1342.

binomial (*see* **III** 1306): **II** 757–759, 766–772, 774–775, 801 (2), 1081   **III** 1303, 1306, 1310–1315, 1316, 1326, 1327, 1332, 1333–1339, 1341–1342, 1373.

Birkenmajer, A.: **I** xxxiii*n*.

al-Bīrūnī (*c.* 1020): **I** xv*n*   **III** 1670, 1672.

Boëtius (*c.* 500): **I** xiii, xxxi, lxi*n*.

Bombelli, R. (*c.* 1560): **III** 1374.

Boncompagni, B.: **I** xiv*n*   **II** 583₁₅.

borrowing: (chap. B–X) **II** 892–894   **III** 1492–1494.

Bubnov, N.: **III** 1693*n*.

bundles: (chap. B–XXI) **II** 1066–1067   **III** 1698–1699.

Busard, H.: **I** xxi*n*.

buying and selling: (chap. B–I) **II** 780–816   **III** 1345–1394 ◊ (mentioned) **II** 778, 844, 892–896, 925, 937, 972, 997, 998.

buying a horse: **II** 1072–1074   **III** 1705–1707, 1709–1714.

Byzantine mathematics: **I** xiii, xiv **III** 1362n.

Caetani, L.: **III** 1428n.

caficius (metrology): **I** lxxviii–lxxix **II** 820$_{1048}$, 822–842, 859, 871–880, 988–998, 1000–1001, 1082.

Calandri, P. M. (c. 1470): **I** lxi **III** 1479.

camel problem: **III** 1434, 1436.

Camerarius, G. (c. 1640): **I** lx **III** 1140n.

Campanus, J. (c. 1260): **III** 1145n.

du Cange, Ch.: **II** 1075, 1125.

Cantor, M.: **III** 1443n, 1693n.

Carmen de ponderibus: **III** 1444, 1445.

census (square, x^2): **I** lxviii **II** 877$_{1260}$, 995$_{1736}$, 1083. See also squares (unknown, x^2).

Chinese mathematics: **III** 1278, 1670.

Chuquet, N. (c. 1484): **I** lxii **III** 1374, 1479, 1516.

circle: **I** lxxviii **II** 865, 867–870, 1045, 1057–1058, 1060–1061, 1083 ◇ (circumference) **II** 868, 869, 1045, 1061 ◇ (area) **II** 865, 869, 977, 1045.
See also rectification; squaring.

cistern problems: (chap. B–XIX) **II** 1039–1045 **III** 1663–1669.

commensurable: **II** 753, 758, 761–764, 767–768, 771, 1084 ◇ (in length) **II** 761 ◇ (in square) **II** 761–762.

common: (number, see **II** 658$_{335}$) **II** 658–662, 663$_{361}$, 664–665, 674, 733, 850, 1084–1085 ◇ (denominator) **III** 1233, 1234, 1287 ◇ (multiple) **II** 936$_{1502}$, 1030$_{1873}$ **III** 1301–1302 ◇ (add in common) **II** 602, 604, 635, 724, 916 (2), 931, 933, 934 (3), 1013, 1037, 1052, 1053 ◇ (subtract a common term) **II** 725, 889, 973, 1048, 1051, 1052, 1074 ◇ (multiply in common) **II** 957 (2).

commutativity: **II** 589$_{36}$, 597$_{73}$, 607$_{120}$, 957$_{1579}$ **III** 1146–1147, 1159, 1164, 1570, 1587.

consumption: (by animals, chap. B–XVI) **II** 988–997 **III** 1603–1616 ◇ (of bread, chap. B–XVII) **II** 999–1005 **III** 1618–1628 ◇ (of lamp-oil, chap. B–XV) **II** 979–987 **III** 1592–1602 ◇ (mentioned) **II** 988, 996.

continued proportion: **I** lxvii **II** 616 (3), 819, 820, 915$_{1418}$, 954$_{1568}$, 968$_{1623}$ **III** 1168, 1250, 1582.

conversion: **II** 826, 1001, 1004, 1006–1008, 1011, 1012, 1034$_{1890}$ ◇ (of fractions) see fractions ◇ (of ratios) see ratio.

crown problem: **III** 1443–1445.

Crusafont y Sabater, M.: **III** 1661.

cubes: (unknown, x^3) **I** lxviii **II** 965–966, 996, 1088 **III** 1566, 1577, 1615 ◇ (sum of consecutive) **II** 709–710 **III** 1237–1238, 1253.

cubit (metrology): **I** lxxix **II** 856–858, 860–870, 976–978, 1040–1050, 1052, 1054–1057, 1059–1067, 1070, 1088.

Curtze, M.: **I** xxi$n$.

decimal positional system: **I** lxii, lxiv **II** 584–586, 611$_{137}$, 622$_{193}$, 630$_{229}$, 1090 **III** 1139, 1142–1143.

Delisle, L.: **I** xxxiii$n$, xxxvi$n$.

demonstrations: *See* proofs.

'denominate': **II** 612, 613, 640 (3), 646 (2), 646$_{289}$, 648 (2), 651 (9), (...), 1089.

'denomination': **II** 609, 639,

640_{261}, 646, 650, 653, 654, 656_{328}, (...), 1089 **III** 1162–1163, 1181, 1182, 1188–1196, 1275, 1282–1283, 1285, 1286.

denominator: **II** 651_{306}, 652, 655, 656 (2), 656_{328}, 658, 659 (5), 660 (8), (...) ◇ (common denominator) see common.

'determination' (see **III** 1632): **II** 1020, 1021, 1024, 1025, 1090 **III** 1632–1635, 1644–1652, 1661.

'digits': **II** 582–587, 584_{23}, 605 (4), 607–609, 613–615, 619–626, 630, 648–651, 653, 655, 658, 676, 700 (4), 1090.

Dijksterhuis, E.: **III** $1443n$.

Diophantos (c. 250): **I** xiii, lxviii, lxix.

dirham: **I** lxviii **II** 841_{1120} **III** 1383, 1428, 1635, 1661, 1712.

distributivity: **II** 676_{405} ('multiplication term by term', **III** 1199) **III** 1145, 1161, 1171, 1174, 1178, 1217.

divisibility: **II** 646–651 **III** 1182, 1188–1191, 1302.

'divisible': **II** 582 (2), 611 (2), 613, 640, 647–650, 746, 812, 936–938.

division: (aspects) **III** 1181, 1183 ◇ (of integers) **I** lxv **II** 635–636, 639–654 **III** 1178–1179, 1181–1196 ◇ (written arrangement) **II** 651–652 ◇ (of fractions) **II** 726–744 **III** 1162–1163, 1275–1302 ◇ (of roots) **II** 756–760, 766–770 **III** 1306–1307, 1309–1310, 1323–1327, 1332–1336.

Djebbar, A.: **I** v, lxiv$n$, lxxiii$n$ **III** 1460.

Domenico d'Agostino Vaiaio (c. 1410): **I** xviii, lxi **III** 1479, 1488, 1491.

Dozy, R.: **III** $1670n$.

'dragma': **III** $1629n$, $1661n$ ◇ (second unknown) **I** lxviii **II** 809 (8), 811 (5), 1093 **III** 1383, 1386.

drapery: (chap. B–VI) **II** 856–862 **III** 1448–1456 ◇ (mentioned) **II** 864, 866, 1042.

duplication: **II** 582, 584_{22}, 605.

elementary (fraction, number): See fractions.

Elements: (knowledge required) **I** xvii, lxii–lxiii, lxv–lxvii **II** 587_{33}, 591_{50}, 598_{80}, 604_{104} **III** 1144, 1151, 1303, 1345, 1362, 1365.
See also Euclid.

emina (metrology; see **III** 1626): **I** lxxix **II** 1003–1005, 1094.

equations: (six standard forms) **I** lxvii **III** 1490 ◇ (reduction to the standard forms) **I** lxviii–lxix ◇ ('side' of an equation) **II** 794–795, 903, 911, 1074, 1103 ◇ (single linear) **I** lxix–lxx **II** 703_{491}, 792_{911}, 793_{915}, 794_{920}, 794_{923}, 795_{925}, 795_{927}, 795_{929}, 884_{1281}, 897_{1340}, 897_{1342}, 898_{1343}, 898_{1344}, 899_{1345}, 899_{1347}, 900_{1348}, 900_{1352}, 901_{1353}, 902_{1360}, 906_{1378}, 907_{1379}, 907_{1380}, 908_{1382}, 908_{1383}, 908_{1384}, 909_{1385}, 909_{1389}, 910_{1391}, 910_{1393}, 911_{1395}, 911_{1397}, 911_{1398}, 911_{1399}, 911_{1401}, 912_{1403}, 912_{1404}, 912_{1405}, 913_{1407}, 913_{1409}, 913_{1410}, 914_{1411}, 917_{1424}, 919_{1428}, 920_{1432}, 921_{1438}, 922_{1440}, 922_{1444}, 925_{1457}, 925_{1461}, 928_{1472}, 929_{1474}, 941_{1517}, 941_{1518}, 941_{1519}, 946_{1528}, 947_{1533}, 947_{1534}, 948_{1538}, 960_{1594}, 961_{1598}, 962_{1605}, 963_{1606}, 975_{1651}, 975_{1652}, 1014_{1807}, 1016_{1818}, 1016_{1820}, 1018_{1827}, 1029_{1870}, 1038_{1904}, 1047_{1944}, 1048_{1948}, 1059_{1982}, 1069_{2011} **III** (e.g.) 1235, 1247, 1345, 1347–1349, 1383, 1490,

1495–1496 ◇ (indeterminate) **III** 1251, 1302 ◇ (pairs of linear) **I** lxx–lxxiii ◇ (indeterminate) **I** lxxiii **III** 1393, 1632–1635, 1644–1652 ◇ (solvability condition) **III** 1389–1391, 1393, 1631–1632, 1640–1641, 1660 ◇ (three or more simultaneous linear) **III** 1705–1707, 1709–1714 ◇ (single quadratic) **I** lxvii$n$, lxxiii–lxxvi **II** 725_{590}, 725_{593}, 796_{935}, 797_{940}, 806_{987}, 806_{990}, 809_{1005}, 810_{1008}, 811_{1016}, 837_{1104}, 841_{1120}, 841_{1121}, 858_{1161}, 862_{1182}, 887_{1292}, 888_{1301}, 890_{1310}, 903_{1364}, 903_{1365}, 904_{1366}, 904_{1370}, 905_{1372}, 905_{1373}, 919_{1430}, 950_{1549}, 958_{1586}, 964_{1612}, 965_{1617}, 995_{1737}, 1032_{1884}, 1047_{1942} **III** 1272, 1300, 1301, 1337, 1340, 1359, 1361, 1365–1366, 1368, 1377, 1379, 1384, 1419, 1425, 1449, 1455, 1461, 1478, 1485, 1488, 1509, 1558, 1560, 1657, 1675, 1679, 1682. See also solution(s).

Euclid (and the *Elements*): **I** xiii, xvi, xxi, xxiv, xxvi–xxviii, xxxi, lviii, lix$n$, lxii–lxiii, lxv, lxxiv–lxxvi **II** 587_{33}, 595_{64}, 718_{553} **III** (e.g.) 1144–1145, 1151, 1153–1158, 1303, 1306, 1311–1315, 1320, 1323n, 1329, 1362, 1716 ◇ (Euclid mentioned) **II** 589 (2), 595, 596, 597 (3), 598 (3), 599, 604, 616 (2), 628, 641, 652, 673, 677, 679, 682, 702, 748 (2), 752, 753, 755, 761, 763 (5), 764 (2), 765 (3), 769, 770, 772 (2), 773, 774 (2), 796, 887, 924, 928, 943, 959, 1043, 1047, 1048, 1050, 1055, 1057, 1058 (2), 1067, 1075, 1095, 1101, 1103. See also '*Elements*'.

Euler, L. (*c.* 1750): **I** v.

exchange (money): (chap. B–XVIII) **II** 1006–1038 **III** 1629–1662.

false position: *See* rule of false.

al-Fārābī (*c.* 920): **I** lx **III** 1140n.

Fibonacci, L. (*c.* 1220): **I** xiv, xv, xxv **III** 1427–1428, 1479, 1708n.

figures (*figure*): (significant figures, thus digits; *see* **II** 620_{185}) **II** 607–609, 613 ff. **III** 1143, 1163–1175 ◇ (figures in demonstrations) **I** lxxxi **II** 589–604, 617, 628, 633, 635, 663, 675, 676, 699, 713, 715, 723, 724, 740, 751, 754–756, 758, 760, 763, 767, 769, 771, 773, 797, 798, 800, 803, 806–808, 810, 814, 818, 820, 835, 837, 838, 859, 862, 866, 870, 884, 888, 889, 915–919, 924, 928, 931, 933–935, 938, 944, 949, 951, 955, 960, 962, 969, 981, 1013, 1015–1017, 1019, 1022, 1023, 1029, 1031, 1033, 1035, 1037, 1046, 1047, 1049–1056, 1058–1061, 1063–1065, 1068 ◇ (mentioned in the text) **II** 617, 914–915, 917, 924, 946, 1018, 1046, 1047, 1049–1051, 1053–1056, 1059, 1063, 1064 ◇ (tables or illustrations; *see* **II** 627_{217}) **II** 586, 623, 627, 629–631, 647, 652, 655 to 701 (*but see* **II** 655_{326} & **III** 1200–1201), 777–779, 781–783, 852, 854 **III** 1228n, 1240n, 1297 ◇ (geometrical figures) **II** 865, 869–870, 1061.

finger reckoning: **II** 614–615, 623, 625–626 **III** 1161, 1171–1172.

Folkerts, M.: **I** xiv$n$.

forward and retrograde motion: **III** 1673–1674, 1691–1693, 1700, 1701–1703.

Fournival, R. de (*c.* 1240): **I** xxxiii.

fractions: (kinds of) **II** 605, 682 **III** 1197 ◊ (expression in Arabic) **I** xviii **III** 1182 ◊ (aliquot) **II** 605$_{109}$ ◊ (ascending) **III** 1193 ◊ ('composite') **II** 657 ◊ (compound) **III** 1159n ◊ ('different'; see **III** 1198n) **II** 682–683, 689, 698–699, 713–714, 717 **III** 1223–1224, 1229, 1241–1243, 1259–1260, 1263–1264 ◊ (elementary fractions —and numbers) **I** xviii **III** 1182, 1188–1189 ◊ (fraction of a fraction) **II** 605 **III** 1159 ◊ ('irregular'; see **III** 1199) **II** 683–688, 700–701, 717–719, 733–735 **III** 1199–1200, 1224–1229, 1243–1244, 1255–1256, 1264–1266, 1276, 1286–1288 ◊ (addition of) **II** 694–703, 743 **III** 1162–1163, 1234–1245, 1295–1297 ◊ (conversion) **II** 667–673, 676, 677, 679–682, 694–699, 711–714, 716–717, 741–742 **III** 1162–1163, 1199, 1212–1216, 1219, 1229, 1234, 1255 ◊ (denomination) **II** 729–730 **III** 1162–1163, 1275, 1282–1283, 1285–1286 ◊ (division) **II** 726–744 **III** 1162–1163, 1275–1299 ◊ (inversion) **II** 726–729 **III** 1250, 1275, 1279–1282 ◊ (multiplication) **II** 655–693, 743 **III** 1162–1163, 1197–1233, 1295–1297 ◊ (written arrangement) **II** 663$_{360}$, 743 ◊ (redintegration) **II** 726–729, 736 **III** 1275, 1279–1280. ◊ (subtraction) **II** 711–719, 743 **III** 1162–1163, 1255–1266, 1295–1297 ◊ (taking fractions of integers with repetitions) **II** 635–638 **III** 1162, 1178–1180.

Gaeta, F.: **I** xxix$n$.

Gautier Dalché, J.: **III** 1661n.

geometric solving: *See* solution.

geometric progressions: *See* progressions.

geometry (practical): **I** lxiv, lxxviii **II** 582, 865$_{1197}$, 976$_{1657}$.

Gerz-von Büren, V.: **I** xxxvii$n$.

Govi, E.: **I** xxvii$n$.

Grand, P.: **I** xxxiii$n$.

Grazia de' Castellani (*c.* 1380): **I** xviii, xix$n$, lxi **III** 1479, 1488.

grinding: (chap. B–VIII) **II** 871–881 **III** 1466–1475.

Guillelmus (*c.* 1370): **I** xxiv, lxiii$n$.

Gundissalinus, D. (*c.* 1150): **I** lx **II** 581$_5$ **III** 1140n, 1141n.

al-Ḥaṣṣār (*c.* 1150): **I** lxiv, lxv **II** 662$_{357}$ **III** 1200, 1236, 1253n, 1275.

Heath, T.L.: **II** 608$_{122}$ **III** 1141n, 1151n, 1320n, 1323n, 1443n.

Hebeisen, C.: **I** v.

height determination: **II** 1065 **III** 1675, 1696–1697.

Heron (*c.* 100): **III** 1304.

hiring workers: (chap. B–XI to B–XIV) **II** 895–978 **III** 1495–1591 ◊ (carpenter) **II** 978 **III** 1590–1591 ◊ (carrier) **II** 952–966 (mentioned) 967, 972, 973, 974, 1000 **III** 1566–1577 ◊ (digger) **II** 976–978 **III** 1589–1590 ◊ (shepherd) **II** 975–976 **III** 1588–1589 ◊ (stonecutter) **II** 967–971 **III** 1579, 1580–1585.

Hochheim, A.: **III** 1362n.

Hughes, B.: **I** xxi$n$.

Hultsch, F.: **III** 1444n.

Ibn 'Abdūn (*c.* 970): **I** lxxiii$n$ **III** 1460n.

Ibn Sīnā: *See* Avicenna.

incommensurable: **II** 771 ◊ (incommensurable in length) **II** 761–762

indeterminate: *See* equations; problems.

Indian mathematics: **I** xiv, lxiv

infinite (infinity, indefinitely, innumerable): **II** 583 (4), 584, 585, 587 (2), 605 (2), 609 **III** 1142 ⋄ (number of possibilities, problems, solutions; see **III** 1634) **II** 842, 847, 1024, 1072 ⋄ (infinite geometric series) **I** lxxvii–lxxviii **II** 848 **III** 1435, 1438.

integers: *See* numbers.

'intermediate result': **II** 613_{147}, 743 (9), 1126 (*servatum*) **III** 1296–1297.

interpolations: **I** lvii–lx.

inversion: (of a fraction) **II** 726–729 **III** 1250, 1275, 1279–1282 ⋄ (of a ratio) *see* ratio.

al-jabr: **I** lxix–lxx **II** 794_{921}, 795_{930}, 890_{1309}, 904_{1369} **III** 1347–1349, 1363.
  See also algebra (algebraic operations).

Johannes Hispalensis (John of Seville, *c.* 1140): **I** xiv, xviii–xix, xxxiii, lxiv **III** 1660.
  See also '*Liber algorismi*'.

Junge, G.: **I** xxi$n$.

al-Karajī (*c.* 1010): **III** $1362n$, $1708n$.

Kazimirski, A.: **III** $1670n$.

Kennedy, E.: **I** v, xv$n$ **III** $1670n$.

al-Khwārizmī (*c.* 820): **I** xiv, xv, xxi, xxxvii, lxiv, lxv **II** 581, 748_{688}, 1077, 1080 **III** 1141, 1434.

Klamroth, M.: **II** 1075.

ladder problems: (chap. B–XX.1) **II** 1046–1055 **III** 1670–1672, 1675–1686.

legacies: **III** 1434.

lending: (chap. B–XXIII) **II** 1071–1074 **III** 1704–1714.

Lévy-Provençal, E.: **I** xix$n$, lxxix$n$ **III** $1466n$.

Liber algorismi: **I** xiv$n$, xix, xxxiii, lix, lx, lxiv, lxv **II** 583_{15}, 583_{18}, 584_{24}, 585_{28}, 588_{34}, 607_{118}, 607_{119}, 608_{121}, 608_{123}, $609_{124-126}$, 610_{135}, 612_{138}, 612_{141}, 613_{142}, 619_{182}, 623_{198}, 625_{206}, 627_{215}, 660_{345}, 703_{492}, 707_{521}, 743_{670}, 748_{688}, 749_{691}, 749_{694}, 777_{845} **III** 1303, $1304n$, $1629n$.

Liber augmenti et diminutionis: **I** xxi, xxxvii–xxxviii **III** $1629n$, $1708n$.

Liber mahameleth: (sources) **I** xvii–xviii ⋄ (Hispano-Arabic background) **I** xvii–xix ⋄ (authorship) **I** xviii–xix, lxi ⋄ (division into two books) **I** xv **II** 582_6 **III** 1141, 1597 ⋄ (contents) **I** xv–xvii ⋄ ('*mahameleth*' mentioned) **II** 581 (2), 597, 598, 604 ⋄ (clarity of the Latin text) **I** lv ⋄ (manuscripts) **I** xix–lv ⋄ (disordered text in the manuscripts) **I** xxxviii–lv ⋄ (incompleteness) **I** lv–lvii, lxiv–lxv ⋄ (early readers) **I** lvii–lx ⋄ (later mediaeval readers) **I** xxiv–xxvi, xxix–xxxiii, xxxiv–xxxvi, lx–lxii ⋄ (mathematical prerequisites) **I** lxii–lxvii ⋄ (mathematics in) **I** lxii–lxxviii ⋄ (metrology) **I** lxxviii–lxxix.

Libri, G.: **I** xxi$n$, xxxvii$n$ **III** $1629n$, $1708n$.

'limit' (*see* **III** 1142–1143): **II** 582–586, 605, 607, 609, 615_{154}, 615_{155}, 617–619, 624–626, 648–650.

linen cloth: (chap. B–VII) **II** 863–870 **III** 1457–1465.

Lorch, R.: **I** xxi$n$, xxv$n$.

'major line' (*see* **III** 1314): **II** 770, 775 (2), 1104.

māl: **I** lxviii **III** 1185, 1235, 1383.

See also squares (unknown, x^2).

Malaga: **III** 1660.

Maldura, F.: **I** xxvii$n$.

Marmura, M.: **III** 1139n.

Marre, A.: **I** lxii$n$ **III** 1479n, 1516n.

masses, metallic: (chap. B–V) **II** 852–855 **III** 1442–1447.

McLennan, J.: **I** v.

'mathematicians': **II** 683 **III** 1228, 1712.

to measure ($\mu\epsilon\tau\rho\epsilon\tilde{\iota}\nu$): **II** 595, 596 (2), 616 (2), 740 (2), 856.

'measure' (metrology): **I** lxxviii–lxxix **II** 882–891, 1040–1045, 1105–1106.

measuring vessels: **I** lxxviii–lxxix **II** 892–894, 997–998, 1003–1005 **III** 1492–1494, 1603, 1616–1617, 1619, 1625–1628.

mediaeval mathematics: **I** xiii–xv.

'medial': **II** 761–765, 771, 772, 775, 1105 **III** 1313–1315, 1329–1331, 1336, 1339–1340.

Mesopotamian mathematics: **I** lxxiii **II** 613$_{147}$ **III** 1145, 1304, 1350, 1381, 1670.

metrology: **I** lxxviii–lxxix.
See also almodi; arrova; caficius; cubit; emina; measure; measuring vessels; mile; modius; ounce; palm; sextarius.

mile (metrology): **I** lxxix **II** 952–965, 967, 974, 1000, 1068–1069.

'minor line' (*see* **III** 1314): **II** 772, 1106.

Mittellateinisches Wörterbuch: **II** 863$_{1186}$, 1075.

modius (metrology): **I** lxxviii–lxxix **II** 784, 788–790, 792–795, 800$_{956}$, 807, 811–813, 819–821, 823, 892–894, 895, 896, 997–998, 1082, 1107.

'money-bag' (*sacellus*): **II** 643–647 **III** 1181–1182, 1185–1188.

morabitinus (coin; *see* **III** 1185): **I** lxxviii–lxxix **II** 643–647, 1006–1038, 1107.
See also exchange.

Moschos (*c*. 1200): **I** xiv.

motion: **II** 581 **III** 1139–1141.
See also forward and retrograde motion; pursuit.

muʿāmalāt: **I** xv.

multiplication: (cases of) **II** 606 **III** 1162–1163 ◊ (of integers) **I** lxiv–lxv **II** 605–635 **III** 1159–1178 ◊ (written arrangement) **II** 625–626 & 625$_{206}$, 627, 630$_{227}$ **III** 1161, 1172 ◊ (table) **II** 583$_{17}$, 607–608, 609$_{129}$ **III** 1164, 1167 ◊ (of fractions) **II** 606, 655–693, 743 **III** 1162–1163, 1197–1233, 1295–1297 ◊ (written arrangement) **II** 663$_{360}$ **III** 1220 ◊ (of roots) **II** 751–753, 760–763 **III** 1305, 1307–1308, 1319–1320, 1327–1329.

al-muqābala (reduction): **I** lvi, lxix–lxx **II** 1074, 1077, 1107 **III** 1347–1349, 1363.
See also algebra (algebraic operations).

Murdoch, J.: **I** xiii$n$.

must (reduction): (chap. B–IX) **II** 882–891 **III** 1476–1491.

mutequefia: *See* reciprocally proportional.

Nagl, A.: **I** xxxvii.

al-Nairīzī (*c*. 880): **I** xxi **III** 1313n.

Nicomachos (*c*. 100): **I** xiii **II** 581, 1080, 1109 **III** 1141.

Niermeyer, J.: **II** 1075, 1105, 1125.

'note' (*see* **III** 1160): **II** 586–587, 609–619 **III** 1143, 1160–1161,

1163, 1165–1169.

numbers (natural): (place among all things) **II** 581 **III** 1139–1141 ◇ (two aspects of, thus two kinds of arithmetic) **II** 581–582 **III** 1139, 1141 ◇ (composition of) **II** 582–587 **III** 1141–1143 ◇ (infinity of) **II** 583, 609 **III** 1141–1142 ◇ ('kinds of') **II** 605, 607 **III** 1159, 1162 ◇ (addition of) **I** lxiv **II** 626 & 626_{210}, 630–631, 707–710 **III** 1159, 1236–1238, 1251–1253 ◇ (division) **II** 635–636, 639–654 **III** 1181–1196 ◇ (multiplication) **II** 605–635 **III** 1159–1178 ◇ (subtraction) **I** lxiv **II** 624–626, 631–635 **III** 1161, 1171–1172, 1176–1177 ◇ (composite, i.e. more than one digit) **II** 582, 586, 605, 607, 613, 623–626, 648, 654, 656–661 **III** 1161, 1171–1172, 1175, 1194, 1195, 1202–1203 — (composite, i.e. not prime) **II** 618 ◇ (divisibility) **II** 646–651 **III** 1182, 1188–1191, 1301–1302 ◇ (negative) **I** lxviii **II** 624_{200}, 704_{497}, 722_{575} **III** 1498, 1560, 1707 ◇ (perfect) **I** xxxi **III** 1141n ◇ (prime) **II** 648_{292} **III** 1189, 1194 ◇ (similar plane) **II** 752 (2) **III** 1320 ◇ (sum of consecutive natural) **III** 1236–1237, 1251–1252 — see also progressions ◇ (finger symbolism) **II** 614–615, 615_{152}, 623, 625–626, 625_{205} **III** 1161, 1171–1172 ◇ (verbal expression) **I** xx **II** 583, 583_{18}, 610 **III** 1139, 1142–1143, 1160, 1165, 1167 ◇ (written expression) **II** 584_{25}, 611_{137}, 622_{193} **III** 1143, 1169, 1172 ◇ (symbols) **I** xx, xxiii, lxiv **II** 585–586, 585_{28}, 625 & 625_{205}, 630_{227}, 652 — see also figures (tables) ◇ (Roman numerals) **I** xxiii, xxiiin.

numerator: **II** (e.g.) 655_{328}, 656–660, 661_{351}, (...).

nummus: (coin) **I** lxviii, lxxviii–lxxix **II** 582, 582_{14}, 639 (4), 704–706, 721–724, 744–746, 780, (...), 1110 ◇ (weight) **II** 1037–1038 ◇ (designation of the constant term in an equation) **I** lxviii **II** (e.g.) 841_{1120}, 925_{1456}, 945_{1527}, 960_{1593}, 995_{1738}, 1018_{1827}.

obol (coin): **I** lxxix **II** 784_{877}.

orders (of numbers): **II** 583–587, 609–611, 613–619, 1112 **III** 1142–1143, 1159–1161, 1163, 1165, 1167, 1169, 1176.

ounce (metrology): **I** lxxix **II** 852–858, 860–862, 1132.

Ouy, G.: **I** xxxviin.

Pacioli, L. (c. 1490): **I** lxi.

palm (metrology): **I** lxxix **II** 852–853, 1066, 1112.

Pappus (c. 350): **I** xxi.

partnership: (chap. B–III) **II** 843–847 **III** 1426–1433 ◇ (mentioned) **II** 812, 852, 878, 935–936 **III** 1388, 1445, 1472, 1542–1543, 1545–1546.

paving floors: **II** 869–870 **III** 1464–1465.

π: See circle.

Picutti, E.: **I** xviiin.

plane numbers, similar: See similar plane numbers.

portions (prescribed): (chap. B–IV) **II** 848–851 **III** 1434–1441 ◇ (mentioned) **II** 1030.

'principal' (see **II** 663_{361}, **III** 1296–1297): **II** 663–667, 673–674, 678 (2), 680–688, 690–692, (...); 703_{493}, 731_{610}, 786_{882}, 865_{1202}, 872_{1233}, 1115–1116.

problems: (the five fundamental problems, see **I** lxv) **III** (e.g.) 1346–1347, 1395, 1492, 1563, 1570 (= sixth problem) ◇ (irrational results) **III** 1373, 1376–

1377, 1406, 1408, 1576, 1687, 1694 ◊ (recreational type) **III** 1670–1696, 1700–1714 ◊ (with 'things' in the data) **III** 1347–1349 & 1350, 1495–1497, 1550, 1566 ◊ ('complicated') **II** 790 ◊ ('foreign') **II** 836 ◊ ('impossible') **II** 751, 768 (2), 832, 856, 886 (3), 888, 888–890, 1012–1013 ◊ (answer indefinite) **III** 1490, 1491 ◊ (indeterminate) **I** lxviii **II** 707, 746, 747, 815, 890 (2), 1012, 1016, 1021, 1037, 1072_{2015}, 1072, 1073 (2) **III** 1251, 1295, 1301–1302, 1393, 1632–1635, 1639, 1644–1652, 1704–1708, 1709–1714 ◊ ('not admissible') **II** 720 ◊ (not applicable) **III** 1450, 1457–1458, 1459, 1464, 1465 ◊ ('not solvable') **II** 702–703, 720, 812, 813, 815, 825, 829, 832, 886, 890, 920, 926, 932, 1012, 1013, 1019–1020, 1021, 1025, 1036, 1050, 1053, 1059, 1065 — (not solvable exactly in numbers) **II** 868, 869 ◊ ('solvable') **II** 812, 813, 825, 857, 870, 890, 917, 920, 932, 1017, 1020, 1021, 1022, 1023, 1024, 1025, 1026 ◊ ('straightforward') **II** 814 ◊ (solvability condition) **II** 812–813, 813, 815, 825, 829, 832, 856–857, 890, 917, 920, 926, 932, 1011–1012, 1012, 1012–1013, 1016–1017, 1019–1020, 1021–1022, 1024, 1036, 1050, 1053, 1059, 1060, 1065 **III** 1389–1390, 1390–1391, 1393, 1406, 1409, 1412, 1450, 1490, 1522, 1524, 1530, 1537, 1639, 1640–1641, 1644–1645, 1647, 1649, 1651, 1660, 1680, 1684, 1690, 1691, 1696.

See also equations; proofs; solution(s); verification.

profit and loss : (chap. B–II) **II** 817–842 **III** 1345, 1395–1425 ◊ (mentioned) **II** 846, 859 ◊ (loss) **II** 831, 935 **III** 1398, 1495, 1498.

See also partnership.

progressions : (arithmetic) **I** lxxvii **III** 1144, 1145–1146, 1168, 1236–1237, 1251–1252, 1278, 1299–1300, 1393–1394, 1503, 1551–1562, 1604, 1615–1616 ◊ (geometric) **I** lxxvii–lxxviii **III** 1168, 1435–1436, 1437–1438.

proofs : (formal, using geometrical figures) **II** 588–604, 615–618, 627–628, 633–634, 635, 663, 674–676, 698–699, 713, 715, 723–724, 724–725, 740, 751–752, 754, 755, 756, 757–758, 758, 760, 763–764, 767, 769, 771–772, 773, 796–797, 797–798, 799–800, 803, 806, 807, 808–809, 810–811, 814, 818, 819–820, 834–835, 836–837, 837–838, 859, 862, 866, 869–870, 884, 887–888, 888–889, 914–915, 916, 917, 918, 918–919, 924, 927–928, 931, 932–933, 933–934, 934–935, 938–939, 943–945, 946, 949, 950–951, 953–954, 959–960, 961–962, 967–969, 980–981, 1012–1013, 1014–1015, 1015–1016, 1017, 1018–1019, 1022–1023, 1023, 1028–1029, 1031, 1032–1033, 1035, 1036–1037, 1043–1044, 1046–1047, 1047, 1047–1048, 1048 (2), 1049–1050, 1050–1051, 1051–1052, 1053, 1054, 1054–1055, 1055, 1056, 1057, 1057–1058, 1059 (2), 1060, 1060–1061, 1062, 1063, 1064, 1064–1065, 1068 ◊ (said to be 'easier' or 'simpler') **II** 591, 675 **III** 1147–1148, 1218 ◊ (justifications) **II** 612, 629_{222}, 636, 637_{255}, 640, 641, 643, 644, 655–656, 658, 659, 663, 673, 683, 684, 686–687, 687, 700, 702, 706–707, 718, 719, 734, 752–753, 753, 761, 762, 762–763,

766, 785, 786–787, 789, 790, 817–818, 819, 830, 843, 844, 849, 856, 857, 860, 863–864, 864, 868, 872, 873, 874, 879, 880, 882, 883, 893, 896, 897, 898, 901, 909, 923–924, 925, 929, 943, 944–945, 945, 946, 947, 948, 951, 952, 953, 955, 956 (2), 957, 963, 963–964, 964, 967, 969 (2), 970, 970–971, 971, 972, 973–974, 975, 976, 976–977, 979, 980, 980–981, 981–982, 982, 983 (2), 984, 985, 986, 987–988, 988 (2), 989 (2), 990 (2), 991, 992, 996–997, 997, 999, 1000, 1002 (2), 1003 (2), 1006 (2), 1007 (2), 1008, 1009, 1010, 1011, 1027, 1034, 1040, 1041, 1043–1044, 1066, 1066–1067.
See also solution, geometrical; verification.

proportion: **I** lxv–lxvii **III** 1235, 1256, 1343, 1345, 1395–1396, 1426, 1436, 1442, 1448, 1457, 1466–1468, 1476, 1492, 1495, 1563, 1578–1580, 1592, 1603, 1618, 1626, 1663, 1697, 1698 ◇ (one or two terms unknown) see problems (five fundamental).
See also continued proportion; ratio; reciprocally proportional.

Ptolemy (c. 150): **I** lxi$n$.

pursuit and distance covered: (chap. B-XXII.1) **II** 1068–1069 **III** 1700–1702.

Pythagorean theorem: **III** 1670–1672, 1674, 1675–1691, 1693–1697.

rank (*locus*): **II** 613$_{144}$, 617 **III** 1160, 1167.

ratio: **I** lxv–lxvii **II** 639 ◇ (compounded) **II** 618, 954 (4), 968 (4) ◇ (double) **II** 608$_{122}$, 795 ◇ (duplicate) **I** lxvii **II** 608$_{122}$, 618, 706, 799 (2), 803, 859 ◇ (*ex æquali*) **I** lxvii **II** 594 (2), 835, 939 ◇ (amount) **II** 882 (2), 1106

◇ (transformations of ratios) **I** lxvi–lxvii ◇ (alternation, *alternando*) **II** 799, 837, 882, 884, 914, 919 ◇ (composition, *componendo*) **II** 818, 819, 839, 840, 849, 873, 887, 901 (2), 910 (2), 917$_{1422}$, 923, 924, 928, 932, 939, 959, 960, 962, 1028, 1044 ◇ (conversion, *convertendo*) **II** 594, 819, 839, 933 ◇ (inversion, *invertendo*) **II** 594–596, 926, 939 (2), 958, 962 (2) ◇ (separation, *separando*) **II** 594, 836, 887$_{1294}$, 897, 911 (2), 917, 918 (2), 919 (2), 962.

rational: **II** 748, 749, 752 (3), 753, 754, 757–759, 761 (3), 764, 768–772, 774, 775 ◇ (rational in square) **II** 753, 768, 769 (4).

Rebstock, U.: **II** 1076 **III** 1362n.

reciprocally proportional (*mutequefia*): **II** 914$_{1414}$, 924 (2), 959 **III** 1518, 1528, 1572.

rectification of the circle: **II** 868 **III** 1463.

redintegration: *See* fraction.

reduction: (*al-muqābala*) see muqābala; algebra (algebraic operations) ◇ (*al-radd*; *al-ikmāl*, 'completion') **I** lxix **II** 841, 890$_{1309}$, 903–905, 958, 995.

'repetition' (*see* **III**1155): **II** 584, 610–615, 620–622, 626–629, 631–638, 640–649, 653–654, 673 **III** 1143, 1160–1162, 1165–1170, 1172–1180, 1181–1182, 1184–1188, 1194–1196.

res: *See* thing; 'thing'.

restoration (*al-jabr*): *See* jabr.

Roman mathematics: **III** 1693.

roots: **I** lxv **II** 582 (2), 706, 710 (3), 724, 746 (5), 748–775, 790, (…) **III** 1141, 1238, 1250, 1253–1254, 1278, 1299–1301, 1303–1342, 1347, 1349–1350, 1358–

1361, 1366–1385, 1399, 1419, 1422–1425, 1427, 1430–1432, 1449–1456, 1459, 1461–1463, 1468, 1471, 1484–1489, 1493, 1497, 1508–1510, 1515–1520, 1523, 1565–1566, 1570, 1575–1576, 1579–1580, 1587, 1590, 1614, 1656–1658, 1671–1672, 1674–1683, 1685–1687, 1690–1691, 1694–1696 ◊ (in the data) III 1358–1361, 1375–1381, 1422–1425, 1430–1432, 1508–1510, 1614, 1656–1658 ◊ (approximation) II 748–751, 790, 858, 964, 1056, 1062 III 1303–1304, 1316–1319, 1358, 1452, 1576, 1687, 1694 ◊ ('irrational') II 754 ◊ ('number without root') II 751, 790, 858 ◊ ('true' root) II 751 ◊ (addition) II 753–754, 763–765 III 1305–1306, 1308–1309, 1320–1322, 1329–1331 ◊ (division) II 756–760, 766–770 III 1306–1307, 1309–1310, 1323–1327, 1332–1336 ◊ (multiplication) II 751–753, 760–763 III 1305, 1307–1308, 1319–1320, 1327–1329 ◊ (subtraction) II 754–756, 765–766 III 1306, 1309, 1322–1323, 1331–1332 ◊ (square root extraction of a binomial or an apotome) I lxxvi–lxxvii II 770–775, 801 III 1310–1315, 1336–1342.

Rouse, R.: **I** xxxiii*n*.

'rule' (*regula*): (mentions in the *Liber mahameleth*) **II** 587, 591, 592, 608–611, 615, 617, 619, 622, 623 (2), 636, 637, 639, 646, 650, 651, 652 (3), 656 (2), 659, 667, 686, 688, 692, 695, 697, 702, 704 (3), 707, 719, 728–730, 743, 745 (2), 754, 755, 763, 785, 793, 795 (4), 800, 817, 818, 831, 834, 838, 843, 844 (2), 856, 857, 860, 864, 868, 872, 873, 874, 897, 944, 1071, 1122–1123 ◊ (termed 'valid' or 'applicable') **II** 623, 745, 793, 795 (4), 868 ◊ (termed 'general') **II** 707, 795 (2), 868, 897.

rule of false: **I** xvi **III** (e.g.) 1235, 1256, 1637.

rule of signs: **II** 622_{193}, 623 **III** 1161, 1171.

rule of three: (simple) **I** xvi **II** (e.g.) 702–703, 719, 777 **III** (e.g.) 1245, 1343, 1356, 1674 ◊ (compound) **III** 1563.

Sacrobosco, J. de (*c.* 1230): **I** xv, xxxvii, lxi*n*.

Saʿīd Abū ʿUthmān (*c.* 1020): **I** xxi.

Saidan, A.: **III** 1362*n*, 1476*n*.

Segovia: **I** lxxix **II** 997–998, 1003–1005, 1125 **III** 1140*n*, 1616–1617, 1626–1628.

selling: See profit and loss.

series: **I** lxxvii–lxxviii **III** 1236–1238, 1251–1254, 1278, 1435–1436, 1437–1438.
See also progressions.

Seville: **I** xix, lxxix*n* **III** 1466*n*, 1660.

sextarius (metrology): **I** lxxix **II** 582, 582_{14}, 639 (5), 780–788, 790, (...), 1127.

sharing: **II** 639, 744–747, 843–851 **III** 1181, 1183, 1277–1278, 1299–1302, 1426–1441.
See also partnership.

shay': **I** lxviii **III** 1347.
See also 'thing'.

side: (of a figure or a composite number) **II** 618, 799, 859, 914, 924, 943, 959, 978, 1102 ◊ (of an equation) *see* equation.

similar plane: (figures) **II** 799, 857, 858, 861, 864, 867, 1041 **III** 1371, 1448, 1451, 1455, 1461, 1665, 1666, 1697 ◊ (numbers) **II** 752 **III** 1320.

Simson, R. (*c.* 1750): **III** 1151.

Smith, D.E.: **II** 782_{861} **III** $1434n$.

solidus (coin): **I** lxxviii–lxxix **II** 821, 823, 826, 1006–1010, 1035–1036, 1128.

solution (solving method): (geometrical computation) **I** lxx–lxxvi **II** 723–724, 757–758, 771–772, 773, 862, 884, 888–889, 949, 950–951, 1013, 1032–1033, 1059, 1060, 1068 **III** 1271–1272, 1325–1326, 1336–1339, 1340–1341, 1455–1456, 1482, 1488–1489, 1557–1559, 1560–1562, 1641, 1657–1658, 1689, 1690–1691, 1701 ◊ (backwards computation) **III** 1704–1705, 1708–1709 ◊ (termed 'general') **II** 707, 897, 899 ◊ ('short solution') **II** 995.

solutions, numerical: ◊ (indeterminate) **I** lxviii **II** 890_{1313}, 1060_{1987} **III** 1490, 1691 ◊ (irrational) **III** 1373, 1376–1377, 1406, 1408, 1576, 1687, 1694 ◊ (negative) **I** lxviii **III** 1707 ◊ (two positive) **I** lxxiii, lxxv–lxxvi **III** 1300–1301, 1419–1422, 1488, 1510, 1560, 1675–1676, 1682–1683 ◊ (zero) **I** lxviii **II** 812_{1021}, 815_{1030}, 1013, 1096 **III** 1412, 1639, 1647 ◊ (verification of the solution) *see* verification.

Sorbon, R. de (*c.* 1240): **I** xxxiii.

Spain: **I** xiii–xv, xvii–xix, lx **III** 1185, 1460, 1592, 1629, 1660.
See also Baeza; Segovia; Seville; Toledo.

specific weights: **III** 1447.

sphere: **I** lxxviii **II** 852–853 **III** 1442, 1446.

squares: (shape) **II** 581, 771–773, 858, 861, 864, 868, 977, 1041–1043, 1045 ◊ (power) **II** 603–604, 618, 708–709, 748–749, 751–756, 758, 761–766, 768–769, 771, 773, 806, 820, 868, 869, 1032, 1060, 1066–1067 ◊ (sum of consecutive) **II** 708–709 **III** 1237, 1252–1253 ◊ (unknown, x^2) **I** lxviii **II** 809, 809_{1005}, 841, 842, 845, 876, 878, 889–890, 903–905, 919, 950, 958, 959, 964–966, 995, 996, 1046–1048, 1083.

squaring the circle: **II** 868 **III** 1458, 1463.

subtraction: (of integers) **I** lxiv **II** 624–626, 631–635 **III** 1161, 1171–1172, 1176–1177 ◊ (of fractions): **II** 711–720, 743 **III** 1162–1163, 1255–1266, 1295–1297 ◊ (of roots) **II** 754–756, 765–766 **III** 1306, 1309, 1322–1323, 1331–132.

Suter, H.: **I** lxiv$n$ **II** 662_{357} **III** $1200n$, $1236n$, $1253n$, $1313n$, $1670n$.

'taccir': **I** xviii, lxiv, lxxviii **II** 865 **III** 1460.

Thesaurus linguae latinae: **II** 871_{1227}.

thing: **II** 581 **III** 1139–1141.
See also 'thing'.

'thing' (*res*, unknown x): **I** lxviii **II** 809_{1001}, 876_{1250}, 877_{1260}, 878_{1266} **III** 1249, 1347, 1362, 1383, 1386 ◊ ('thing' mentioned in the *Liber mahameleth*) **II** 792–796, 807–811, 876–878, 880, 888–890, 892–894, 896–922, 924–926, 928–932, 936–937, 940–942, 945–948, 950–951, 955–956, 958–965, 974–975, 983–984, 994–996, 1013–1014, 1016, 1018, 1029–1030, 1033, 1035, 1038, 1046–1048, 1059, 1069, 1073–1074, 1123–1124.
See also algebra; equations; unknowns.

Thorndike, L.: **I** xiv$n$.

Toledo: **I** xix, lx, lxxix **II** 997–998, 1003–1005, 1130 **III** 1616–

1617, 1626–1628.

Toomer, G.: **I** v.

'treasury' (*domus pecunie*): **II** 644–647, 1113 **III** 1181–1182, 1185–1188.

trees: (chap. B-XX.2&3) **II** 1055 –1062 **III** 1672–1674, 1686–1693.

triangle: **II** 865, 1046, 1047, 1056, 1062, 1131.

Tropfke, J.: **II** 622_{193} **III** $1434n$, $1704n$, $1708n$.

two towers: (chap. B–XX.5) **II** 1062–1065 **III** 1674–1675, 1693–1696.

unit: (origin of number) **II** 582–583 **III** 1141 ◇ (not a number) **II** 582, 653_{316} **III** 1141.

unknowns (designation in the *Liber mahameleth*): **I** lxviii **II** 809_{1001} **III** $1345n$; *see also* 'thing' ◇ (two unknowns) **II** 809 & 809_{1002}, 811 **III** 1383, 1386 ◇ (other mediaeval symbol and designation) **II** 782_{861}, 852_{1146} **III** 1249, 1712.

verbal mathematics: **I** lv **II** 583_{18}, 613_{147}, 659_{340}, 723_{583}, 743_{670}, 744_{677}, 772_{824}, 801_{962}, 801_{964} **III** 1139, 1228, $1233n$, 1250, 1322–1323, 1373–1374, $1376n$, 1527, 1631, 1713.
See also ambiguity.

verification (of the numerical solution): **II** 640, 824, 831, 833, 834, 836, 876, 877, 898, 900_{1351}, 904, 906 (2), 907, 908, 910, 913, 921, 922, 926, 929, 929–930, 936, 1030, 1038, 1069; *see also* 793, 794, 826, 829, 833.

Villa Dei, A. de (*c.* 1125): **I** xv.

Vitrac, B.: **III** $1323n$.

Vitruvius (*c.* 20 B.C.): **III** 1444.

Vlasschaert, A.-M.: **I** xvn.

Vogel, K.: **I** $xivn$ **III** $1362n$, $1670n$.

volumes: **I** lxxviii **II** 582, 853, 855, 976–978, 1040–1043, 1104, 1128 **III** 1442, 1446, 1589–1590, 1663–1669.

wages: *See* hiring ◇ (wages in arithmetic progression) (chap. B–XII) **II** 943–951 **III** 1551–1562 ◇ (mentioned) **II** 997.

ways (the three banal, *see* **III** 1344): **II** 778, 778_{848}, 780_{852}, 782_{859}, 784_{876}, 789_{899}, 817_{1035}, 819_{1043}, 872_{1231}, 872_{1232}, 882_{1277}, (...), 968_{1625}, (...) **III** (e.g.) 1245, 1266, 1344, 1346, 1351, 1352, 1354, 1355, 1357, 1399, 1428, 1450, 1458, 1469, 1481, 1502, 1503, 1532, 1534, 1567, 1575, 1636.

weights and measures: *See* metrology; specific weights.

Wiedemann, E.: **I** lxn **III** $1670n$.

Woepcke, F.: **III** $1362n$.

workers: *See* hiring.

zero: **II** 630_{227}, 648 & 648_{293}, 649 & 649_{296}, 650 & 650_{303} **III** 1175.